# Seifert Fiberings

Mathematical
Surveys
and
Monographs

Volume 166

# Seifert Fiberings

**Kyung Bai Lee**
**Frank Raymond**

**American Mathematical Society**
Providence, Rhode Island

EDITORIAL COMMITTEE

Ralph L. Cohen, Chair        Michael A. Singer
Eric M. Friedlander          Benjamin Sudakov
Michael I. Weinstein

2000 *Mathematics Subject Classification.* Primary 55R55, 57S30, 57–XX;
Secondary 53C30, 55R91, 58E40, 58D19, 57N16.

For additional information and updates on this book, visit
www.ams.org/bookpages/surv-166

Library of Congress Cataloging-in-Publication Data
Lee, Kyung Bai, 1943–
  Seifert fiberings / Kyung Bai Lee, Frank Raymond.
    p. cm. — (Mathematical surveys and monographs : v. 166)
  Includes bibliographical references and index.
  ISBN 978-0-8218-5231-6 (alk. paper)
  1. Fiberings (Mathematics).  2. Complex manifolds.  I. Raymond, Frank, 1932–  II. Title.

QA612.6.L44  2010
514'.2—22
                                                                    2010022528

**Copying and reprinting.** Individual readers of this publication, and nonprofit libraries acting for them, are permitted to make fair use of the material, such as to copy a chapter for use in teaching or research. Permission is granted to quote brief passages from this publication in reviews, provided the customary acknowledgment of the source is given.

Republication, systematic copying, or multiple reproduction of any material in this publication is permitted only under license from the American Mathematical Society. Requests for such permission should be addressed to the Acquisitions Department, American Mathematical Society, 201 Charles Street, Providence, Rhode Island 02904-2294 USA. Requests can also be made by e-mail to reprint-permission@ams.org.

© 2010 by the American Mathematical Society. All rights reserved.
The American Mathematical Society retains all rights
except those granted to the United States Government.
Printed in the United States of America.

∞ The paper used in this book is acid-free and falls within the guidelines
established to ensure permanence and durability.
Visit the AMS home page at http://www.ams.org/

10 9 8 7 6 5 4 3 2 1      15 14 13 12 11 10

To Hyang, Myung, Kwang, Chung, Natasha, Nyra and Rob

# Acknowledgments

We wish to acknowledge our debt to our many collaborators, teachers, students, and colleagues who have, over the years, contributed so much to our work and enjoyment of mathematics. We are also grateful for past financial support from University of Oklahoma and University of Michigan as well as from the National Science Foundation and RIP of the Mathematisches Institut Oberwolfach.

# Contents

| | |
|---|---|
| Preface | xiii |
| **Chapter 1. Transformation Groups** | 1 |
| 1.1. Introduction | 1 |
| 1.2. (Locally) Proper $G$-spaces | 4 |
| 1.3. Fiber bundles | 6 |
| 1.4. Classifying spaces | 9 |
| 1.5. Borel spaces and classifying spaces | 10 |
| 1.6. Tubular neighborhoods and slices | 11 |
| 1.7. Existence of slices | 15 |
| 1.8. Cohomology manifolds and the Smith theorems | 18 |
| 1.9. Actions of $G \cdot \Pi$ ($G$ Lie group, $\Pi$ discrete) | 20 |
| **Chapter 2. Group Actions and the Fundamental Group** | 23 |
| 2.1. Covering spaces | 23 |
| 2.2. Lifting group actions to covering spaces | 26 |
| 2.3. Lifting an action of $G$ when $G$ has a fixed point | 28 |
| 2.4. Evaluation homomorphism | 31 |
| 2.5. Lifting connected group actions | 32 |
| 2.6. Example (Semi-free $S^1$-actions on 3-manifolds) | 35 |
| 2.7. Lifting the slice representation | 39 |
| 2.8. Locally injective actions | 42 |
| **Chapter 3. Actions of Compact Lie Groups on Manifolds** | 47 |
| 3.1. Actions of compact Lie groups on aspherical manifolds | 47 |
| 3.2. Actions of compact Lie groups on admissible manifolds | 54 |
| 3.3. Compact Lie group actions on spaces which map into $K(\Gamma, 1)$ | 59 |
| 3.4. Manifolds with few or no periodic homeomorphisms | 65 |
| 3.5. Injective torus actions | 66 |
| **Chapter 4. Definition of Seifert Fibering** | 69 |
| 4.1. Examples | 69 |
| 4.2. $\mathrm{TOP}_G(P)$, the group of weak $G$-equivalences | 73 |
| 4.3. Seifert fiberings modeled on a principal $G$-bundle | 78 |
| 4.4. The topology and geometry of the fibers | 81 |
| 4.5. Examples with $\Pi$ discrete | 83 |
| 4.6. The Seifert Construction | 91 |
| **Chapter 5. Group Cohomology** | 95 |
| 5.1. Introduction | 95 |
| 5.2. Group extensions | 95 |

| | | |
|---|---|---|
| 5.3. | Pullback and pushout of short exact sequences | 97 |
| 5.4. | Extensions with Abelian kernel $A$ and $H^2(Q;A)$ | 100 |
| 5.5. | Central extensions | 102 |
| 5.6. | Extensions with non-Abelian kernel $G$ and $H^2(Q;\mathcal{Z}(G))$ | 104 |
| 5.7. | $H^1(Q;C)$ with a non-Abelian $C$ | 105 |

Chapter 6. Lie Groups — 109
- 6.1. Introduction — 109
- 6.2. Nilpotent Lie groups — 111
- 6.3. Solvable Lie groups — 114
- 6.4. Semisimple Lie groups — 115

Chapter 7. Seifert Fiber Space Construction for $G \times W$ — 119
- 7.1. Introduction — 119
- 7.2. Cohomological criteria — 120
- 7.3. Main Construction Theorem for $\mathrm{TOP}_G(G \times W)$ — 122
- 7.4. The meaning of existence, uniqueness and rigidity — 123
- 7.5. $H^p(Q;\mathrm{M}(W,\mathbb{R}^k)) = 0$, $p > 0$ — 127
- 7.6. Proof of the Construction Theorem — 131
- 7.7. When is $\theta$ injective? — 135
- 7.8. Smooth case — 137

Chapter 8. Generalization of Bieberbach's Theorems — 139
- 8.1. Bieberbach's theorems — 139
- 8.2. Proof of Bieberbach's theorems — 139
- 8.3. The First Bieberbach Theorem — 143
- 8.4. The Second Bieberbach Theorem — 151
- 8.5. The Third Bieberbach Theorem — 156

Chapter 9. Seifert Manifolds with $\Gamma\backslash G/K$-Fiber — 159
- 9.1. Introduction — 159
- 9.2. The group $\mathrm{TOP}_{G,K}(G/K \times W)$ — 160
- 9.3. When $N_G(K) = K$ and $\mathrm{Aut}^0(G,K) = 1$ — 164
- 9.4. Symmetric spaces of noncompact type — 166
- 9.5. Solvmanifolds — 172

Chapter 10. Locally Injective Seifert Fiberings with Torus Fibers — 179
- 10.1. Introduction — 179
- 10.2. When does $E(P,Q)$ split? — 187
- 10.3. From local to global: $H^*(Q;\mathcal{T}^k)$ and $H^*(Q;\mathcal{Z}^k)$ — 190
- 10.4. The product case, $\mathbb{R}^k \times W$ — 198
- 10.5. Some aspects of the $'E^{p,q}$-spectral sequence — 201

Chapter 11. Applications — 205
- 11.1. Existence of closed $K(\Pi,1)$-manifolds — 206
- 11.2. Rigidity of Seifert fibering — 210
- 11.3. Lifting problem for homotopy classes — 216
- 11.4. Polynomial structures for solvmanifolds — 227
- 11.5. Applications to fixed-point theory — 231

| | | |
|---|---|---|
| 11.6. | Homologically injective torus operations | 236 |
| 11.7. | Maximal torus actions | 248 |
| 11.8. | Toral rank of spherical space forms | 265 |

Chapter 12. Seifert Fiberings with Compact Connected $Q$ — 269
- 12.1. Introduction — 269
- 12.2. Lifting $Q$-actions — 269
- 12.3. Lifting $Q$-actions (for connected $Q$) — 271
- 12.4. Examples — 278

Chapter 13. Deformation Spaces — 283
- 13.1. Uniformizing groups — 283
- 13.2. $\widetilde{\mathrm{PSL}}(2,\mathbb{R})$-geometry — 284
- 13.3. Lorentz structures and $\widetilde{\mathrm{PSL}}(2,\mathbb{R})$-geometry — 288
- 13.4. Deformation spaces for $\widetilde{\mathrm{PSL}}(2,\mathbb{R})$-geometry — 290
- 13.5. Deformation spaces for Nil-geometry — 296

Chapter 14. $S^1$-actions on 3-dimensional Manifolds — 299
- 14.1. Introduction — 299
- 14.2. $S^1$-actions on 3-manifolds — 302
- 14.3. $S^1$-actions on 3-manifolds as Seifert fiberings — 304
- 14.4. The classification of $S^1$-actions on closed aspherical 3-manifolds — 309
- 14.5. The classification of $S^1$-actions with fixed points on 3-manifolds — 312
- 14.6. A complete set of invariants — 314
- 14.7. The Euler number — 318
- 14.8. $\Gamma\backslash G$ as 3-dimensional Seifert manifolds — 322
- 14.9. $H_1(M;\mathbb{Z})$ — 329
- 14.10. Injective holomorphic Seifert fiberings — 331
- 14.11. Brieskorn complete intersections — 335
- 14.12. Generalized Seifert 3-manifolds as local SO(2)-actions — 338
- 14.13. A complete set of invariants for the 3-dimensional Seifert fiberings — 344
- 14.14. Historical remarks — 348

Chapter 15. Classification of Seifert 3-manifolds via equivariant cohomology — 353
- 15.1. $H^2(Q;\mathbb{Z}^k)$ and codimension-2 injective actions — 353
- 15.2. A presentation for $H^2(Q;\mathcal{Z})$ — 356
- 15.3. 3-dimensional spherical space forms — 366
- 15.4. Seifert fiberings with $Q$ Euclidean crystallographic — 374
- 15.5. Seifert fiberings with $Q$ hyperbolic — 376
- 15.6. Equivariant classification — 377
- 15.7. An Illustration — 377

Bibliography — 383

Index — 393

# Preface

This book is an exploration of Seifert fiberings. These are mappings which extend the notion of fiber bundle mappings by allowing some of the fibers to be singular. Seifert fiberings are mappings whose typical fibers are homeomorphic to a fixed homogeneous space. The singular fibers are quotients of the homogeneous space by distinguished groups of homeomorphisms.

In a remarkable paper, in 1933, Herbert Seifert, introduced a class of 3-manifolds which became known as Seifert manifolds. They play a very significant role in low dimensional topology and remain under intense scrutiny today. A Seifert manifold maps onto a 2-dimensional surface such that the inverse image of each point on the surface is homeomorphic to a circle. The set of singular fibers are isolated from each other and the typical fibers wind nontrivially around the singular fibers. We will describe in detail the Seifert 3-manifolds as a special case of the general construction of Seifert fiberings. Our major focus, however, is on higher dimensional phenomenon where the typical fiber is a homogeneous space.

A major inspiration for a generalization to higher dimensions comes from transformation groups. Let $(G, X)$ be a proper action of the connected Lie group $G$ on a path-connected $X$ and examine the homomorphism $\text{ev}_*^x : \pi_1(G, e) \to \pi_1(X, x)$ induced by the evaluation map $\text{ev}^x : g \mapsto gx$, $g \in G$. The image $H$ of $\text{ev}_*^x$ is a central subgroup of $\pi_1(X, x)$ independent of the base point $x$.

The $G$-action on $X$ can be lifted to the covering space $X_H$ of $X$ associated with the subgroup $H$ of $\pi_1(X, x)$ so that $\pi_1(X_H, \widehat{x}) = H$. The covering transformations, $Q = \pi_1(X, x)/H$, commute with the lifted $G$-action on $X_H$ and so induce a proper $Q$-action on $W = G\backslash X_H$. We get the following commutative diagram of orbit mappings:

$$\begin{array}{ccc} (G, X_H) & \xrightarrow{G\backslash}_{\tau'} & (Q, W) \\ \nu' \downarrow Q\backslash & & \nu \downarrow Q\backslash \\ (G, X) & \xrightarrow{G\backslash}_{\tau} & G\backslash X = Q\backslash W \end{array}$$

What we discover is that the lifted action of $G$ on $X_H$ is usually simpler than on the original $X$. The discrete action of $Q$ on $W$ can be used to describe the action of $G$ on $X$ locally. In fact, under the appropriate circumstances, the $Q$-action on $W$ can be used to construct all the possible $G$-actions on $X$ whose orbit space is $Q\backslash W$.

For example, if $G = T^k$, the $k$-dimensional torus, and $\text{ev}_*^x$ is injective, then $(T^k, X)$ is called an injective torus action. For this, $X_H$ splits into a product

$T^k \times W$, where $T^k$ acts as translations on the first factor. In this case, the elements of $H^2(Q; \mathbb{Z}^k)$ completely determine all the injective torus actions, up to equivalence, where the $T^k$-orbit space will be $Q \backslash W$.

For a general Seifert fibering, the covering space $X_H$ is replaced by a principal $G$-bundle $P$ over the base of the bundle $W$. Acting properly on $P$ is a group, $\Pi$, normalizing $\ell(G)$, the left translational action of $G$ on $P$.

Put $\Gamma = \ell(G) \cap \Pi$. On $W$ there is induced an action of $Q = \Pi/\Gamma$.

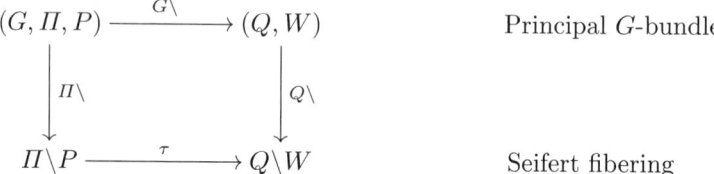

The induced mapping $\tau$ is our Seifert fibering. The typical fiber is the homogeneous space $\Gamma \backslash G$ where $\Gamma = \Pi \cap \ell(G)$, and $Q = \Pi/\Gamma$. The singular fibers are quotients of $\Gamma \backslash G$.

Even though we formulate Seifert fiberings for general spaces, our interest is directed towards geometric applications. Consequently, almost all of our illustrations and applications are devoted to manifolds or manifolds with singularities. We also focus on the Seifert fiberings where $\Pi$ and $Q$ are discrete, and it is only in Chapter 12 that we consider Seifert fiberings where $Q$ is a compact connected Lie group.

The many different topics covered in this book show the broad range of applicability at many levels. From this, it is clear that a mathematician studying geometric problems will often have to analyze singular fiberings, and we believe that this book provides some good tools for attacking interesting geometric phenomena.

Our interest is to engage the reader who has a modest background in topology, geometry and algebra as found in the second year of graduate school. We set language and notation and add some background material to fill in things that may lie outside the standard courses. Occasionally, we use and quote results that are readily available in good sources elsewhere. But on the whole, we have tried to be reasonably complete in our presentations. Examples are given to familiarize the reader with definitions and to illustrate special cases of the theorems. Similarly, the exercises are designed to enhance understanding of the text. Most of them are not difficult, and the reader is encouraged to do them.

Chapter 1 is an introductory chapter which establishes notation and fills in needed facts about proper actions of noncompact Lie groups.

In Chapter 2, covering spaces and lifting group actions to covering spaces are discussed. While much may be familiar to the reader, our approach is very explicit for ease in later computations.

Chapter 3 continues the theme that lifting a group action to a covering space simplifies the action. For example, locally injective actions lift to free actions and injective torus actions lift to product actions. These lifting techniques are then used to study actions of compact Lie groups $G$ on closed aspherical manifolds and their generalizations, the admissible manifolds. It is shown that the effective action of the connected component of $G$ on an admissible manifold $M$ is a torus $T^k$ which acts injectively and with $k \leq$ the rank of the center of $\pi_1(M)$. If the center of $\pi_1(M)$ is finite, then the finite $G$ injects into the outer automorphism group of $\pi_1(M)$, $\mathrm{Out}(\pi_1(M))$. These results lead to constructions of closed manifolds that admit no effective action of any finite group.

In Chapter 4, a formal definition of a Seifert fibering is given. It is motivated by some new examples and the actions studied in Chapter 3. Let $P$ be a principal $G$-bundle over a space $W$. Let $\Pi$ act properly on $P$ and normalize the left principal $G$-action $\ell(G)$ on $P$. Then there exists a commutative diagram

$$\begin{array}{ccc} (G, \Pi, P) & \xrightarrow{G\backslash} & (Q, W) \\ \downarrow{\Pi\backslash} & & \downarrow{Q\backslash} \\ X = \Pi\backslash P & \xrightarrow{\tau} & B = Q\backslash W \end{array} \quad \text{where } Q = \Pi/\Pi \cap \ell(G), \; W = G\backslash P.$$

Assume the induced action of $Q$ on $W$ is proper. The map $\tau$, induced by the three other orbit mappings, is by definition a Seifert fibering modeled on the principal $G$-bundle $P$. In our definition, $G$ and $\Pi$ are Lie groups with $G$ and $P$ usually connected. The inverse image $\tau^{-1}(b)$, $b \in B$, is called a fiber. If $\tau^{-1}(b)$ is the homogeneous space $\Gamma\backslash G$, where $\Gamma = \Pi \cap \ell(G)$, it is called a typical fiber; otherwise, it is called a singular fiber. The singular fibers turn out to be quotients of the typical fiber by the action of a compact group of affine diffeomorphism of $\Gamma\backslash G$. More precisely, a singular fiber is a quotient of $G$ by a group $\Gamma' \subset \mathrm{Aff}(G) = \ell(G) \rtimes \mathrm{Aut}(G)$, where $\Gamma' \cap \ell(G) = \Gamma$, and $\mathrm{Aut}(G)$ is the group of continuous automorphisms of $G$.

Let $\mathrm{TOP}_G(P)$ be the normalizer of $\ell(G)$ in $\mathrm{TOP}(P)$, the group of homeomorphisms of $P$ with the compact-open topology. The former is also the same as both the weak bundle automorphisms of $P$ and the weak $\ell(G)$-equivalences of $P$. Thus, a Seifert mapping $\tau$ arises from an embedding of a group $\Pi$ in $\mathrm{TOP}_G(P)$. The imposed condition of properness excludes pathological situations.

One goal of the later chapters is to emulate fiber bundle theory and determine the various spaces $X$ which fiber over a fixed $B$ with typical fiber $\Gamma\backslash G$. Obviously, we need to find all the $\Pi$ that embed into $\mathrm{TOP}_G(P)$ satisfying the properness conditions. In particular, each $\Pi$ will be a (topological) extension of the Lie group $\Gamma$ by $Q$.

Locally injective and injective actions on $X$ and their orbit mappings $\tau$ give rise to a large and important class of Seifert fiberings. In most cases, $Q$ is discrete.

If $\mathcal{U}$ is a geometrically interesting subgroup of $\mathrm{TOP}_G(P)$ (for example, $\mathcal{U} = \mathrm{Isom}(P)$, the group of isometries of $P$) and $\Pi \subset \mathrm{TOP}_G(P)$, then determining when $\Pi$ can be mapped into $\mathcal{U}$, or at least can be deformed into $\mathcal{U}$, is another important goal investigated in later chapters.

Chapter 5 fills in what is needed from the cohomology of groups. The emphasis here is on the low dimensional cohomology and cohomology sets for discrete groups with non-Abelian coefficients.

Chapter 6 discusses facts needed from Lie group theory, especially for nilpotent, completely solvable, and semisimple Lie groups.

Chapter 7 answers the questions concerning the existence and construction of Seifert fiberings. Treated also is the uniqueness and rigidity of the construction for simply connected Abelian, nilpotent, and completely solvable $G$ as well as for semisimple $G$ in adjoint form. They possess the property that an isomorphism of a lattice $\Gamma$ in a $G$ with another lattice $\Gamma'$ in $G'$ can be uniquely extended to an isomorphism of $G$ into $G'$. This is called the ULIEP (Unique Lattice Isomorphism Extension Property).

The theorems of this chapter are crucial for later chapters; Chapters 8, 9, 11, 13, and 14.

In Chapter 8, we investigate the significance and geometric meaning of a very important special case of the existence, uniqueness, and rigidity theorems of Chapter 7. For example, if $P = \mathbb{R}^n \times$ point, then $\mathrm{TOP}_{\mathbb{R}^n}(P) = \ell(\mathbb{R}^n) \rtimes \mathrm{Aut}(\mathbb{R}^n) = \mathbb{R}^n \rtimes \mathrm{GL}(n,\mathbb{R}) = \mathrm{Aff}(\mathbb{R}^n)$, the group of affine diffeomorphisms of Euclidean affine space. Now let $\Pi$ be any extension $1 \to \mathbb{Z}^n \to \Pi \to Q \to 1$, where $Q$ is a finite group and the action of $Q$ on $\mathbb{Z}^n$, induced by conjugation by elements of $\Pi$, is faithful. Then, and only then, does there exist an injection $\theta : \Pi \to \mathrm{Aff}(\mathbb{R}^n)$ such that $\theta(\Pi) \cap \ell(\mathbb{R}^n) = \mathbb{Z}^n$. This, unless $\Pi$ is $\mathbb{Z}^n$, results in a Seifert fibering $\tau : \theta(\Pi)\backslash\mathbb{R}^n \to$ point, where the only fiber is a singular fiber. The typical fiber $\theta(\mathbb{Z}^n)\backslash\mathbb{R}^n$, an $n$-torus, only appears as a covering of the singular fiber.

Because $Q$ is finite and is mapped injectively into $\mathrm{GL}(n,\mathbb{R})$, the uniqueness theorem of Chapter 7 implies that $\theta(\Pi)$ can be conjugated in $\mathrm{Aff}(\mathbb{R}^n)$ so that the image $\theta'(\Pi)$ now lies in $E(n) = \ell(\mathbb{R}^n) \rtimes \mathrm{O}(n,\mathbb{R})$. Since $E(n)$ is the full group of isometries of Euclidean space, $\theta'(\Pi)$ is a Euclidean crystallographic group, and $\theta'(\Pi)\backslash\mathbb{R}^n$ is a Euclidean crystal. If $\Pi$ is torsion free, $\theta'(\Pi)\backslash\mathbb{R}^n$ is a compact flat Riemannian manifold. The so-called first Bieberbach theorem states that if $\Delta$ is a discrete subgroup of $E(n)$ with $\Delta\backslash E(n)$ compact, then $\Delta \cap \ell(\mathbb{R}^n) \cong \mathbb{Z}^n$, a lattice of $\ell(\mathbb{R}^n)$, with $\mathbb{Z}^n$ having finite index in $\Delta$. Thus $\Delta$ must be one of the crystallographic groups $\theta'(\Pi)$ constructed above. Now the uniqueness and rigidity theorem of Chapter 7 asserts if $\theta'(\Pi)$ and $\theta''(\Pi)$ are two embeddings in $E(n)$, then they are conjugate in $\mathrm{Aff}(\mathbb{R}^n)$. Consequently, if $\Pi$ is torsion free, then the flat Riemannian manifold $\theta'(\Pi)\backslash\mathbb{R}^n$ is diffeomorphic to $\theta''(\Pi)\backslash\mathbb{R}^n$ by an affine diffeomorphism. This becomes the content of the second theorem of Bieberbach. (Furthermore, the classification of all crystallographic groups up to conjugacy in $\mathrm{Aff}(\mathbb{R}^n)$ can now be determined by cohomological means.)

These results of Bieberbach are also extended, in Chapter 8, to simply connected nilpotent Lie groups and certain completely solvable Lie groups as well as to the corresponding infra-nilmanifolds and infra-solvmanifolds resulting from the construction.

Chapter 9 extends the definition of a Seifert fibering to one modeled on the product fiber bundle $G/K \times W$ over $W$. The group $K$ is a closed subgroup of the Lie group $G$, the typical fiber is the double coset space $\Gamma\backslash G/K$, and singular fibers are finite quotients of the typical fiber. When $G$ has finitely many connected components and $K$ is a maximal compact subgroup, then $G/K$ is diffeomorphic to $\mathbb{R}^n$. In this circumstance, if $\Gamma$ is a torsion-free lattice in $G$ with $\Gamma\backslash G$ compact, then the double coset space $\Gamma\backslash G/K$ is a closed aspherical manifold. Especially interesting are the Riemannian symmetric spaces of noncompact type $G/K$ and the corresponding locally symmetric spaces $\Gamma\backslash G/K$. The existence, uniqueness, and rigidity theorems have analogues for these new types of fiberings.

Chapter 10 turns to Seifert fiberings modeled on nontrivial principal $G$-bundles $P$ over $W$ with $G$ a $k$-dimensional torus. This chapter is independent of the previous three chapters and also recaptures the results of Chapter 7 for $G = \mathbb{R}^k$. Let $\mathrm{M}(W, T^k)$ denote the space of continuous maps from $W$ into $T^k$. It is an Abelian group. The structure of $\mathrm{TOP}_{T^k}(P)$ (weakly $T^k$-equivariant homeomorphisms of $P$) is given by the exact sequence

$$1 \longrightarrow \mathrm{M}(W, T^k) \xrightarrow{\psi} \mathrm{TOP}_{T^k}(P) \xrightarrow{j} \mathrm{Aut}(T^k) \times \mathrm{TOP}(W).$$

To construct a Seifert fibering modeled on the principal fibering $T^k \to P \to W$, we begin with a proper action $\rho : Q \to \mathrm{TOP}(W)$ with $Q$ discrete, and a homomorphism $\varphi : Q \to \mathrm{Aut}(T^k)$. We seek extensions $1 \to F \to \Pi \to Q \to 1$, and injective homomorphisms $\theta : \Pi \to \mathrm{TOP}_{T^k}(P)$ such that $\theta(\Pi) \cap \ell(G) = F \subset \mathrm{M}(W, T^k)$ and $(\varphi \times \rho)(Q) \subset \mathrm{Aut}(T^k) \times \mathrm{TOP}(W)$. To ensure $(\varphi \times \rho)(Q) \subset \mathrm{Im}(j)$, we show this holds if and only if $P$ is invariant under the action of $Q$. This translates, in cohomological terms, to $[P] \in H^2(W; \mathbb{Z}^k)^Q$, where $[P]$ is the cohomology class representing the principal $T^k$-bundle $P$ over $W$.

The Borel space, $EQ \times_Q W = W_Q$, associated to the $Q$-action on $W$, plays an important role. For example, the $Q$-action on $W$ lifts to a group of weak bundle automorphisms of $P$ if and only if the bundle $P$ is the pullback of a $T^k$-bundle $\widetilde{P}$ over $W_Q$ via the inclusion $W \xrightarrow{i} EQ \times W \xrightarrow{\pi} EQ \times_Q W$. If $W$ is simply connected, we obtain an exact sequence

$$0 \to H^2(Q; \mathbb{Z}^k) \xrightarrow{e_1} H^2(W_Q; \mathbb{Z}^k) \xrightarrow{e_2} H^2(W; \mathbb{Z}^k)^Q$$
$$\xrightarrow{\delta} H^2(Q; \mathrm{M}(W, T^k)) \to H^3(W_Q; \mathbb{Z}^k).$$

The classification of the Seifert fiberings $\tau : X \to B$ reduces to an analysis of the terms of this exact sequence. For $[P] \in H^2(W; \mathbb{Z}^k)^Q$, $\delta[P]$ represents an extension of $\mathrm{M}(W, T^k)$ by $Q$. This extension, $E(P, Q)$, is the group of all weak bundle automorphisms of $P$ that project onto the image of $Q$ in $\mathrm{Aut}(T^k) \times \mathrm{TOP}(W)$. The group $\Pi$ must map into $E(P, Q)$ before it can map into $\mathrm{TOP}_{T^k}(P)$. The group $E(P, Q)$ splits (that is, the $Q$-action on $W$ lifts to a group of weak bundle automorphisms on $P$) if and only if $\delta[P] = 0$. If $e_2[\widetilde{P}] = P$, then $P$ is the pullback of the bundle $\widetilde{P}$ over the Borel space. The cohomology $H^*(W_Q; \mathbb{Z}^k)$ is called the $Q$-equivariant cohomology of $W$, often written as $H^*_Q(W; \mathbb{Z}^k)$. The elements of $H^2(Q; \mathbb{Z}^k)$ then classify all the distinct $\theta : Q \to \mathrm{TOP}_{T^k}(P)$, for a fixed $(Q, W)$ that lifts to $P$.

In Chapter 11, a large group of applications capitalizing on the theorems of Chapter 7 are presented.

• We use the Seifert Construction to create a wide class of closed aspherical manifolds and the theorems of Chapter 7 to topologically classify some of them. The rigidity of some Seifert fiberings is used to topologically classify Seifert fiberings.

• We show how homotopy and algebraic data lead to fiber preserving group actions on Seifert manifolds.

• A torsion-free polycyclic-by-finite group $\Gamma$ sometimes fails to be the fundamental group of a complete affinely flat manifold. However, by using an iteration of the

Seifert fiber construction, it is shown that $\Gamma$ is the fundamental group of a compact solvmanifold with a polynomial structure that generalizes a complete affinely flat structure.

- A generalization of the second Bieberbach theorem for nilpotent groups in Chapter 8 is proved and used to show that the Nielsen number equals the Lefschetz number for homotopically periodic self-homeomorphisms on infra-solvmanifolds.

- A torus action $(T^k, X)$ is homologically injective if the evaluation homomorphism $\mathrm{ev}_*^x : H_1(T^k, \mathbb{Z}) \to H_1(X; \mathbb{Z})$ is injective. We show this type of torus action can be written as $(T^k, T^k \times_\Delta Y)$. That is, $X$ is finitely covered by the product $T^k \times Y$, the left translation of $T^k$ on $T^k \times Y$ descends to $X$ via the commuting covering transformations. The finite Abelian covering group $\Delta$ acts freely as translations on $T^k$ while acting diagonally on the product. These splittings, which are not necessarily unique, are classified. With the appropriate definition of homologically injective Seifert fiberings, a similar splitting theorem for completely solvable $G$ is also obtained.

- An effective torus action $(T^k, M)$ on a closed aspherical manifold $M$ with center of $\pi_1(M)$ isomorphic to $\mathbb{Z}^k$ is called a maximal torus action of $M$. Smooth maximal torus actions are shown to exist for infra-nilmanifolds as well as for many other Seifert fiberings. For an infra-nilmanifold, the connected component of the affine diffeomorphisms contains a maximal torus action. In particular, for a compact flat Riemannian manifold, a maximal torus action is the connected component of the full isometry group of $M$.

- This Chapter concludes with a determination of the dimension of the largest torus that acts effectively on most spherical space forms.

Chapter 12 investigates the Seifert Construction for connected $Q$. If the group $Q$ acting on $W$ is a *compact connected* Lie group, then the Seifert Construction to produce Seifert fiberings $\tau : X \to Q \backslash W = B$ modeled on principal $T^k$-bundles $P$ over $W$ requires that we classify the liftings of $Q$ to groups of bundle automorphisms of $P$. While requiring additional techniques than those used so far for discrete $Q$, the reader will find that the first two sections of Chapter 10 provides a good introduction to what must be overcome to accomplish this classification. This classification with applications is presented in Chapter 12. Some of these results closely resemble those obtained in Chapter 10. Chapter 12, however, is formally independent of the other chapters and can be studied independently of the other chapters.

Chapter 13 studies deformation spaces for 3-dimensional Seifert spaces. There are three classical 2-dimensional Riemannian geometries: spherical, Euclidean, and hyperbolic. For each closed 2-manifold $M$, one can impose metrics so that the metric universal covering of $M$ is one of these classical 2-dimensional geometries, the standard sphere, the Euclidean plane, or the hyperbolic plane. That is, one can embed the fundamental group $\Pi$ of $M$ into the full isometry group of the sphere, the Euclidean plane, or the hyperbolic plane so that $\Pi$ acts as covering transformations to get a 2-dimensional geometric structure on $M$.

The Thurston geometrization conjecture, now a theorem due to Perelman, states that a closed 3-manifold can be split into pieces such that each piece has a 3-dimensional geometric structure. There are eight so-called 3-dimensional geometries to consider.

As it turns out, each closed Seifert 3-manifold admits exactly one of six of the eight Riemannian geometries.

Any closed Seifert 3-manifold $M$ not covered by the 3-sphere has a finite covering $M'$ which is a principal circle bundle over a closed surface $B'$. Since $B'$ admits a 2-dimensional geometric structure, one expects $M'$ will admit either a product geometric structure or a 1-dimensional structure coming from the fiber twisted by a 2-dimensional structure coming from the base.

We classify up to isometry the Seifert 3-manifolds for the two twisted geometries. For the Seifert manifold $M$ with $B$ a hyperbolic orbifold and with twisted geometry, the moduli space of geometric structures on $M$ up to isometry is itself a Seifert fibering over the moduli space of the hyperbolic orbifold $B = Q\backslash W$ with typical fiber a torus $T^{2g}$ where $g$ is the genus of $B$. We will use the term orbifold to simply mean the orbit space of a locally proper action of a discrete group on a space $X$. This usage does not conform with the accepted meaning and usage of the term "orbifold" as, say in [**Thu97**], but when we use the term in our discussions concerning manifolds, the two meanings will usually coincide.

It also turns out that the manifolds with the most interesting of these geometries also has a complete Lorentz structure whose classification is very similar to the Riemannian case.

Chapter 14 begins the analysis of closed 3-dimensional manifolds that admit an effective $S^1$-action. The classification is given up to $S^1$-equivalence and for those with infinite $\pi_1(M)$ up to topological equivalence. For the latter case, the rigidity result of Chapter 7 is employed.

When the $S^1$-action has no fixed points and the fundamental group is not finite Abelian, the orbit mapping is always a Seifert fibering modeled on a principal $\mathbb{R}^1$ or $S^1$-bundle over $W = S^2$ or $\mathbb{R}^2$. Any 3-dimensional Seifert fibering always has a two-fold Seifert covering which is the orbit mapping of an $S^1$-action. A large number of algebraic, topological and geometric properties for the orientable Seifert fiberings over an orientable base are derived. A short excursion into holomorphic Seifert fiberings is also given. The chapter concludes with the classification of all the 3-dimensional Seifert fiberings up to fiber preserving homeomorphism in the spirit of Seifert's original methods as well as the topological classification using the rigidity theorem of Chapter 7. Actually, Chapter 14 can be read independently of Chapter 7 provided that one just accepts the rigidity result of Chapter 7 for $G = \mathbb{R}^1$.

Chapter 15 is a continuation of Chapter 10 in that we use the methods and results of Chapter 10 to classify the 3-dimensional Seifert fiberings purely in terms of the equivariant cohomology $H^2(W_Q; \mathbb{Z})$, where $W = S^2$ or $\mathbb{R}^2$. When $W = \mathbb{R}^2$, $H^2(Q; \mathbb{Z})$ is isomorphic to $H^2(W_Q; \mathbb{Z})$. However, the spherical space form case ($W = S^2$) turns out to be especially interesting, and we give a full classification of the 3-dimensional space forms up to topological equivalence by this method. These results and methods provide an alternative to those employed in Chapter 14.

If interest is mainly in later chapters, then the introductory Chapters 1, 2, 3 can be quickly reviewed and Chapter 4 should be understood on a conceptual level—not all the proofs being necessary. The examples of Chapter 7 should be studied and the statement of Theorem 7.3.2 clearly understood. The proofs in Chapter 7 are not needed for the applications. An abbreviated version of some of the topics in this book appeared in [**LR02**].

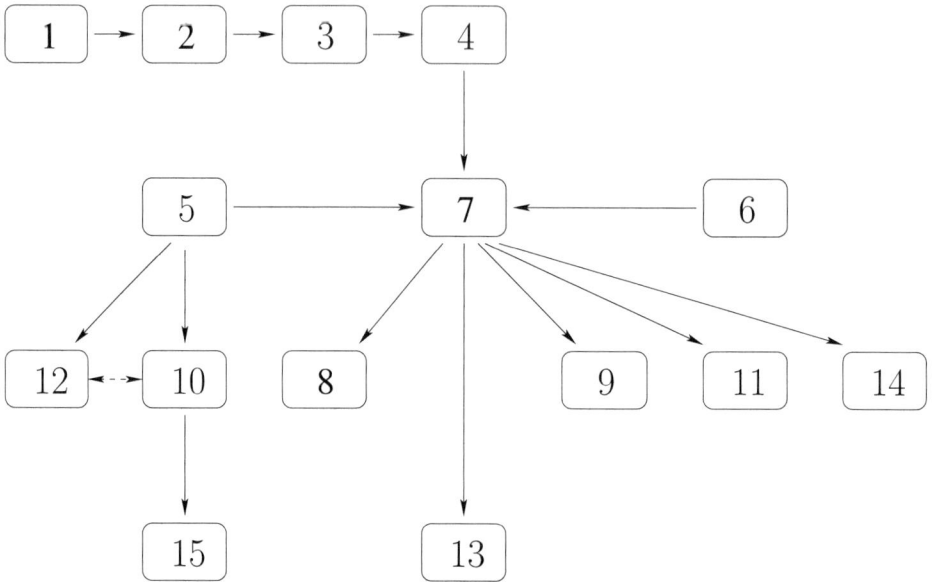

Flow diagram in chapters

CHAPTER 1

# Transformation Groups

This is an introductory chapter. It is used to set notation and recount basic ideas of group actions on general spaces as well as related material on fiber bundles, classifying spaces, and the Borel construction. Locally compact Lie groups are significant players in Seifert fiberings, and the notion of proper $G$-spaces plays an important role. So a background on proper actions is included leading to Palais' proof of the existence of a slice for proper $G$-spaces. The final section sets the stage for how the preliminary work will be applied in the Seifert Constructions.

## 1.1. Introduction

1.1.1. Our spaces, in general, will be path-connected, completely regular and Hausdorff. When using covering space theory, we shall also assume our spaces locally path-connected and semilocally simply connected (that is, every point $x$ in $X$ has a neighborhood $U$ such that the homomorphism from the fundamental group of $U$ to the fundamental group of $X$, induced by the inclusion map of $U$ into $X$, is trivial).

NOTATION 1.1.2. Let $G$ be a Lie group, and let $K$ be a subgroup of $G$.

$$
\begin{aligned}
N_G(K) &= \text{Normalizer of } K \text{ in } G \\
C_G(K) &= \text{Centralizer of } K \text{ in } G \\
\text{Aut}(G) &= \text{The group of continuous automorphisms of } G \\
\text{Inn}(G) &= \text{The group of inner automorphisms of } G \\
\text{Aut}(G,K) &= \{\alpha \in \text{Aut}(G) : \alpha|_K \in \text{Aut}(K)\} \\
\text{Inn}(G,K) &= \text{Inn}(G) \cap \text{Aut}(G,K) \\
\text{Out}(G,K) &= \text{Aut}(G,K)/\text{Inn}(G,K) \\
\mu(a) &= \text{Conjugation by } a; \text{ so } \mu(a)(x) = axa^{-1} \text{ for } x \in G
\end{aligned}
$$

1.1.3. Let $\mathcal{P}$ be a property on a group. We say a group is *virtually* $\mathcal{P}$ if it contains a normal subgroup of finite index which is $\mathcal{P}$. For example, a crystallographic group is virtually free Abelian; see Theorem 8.1.2.

1.1.4. A *left action* of a topological group $G$ on a topological space $X$ is a continuous function

$$\varphi : G \times X \longrightarrow X$$

such that
  (i) $\varphi(gh, x) = \varphi(g, \varphi(h, x))$ for all $g, h \in G$ and $x \in X$, and
  (ii) $\varphi(1, x) = x$, for all $x \in X$, where 1 is the identity element of $G$.

The point $\varphi(g, x) \in X$ is usually written simply as $gx$, $g(x)$, or sometimes $g \cdot x$. Clearly, each element $g \in G$ can be viewed as a homeomorphism of $X$ onto itself. We may denote this action by $(G, X, \varphi)$, or more simply suppress the $\varphi$, $(G, X)$, and call $X$ a $G$-space. If $X$ and $Y$ are $G$-spaces, then a $G$-map is a continuous function

$f: X \to Y$ which is *equivariant*; i.e., $f(gx) = gf(x)$ for all $g \in G$ and $x \in X$. If $f$ is a $G$-map and a homeomorphism, then $f$ is called a *G-equivalence* or *G-isomorphism* (in the relevant category). A map $f: X \to Y$ is *weakly G-equivariant* if there exists a continuous automorphism $\alpha_f$ of $G$ such that $f(gx) = \alpha_f(g)f(x)$, for all $g \in G$, $x \in X$. If $f$ is a homeomorphism, then $f$ is a *weak G-equivalence* or a *weak G-isomorphism*. Similarly, if there is a homeomorphism $f$, we say that $(G, X)$ and $(G, Y)$ are $G$-equivalent (respectively, are weakly $G$-equivalent). When the context is clearly understood, we can drop the "$G$".

There is an analogous notion of a *right action*,

$$\psi: X \times G \longrightarrow X$$

which we denote by $\psi(x, g) = xg$ or $x \cdot g$. Then $\psi(x, gh) = (xg)h = xgh$. Any right $G$-action $\psi(x, g)$ can be converted to a left $G$-action $\varphi(g, x)$ by $\varphi(g, x) = \psi(x, g^{-1})$ and vice versa. Note that, for a reasonably nice space $X$ (e.g., locally compact Hausdorff which is either connected or locally connected), the set of self-homeomorphisms of $X$ becomes a topological group $\text{TOP}(X)$ with respect to the compact-open topology on $\text{TOP}(X)$ (see Subsection 1.2.6), and a left $G$-action is equivalent to having a group homomorphism $G \to \text{TOP}(X)$. A right $G$-action becomes an antihomomorphism. We will always assume we have a left action unless we specify otherwise.

An action $(G, X, \varphi)$ is called a *smooth action* if $G$ is a Lie group, $X$ is a smooth manifold, and the function $\varphi$ is smooth.

1.1.5. If $X$ is a $G$-space and $x \in X$, then

$$Gx = \{y \in X : y = gx \text{ for some } g \in G\}$$

is called the *orbit of $G$ through $x$*. It can be denoted by $Gx$, $G \cdot x$, or $G(x)$.

The collection of all orbits of $X$ forms a partition of $X$ into disjoint sets. The collection of orbits with the quotient topology (identification topology) is called the *orbit space* of the $G$-action on $X$, and is denoted by $G \backslash X$ (for a right action, $X/G$ is used). The quotient map $\nu: X \to G \backslash X$ is called the *orbit map*. It is an open mapping, for if $U$ is open in $X$, then $\nu^{-1}(\nu(U)) = \bigcup_{g \in G} gU$, and each $gU$ is open because $g$ is a homeomorphism. Thus $\nu(U)$ is open in the quotient topology.

A map $f: X \to Y$ is called a *proper mapping* if the preimage of a compact set is compact. For a general $G$-space, the orbits may fail to be closed subsets of $X$, and consequently, $G \backslash X$ would not be $T_1$ (i.e., points in $G \backslash X$ may not be closed). However, when $G$ is compact and $X$ is Hausdorff, we have:

THEOREM 1.1.6 ([**Bre72**, 3.1]). *If $X$ is a Hausdorff $G$-space with $G$ compact, then*

(1) *$G \backslash X$ is Hausdorff,*
(2) *the orbit map $X \to G \backslash X$ is a closed mapping,*
(3) *the orbit map $X \to G \backslash X$ is a proper mapping,*
(4) *$X$ is compact if and only if $G \backslash X$ is compact,*
(5) *$X$ is locally compact if and only if $G \backslash X$ is locally compact.*

These facts are easy to prove. To obtain similar properties when $G$ is a non-compact Lie group, we must impose the notion of a proper action (see Section 1.2 below).

## 1.1.7. Let
$$G_x = \{g \in G : gx = x\}.$$
This subgroup of $G$ is a closed subgroup of $G$ if points of $X$ are closed (i.e., $X$ is $T_1$). It is called the *isotropy* or *isotropy subgroup* of $G$ at $x$. The set
$$X^G = \{x \in X : gx = x \text{ for all } g \in G\}$$
is called the *fixed-point set* of the action of $G$ on $X$. It is a *closed subset of $X$ if $X$ is Hausdorff*. We sometimes write $X^G$ as $\operatorname{Fix}(G, X)$. Clearly,
$$G_{gx} = gG_x g^{-1}$$
so that, if $y \in Gx$, then $G_y = gG_x g^{-1}$ for any $g$ such that $gx = y$. By $(G_x)$ we mean the conjugates of $G_x$. This set of conjugates is called the *orbit type* of the orbit $Gx$.

If $K$ is a normal subgroup of $G$ and $X$ is a $G$-space, then there is induced a natural action of $G/K$ on $K\backslash X$, and the orbit mapping $\nu : X \to G\backslash X$ factors through $X \xrightarrow{K\backslash} K\backslash X \xrightarrow{G/K\backslash} G\backslash X$. The induced natural action
$$G/K \times K\backslash X \longrightarrow K\backslash X$$
is given by $gK \cdot (Kx) = g(Kx)$. (Strictly speaking, if $K$ is not a closed subgroup of $G$, then the topological group $G/K$ is not a Hausdorff topological group. It is often assumed that a topological group is Hausdorff which will automatically imply that it is completely regular. Of course, our main concern is with Lie groups and their closed subgroups).

If there exists a subgroup $K$ of $G$, $K \neq 1$, such that $X^K = X$, we say that $G$ acts *ineffectively*; otherwise, $G$ acts *effectively*. Therefore, the action is effective if and only if $\bigcap_{x \in X} G_x = 1$. We call the largest subgroup $K$ of $G$ that fixes all of $X$, the *ineffective part* of the action. It is a closed (assuming $X$ is $T_1$) normal subgroup, and $G/K$ acts effectively on $X$.

## 1.1.8. 
An action of $G$ on $X$ is *transitive* if the orbit through some point of $X$ consists of all of $X$. That is to say, given any $x$ and $y$ in $X$, there exists $g \in G$, such that $gx = y$. Such an action is *simply transitive* if such $g$ is unique. If $G$ acts transitively on $X$, then the function $G/G_x \to G(x) \hookrightarrow X$ given by $gG_x \to gx$ is *onto*. Clearly the map is one-to-one. Moreover, the map is continuous. This is an immediate consequence of the universal properties of quotient mappings. *If $G/G_x$ is compact and $X$ is Hausdorff, then the mapping is a homeomorphism.* In general, however, the inverse mapping may fail to be continuous. See Subsection 1.2.4 for a condition guaranteeing continuity of the inverse.

EXAMPLE 1.1.9. Let $G(\cong \mathbb{R})$ be a linear subspace of $\mathbb{R}^2$ which consists of the points on a line through the origin with irrational slope. Reducing the coordinates in $\mathbb{R}^2$ modulo 1 induces the standard covering projection,
$$p : \mathbb{R}^2 \longrightarrow \mathbb{Z}^2 \backslash \mathbb{R}^2 = T^2,$$
of $\mathbb{R}^2$ onto the 2-torus $T^2$. This is a homomorphism of the additive group $\mathbb{R}^2$ onto $T^2$ with kernel the standard integral lattice subgroup $\mathbb{Z}^2$. Since $G \cap \mathbb{Z}^2 = \{0\}$ is the trivial group, $G$ descends to $T^2$ as a one-to-one continuous homomorphism onto its image. Let $X = p(G)$ with the relative topology of $T^2$. Then $G$ acts on $G$ by left translations, and this descends to a transitive $G$-action on $X$, with the isotropy $G_0 = 0$. Now note that this topology on $X$ is strictly weaker than the

topology of the original $G$. Consequently, the natural map $G/G_x \to G(x)$ is not a homeomorphism.

EXERCISE 1.1.10. (a) Show that the two $\mathbb{Z}_5 = \langle \lambda \rangle$ actions on the unit disk $D$ given by $(\lambda, z) \mapsto e^{\frac{2\pi}{5}i}z$ and $(\lambda, z) \mapsto e^{\frac{4\pi}{5}i}z$ are not $G$-equivalent but are weakly $G$-equivalent.

(b) Show that there are an infinite number of different $G$-maps between the two actions in (a).

(c) Show that $(\lambda, z) \mapsto e^{\frac{2\pi}{5}i}z$ and $(\lambda, z) \mapsto e^{\frac{8\pi}{5}i}z$ are $G$-equivalent.

(d) Show that if $f : (G, X) \to (G, Y)$ is a weak $G$-equivalence, then $f$ induces a homeomorphism $G\backslash X \to G\backslash Y$ which sends orbits of type $(H)$ to orbits of type $(\alpha_f(H))$.

DEFINITION 1.1.11. $G$ acts *freely* on $X$ if $G_x = 1$ for all $x \in X$.

Examples of free actions are groups of covering transformations, and the left translations in a principal $G$-bundle. See Subsection 1.3.5 for a definition. However, free actions are more general than the left translations in a principal $G$-bundle. In Example 1.1.9, $\mathbb{R}$ acts freely on $p(G)$ but $p(G) \to \mathbb{R}\backslash p(G)$ (a point) is not a principal $\mathbb{R}$-bundle projection. The reason is that the action fails to be *locally proper* (see the next section). See Example 1.7.7 for another such occurrence.

DEFINITION 1.1.12. An action $(G, X)$ is *semifree* if the action is free away from a nonempty fixed-point set. The two actions in Exercise 1.1.10(a) are semifree, because they are free on $D - \{0\}$.

## 1.2. (Locally) Proper $G$-spaces

1.2.1. We shall be largely dealing with actions of Lie groups on spaces. Because our group may not be compact, we need to recall the notion of a *proper action* of a locally compact topological group on a topological space. Compact Lie group actions tend to leave important geometrical structures of spaces invariant whereas noncompact Lie groups often do not.

Properness is the concept that enables properties of the actions of noncompact Lie groups to resemble those of compact groups. There are good sources for properties of proper actions; e.g., R. Palais [**Pal61**], R. Kulkarni [**Kul81**]. However, we caution the reader that there is no uniformity in terminology for this concept. We shall, for our convenience, recall what we shall need and refer to Palais for some of the proofs.

In Sections 1.2 through 1.6, with the exception of 1.3, *$G$ is a locally compact topological group and $X$ is a completely regular Hausdorff space* and neighborhoods will be open sets unless specified or commented differently. The point of complete regularity is to help ensure that $G\backslash X$ will have nice separation properties, and this coupled with the notion of (local) *properness* enables us to have a *slice theorem*, (cf. Subsection 1.7.1).

DEFINITION 1.2.2. An action of $G$ on $X$ is called *locally proper* if for each $x \in X$, there exists a neighborhood $U$ of $x$ such that

$$\{g \in G : gU \cap U \neq \emptyset\}$$

has compact closure. In particular, $G_x$, being a closed subset of the above set (since $X$ is $T_1$), is compact . If $G$ is discrete, the above set is finite. (In Palais [**Pal61**],

a locally proper $G$-space is called a *Cartan $G$-space*). The action is called *proper* if for each $x$, there exists a neighborhood $U$ of $x$ such that, for each $y \in X$, there exists a neighborhood $V$ of $y$ for which the closure of
$$\{g \in G : gV \cap U \neq \emptyset\}$$
is compact. Observe that if closure of $\{g \in G : gV \cap U \neq \emptyset\}$ is compact, then closure of $\{g \in G : V \cap gU \neq \emptyset\}$ is compact: for, if $C$ is compact, then $C^{-1}$ is compact also.

1.2.3. The following properties of locally proper and proper actions are proved in Palais [**Pal61**]. The proofs are easier for compact or discrete $G$. The use of nets can be avoided if one assumes that $X$ is first countable. We will use these properties mostly for compact or discrete $G$ and some noncompact connected Lie groups. We suggest that the reader furnish his/her own proofs for discrete $G$.

PROPOSITION 1.2.4. *For locally proper actions $(G, X)$, we have:*
(1) *Each orbit is closed in $X$. Hence, $G\backslash X$ is $T_1$. In fact, $G\backslash X$ is locally completely regular. However, $G\backslash X$ may fail to be Hausdorff.*
(2) *The map $gG_x \to gx$ is a homeomorphism of $G/G_x$ onto $Gx$.*
(3) *If $K$ is the ineffective part of $G$, then $G/K$ acts locally properly. (Similarly, for proper actions.)*
(4) *If an $x$ has a neighborhood $U$ such that $\{g \in G : U \cap gU \neq \emptyset\}$ is finite, then $Gx$ is discrete in $X$, and in fact, $G$ itself is discrete.*
(5) *$(G, X)$ is proper if and only if $G\backslash X$ is regular. In fact, $G\backslash X$ is completely regular when $(G, X)$ is proper.*

Here are some additional facts for $G$-spaces.
(6) *If $X$ and $Y$ are $G$-spaces, and if $X$ is a locally proper (respectively, proper) $G$-space, then so is $X \times Y$.*
(7) *Let $Y$ be a locally proper (respectively, proper) $G$-space and $X$ a (respectively, completely regular) space. If $f : X \longrightarrow G\backslash Y$ is a map, then the pullback $\widetilde{f} : \widetilde{X} \longrightarrow Y$ (see Subsection 1.3.7) is a locally proper (respectively, proper) $G$-space whose orbit space is $X$.*
(8) *If $X$ is a locally proper (respectively, proper) $G$-space, $H$ a closed subgroup of $G$, and $Y$ an $H$-invariant subspace of $X$, then $Y$ is locally proper (respectively, proper) $H$-space. (Note, the Example 1.1.9 shows that $H$ must be closed, for we may take $H = p(G) \subset T^2$).*
(9) *If $X$ is a locally compact $G$-space, the following are equivalent:*
  (a) *$(G, X)$ is locally proper and $G\backslash X$ is Hausdorff.*
  (b) *$(G, X)$ is proper.*
  (c) *For each compact subset $C$ of $X$, the closure of $\{g \in G : gC \cap C \neq \emptyset\}$ is compact.*
(10) *Let $X$ be a proper $G$-space and $H$ a closed normal subgroup of $G$. Then $H\backslash X$ is a proper $G/H$-space.*

**Note:** All citations are from Palais' paper [**Pal61**]: (1) follows from (1.1.4) and Corollary 2 of (1.2.8); (2) is (1.1.5); (3) is (1.1.6); for (5), (1.2.5) says that locally proper and $G\backslash X$ completely regular imply properness, and (1.2.8) states that properness implies $G\backslash X$ regular; (6a) is (1.3.3), (6b) is (1.3.4); and (6c) is (1.3.1); and (7) is (1.2.9). The item (7) above is a common criterion for proper action. We state it as a corollary.

COROLLARY 1.2.5. *Let $X$ be a completely regular, locally compact Hausdorff space. A $G$-action on $X$ is proper if and only if, for each compact subset $C$ of $X$, the closure of $\{g \in G : gC \cap C \neq \emptyset\}$ is compact.*

REMARK 1.2.6. Even if $(G, X, \varphi)$ is not necessarily a proper $G$-space, there is a natural homomorphism $\widetilde{\varphi} : G \to \mathrm{TOP}(X)$, where $\mathrm{TOP}(X)$ is the group of all self-homeomorphisms of $X$. We may topologize $\mathrm{TOP}(X)$ so that $\widetilde{\varphi}$ becomes continuous if we choose to do so. For example, $\widetilde{\varphi}$ will be continuous if we take the compact-open topology on $\mathrm{TOP}(X)$ (see [**Hu59**, §9.4, p.75]) and, consequently, also continuous if we take the smaller point-open topology (i.e., the topology of pointwise convergence). We shall often denote $\widetilde{\varphi}$ by $\rho : G \to \mathrm{TOP}(X)$, thinking of $\rho$ as a representation.

Under the compact-open topology, $\mathrm{TOP}(X)$ is almost a topological group but fails only in that inversion may not be continuous. However, if $X$ is assumed to be locally compact Hausdorff and either connected or locally connected, then $\mathrm{TOP}(X)$ under the compact-open topology is a topological group. (See [**Die48**] for $X$ connected, and [**Are46**] for $X$ locally connected.) If and when the topology on $\mathrm{TOP}(X)$ becomes an issue, we shall be explicit about it.

PROPOSITION 1.2.7 ([**Pal61**, (1.1.7) and (1.1.8)]). *Suppose $G$ is a locally compact group and $X$ is a Hausdorff space. If $\rho : G \to \mathrm{TOP}(X)$ a locally proper action, then $\rho$ is a continuous and relatively open map of $G$ when $\mathrm{TOP}(X)$ is given the point-open topology or the compact-open topology. Thus, if $\rho$ is injective (i.e., the $G$-action is effective), $\rho$ is a topological isomorphism onto its image. The image $\rho(G)$ is a closed subgroup of $\mathrm{TOP}(X)$ if the latter is a topological group under the point-open topology or the compact-open topology.*

Palais proves these statements with the point-open topology on $\mathrm{TOP}(X)$, but it is easy to see that they are true with the compact-open topology on $\mathrm{TOP}(X)$ as well.

## 1.3. Fiber bundles

We assume the reader is familiar with the notion and elementary properties of fiber bundles. Some references are [**Ste51**], [**Hus94**], [**Hu59**], and [**Bre72**].

1.3.1. A *fiber bundle* consists of the data $(E, B, \pi, F)$, where $E$, $B$, and $F$ are topological spaces and $\pi : E \to B$ is a continuous surjection satisfying a local triviality condition below. $B$ is called the *base space* of the bundle, $E$ the *total space*, and $F$ the *fiber*. The map $\pi$ is called the projection map.

We require the *local triviality condition* that for any $x \in B$, there is an open neighborhood $U$ of $x$ (which will be called a trivializing neighborhood) and a homeomorphism $\phi : \pi^{-1}(U) \to U \times F$ so that the diagram

$$\begin{array}{ccc} \pi^{-1}(U) & \xrightarrow{\phi} & U \times F \\ \pi \downarrow & & \downarrow \mathrm{proj} \\ U & \xrightarrow{=} & U \end{array}$$

is commutative. The set of all $\{(U_i, \phi_i)\}$ is called a local trivialization of the bundle.

For any $x \in B$, the preimage $\pi^{-1}(x)$ is homeomorphic to $F$ and is called the *fiber over $x$*. A fiber bundle $(E, B, \pi, F)$ is often denoted by $F \to E \to B$.

A smooth fiber bundle is a fiber bundle in the category of smooth manifolds. That is, $E$, $B$, and $F$ are required to be smooth manifolds and all the functions above are required to be smooth maps.

EXAMPLE 1.3.2. Let $E = B \times F$, and let $\pi : E \to B$ be the projection onto the first factor. Then $E$ is a fiber bundle (of $F$) over $B$. Here $E$ is not just locally a product but globally one. Any such fiber bundle is called a *trivial bundle*.

Perhaps the simplest example of a nontrivial bundle $E$ is the Möbius strip. It is obtained from a square by identifying one pair of opposite edges by flipping. The Möbius strip has a circle for a base $B$ and a line segment for the fiber $F$, so the Möbius strip is a bundle of the line segment over the circle. It is not a trivial bundle.

THEOREM 1.3.3. *Let $F \xrightarrow{i} E \xrightarrow{p} B$ be a fiber bundle. Choose base points $b_0 \in B$, $x_0 \in F = p^{-1}(b_0) \subset E$. If $B$ is path-connected, there is a long exact sequence of homotopy groups where the last three are just pointed sets*

$$\cdots \to \pi_n(F, x_0) \xrightarrow{i_*} \pi_n(E, x_0) \xrightarrow{p_*} \pi_n(B, b_0) \to \pi_{n-1}(F, x_0) \to \cdots \to \pi_0(B, b_0).$$

1.3.4. Most of the time we will be dealing with fiber bundles with structure group a topological group $G$. let $U_i$ and $U_j$ be trivializing neighborhoods with $\phi_i$ and $\phi_j$ local trivializations. For $U_i \cap U_j \neq \emptyset$, we have the homeomorphism $\phi_j \circ \phi_i^{-1} : (U_i \cap U_j) \times F \to \pi^{-1}(U_i \cap U_j) \to (U_i \cap U_j) \times F$. This can be written as the identity on the first factor and $g_{ji}(x) : F \to F$ on the second factor. The functions $g_{ji}$ are called *transition functions*. That is, $g_{ji} : U_i \cap U_j \to \text{TOP}(F)$ is a continuous map in the compact open topology. If $U_i \cap U_j \cap U_k \neq \emptyset$, then we necessarily have $g_{ki} = g_{kj} g_{ji}$ (1-cocycle condition) for all $x \in U_i \cap U_j \cap U_k$.

If the range of all the $g_{ji}$'s, for all possible $ji$'s (it suffices to have a covering of $B$ by the $U_i$'s), all lie in a topological group $G$ contained in $\text{TOP}(F)$ (with induced topology from $\text{TOP}(F)$), then we say that the bundle $E$ has *structure group $G$*. Note this implies that $G$ acts effectively on $E$.

1.3.5. A *principal $G$-bundle* is a fiber bundle $\pi : P \to X$ together with a continuous *right* action $P \times G \to P$ by a topological group $G$ such that $G$ preserves the fibers of $P$ and acts freely and transitively on them. The abstract fiber of the bundle is taken to be $G$ itself. (One often requires the base space $X$ to be a Hausdorff space and possibly paracompact.) It follows that the orbits of the $G$-action are precisely the fibers of $\pi : P \to X$ and the orbit space $P/G$ is homeomorphic to the base space $X$.

It is convenient, in later chapters, to usually think of our group as acting on the *left*. That is, a *principal $G$-bundle* is a fiber bundle $\pi : P \to X$ together with a free locally proper *left* action $G \times P \to P$ by a topological group $G$. In this way, our principal $G$-bundles differ from the usual conventions where $G$ is assumed to be acting (freely and locally properly) on the right. The main reason for using the left action is that an action then becomes a group homomorphism (rather than antihomomorphism). Of course, these two can be converted to the other by the conversion used in Subsection 1.1.4, and there is a one-to-one correspondence between the usual structural formulas. It is obvious that a principal $G$-bundle is a fiber bundle with fiber $G$ and structure group $G$ where the $G$-action on $G$ can be thought as translation along the fiber.

One can also define principal $G$-bundles in the category of smooth manifolds. Here $\pi : P \to X$ is required to be a smooth map between smooth manifolds, $G$ is required to be a Lie group, and the corresponding action on $P$ should be smooth. A smooth fiber bundle is a fiber bundle in the category of smooth manifolds. That is, $E$, $B$, and $F$ are required to be smooth manifolds and all the functions above are required to be smooth maps.

1.3.6. If a topological group $G$ acts freely and locally properly on a completely regular space $E$ with orbit space $B$, there is a local cross section to the orbit mapping (see the next section). Therefore, $(G, E, B = G\backslash E)$ is a principal $G$-bundle over the base $B$.

1.3.7. Let $E \xrightarrow{\pi} B$ and $E' \xrightarrow{\pi'} B'$ be principal $G$-bundles. A $G$-map $f : E \to E'$ is called a *principal bundle map*. Obviously, $f$ induces a map $\overline{f} : B \to B'$ such that $\pi' \circ f = \overline{f} \circ \pi$. If $E = E'$ and $\overline{f}$ is the identity on $B = B'$, then $f$ is an *equivariant map covering the identity*. In this case, $f$ is called an *equivalence*, and $E$ and $E'$ are *equivalent*

Let $E' \xrightarrow{\pi'} B'$ be a principal $G$-bundle, and $\overline{f} : B \to B'$ a map. Then
$$\overline{f}^*(E') = \{(b, e') \in B \times E' : \overline{f}(b) = \pi'(e')\}$$
is a bundle over $B$ by defining the projection $\pi : \overline{f}^*(E') \to B$ by $\pi(b, e') = b$ and a $G$-action by $(b, e') \cdot g = (b, e' \cdot g)$. This bundle $\overline{f}^*(E') \xrightarrow{\pi} B$ is called the *inverse image bundle* or *pullback bundle* over $B$.

Now suppose there is a principal bundle map $f : E \to E'$ inducing $\overline{f} : B \to B'$. Then $\overline{f}^*(E')$ is equivalent to $E$.

The equivalence $h : \overline{f}^*(E') \to E$ is given by
$$h(b, e') = \text{the unique } e \in \pi^{-1}(b) \text{ such that } f(e) = e'.$$
The natural $G$-map $\overline{f}^*(E') \to E'$ is given by $(b, e') \mapsto e'$, and is the same as $f \circ h$. If $E$ and $E'$ are principal $G$-bundles over $B = B'$ and $f$ is a $G$-map such that $\overline{f}$ is a homeomorphism, then it is reasonable to say that $f$ is a *bundle isomorphism* and $E$ and $E'$ are *isomorphic $G$-bundles*.

1.3.8. If $E \to B$ is a fiber bundle with fiber $F$ and structure group $G$, there is a principal $G$-bundle $P \to B$ over $B$ associated to it. The principal bundle $P$ may be constructed, up to equivalence, from $E$ by taking all of the local trivializations $\{(U_i, \phi_i)\}$ of the bundle $E$ and replacing them by $\{(U_i, \phi'_i)\} = \{\pi'^{-1}(U_i) \xrightarrow{\phi'_i} U_i \times G\}$, where $\phi'_j \circ {\phi'_i}^{-1}(u \times g) = (u, g_{ji}(u) \cdot g)$, $u \in U_i \cap U_j$ and $\phi_j \circ \phi_i^{-1}(u, f) = (u, g_{ji}(u)(f))$. Since $G$ acts effectively on $F$, $g_{ji}(u)$ is unique in $G \subset \text{TOP}(F)$. Then form $P$ by gluing all the disjoint $U_i \times G$ together according to the relation $(u, g) \sim (u, g_{ji}(u)g)$ with $(u, g) \in U_i \times G$ and $(u, g_{ji}(u)g) \in U_j \times G$ and $u \in U_i \times U_j$.

Conversely, from a principal $G$-bundle $P$ and a space $F$ with (left) $G$-action, one can define a fiber bundle with fiber $F$ and structure group $G$. (We use principal *right* action on $P$ here).
$$P \times_G F = (P \times F)/\sim$$
where
$$(p, f) \sim (p \cdot g^{-1}, g \cdot f) \text{ for } (p, f) \in P \times F \text{ and } g \in G.$$
With this notation, the fiber bundle is equivalent to $P \times_G F \to B$. That is, the $G$-bundles over $B$ with the same effective action of $G$ on $F$ are equivalent if and only if their associated principal $G$-bundles over $B$ are equivalent. Thus the

classification of fiber bundles over $B$ with structure group $G$ and the fiber $F$ (the effective $G$-space) is reduced to the classification of principal $G$-bundles $P$ over $B$.

## 1.4. Classifying spaces

1.4.1. Recall that a principal $G$-bundle is universal if the total space $EG$ is contractible. The base space, $BG$, is called the *universal classifying space*. $BG$ itself is unique up to homotopy. The principal $G$-bundles over a paracompact base $B$ (or more generally, the numerable bundles over an arbitrary space $B$), are, up to bundle equivalence, in one-to-one correspondence with $[B, BG]$, the homotopy classes of maps of $B$ into $BG$. In fact, for each principal bundle $P$ over $B$, there is a map $f : B \to BG$ such that the pullback $f^*(EG)$ is a principal $G$-bundle equivalent to $P$. Any two pullbacks $f^*(EG)$ and $g^*(EG)$ are equivalent principal $G$-bundles if and only if $f$ is homotopic to $g$. Certainly $[B, BG]$ is a worthy set to compute. For example, if $B$ is contractible, or if $G$ is contractible, then $[B, BG]$ is a singleton and $P$ must be equivalent to the product bundle $B \times G$.

1.4.2. To construct $EG$, when $G$ is a Lie group, one can construct the $n$-fold join $G^{(n)}$ of $G$, with $G$ acting diagonally and hence freely. This is an approximation of $EG$ in the sense that $G^{(n)}$ is an $(n-1)$-connected simplicial complex if $G$ is connected. $EG$ is just the infinite join of $G$ with itself, and under a suitable topology, it is contractible. It is often a technical convenience to replace $EG$ by $G^{(n)} \subset EG$, for $n$ very large. This will be satisfactory for classifying purposes of bundles over $X$, if the dimension of $X$ is finite.

The convenience arises in that we can use Čech cohomology with various supports without worrying if things such as the universal coefficient formulas remain valid in its most abstract setting. We will not mention this minor technicality any further, and we refer to the reader A. Borel et al., "Seminar on Transformation Groups" [**Bor60**], for the various ways one gets around these technicalities. Since our interest is mainly in finite dimensional geometric situations, any problem can usually be avoided by dealing with $G^{(n)}$, $n$ large, instead of $EG$.

1.4.3. If $G$ is discrete, then $\pi_i(BG) = 0$, for $i \geq 2$, and so $BG$ is a $K(G, 1)$. Then $[B, BG]$ is in one-to-one correspondence with the conjugacy classes of homomorphisms of $\pi_1(B) \to \pi_1(BG)$ (cf. Spanier [**Spa66**, Chapter 8, Theorem 11, p.428] or Hu [**Hu59**, p.198]). (Choose base points $b \in B$, $b' \in BG$. Two homomorphisms $f_*, g_* : \pi_1(B, b) \to \pi_1(BG, b') = G$ are conjugate if there is an element $\gamma \in \pi_1(BG, b')$ such that $f_*(\alpha) = \gamma g_*(\alpha) \gamma^{-1}$, for all $\alpha \in \pi_1(B, b)$.) Note if $G$ is Abelian, then $[B, BG] = \operatorname{Hom}(\pi_1(B), G)$.

1.4.4. If $G = S^1$, then $G^{(2)} = S^1 \circ S^1$ (the 3-sphere), where $\circ$ denotes join and $S^1 \backslash G^{(2)} = (BG)^{(2)} = S^2$, and $[B, S^2] \to [B, BS^1]$ is bijective for all $B$ of dimension $\leq 2$.

If $G = T^k$, then $BG = \mathbb{C}P_\infty \times \cdots \times \mathbb{C}P_\infty$, $k$ copies. This is a $K(\mathbb{Z}^k, 2)$. When $k = 1$, $EG = S^\infty$, the infinite join of circles and $(T^1)^{(n)} = S^{2n-1}$.

$$[B, BT^k] = [B, K(\mathbb{Z}^k, 2)] \cong H^2(B, \mathbb{Z}^k).$$

If $G = \mathbb{Z}_{n_1} \times \cdots \times \mathbb{Z}_{n_k}$, where each $\mathbb{Z}_{n_i}$ is a finite cyclic group of order $n_i$, then $BG = \mathbb{Z}_{n_1} \times \cdots \times \mathbb{Z}_{n_k} \backslash S_1^\infty \times \cdots \times S_k^\infty$, where each $\mathbb{Z}_{n_i}$ acts freely as covering transformations on $S_i^\infty$ and as a subgroup of $T^1$. Then, $[B, BG] = [B, K(G, 1)] \cong H^1(B; G)$. Note, there is a principal $T^k$ fibering over $BT^k$ with total space $BG$.

$(BG)^{(n)}$ is a product of $(2n-1)$-dimensional lens spaces with the $i$th factor being $\mathbb{Z}_{n_i}\backslash S^{2n-1}$.

So all the principal $T^k$-(respectively, $\mathbb{Z}_{n_1}\times\cdots\times\mathbb{Z}_{n_k}$-)bundles over $B$, up to bundle equivalence, are in one-to-one correspondence with the elements of $H^2(B;\mathbb{Z}^k)$ (respectively, $H^1(B;\mathbb{Z}_{n_1}\times\cdots\times\mathbb{Z}_{n_k})$). The element of $H^2(B;\mathbb{Z}^k)$ (respectively, $H^1(B;\mathbb{Z}_{n_1}\times\cdots\times\mathbb{Z}_{n_k})$) representing a principal bundle is called the characteristic class of that bundle.

When $k=1$, the element of $H^2(B;\mathbb{Z})$ representing the principal $S^1$-bundle over $B$ is called the *Euler class* of this principal $S^1$-bundle.

The cohomology used here is the Čech or, equivalently, the Alexander-Spanier-Wallace cohomology groups when $B$ is paracompact. If $B$ has the homotopy type of a CW-complex, then the Čech cohomology groups are isomorphic to the singular cohomology groups; see [**Spa66**] for example.

## 1.5. Borel spaces and classifying spaces

**1.5.1.** Let $(G,X)$ be a $G$-space and $EG$ a contractible space on which $G$ acts freely and properly. A technique due to A. Borel for studying $G$-actions is the so-called *Borel space* $EG\times_G X$ associated to the $G$-space $X$. On $EG\times X$, there is the diagonal $G$-action given by

$$g(e,x) = (ge, gx).$$

We define the Borel space to be the quotient space $G\backslash(EG\times X)$. This is usually written as either $EG\times_G X$ or $X_G$. Thus we get a fiber bundle

$$X \to EG\times_G X \to BG.$$

More precisely, we have the commutative diagram

$$\begin{array}{ccccc} X & \xleftarrow{\pi_2} & EG\times X & \xrightarrow{\pi_1} & EG \\ \downarrow & & \downarrow{\scriptstyle G\backslash} & & \downarrow{\scriptstyle G\backslash} \\ G\backslash X & \xleftarrow{\overline{\pi}_2} & EG\times_G X = X_G & \xrightarrow{\overline{\pi}_1} & BG \end{array}$$

where $\overline{\pi}_1$ is a fiber bundle mapping with fiber $X$ and structure group $G/K$ where $K$ is the ineffective part of the $G$-action on $X$, $\overline{\pi}_2$ is a mapping such that $\overline{\pi}_2^{-1}(x^*) = BG_x$, or $G_x\backslash EG$, the classifying space for $G_x$, where $x^* \in G\backslash X$, and $x \in x^*$.

Therefore, if $F$ is the set of fixed points in $G\backslash X$, then we claim $\overline{\pi}_2^{-1}(F) = F\times BG \subset X_G$. For, over each point $b \in BG$, we have the fiber $X$, and if $x \in X^G \neq \emptyset$, then there is a unique point $x_b$ in the fiber over $b$ corresponding to $x$. Since $G$ acts on the fiber and fixes $x_b$, then $b \mapsto x_b$ defines a cross section of $X_G \to BG$. Then, for any principal ideal domain $L$, $\overline{\pi}_1^* : H^*(BG;L) \to H^*(X_G;L)$ is a *direct summand*. If we view $x \in X^G$ as $x^* \in G\backslash X$, then $\overline{\pi}_2^{-1}(x^*) = BG$ and so $\overline{\pi}_2^{-1}(F) = F\times BG = F_G \subset X_G$.

**1.5.2.** It should be clear that the Borel construction is functorial. The cohomology group $H^*(X_G;A)$, where $A$ is a $G$-module, is called the equivariant cohomology of $X$ with coefficients in $A$, and is often denoted by $H_G^*(X;A)$. Relative groups are defined in the usual way. If we fix $e \in EG$, then the inclusion $X \hookrightarrow e\times X \to EG\times_G X$ induces a homomorphism $H^*(X_G;A) \to H^*(X;A)$ which coincides with the edge homomorphism derived from the Leray spectral sequence associated to the bundle mapping $\overline{\pi}_1$.

EXERCISE 1.5.3. Let $G$ act on $X$ and $Y$, and diagonally on $X \times Y$. We have the commutative diagram

$$\begin{array}{ccccc} X & \xleftarrow{p_1} & X \times Y & \xrightarrow{p_2} & Y \\ G\backslash \downarrow \nu_1 & & G\backslash \downarrow & & G\backslash \downarrow \nu_2 \\ G\backslash X & \xleftarrow{\bar{p}_1} & X \times_G Y & \xrightarrow{\bar{p}_2} & G\backslash Y \end{array}$$

where $p_i$ are the obvious equivariant projections and $\nu_i$ the orbit mappings. Show that if $x \in X$, then $\bar{p}_1^{-1}(\nu_1(x)) \subset X \times_G Y$ is homeomorphic to $G_x\backslash Y$ and $y \in Y$, then $\bar{p}_2^{-1}(\nu_2(y))$ is homeomorphic to $G_y\backslash X$.

## 1.6. Tubular neighborhoods and slices

The most important tool for analyzing Lie group actions is the existence of a *slice*. Slices give us the complete equivariant structure of an invariant tubular neighborhood of each orbit.

DEFINITION 1.6.1. Let $X$ be a $G$-space and $H$ a closed subgroup of $G$ with a local cross section. A subset $S$ of $X$ is an *$H$-kernel* if there exists an equivariant map $f : GS \to G/H$ such that $f^{-1}(H) = S$. If, in addition, $GS$ is open in $X$, we call $S$ an *$H$-slice* in $X$. If $GS = X$, we call $S$ a *global $H$-slice* in $X$. For $x \in X$, by a *slice at $x$*, we mean a $G_x$-slice in $X$ which contains $x$.

EXAMPLE 1.6.2. Let $G = \mathrm{SO}(3)$ acting on $X = \mathbb{R}^3$ as the full group of rotations, and $x = (0, 0, 1)$. Then $G_x = \mathrm{SO}(2)$ and $G_x\backslash G = S^2$, the 2-sphere; $S = \{(0, 0, t) : 0 < t < \infty\}$. Let

$$f : GS = \mathrm{SO}(3) \cdot S \to S^2 = \mathrm{SO}(2)\backslash \mathrm{SO}(3)$$

given by $f(y) = \frac{y}{\|y\|}$. Then $S$ is a local slice at $x$. If $X = \mathbb{R}^3 - 0$, then $S$ is a global slice.

1.6.3. The following is a prototypical example of a global $H$-slice in $X$. In fact, it follows from Proposition 1.6.4 that *any two spaces with the same global $H$-slices are $G$-isomorphic*.

**Prototypical Example**: Let $H$ be a closed subgroup of $G$ with a local cross section, and suppose $H$ acts on $S$. On $G \times S$, define an action of $G \times H$ by

$$(g, h)(\bar{g}, x) = (g\bar{g}h^{-1}, hs).$$

Denote the quotient of the *diagonal $H$-action* by $G \times_H S$, and the image of $(g, s)$ by $\langle g, s \rangle \in G \times_H S$. Since $H$ commutes with the $G$-action on $G \times S$, the $G$-action descends to $G \times_H S$ and is given by $(g, \langle \bar{g}, s \rangle) \to \langle g\bar{g}, s \rangle$. We get a commutative diagram of projections and orbit mappings:

$$\begin{array}{ccccc} (G \times H, G) & \xleftarrow{\pi_1} & (G \times H, G \times S) & \xrightarrow{G\backslash} & (H, S) \\ \downarrow H\backslash & & \downarrow H\backslash & & \downarrow H\backslash \\ (G, G/H) & \xleftarrow{\bar{\pi}_1} & (G, G \times_H S) & \xrightarrow{G\backslash} & H\backslash S = G\backslash(G \times_H S). \end{array}$$

Clearly, $G \times_H S$ is a $G$-space whose orbit space is $H\backslash S$ and also a fiber bundle over $G/H$ with fiber $S$ and structure group $H/K$, where $K$ is the ineffective part of the action of $H$ on $S$. The associated principal bundle has total space $G/K$ and structure group $H/K$. There is the obvious $G$-isomorphism between $(G, G \times_H S)$

and $(G, G/K \times_{H/K} S)$. The map $\overline{\pi}_1$ induced from $\pi_1$, the projection onto the first factor, is $G$-equivariant such that $\overline{\pi}_1^{-1}(H) = S = \langle e, S \rangle$. Thus, the $G$-space $G \times_H S$ has a global slice $\langle e, S \rangle$. We have the following converse to this Prototypical Example.

PROPOSITION 1.6.4. *Suppose $H$ is a closed subgroup of $G$ with a local cross section, and $X$ is a $G$-space with a $G$-equivariant map $f : X \to G/H$. Let $S = f^{-1}(H)$. Then there exists a $G$-isomorphism $\varphi : (G, G \times_H S) \longrightarrow (G, X)$.*

PROOF. We first show that there exists a $G$-map $\varphi$ which is continuous, one-to-one and onto. The set $S$ is $H$-invariant because $f(hs) = hf(s) \in H$ for all $h \in H$ and $s \in S$. Define an action of $G \times H$ on $G \times S$ as above and a $G$-map $\widetilde{\varphi} : G \times S \to X$ by $\widetilde{\varphi}(g, s) = gs$. The map is easily seen to be continuous. We show that $\widetilde{\varphi}$ is also onto. For any $x \in X$, $f(x) = gH \in G/H$ for some $g \in G$. Then $f(g^{-1}x) = H$. Therefore, $s = g^{-1}x \in S$ for some $s \in S$, and $x = gs = \widetilde{\varphi}(g, s)$.

If $\widetilde{\varphi}(g, s) = \widetilde{\varphi}(g', s')$, then $gs = g's'$, hence $s' = g'^{-1}gs$. But as $f(s') = g'^{-1}gf(s) = H$, we have $g'^{-1}g \in H$. Thus, $g' = gh^{-1}$, $s' = hs$ for $h = g'^{-1}g$. Consequently, $\widetilde{\varphi}(g', s') = \widetilde{\varphi}(gh^{-1}, hs)$ and $\widetilde{\varphi}$ factors through $\varphi : (G, G \times_H S) \longrightarrow (G, X)$.

In fact, we have actually shown that $\varphi$ is continuous one-to-one and onto. Furthermore, it is $G$-equivariant. To show $\varphi^{-1}$ is continuous when $G$ is compact, observe that $G \times S \longrightarrow GS$ is a *closed* mapping since $S$ is closed in $X$. We defer the continuity of $\varphi^{-1}$ in the noncompact case until Corollary 1.6.12. □

The proposition suggests the following.

DEFINITION 1.6.5. If a $G$-space has a slice $S_x$ at $x$, and $G_x$ has a local cross section in $G$, then $GS_x$, which is $G$-isomorphic to $G \times_{G_x} S_x$, by Proposition 1.6.4, is a fiber bundle over the orbit $Gx$ with structure group $G_x / \bigcap_{s \in S} G_s$ ($\bigcap_{s \in S} G_s$ is the ineffective part of the action of $G_x$ on the slice $S_x$) and fiber $S_x$.

The set $GS_x$ is called a *$G$-invariant tube about the orbit $Gx$* or a *$G$-invariant tubular neighborhood of $Gx$*. The fiber over $gx$ is $gS_x$.

1.6.6. Proposition 1.6.4 says that if a $G$-action has a slice at $x$, then there exists a $G$-invariant tubular neighborhood about the orbit $G(x)$. Conversely, the $G$-invariant map $\overline{\pi}_1$ in Subsection 1.6.3 shows that if there exists a $G$-invariant neighborhood of $G(x)$ in $X$, of the type $(G, G \times_{G_x} S)$ with $S$ containing $x$, then $S$ is a slice at $x$. Therefore *the existence of a slice at $x$ is equivalent to the existence of a $G$-invariant tubular neighborhood at $x$.*

EXAMPLE 1.6.7. Let $(G, X)$ be a group of regular covering transformations, $\nu : X \longrightarrow G \backslash X$, the covering projection. For $\nu(x) = x^* \in G \backslash X$, let $U^*$ be a neighborhood of $x^*$ which is evenly covered. That is, $\nu^{-1}(U^*)$ is the disjoint union of copies of open sets homeomorphic to $U^*$. If $U$ denotes the copy containing $x$, then $\nu^{-1}(U^*) = GU$, which is isomorphic to $G \times U$ and forms a $G$-tubular neighborhood of $\nu^{-1}(x^*)$.

EXAMPLE 1.6.8. Consider the affine transformations of $\mathbb{Z} \rtimes \mathbb{Z}_2 = G$ on the real line as given by
$$(n, \epsilon)x = \epsilon x + n, \qquad \epsilon = \pm 1.$$
The isotropy at $x$ is trivial if $x$ is not an integer or a half integer. If $x = \frac{m}{2}$ for some integer $m$, then $G_x = \{(0, 1), (m, -1)\} \cong \mathbb{Z}_2$. For a slice at 0, one can choose the

set $S = (-\frac{1}{2}, \frac{1}{2}) \subset \mathbb{R}$. One has that $(\mathbb{Z} \rtimes \mathbb{Z}_2) \times_{\mathbb{Z}_2} (-\frac{1}{2}, \frac{1}{2})$ is a tubular neighborhood $V$ of the orbit. It consists of all of $\mathbb{R}$ except for the orbit through $\frac{1}{2}$. That is, $G \times_{\mathbb{Z}_2} S = V = GS = \mathbb{R} - (\frac{1}{2} + \mathbb{Z})$. We can define a $(\mathbb{Z} \rtimes \mathbb{Z}_2)$-equivariant map
$$f : G \times_{\mathbb{Z}_2} S = V = GS \longrightarrow G/G_x = \mathbb{Z} \rtimes \mathbb{Z}_2/\mathbb{Z}_2$$
by $v = (n, \epsilon)(s) = (n, -\epsilon)(-s) \xrightarrow{f} f(\epsilon s + n) = n = \{(n, 1) \cup (n, -1)\} \in (\mathbb{Z} \rtimes \mathbb{Z}_2)/\mathbb{Z}_2$. Then $f^{-1}(0) = f^{-1}\{(0, 1) \cup (0, -1)\} = f^{-1}(\mathbb{Z}_2) = f^{-1}(G_0) = S$. (Note that the equivariant mapping $f$ cannot be extended to all of $\mathbb{R}$).

EXAMPLE 1.6.9 (Brieskorn varieties). Consider an action $\mathbb{C}^* \times \mathbb{C}^n \to \mathbb{C}^n$ given by
$$z \times (z_1, \ldots, z_n) \to (z^{b_1} z_1, \ldots, z^{b_n} z_n),$$
where $b_i$ are positive integers $\geq 1$.

Notice the restriction to $S^1 \times S^{2n-1} \to S^{2n-1}$ where $z \in S^1 \subset \mathbb{C}^*$, and $(z_1, \ldots, z_n) \in S^{2n-1} \subset \mathbb{C}^n$ is well defined.

The $b_i$'s that we shall use are arrived at as follows: Define a set
$$V(a_1, a_2, \ldots, a_n) = \{(z_1, \ldots, z_n) \mid z_1^{a_1} + \cdots + z_n^{a_n} = 0\},$$
where $a_i$ are integers $\geq 2$. Put $a = \mathrm{lcm}\{a_1, \ldots, a_n\}$ and define $b_i = a/a_i$. Then $V$ is invariant under the $\mathbb{C}^*$-action for
$$(z^{b_1} z_1)^{a_1} + \cdots + (z^{b_n} z_n)^{a_n} = z^a(z_1^{a_1} + \cdots + z_n^{a_n}) = 0$$
if $(z_1, \ldots, z_n) \in V$. Note that the set
$$K(a_1, \ldots, a_n) = V(a_1, \ldots, a_n) \cap S^{2n-1}$$
is also $S^1$ invariant. Then
$$K(a_1, \ldots, a_n) = \{(z_1, \ldots, z_n) \mid z_1 \bar{z}_1 + \cdots + z_n \bar{z}_n = 1 \quad \text{and} \quad z_1^{a_1} + \cdots + z_n^{a_n} = 0\}.$$
Let
$$p(z_1, \ldots, z_n) = z_1^{a_1} + \cdots + z_n^{a_n}.$$
The polynomial function $p : \mathbb{C}^n - \mathbf{0} \to \mathbb{C}$ has 0 as a regular value. Therefore, $p^{-1}(0) = V(a_1, \ldots, a_n) - \mathbf{0}$ is a *complex manifold* of dimension $n-1$ and $K(a_1, \ldots, a_n)$ is a *real analytic manifold* of dimension $2n-3$. It is not difficult to see that $K(a_1, \ldots, a_n) \times \mathbb{R}^1 \approx V(a_1, \ldots, a_n) - \mathbf{0}$.

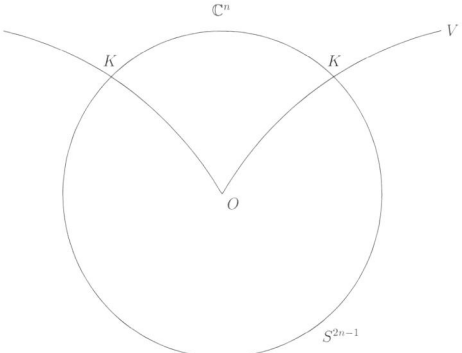

Define $\varphi : S^{2n-1} - K \to S^1$ by
$$\varphi(\vec{z}) = \frac{p(\vec{z})}{|p(\vec{z})|} \in S^1.$$

If we define an $S^1$-action on this image by
$$z \times \frac{p(\vec{z})}{|p(\vec{z})|} \mapsto z^a \frac{p(\vec{z})}{|p(\vec{z})|},$$
we see that the map $\varphi$ is $S^1$-equivariant. Therefore there exists a global $S^1$-slice $Y = \varphi^{-1}(1)$, $1 \in S^1$ for the action $(S^1, S^{2n-1} - K)$. That is,
$$(S^1, S^{2n-1} - K) \xrightarrow{\approx} (S^1, S^1 \times_{\mathbb{Z}_a} Y)$$
an $S^1$-equivariant homeomorphism. Thus, $S^{2n-1} - K$ fibers over $S^1/\mathbb{Z}_a$ equivariantly with fiber $Y = \varphi^{-1}(1)$ and with structure group $\mathbb{Z}_a$.

It can be shown ([**Mil68**, p.55]) that $Y \cup K$ is a compact manifold with boundary. $Y$ is parallelizable of dimension $2(n-1)$ and has the homotopy type of a wedge of $\prod_{i=1}^n (a_i - 1)$ spheres of dimension $n-1$. In many cases, $K$ is a smoothly embedded topological sphere in $\mathbb{C}^n$ not diffeomorphic to the standard sphere.

As a special case, take $\mathbb{C}^2$ and the action $z \times (z_1, z_2) \to (z^2 z_1, z^3 z_2)$. On $S^3$ this action results in a fixed point free action. The isotropy group at each point $(z_1, 0)$ in $S^3$ is $\mathbb{Z}_2$, and the isotropy group at each point $(0, z_2)$ in $S^3$ is $\mathbb{Z}_3$. The 1-manifold $K = \{(z_1, z_2) | (z_1)^3 + (z_2)^2 = 0 \text{ and } z_1 \bar{z}_1 + z_2 \bar{z}_2 = 1\}$ is a free orbit of the $S^1$-action.

From above, $S^3 - K$ fibers over the circle with fiber a 2-manifold having the homotopy type of $S^1 \vee S^1$. Since its boundary is $K$, $F$ is a punctured torus $T'$, so $S^3 - K = S^1 \times_{\mathbb{Z}_6} T'$, which gives the smooth fibered structure of the complement of the $(3,2)$ torus knot $K$.

LEMMA 1.6.10 ([**Pal61**, 2.1.2]). *Let $S$ be an $H$-kernel in the $G$-space $X$, and let $\eta : U \to G$ be a local cross section from $G/H$ to $G$, $(\eta(H) = 1)$. Then if $g_0 \in G$, the map $F : (u, s) \mapsto g_0 \eta(g_0^{-1} u) s$ is a homeomorphism of $g_0 U \times S$ onto an open neighborhood of $g_0 S$ in $GS$. Moreover, $f(F(u, s)) = u$ when $f$ is the equivariant map defining the $H$-kernel.*

PROOF. $f(F(u,s)) = g_0(\eta(g_0^{-1}u)H) = g_0(g_0^{-1}u) = u$. Therefore, $F(g_0 U \times S) = f^{-1}(g_0 U)$, which is an open neighborhood of $g_0 S$ in $GS$. Note that $F$ is one-to-one and continuous. We claim that $F^{-1}$ is continuous by showing that if $F(u_\alpha, s_\alpha)$ converges to $F(u, s)$, then $u_\alpha \to u$ and $s_\alpha \to s$, where we use nets if there is no countable neighborhood base. Now because $f$ is continuous, $u_\alpha = f(F(u_\alpha, s))$ converges to $u = f(F(u,s))$. Therefore, $\eta(g_0^{-1} u_\alpha)^{-1}$ converges to $\eta(g_0^{-1} u)^{-1}$ because $\eta$ is continuous. Now, $\eta(g_0^{-1} u_\alpha) s_\alpha = g_0^{-1} F(u, s_\alpha)$ which converges to $g_0^{-1} F(u,s) = \eta(g_0^{-1} u) s$, which implies $s_\alpha$ converges to $s$.

Note, taking $g_0 = 1$, the argument shows that if $W$ is open in $S$, then $G(W)$ is open in $GS$. □

PROPOSITION 1.6.11 ([**Pal61**, 2.1.3]). *Let $S_1$ and $S_2$ be $H$-kernels in $G$-spaces $X_1$ and $X_2$, respectively, and let $f_0$ be an $H$-equivariant map of $S_1$ into $S_2$. Assume $H$ has a local cross section from $G/H$ to $G$. Then there exists a unique $G$-equivariant map $f$ of $GS_1$ onto $GS_2$ such that $f|S_1 = f_0$; namely, $f(gs) = gf_0(s)$ for $g \in G$, $s \in S$. Moreover, if $f_0$ embeds $S_1$ into $S_2$, then $f$ embeds $GS_1$ into $GS_2$.*

PROOF. We are able to extend $f_0$ to $f$ because $f_0$ being $H$-equivariant implies $H_s \subset H_{f_0(s)}$, $s \in S$, and so, $G_s \subset G_{f(s)}$. To check continuity of $f$ we use the previous lemma. Let $\eta : U \to G$ be a local cross section, where $U$ is a neighborhood of $H$ in $G/H$. Now $F_i : (u, s) \mapsto g_0 \eta(g_0^{-1} u) s$ is a homeomorphism of $g_0 U \times S_i$

onto a neighborhood of $g_0 S_i$ in $GS_i$, $i = 1, 2$, and $g_0 \in G$. Since $f(F_1(u, s)) = f(g_0(\eta(g_0^{-1}(u))s)) = g_0\eta(g_0^{-1}u)f_0(s) = F_2(u, f_0(s))$, the continuity of $f$ follows. Also if $f_0^{-1}$ exists and is continuous, then $f^{-1}$ is continuous by symmetry. $\square$

1.6.12 (Proof of Proposition 1.6.4). Suppose $(G, X)$ and $f : X \to G/H$ and $S = f^{-1}(H)$ are given as in Proposition 1.6.4. Choose the $H$-equivariant identity map id $: S \to S$ and extend it to $\varphi : G \times_H S \to (G, X)$. This extension is unique. Therefore, $\varphi$ is a $G$-isomorphism. $\square$

COROLLARY 1.6.13. *If $(G, X)$ is a completely regular proper $G$-space, $G$ has a slice $S$ at $x$, and $G_x$ has a local cross section on $G$, then $S$ can be chosen so small such that each $y \notin G(x)$ has a neighborhood $V_y$ for which the closure of $\{g \in G : GS \cap V_y \neq \emptyset\}$ is compact.*

PROOF. By Lemma 1.6.10, there is a homeomorphism of $U \times S$ onto a neighborhood of $S$ in $GS$ where $U$ is a neighborhood of $G_x$ in $G/G_x$. Therefore, $\nu : S \to G_x \backslash S$ has an open image on $G \backslash X$. Take a sufficiently small neighborhood $W$ of $\nu(x)$ and put $\nu^{-1}(W) \cap S = S'$. Then $\eta(U)S'$ is a neighborhood of $x$. If $U$ is also to be sufficiently small, then the definition of proper will imply that for each $y \in X$, there exists $V_y$ so that the closure of $\{g \in G : GS' \cap V_y \neq \emptyset\}$ is compact. $\square$

## 1.7. Existence of slices

As one can see from Proposition 1.6.4, the existence of a slice at $x$ means there exists a $G$-equivariant open tube about the orbit $Gx$ which is $G$-isomorphic to $(G, G \times_{G_x} S_x)$, where $S_x$ is the slice at $x$. Therefore, the global structure of the $G$-action can be achieved by successfully piecing together the tubes about the orbits. But of course, slices do not exist without restrictions on $G$ as Example 1.1.9 shows. Showing that slices exist for various situations has had an illustrious history with significant contributions having been made by Montgomery, Zippin, Gleason, Yang, Kozul, Mostow, Palais [**Pal61**] and others. Excellent expositions for the existence of a slice for compact Lie group actions can be found in [**Bor60**] or in [**Bre72**]. We verify, in Theorem 1.7.5, the existence of slices for discrete $G$ acting locally properly. We also sketch, in Theorem 1.7.8, the existence for compact $G$ acting smoothly.

For the general Lie groups, we have

THEOREM 1.7.1 ([**Pal61**]). *If $(G, X)$ is a locally proper Lie group action on a completely regular space $X$, then $G$ has a slice at each $x \in X$.*

COROLLARY 1.7.2. *If $G$ is a Lie group acting properly on a completely regular space $X$ and $x$ and $y$ lie on different orbits of $G$, then there exist slices $S_x$ and $S_y$ at $x$ and $y$, respectively, such that the closures of $GS_x$ and $GS_y$ are disjoint.*

PROOF. Since $G$ is a Lie group, $G/H$ has a local cross section to $G$ for any closed subgroup $H$. By Theorem 1.7.1, $G$ has a slice at each $x \in X$. By Proposition 1.2.4(5), $G \backslash X$ is completely regular. We can separate $\overline{x}, \overline{y} \in G \backslash X$ by open sets whose closures are disjoint. Now we apply the procedure used in the proof of Corollary 1.6.13 to obtain the desired conclusion. $\square$

COROLLARY 1.7.3. *For a completely regular $G$-space where $G$ is a Lie group, the following are equivalent:*

(i) $G$ acts locally properly;
   (ii) $G$ has a $G_x$-slice at each $x \in X$, and $G_x$ is compact;
   (iii) The closure of $\{g \in G : gS_x \cap S_x \neq \emptyset\}$ is compact for each $x \in X$.

For our purposes, we shall deal mostly with compact Lie group actions and two types of noncompact Lie group actions. If $G$ is not compact and not discrete, then $G$ will usually be acting freely as left translations on a principal $G$-bundle. In this case, the existence of a slice is built into the definition of a principal $G$-bundle. If $G$ is discrete, we shall give a direct elementary proof of the existence of a slice. Let us begin with this latter case.

Again, Example 1.1.9 shows that if we take the usual copy of $\mathbb{Z}$ in $\mathbb{R}$, the free smooth action on the descent of the irrational line is *not* locally proper and no slices exist for this action.

PROPOSITION 1.7.4. *Let $(G, X)$ be a locally proper $G$-space with $G$ discrete and $X$ Hausdorff. For each $x \in X$, there exists a $G_x$-invariant neighborhood $S$ such that $\{g \in G : gS \cap S \neq \emptyset\} = G_x$ and $gS \cap \bar{g}S \neq \emptyset$ if and only if $g$ and $\bar{g}$ belong to the same $G_x$-coset.*

PROOF. Since X is Hausdorff, properties (1), (2), and (3) of Proposition 1.2.4 imply, without assuming complete regularity, that the G-orbits are closed subsets of X. Because the action is locally proper, $G$ discrete, then all orbits are discrete and all their isotropy groups are finite. Let $x \in X$. Then there exists $U$ such that $F = \{g \in G : gU \cap U \neq \emptyset\}$ is a finite set. If $g \notin G_x$ and $g \in F$, then $gx \neq x$. There exist neighborhoods $V_x{}^g$ of $x$ and $V_{gx}$ of $gx$ such that $V_x{}^g \cap V_{gx} = \emptyset$, $V_x{}^g \subset U$, and $gV_x{}^g \subset V_{gx}$. Put $V = \bigcap_{g \in F - G_x} V_x{}^g$. Then we have $\{g \in G - G_x : gV \cap V \neq \emptyset\} = \emptyset$. Take $W$ a neighborhood of $x$, $W \subset V$ and such that $gW \subset V$ for all $g \in G_x$. Put $S = G_x(W)$, then $S \subset V$ and $S$ is $G_x$-invariant, and $gS \cap S = \emptyset$ if and only if $g \in G - G_x$. Furthermore, observe that $gG_x \cap \bar{g}G_x \neq \emptyset$ if and only if $g$ and $\bar{g}$ belong to the same $G_x$-coset. Thus, $\{gG_xS\}$, as $gG_x$ runs through the disjoint $G_x$-cosets, forms an invariant neighborhood of the orbit with each distinct $gG_xS$ being disjoint from all the others. □

THEOREM 1.7.5. *Let $(G, X)$ be a locally proper $G$-space with $G$ discrete and $X$ Hausdorff. Then for each $x \in X$, there exists a slice $S$, and $(G, GS)$, a $G$-invariant neighborhood of $Gx$, is $G$-isomorphic to $(G, G \times_{G_x} S)$ and homeomorphic to $G/G_x \times S$.*

PROOF. For $S$, we choose the $S$ from Proposition 1.7.4. Define $f : GS \to G/G_x$ by
$$f(gs) = gG_x, \quad g \in G, \quad s \in S.$$
If $gs = g's'$, then $g'^{-1}g \in G_x$ by Proposition 1.7.4. Therefore, $f(gs) = f(g's')$ and so $f$ is well defined. It is continuous since $G/G_x$ is discrete, $GS$ is a product space homeomorphic to $G/G_x \times S$ since $GS$ is a disjoint collection of open sets each homeomorphic to $S$ and indexed by $G/G_x$. The $G$-map
$$\varphi : (G, G \times_{G_x} S) \longrightarrow (G, GS)$$
defined by $(g, \langle \bar{g}, s \rangle) \xrightarrow{\varphi} g\bar{g}s$ is clearly one-to-one, continuous, $G$-invariant, and onto. The inverse is also clearly continuous. □

## 1.7. EXISTENCE OF SLICES

REMARK 1.7.6. Note this form of the slice theorem does not require that $X$ be completely regular. Even when $X$ is a smooth manifold and the action is smooth, the orbit space can fail to be Hausdorff. However if the action is *proper*, $G\backslash X$ will be completely regular and, in particular, Hausdorff.

EXAMPLE 1.7.7. Let the free $\mathbb{R}$-action on the strip $Z = \{(x,y) : -1 \leq y \leq 1\}$ be that whose orbits are pictured below:

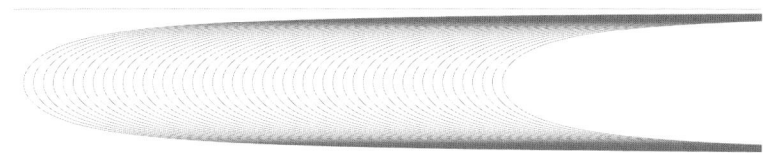

FIGURE 1. Flow

This action is locally proper for if we delete either $y = 1$ or $y = -1$, then the deleted strip is a (locally trivial) principal $\mathbb{R}$-bundle with a cross section (i.e., a global slice). However, there is no global cross section. The action is proper on $\{(x,y) : -1 < y < 1\}$ but properness on the strip is violated for if $U$ is a neighborhood of $(x,1)$ and $V$ is a neighborhood of $(x',-1)$, then closure $\{g \in G : gU \cap V \neq \emptyset\}$ is not compact. The orbit space is a half open interval with a double point at the (closed) end point, corresponding to the two lines $y = \pm 1$. This is a non-Hausdorff 1-manifold with boundary. For a discrete example, just choose $\mathbb{Z} \subset \mathbb{R}$ in this example.

THEOREM 1.7.8. *Let $(G, M)$ be a smooth action of a compact Lie group on a smooth manifold. Then each point $x \in M$ has a smooth $G_x$-slice.*

PROOF. [Sketch] (cf. [**Bor60**, p.108]) Introduce a Riemannian metric $\rho$ on $M$. Then average this metric over the compact Lie group $G$:

$$\rho'(v, v')_x = \int_G \rho(gv, gv')_{gx} d\mu(g)$$

to get a new $G$-invariant Riemannian metric $\rho'$. The $G$-action via the differential on the tangent bundle $T(M)$ now acts as isometries with respect to the new metric $\rho'$. The exponential map $\exp : T(M) \to M$ is $G$-equivariant. That is,

$$g \exp(v) = \exp(g_* v),$$

where $g_*$ denotes the differential of the diffeomorphism $g \in G$. For a (compact) orbit $Gx$, the tangent bundle along the orbit splits equivariantly into the tangent vectors along the orbit and those normal to the orbit. For $\epsilon > 0$, sufficiently small, the exponential map sends the vectors of length less than $\epsilon$ of the normal bundle of the compact orbit $Gx$ diffeomorphically onto an open neighborhood consisting of those points of $M$ whose distance from $Gx$ is less than $\epsilon$. The normal vectors at $x \in Gx$ are invariant under the action of $G_x$ and so in any normal coordinate system about $x$, $G_x$ acts orthogonally. For a slice $S$ at $x$, take the exponential of the normal vectors of length less than $\epsilon$ to $Gx$ at $x$. Then $S$ is an open disk centered at $x$. Clearly, $G_x S = S$ and $GS$ is an open tubular neighborhood of $Gx$. We claim

$$gS \cap S \neq \emptyset \implies g \in G_x.$$

Suppose $g(\exp v) = \exp w$, with $v, w \in \exp_x^{-1}(S)$. Because the restricted exponential is a diffeomorphism on $GS$, $\exp(g_* v) = g \exp(v) = \exp(w)$, and so $g_*(v) = w$. But $w \in \exp^{-1}(S)$, hence $g_*(v) \in \exp^{-1}(S)$ which implies that $g \in G_x$.

Next, we claim there is a $G$-isomorphism between $(G, G \times_{G_x} S)$ and $(G, GS)$. Define a smooth $G$-equivariant map $(G, G \times S) \xrightarrow{f} (G, GS)$ by $(\overline{g}, (g, s)) = (\overline{g}g, s) \mapsto \overline{g}(gs) = \overline{g}g(s)$. This map is smooth, $G$-equivariant, and onto. It is not one-to-one in general. In fact, $f$ factors through $(G, G \times_{G_x} S)$: Suppose $(g, s) \mapsto gs$ and $(g', s') \mapsto g's' = gs$. Then $g'^{-1}g = h \in G_x$. Consequently, $(g', s') = (gh^{-1}, s') = (gh^{-1}, g'^{-1}gs) = (gh^{-1}, hs)$. Conversely, $(gh^{-1}, hs) \mapsto gs$ for all $h \in G_x$. Thus, the map $f$ factors through $(G, G \times_{G_x} S)$. Note that the $G$-action descends to $G \times_{G_x} S$ since the diagonal $G_x$-action commutes with the $G$-action on $G \times S$. If we denote the image of $(g, s)$ by $\langle g, s \rangle \in G \times_{G_x} S$, then $\overline{g}\langle g, s \rangle = \langle \overline{g}g, s \rangle$ and is well defined. It is clear that the induced $\overline{f} : G \times_{G_x} S \longrightarrow GS$ is now one-to-one. It suffices to show $\overline{f}$ is open to conclude that $\overline{f}$ is a diffeomorphism. Let $t : U \longrightarrow G$ be a local cross section in $G/G_x$ and define $K : (u, v) \mapsto K_*(u)(v)$ on $U \times \exp^{-1}(S)$ diffeomorphically onto an open set in the normal bundle to $G(x)$. Then $\widetilde{K} = \exp \circ K$ is a diffeomorphism onto an open set in $X$. Since $\exp(t_*(u)v) = t(u)\exp(v)$, then $(u, s) \mapsto t(u)s$ is a diffeomorphism of $U \times S$ onto an open set in $GS$. □

REMARK 1.7.9. Let $(G, X)$ be a locally proper action with $X$ connected and locally connected. Let $V$ be the set of points lying on orbits of maximal dimension and with fewest components in their isotropy groups. Then, $V$ is a $G$-invariant open subset of $X$. More generally, if $k$ is an integer greater than $-1$, the set $U$ of points which lie on orbits of dimension greater than $k$ is a $G$-invariant open subset of $X$. These facts are an immediate consequence of the slice theorems. For, if $S_x$ is a slice at $x$, then $G_y \subset G_x$, $y \in S_x$, and $G \times_{G_x} S_x = GS_x$ is an open $G$-invariant subset of $X$. An orbit in $V$ is called a *principal orbit* for $(G, X)$. If $X$ is a manifold (or $\mathbb{Z}$-cohomology manifold), it can be shown that $V$ is dense in $X$. (See [**Bor60**, Chap 10] for the general case or [**Bre72**, Chap IV] for the locally smooth case.)

## 1.8. Cohomology manifolds and the Smith theorems

Lemma 3.1.6 is an example of one of the important Smith theorems. The technique of proof employed there can be expanded to prove much stronger results of Smith. These Smith theorems play a very important role in transformation groups. We shall now record some of them. For a full account, one can consult the original papers of P. A. Smith or more recent proofs and/or expositions found in [**Bor60**] or [**Bre72**], [**Bre97**]

1.8.1. Let $(G, M)$ be a nontrivial action on a connected $m$-manifold $M$, where $G = T^k$ or a finite $p$-group, $p$ a prime. Then

(1) $F = M^G$ is a cohomology manifold over $\mathbb{Z}$ for $G = T^k$ (respectively, over $\mathbb{Z}_p$, for $G$ a $p$-group) of dimension $\leq m - 2$ (respectively, $\leq m - 1$). $F$, itself, could be empty, in which case, we say $\dim F = -1$.
(2) $\dim F \equiv m \pmod 2$ if $G = T^k$, or $p$-group, $p \neq 2$.
(3) If the action is effective, then $\dim F \leq m - 2k$ for $G = T^k$ or $G = \mathbb{Z}_{p^k}$ with $k \neq 1$, if $p = 2$. If $p = 2$, $\dim(F) \leq m - 1$.
(4) $\chi(F) = \chi(M)$ if $G = T^k$ (respectively, $\chi(F) \equiv \chi(M)$ mod $p$ if $G = \mathbb{Z}_p$).
(5) If $M$ is oriented and $G$, a 2-group, preserves orientation, then $\dim F \equiv m \pmod 2$.

(6) If the action is smooth, then $F$ is a smooth submanifold of $M$.
(7) If $M$ is $\mathbb{Z}$-acyclic (respectively, $\mathbb{Z}_p$-acyclic), then $M^{T^k}$ (respectively, $M^G$, $G$ a $p$-group) is $\mathbb{Z}$-acyclic (respectively, $\mathbb{Z}_p$-acyclic).
(8) If $M$ has the $\mathbb{Z}$-homology (respectively, the $\mathbb{Z}_p$-homology) of the $m$-sphere, then $M^{T^k}$ (respectively, $M^G$, $G$ a $p$-group) has the $\mathbb{Z}$-homology (respectively, the $\mathbb{Z}_p$-homology) of an $(m-r)$-sphere, $r \geq 0$.

Statement (7) implies that a finitely generated group which acts freely and properly on a contractible manifold is torsion free. See Corollary 3.1.9 for a proof. The hypothesis of the theorem can be weakened to replacing the manifold $M$ by a cohomology manifold over $\mathbb{Z}$ or over $\mathbb{Q}$ for $G = T^k$, or over $\mathbb{Z}_p$ for $G$ a $p$-group.

1.8.2. A definition of a *cohomology $m$-manifold* $X$ over a PID (Principal Ideal Domain) $L$ can be given as:

(1) $X$ is a locally compact space with a countable local basis.
(2) $H^{m+1}(X, A; L) = 0$ for all closed subsets $A$ of $X$, (i.e., $\dim_L(X) \leq m$).
(3) If $U$ is a neighborhood of $x$, then there exists a neighborhood $V$ of $x$, contained in $U$ and $H^*(U; L) \to H^*(V; L)$ is trivial, (i.e., *clc = cohomologically locally connected*).
(4) $H^p(X, X - x; L) = L$ if $p = m$, and 0 otherwise, for all $x \in X$. ($H^p(X, X - x; L)$ is called the *local cohomology group* in dimension $p$ at $x$. For a finite dimensional simplicial complex, this reduces to the condition

$$H^p(\text{St}(v), \partial \text{St}(v); L) \cong H^p(D^m, S^{m-1}; L) \ (\cong H^p(X; X - v; L))$$

for each vertex $v$, where $\text{St}(v)$ is the star of $v$.)

Each $U$ open, connected, and with compact closure satisfies $H^m(X, X - U; L) = L$ or $L/2L$. If it is always $L$, then $X$ is called *orientable* over $L$. If not, then $X$ is *nonorientable*. In particular, if $X$ is compact and connected, then $H^m(X; L) \cong L$ if orientable and $H^m(X; L) \cong L/2L$ if not. Furthermore, if $V \subset U$, then the homomorphism induced by the inclusion $H^m(X, X - V) \to H^m(X, X - U)$ is an isomorphism, when $U$ and $V$ are as above. If $A$ is closed in $X$ and $A \neq X$, $X$ connected, then $H^m(A; L) = 0$.

Cohomology manifolds cannot be avoided if one wishes to study nonsmooth actions since the fixed-point set of a $p$-group or torus group acting nonsmoothly may fail to be locally Euclidean. However, it will be a cohomology manifold over the appropriate $L$. These cohomology manifolds behave homologically just as manifolds. They satisfy Poincaré duality, both globally and locally (i.e., relative duality). They have a fundamental class when orientable and a twisted fundamental class when not orientable. While these cohomology manifolds will appear in creating topological actions, all that we will actually use are the facts of Subsection 1.8.1. More details about cohomology manifolds can be found in [**Bor60**], [**Bre97**], and [**Wil79**].

Cohomology manifolds are not necessarily Absolute Neighborhood Retracts (ANR's), and so the convenient cohomology theory to use is the Čech, or equivalently, the Alexander-Spanier theory. This agrees with the singular theory if $X$ is (*homologically locally connected*) over $L$; that is, if every neighborhood $U$ of $x \in X$ has a neighborhood $V$, $x \in V \subset U$, so that $i_* : H_*(V; L) \to H_*(U; L)$ is trivial with respect to singular homology (for example, if $X$ is locally contractible). The appropriate homology to use is the Borel-Moore homology (or, equivalently, the Čech homology whenever $L$ is a field). Again we can substitute the singular homology if the *homologically locally connected* over $L$ condition holds.

1.8.3. Examples of simplicial cohomology manifolds that arise as fixed points in nonsmooth actions on smooth manifolds are (a) Suspensions of smooth manifolds having the integral homology type of $S^{n-1}$ but are not $S^{n-1}$ and (b) Suspension of $\mathbb{R}P_n$ when $L = \mathbb{Z}_p$ or $\mathbb{Q}$, $p \neq 2$ and $n$ odd.

A separable metric cohomology $n$-manifold over a PID $L$, $n \leq 2$, is always locally Euclidean [**Wil79**]. If $M$ is a simplicial cohomology 3-manifold over $\mathbb{Z}$, then $M$ is locally Euclidean, but a simplicial cohomology $n$-manifold over $\mathbb{Z}$ is not necessarily locally Euclidean for $n > 3$.

For nonsimplicial examples, one can take a badly embedded (i.e., wild) arc $A$ in a smooth manifold $M$ and collapse $A$ to a point. Then $M/A$ is a cohomology manifold of the same homotopy type as $M$, but is not locally Euclidean at the point $\{A/A\}$.

1.8.4. A way to avoid the use of cohomology manifolds is to use the concept of "locally smooth actions", which was introduced by Bredon in [**Bre72**]. Let $(G, M)$ be a $G$-space. The action is called *locally smooth* if for each $x \in M$, there is a $G_x$-slice $S$ at $x$ such that the $G_x$-space $S$ is equivalent to an orthogonal $G_x$-space. Consequently, the $G$-invariant tube $G \times_{G_x} S$ is a *linear* tube. That is, $S$ is homeomorphic to $\mathbb{R}^n$ and the action of $G_x$ on $S$ is topologically equivalent to an orthogonal action of $G_x$ on $\mathbb{R}^n$. Thus, $M$ is a topological manifold. It is not necessarily a smooth manifold, but each $G_x$-invariant tube is a smooth $G$-space. Moreover, if $H \subset G$ is a closed subgroup, then $M$ is locally smooth as an $H$-space. A consequence of this is $M^H$ is a (topological) submanifold of $M$, for each closed $H \subset G$. Of course, smooth actions are locally smooth. For further information, the reader should consult [**Bre72**].

## 1.9. Actions of $G \cdot \Pi$ ($G$ Lie group, $\Pi$ discrete)

DEFINITION 1.9.1. A closed subgroup $\Gamma$ of a Lie group $G$ is called *cocompact* if the coset space $G/\Gamma$ is compact.

The following will be extremely useful in our study of Seifert fiberings.

THEOREM 1.9.2. *Let $G$ be a Lie group acting effectively and properly on a locally compact Hausdorff space $X$. Suppose there also exists a discrete group $\Pi$ acting effectively on $X$. Assume*

(1) *$\Pi$ normalizes $G$ in $\mathrm{TOP}(X)$, and*
(2) *$\Gamma = G \cap \Pi$ is a closed discrete subgroup of $G$.*

*Put $Q = \Pi/\Gamma$. Then there exists an induced action of $Q$ on $G \backslash X = W$. If $\Pi$ acts properly and $\Gamma$ is cocompact in $G$, then the $Q$ action on $W$ is proper. Conversely, if the $Q$-action on $W$ is proper, then the group $G \cdot \Pi$, generated by $G$ and $\Pi$, acts properly on $X$, and consequently, $\Pi$ acts properly on $X$.*

PROOF. Let $\nu : X \to W$ denote the $G$-orbit mapping. Since $\Pi$ normalizes $G$, $\Pi$ acts on $W$ as a $Q$-action. Since $G$ acts properly on $X$, the orbit space $W$, by Proposition 1.2.4(5), is completely regular since $X$ is locally compact. The map $\nu$ is $\Pi$-equivariant. We shall prove $(Q, W)$ is proper.

Let $K \subset W$ be compact. There is a compact set $K' \subset \nu^{-1}(K)$ such that $\nu(K') = K$. To find such a $K'$, take for each $w \in K$ and $x \in \nu^{-1}(K)$, a $G_x$-slice at

$x$, with $\nu(x) = w$. The slice $S_x$ projects to an open subset of $W$. Without loss of generality, we can take the slice $S_x$ to be a compact set whose interior contains $x$. Because $K$ is compact, choose a finite number, $S_{x_1}, \ldots, S_{x_n}$, of slices for which the union of the projections of the interior covers $K$. Thus $K' = (\bigcup_{i=1}^{n} S_{x_i}) \cap \nu^{-1}(K)$ is compact and $\nu(K') = K$.

Let $e \in F \subset G$ be a compact set such that $\Gamma \cdot F = G$ (say, a fundamental domain). Then $F \cdot K'$ is compact. Suppose $\alpha \in Q$ satisfies $\alpha(K) \cap K \neq \emptyset$. Then, by adjusting by an element of $\Gamma$ if necessary, we can find $\widetilde{\alpha} \in \Pi$ such that $\widetilde{\alpha}(F \cdot K') \cap (F \cdot K') \neq \emptyset$. Since the $\Pi$-action is proper, there are only finitely many such $\widetilde{\alpha}$'s in $\Pi$, and hence there are only finitely many such $\alpha$'s in $Q$.

For the converse, let $C \subset X$ be a compact set. Then $\nu(C) \subset W$ is compact. Since $(Q, W)$ is proper, there exist only finitely many elements $\alpha_1, \alpha_2, \ldots, \alpha_n \in Q$ such that $\alpha_i(\nu(C)) \cap \nu(C) \neq \emptyset$. For each $i$, pick a preimage $\widetilde{\alpha}_i \in \Pi$ of $\alpha_i$, and let

$$D_i = \widetilde{\alpha}_i(C) \cup C.$$

Then $D_i$ is a compact subset of $X$.

Suppose $f \in G \cdot \Pi$ satisfies $f(C) \cap C \neq \emptyset$. Let $\overline{f} \in Q$ denote the image of $f$. Then

$$\overline{f}(\nu(C)) \cap \nu(C) = \nu(f(C)) \cap \nu(C) \supset \nu(f(C) \cap C) \neq \emptyset.$$

This implies that $\overline{f}$ is one of the $\alpha_i$'s; that is, $f = g\widetilde{\alpha}_i$ for some $g \in G$. Then $g\widetilde{\alpha}_i(C) \cap C \neq \emptyset$. This implies

$$g(D_i) \cap D_i = g(\widetilde{\alpha}_i(C) \cup C) \cap (\widetilde{\alpha}_i(C) \cup C) \supset g\widetilde{\alpha}_i(C) \cap C \neq \emptyset.$$

Consequently, we have

$$\{f \in G \cdot \Pi : f(C) \cap C \neq \emptyset\} \subset \bigcup_i \{g\widetilde{\alpha}_i \in G \cdot \Pi : g(D_i) \cap D_i \neq \emptyset\}$$
$$= \bigcup_i \{g \in G : g(D_i) \cap D_i \neq \emptyset\}\widetilde{\alpha}_i.$$

Since $(G, X)$ is proper, the last term is compact, and the proof is complete. $\square$

The next theorem is a stronger version of the converse of the preceding theorem and is needed for the Seifert Constructions when $X$ is not assumed to be locally compact; see Chapter 7.

THEOREM 1.9.3. *Let $1 \to G \to E \xrightarrow{\lambda} Q \to 1$ be an exact sequence of Lie groups with $Q$ discrete. Suppose $E$ acts on a completely regular $X$ and the subgroup $G$ acts properly. On $W = G \backslash X$, assume that $Q$ acts properly. Then $E$ acts properly.*

PROOF. It suffices to show $E$ acts locally properly on $X$ since $E \backslash X = Q \backslash W$ is completely regular ($Q$ acts properly on $W$, which is completely regular because $G$ acts properly on $X$).

First, we assert that $E$ acts properly when $Q$ is finite. To see this, let $x \in X$, and $\{e_1, e_2, \ldots, e_n\}$ be a set of elements of $E$ representing distinct $G$-cosets. Thus, each $e \in E$ can be uniquely represented by $e = ge_i$, for some $g \in G$, and some $e_i$. Since $G$ acts properly, there is a neighborhood $U_x$ such that for $y \in X$, there is a neighborhood $U_y$ such that the closure of $\{g \in G : gU_y \cap U_x \neq \emptyset\}$ is compact. In particular, there are neighborhoods $U_{e_i \cdot x}$ for which we can take $e_i \cdot U_x$ such that the closure of $\{g \in G : gU_{e_i \cdot x} \cap U_x \neq \emptyset\}$ is compact, for each $i$. Therefore,

$$\{e \in E : eU_x \cap U_x \neq \emptyset\} = \bigcup_{i=1}^{n}\{g \in G : ge_iU_x \cap U_x \neq \emptyset\}$$
$$= \bigcup_{i=1}^{n}\{g \in G : gU_{e_i \cdot x} \cap U_x \neq \emptyset\}.$$

As the closure of each of the unions is compact, $E$ acts locally properly on $X$ when $Q$ is finite.

Now put $w = \nu(x)$, where $\nu : X \to G\backslash X = W$ is the $G$-orbit mapping. Let $S'_x$ be a $G_x$-slice at $x$. The image $G_x\backslash S'_x$ is an open neighborhood of $w$. Choose $V_w$, a neighborhood of $w$ such that $V_w$ is a $Q_w$-slice at $w$, and such that $V_w \subset G_x\backslash S'_x$. Put $S_x = \nu^{-1}(V_w) \cap S'_x$. Then $S_x$ is a $G_x$-slice at $x$. Note that $qV_w \cap V_w \neq \emptyset$ implies $q \in Q_w$, a finite subgroup of $Q$. The $Q$-tube about $w$, $Q \times_{Q_w} V_w$, is a collection of homeomorphic copies of $V_w$, the closures of which are disjoint from each other ($qV_w$ and $q'V_w$ are disjoint if $q^{-1}q' \notin Q_w$, and are equal if $q^{-1}q' \in Q_w$). Let $E^{(w)}$ be the subgroup of $E$ fitting into the exact sequence of (Lie) groups $1 \to G \to E^{(w)} \to Q_w \to 1$. Then the action of $E^{(w)}$ is proper on the $E^{(w)}$-invariant set $\nu^{-1}(V_w)$.

There is a neighborhood $U_x \subset \nu^{-1}(V_w)$ such that the closure of $\{e \in E^{(w)} : eU_x \cap U_x \neq \emptyset\}$ is compact. Now for any $e \notin E^{(w)}$, $\lambda(e) \notin Q_w$ so that $eU_x \cap U_x = \emptyset$. Therefore the $E$-action on $X$ is locally proper. □

REMARK 1.9.4. Theorem 1.9.3 can be applied as follows: Suppose $G$ is a Lie group acting effectively and properly on $X$. Let $\Phi$ be a group acting effectively on $X$ and normalizing the $G$-action and such that $(G \cdot \Phi)/G = Q$ is discrete. If $Q$ acts properly on $W = G\backslash X$, then $G \cdot \Phi$ acts properly on $X$.

Observe that $\widetilde{\varphi} : G \to \mathrm{TOP}(X)$ is a topological isomorphism of $G$ onto $\widetilde{\varphi}(G)$. Thus, identifying $G$ with $\widetilde{\varphi}(G)$, we have that $G$ is a closed subgroup of the topological group $G \cdot \Phi$ (group generated by $\widetilde{\varphi}(G)$ and $\Phi$ in $\mathrm{TOP}(X)$). Thus, $Q = (G \cdot \Phi)/G$ acts (properly by assumption) on $W$. When $Q$ is finite, it is clear that $G \cdot \Phi$ is a Lie group and the argument in Theorem 1.9.3 also shows under the hypothesis above that $G \cdot \Phi$ is also a Lie group. In fact, the assumption that $E$ is a Lie group was not actually used in the proof of Theorem 1.9.3.

EXAMPLE 1.9.5. Let $G = \mathbb{R}$ act on $X = \mathbb{R}^2$ by translation on the first factor. Let $\Pi = \mathbb{Z}^2$ with generators $(1,1)$ and $(1,\sqrt{2}) \in \mathbb{R} \times \mathbb{R}$ and act on $\mathbb{R}^2$ by translating as a subgroup of $\mathbb{R}^2$. The actions commute and $\Pi \cap G = 0 \in \mathbb{R}$. Both $G$ and $\Pi$ act properly on $\mathbb{R}^2$ with quotients $W = G\backslash X = \mathbb{R}$ and $\Pi\backslash\mathbb{R}^2 = T^2$, respectively. Neither of the induced actions $(\frac{G\times \Pi}{G}, G\backslash X) = (\mathbb{Z}^2, W) = (\mathbb{Z}^2, \mathbb{R})$ or $(G, \Pi\backslash\mathbb{R}^2) = (\mathbb{R}, T^2)$ are locally proper (cf. Example 1.1.9). If we replace $\pi = \mathbb{Z}^2$ by a new $\pi = \mathbb{Z}^2$ generated by $(1,0)$ and $(1,\sqrt{2})$, then $\pi \cap \mathbb{R} = \mathbb{Z}$ and the induced actions are now proper.

CHAPTER 2

# Group Actions and the Fundamental Group

## 2.1. Covering spaces

Covering spaces play a crucial role in this book. The subject is very familiar and, to set notation, we adopt the terminology of covering projections as in [**Spa66**, §2.5 and §2.6]. Unless specified otherwise, a *covering projection* will be a map

$$\nu : (Y, y_0) \longrightarrow (X, x_0),$$

where $Y$ is path-connected, $X$ is locally path-connected, and *semilocally simply connected* (each $x$ has a neighborhood $U$ such that $i_* : \pi_1(U, x) \to \pi_1(X, x)$ is trivial) such that

(i) $\nu$ is onto,
(ii) each $x \in X$ has a neighborhood $U$ such that $\nu^{-1}(U)$ is a disjoint collection of open sets each of which maps, by $\nu$, homeomorphically onto $U$ (i.e., $U$ is *evenly covered*).

The category of covering spaces has objects which are covering projections $\nu : Y \to X$ and morphisms which are commutative triangles

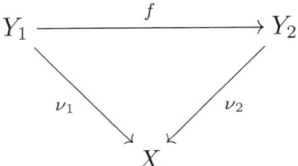

where $\nu_1$ and $\nu_2$ are covering projections. In this category, *every morphism is also a covering projection*. If $f$ is a homeomorphism, the covering spaces $Y_1$ and $Y_2$ are called *equivalent*. If $Y_1 = Y_2$ and $f$ is a homeomorphism, we call $f$ a *covering transformation*. It is well known that such an $f$ is determined by what it does to a single point.

2.1.1 (Construction of covering spaces). Let $K$ be a subgroup of $\pi_1(X, x)$. We define

$$X_K = \text{ the covering space of } X \text{ associated with the subgroup } K$$

so that $\pi_1(X_K, \widehat{x}) = K$. It can be constructed as follows.

Denote by $P(X, x)$ the set of all paths in $X$ with initial point $x$. Define an equivalence relation on $P(X, x)$ by $p_1 \sim_K p_2$ if and only if $p_1(1) = p_2(1)$ and the closed loop

$$p_1 * \overline{p_2} = \begin{cases} p_1(2t), & 0 \leq t \leq 1/2, \\ p_2(2 - 2t), & 1/2 \leq t \leq 1 \end{cases}$$

represents an element of $K$. Finally let
$$X_K = P(X,x)/\sim_K .$$
A basis for the topology of $X_K$ consists of the collection $\{\langle p, U\rangle\}$, where $U$ is open in $X$, $p$ is a path such that $p(0) = x$, $p(1) \in U$, and $\langle p, U\rangle$ denotes all equivalence classes of paths having a representative of the form $p * p'$, where $p'(0) = p(1)$ and $p'(t) \in U$. (If $X$ is Hausdorff, this topology is the same as that induced by the quotient topology, where $P(X,x)$ is given the compact-open topology.)

The map $p \mapsto p(1)$ induces the projection map $\nu : X_K \to X$. If $\widehat{x} \in X_K$ denotes the equivalence class of the constant path at $x$, then $\nu_* : \pi_1(X_K, \widehat{x}) \to \pi_1(X,x)$ has image $K$.

If $K$ is a *normal* subgroup, then a free and locally proper action of the quotient group $\pi_1(X,x)/K$ on $X_K$ can be defined to show that $\nu : X_K \to X$ is a regular covering: Recall that, the group operation of $\pi_1(X,x)$ is defined by juxtaposition. That is, given loops $\ell_1(t), \ell_2(t)$ at $x$, $[\ell_1 \ell_2]$ is the homotopy class of
$$(\ell_1 * \ell_2)(t) = \begin{cases} \ell_1(2t), & 0 \le t \le 1/2, \\ \ell_2(2t-1), & 1/2 \le t \le 1. \end{cases}$$

Let $\alpha \in \pi_1(X,x)/K$ and $\widehat{y} \in X_K$. Take a path $p$ in $P(X,x)$ representing $\widehat{y}$, and a loop $\ell$ in $P(X,x)$ representing $\alpha$. Define
$$\alpha \cdot \widehat{y} = \text{the ``}\sim_K\text{'' equivalence class represented by } \ell * p \in P(X,x).$$

Suppose $p' \in P(X,x)$ and $\ell' \in P(X,x)$ are other elements representing $\widehat{y}$ and $\alpha$, respectively. Then $p * \overline{p'}$ and $\ell * \overline{\ell'}$ represent elements of $K$ since $[\ell] = \alpha = [\ell'] \in \pi_1(X,x)/K$. Thus, $(\ell * p)*(\overline{\ell' * p'}) \simeq \ell*(p*\overline{p'})*\overline{\ell'}$. Since $[p*\overline{p'}] \in K$ and $K$ is normal in $\pi_1(X,x)$, $[\ell * (p*\overline{p'}) * \overline{\ell'}] = [\ell] * [p*\overline{p'}] * [\overline{\ell'}] = ([\ell] * [p*\overline{p'}] * [\ell]^{-1}) \cdot ([\ell * \overline{\ell'}]) \in K$. Consequently, the two paths $\ell * p$ and $\ell' * p'$ represent the same point of $X_K$. Finally, $K$ acts trivially on all of $X_K$ so that $\pi_1(X,x)/K$ is the group of covering transformations.

2.1.2 (Universal covering space). If, in particular, $K$ is the trivial group, $X_K$ is the *universal covering space* $\widetilde{X}$. In general, we have a commuting diagram of spaces with free, locally proper actions:

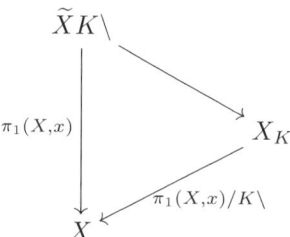

Let $\nu' : (X', x') \to (X, x)$ be a covering projection such that $\nu'_*(\pi_1(X', x')) = K \subset \pi_1(X,x)$. Let $X_K$ be the covering space of $X$ associated with the space with base point $(X,x)$. If $p \in P(X,x)$, then its "$\sim_K$" equivalence class represents a point $\widehat{y} \in X_K$. We may lift the path $p$ to a path $\widetilde{p}$ in $X'$ with initial point $x'$ and terminal point $\widetilde{p}(1)$. The projections $\nu(\widehat{y}) = p(1)$ and $\nu'(\widetilde{p}(1))$ are equal.

EXERCISE 2.1.3 ($X_K$ is unique). Show that the assignment $\widehat{y} \xrightarrow{f} \widetilde{p}(1)$ defines an equivalence

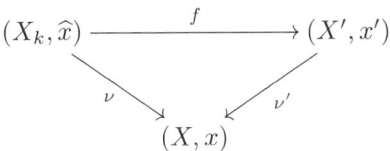

We may now transfer all of our constructions with $X_K$ to $X'$ by just lifting the paths used in the construction of $X_K$ to $X'$.

2.1.4 (Induced homomorphism on fundamental groups). A continuous map
$$f : (X, x_0) \to (Y, y_0)$$
induces a map on the spaces of paths
$$\mathrm{P}f : P(X, x_0) \longrightarrow P(Y, y_0),$$
defined by $p(t) \mapsto (f \circ p)(t)$. In particular, this restricts to a map on the loop spaces
$$\Omega(X, x_0) \longrightarrow \Omega(Y, y_0).$$
Since $p \simeq q$ implies $f \circ p \simeq f \circ q$, this yields a homomorphism $f_*$ by
$$f_* : \pi_1(X, x_0) \longrightarrow \pi_1(Y, y_0),$$
$$\alpha \longrightarrow f \circ \alpha,$$
where $\alpha \in \pi_1(X, x_0)$ is viewed as a path $(I, \partial I) \to (X, x_0)$.

We shall describe how $f_*$ is related to a *lifting* of $f$ to the universal covering spaces. Let $\nu : (\widetilde{X}, \widetilde{x}_0) \longrightarrow (X, x_0)$, $\nu' : (\widetilde{Y}, \widetilde{y}_0) \longrightarrow (Y, y_0)$ be the universal covering projections, where $\widetilde{x}_0$ and $\widetilde{y}_0$ are the classes of constant paths. Recall $\widetilde{X}$ and $\widetilde{Y}$ are quotient spaces of $P(X, x_0)$ and $P(Y, y_0)$. Naturally, the map $\mathrm{P}f$ induces a map $\widetilde{f} : (\widetilde{X}, \widetilde{x}_0) \longrightarrow (\widetilde{Y}, \widetilde{y}_0)$, a continuous lifting of $f$, and the diagram

$$\begin{array}{ccc} (\widetilde{X}, \widetilde{x}_0) & \xrightarrow{\widetilde{f}} & (\widetilde{Y}, \widetilde{y}_0) \\ \nu \downarrow & & \downarrow \nu' \\ (X, x_0) & \xrightarrow{f} & (Y, y_0) \end{array}$$

is commutative. Note that $\widetilde{f}(\widetilde{x}) = f \circ \widetilde{x}$ ($x$ as a path class).

Now let $\ell(t)$ be a loop at $x_0$ (representing $\alpha$). Then
$$(f \circ (\ell * p))(t) = \begin{cases} f(\ell(2t)), & 0 \leq t \leq 1/2, \\ f(p(2t-1)), & 1/2 \leq t \leq 1 \end{cases}$$
$$= (f \circ \ell * f \circ p)(t),$$
which yields $\mathrm{P}f(\ell * p) = \mathrm{P}f(\ell) * \mathrm{P}f(p)$. This in turn, gives rise to $\widetilde{f}(\alpha(\widetilde{x})) = \widetilde{f}(\alpha) * \widetilde{f}(\widetilde{x}) = f_*(\alpha) \cdot \widetilde{f}(\widetilde{x})$. We have shown

LEMMA 2.1.5. *Let $f : (X, x_0) \longrightarrow (Y, y_0)$ be a continuous map. Suppose $\widetilde{f} : (\widetilde{X}, \widetilde{x}_0) \longrightarrow (\widetilde{Y}, \widetilde{y}_0)$ is a lifting of $f$, and $f_* : \pi_1(X, x_0) \longrightarrow \pi_1(Y, y_0)$ is the induced homomorphism. Then $\widetilde{f} \circ \alpha = f_*(\alpha) \circ \widetilde{f}$ for $\alpha \in \pi_1(X, x_0)$.*

2.1.6 (Change of base points). We shall sketch the well-known procedure for changing base points. Let $a \in P(X, x_0, y_0)$, a path from $x_0$ to $y_0$. We define

$$T_a : P(X, x_0) \to P(X, y_0) \quad \text{and} \quad L_a : \Omega(X, x_0) \to \Omega(X, y_0)$$

by

$$T_a(p) = \bar{a} * p, \quad L_a(\ell) = \bar{a} * \ell * a,$$

for $p \in P(X, x_0)$ and $\ell \in \Omega(X, x_0)$. Similarly, we define $T_{\bar{a}} : P(X, y_0) \to P(X, x_0)$ and $L_{\bar{a}} : \Omega(X, y_0) \to \Omega(X, x_0)$. We have the following

$$T_{\bar{a}} \circ T_a(p) \simeq p, \quad T_a \circ T_{\bar{a}}(q) \simeq q,$$
$$L_{\bar{a}} \circ L_a(\ell) \simeq \ell, \quad L_a \circ L_{\bar{a}}(h) \simeq h,$$

with all homotopies keeping the end points fixed. Furthermore, the homotopy classes of the images of $T_a$, $T_{\bar{a}}$, $L_a$ and $L_{\bar{a}}$ depend only upon the homotopy class of $a$ and the variables $p$, $\ell$, etc. We also have that

$$[L_a(\ell_1 * \ell_2)] = [L_a(\ell_1)][L_a(\ell_2)],$$

where $[\ell]$ denotes the homotopy class of $\ell$. Similarly for $L_{\bar{a}}$.

Thus, we have shown that $L_a$ induces an isomorphism of $\pi_1(X, x_0)$ to $\pi_1(X, y_0)$ via $[\ell] \mapsto [L_a(\ell)]$. If $K$ is a subgroup (respectively, a normal subgroup) of $\pi_1(X, x_0)$, then $L_a : K \to \bar{a}Ka$ is a subgroup (respectively, a normal subgroup) of $\pi_1(X, y_0)$. We may now establish an isomorphism between the covering space $(X_K, \hat{x}_0)$ over $(X, x_0)$ and the covering space $(X_{\bar{a}Ka}, \hat{y}_0)$ over $(X, y_0)$. Consider $p \in P(X, x_0, x)$ and $b \in P(X, y_0, x)$. Suppose $p \simeq_K a * b$. Then $\bar{a} * p \simeq_{\bar{a}Ka} b$, for, $p * \bar{b} * \bar{a} \in K$ and $\bar{a} * (p * \bar{b} * \bar{a}) * a \in \bar{a}Ka$.

Let us also observe that $T_a(p) \simeq T_c(p)$ and $L_a(\ell) \simeq L_c(\ell)$ if $a * \bar{c} \in \mathcal{Z}(\pi_1(X, x_0))$, the center of $\pi_1(X, x_0)$. More importantly, if $[\ell] \in \mathcal{Z}(\pi_1(X, x_0))$, then $L_a[\ell]$ is independent of the homotopy class of $a$ and will be a central element of $\pi_1(X, y_0)$.

## 2.2. Lifting group actions to covering spaces

In this section, we shall describe how we may lift actions on a space $X$ to actions on covering spaces of $X$.

2.2.1 (Universal lifting sequence). Let $\nu : \widetilde{X} \longrightarrow X$ be the universal covering projection, and $f : X \longrightarrow X$ a homeomorphism. Then $f$ lifts to a homeomorphism $\widetilde{f} : \widetilde{X} \longrightarrow \widetilde{X}$, making the diagram

$$\begin{array}{ccc} \widetilde{X} & \xrightarrow{\widetilde{f}} & \widetilde{X} \\ \downarrow & & \downarrow \\ X & \xrightarrow{f} & X \end{array}$$

commutative. Since $\pi_1(X, x)$ acts on $\widetilde{X}$ effectively as the group of covering transformations, we view $\pi_1(X, x)$ as a subgroup of $\text{TOP}(\widetilde{X})$, the group of all self-homeomorphisms of $\widetilde{X}$.

By elementary covering space theory, every homeomorphism of $X$ lifts to a homeomorphism of $\widetilde{X}$. Moreover, since each element $\alpha \in \pi_1(X, x)$ induces the identity map on $X$, there are many such lifts $\widetilde{f}$; namely, $\widetilde{f} \circ \alpha$, for all $\alpha \in \pi_1(X, x)$, are lifts of $f$. Consequently, the totality of the liftings of the group $\text{TOP}(X)$ to

## 2.2. LIFTING GROUP ACTIONS TO COVERING SPACES

$\mathrm{TOP}(\widetilde{X})$ is $N(\pi_1(X,x))$, the normalizer of $\pi_1(X,x)$ in $\mathrm{TOP}(\widetilde{X})$. Thus we have an exact sequence of groups:

$$1 \longrightarrow \pi_1(X,x) \longrightarrow N(\pi_1(X,x)) \longrightarrow \mathrm{TOP}(X) \longrightarrow 1.$$

2.2.2 (Lifting sequence). Let $\rho : G \to \mathrm{TOP}(X)$ be an action of $G$ on $X$, and let $K \subset G$ be the kernel of $\rho$. That is, $K$ is the ineffective part of the action. The group of all liftings of elements of $G$ is denoted by $G^*$, and fits into the following commuting diagram:

$$\begin{array}{ccccccccc}
1 & \longrightarrow & \pi_1(X,x) & \longrightarrow & \pi_1(X,x) \times K & \longrightarrow & K & \longrightarrow & 1 \\
& & {\scriptstyle =}\downarrow & & \downarrow & & \downarrow & & \\
1 & \longrightarrow & \pi_1(X,x) & \longrightarrow & G^* & \longrightarrow & G & \longrightarrow & 1 \\
& & & & {\scriptstyle \rho^*}\downarrow & & {\scriptstyle \rho}\downarrow & & \\
& & & & \mathrm{TOP}(\widetilde{X}) & \longrightarrow & \mathrm{TOP}(X). & &
\end{array}$$

In fact, the lifts of all of $\mathrm{TOP}(X)$ is $N(\pi_1(X,x)) = N_{\mathrm{TOP}(\widetilde{X})}(\pi_1(X,x))$, the normalizer of $\pi_1(X,x)$ in $\mathrm{TOP}(\widetilde{X})$. Note that $N(\pi_1(X,x)) \subset \mathrm{TOP}(\widetilde{X})$ is the group of all homeomorphism of $\widetilde{X}$ which induce homeomorphisms on $X$. The normalizer acts on $\pi_1(X,x)$ by conjugation yielding a homomorphism $\theta : N(\pi_1(X,x)) \longrightarrow \mathrm{Aut}(\pi_1(X,x))$. This induces a homomorphism into the outer automorphism group $\psi : \mathrm{TOP}(X) \longrightarrow \mathrm{Out}(\pi_1(X,x))$ yielding the commutative diagram of exact sequences:

$$\begin{array}{ccccccccc}
1 & \longrightarrow & \pi_1(X,x) & \longrightarrow & N(\pi_1(X,x)) & \longrightarrow & \mathrm{TOP}(X) & \longrightarrow & 1 \\
& & {\scriptstyle \theta}\downarrow & & {\scriptstyle \theta}\downarrow & & {\scriptstyle \psi}\downarrow & & \\
1 & \longrightarrow & \mathrm{Inn}(\pi_1(X,x)) & \longrightarrow & \mathrm{Aut}(\pi_1(X,x)) & \longrightarrow & \mathrm{Out}(\pi_1(X,x)) & \longrightarrow & 1.
\end{array}$$

Now with the action $\rho : G \to \mathrm{TOP}(X)$, we pull back the top exact sequence to get (see Section 5.3.1):

$$\begin{array}{ccccccccc}
1 & \longrightarrow & \pi_1(X,x) & \longrightarrow & G^* & \longrightarrow & G & \longrightarrow & 1 \\
& & {\scriptstyle =}\downarrow & & {\scriptstyle \rho^*}\downarrow & & {\scriptstyle \rho}\downarrow & & \\
1 & \longrightarrow & \pi_1(X,x) & \longrightarrow & N(\pi_1(X,x)) & \longrightarrow & \mathrm{TOP}(X) & \longrightarrow & 1.
\end{array}$$

The top exact sequence is called the *lifting exact sequence* of the group action $(G,X)$; the action $(G^*, \widetilde{X})$ is called the *extended lifting* of $(G,X)$. When the lifting exact sequence splits (i.e., $G^* = \pi_1(X,x) \rtimes G$), the action $(G, \widetilde{X})$ is called a *lifting* of $(G,X)$. Note that $(G^*, \widetilde{X})$ always exists, but $(G, \widetilde{X})$ may not. See the next example (and the next three sections).

EXERCISE 2.2.3. (1) Show that if $G = \mathbb{Z}_2$ acts on the circle group $X = S^1 = \{z \in \mathbb{C} : |z| = 1\}$ by $z \mapsto \bar{z}$, then $G^* = \mathbb{Z} \rtimes \mathbb{Z}_2$ acts on $\widetilde{X} = \mathbb{R}$ as in Example 1.6.8. In this case, the $(\mathbb{Z}_2, S^1)$-action lifts to $(\mathbb{Z}_2, \mathbb{R}^1)$.

(2) Consider $\eta(z) = ze^{\pi i}$. Since $\eta^2(z) = ze^{2\pi i} = z$, $\eta$ induces an action of $\mathbb{Z}_2$ on $S^1$. The lifting sequence is

$$0 \to \mathbb{Z} \to \mathbb{Z} \to \mathbb{Z}_2 \to 1,$$

which does not split. So, the $(\mathbb{Z}_2, S^1)$-action does not have a lift $(\mathbb{Z}_2, \mathbb{R})$. Instead, there is an extended lifting $(\mathbb{Z}, \mathbb{R}^1)$.

PROPOSITION 2.2.4. *Suppose $G$ is a discrete group, acting properly on a completely regular space $X$. Then the extended-lifting $G^*$ on the universal covering space $\widetilde{X}$ is proper.*

PROOF. The proposition follows from Theorem 1.9.3. □

EXERCISE 2.2.5. Let $G^*$ be an extended-lifting of $(G, X)$ to the universal covering $\widetilde{X}$. Suppose $K$ is a normal subgroup of $\pi_1(X, x)$ and conjugation by $G^*$ on $\pi_1(X, x)$ leaves $K$ invariant. Then there is an induced action of $G^*/K$ on $K\backslash\widetilde{X}$. This action is proper if $G$ is discrete, acting properly, and $X$ is completely regular. A proof follows from Theorem 1.9.3 and Proposition 1.2.4(8).

REMARK 2.2.6. The lifting exact sequence can be made very explicit. Take $x$ as a base point, and for each $g \in G$, take a path $p_g \in P(X, x)$ so that $p_g(1) = gx$. For a point in $b \in \widetilde{X}$, take a path $\tilde{b} \in P(X, x)$ representing $b$. A lift of $g$ can be given as follows: Define a path $p_g * (g \cdot \tilde{b}) \in P(X, x)$. Then the map $\tilde{b} \mapsto p_g * (g \cdot \tilde{b})$ defines a map $\widetilde{X} \to \widetilde{X}$, which is a lift of $g : X \to X$.

A careful description and construction of an explicit lifting exact sequence as an extension of $\pi_1(X, x)/K$ and $G$ can be found in [**CR72d**, §2]. The explicit and rather technical construction enables one to give a detailed analysis of the group structure of $G^*$ in terms of the choice of the paths $p_g$. The explicitness then leads to several interesting applications.

If we assume that $G$ has a fixed point at $x$, then $p_g$ can be chosen to be the trivial path and one constructs the extended-lifting $G^*$ as $\pi_1(X, x) \rtimes_\varphi G$. The automorphism $\varphi : G \longrightarrow \mathrm{Aut}(\pi_1(X, x))$ is induced from sending a loop class $[\ell(t)] \in \pi_1(X, x)$ to the loop class $[g(\ell(t))] \in \pi_1(X, x)$. Thus, $\varphi(g)([\ell(t)]) = [g(\ell(t))]$. We examine this type of lifting in the next section.

## 2.3. Lifting an action of $G$ when $G$ has a fixed point

Throughout this section, $G$ is a topological group acting on $X$ with a *base point $x$ which is fixed under the action of $G$*. Each element $g \in G$, considered as a continuous map $g : (X, x) \to (X, x)$, induces an automorphism $g_* : \pi_1(X, x) \longrightarrow \pi_1(X, x)$ by $g_*[\ell(t)] = [g \cdot \ell(t)]$. Clearly, $g \mapsto g_*$ defines a homomorphism

$$G \longrightarrow \mathrm{Aut}(\pi_1(X, x)).$$

Observe that this homomorphism factors through $G \longrightarrow G/G_0$, where $G_0$ is the path-component of $G$ containing the identity element.

Furthermore, since $x \in X^G = \mathrm{Fix}(G, X)$, $G$ operates naturally on the space of paths issuing out of $x$, $P(X, x)$. We wish to explicitly describe the liftings of this action to covering spaces. We shall always denote the base point of a covering space by the equivalence class of the trivial path and denote it by $\hat{x}$. We will denote $\varphi(g)y$ by $g \cdot y$, for every $y \in X$.

THEOREM 2.3.1 ([**CR69**, 3.1]). *Suppose $G$ is a topological group acting on $X$ with $x \in \mathrm{Fix}(G, X)$. If $K \subset \pi_1(X, x)$ is invariant under the action of $G$ on $\pi_1(X, x)$, then there is a covering action*

$$(G, (X_K, \hat{x})) \xrightarrow{\nu} (G, (X, x))$$

*for which $\nu$ is $G$-equivariant.* If $K$ *is normal in* $\pi_1(X, x)$*, then*

$$g \cdot (\alpha \hat{y}) = g_*(\alpha)(g \cdot \hat{y})$$

*for all* $g \in G$, $\hat{y} \in X_K$, *and* $\alpha \in \pi_1(X, x)$. *The extended-lifting on* $X_K$ *is a semidirect product* $(\pi_1(X, x)/K) \rtimes G$ *which operates as*

$$(\alpha, g)(\hat{y}) = \alpha(g \cdot \hat{y}).$$

*Thus,*

$$1 \longrightarrow \pi_1(X, x)/K \longrightarrow (\pi_1(X, x)/K) \rtimes G \longrightarrow G \longrightarrow 1$$

*is exact, where* $\pi_1(X, x)/K$ *is the deck transformation group of the covering space* $X_K \longrightarrow X$.

PROOF. $G$ acts on $P(X, x)$ leaving the trivial path fixed. The $G$-invariance of $K$ allows us to introduce an action of $G$ on $(X_K, \hat{x})$: If $p(t)$ represents the equivalence class of a point $\hat{y}$ in $X_K$ (a path issuing from $x \in X$), then $g \cdot \hat{y}$ is the equivalence class of $g \cdot p(t)$.

To show that this is well defined, suppose $q(t)$ represents the same equivalence class of the point $\hat{y}$ in $X_K$; that is, $[p(t) * \bar{q}(t)] \in K$. Then,

$$[g \cdot p(t) * \overline{g \cdot q(t)}] = [g \cdot (p(t) * \bar{q}(t))] \in g_*[(p(t) * \bar{q}(t))] \in g_*(K) = K,$$

which shows $g \cdot p(t)$ and $g \cdot q(t)$ represent the same point in $X_K$.

Now let $\ell(t)$ be a loop at $x$ (representing $\alpha$), then

$$g \cdot (\ell * p)(t) = \begin{cases} g \cdot \ell(2t), & 0 \le t \le 1/2, \\ g \cdot p(2t-1), & 1/2 \le t \le 1 \end{cases}$$
$$= (g \cdot \ell * g \cdot p)(t),$$

which yields the formula $g \cdot (\alpha \hat{y}) = g_*(\alpha)(g \cdot \hat{y})$.

Let us now define an action of $(\pi_1(X, x)/K) \rtimes G$ on $X_K$ by

$$(\alpha, g)\hat{y} = \alpha(g \cdot \hat{y}).$$

This defines an action because

$$\begin{aligned}(\beta, h)(\alpha, g)(\hat{y}) &= (\beta, h)(\alpha(g \cdot \hat{y})) \\ &= \beta(h \cdot (\alpha(g \cdot \hat{y}))) \\ &= \beta(h_*(\alpha)(h \cdot (g \cdot \hat{y}))) \\ &= (\beta h_*(\alpha))((hg) \cdot \hat{y}) \\ &= (\beta h_*(\alpha), hg)(\hat{y}).\end{aligned}$$

Also

$$\nu((\alpha, g)\hat{y}) = \nu(\alpha(g \cdot \hat{y})) = \nu(g \cdot \hat{y}) = g \cdot \nu(\hat{y}).$$

The map $\nu$ is equivariant under the homomorphism which projects $(\pi_1(X, x)/K) \rtimes G$ onto its second coordinate. The lifting of $G$ described above is just the restriction to $1 \rtimes G \subset (\pi_1(X, x)/K) \rtimes G$. This is clearly the full lifting sequence of all lifts of $G$-actions covering the given $G$-action. □

COROLLARY 2.3.2. *Suppose a topological group $G$ acts on $X$ with $x \in \mathrm{Fix}(G, X)$. Then the lifting sequence*

$$1 \longrightarrow \pi_1(X, x) \longrightarrow G^* \longrightarrow G \longrightarrow 1$$

*splits so that* $G^* = \pi_1(X, x) \rtimes G$.

PROPOSITION 2.3.3. *Suppose $G$ acts on $X$, leaving a normal subgroup $K$ of $\pi_1(X, x)$ invariant. Put $E = (X_K)^G$. Let $\alpha \in \pi_1(X,x)/K$, $\alpha \neq 1$. Then $g_*(\alpha) = \alpha$ for all $g \in G$ if and only if $\alpha E \cap E \neq \emptyset$ if and only if $\alpha E = E$.*

PROOF. Suppose $g_*(\alpha) = \alpha$. Let $\hat{y} \in E$. Then $g \cdot (\alpha \hat{y}) = g_*(\alpha)(g \cdot \hat{y}) = g_*(\alpha)(\hat{y}) = \alpha \hat{y}$. This implies $\alpha E \subset E$. For $\hat{y} \in E$, consider $\alpha^{-1}\hat{y}$. Then $g \cdot (\alpha^{-1}\hat{y}) = g_*(\alpha^{-1})(g \cdot \hat{y}) = \alpha^{-1}\hat{y}$ so that $\alpha^{-1}E \subset E$. Consequently, $\alpha E = E$.

Conversely, suppose there exists $\hat{y} = \alpha\hat{z}$, with $\hat{y} \in E$ and $\hat{z} \in E$. Then $\alpha(\hat{z}) = \hat{y} = g \cdot (\hat{y}) = g \cdot (\alpha\hat{z}) = g_*(\alpha)(g \cdot \hat{z}) = g_*(\alpha)(\hat{z})$. Since $\alpha, g_*(\alpha)$ are elements of deck transformation, $g_*(\alpha) = \alpha$. □

COROLLARY 2.3.4. *Suppose $G$ acts on $X$, leaving the normal subgroup $K$ of $\pi_1(X, x)$ invariant. Put $F = X^G$, and $E = (X_K)^G$. If $\Gamma = \{\alpha \in \pi_1(X,x)/K : g_*(\alpha) = \alpha$, for all $g \in G\}$, then $\Gamma \backslash E = \nu(E) \subset X^G$.*

LEMMA 2.3.5. *Suppose $G$ acts on $X$, leaving the normal subgroup $K$ of $\pi_1(X, x)$ invariant. Put $F = X^G$, and $E = (X_K)^G$. Let $E_{\hat{x}}$ denote the path component of $E$ that contains $\hat{x}$ and $F_x$ the path component of $F = X^G$ that contains $x$. Then $\nu(E_{\hat{x}}) = F_x$.*

PROOF. Clearly $\nu(E_{\hat{x}}) \subset F_x$. Suppose $y \in F_x$ and $p$ is a path in $F_x$ from $x$ to $y$. Then the lift of this path is a path starting at $\hat{x}$ and ending at $\hat{y}$, where $\nu\hat{y} = y$. Since $g \circ p(t) = p(t)$ for each $t$, this lift is in $E_{\hat{x}}$. So $\nu$ maps $E_{\hat{x}}$ onto $F_x$. □

COROLLARY 2.3.6. *Suppose $G$ acts on $X$, leaving the normal subgroup $K$ of $\pi_1(X, x)$ invariant. Put $F = X^G$, and $E = (X_K)^G$. If $E$ is path-connected, then the image of $\pi_1(F, x) \longrightarrow \pi_1(X,x)/K$ is the subgroup*

$$\Gamma = \{\alpha \in \pi_1(X,x)/K : g_*(\alpha) = \alpha \text{ for all } g \in G\} = (\pi_1(X,x)/K)^G.$$

*The following diagram of exact rows*

$$\begin{array}{ccccccccc} 1 & \longrightarrow & \pi_1(E, \hat{x}) & \longrightarrow & \pi_1(F, x) & \longrightarrow & \Gamma & \longrightarrow & 1 \\ & & \downarrow & & \downarrow & & \downarrow & & \\ 1 & \longrightarrow & K & \longrightarrow & \pi_1(X, x) & \longrightarrow & \pi_1(X,x)/K & \longrightarrow & 1 \end{array}$$

*is commutative. If, in addition, $\Gamma = \pi_1(X,x)/K$, then $\nu^{-1}(F_x) = E$.*

PROOF. We have $\Gamma \backslash E = F_x$, the path component of $F$ containing $x$. For $\alpha \in \Gamma$, if $\alpha\hat{x} = \hat{y}$, then $\hat{y} \in E$ and there is a path $\alpha(t)$ in $E$ starting at $\hat{x}$ and ending at $\hat{y}$ and representing $\alpha$. Then $\nu(\alpha(t))$ is a loop in $F_x$ based at $x$. This represents a nontrivial element of $\pi_1(X,x)/K$ if and only if the image of this class does not lie in $K$, or equivalently, $\hat{x} \neq \hat{y}$. Conversely, a loop $\ell$ in $F_x$ based at $x$, lifts to a path in $E$ starting at $\hat{x}$ and ending at $\alpha(\hat{x})$, where $\alpha \in \text{Image}(\pi_1(F_x, x))$ in $\pi_1(X,x)/K$ is represented by $\ell$. If $\Gamma = \pi_1(X,x)/K$, then $\nu^{-1}(F_x) = E$ for $\alpha(\hat{x})$ is in $E$, for each $\alpha \in \Gamma$. □

PROPOSITION 2.3.7. *Suppose $G$ is compact and acts on the locally compact Hausdorff $X$ with fixed point, leaving the normal subgroup $K$ of $\pi_1(X, x)$ invariant. Then $(\pi_1(X,x)/K) \rtimes G = G^*$ acts properly on $X_K$.*

PROOF. Let $C$ be a compact subset of $X_K$. Then we want to show that closure of $\{(\alpha, g) \in G^* : (\alpha, g)C \cap C \neq \emptyset\}$ is compact. We know $\{\alpha \in \pi_1(X,x)/K : \alpha(G \cdot C) \cap (G \cdot C) \neq \emptyset\}$ is finite, say, $\alpha_1, \ldots, \alpha_k$, because $G \cdot C$ is compact, and

$\pi_1(X,x)/K$ acts locally properly on $X_K$ (and hence properly as $X$ is Hausdorff). Now $(\alpha_i, g)C \cap C \neq \emptyset$ if and only if $gC \cap \alpha_i^{-1}C \neq \emptyset$. Since $G$ is compact, the closure of $\{g \in G : gC \cap \alpha_i^{-1}C \neq \emptyset\}$ is compact. The conclusion follows because a finite union of a compact set is compact. $\square$

EXAMPLE 2.3.8 ([**CR69**, 3.7]). Let $X$ be a 2-sphere $S^2$ and consider the rotation of 180 degrees around the north-south axis. This action is equivariant with respect to the antipodal map and so induces an action of $\mathbb{Z}_2$ on the projective plane $P_2$ with fixed point equal to an isolated point (the image of the poles) and a circle which generates the fundamental group (the image of the equator). If we now lift this induced action on $P_2$ to its universal covering space, using the isolated fixed point as a base point, we obtain an action equivalent to the original rotation about the polar axis. Notice that $E \neq \nu^{-1}(F)$ and $E$ consists of only two points; $\pi_1(\text{pt}) \to \pi_1(P_2)$ is surely not onto. On the other hand, if we were to lift the action but with base point at one of the points on the circle, we would induce an action on $S^2$ equivalent to reflecting across the equator. In this case, since $E$ projects onto the component containing the base point, and $\pi_1(F_x) \to \pi_1(P_2)$ is onto.

## 2.4. Evaluation homomorphism

We are also interested in getting a more precise description of lifting group actions when $G$ is path-connected.

Let $G$ be a path-connected group acting on a path-connected Hausdorff space $X$. Fix $x \in X$. The evaluation map $\text{ev}^x : (G, e) \to (X, x)$ is defined by

$$\text{ev}^x(g) = g \cdot x.$$

This induces the evaluation homomorphism

$$\text{ev}^x_* : \pi_1(G, e) \to \pi_1(X, x).$$

The main reference for the following is [**CR69**, §4]. We will see that in general the image of $\pi_1(G, e)$ in $\pi_1(X, x)$ is a central subgroup of $\pi_1(X, x)$ and independent of the base point $x$.

LEMMA 2.4.1. *Let $g \in P(G, e)$ and $p \in P(X, x)$ (paths emanating from $e$ and $x$, respectively). Then the three paths*

(A) $\begin{cases} g(2t) \cdot x, & 0 \leq t \leq 1/2, \\ g(1) \cdot p(2t-1), & 1/2 \leq t \leq 1, \end{cases}$

(B) $\quad g(t) \cdot p(t), \quad 0 \leq t \leq 1,$

(C) $\begin{cases} p(2t), & 0 \leq t \leq 1/2, \\ g(2t-1) \cdot p(1), & 1/2 \leq t \leq 1 \end{cases}$

*are homotopic by fixed endpoint homotopies.*

PROOF. Schematically, we will have a diagram below. Introduce a path in $P(G, e)$ by

$$(g * c_{g(1)})(t) = \begin{cases} g(2t), & 0 \leq t \leq 1/2, \\ g(1), & 1/2 \leq t \leq 1, \end{cases}$$

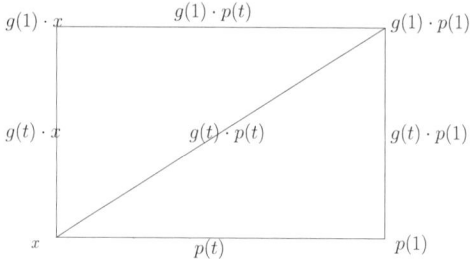

and a path in $P(X,x)$ by

$$(c_x * p)(t) = \begin{cases} x, & 0 \leq t \leq 1/2, \\ p(2t-1), & 1/2 \leq t \leq 1 \end{cases}$$

($c$ denotes the constant path). Now $g * c_{g(1)} \simeq g$ and $(c_x * p) \simeq p$ implies $(g * c_{g(1)}) \cdot (c_x * p) \simeq g \cdot p$ relative $\partial I$. But $(g * c_{g(1)}) \cdot (c_x * p) = (g \cdot x) * g(1) \cdot p$ is the path labeled (A), while $g \cdot p$ is the path labeled (B).

Thus the path (A) is homotopic to (B) via a fixed endpoint homotopy. A similar argument relates (C) to (B). $\square$

THEOREM 2.4.2. *The image of $\mathrm{ev}_*^x : \pi_1(G, e) \to \pi_1(X, x)$ is a central subgroup of $\pi_1(X, x)$ which is independent of choices of $x$.*

PROOF. Suppose $g \in P(G, e)$ and $p \in P(X, x)$ are closed loops. Then $\mathrm{ev}_*^x$ is induced by

$$g(t) \mapsto g(t) \cdot x.$$

That is, $\mathrm{ev}_*^x[g] = [g(t) \cdot x]$. Noting that $g(1) = e$ and $p(1) = x$, we have

$$g(t) \cdot x * p(t) = \begin{cases} g(2t) \cdot x, & 0 \leq t \leq 1/2, \\ g(1) \cdot p(2t-1), & 1/2 \leq t \leq 1, \end{cases}$$

while

$$p(t) * (g(t) \cdot x) = \begin{cases} p(2t), & 0 \leq t \leq 1/2, \\ g(2t-1) \cdot p(1), & 1/2 \leq t \leq 1. \end{cases}$$

In view of Lemma 2.4.1, these two loops represent the same element of $\pi_1(X, x)$,

$$\mathrm{ev}_*^x([g]) \cdot [p] = [p] \cdot \mathrm{ev}_*^x([g]),$$

thus the image of $\mathrm{ev}_*^x$ lies in the center of $\pi_1(X, x)$.

Let $y$ be another point in $X$ and $C$ a path from $x$ to $y$. Then $C$ induces an isomorphism $C_\# : \pi_1(X, y) \longrightarrow \pi_1(X, x)$ by sending a loop $\ell$ based at $y$ to a loop $C * \ell * \overline{C}$ based at $x$. Let $\ell = g(t) \cdot y$, and $\ell' = g(t) \cdot x$. Then $C * \ell * \overline{C} \simeq \ell'$ (for the homotopy just moves along $C$). Choosing a different path $C'$ will send $\ell$ to a loop homotopic to $\ell'$. Therefore, $\mathrm{ev}_*^x(\pi_1(G, e))$ is independent of choice. $\square$

## 2.5. Lifting connected group actions

THEOREM 2.5.1 ([**CR69**, §4], [**Bre72**, Chapter I, §9]). *Let $G$ be a path-connected topological group acting on a space $X$ which admits covering space theory. Let $K$ be a normal subgroup of $\pi_1(X, x)$ containing the image of $\mathrm{ev}_*^x : \pi_1(G, e) \to \pi_1(X, x)$ and put $Q = \pi_1(X, x)/K$. Then*

(1) The $G$-action on $X$ lifts to a $G$-action on $X_K$ which commutes with the covering $Q$-action.
(2) The lifted $G$-action does not depend upon the choice of base points in $X$ or in $X_K$.
(3) If, in addition, $X$ is completely regular and $G$ is a connected Lie group acting properly on $X$, then the $G$-action on $X_K$ is proper and the induced action of $Q$ on $W = G\backslash X_K$ is also proper.

PROOF. (1) The $G$-action on $X_K$ is described as follows. Given $u \in G$ and $\widehat{y} \in X_K$, select a path $g \in P(G,e)$ with $g(1) = u$, and choose a path $p$ which represents $\widehat{y}$. We define $u \cdot \widehat{y}$ to be the common equivalence class of the three paths listed in Lemma 2.4.1. In particular, $u \cdot \widehat{y}$ is represented by a path $(g \cdot x) * (u \cdot p)$ connecting $x$ and $u \cdot y$, where $y = p(1)$.

(Well-definedness). Suppose $g' \in P(G,e)$ also has $g'(1) = u$, and $p'$ also represents $\widehat{y}$. Then

$$((g \cdot x) * (u \cdot p)) * \overline{(g' \cdot x) * (u \cdot p')}$$
$$\simeq (g \cdot x) * ((\overline{g'} \cdot x) * (g' \cdot x)) * (u \cdot p) * (u \cdot \overline{p'}) * (\overline{g'} \cdot x)$$
$$\simeq ((g * \overline{g'}) \cdot x) * ((g' \cdot x) * u \cdot (p * \overline{p'}) * (\overline{g'} \cdot x)).$$

Since $K$ contains $\operatorname{Im}\{\operatorname{ev}^x_* : \pi_1(G,e) \to \pi_1(X,x)\}$, $(g * \overline{g'}) \cdot x$ represents an element of $K$.

Now consider the path $(g' \cdot x) * u \cdot (p * \overline{p'}) * (\overline{g'} \cdot x) = (g' \cdot x) * g'(1) \cdot (p * \overline{p'}) * (\overline{g'} \cdot x)$. The map
$$F(t,s) = (g'(st) \cdot x) * g'(s) \cdot (p(t) * \overline{p'}(t)) * (\overline{g'}(st) \cdot x)$$
gives a homotopy from $F(t,0) = p(t) * \overline{p'}(t)$ to $F(t,1) = (g' \cdot x) * g'(1) \cdot (p * \overline{p'}) * (\overline{g'} \cdot x)$. Also, since $p * \overline{p'}$ represents an element of $K$, the definition of $u \cdot \widehat{y}$ does not depend on the choices of $g$ and $p$.

Recall that the covering action of $Q$ on $X_K$ is induced from the action of $\pi_1(X,x)$ on $\widetilde{X}$. Suppose $\alpha \in \pi_1(X,x)$ is represented by a closed loop $\ell$ at $x$. By Lemma 2.4.1, the part $0 \le t \le 3/4$ of the two paths

$$\begin{cases} g(2t) \cdot x, & 0 \le t \le 1/2, \\ u \cdot \ell(4t-2), & 1/2 \le t \le 3/4, \\ u \cdot p(4t-3), & 3/4 \le t \le 1, \end{cases}$$

and

$$\begin{cases} \ell(2t), & 0 \le t \le 1/2, \\ g(4t-2) \cdot x, & 1/2 \le t \le 3/4, \\ u \cdot p(4t-3), & 3/4 \le t \le 1 \end{cases}$$

are equivalent so that $u \cdot (\alpha \cdot \widehat{y}) = \alpha \cdot (u \cdot \widehat{y})$ in $X_K$. This shows the $G$-action and $Q$-action on $X_K$ commute with each other. We have not used complete regularity or the fact that $G$ is a Lie group so far, but these assumptions are needed for (3).

(2) If $\nu : (X', x') \to (X, x)$ is a covering projection of the $G$-space $X$ with $\nu_*(\pi_1(X', x')) = K \supset \operatorname{ev}^x_*(\pi_1(G, e))$, then we can construct the lift of the $G$-action on $X'$ by lifting the paths used in the construction of $(G, X_K)$ to $X'$. We may then easily see what happens if we choose a different base point to lift paths.

Choose base points $y' \in X'$ such that $\nu(y') = y \in X$. Take a path $\beta : (I, 0, 1) \longrightarrow (X', y', x')$. If $b$ is a point in $X'$, then let $\gamma : (I, 0, 1) \longrightarrow (X', x', b)$.

For $g \in G$, let $g(t)$ be a path in $G$ from $e$ to $g$. Then $gb$ is given by the lift of the path $\nu(\gamma(t))*g(t)\cdot\nu(\gamma(1))$ with initial point $x'$. The point $b$ can also be represented by the end point of $\beta(t)*\gamma(t)$. Then $g(b)$ using $y'$ and $y$ as base points is represented by the lift of the paths $\nu(\beta(t)*\gamma(t))*g(t)\cdot\nu(\gamma(1))$ with initial point $y'$. Thus we see that we have the same image $gb$. So our lifting construction can be described at any base point *and the lifted $G$-action is independent of the choice of the base point.*

(3) Let the equivalence class of $\beta : (I, 0, 1) \to (X, x, y)$ represent a point $\hat{y}$ in $X_K$. We will show that the $G$-action on $X_K$ is locally proper at $\hat{y}$. Since the $G$-action on $X$ is proper, we can choose a slice $S_y$ at $y$ which we can assume to be path-connected and small enough so that when we lift a neighborhood of $y$ to $X_K$ at $\hat{y}$, the slice lifts homeomorphically to $\hat{S}$. The isotropy of $\hat{y}$ is $G_{\hat{y}} \subset G_y$. By continuity and the $G$-equivariance of $\nu$, $\hat{S}$ is $G_{\hat{y}}$-invariant. In fact, the map $G\hat{S} \xrightarrow{\nu'} GS_y \xrightarrow{f} G/G_y$ is $G$-invariant and $(f \circ \nu')^{-1}(G_y)$ gives a $G_y$-kernel. (Here $f$ is the $G$-invariant map defining the $G_y$-slice at $y$ and $\nu'$ denotes the restriction of $\nu$ to $G\hat{S}$.) $(f \circ \nu')^{-1}(G_y)$ will be disjoint copies of $\hat{S}$, one for each element of $G_y/G_{\hat{y}}$ which is a finite group since $G_y$ is compact, $G_{\hat{y}}$ is a closed normal subgroup and $G_y/G_{\hat{y}}$ is isomorphic to a subgroup of the discrete group $Q$.

(Let $p(t)$ be a path class representing $\hat{y} \in X_K$ where $\nu(\hat{y}) = y$. For $u \in G_y$, let $h \in P(G, e)$ with $h(1) = u$. Then $p(t)*h(t)\cdot y$ is the path class $u\cdot\hat{y}$. This image, $u\cdot\hat{y}$, is the same as the point represented by $p(t)*h(t)\cdot y*\overline{p(t)}*p(t)$. The loop $p(t)*h(t)\cdot y*\overline{p(t)}$ defines an element $\alpha \in \pi_1(X, x)$. This is well defined up to an element of $\text{Im}(\text{ev}_*^x)$. Denote its image in $\pi_1(X, x)/\text{Im}(\text{ev}_*^x) \to \pi_1(X, x)/K$ by $[\alpha]$. (Note that $u \cdot \hat{y}$ is the same as $[\alpha]\hat{y}$, where $[\alpha]$ acts as a covering transformation on $X_K$.) It is not difficult to check that the assignment $u \mapsto [\alpha]$ is an antihomomorphism of $G_y$ into $\pi_1(X, x)/K$, with kernel $G_{\hat{y}}$. In fact, $G_y/G_{\hat{y}}$ is anti-isomorphic to $Q_w \subset Q$, where $w$ is the orbit $G(\hat{y})$ in $W$ and $Q_w$ is the isotropy of the induced $Q$-action on $W$. (The reader may wish to compare this with Corollary 2.7.7 where the anti-isomorphism is converted to an isomorphism.))

In other words, we get an $H$-slice, $H = G_y$, $\bigcup_{g \in G_y/G_{\hat{y}}} g\hat{S}$, in $X_K$. This slice, $(G_y/G_{\hat{y}})(\hat{S})$, is a disjoint covering of the slice $S_y$ at $y \in X$.

Now since $X$ is completely regular, we need only show that $(G, X_K)$ is locally proper at the arbitrary point $\hat{y}$. But this is clear as $G \times_{G_{\hat{y}}} \hat{S}$ is an open path-connected $G$-invariant tube about the orbit $G\hat{y}$ which covers the tube $G \times_{G_y} S_y$. Therefore we have $\hat{S}$ is a $G_{\hat{y}}$-slice at $\hat{y}$.

The action of $G^*$, the group of homeomorphisms generated by the commuting groups $G$ and the covering transformations $Q$ is also locally proper because $(f \circ \nu)^{-1}(G \times_{G_y} S_y)$ is a $G^*$-invariant disjoint union of components each homeomorphic to the component $GS_{\hat{y}} = G\hat{S}$. Consequently, the action of $Q$ on $W = G \backslash X_K$ is also proper by Proposition 1.2.4(8). □

EXERCISE 2.5.2. Show that $G_{\hat{y}}$ is normal subgroup of $G_y$.

REMARK 2.5.3. Suppose $G$ is path-connected and acts on $X$ with fixed point $x$, and on $\pi_1(X, x)$ leaving the normal subgroup $K$ invariant. Then the lifted action of $G$ described in Sections 2.4 and 2.5 are identical. In particular, $G^*$ is just $G \times Q$, since it is a commuting semidirect product.

2.5.4 (Examples of lifting actions (cf. Example 2.3.8)). (1) Consider the standard action of $S^1$ on the 2-sphere $S^2$ by rotation around the axis joining the north and south poles. The action of $\mathbb{Z}_2$ by antipodal map commutes with the $S^1$-action. Thus, the $S^1$-action induces an action on $\mathbb{R}P_2 = S^2/\mathbb{Z}_2$.

For the action $(S^1, \mathbb{R}P_2)$, the evaluation homomorphism
$$\mathrm{ev}_*^{x*} : \pi_1(S^1) = \mathbb{Z} \to \mathbb{Z}_2 = \pi_1(\mathbb{R}P_2)$$
is trivial. (So it lifts to the 2-sphere $S^2$, and it lifts to the standard rotations about the polar axis.) The $N$ and $S$ poles project to the fixed-point set.

The projective line corresponding to the *boundary* of identification of the disk lifts to a semicircular arc. On $\mathbb{R}P_2$ this orbit has $\mathbb{Z}/2$ isotropy and so on $\mathbb{R}P_2$ when $e^{2\pi i t}(a)$ has gone from $a$ to $a$ as $0 \leq t \leq \frac{1}{2}$ the lifted action has gone half way around the equator. On $\mathbb{R}P_2$ we have a loop (going from 0 to $a = -a$ to 0) which lifts to an arc on $S^2$ going from $N$ to $S$ which *covers* the simple closed curve on $\mathbb{R}P_2$. Check explicitly that the lifts are independent of the base points in $\mathbb{R}P_2$.

(2) SO(3)-action on $\mathbb{R}P_3$. Since $\mathbb{R}P_3 \approx \mathrm{SO}(3)$, $\mathrm{SO}(3)$ acts on $\mathbb{R}P_3$ as just left translation. Moreover, $\mathrm{ev}_*^x : \pi_1(\mathrm{SO}(3)) = \mathbb{Z}_2 \to \mathbb{Z}_2 = \pi_1(\mathbb{R}P_3)$ is an isomorphism, so this action *cannot* be lifted to an action on $S^3$.

However, $S^3 = \mathrm{Spin}(3)$ doubly covers as a group $\mathrm{SO}(3)$, and $\mathrm{Spin}(3)$ is simply connected. So the noneffective action of $S^3$ on $\mathbb{R}P_3$ via $S^3 \to \mathrm{SO}(3)$ can be lifted to $S^3$. On $S^3$ it is a free transitive action.

(3) $S^1$-action on the Klein bottle $K$. Write the Möbius band as $M = S^1 \times_{\mathbb{Z}_2} I$, where $\mathbb{Z}_2$ acts by $(z, t) \mapsto (-z, -t)$, with $z \in S^1$, $t \in [-1, 1]$. Represent a point of $M$ by $\langle z, t \rangle$. Then $\langle z, t \rangle = \langle -z, -t \rangle$. There is an effective $S^1$-action on $M$ given by $\bar{z} \times \langle z, t \rangle \mapsto \langle \bar{z}z, t \rangle$. The core orbit $\langle \bar{z}, 0 \rangle$ has isotropy group $\mathbb{Z}_2 \subset S^1$ and is doubly covered by all the other (free) orbits $\langle \bar{z}, t \rangle$, $t \neq 0$. Attach two identical Möbius bands by the identity homeomorphism on their boundaries. The $S^1$-action extends obviously to the union which is the Klein bottle. Let $\pi_1$ of the two core orbits be denoted by $a$ and $b$, and $\pi_1$ of a free orbit by $t$. By van Kampen's theorem, $\{a, b, t \mid t = a^2 = b^{-2}\} = \{a, b \mid a^2b^2 = 1\}$. Now the center is generated by $t$ which is also a generator of $\mathrm{ev}_*^x(\pi_1(S^1))$. We can lift now all the way up to $K_\mathbb{Z}$ with $\pi_1(K_\mathbb{Z}) \cong \mathbb{Z}$, but no further. This is an annulus $K_\mathbb{Z} = S^1 \times \mathbb{R}$ and the $S^1$-action is free and just translates along the first factor.

The orbit space of this $S^1$-action is $\mathbb{R}$ and there is an action of $\pi_1(K)/\mathrm{Im}(\mathrm{ev}_*^x) \cong \mathbb{Z}_2 * \mathbb{Z}_2 \cong \mathbb{Z} \rtimes \mathbb{Z}_2$ on $\mathbb{R}$. This induced action on $\mathbb{R}$ is just the action of the infinite dihedral group: $(n, \epsilon)(x) = \epsilon x + n, \epsilon = \pm 1$. The orbit space is an interval which is identical with the orbit space of the $S^1$-action on the Klein bottle. The covering action on $S^1 \times \mathbb{R}^1$ is given by the isotropy groups on the Klein bottle $(n, \epsilon)(z, x) = (\epsilon z, \epsilon x + n)$; cf. Example 1.6.8.

## 2.6. Example (Semi-free $S^1$-actions on 3-manifolds)

2.6.1. The lift of a semifree $S^1$-action on a connected 3-manifold $M$ to its universal covering $\widetilde{M}$ is tractable. The action of $S^1$ on $S^2 \times S^1$ and its lift to the universal covering is a good guide for the analysis of semifree $S^1$-actions on connected oriented closed 3-manifolds. For example, there is an action of $S^1$ on $S^2 \times S^1$ obtained by rotating $S^2$ about its north and south pole and acting trivially on the second factor. Clearly, the orbit space is $\{\mathrm{arc}\} \times S^1$, an annulus with the boundary corresponding to the two circles of fixed points. The universal covering is $S^2 \times \mathbb{R}^1$

which can be thought of $\mathbb{R}^3 - \{\text{origin}\}$. The lift of this action is just rotation on the $S^2$-factor and trivial on $\mathbb{R}^1$. A fundamental domain is $yz$-plane $- \{\text{origin}\}$, $y \geq 0$, and the action is equivalent to rotating this domain about the $z$-axis. The two components of ($z$-axis $-$ origin) are the lifts of the two circles of the fixed points on $S^2 \times S^1$. The covering transformations $\pi_1(S^2 \times S^1) = \mathbb{Z}$ acts on the quotient $S^1 \backslash (S^2 \times \mathbb{R}^1) \approx I \times \mathbb{R}$ by translating freely along the $\mathbb{R}$-factor yielding the annulus as quotient.

THEOREM 2.6.2 (cf. [**Ray68**]). *If $(S^1, M)$ is a semifree $S^1$-action on a connected closed oriented 3-manifold, then*

(1) *The orbit space $S^1 \backslash M$ is a compact orientable surface of genus $g$ with $k(>0)$ boundary components.*
(2) *There is a global cross section to the orbit mapping with the $k$-boundary components mapping homeomorphically onto the fixed set. Thus the action, up to $S^1$-equivalence, is completely determined by $g$ and $k$.*
(3) *$M$ is homeomorphic to $S^3 \# (S^2 \times S^1) \# \cdots \# (S^2 \times S^1)$ with $(2g + k - 1)$ $S^2 \times S^1$-factors in the connected sum.*
(4) *The universal covering $\widetilde{M}$ is homeomorphic to $\mathbb{R}^3 - C$, where $C$ is a closed subset of the $z$-axis and the lifted $S^1$-action to $\widetilde{M}$ is $S^1$-equivalent to rotation in $\mathbb{R}^3 - C$ about the $z$-axis. The ($z$-axis $- C$) is the inverse image of the fixed set $M^{S^1}$.*
(5) *The orbit mapping $\widetilde{M} \to W = S^1 \backslash \widetilde{M}$ has a cross section whose image is homeomorphic to $(yz$-plane $- C)$, $y \geq 0$. The action of $\pi_1(M)$, free product of $2g + k - 1$ copies of $\mathbb{Z}$, is free on $W$ and $\pi_1(M) \backslash W$ is the same as $S^1 \backslash M$.*

Schematically we have

$$\begin{array}{ccc} (S^1, \mathbb{R}^3 - C) & \xrightarrow{S^1 \backslash} & (*^n \mathbb{Z}, W) \\ \nu \downarrow *^n \mathbb{Z} \backslash & & \downarrow *^n \mathbb{Z} \backslash \\ (S^1, M) & \xrightarrow{S^1 \backslash} & S^1 \backslash M \end{array}$$

as a commutative diagram of orbit mappings.

2.6.3. Much of this theorem can be proven with the material already developed. However, this will also follow from a general classification of $S^1$-actions, with fixed points, on 3-manifolds to be discussed in Section 14.5. Instead, we shall for each orientable surface of genus $g$ with $k(\geq 1)$ boundary components, construct a semifree action on a connected sum of $2g + k - 1$ copies of $S^2 \times S^1$'s whose orbit space is a given surface. Because of (2) in Theorem 2.6.2, this will be the only possible $S^1$-action with invariants $g$ and $k$ up to $S^1$-equivalence. Given that (1) and (2) hold, we shall be able to verify that (3), (4), and (5) hold.

The reader may easily check that the case where $g = 0$ and $k = 1$ corresponds to the standard $S^1$-rotations in $\mathbb{R}^3 \cup \infty = S^3$, with ($z$-axis $\cup \infty$) as fixed set and the closed 2-disk as orbit space. We have already discussed the case where $g = 0$ and $k = 2$ so that $\#^{2g+k-1}(S^2 \times S^1) = S^2 \times S^1$. On $(S^1, S^2 \times S^1)$, there is a 3-cell, which is an invariant tubular neighborhood about a point on the fixed set. This can be represented as the inverse image of the orbit mapping of the shaded region in the orbit space as shown.

## 2.6. EXAMPLE (SEMI-FREE $S^1$-ACTIONS ON 3-MANIFOLDS)

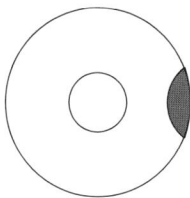

Let us take two copies of $S^2 \times S^1$, remove the interiors of an invariant 3-ball and match equivariantly their boundaries by an orientation reversing homeomorphism as shown. This gives us an $S^1$-equivariant connected sum of $S^2 \times S^1$ with $S^2 \times S^1$.

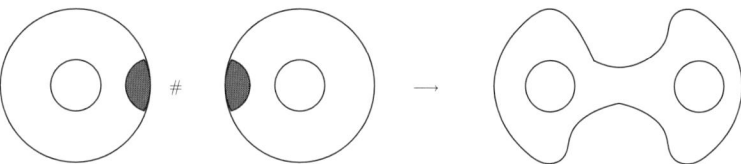

We can do this any number of times, say $n > 1$, getting $M_n = (S^2 \times S^1) \# \cdots \# (S^2 \times S^1)$ whose orbit space is a disk with $n$-holes and $n+1$ boundary components, corresponding to the $n+1$ circles of fixed points.

We may describe *all* the $S^1$-actions on $M_n$, $n \geq 0$, as follows. Take an oriented compact surface $A$ of genus $g$ with $k(> 0)$ boundary components such that $2g + k - 1 = n$. Form $S^1 \times A$, and let $S^1$ act as standard rotations on the first factor and trivially on the second factor. The boundary consists of $S^1 \times \partial A$. Now if we take each orbit $S^1 \cdot a$, where $a \in \partial A$, and collapse it to a distinct point, we obtain the quotient space with an $S^1$-action with exactly $k$ connected components of fixed points homeomorphic to the boundary curves on $A$. The action is free elsewhere. To see that this space is homeomorphic to $M_n$, let us take a punctured torus for $A$ written as follows.

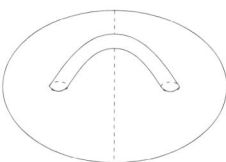

Figure 1. E1–3

Make the construction above to create $M$. Above the dotted line we have an $S^1$-invariant 2-sphere. Cut open the 3-manifold along this 2-sphere and add two $S^1$-invariant 3-balls. The action on each of the 3-balls being just the cone of the $S^1$-action on the boundary 2-sphere. The $S^1$ action now extends to this new 3-manifold. It now has two boundary components of fixed points and the orbit space looks like

FIGURE 2. E1-4

which is an annulus. Then, as we have seen, the $S^1$-space above this annulus is $S^2 \times S^1$. Now we can reconstruct our original $M$ by just removing the interiors of the $S^1$-invariant 3-balls and identifying equivariantly the boundaries. This operation is a connected sum with $S^2 \times S^1$. (Observe that $(S^2 \times S^1) \# S^3 = S^2 \times S^1$ and if one removes the interior of two disjoint 3-cells from $S^3$ and identifies the two boundary 2-spheres, it yields $S^2 \times S^1$.) So our $M' = M_2 = S^2 \times S^1 \# S^2 \times S^1$. This action on $M_2$ is different from the one obtained using $A$, a twice punctured disk, since the orbit spaces are different. To identify $M$ with $M_n$, we just observe that any $A$ can be obtained from doing boundary connected sums with an annulus or a punctured torus with one boundary component. The corresponding $M_n$ is obtained by doing the corresponding $S^1$-equivariant connected sums.

For $n \geq 1$, we get $\pi_1(M_n) = \mathbb{Z} * \cdots * \mathbb{Z}$, a free product of $n$ copies of $\mathbb{Z}$. $\widetilde{M_n}$ is a simply connected 3-manifold with a semifree $S^1$-action.

Since the $S^1$-action on $M$ has fixed points, we may lift the action to the universal covering $\widetilde{M}$. The lifted action is also semifree with the fixed set of $\widetilde{M}$ over the fixed set of $M$ and the free orbits of $\widetilde{M}$ over those of $M$. Let $\chi : A \to \chi(A) \subset M$ be a section to the orbit mapping $M \to S^1 \backslash M = A$. Because $\pi_1(\chi(A)) \xrightarrow{i_*} \pi_1(M)$ is an isomorphism, $\nu^{-1}(\chi(A))$ is the universal covering of $A$ and it too is a section of the orbit mapping $\widetilde{M} \to S^1 \backslash \widetilde{M} = W$. Thus $W$ is the universal cover of $A$ with boundary points projecting to the fixed orbits in $A$. Furthermore, the induced action of $\pi_1(M)$ on $W$ is free. Since $W$ is a simply connected 2-manifold, it is homeomorphic to the part of the $yz$-plane in $\mathbb{R}^3$ with $y \geq 0$ and a closed subset $D$ of the $z$-axis removed. The lifted $S^1$-action on $\widetilde{M}$ is equivalent to translation on the $S^1$-factor of $S^1 \times W$ and with each orbit over $S^1 \times \partial W$ collapsed to a point. This is equivalent to a rotation of the part of the $yz$-plane about the $z$-axis described above. □

2.6.4. In Section 14.5 where this is treated in full detail, finite isotropy also is allowed. Each component of a finite cyclic isotropy adds an equivariant connected sum of a lens space to $M$. $W$ will still be a simply connected 2-manifold and the fundamental group, $\mathbb{Z} * \cdots * \mathbb{Z} * \mathbb{Z}_{p_1} * \cdots * \mathbb{Z}_{p_n}$ will act properly on $W$ but will no longer be free if there are any finite isotropy subgroups. The orbit space $A$ will still be a surface of genus $g$ with $k$ boundary components corresponding to the fixed set, and the orbits with finite cyclic isotropy are isolated and in the interior of $A$.

Once again, one can show by similar methods that the lift of an orientable $(S^1, M)$ to $(S^1, \widetilde{M})$ is semifree and has a global cross section. The orbit space of the lift is a simply connected 2-manifold with boundary. The boundary $F'$ is the homeomorphic image of the fixed points $\widetilde{F}$ of $\widetilde{M}$ and these are lifts of the fixed points $F$ from $M$. Therefore, $S^1 \backslash \widetilde{M} = W$ is diffeomorphic to a closed 2-cell with a closed set $D'$ removed from the boundary. $\pi_1(M)$ acts properly on $S^1 \backslash \widetilde{M}$ with $S^1 \backslash M$ as quotient. The action of $\pi_1(M)$ on $W$ is free except for the preimages of the singular orbits in $S^1 \backslash M$. Because $\widetilde{M} \to S^1 \backslash \widetilde{M}$ has a cross section, the action $(S^1, \widetilde{M})$ is smoothly (and/or topologically) equivalent to the restriction of the linear $(S^1, U) \subset (S^1, S^3)$ where $U = (\mathbb{R}^3 \cup \infty = S^3) - D$. Here $D \cup \widetilde{F}$ is the $z$-axis $\cup \infty$ with the $S^1$-action equivalent to a rotation about the axis.

Often underlying a topological or differentiable classification is a finer and deeper geometric structure. The topological classification can guide a geometric classification. In [**Kam88**], Kamishima has put this topological classification into geometric form. He shows that each of the $S^1$-manifolds discussed admit a conformally flat structure whose automorphism group contains $SO(2)$ acting with fixed points. The moduli space of conformally flat structures with an invariant $SO(2)$-action with fixed points turns out to be a Seifert fiber space over the moduli space of representations of $\pi_1(M)$ into $PSL(2, \mathbb{R})$ with the typical fiber a torus of dimension $2g + k - 1$; see [**Kam88**, 3.1.4]. This comes about as follows. Let $S^1_\infty$ be a geometric circle in conformally flat $S^3$. Then $S^3 - S^1_\infty$ is conformally equivalent to the Riemannian product $\mathbb{H}^2 \times S^1$. Thus, the group of conformal transformations of $S^3$ preserving $S^1_\infty$, the group of conformal maps of $S^3 - S^1_\infty$, is isomorphic to $SO(2,1) \times O(2)$. This implies a representation of $\Pi$ into Conformal$(S^3 - S^1_\infty)$ gives rise to representations of $\Pi$ into $SO(2)$ and into $SO(2,1)$. The maps of $\Pi = \pi_1(M)$ into $SO(2)$ yield the fiber torus $T^{2g+k-1}$, of course. The typical fiber of this Seifert fibering is a torus of dimension $2g + k - 1$; see [**Kam88**, 3.1.4].

In Chapter 13, the analogous moduli of geometric structures on closed 3-manifolds admitting an $SO(2)$-action *without* fixed points and over a hyperbolic base is explained in detail.

## 2.7. Lifting the slice representation

In Section 2.3 we discussed lifting actions when the action had fixed points and in Section 2.5 when the group was path-connected. If $G$ is path-connected and a closed subgroup $H \subset G$ fixes $x \in X$, then $H$ may be lifted by the method of Section 2.3 and also by the method of Section 2.5. Exactly how these liftings may be compared is the subject of this section.

LEMMA 2.7.1 ([**CR69**, Lemma 4.5]). *Let $X$ be a path-connected space; $G$ a path-connected topological group acting on $X$. Let $H$ be a closed subgroup of $G$ fixing $x \in X$. Then there is a natural homomorphism $\psi : H \longrightarrow \pi_1(X, x)/\mathrm{Im}(\mathrm{ev}^x_*)$ such that*

$$h_*(\alpha) = \psi(h) \; \alpha \; \psi(h)^{-1} \text{ for all } h \in H, \text{ and } \alpha \in \pi_1(X, x),$$

*where $h_* \in \mathrm{Aut}(\pi_1(X,x)/\mathrm{Im}(\mathrm{ev}^x_*))$ is induced by $h_* \in \mathrm{Aut}(\pi_1(X,x))$ given by $h_*([\ell(t)]) = [h \cdot \ell(t)]$.*

PROOF. Let $h(t) \in P(G,e)$ such that $h(1) = h \in H \subset G$, and let $\ell(t)$, a loop in $P(X,x)$ represent $\alpha \in \pi_1(X,x)$. Then by examining the homotopy diagram as in Subsection 2.4.1

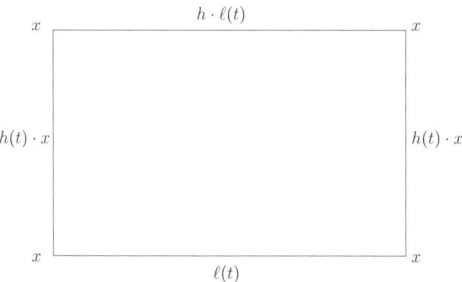

we have $\ell(t) * h(t) \cdot x \simeq h(t) \cdot x * h(\ell(t))$. Consequently, $\overline{h(t) \cdot x} * \ell(t) * h(t) \cdot x \simeq h(\ell(t))$ represents $h_*(\alpha)$. If $h(t) \cdot x$ represents $\beta_h \in \pi_1(X,x)$, then $h_*(\alpha) = \beta_h^{-1} \alpha \beta_h$. To see that this is well defined, let $h'(t) \in P(G,e)$ with $h'(1) = h$. Then $h'(t) \cdot x$ represents $\beta'_h \in \pi_1(X,x)$ and $\beta_h {\beta'_h}^{-1} \in \mathrm{Im}(\mathrm{ev}^x_*) \subset \mathcal{Z}(\pi_1(X,x))$. Therefore ${\beta'_h}^{-1} \alpha \beta'_h = \beta_h^{-1} \alpha \beta_h$ and so $h_*(\alpha) = \beta_h^{-1} \alpha \beta_h$ is a well-defined formula.

The function $h \mapsto \beta_h$ is not well defined as an element of $\pi_1(X,x)$ but it is well defined by taking the $\pi_1(X,x)/\mathrm{Im}(\mathrm{ev}^x_*)$-path class of $h(t) \cdot x$. Therefore, we define $\psi : H \to \pi_1(X,x)/\mathrm{Im}(\mathrm{ev}^x_*)$ by

$$\psi(h) = [\beta_h^{-1}].$$

Examine the homotopy diagram

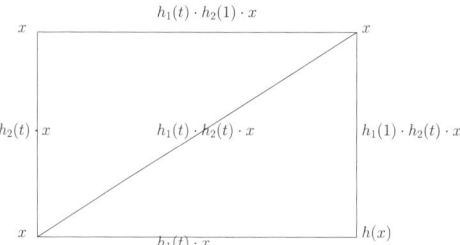

which is obtained from Lemma 2.4.1 by $g(t) = h_1(t)$ and $p(t) = h_2(t) \cdot x$. Note that $h_1(t) \cdot h_2(1) \cdot x = h_1(t) \cdot x$. This shows that $\psi$ is a homomorphism. The formula $h_*(\alpha) = \psi(h) \, \alpha \, \psi(h)^{-1}$ is clear from the construction. $\square$

COROLLARY 2.7.2. *Let $(G,X)$ be an action of a path-connected topological group on a path-connected space with $H$, a closed subgroup of $G$, fixing $x \in X$. Let $K$ be a normal subgroup of $\pi_1(X,x)$ containing the image of $\mathrm{ev}^x_* : \pi_1(G,e) \to \pi_1(X,x)$. Let $Q = \pi_1(X,x)/K$. Then there is a natural homomorphism $\psi : H \longrightarrow Q$ such that*

$$h_*(\alpha) = \psi(h) \, \alpha \, \psi(h)^{-1} \text{ for all } h \in H, \text{ and } \alpha \in \pi_1(X,x),$$

*where $h_* \in \mathrm{Aut}(Q)$ is induced by $h_* \in \mathrm{Aut}(\pi_1(X,x))$ given by $h_*([\ell(t)]) = [h \cdot \ell(t)]$.*

Lemma 2.7.1 and Corollary 2.7.2 are especially applicable to locally injective actions (Section 2.8); see Subsection 2.8.9 for an illustration.

2.7.3. Since $H$ leaves $K$ invariant and $K \supset \mathrm{Im}(\mathrm{ev}^x_*)$, we may lift the $H$-action on $X$ to $X_K$ by the method of Section 2.3. Denote this lifted $H$-action by
$$h \times \hat{y} \mapsto h \diamond \hat{y}.$$
$H$ also lifts by the method of Section 2.5, to the restriction of the lifted $G$-action to $H$. This will be denoted by
$$h \times \hat{y} \mapsto h \bullet \hat{y}.$$
Let $p(t) \in P(X, x)$ represent $\hat{y}$; $h(t) \in P(G, e)$ is a path in $G$ from $e$ to $h$. Then
$$\begin{aligned} h \diamond \hat{y} & \text{ is represented by } & h \cdot p(t), \\ h \bullet \hat{y} & \text{ is represented by } & h(t) \cdot p(t). \end{aligned}$$
For a simple illustration, take the $\mathbb{Z}_2 \subset S^1$ acting on $\mathbb{R}P^2$ and use a point on the fixed circle as base point. We get three $\mathbb{Z}_2$-actions on $S^2$ in Example 2.3.8. Then $\psi(h)$ is the antipodal map, $(h, \diamond)$ is the reflection across the equator, and $(h, \bullet)$ is the rotation about the polar axis.

LEMMA 2.7.4. $h \diamond \hat{y} = h \bullet (\psi(h) \cdot \hat{y})$.

Consider the following homotopy diagram.

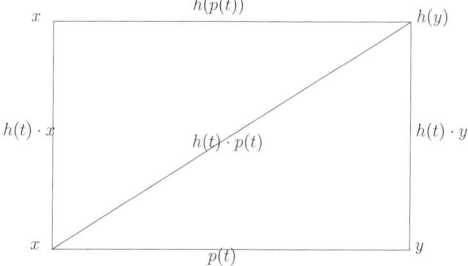

This shows
$$(h(t) \cdot x) * (h \cdot p(t)) \simeq h(t) \cdot p(t),$$
which implies $\psi(h)^{-1} \cdot (h \diamond \hat{y}) = h \bullet \hat{y}$. Thus, $h \diamond \hat{y} = \psi(h) \cdot (h \bullet \hat{y}) = h \bullet (\psi(h) \cdot \hat{y})$.

2.7.5. Let $\rho : (G, X_K) \xrightarrow{G \backslash} (G \backslash X_K) = W$ denote the orbit mapping. If $\hat{y} \in X_K$, put $\rho(\hat{y}) = w$. Since the covering transformations $Q = \pi_1(X, x)/K$ commutes with the lifted $G$-action, there is induced a $Q$-action on $W$, denoted by $\alpha \times w \mapsto \alpha \cdot w$, and $\rho$ is $Q$-equivariant as well as $G$-equivariant (using trivial $G$-action on $W$). $H$ also acts on $W$ via $\psi : H \to \psi(H) \subset Q$.

PROPOSITION 2.7.6. $\rho : (H, X_K, \diamond) \longrightarrow (H, W, \psi)$ is an equivariant map.

This is true because $\rho(h \diamond \hat{y}) = \rho(h \bullet (\psi(h) \cdot \hat{y})) = \rho(\psi(h) \cdot (h \bullet \hat{y})) = \psi(h) \cdot w = \psi(h) \cdot \rho(\hat{y})$. $\square$

COROLLARY 2.7.7. Let $(G, X)$ be as above with $G$ a connected Lie group acting properly on a completely regular $X$ which admits covering space theory. Let $x \in X$, and $K \supset \mathrm{Im}(\mathrm{ev}^x_*)$. Assume $G_{\hat{x}} = 1$, $\hat{x} \in X_K$, $\nu(\hat{x}) = x$, and $w = \langle \hat{x} \rangle \in W = G \backslash X_K$. Then $G_x \xrightarrow{\psi} Q_w$ is an isomorphism and there exists slices $S_x$ at $x$ and $\Sigma_w$ at $w$ such that the slice actions are $G_x$-equivalent.

PROOF. Let $S_x$ be a slice at $x$ and assume it is path-connected and sufficiently small as to be included in an evenly covered neighborhood of $x$. Then $S_x$ lifts homeomorphically to $S_{\hat{x}}$. In Theorem 2.5.1, we saw that $S_{\hat{x}}$ is a $(G_{\hat{x}} = 1)$-slice at $\hat{x}$. Thus, $S_{\hat{x}}$ projects by $\rho$ homeomorphically into $W$. Note that the $G_x$-action on $S_x$ in $X$ is equivalent to the $(G_x, \diamond)$ action on $S_{\hat{x}}$ in $X_K$. (Since $\rho : h \diamond \hat{y} \to \phi(h) \cdot \rho(\hat{y})$ is an $H$-equivariant homeomorphism from $(H, S_x) = (H, S_{\hat{x}})$ to $(Q_w, \Sigma_w)$, the slices $S_x$ at $x$ and $\Sigma_w$ at $w$ are $\Psi(H)$-equivalent.) Put $= \Sigma_w = \rho(S_{\hat{x}})$, and $S_{\hat{x}}$ can be regarded as a section over $\Sigma_w$. Since $\rho(h \diamond \hat{y}) = h \cdot \rho(\hat{y})$ is an $H$-equivariant homeomorphism, the slices $S_x$ at $x$ and $\Sigma_w$ at $w$ are $\psi(H)$-equivalent. Therefore, the slice action $(G_x, S_x)$ is equivalent to the slice action of $Q_w$ at $w$ via $\rho : (G_x, S_{\hat{x}}, \diamond) \to (\psi(G_x), \rho(S_{\hat{x}}))$. □

REMARK 2.7.8. Note also that if the $(G, X)$-action is smooth or holomorphic, then the slice actions

$$(G_x, S_x), \ (G_x, S_{\hat{x}}, \diamond), \ \text{and} \ (\psi(G_x), \rho(S_{\hat{x}}))$$

are also smoothly or holomorphically equivalent.

In the smooth case, the slice representations ($S_x$ is a cell, normal to the orbit at $x$) are smoothly equivalent to linear representations, and it can easily be shown that as linear representations, they are linearly equivalent.

If $G_x \neq 1$, then $Q_w = \psi(G_x) \cong G_x/G_{\hat{x}}$ and $(G_x, S_{\hat{x}})$ factors as

$$
\begin{array}{ccc}
(H, S_{\hat{x}}) = (G_x, S_{\hat{x}}) & \xrightarrow{G_{\hat{x}} \backslash} & (G_x/G_{\hat{x}}, G_{\hat{x}} \backslash S_{\hat{x}}) = (\psi(G_x), \Sigma_w) \\
{\scriptstyle \nu} \downarrow & & \downarrow {\scriptstyle \psi(G_x) \backslash} \\
(G_x, S_x) & \xrightarrow{G_x} & G_x \backslash S_x.
\end{array}
$$

It is significant to note that when $(G, X_K)$ is a principal action, then the isotropy subgroups of the $Q$-action on $W$ encodes all the slice information of $(G, X)$. We shall address the reverse of this procedure in a later chapter. Namely, if we are given $(Q, W)$ acting properly with $Q$ discrete, and $P \to W$ a fixed principal $G$-bundle, then we can ask how may we construct a covering $Q$-action on $P$ which commutes with the principal $G$-action. This reverse process may or may not be possible, and if possible, it may not be unique.

EXERCISE 2.7.9. The two $H$-actions on $X_K$ in Subsection 2.7.3 are subgroups of the group, $(\pi_1(X, x)/K) \rtimes H$, generated by all lifts of the $H$-action on $X$. Determine explicitly these two subgroups in $(\pi_1(X, x)/K) \rtimes H$.

EXERCISE 2.7.10. Show that $\psi$ induces the exact sequence $1 \to G_{\hat{x}} \to G_x \to Q_w \to 1$. (Hint: For $q \in Q_w$, there is a unique $h \in G_x/G_{\hat{x}}$ so that $q\hat{x} = h^{-1} \circ \hat{x}$.)

## 2.8. Locally injective actions

Let $X$ be a path-connected space, and let $G$ be a path-connected topological group acting on $X$. Let $H$ be a closed subgroup of $G$ fixing $x \in X$. In Lemma 2.7.1, the homomorphism $\psi : H \to \pi_1(X, x)/\text{Im}(\text{ev}_*^x)$ was defined as follows: For $h \in H$, pick a path $h(t)$ from $e$ to $h$. Then $\beta_h = h(t) \cdot x$ is a loop at $x$. Then $\psi(h)$ is defined by $\psi(h) = [\beta_h^{-1}]$. It has the property

$$h_*(\alpha) = \psi(h) \ \alpha \ \psi(h)^{-1} \text{ for all } h \in H, \text{ and } \alpha \in \pi_1(X, x),$$

where $h_* \in \text{Aut}(\pi_1(X, x)/\text{Im}(\text{ev}_*^x))$ is induced by $h_* \in \text{Aut}(\pi_1(X, x))$ given by $h_*([\alpha(t)]) = [h \cdot \alpha(t)]$.

## 2.8. LOCALLY INJECTIVE ACTIONS

DEFINITION 2.8.1. Suppose a path-connected $G$ acts on $X$ admitting covering space theory. If, at each $x \in X$, $\psi : G_x \to \pi_1(X,x)/\text{Im}(\text{ev}_*^x)$ is a monomorphism, then $(G, X)$ is called a *locally injective $G$-action*.

EXAMPLE 2.8.2. All free actions $(G, X)$ are locally injective. This follows immediately from the definition.

PROPOSITION 2.8.3. *Let $(G, X)$ be a locally proper and locally injective action of a connected Lie group on a space $X$. Suppose that*
$$Q = \pi_1(X,x)/\text{Im}(\text{ev}_*^x(\pi_1(G,e)))$$
*is torsion free. Then the $G$ action must be free.*

PROOF. Let $y \in X$. Then $Q' = \pi_1(X,y)/\text{Im}(\text{ev}_*^y(\pi_1(G,e)))$ is isomorphic to $Q$ and so $Q'$ is torsion free. Since $(G, X)$ is locally injective, $\psi_y : G_y \to Q'$ must be injective. Since $(G, X)$ is locally proper, $G_y$ is compact; see Definition 1.2.2. Therefore, $G_y = 1_G$. □

PROPOSITION 2.8.4. *If $(G, X)$ is a locally injective $G$-action, then the lifted $G$-action $(G, X_{\text{Im}(\text{ev}_*^x)})$ is a free action and conversely. Moreover, if $G$ is a Lie group acting properly on $X$, then $(G, X_{\text{Im}(\text{ev}_*^x)})$ is proper so the action is a principal $G$-action on $X_{\text{Im}(\text{ev}_*^x)}$.*

PROOF. Assume $X$ is locally injective. Pick a base point $x \in X$, and let $H = \text{Im}(\text{ev}_*^x)$. Then $(G, X)$ lifts to $(G, X_H)$, and let $\widehat{x}$ to be the path class of the constant loop at $x$. We have seen $\psi(G_x) = Q_{\rho(\widehat{x})} \subset \pi_1(X, x)/H$. The kernel of $\psi$ is $G_{\widehat{x}}$ (see Exercise 2.7.10), and this, by hypothesis, is trivial. By Theorem 2.5.1(2), the $G$-action is independent of choice of base points, so we have $\psi(G_y) = Q_{\rho(\widehat{y})} \subset \pi_1(X,y)/\text{Im}(\text{ev}_*^y)$. Thus $G_{\widehat{y}} = 1$, so the action $(G, X_H)$ is free.

Conversely, if the lifted action $(G, X_H)$ is free, then $G_{\widehat{y}} = 1$, for all $\widehat{y} \in X_H$. But $\psi : G_y \to Q_{\rho(\widehat{y})} \subset \pi_1(X,y)/\text{Im}(\text{ev}_*^y)$ is onto with kernel $G_{\widehat{y}} = 1$. So using $y$ and $\widehat{y}$ as base points for $X$ and $X_H$, we have that $\psi : G_y \to \pi_1(X,y)/\text{Im}(\text{ev}_*^y)$ is monic and so $(G, X)$ is locally injective.

The argument, so far, has not required that $G$ be a Lie group. In order that $(G, X_H)$ be a principal $G$-action, we need to show that the action is locally proper. If $G$ is a Lie group and the action $(G, X)$ is proper, then by Theorem in 2.5.1(3), the action will be proper. □

We remark that injective torus actions (see Definition 3.1.10 and Section 3.5 for a definition) are locally injective, but locally injective torus actions are not necessarily injective torus actions. For example, the Hopf fibration $S^1 \to S^3 \to S^2$ is locally injective, but not injective. Any $T^k$-action with a global $H$-slice where $H$ is a finite subgroup of $T^k$ is an injective $T^k$-action. For example, the $\mathbb{C}^*$ and $S^1$-actions $\mathbb{C}^* \times \mathbb{C}^n - V \longrightarrow \mathbb{C}^n - V$, $S^1 \times S^{2n-1} - K \longrightarrow S^{2n-1} - K$, from Example 1.6.9, have global $\mathbb{Z}_a$ slices. However, the extended action $(S^1, S^{2n-1})$ is not locally injective since $S^{2n-1}$ is simply connected and the action has nontrivial isotropy.

EXAMPLE 2.8.5. Many spherical space forms have automorphisms of their geometry which are locally injective actions. See Subsection 4.5.18 and Chapter 15 for specific descriptions.

EXAMPLE 2.8.6. Suppose we take any smooth orientable $n$-manifold $W$, and let $Q$ be either a connected Lie group or a discrete group acting (on the left) effectively, smoothly properly, and preserving orientation on $W$. Let $P$ be the bundle of oriented frames over $W$. It is a $\mathrm{GL}^+(n,\mathbb{R})$ principal bundle (with $\mathrm{GL}^+(n,\mathbb{R})$ acting on $W$ on the right). Then the $Q$-action on $W$ lifts to a $Q$-action on $P$ and commutes with $G = \mathrm{GL}^+(n,\mathbb{R})$. In fact, $Q$ acts freely on $P$. We get

$$\begin{array}{ccc} (G\times Q, P) & \xrightarrow{G\backslash} & (Q, W) \\ \downarrow & & \downarrow \\ (G, Q\backslash P) & \xrightarrow{G\backslash} & Q\backslash W, \end{array}$$

where $P \to Q\backslash P$ is a principal $Q$-fibering. Therefore, if $Q$ is discrete, we have constructed for each $(Q, W)$ a *locally injective* $G$-action on a manifold $Q\backslash P$.

PROPOSITION 2.8.7. *Suppose $P$ is a principal $G$-bundle where $G$ is a connected Lie group, and $\Pi \subset \mathrm{TOP}_G(P)$ is a group of covering transformations of $P$ acting properly, that centralizes $\ell(G)$ and $\ell(G) \cap \Pi = 1$. Then the induced $G$-action on $\Pi \backslash P = X$ is locally injective.*

PROOF. Since $\Pi$ commutes with $\ell(G)$, there is induced a $G$-action on $\Pi\backslash P = X$, which is covered by $\ell(G)$ on $P$. Because $G$ is connected, this lift is the unique lift to $P$ covering the induced $G$-action on $X$. Therefore, $\pi_1(G, e) \to \mathrm{ev}_*^x(\pi_1(G,e)) \subset \pi_1(P, \widehat{x})$. We have seen that $G_x \to \pi_1(X, x)/\mathrm{Im}(\mathrm{ev}_*^x)$ is a monomorphism in Corollary 2.7.7. Since the choice of $x$ is arbitrary, the $G$-action must be locally injective. $\square$

EXERCISE 2.8.8. Examine the slice representations in light of Theorem 2.5.1(3) and Section 2.7 for the actions in Sections 2.6 and 2.8

2.8.9 (Illustration of Lemma 2.7.1 and Corollary 2.7.2). In Subsection 1.6.9, we have an $S^1$-action on $S^3$ given by $z \times (z_1, z_2) \mapsto (z^2 z_1, z^3 z_2)$. On $S^3$, the orbits $(z_1, 0)$ and $(0, z_2)$ have isotropy subgroups $\mathbb{Z}_2$ and $\mathbb{Z}_3$, respectively. All other orbits are free and the orbit space is the 2-sphere. If we remove the free orbit $K = \{(z_1, z_2) : z_1^3 + z_2^2 = 0 \text{ and } z_1 \bar{z}_1 + z_2 \bar{z}_2 = 1\}$, then $(S^1, S^3 - K)$ can be written equivariantly as $(S^1, S^1 \times_{\mathbb{Z}_6} T')$ which fibers equivariantly over $S^1/\mathbb{Z}_6$ with fiber the once punctured torus $T'$. The orbit space now is the once punctured sphere with the two singular orbits $(z_1, 0)$ and $(0, z_2)$. The action is injective and therefore locally injective since it lifts to $(S^1, S^1 \times T')$.

If $x \in (z_1, 0)$, then $h(t) \cdot (x) = e^{2\pi i \frac{t}{2}}(x)$, $0 \le t \le 1$, is the orbit $(z_1, 0)$ and this loop in $S^3 - K$ generates the image of $\mathbb{Z}_2$ in $\pi_1(S^3 - K, x)/\mathrm{Im}(\mathrm{ev}_*^x) = Q$. In Chapter 14, we see that $Q \cong \mathbb{Z}_2 * \mathbb{Z}_3$ where the generator of $\mathbb{Z}_2$ in $\mathbb{Z}_2 * \mathbb{Z}_3$ can be taken to be the image of $h(t)$ in $Q$. The group element $h = h(1)$ in $S^1$ fixes $x$ and induces an automorphism of $\pi_1(S^3 - K, x)$. It also induces an involution on $Q$ where it is evaluated as $h_*(\alpha) = \psi(h) \, \alpha \, \psi(h)^{-1}$, for $\alpha \in Q$ and where $\psi(h)$ is the element of $Q$ generated by the loop $\overline{h(t) \cdot (x)}$. To illustrate Corollary 2.7.2, take the covering $(S^1, S^1 \times T')$. Conjugation is now trivial as the new $Q$ is just $\mathbb{Z}_6$.

Analogously, an effective $S^1$-action on $S^3$ without fixed points can be written as $z \times (z_1, z_2) \mapsto (z^a z_1, z^b z_2)$, where $\gcd(a, b) = 1$; see Chapter 14. The $S^1$-invariant 1-manifold $K_{(a,b)} = \{(z_1, z_2) : z_1^b + z_2^a = 0 \text{ and } z_1 \bar{z}_1 + z_2 \bar{z}_2 = 1\}$ is the $(b, a)$-torus knot. Then, $(S^1, S^3 - K_{(a,b)})$ can be written as $(S^1, S^1 \times_{\mathbb{Z}_{ab}} Y)$, where $Y$ is a once punctured surface of genus $((a-1)(b-1))/2$. ($Y$ is the spanning surface of this torus knot.) Moreover, $\pi_1(S^3 - K_{(a,b)}, x)/\mathrm{Im}(\mathrm{ev}_*^x) = Q$ is $\mathbb{Z}_a * \mathbb{Z}_b$ and $e^{2\pi i t} \cdot x = h(t)$,

$0 \leq t \leq \frac{1}{b}$, $x \in (z_1, 0)$, generates $\mathbb{Z}_a$ in $\mathbb{Z}_a * \mathbb{Z}_b$. Once again, $h_*$ on $Q$ is given by conjugation by $\psi(h)$.

CHAPTER 3

# Actions of Compact Lie Groups on Manifolds

## 3.1. Actions of compact Lie groups on aspherical manifolds

3.1.1. A path-connected space $X$, admitting covering space theory and which is either locally contractible or has the homotopy type of a CW-complex, is called *aspherical* if $\pi_i(X,x) = 0$ for all $i > 1$. Note that the universal covering of $X$ is contractible and that $X$ is a $K(\pi_1(X,x),1)$. Such spaces are unique up to homotopy type.

Manifolds which are aspherical enjoy many special properties and are extensively studied in topology and geometry. For example, if $M$ is a Riemannian manifold whose sectional curvature is nonpositive, then $M$ is aspherical. Moreover, if $M$ is closed and the sectional curvature is negative, the center of the fundamental group is trivial.

In this section we will investigate actions of compact Lie groups $G$ on closed aspherical manifolds $M$. It will follow that if $G$ is connected and acts effectively, then $G$ must be a torus and the torus must act injectively. Therefore, the isometry group of such a Riemannian manifold $M$ will have its connected component of the identity equal to the $k$-torus where $k$ is less than or equal to the rank of the center of $\pi_1(M)$.

3.1.2. Much can also be said about the effective actions of finite groups on a closed aspherical manifold $M$. For example, if the center of $\pi_1(M)$ is trivial and a finite group $H$ acts effectively on $M$, then we shall show that $H$ injects into $\mathrm{Out}(\pi_1(M))$. It is known that if $M$ is Riemannian, with constant negative sectional curvature, then $M$ is aspherical, the center of $\pi_1(M)$ is trivial, and $\mathrm{Out}(\pi_1(M))$ is finite and isomorphic to the full group of isometries of $M$.

Consequently, if we can show that $\mathcal{Z}(\pi_1(M)) = 1$, and $\mathrm{Out}(\pi_1(M))$ has no torsion, then the closed aspherical manifold admits no action of (not necessarily connected) compact Lie groups whatsoever (totally rigid). Such manifolds exist even in dimension 3, and some of them fiber over the circle. In fact, in every cobordism class, there exists a closed manifold without any finite group actions if dimension is greater than 2.

LEMMA 3.1.3. *Let $\gamma : G \times X \to X$ be an action of a path-connected group $G$ on a path-connected space $X$. Suppose the $G$-action lifts to a regular covering space $p : \widehat{X} \to X$. Suppose $F = X^G$, the set of points of $X$ fixed by $G$, is nonempty and $F'$ is a path component of $F$. Similarly, let $E = \widehat{X}^G$. Then $p^{-1}(F) = E$. The path components of $E$ which project into $F'$ (and hence onto) are exactly the path components of $p^{-1}(F')$.*

PROOF. This lemma follows easily from Lemma 2.3.5 but we shall also give a direct proof. Let $x \in F$ be a base point, and choose any $\hat{x} \in \widehat{X}$ such that

$p(\hat{x}) = x$ as the base point for $\widehat{X}$. We claim that $g(\hat{x}) = \hat{x}$ for all $g \in G$. First $p(g\hat{x}) = g(p(\hat{x})) = x$, (see Theorem 2.5.1(1)). But the orbit of $\hat{x}$ is connected and $g\hat{x} \in p^{-1}(x)$, a discrete set. So $g\hat{x} = \hat{x}$. Since the choice of base point does not alter the lifted $G$-action, we see that $p(E) \supset F$. Since $p(E) \subset F$, we have that $p(E) = F$ and $p^{-1}(F) = E$.

Let $F'$ be a path component of $F$, and for $x \in F'$, choose $\hat{x}$ such that $p(\hat{x}) = x$ as base points. Let $E'$ be the path component of $E$ containing $\hat{x}$. Then $p(E') \subset F'$. Moreover, if we lift a path in $F'$ starting at $x$ and lift to a path in $\widetilde{X}$ starting at $\hat{x}$, then it lifts to a path in $E'$. Hence $p(E') \supset F'$ and $p(E') = F'$. □

EXAMPLE 3.1.4 (cf. Subsection 2.6.1). Let $S^1$ act on $S^2 \times S^1$ by the rotation about the two poles of the first factor and trivially on the second. Then the fixed-point set is two circles. The lifted $S^1$-action to $S^2 \times \mathbb{R}^1$ has two components of fixed points each homeomorphic to $\mathbb{R}^1$ and which project to the two circles. However, if we take an $S^1$-action on $S^2 \times S^1 \# S^2 \times S^1$ with three components of fixed points (3-circles), on the universal covering there is an infinite number of fixed components (each is line), each projecting onto one of the fixed circles. (Recall from Theorem 2.6.2 and Subsection 2.6.4 that the universal cover $\widetilde{M}$ of $M = (S^2 \times S^1 \# S^2 \times S^1)$ is homeomorphic to $\mathbb{R}^3 - D$, where $D$ is a closed set deleted from the $z$-axis. Then $F = \widetilde{M}^{S^1} = (z\text{-axis}) - D$ projects homeomorphically onto the boundary $F'$ of $S^1 \backslash \widetilde{M}$. The induced action of $\pi_1(M) \cong \mathbb{Z} * \mathbb{Z}$ acts freely on $F'$ (an open subset of the line) and has quotient the image of $F$ in $S^1 \backslash M$ which is the three circle boundary components of the 2-disk with two holes).

The next lemma is a weak version of one of the Smith theorems. It has a comparatively elementary proof. It plays a crucial role in proving Lemmas 3.1.11 and 3.1.13.

DEFINITION 3.1.5. Let $A$ be a commutative ring. A space $X$ is $A$-*acyclic* if $\widetilde{H}^i(X; A) = 0$ for all $i$.

LEMMA 3.1.6. (1) *If $S^1$ acts on a $\mathbb{Q}$-acyclic paracompact finite dimensional, locally connected space $X$, then $F = X^{S^1} \neq \emptyset$ and is also $\mathbb{Q}$-acyclic.*
(2) *If a finite $p$-group ($p$ prime) $G$ acts on a $\mathbb{Z}_p$-acyclic paracompact finite dimensional space $X$, then $F = X^G \neq \emptyset$ and is also $\mathbb{Z}_p$-acyclic.*

PROOF. Form the Borel space and the diagram from Section 1.5:

$$\begin{array}{ccccc}
X & \xleftarrow{\pi_2} & S^\infty \times X & \xrightarrow{\pi_1} & S^\infty = ES^1 \\
\downarrow S^1 \backslash & & \downarrow S^1 \backslash & & \downarrow S^1 \backslash \\
S^1 \backslash X & \xleftarrow{\bar{\pi}_2} & X_{S^1} = S^\infty \times_{S^1} X & \xrightarrow{\bar{\pi}_1} & \mathbb{C}P_\infty = BS^1.
\end{array}$$

$\bar{\pi}_1$ is a fibering with $X$ as fiber and structure group $S^1$. Since $X$ is $\mathbb{Q}$-acyclic, $\bar{\pi}_1^*$ is an isomorphism on cohomology; see [**Bor60**, XVI; 4.4] for a proof in Čech cohomology. Suppose $F = \emptyset$, then $\bar{\pi}_2^{-1}(x^*) = S_x^1 \backslash S^\infty = BS_x^1$, and $S_x^1$ is trivial or finite cyclic. But then $BS_x^1$ is also $\mathbb{Q}$-acyclic and so $\bar{\pi}_2^*$ is an isomorphism in rational cohomology by the Vietoris mapping theorem; see Remark 3.1.7.

Since $S^1 \backslash X$ is finite dimensional, its $\mathbb{Q}$-cohomology vanishes above the dimension of $S^1 \backslash X$. This contradicts that $\bar{\pi}_2^*$ is an isomorphism, for $H^*(X_{S^1}; \mathbb{Q}) \cong H^*(\mathbb{C}P_\infty; \mathbb{Q})$. Therefore $F \neq \emptyset$.

For $x^* \in S^1\backslash X - F$, $\bar{\pi}_2^{-1}(x^*)$ is still $\mathbb{Q}$-acyclic. This again implies that $\bar{\pi}_2^* : H^q(S^1\backslash X, F; \mathbb{Q}) \to H^*(X_{S^1}, F_{S^1}; \mathbb{Q})$ is an isomorphism in Čech rational cohomology. Observe $F_{S^1} \cong F \times \mathbb{C}P_\infty$. So we examine

$$\xrightarrow{\delta} H^q(X_{S^1}, F_{S^1}; \mathbb{Q}) \xrightarrow{j^*} H^q(X_{S^1}; \mathbb{Q}) \xrightarrow{i^*} H^q(F \times \mathbb{C}P_\infty; \mathbb{Q}) \xrightarrow{\delta}.$$

Since $H^q(S_1\backslash X, F; \mathbb{Q})$ vanishes for $q > \dim(S^1\backslash X)$, $H^q(X_{S^1}, F_{S^1}; \mathbb{Q}) = 0$ and $i^*$ is an isomorphism for $q > \dim(S^1\backslash X)$. We have seen $\bar{\pi}_1^* : H^q(X_{S^1}; \mathbb{Q}) \to H^q(\mathbb{C}P_\infty; \mathbb{Q})$ is an isomorphism and so, by the Künneth rule, we have

$$H^q(F_{S^1}; \mathbb{Q}) = H^q(F \times \mathbb{C}P_\infty; \mathbb{Q}) = \sum_{i+j=q} H^i(F; \mathbb{Q}) \otimes H^j(\mathbb{C}P_\infty; \mathbb{Q}).$$

Recall that

$$H^q(X_{S^1}; \mathbb{Q}) \xrightarrow{i^*} H^q(F \times \mathbb{C}P_\infty; \mathbb{Q})$$

is an isomorphism for $q > \dim(S^1\backslash X)$, and

$$H^*(X_{S^1}; \mathbb{Q}) \xleftarrow{\bar{\pi}_1^*} H^*(\mathbb{C}P_\infty; \mathbb{Q})$$

are isomorphisms. Consequently, for big $q$, we have

$$H^q(\mathbb{C}P_\infty; \mathbb{Q}) \cong \sum_{i+j=q} H^i(F; \mathbb{Q}) \otimes H^j(\mathbb{C}P_\infty; \mathbb{Q}).$$

This implies that $F$ must be $\mathbb{Q}$-acyclic.

Essentially the same argument used for $S^1$ now holds for $\mathbb{Z}_p$ if $S^1$, $\mathbb{Q}$, and $BS^1$ are replaced by $\mathbb{Z}_p$, $\mathbb{Z}_p$, and $B\mathbb{Z}_p$. Note $(\mathbb{Z}_p)_x = 1$ or $\mathbb{Z}_p$. For the general case of a finite $p$-group, use the fact that $G$ is solvable and contains a nontrivial normal subgroup $H \subset G$. Consequently, $X^G = (X^H)^{G/H}$. Now, reduce this case to $G = \mathbb{Z}_p$ by induction. □

REMARK 3.1.7. The *Vietoris mapping theorem* states that *under a closed mapping for which inverse images of points are acyclic in Čech cohomology, then the mapping induces an isomorphism in cohomology*; see [**Bre97**, II, 11.1 and/or IV, 6.1] for a modern proof of the Vietoris mapping theorem. To satisfy the technical requirements of the theorem, we replace $ES^1 = S^\infty$ by $S^{2n+1}$ and thus $\mathbb{C}P_\infty$ by $\mathbb{C}P_n$ which is $2n$-dimensional. Then $S^{2n+1}$ is compact and $S_x^1\backslash S^{2n+1}$ is $\mathbb{Q}$-acyclic up through dimension $2n$, where $n$ is very large. The argument is unchanged for all $q < 2n$. As $n$ can be arbitrarily large, $F$ is $\mathbb{Q}$-acyclic.

One must also justify the validity of the Künneth theorem for Čech cohomology (with closed supports). This is not a problem if $F$ has the homotopy type of a CW-complex or if $F$ is compact. In general, of course, this may not be the case but the Künneth rule still applies without restrictions on $F$ because $\mathbb{C}P_\infty$, or its approximation $\mathbb{C}P_n$, is a very nice space (cf. [**Bor60**, XVI, §5] or [**Bre97**, IV, 7.6]). Similarly, in showing $\bar{\pi}_1^*$ is an isomorphism on cohomology, one uses the approximation by $\mathbb{C}P_n$ for $\mathbb{C}P_\infty$ so that the conditions that Borel requires in [**Bor60**, IV, §3.8] are met.

A different proof of the lemma, using Smith theory, can be found in [**Bre72**, III,§10]. The lemma is also valid in cohomology with integral coefficients provided that $(S^1, X)$ has only a finite number of distinct isotropy groups.

EXERCISE 3.1.8. Show that the orbit space $S^1\backslash X$ in Remark 3.1.7 (respectively, $\mathbb{Z}_p\backslash X$) is $\mathbb{Q}$-acyclic (respectively, $\mathbb{Z}_p$-acyclic). (Hint: Show that $H^q(X_{S^1}, F_{S^1}; \mathbb{Q}) =$

0 for all $q$ using the fact that $F$ is acyclic and that $\bar{\pi}_1$ has a cross section.) Also show that Lemma 3.1.6 remains true if $S^1$ is replaced by the $k$-torus $T^k$.

COROLLARY 3.1.9. *If $X$ is a finite dimensional aspherical space, then $\pi_1(X,x)$ is torsion free.*

PROOF. The universal covering $\widetilde{X}$ of $X$ is contractible and finite dimensional. If the covering transformations of $\widetilde{X}$ contained a cyclic group of prime order, then it would contradict Lemma 3.1.6(2). □

DEFINITION 3.1.10. A torus action $(T^k, X)$ on a path-connected space is said to be *injective* if $\text{ev}_*^x : \pi_1(T^k, e) \to \pi_1(X, x)$, where $\text{ev}^x(t) = t \cdot x$, is injective. Note, if $(T^k, X)$ is injective, then $T_x^k$ is finite for each $x \in X$; see Section 3.5 for additional properties of injective actions.

LEMMA 3.1.11. *If $(S^1, M)$ is a nontrivial action where $M$ is a closed connected aspherical manifold, then the action must be injective.*

PROOF. Assume that the kernel of $\text{ev}_*^x$ is not trivial. Then there exists a finite covering group $'S^1 \to S^1$ which acts nontrivially on $M$, and for which the image of $\text{ev}_*^x$ is trivial. Then this action of $'S^1$ lifts to the universal cover $p : \widetilde{M} \to M$. Since $\widetilde{M}$ is contractible, it is $\mathbb{Q}$-acyclic.

By Lemma 3.1.6, $E = \widetilde{M}^{\,'S^1}$ must be $\mathbb{Q}$-acyclic and by Lemma 3.1.3, $p^{-1}(F) = E$, and $\pi_1(M)\backslash E = F = M^{\,'S^1}$. Since $E$ is $\mathbb{Q}$-acyclic, then $H^*(\pi_1(M)\backslash E; \mathbb{Q}) = H^*(F; \mathbb{Q})$ is the same as the group cohomology $H^*(\pi_1(M); \mathbb{Q})$, which is the same as $H^*(M; \mathbb{Q})$ because $M$ is aspherical. (Alternatively, the inclusion map $\tilde{i} : p^{-1}(F) \hookrightarrow \widetilde{M}$ is $\pi_1$-equivariant and induces an isomorphism on rational cohomology. As the actions are free, the inclusion $i : F \hookrightarrow M$ also induces an isomorphism on rational cohomology.)

If $M$ is orientable, we have $H^n(M; \mathbb{Q}) = \mathbb{Q}$. But $F \neq M$, since the action $(S^1, M)$ was assumed nontrivial, and $H^n(F; \mathbb{Q}) = 0$ because $F$ is a proper closed subset of $M$. This is a contradiction, so the $S^1$-action is injective.

If $M$ is not orientable, we can lift the $S^1$-action to the orientable double cover $\widehat{M}$, because the elements of $\pi_1(M)$ which preserve the orientation of $\widehat{M}$ are left invariant by $S^1$. Thus the $S^1$-action lifts to orientable double cover $\widehat{M}$ of $M$. Then the $S^1$-action on $\widehat{M}$ is injective, and consequently it is injective on $M$. □

COROLLARY 3.1.12 ([**CR69**, Theorem 5.6]). *If $G$ is a nontrivial compact connected Lie group acting effectively on a closed aspherical manifold $M$, then $G$ is a torus group and acts injectively on $M$.*

PROOF. Let $T^k$ be a maximal torus in $G$. If the restriction of the evaluation homomorphism $\text{ev}_*^x : \pi_1(G, e) \to \pi_1(X, x)$ to the torus $T^k$ is not an injection, then there exists a circle subgroup $C$ in $T^k$, for which the restriction of the evaluation homomorphism is not injective. This contradicts Lemma 3.1.11. Consequently, the evaluation homomorphism $\text{ev}_*^x : \pi_1(T^k, e) \to \pi_1(M, x)$, which factors through $\pi_1(G, e)$, is injective. But, $\pi_1(T^k, e) \to \pi_1(G, e)$ (under the homomorphism induced by inclusion) is injective if and only if $G$ itself is $T^k$; see Proposition 6.4.4. □

LEMMA 3.1.13. *Let $M$ be a closed aspherical $n$-manifold and $p : \widetilde{M} \to M$ its universal covering. Then no finite group can act effectively on $\widetilde{M}$ and commute with the covering transformations.*

PROOF. Suppose the lemma is false. Then there exists a cyclic subgroup $\mathbb{Z}_p$, of prime order, which acts effectively on $\widetilde{M}$ and commutes with $\Pi = \pi_1(M)$. Since $M$ is contractible, it is $\mathbb{Z}_p$-acyclic and therefore, by Lemma 3.1.6, $E = \widetilde{M}^{\mathbb{Z}_p}$ is a nonempty proper closed subset of $\widetilde{M}$ and is also $\mathbb{Z}_p$-acyclic. The set $E$ is also $\Pi$-invariant because $\Pi$ and $\mathbb{Z}_p$ commute. Thus, $\Pi$ acts freely and properly on $E$ with quotient $F \subset M$. Now $F$ is a proper closed subset of $M$, for otherwise, $E$ would not be a proper closed subset of $\widetilde{M}$.

Because $E$ is $\mathbb{Z}_p$-acyclic and $\Pi = \pi_1(M)$ acts freely and properly on $E$, $H^*(F;\mathbb{Z}_p) \cong H^*(\Pi;\mathbb{Z}_p) \cong H^*(M;\mathbb{Z}_p)$. Suppose $M$ is $\mathbb{Z}_p$-orientable (this holds if $p = 2$, or if $M$ is orientable). Thus, $H^n(M;\mathbb{Z}_p) = \mathbb{Z}_p$. But $H^n(F;\mathbb{Z}_p) = 0$ since $F$ is a proper subset of $M$. This yields a contradiction.

If $M$ is not orientable, the $\mathbb{Z}_p$-action on $\widetilde{M}$ commutes with the subgroup of orientation preserving covering transformations (a subgroup of index 2), and we reach a similar contradiction on the orientable double covering of $M$. $\square$

LEMMA 3.1.14. *Let a finite group $G$ act effectively on a connected space $X$ (admitting covering space theory). Let $1 \to \pi_1(X) \to G^* \to G \to 1$ be the lifting exact sequence of Subsection 2.2.2. Recall the induced homomorphisms $\theta : G^* \to \mathrm{Aut}(\pi_1(X,x))$ and $\psi : G \to \mathrm{Out}(\pi_1(X,x))$. Let $K$ be the kernel of $\psi$. Then, there is induced commutative diagram*

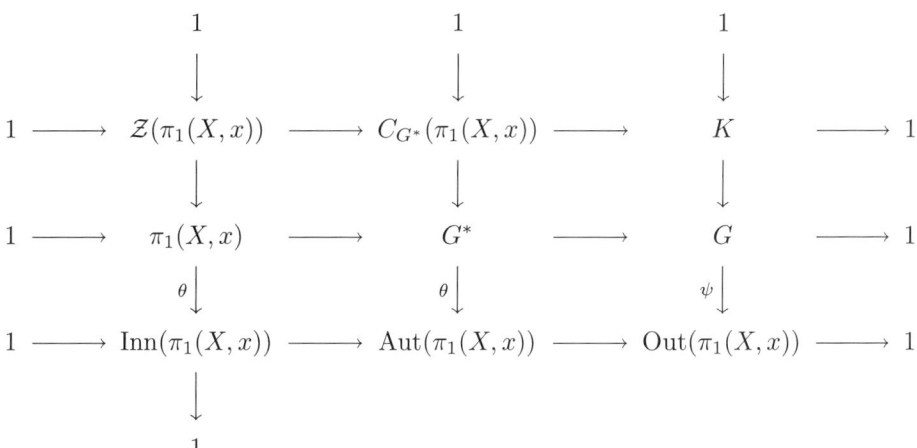

*with exact rows and columns.*

Here $C_{G^*}(\pi_1(X,x))$ is the centralizer of $\pi_1(X,x)$ in $G^*$, which is the kernel of $\theta$. The extended-lifting $G^*$ acts on the universal covering space and the elements of $C_{G^*}(\pi_1(X,x))$ commute with the covering transformations of $\widetilde{X}$, the universal covering of $X$. The exactness of the top row is easily checked by a diagram chase.

EXERCISE 3.1.15 ([**GLÖ85**]). The diagram in Lemma 3.1.14 can be improved. Let $H$ be a normal subgroup of $\Pi = \pi_1(X,x)$, and let $(X_H, \overline{x})$ be the covering space of $(X,x)$ corresponding to the subgroup $H$. Let $G^*$ be the group of homeomorphisms of $\widetilde{X}$ covering those in $G$, and let $\widehat{G}$ be the group of homeomorphisms of $X_H$ covering those in $G$. Unfortunately, the map $\widehat{G} \to G$ may not be onto if $H$ is not a normal subgroup of $G^*$. Show that $G^*/H \cong \widehat{G}$ and hence $G^* \to \widehat{G}$ is onto if $H$ is normal in $G^*$.

If $H$ is normal in $G^*$, conjugation in $G^*$ gives a homomorphism from $G^*$ into $\mathrm{Aut}(\Pi, H)$, the automorphisms of $\Pi$ leaving $H$ invariant. There are natural homomorphisms from $\mathrm{Aut}(\Pi, H)$ into $\mathrm{Aut}(\Pi/H)$ and from $\mathrm{Out}(\Pi, H)$ into $\mathrm{Out}(\Pi/H)$. We have the exact commutative diagram:

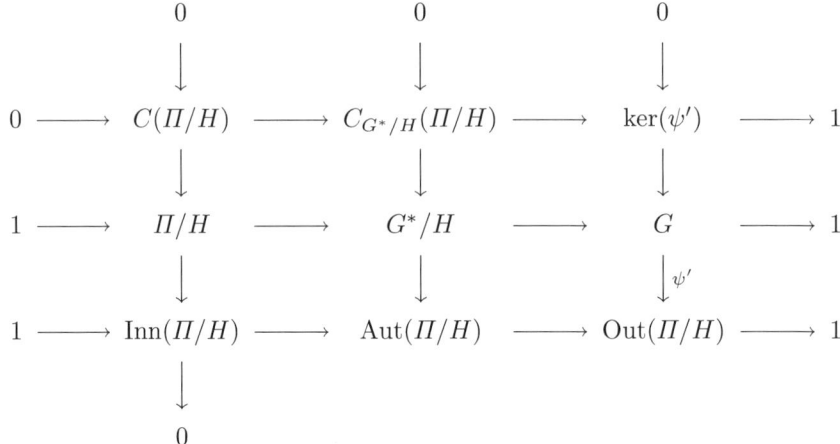

THEOREM 3.1.16. *Let a finite group $G$ act effectively on a closed aspherical manifold $M$. Then, with the notation of Lemma 3.1.14,*

(1) $C_{G^*}(\pi_1(M,x))$ *is torsion free.*
(2) *The torsion of $G^*$ injects into* $\mathrm{Aut}(\pi_1(M,x))$.
(3) *The subgroup $K$, the kernel of $\psi$, is Abelian.*

*In addition, if the center of $\pi_1(M,x)$ is finitely generated of rank $k$, then $K$ is isomorphic to a subgroup of the $k$-torus $T^k$.*

PROOF. If $C_{G^*}(\pi_1(M,x))$ is not torsion free, let $\mathbb{Z}_p$ be one of its cyclic subgroups of prime order $p$. Now $\mathbb{Z}_p$ acts on $\widetilde{M}$ and commutes with $\pi_1(M,x)$. This is impossible by Lemma 3.1.13. Thus, all of the torsion of $G^*$ must inject into $\mathrm{Aut}(\pi_1(M,x))$. Now $C_{G^*}(\pi_1(M,x))$ is a torsion-free central extension of an Abelian group $\mathcal{Z}(\pi_1(M,x))$ by a finite group $K$. By Corollary 5.5.3, the extension is Abelian. If the perhaps-redundant hypothesis that $\mathcal{Z}(\pi_1(M,x))$ is finitely generated, say of rank $k$, holds, then $C_{G^*}(\pi_1(M,x))$ is isomorphic to $\mathbb{Z}^k$. Consequently, $K$ is a subgroup of $T^k$. □

COROLLARY 3.1.17. *Let $G$, a compact Lie group, act effectively on a closed aspherical manifold $M$. Then*

(1) *If $G$ is connected, then $G$ is a $k$-torus and acts injectively.*
(2) *If $x \in M^G$, then $G$ is finite and $\theta : G \to \mathrm{Aut}(\pi_1(M,x))$ is injective.*
(3) *If $\mathcal{Z}(\pi_1(M,x)) = 1$, then $G$ is finite and $\psi : G \to \mathrm{Out}(\pi_1(M,x))$ is injective.*

PROOF. (1) While we have already verified (1) in Corollary 3.1.12, we shall provide yet another argument that uses Corollary 3.1.9 and Lemma 3.1.13. If $G$ is a compact, connected Lie group, let $F$ be any finite nontrivial subgroup of $G$. Since $G$ is connected, every element of $G$ is a homeomorphism of $M$ isotopic to the identity homeomorphism. Therefore, $\psi : F \to \mathrm{Out}(\pi_1(M,x))$ is trivial. Then, by (3) of Theorem 3.1.16, $F$ is Abelian. But the only compact connected Lie group with the property that all of its finite subgroups are Abelian is a torus. Therefore, $G$ must be a torus.

Assume that $G = T^k$ is nontrivial and that the action is not injective. Then there is a circle subgroup $S^1$ for which the induced homomorphism is not injective on $\pi_1(S^1, 1)$. The image must be finite but, by Corollary 3.1.9, $\pi_1(M)$ is torsion free, so the image must be trivial. We lift the action of $S^1$ to an effective action on the universal covering $\widetilde{M}$. Let $\mathbb{Z}_p$ be a cyclic subgroup of $S^1$. The group acts effectively on $\widetilde{M}$ and commutes with $\pi_1(M)$ (because $S^1$ commutes with the covering transformations). This is not possible by Lemma 3.1.13.

(2) The connected component of the identity $G^0$ of $G$ must be trivial, for otherwise (1) is contradicted because $\operatorname{ev}^x_*(\pi_1(G^0, x))$ is trivial. Therefore $G$ lifts to $G^*$ and must inject into $\operatorname{Aut}(\pi_1(M, x))$ by (2) of the theorem.

(3) Since $\operatorname{ev}^x_*(\pi_1(G^0, x))$ injects into $\mathcal{Z}(\pi_1(M, x)) = 1$, $G$ is finite. Therefore, as $C_{G^*}(\pi_1(M, x))$ is torsion free, the kernel of $\psi$ must be trivial. $\square$

REMARK 3.1.18. In Corollary 3.1.17, statements (1) and (2) are due to [**CR69**, §5 and §6]. Statement (3) is due to A. Borel (unpublished) and appears in [**CR72d**]. Actually the two proofs to show that when $G$ is nontrivial and connected, $G$ is a torus, use Lemma 3.1.6 in an essential way. What is different in the two approaches are the statements (a) and (b) used to show $G$ is $T^k$: (a) If every finite subgroup of $G$ is Abelian, then $G$ is a torus. (b) If $\pi_1(T^k) \to \pi_1(G)$ is injective, where $T^k$ is a maximal torus of $G$, then $G$ is a torus.

REMARK 3.1.19. We mention two very interesting unsolved problems.

(1) Let $M$ be a closed aspherical manifold and $\mathcal{Z} = \operatorname{Center} \pi_1(M)$. Is $\mathcal{Z}$ finitely generated? No examples of closed aspherical manifolds with nonfinitely generated $\mathcal{Z}$ are known to us.

(2) Let $M$ be a closed aspherical manifold. Suppose $\mathcal{Z}(\pi_1(M)) \neq 1$. Does $M$ admit a nontrivial action of $S^1$? If $\mathcal{Z}(\pi_1(M))$ is finitely generated of rank $k$, $k > 0$, does $M$ admit an effective $T^k$-action? Again no examples of closed aspherical manifolds that fail to have such action are known to us. However, in the smooth category, the answer is no in general. We shall return to these questions in Section 11.7.

REMARK 3.1.20. There is no shortage of closed aspherical manifolds. For example, if $\Gamma$ is discrete and acts properly and freely on $\mathbb{R}^n$, then $\Gamma \backslash \mathbb{R}^n$ is an aspherical manifold (a $K(\Gamma, 1)$-manifold). If $\Gamma$ acts so that the quotient is compact, then $\Gamma \backslash \mathbb{R}^n$ is a closed aspherical manifold. Typical examples arise by taking $\Gamma$ a torsion-free cocompact discrete subgroup of a Lie group $G$ with finitely many components and $K$ a maximal compact subgroup of $G$. Since $K$ is a maximal compact subgroup, $G/K$ is diffeomorphic to $\mathbb{R}^n$ for some $n$, and $\Gamma$ being torsion free implies $\Gamma \cap K = 1$ so that $\Gamma \backslash \mathbb{R}^n = \Gamma \backslash (G/K) = (\Gamma \backslash G)/K$ is a closed aspherical manifold.

For example, each closed 2-manifold whose Euler characteristic is negative is of the form $\Gamma \backslash \operatorname{PSL}(2, \mathbb{R}) \rtimes \mathbb{Z}_2 / O(2)$ where $\Gamma$ is the fundamental group of the surface and is isomorphic to a torsion-free cocompact subgroup of the full group of isometries of the hyperbolic plane. In dimension 3, each closed 3-manifold can be written as a connected sum of closed aspherical manifolds and manifolds covered by $S^3$ or $S^2 \times S^1$. All of the closed 2- and 3-dimensional closed aspherical manifolds are covered by $\mathbb{R}^2$ or $\mathbb{R}^3$. However, there is an important class of closed aspherical $n$-manifolds $n > 3$, constructed by M. Davis [**Dav83**], whose universal coverings are not homeomorphic to $\mathbb{R}^n$. All of these manifolds have fundamental group with trivial center.

Another source of aspherical manifolds arises from the fact that products of aspherical spaces or, more generally, bundles whose base and fiber are aspherical are again aspherical.

PROPOSITION 3.1.21. *Let $G$ be a nontrivial finite group acting freely on a connected space $X$ such that $\pi_1(X)$ and $\pi_1(G\backslash X)$ are torsion free. Assume $\mathcal{Z}(\pi_1(X))$ is finitely generated and that every element of $G$ is homotopic to the identity. Then*

(1) *$\pi_1(X)$ and $\pi_1(G\backslash X)$ have nontrivial centers of rank $k \geq 1$.*
(2) *$G$ is an Abelian group of rank $\leq k$ (i.e., can be embedded in a $k$-torus).*
(3) *the center of $\pi_1(G\backslash X)$ is the centralizer of $\pi_1(X)$ in $\pi_1(G\backslash X)$ and is an extension of $\mathcal{Z}(\pi_1(X))$ by $G$.*

PROOF. The lifting exact sequence (see Subsection 2.2.2) for the $G$-action is given by
$$1 \to \pi_1(X) \to \pi_1(G\backslash X) \to G \to 1.$$
Let us abbreviate $\pi_1(X)$ by $\Pi$ and $\pi_1(G\backslash X)$ by $G^*$, as in the diagram of Lemma 3.1.14. Since $G$ acts homotopically trivially, $G$ goes trivially into $\mathrm{Out}(\Pi)$, $G = K$ and $C_{G^*}(\Pi)$ is mapped onto $G$. Since $G^*$ is torsion free, $C_{G^*}(\Pi)$ is also torsion free. Since $G \neq 1$, $G^*$ is not finite. Therefore, $\mathcal{Z}(\Pi)$ has rank $k$ for some $k \geq 1$. (Otherwise, $C_{G^*}(\Pi)$, which is torsion free, could not map onto $G$.) The top row in diagram of Lemma 3.1.14 is now a central extension and so $C_{G^*}(\Pi) \cong \mathbb{Z}^k$ and $\mathcal{Z}(\Pi)$ is a sublattice of $C_{G^*}(\Pi)$. The quotient $G$ is isomorphic to a subgroup of $T^k$.

Now, for $\alpha \in G^*$, $\theta(\alpha)$ goes trivially into $\mathrm{Out}(\Pi)$. Consequently, $\theta(\alpha)$ is the conjugation $\mu(\gamma)$ by some element $\gamma \in \Pi$. Therefore, $\alpha c \alpha^{-1} = \gamma c \gamma^{-1} = c$, for $c \in \mathcal{Z}(\Pi)$. So, $\alpha$ acts trivially on $\mathcal{Z}(\Pi)$ and because $\mathcal{Z}(\Pi)$ is a sublattice of $C_{G^*}(\Pi)$, $G^*$ also acts trivially on $C_{G^*}(\Pi)$ implying that $C_{G^*}(\Pi)$ is the full center of $G^*$. □

COROLLARY 3.1.22. *Let $X$ be closed aspherical manifold with $\mathcal{Z}(\pi_1(X))$ finitely generated. Let $G$, a finite group, act effectively and homotopically trivially on $X$. Then*

(1) *$\pi_1(X)$ has a nontrivial center of rank $k \geq 1$.*
(2) *$G$ is an Abelian group of rank less than or equal to $k$ (i.e., can be embedded into $k$-torus).*
(3) *In the lifting sequence $1 \to \pi_1(X) \to G^* \to G \to 1$ for $(G, X)$, the center of $G^*$ is $C_{G^*}(\Pi)$ and is a central extension of $\mathcal{Z}(\pi_1(X))$ by $G$.*

PROOF. Effectiveness of the $G$-action guarantees that $C_{G^*}(\Pi)$ is torsion free. We observe that $\mathcal{Z}(\pi_1(X)) \neq 1$, for otherwise $G$ would have to inject into $\mathrm{Out}(\Pi)$. The rest of the argument proceeds as in the Proposition. □

## 3.2. Actions of compact Lie groups on admissible manifolds

DEFINITION 3.2.1. A closed manifold $M$ is called *admissible* [**LR87**] if the only periodic self-homeomorphisms of $\widetilde{M}$ commuting with the deck transformation group $\pi_1(M)$ are elements of the center of $\pi_1(M)$.

In light of Lemma 3.1.14 and Subsection 2.2.2, this is equivalent to saying that the torsion subgroup of $\mathcal{Z}(\pi_1(M,x))$ injects isomorphically onto the torsion subgroup of $C_{G^*}(\pi_1(M,x))$ for any lifting exact sequence $1 \to \pi_1(M,x) \to G^* \to G \to 1$. Note also, by Corollary 3.1.9 and Lemma 3.1.13, that *closed aspherical manifolds are admissible*. We adapt these arguments to derive similar results for admissible manifolds.

The following extends well-known results of [**CR69**], [**CR71**], [**CR72d**], [**SY79**], [**DS82**], [**GLÖ85**], and an unpublished result of A. Borel.

THEOREM 3.2.2. *Let $G$ be a compact Lie group acting effectively on an admissible manifold $M$.*

(1) *If $G$ is connected, then $G$ is a $k$-torus and acts injectively.*
(2) *If $G$ is finite and fixes $x \in M$, then $\theta : G \to \mathrm{Aut}(\pi_1(M,x))$ is injective.*
(3) *If $G$ is finite, then $K$, the kernel of $\psi : G \to \mathrm{Out}(\pi_1(M,x))$, is Abelian.*
(4) *If $G$ is finite and $\mathcal{Z}(\pi_1(M,x))/\mathrm{Torsion}(\mathcal{Z}(\pi_1(M,x)))$ is finitely generated of rank $k$, then $K$ is a subgroup of $T^k$. In particular, if $\mathcal{Z}(\pi_1(M,x))$ is finite, then $\psi$ is injective.*

PROOF. (1) (cf. Lemma 3.1.11). Let $T^k$ be a maximal torus of $G$. Suppose there exists a subgroup $S^1 \subset T^k$ which acts non-injectively. Therefore, $\mathrm{Im}(\mathrm{ev}_*^x(\pi_1(S^1,1))) = \mathbb{Z}_n$, $n \geq 1$. Then the $n$-fold covering $f : {}'S^1 \to S^1$ acts on the universal covering $\widetilde{M}$ effectively, covering the action $(S^1, M)$. Now for any non-trivial finite subgroup $\mathbb{Z}_p \subset {}'S^1$, $\mathbb{Z}_p$ commutes with $\pi_1(M)$ and so must be in the center of $\pi_1(M)$. If we choose $p$, coprime to $n$, then $\mathbb{Z}_p$ maps injectively onto a cyclic subgroup of $S^1$. Let $\nu : \widetilde{M} \to M$ be the covering projection and $g \in \mathbb{Z}_p$. Since, for all $\widetilde{x} \in \widetilde{M}$, $f(g) \cdot \nu(\widetilde{x}) = \nu(g\widetilde{x}) = \nu(\widetilde{x})$, $f(g)$ must be the identity. This is a contradiction. So the maximal torus acts injectively. But, $\mathrm{ev}_*^x(\pi_1(T^k,1)) \to \pi_1(M,x)$ can be factored as $\pi_1(T^k,1) \xrightarrow{i_*} \pi_1(G,1) \xrightarrow{\mathrm{ev}_*^x} \pi_1(M,x)$. Since $i_*$ must be an injection, $G = T^k$, see Proposition 6.4.4.

(2) Let $L$ be the kernel of $\theta : G \to \mathrm{Aut}(\pi_1(M,x))$ where the finite $G$ fixes $x \in M$. Let $\mathbb{Z}_p$ be a cyclic subgroup of $L$. The lifting sequence of $G$ restricted to $\mathbb{Z}_p$ yields $\pi_1(M) \times \mathbb{Z}_p$. By hypothesis, $\mathbb{Z}_p$ must be in the center of $\pi_1(M)$. But then $\mathbb{Z}_p$ would have to act trivially on $M$ contradicting the effectiveness of $G$.

(3) From Lemma 3.1.14 (the lifting sequence for $K$), we have the following commutative diagram.

$$\begin{array}{ccccccccc} 0 & \longrightarrow & \mathcal{Z}(\pi_1(M)) & \longrightarrow & C_{K^*}(\pi_1(M)) & \longrightarrow & K & \longrightarrow & 1 \\ & & \downarrow & & \downarrow & & \downarrow & & \\ 1 & \longrightarrow & \pi_1(M) & \longrightarrow & K^* & \longrightarrow & K & \longrightarrow & 1 \end{array}$$

As the torsion of $\mathcal{Z}(\pi_1(M))$ maps isomorphically onto the torsion of $C_{K^*}(\pi_1(M))$, no nontrivial element of finite order of $C_{K^*}(\pi_1(M))$ maps onto a nontrivial element of $K$. Dividing out by the torsion subgroup, $\tau$, of $\mathcal{Z}(\pi_1(M))$, we get the central extension

$$0 \to \mathcal{Z}(\pi_1(M))/\tau \to C_{K^*}(\pi_1(M))/\tau \to K \to 1.$$

The first two groups are torsion free and by Corollary 5.5.3, the middle group and hence the last is Abelian.

(4) The first part follows as in Theorem 3.1.16 using $\mathcal{Z}(\pi_1(M))/\tau$. For the second part, if $\mathcal{Z}(\pi_1(M))$ is finite, then $M$ is admissible and $K$ finite implies that $K$ is trivial. □

COROLLARY 3.2.3. *If $M$ is admissible and $G$ is a compact Lie group which acts effectively, then*
  (i) *$\mathcal{Z}(\pi_1(M)) = 1$ implies $G$ is finite and $\Psi : G \to \mathrm{Out}(\pi_1(M))$ is injective.*
  (ii) *$\mathrm{Fix}(G, M) \neq \emptyset$ implies $G$ is finite and $\theta : G \to \mathrm{Aut}(\pi_1(M))$ is injective.*
*Furthermore, if $M$ is a nonorientable manifold whose orientable double cover $M'$ is admissible, then statements* (i) *and* (ii) *also hold for $M$.*

The reader has probably observed that admissibility is a stronger condition than necessary to obtain some of the conclusions of the above theorem. In particular, we did not use that $M$ was a manifold. Let us examine this more carefully.

DEFINITION 3.2.4. Let $X$ be a path-connected space admitting covering space theory with $\widetilde{X}$ its universal covering. Let $\mathcal{A}$ be a set of primes. We say that $X$ is an $\mathcal{A}$-*admissible* space if $p \in \mathcal{A}$ implies that $\Phi \times \mathbb{Z}_p$ does not act effectively on $\widetilde{X}$. Obviously, an admissible manifold (or space) is an $\mathcal{A}$-admissible manifold (or space) for any $\mathcal{A}$.

THEOREM 3.2.5. *Let $M$ be a path-connected space admitting covering space theory with $\widetilde{M}$ the universal covering of $M$ and $\pi_1(M) = \Phi$, the group of covering transformations. Suppose $X$ is $\mathcal{A}$-admissible for a subset $\mathcal{A}$ of primes. Then*
  (1) *Suppose $\mathcal{A}$ is an infinite set. If $G$ is connected, then $G$ is a $k$-torus and acts injectively.*
  (2) *If $p \in \mathcal{A}$ and $G$ is finite, then $L = \ker\{\theta : G \to \mathrm{Aut}(\Phi)\}$ has order not divisible by $p$. In general, if every prime $p$ that divides the order of $G$ is in $\mathcal{A}$, then $\theta : G \to \mathrm{Aut}(\Phi)$ is injective.*
  (3) *If $p \in \mathcal{A}$, $G$ is finite, and $\mathcal{Z}(\Phi) = 1$, then $K = \ker(\psi)$ has order not divisible by $p$. In general, if every prime $p$ that divides the order of $G$ is in $\mathcal{A}$, then $\psi : G \to \mathrm{Out}(\Phi)$ is injective.*

For (1), as in the proof of (1) of the previous theorem, we may choose $p$ coprime to $n$. Then $\mathbb{Z}_p \subset {}'S^1$ commutes with $\pi_1(M)$ and, as $\frac{\Phi \cdot \mathbb{Z}_p}{\Phi} = \mathbb{Z}_p/(\mathbb{Z}_p \cap \Phi)$, we have that $\mathbb{Z}_p \cap \Phi = \mathbb{Z}_p$ if $p \in \mathcal{A}$. Thus, $f(\mathbb{Z}_p) \cong \mathbb{Z}_p \subset S^1$ acts trivially on $M$, contradicting that $G$ acts effectively on $M$.

3.2.6. We have extended the major results of compact group actions on closed aspherical manifolds to admissible manifolds. We shall show that this is a much wider class of manifolds than asphericals. This material is taken from [**LR87**]. Let us recall the following definitions:

DEFINITION 3.2.7. A connected, closed, oriented $m$-manifold $M$ with $\Pi = \pi_1(M)$ is called:
  (1) *Aspherical* if $\pi_i(M) = 0$, for all $i > 1$; $M$ is therefore a $K(\Pi, 1)$.
  (2) *Hyper-aspherical* [**DS82**] if there exists a closed aspherical $m$-manifold $N$ and a map $f : M \to N$ of degree 1. This is equivalent to saying $f^* : H^m(N; \mathbb{Z}) \to H^m(M; \mathbb{Z})$ is onto.
  (3) *$K$-manifold* [**GLÖ85**] if there exists a *torsion-free* group $\Gamma$ and a map $f : M \to K(\Gamma, 1)$ so that $f^* : H^m(K(\Gamma, 1), \mathbb{Z}) \to H^m(M, \mathbb{Z})$ is onto.

(4) *Admissible* [**LR81**] if the only periodic self-homeomorphisms of $\widetilde{M}$ commuting with the deck transformation group $\pi_1(M)$ are elements of the center of $\pi_1(M)$.

THEOREM 3.2.8. *Aspherical* $\Longrightarrow$ *Hyper-aspherical* $\Longrightarrow$ *K-manifold* $\Longrightarrow$ *Admissible*.

3.2.9. Each of the implications of Theorem 3.2.8 cannot be reversed; See [**LR87**] for a complete discussion. For example, the connected sum of two nonhomeomorphic 3-dimensional spherical space forms, which are also not lens spaces, is admissible but is not a $K$-manifold.

Hyper-aspherical manifolds which are not aspherical are easily obtained by taking any closed oriented aspherical $n$-manifolds $N$ and forming the orientable connected sum with any other closed oriented $n$-manifold $P$. Then $M = P \# N$ maps onto $N$ with a map of degree 1 by collapsing the portion of $P$ of $M$ to a point.

LEMMA 3.2.10 ([**DS82**, Lemma 2.5]). *Let $G = \mathbb{Z}_q$ ($q$ a prime) act nontrivially on a closed connected oriented $n$-manifold $M$. Let $p : M \to \mathbb{Z}_q \backslash M$ be the natural projection. Then the map $p^* : H^n(\mathbb{Z}_q \backslash M; \mathbb{Z}) \to H^n(M; \mathbb{Z})$ is not surjective.*

PROOF. Let $F$ be the fixed-point set of $G$. From the general Smith theory, which we recalled in Subsection 1.8.1, it is known that $F$ is nowhere dense (since the dimension of $F$ is lower than that of $M$) and the induced orbit map $p_1 : M - F \longrightarrow G \backslash M - F$ is a finite covering of topological manifolds. In particular, both $(M, F)$ and $(G \backslash M, F)$ are relative topological $n$-manifolds; i.e., their differences are $n$-manifolds. Also if $G$ acts orientation preservingly, then $\dim F \leq n - 2$, and $M - F$ is connected. Hence the horizontal arrows in the diagram below

$$\begin{array}{ccc} H^n(G \backslash M, F) & \xrightarrow{j_0^*} & H^n(G \backslash M) \\ \downarrow p_0^* & & \downarrow p^* \\ H^n(M, F) & \xrightarrow{j^*} & H^n(M) \end{array}$$

are isomorphisms. Therefore it suffices to show that $p_0^*$ is not surjective. But it is clear that $M - F$ and $G \backslash M - F$ are connected topological manifolds and $p_0$ has degree $q$. A diagram chase then shows that $H^n(G \backslash M, F) \cong \mathbb{Z} \cong H^n(M, F)$ and $p_0^*$ corresponds to multiplication by $q$. Hence $p^*$ is not surjective.

If $G$ does not preserve orientation, then $G = \mathbb{Z}_2$. In this case the invariant cohomology group is $H^n(M; \mathbb{Q})^{\mathbb{Z}_2} = 0$ since $H^n(M; \mathbb{Q}) = H^n(M; \mathbb{Z}) \otimes \mathbb{Q}$. Consider the following diagram

$$\begin{array}{ccccc} H^n(M; \mathbb{Z}) & \xrightarrow{\alpha} & H^n(M; \mathbb{Z}) \otimes \mathbb{Q} & \xrightarrow{=} & H^n(M; \mathbb{Q}) \\ \uparrow p^* & & \uparrow p^* \otimes 1 & & \uparrow \\ H^n(G \backslash M; \mathbb{Z}) & \xrightarrow{\beta} & H^n(G \backslash M; \mathbb{Z}) \otimes \mathbb{Q} & \xrightarrow{=} & H^n(G \backslash M; \mathbb{Q}) \end{array}$$

where the last map $H^n(G \backslash M; \mathbb{Q}) \to H^n(M; \mathbb{Q})$ factors through

$$H^n(G \backslash M; \mathbb{Q}) \xrightarrow{p_1^*} H^n(M; \mathbb{Q})^{\mathbb{Z}_2} \xrightarrow{\text{incl}} H^n(M; \mathbb{Q}).$$

The map $\alpha$ is injective, and $\beta$ is injective on the free part of $H^n(G \backslash M; \mathbb{Z})$. Furthermore $p_1^*$ is an isomorphism by a standard transfer argument. Compare [**Bor60**,

Chapter 2] or [**Bre72**, §7.2]. Therefore $H^n(G\backslash M; \mathbb{Q}) = 0$, so that $H^n(G\backslash M, \mathbb{Z})$ is a torsion group. In fact, similar considerations also show that the latter is at most a 2-torsion group. In any case, we have shown enough to guarantee that $p^*$ is zero if $G$ does not preserve orientation. □

REMARK 3.2.11. The lemma is also valid for a closed, connected, oriented $\mathbb{Z}$-cohomology $n$-manifold $M$. The proof applies without changes.

3.2.12 (Proof of Theorem 3.2.8). We only need to prove $K$-manifold $\Longrightarrow$ Admissible. Let $\Pi = \pi_1(X)$. Suppose $M$ is a $K$-manifold which is not admissible. Then there exists a homeomorphism $h$ of $\widetilde{M}$ so that

(i) $h$ commutes with $\Pi$,
(ii) $h^k = \mathrm{id}$, for some $k > 1$,
(iii) $h \notin \mathcal{Z}(\Pi)$, the center of $\Pi$.

Let $p$ be the smallest integer so that $h^p \in \mathcal{Z}(\Pi)$, $1 < p \leq k$. Let $k = d \cdot p$. We may assume $p$ is a prime by choosing a power of $h$ if necessary. Then
$$\mathbb{Z}_k = \{h, h^2, \ldots, h^k\} \subset C_{\mathrm{TOP}(\widetilde{M})}(\Pi),$$
the centralizer of $\Pi$ in $\mathrm{TOP}(\widetilde{M})$,
$$\mathbb{Z}_d = \{h^p, h^{2p}, \ldots, h^{dp}\} = \mathbb{Z}_k \cap \Pi = \mathbb{Z}_k \cap \mathcal{Z}(\Pi).$$
Such an $h$ defines an action of $\mathbb{Z}_p \cong \mathbb{Z}_k/\mathbb{Z}_d$ on $M$. The lifting sequence (see Subsection 2.2.2) of $(\mathbb{Z}_p, M)$ is $1 \to \Pi \xrightarrow{i} E \to \mathbb{Z}_p \to 1$ and $\Pi \supset \mathcal{Z}(\Pi) \supset \mathbb{Z}_d$, $E \supset C_E(\Pi) \supset \mathbb{Z}_k$ so that $1 \to \mathbb{Z}_d \to \mathbb{Z}_k \to \mathbb{Z}_p \to 1$ is exact.

Assume $\mathrm{Fix}(\mathbb{Z}_p, M) = \emptyset$. Then $\Pi_1(\mathbb{Z}_p\backslash M) = E$. The set of torsion elements of $C_E(\Pi)$ forms a fully invariant subgroup of $C_E(\Pi)$ coinciding with $tC_E(\Pi)$, the smallest normal subgroup containing all torsion elements, and
$$1 \to t(\mathcal{Z}(\Pi)) \to t(C_E(\Pi)) \to \mathbb{Z}_p \to 1$$
is exact; see Corollary 5.5.2. Then $\Pi/t(\mathcal{Z}(\Pi)) \cong E/t(C_E(\Pi))$. The kernel of the homomorphism $f_* : \Pi \longrightarrow \Gamma$, induced from $f : M \to K(\Gamma, 1)$, contains the smallest normal subgroup containing all the torsion of $\Pi$. Therefore $\Pi \to \Gamma$ factors through $\Pi/t(\mathcal{Z}(\Pi))$. Consequently, we may extend the homomorphism $\Pi \to \Gamma$ to $E \to \Gamma$ via

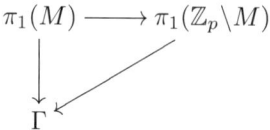

If $\mathrm{Fix}(\mathbb{Z}_p, M) \neq \emptyset$, then $E \cong \Pi \times \mathbb{Z}_p$, where $\mathbb{Z}_p$ is the isotropy of $E$ at a preimage of a fixed point. Now $\pi_1(\mathbb{Z}_p\backslash M) \cong E/N$, where $N$ is the smallest normal subgroup containing all the isotropy subgroups [**Arm68**]. Since $\mathbb{Z}_p \subset N$ already, $\pi_1(\mathbb{Z}_p\backslash M)$ is a quotient of $\Pi$ by a normal subgroup of $\Pi$ generated by torsion elements. Thus the homomorphism $\Pi \to \Gamma$ again factors through $\pi_1(\mathbb{Z}_p\backslash M) \to \Gamma$.

In either case, we have the following.

This induces a homotopy commutative diagram

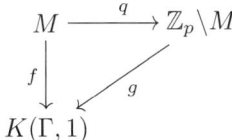

where $q$ is the orbit mapping. The map $g$ can be constructed [**Spa66**, p.428] because $\mathbb{Z}_p\backslash M$ has the homotopy type of a CW-complex since Floyd has shown that $\mathbb{Z}_p\backslash M$ is an ANR [**Flo55**]. The induced diagram on cohomology $f^* = q^* \circ g^*$ in dimension $m$ leads to a contradiction, for it was assumed that $f^*$ was onto, but $q^* : H^m(\mathbb{Z}_p\backslash M; \mathbb{Z}) \longrightarrow H^m(M; \mathbb{Z})$ is never onto, by Lemma 3.2.10. □

REMARK 3.2.13. Hyper-aspherical and $K$-manifolds, by their nature, are orientable. The following ensures that manifolds (perhaps nonorientable), covered by these manifolds, are admissible. Consequently, they will satisfy the hypothesis of Theorem 3.2.8.

PROPOSITION 3.2.14. *If $M$ is admissible and covers $N$, then $N$ is admissible.*

PROOF. Here we abuse the definition of admissible (the only periodic self-homeomorphisms of $\widetilde{N}$ commuting with $\pi_1(N)$ are elements of $\mathcal{Z}(\pi_1(N))$) by allowing $N$ be nonorientable. Let $\mathbb{Z}_n$ be a cyclic group acting effectively on $\widetilde{N}$, the universal covering of $N$, and commuting with $\pi_1(N)$. Since $\pi_1(M) \subset \pi_1(N)$, $M$ is admissible and $\widetilde{M} = \widetilde{N}$, $\mathbb{Z}_n$ is in the center of $\pi_1(M)$. Hence it must also be in the center of $\pi_1(N)$. □

REMARK 3.2.15. Because of the Remark 3.2.11, Theorem 3.2.8 will also be valid for a connected ANR closed orientable $\mathbb{Z}$-cohomology manifold. Again no change is needed in the argument.

## 3.3. Compact Lie group actions on spaces which map into $K(\Gamma, 1)$

Similarly to Theorem 3.2.5, we are able to show that generalizations of $K$-manifolds are certain $\mathcal{A}$-admissible spaces.

THEOREM 3.3.1. *Let $M$ be a connected, closed orientable (respectively, an ANR $\mathbb{Z}$-cohomology) $m$-manifold and $f : M \to K(\Gamma, 1)$, a map. Suppose that $f^*(H^m(K(\Gamma, 1); \mathbb{Z})) = d\mathbb{Z} \subset \mathbb{Z} = H^m(M; \mathbb{Z})$, with $|d| \geq 1$. Let $\mathcal{P}$ be the set of primes that do not divide the orders of the torsion elements of $\Gamma$, and let $\mathcal{D}$ be the set of primes that do not divide the integer $d$. Put $\mathcal{A} = \mathcal{P} \cap \mathcal{D}$. Then, $M$ is $\mathcal{A}$-admissible and consequently (1), (2) and (3) of Theorem 3.2.5 hold.*

PROOF. Suppose to the contrary that there exists a prime $p \in \mathcal{A}$ such that $\Phi \times \mathbb{Z}_p$ acts effectively on $\widetilde{M}$. Since $\Phi \cap \mathbb{Z}_p = 1$, there is an effective action of $\mathbb{Z}_p$ on $M$. If the action of $\mathbb{Z}_p$ on $M$ is free, then the original action of $\Phi \times \mathbb{Z}_p$ is free and $\pi_1(\mathbb{Z}_p\backslash M) = \Phi \times \mathbb{Z}_p$. The homomorphism $f_* : \Phi \to \Gamma$ can be factored as $\Phi \xrightarrow{q_*} \Phi \times \mathbb{Z}_p \longrightarrow \Phi \xrightarrow{f} \Gamma$ where $q_*$ is the isomorphism of $\Phi$ onto the first factor of $\Phi \times \mathbb{Z}_p$ induced by the quotient mapping $q : M \to \mathbb{Z}_p\backslash M$. The second map is projection back to $\Phi$. The composite is again equal to $f$. As $\mathbb{Z}_p\backslash M$ is an ANR, there is a map $g : \mathbb{Z}_p\backslash M \to K(\Gamma, 1)$ such that $f \simeq g \circ q$. If $\mathbb{Z}_p$ preserves orientation of $M$ (which holds in particular if $p$ is odd), then $H^m(\mathbb{Z}_p\backslash M; \mathbb{Z}) = \mathbb{Z}$

and $q^*(H^m(\mathbb{Z}_p\backslash M;\mathbb{Z})) = p\mathbb{Z} \subset \mathbb{Z} = H^m(M;\mathbb{Z})$. Now $g^*(H^m(K(\Gamma,1);\mathbb{Z})) = a\mathbb{Z} \subset H^m(\mathbb{Z}_p\backslash M;\mathbb{Z})$ and so $d\mathbb{Z} = p \cdot a\mathbb{Z}$. Thus, $p$ divides $d$, contradicting that $p \in \mathcal{D}$. If $p = 2$ and reverses orientation, then $H^m(\mathbb{Z}_2\backslash M;\mathbb{Z}) = \mathbb{Z}_2$ and $d\mathbb{Z} = 0\mathbb{Z}$, a contradiction.

On the other hand, if $x \in M^{\mathbb{Z}_p}$, then by lifting the $\mathbb{Z}_p$-action back up to $\widetilde{M}$ using $x$ as base point on $M$ and $\widetilde{x}$ on $\widetilde{M}$, the lifting sequence becomes $\Phi \rtimes \mathbb{Z}_p$ which is congruent to the original $\Phi \times \mathbb{Z}_p$. The fundamental group of $\mathbb{Z}_p\backslash M$ is given by $(\Phi \rtimes \mathbb{Z}_p)/N$ where $N$ is the smallest normal subgroup of $\Phi \rtimes \mathbb{Z}_p$ containing all the isotropy subgroups of the $\Phi \rtimes \mathbb{Z}_p$-action on $\widetilde{M}$. Therefore $(\Phi \rtimes \mathbb{Z}_p)/N = (\Phi \cdot N)/N \cong \Phi/(\Phi \cap N)$. (Note in terms of $\Phi \times \mathbb{Z}_p$, an isotropy subgroup is generated by $(\varphi(g), g)$, where $g$ is a generator of $\mathbb{Z}_p$ and $\varphi : \mathbb{Z}_p \to \Phi$ is a homomorphism. A typical element of $N$ is given by
$$((x_1,1)(\varphi_1^{n_1}(g),g^{n_1})(x_1^{-1},1))\cdots((x_k,1)(\varphi_k^{n_k}(g),g^{n_k})(x_k^{-1},1))$$
$$= ((x_1\varphi_1^{n_1}(g)x_1^{-1})\cdots(x_k\varphi_k^{n_k}(g)x_k^{-1}), g^{n_1+\cdots+n_k}),$$
where $x_i \in \Phi$. The element belongs to $\Phi \cap N$ if $g^{n_1+\cdots+n_k} = 0 \in \mathbb{Z}_p$; in other words, if $n_1 + n_2 + \cdots + n_k$ is a multiple of $p$. Thus $\Phi \cap N$ is generated by elements of order $p$, and if $\Phi$ has no $p$-torsion, $\Phi \cap N = 1$.) Since $\Gamma$ has no $p$-torsion, the kernel $K$ of the map $f_* : \Phi \to \Gamma$ must contain the smallest subgroup of $\Phi$ that contains all the $p$-torsions of $\Phi$. Consequently the homomorphism $f_* : \Phi \to \Gamma$ factors as $\Phi \to \Phi/(\Phi \cap N) \to \pi_1(\mathbb{Z}_p\backslash M) \to \Phi/K \subset \Gamma$. Once again, $f$ factors through $\mathbb{Z}_p\backslash M$. If $\mathbb{Z}_p$ preserves orientation, then $F = M^{\mathbb{Z}_p}$ has dimension less than $m-1$, by the Smith theorems. Consequently, $q^* : \mathbb{Z} = H^m(\mathbb{Z}_p\backslash M, F;\mathbb{Z}) \to H^m(M,F;\mathbb{Z}) \cong H^m(M;\mathbb{Z})$ is multiplication by $p$. Since $pa = d$, $p$ must divide $d$, but $p \in \mathcal{D}$, a contradiction.

Finally, if $\mathbb{Z}_p = \mathbb{Z}_2$ and reverses orientation of $M$, $M^{\mathbb{Z}_2} \neq \emptyset$, then $H^m(\mathbb{Z}_2\backslash M;\mathbb{Z}) \otimes \mathbb{Q} = 0$, so $H^m(\mathbb{Z}_2\backslash M;\mathbb{Z})$ is at most a torsion group. Consequently, $d\mathbb{Z} = 0\mathbb{Z}$, a contradiction. $\square$

REMARK 3.3.2. The proof of the theorem also shows that $M$ is $\mathcal{A}$-admissible if the set of primes $\mathcal{P}$ is replaced by the set of primes $\mathcal{P}'$ that do not divide the order of the torsion elements of $\Phi$.

COROLLARY 3.3.3. *Let $M$ be a connected, closed orientable ANR $\mathbb{Z}$-cohomology $m$-manifold with a map $f : M \to K(\Gamma,1)$ such that $f^*(H^m(K(\Gamma,1);\mathbb{Z})) = d\mathbb{Z} \subset H^m(M;\mathbb{Z})$, $|d| \geq 1$. Suppose $p$, a prime, does not divide $d$ and either $\Gamma$ or $\Phi = \pi_1(M)$ (not necessarily both) has no $p$-torsion. Then $\Phi \times \mathbb{Z}_p$ cannot act effectively on $\widetilde{M}$. That is, $M$ is an $\mathcal{A}$-admissible space whenever $p \in \mathcal{A}$.*

COROLLARY 3.3.4. *If $M$ is a connected, closed orientable $m$-manifold with no $p$-torsion in $\pi_1(M)$ for an infinite number of primes $p$, and if $f : M \to K(\Gamma,1)$ is a map such that $f^* : (H^m(K(\Gamma,1);\mathbb{Q})) \to H^m(M;\mathbb{Q})$ is nontrivial, then any connected compact Lie group that acts on $M$ effectively must be a torus acting injectively.*

Note that the condition here on $f^*$ is equivalent to the condition on $f^*$ in the theorem that $f^* : (H^m(K(\Gamma,1);\mathbb{Z})) \to H^m(M;\mathbb{Z})$ is nontrivial.

The last corollary was proved for smooth actions, without the perhaps redundant torsion restrictions on $\pi_1(M)$, by W. Browder and W.C. Hsiang [**BH82**] by a different method. Their theorem has other interesting applications.

3.3.5. Let $X$ be a connected, locally compact, finite dimensional ANR, and let $f : X \to K(\Gamma,1)$ be a map so that $f_*$ is surjective on the fundamental groups.

## 3.3. COMPACT LIE GROUP ACTIONS ON SPACES WHICH MAP INTO $K(\Gamma, 1)$

Let $G$ be a compact connected Lie group acting effectively on $X$, and let $H$ be the image of $\operatorname{ev}_*^x : \pi_1(G, e) \to \pi_1(X, x)$. Since $H$ lies in the center of $\pi_1(X, x)$ and $f_*$ is onto, $f_*(H)$ lies in Center($\Gamma$). Let $p : X \to G\backslash X$ be the projection, and put $\Gamma' = \Gamma/f_*(H)$ and $\alpha' : \Gamma \to \Gamma'$ be the natural homomorphism. This induces a map $\alpha : K(\Gamma, 1) \to K(\Gamma', 1)$. In addition, assume that the action has locally finite orbit structure. That is, at each $x$, there is a slice $S_x$ with only a finite number of orbit types for the $G_x$-orbit on $S_x$.

THEOREM 3.3.6 (Browder–Hsiang[**BH82**]). *There is a homomorphism $\varphi : H_*(G\backslash X) \to H_*(K(\Gamma', 1))$ in rational homology that makes the following diagram commutative.*

$$\begin{array}{ccc} H_*(X) & \xrightarrow{f_*} & H_*(K(\Gamma, 1)) \\ p_* \downarrow & & \downarrow \alpha_* \\ H_*(G\backslash X) & \xrightarrow{\varphi} & H_*(K(\Gamma', 1)) \end{array}$$

This result was obtained by Browder and Hsiang under the hypothesis that $X$ was a smooth manifold. The proof given here is very different from theirs. Note $X$ automatically has locally finite orbit structure if $X$ is a ($\mathbb{Z}$-cohomology) manifold.

LEMMA 3.3.7. *Let $G$ be a connected Lie group acting properly on a space $X$ which admits covering space theory. Let $H$ be the image of $\operatorname{ev}_*^x : \pi_1(G, e) \to \pi_1(X, x)$ and put $Q = \pi_1(X, x)/H$. Then the $G$-action on $X$ lifts to a $G$-action on $X_H$ (a covering space of $X$ with $\pi_1(X_H) = H$), and $W = G\backslash X_H$ is simply connected.*

PROOF. From Theorem 2.5.1, we know the $G$-action lifts to an action on $X_H$, and it commutes with the covering $Q = \pi_1(X)/H$-action. First, $\operatorname{ev}_*^{x'} : \pi_1(G, e) \to \pi_1(X_H, x')$ is an epimorphism. By [**MY57**, Corollary 1], it is possible to lift paths, $h : I \to W$ to $h' : I \to X_H$

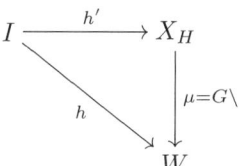

so that $\mu(h'(t)) = h(t)$. Thus, every loop based at $w \in W$ lifts to a path beginning at $x'$ (so $\mu(x') = w$), and ends at $g(x')$ for some $g \in G$. We reparametrize these loops as follows.

$$h_1(t) = \begin{cases} h(2t), & 0 \leq t \leq 1/2, \\ w, & 1/2 \leq t \leq 1, \end{cases}$$

$$h_1'(t) = \begin{cases} h'(2t), & 0 \leq t \leq 1/2, \\ g(t) \cdot x', & 1/2 \leq t \leq 1, \end{cases}$$

with $g(t)$ a path in $G$ that begins at $g \in G$ and ends at $e$ as $t$ goes from $1/2$ to $1$.

Now $h_1$ is homotopic to $h$ and $h_1'(t)$ is a loop covering $h_1$. Thus $\mu_* : \pi_1(X_H) \to \pi_1(W)$ is onto. But any loop in $X_H$ such as $h_1'$ is homotopic to the image of the evaluation map $\operatorname{ev}^{x'}$ of a loop in $G$ based at $e$. Thus, $W$ is simply connected. □

3.3.8 (Proof of Theorem 3.3.6). Lift the action of $G$ on $X$ to $X_H$, the covering space with $\pi_1(X_H) = H$; see Theorem 2.5.1. The lifted $G$-action commutes with the group, $Q = \pi_1(X,x)/H$, of covering transformations, and so induces a $Q$-action on $G\backslash X_H = W$.

Let us form the Borel space $X_H \times_Q EQ$ associated to the $Q$-action on $X_H$. On $X_H \times EQ$ we have a $G$-action on the first factor and a commuting diagonal $Q$-action. Similarly, form the Borel space $W_Q = W \times_Q EQ$ associated with the $Q$-action on $W$.

Now $\alpha' = \alpha_*$ (the homomorphism induced from $\alpha : K(\Gamma,1) \to K(\Gamma',1)$). The map
$$X_H \xrightarrow{\nu} X \xrightarrow{f} K(\Gamma,1)$$
induces
$$H = \pi_1(X_H) \xrightarrow{\nu_*} H \subset \pi_1(X,x) \xrightarrow{f_*} \Gamma,$$
and so $(f \circ \nu)_*(H) = f_*(H) = \ker(\alpha_*) = \ker(\alpha')$. From $1 \to \ker(\alpha_*) \to \Gamma \xrightarrow{\alpha_*} \Gamma' = \Gamma/\ker(\alpha_*) \to 1$, there is a fiber bundle
$$K(\ker(\alpha'),1) \longrightarrow K(\Gamma,1) \xrightarrow{\alpha} K(\Gamma',1).$$
Let $\widehat{f} : X_H \to K(\ker(\alpha_*),1)$ be the geometric realization induced from $(f \circ \nu)_*$. This is possible since $X_H$ is an ANR. We form the following diagram.

$$\begin{array}{ccccccc}
K(\ker\alpha_*,1) & \xleftarrow{\widehat{f}} & (G,X_H) & \xleftarrow{p_1} & X_H \times EQ & \xrightarrow{G\backslash} & W \times EQ \\
\downarrow & & \nu \downarrow Q\backslash & & \downarrow Q\backslash & & \downarrow Q\backslash \\
K(\Gamma,1) & \xleftarrow{f} & (G,X) & \xleftarrow{p'_1} & X_H \times_Q EQ & \xrightarrow{G\backslash} & W \times_Q EQ \\
\alpha \downarrow & & p \downarrow G\backslash & & p' \downarrow G\backslash & & \downarrow = \\
K(\Gamma',1) & & G\backslash X & \xleftarrow{\overline{p'_1}=\beta} & G\backslash(X_H \times_Q EQ) & = & W \times_Q EQ = W_Q
\end{array}$$

The upper left-hand square homotopy commutes, while the entire rest of the diagram commutes. To the right of the second column, we have equivariant maps. The maps $p_1$ and $p'_1$ are homotopy equivalences but the natural homotopy inverses are not equivariant maps.

The idea now is to produce maps $\eta$ and $\overline{f}$ so that the following diagram

(3.3.1)
$$\begin{array}{ccccc}
X & \xleftarrow{=} & X & \xrightarrow{f} & K(\Gamma,1) \\
p \downarrow & & \eta \downarrow & & \downarrow \alpha \\
G\backslash X & \xleftarrow{\beta} & W_Q & \xrightarrow{\overline{f}} & K(\Gamma',1)
\end{array}$$

commutes up to homotopy.

Because $Q$ acts freely on $X_H$, $X_H \times_Q EQ$ is a $Q$-bundle over $X$ with contractible fiber $EQ$. Thus $p'_1$ is a homotopy equivalence. Let $q$ be a homotopy inverse. Put
$$\eta : X \xrightarrow{q} (X_H \times_Q EQ) \xrightarrow{p'} W_Q.$$
Now $\pi_1(W_Q) = Q$, since $W \times EQ$ is simply connected and $Q$ acts freely.

To realize the map $\overline{f}$, we note that $W_Q$ has the homotopy type of a CW-complex because $W$ itself is an ANR. This follows from the fact that $X_H$ is an ANR and $G$ is compact group acting on $X_H$ with locally finite orbit structure. R. Oliver's

## 3.3. COMPACT LIE GROUP ACTIONS ON SPACES WHICH MAP INTO $K(\Gamma, 1)$

solution to the Conner Conjecture [**Oli76**] assures us that $W = G \backslash X_H$ is an ANR. Furthermore, $Q$ acts properly on $W \times EQ$ and so $W_Q$ is an ANR. Note that the natural homomorphism $\overline{f}_* : Q \to \Gamma'$ fits into the following diagram.

$$\begin{array}{ccccccccc}
1 & \longrightarrow & H & \longrightarrow & \pi_1(X, x) & \longrightarrow & Q & \longrightarrow & 1 \\
& & \downarrow f_* & & \downarrow f_* & & \downarrow \overline{f}_* & & \\
1 & \longrightarrow & f_*(H) & \longrightarrow & \Gamma & \longrightarrow & \Gamma' & \longrightarrow & 1
\end{array}$$

Consequently, there is a map $\overline{f} : W_Q \to K(\Gamma', 1)$ so that $\overline{f}$ induces $\overline{f}_*$ on fundamental groups, and makes the right rectangle in the diagram (3.3.1) homotopy commute.

Now, consider the natural map $\beta : W_Q = W \times_Q EQ \longrightarrow Q \backslash W$. If $w^* \in Q \backslash W$, then $\beta^{-1}(w^*) = Q_w \backslash (EQ) = BQ_w$, where $Q_w$ is the isotropy of $Q$ at $w \in W$ and $\overline{\nu}(w) = w^*$, where $\overline{\nu} : W \to Q \backslash W = G \backslash X$.

Observe that $Q_w = G_x / G_{x'}$, where $\nu(x') = x$, and $x'$ is on a $G$-orbit over $w \in W$. But the connected components of the identity $(G_x)_0$ and $(G_{x'})_0$ must be identical. So $Q_w$ is finite and is equal to $\pi_0(G_x)/\pi_0(G_{x'})$.

Now using the Leray spectral sequence of the map $\beta$, we have that $\beta^*$ is an isomorphism in rational cohomology because $H^p(BQ_w; \mathbb{Q}) = 0$ for $p > 0$.

(Technically, to apply the Vietoris mapping theorem, we need to show that the presheaf $V \longrightarrow H^*(\beta^{-1}(V); \mathbb{Q})$ determines $H^*(\beta^{-1}(w^*); \mathbb{Q})$. That is,

$$\mathop{\mathrm{dir\,lim}}_{V} H^*(\beta^{-1}(V); \mathbb{Q}) = H^*(\beta^{-1}(w^*); \mathbb{Q}),$$

where $V$ runs through a fundamental system of neighborhoods of $w^* \in Q \backslash W$. Now $\beta^{-1}(V) = U \times_{Q_w} EQ$ and $\beta^{-1}(w^*) = Q_w \backslash EQ \cong BQ_w$. Here, $U$ is a $Q_w$-slice at $w \in W$ with $V = Q_w \backslash U$. We have the commutative diagram

$$\begin{array}{ccc}
U \times EQ_w & \hookrightarrow & U \times EQ \\
Q_w \backslash \downarrow & & Q_w \backslash \downarrow \\
U \times_{Q_w} EQ_w & \hookrightarrow & U \times_{Q_w} EQ \\
\beta' \downarrow & & \beta \downarrow \\
V & = & V
\end{array}$$

induced by the inclusion $EQ_w \hookrightarrow EQ$. The horizontal maps are homotopy equivalences and the top map is $Q_w$-invariant. The map $\beta'$ is the restriction of $\beta$. We can think of $EQ_w$ and $EQ$ as the infinite join of $Q_w$ and $Q$, respectively. These can be approximated up to $n - 1$ equivalences by taking the $n$-fold joins of $Q_w$ and $Q$, respectively. The approximation of $EQ_w$ is now compact and we can apply the Vietoris mapping theorem to the restriction of $\beta'$ to this approximation. We get that $\beta'^* : H^j(V, \mathbb{Q}) \to H^j(U \times_{Q_w} EQ_w; \mathbb{Q})$ is an isomorphism for $j < n$, where $n$ can be taken arbitrarily large. The presheaf determines $H^*(\beta^{-1}(w^*); \mathbb{Q})$, up through dimension $n - 1$, by taking the direct limit of $H^j(U \times_{Q_w} EQ_w; \mathbb{Q})$ over the neighborhoods $V$ of $w^*$. Thus, we conclude that $\beta^* : H^j(Q \backslash W; \mathbb{Q}) \to H^j(W_Q; \mathbb{Q})$ is an isomorphism for all $j$.)

Consequently, the homomorphism $\beta_* : H_j(W \times_Q EQ; \mathbb{Q}) \to H_j(Q \backslash W; \mathbb{Q})$ is bijective for all $j$. Our homomorphism $\varphi$ is just $\overline{f}_* \circ \beta_*^{-1}$. $\square$

COROLLARY 3.3.9. *Under the same hypothesis of Theorem 3.3.6, the rational coefficients in the conclusion may be replaced by coefficients in $\mathbb{Z}_p$, where $p$ is a prime that does not divide the order of the $\{Q_w : w \in W\}$.*

PROOF. There is no change in argument where we observe that $H^j(BQ_w; \mathbb{Z}_p) = 0$, for $j > 0$. □

REMARK 3.3.10. The proof of Theorem 3.3.6 would be simplified if one could find a map $\varphi : G\backslash X \to K(\Gamma', 1)$ which makes the diagram commute. Unfortunately, this is not always possible. An example is given in [**BH82**, §5] of an $S^1$-closed manifold $M$ and a group $\Gamma$, for which there is no map $\varphi : G\backslash M \to K(\Gamma', 1)$ which makes the diagram commute. To overcome this difficulty, we essentially replaced $(G, X)$ by the homotopically equivalent $(G, X_H \times_Q EQ) = (G, (X_H)_Q)$ and $G\backslash X$ by the rationally homologically equivalent $W_Q = (G\backslash(X_H)_Q)$. (Note $Q\backslash W = G\backslash X$, where $W = G\backslash X_H$.) On the other hand, there are special conditions where $f$ factors through $G\backslash M$. For example,

EXERCISE 3.3.11 (Donnelley-Schultz). Show that if $\Gamma$ is torsion free and $G$ is compact connected semisimple Lie group acting on a manifold $M$, then there is a continuous map $\varphi : G\backslash M \to K(\Gamma', 1)$ which makes the diagram commute.

For the applications in the literature, it is only the factorization via homology or cohomology that is actually needed. Browder and Hsiang succinctly put it as follows:

PROPOSITION 3.3.12. *Let $G$ be a connected, compact Lie group acting on a connected (ANR $\mathbb{Z}$-cohomology) $m$-manifold $M$. Let $k$ be the dimension of the orbit space. Then*
$$(\alpha \circ f)_*(H_j(M; \mathbb{Q})) = 0 \text{ for } j > k.$$
*(Note: If $\ell$ is the dimension of a principal orbit on $M$, then $k = m - \ell$).*

PROOF. Observe that the dimension of $G\backslash M$ is $k$, and so $p_*$ is trivial in dimension greater than $k$. □

COROLLARY 3.3.13. *If $f_*(H) = (f \circ \mathrm{ev}^x)_*(\pi_1(G, e))$ is finite, then $f_*(H_j(M; \mathbb{Q})) = 0$, for $j > k$.*

PROOF. Because $f_*(H) = \ker(\alpha')$ is finite,
$$\alpha_* : H_*(K(\Gamma, 1); \mathbb{Q}) \longrightarrow H_*(K(\Gamma', 1); \mathbb{Q})$$
is an isomorphism. The corollary follows from $(\alpha \circ f)_* = \varphi \circ p_*$. □

When $G$ has a fixed point or when $\pi_1(G)$ is finite are typical situations in which the hypothesis of the corollary is satisfied. The corollary is also used to recapture Corollary 3.3.4.

COROLLARY 3.3.14 (cf. Corollary 3.3.4). *Let $G$ be a connected, compact Lie group acting effectively on a closed connected (ANR cohomology) $m$-manifold $M$, and $f_* : H_m(M; \mathbb{Q}) \to H_m(K(\Gamma, 1); \mathbb{Q})$ is nontrivial for some map $f : M \to K(\Gamma, 1)$, then $G$ is a torus acting injectively on $M$.*

PROOF. For any circle subgroup $S^1$ of $G$, if $f_*(\mathrm{ev}^x_*(\pi_1(S^1))) = H$ is finite, then as $\dim(S^1\backslash M) = m - 1$, $\alpha_* \circ f_* : H_m(M; \mathbb{Q}) \to H_m(K(\Gamma', 1); \mathbb{Q})$ must be trivial. But $\alpha_*$ is an isomorphism implying that $f_*$ is trivial. This is a contradiction. So

$\mathrm{ev}_*^x : \pi_1(S^1) \to \pi_1(M)$ is injective for each circle subgroup $S^1 \subset G$. As shown in Proposition 6.4.4, this implies $G = T^k$ for some $k$ and $T^k$ acts injectively. $\square$

EXAMPLE 3.3.15. The connected sum $M = M_1 \# M_2$ ($M_1 \neq M_2$), where each $M_i$ is a 3-dimensional spherical space form with $\pi_1(M_i)$ non-Abelian, is an admissible manifold, but not a $K$-manifold nor an essential manifold, because $f_*(H_3(M;\mathbb{Q}))$ is always trivial [**LR87**]. So the nonexistence of $S^1$-actions on $M$ is a consequence of Theorem 3.2.2 but not of Corollary 3.3.14.

More applications exploiting Proposition 3.3.12 can be found in [**BH82**], [**KK83**], and [**Ber85**]. Some other papers related to this section are [**Yau77**], [**SY79**] and [**DS82**].

## 3.4. Manifolds with few or no periodic homeomorphisms

3.4.1. Theorem 3.2.2 obviously gives a recipe for constructing manifolds with no nontrivial effective compact Lie group actions. One only needs to find admissible manifolds $M$ for which $\pi_1(M)$ has trivial center and $\mathrm{Out}(\pi_1(M))$ is torsion free.

3.4.2. The first aspherical manifolds with no actions were constructed by Conner-Raymond-Weinberger [**CRW72**] in various dimensions greater than or equal to 7. The examples all fiber over $S^1$. In various dimensions greater than or equal to 7, it is even possible to find aspherical manifolds with $\mathrm{Out}(\pi_1(M)) = 1$, or solvmanifolds with no finite group actions.

About the same time and independently, E. Bloomberg, constructed the first hyper-aspherical manifolds without a group action, [**Blo75**]. The first aspherical 3-manifolds with no finite group actions were constructed by Raymond and Tollefson [**RT76**] (and [**RT82**]). All of the above relied on showing that the torsion in $\mathrm{Out}(\pi_1(M))$ was trivial. However, extra nonalgebraic work was needed to eliminate involutions in the 3-dimensional examples.

In every bordism class (dim $\geq 3$), R. Schultz [**Sch81a**] has shown that there are manifolds with no group actions.

Puppe in [**Pup95**] has shown that there exists 6-dimensional *simply* connected closed manifolds without any effective finite orientation preserving group action. However, many of these manifolds seem to admit orientation reversing involutions; see [**Pup07**] for a recent update.

EXAMPLE 3.4.3. [**CR72d**, p.43, §7.2] Here is a rather elementary example of a nonorientable 3-manifold that admits no action of any finite group except for $\mathbb{Z}_2$. Furthermore, smoothly there is only one such action. Take the matrix $h = \begin{bmatrix} 0 & 1 \\ 1 & 1 \end{bmatrix}$. This linear transformation on $\mathbb{R}^2$ preserves the integral lattice. The matrix $h$ normalizes the $\mathbb{Z} \times \mathbb{Z}$ group of standard translations

$$\begin{bmatrix} x \\ y \end{bmatrix} \mapsto \begin{bmatrix} x+m \\ y+n \end{bmatrix}, \quad (m,n) \in \mathbb{Z} \times \mathbb{Z},$$

and so induces a homeomorphism $h$ on the 2-torus $T = \mathbb{Z}^2 \backslash \mathbb{R}^2$. Form $M = T^2 \times_{\mathbb{Z}} \mathbb{R}$, with $\mathbb{Z} = \langle h \rangle$ acting diagonally. The generator of $\mathbb{Z}$ acts on $T^2$ as $h$ and on $\mathbb{R}$ by sending $r \mapsto r - 1$. $M$ is then an affinely flat 3-manifold which fibers over $S^1$ with fiber $T^2$ and structure group $\mathbb{Z}$. It is not too difficult to show that $\mathcal{Z}(\pi_1(M)) = 1$ and $\mathrm{Out}(\pi_1(M)) = \mathbb{Z}_2$, $h$ and all its powers fix just one point on $T^2$. There is a

$\mathbb{Z}_2$-action on $M$ which leaves each fiber $T^2$ invariant (because $\begin{bmatrix} -1 & 0 \\ 0 & -1 \end{bmatrix}$ centralizes $h$). This action on $M$ has exactly two circles of fixed points.

3.4.4 (Bloomberg's example [**Blo75**]). Bloomberg proved if $A$ and $B$ are two nonisomorphic groups, neither of which is isomorphic to a nontrivial free product or an infinite cyclic group, then $\mathrm{Out}(A*B) = \mathrm{Aut}(A) \times \mathrm{Aut}(B)$. Conner-Raymond [**CR72d**, p.60] had shown that there exist closed aspherical 4-manifolds $B(k)$, $k > 1$, which fiber over $S^1$, with nilmanifolds fiber, with the property that each $\pi_1(B(k))$ is distinct, centerless, and $\mathrm{Aut}(\pi_1(B(k)))$ is torsion free. Consequently, every action of a finite group on $B(k)$ is necessarily free. In particular, the nonclosed $B(k) - \{\mathrm{point}\}$ has no action. Therefore, $\pi_1(B(k)\#B(k')) = \pi_1(B(k)) * \pi_1(B(k'))$, and $\mathrm{Out}(\pi_1(B(k)) * \pi_1(B(k')))$ is torsion free. Since the connected sum is hyperaspherical, Theorem 3.2.8 applies and $B(k)\#B(k')$ has no effective finite group action other than the trivial one. However, at the time, Theorem 3.2.8 was only known for aspherical manifolds and $B(k)\#B(k')$ was not aspherical. By a careful analysis of how a cyclic group might act on $\pi_1(B(k)) * \pi_1(B(k'))$, Bloomberg was able to reach a contradiction. He does mention that the Conner-Raymond examples generalize to all higher dimensions and so one could construct similar examples in all dimensions greater than 4. However, we have not been able to verify this claim.

EXAMPLE 3.4.5. $M = B(k)\#\mathbb{C}P_2$ has no effective nontrivial compact Lie group action.

PROOF. Since $B(k)$ fibers over the circle, $\chi(B(k)) = 0$. Consequently, $\chi(M) = \chi(B(k)) + \chi(\mathbb{C}P_2) - 2 = 0 + 3 - 2 = 1$. Suppose $\mathbb{Z}_p$ ($p$ prime) acts on $M$. By the Smith theorems, $\chi(M^{\mathbb{Z}_p}) \cong \chi(M)$, mod $p$. As $\chi(M) = 1$, $M^{\mathbb{Z}_p} \neq \emptyset$. Since $M$ is hyper-aspherical, $\theta : \mathbb{Z}_p \to \mathrm{Aut}(\pi_1(M))$ is faithful if and only if $\mathbb{Z}_p$ is effective (Theorem 3.2.2). But as $\mathrm{Aut}(\pi_1(M)) = \mathrm{Aut}(\pi_1(B(k)))$ is torsion free, $\theta$ is trivial implying that $\mathbb{Z}_p$ acts trivially. Since this holds for any prime $p$, no nontrivial compact Lie group can act effectively on $M$. □

EXERCISE 3.4.6. Let $G$ be a compact Lie group acting effectively on $M = T^4\#\mathbb{C}P_2$. Show that $G$ must be finite, in fact a subgroup of $\mathrm{GL}(4,\mathbb{Z})$, and each nontrivial element of $G$ acts homotopically nontrivially on $M$.

EXERCISE 3.4.7. Show that in each dimension greater than 3 there exists compact, simply connected manifolds with nonempty connected boundary, admitting no nontrivial compact Lie group actions. (Hint: The restriction of an action to the boundary must be effective.)

## 3.5. Injective torus actions

Recall that (Definition 3.1.10) a torus action $(T^k, X)$ is said to be *injective* If $\mathrm{ev}_*^x : \pi_1(T^k, e) \to \pi_1(X, x)$, where $\mathrm{ev}^x(t) = t \cdot x$, is injective.

An element $a$ of a group $G$ is called *primitive* if $a$ does not have an $n$th root for any $n > 1$; that is, there is no $x \in G$ such that $x^n = a$.

LEMMA 3.5.1. *Suppose $S^1$ acts on a path-connected space $X$ (which admits covering space theory) so that the generator of $\pi_1(S^1, 1)$ is mapped to a primitive element of $\pi_1(X, x)$ by $\mathrm{ev}_*^x : \pi_1(S^1, 1) \to \pi_1(X, x)$. Then the $S^1$-action is free.*

PROOF. Let $x$ be a base point, and let $a$ be represented by the loop $e^{2\pi it}(x)$, $0 \leq t \leq 1$. Let $\alpha : (I, 0, 1) \to (X, x, y)$ be a path. Then the loop $e^{2\pi it}(y)$ represents a primitive element $a' \in \pi_1(X, y)$; see Subsection 2.1.6. In fact, $\bar{\alpha} * e^{2\pi it}(x) * \alpha$, a loop that represents a primitive element of $\pi_1(X, y)$, is homotopic to $e^{2\pi it}(y)$ with the base point $y$ fixed. (The homotopy just pulls $e^{2\pi it}(x)$ back along the path $\alpha$ to $e^{2\pi it}(y)$.) Let $n$ be a positive integer such that $e^{\frac{2\pi i}{n}t}(y)$ is a loop based at $y$. Then $n$ must be 1, for otherwise, it would be a loop representing an $n$th root of $a'$. Therefore, as $y$ is arbitrary, the $S^1$-action is free. $\square$

Suppose $T^k$ acts injectively on a path-connected, locally path-connected, semilocally simply connected paracompact Hausdorff space $X$. Then, by Theorem 2.5.1, the $T^k$-action lifts to an action on the covering space $X_{\mathbb{Z}^k}$, where $\mathbb{Z}^k$ is the image of $\pi_1(T^k) \to \pi_1(M)$ and $X_{\mathbb{Z}^k}$ is the covering space of $X$ with $\pi_1(X_{\mathbb{Z}^k}) = \mathbb{Z}^k$. We show that the lifted action $(T^k, X_{\mathbb{Z}^k})$ is free and splits.

THEOREM 3.5.2 ([**CR71**, §3.1]). *If $T^k$ acts injectively on $X$, then $X_{\mathbb{Z}^k}$ (the covering space of $X$ with $\pi_1(X_{\mathbb{Z}^k}) = \mathbb{Z}^k$) splits into $T^k \times W$ so that $(T^k, X_{\mathbb{Z}^k}) = (T^k, T^k \times W)$, where the $T^k$-action on $T^k \times W$ is via translation on the first factor and trivial on the simply connected $W$ factor.*

PROOF. Lift the $T^k$-action to $X' = X_{\mathbb{Z}^k}$. For any $S^1 \subset T^k$, the generator of $\pi_1(S^1, 1)$ is mapped to a primitive element of $\pi_1(X', x) = \mathbb{Z}^k$ by $\mathrm{ev}_*^x : \pi_1(S^1, 1) \to \pi_1(X', x)$. By Lemma 3.5.1, the $S^1$-action on $X'$ is free. Consequently, the lifted torus action $(T^k, X')$ must be a free action. Since $T^k$ is compact, the action $(T^k, X')$ is proper and therefore $X' \to W = T^k \backslash X'$ is a principal $T^k$ bundle. (Here we used complete regularity implied by paracompactness.)

We shall show that this bundle is trivial. In [**CR71**], this is done using the Hirsh-Leray-Dold theorem. Here we prove this directly. Since $\mathrm{ev}_*^{x'} : \pi_1(T^k, 1) \to \pi_1(X', x')$ is an isomorphism, $\pi_1(W) = 1$. Let $\mathbb{R}^k$ act on $X'$ through the covering projection $\mathbb{R}^k \to T^k$ (so $(\mathbb{R}^k, X')$ has ineffective part $\mathbb{Z}^k$). Then the action of $\mathbb{R}^k$ lifts to the universal covering $\widetilde{X}$ of $X'$. Thus the diagram

$$\begin{array}{ccccc}
\mathbb{R}^k & \longrightarrow & \widetilde{X} & \longrightarrow & T^k \backslash X' = W \\
\downarrow {\scriptstyle /\mathbb{Z}^k} & & \downarrow {\scriptstyle /\mathbb{Z}^k} & & \downarrow = \\
T^k & \longrightarrow & X' & \longrightarrow & T^k \backslash X' = W
\end{array}$$

is commutative. In fact, as $X'$ is a principal $T^k$-bundle, the lift $(\mathbb{R}^k, \widetilde{X})$ is a principal $\mathbb{R}^k$-bundle over $W$. This is seen by observing that $\pi_1(T^k, x') \to \pi_1(X', x')$ is an isomorphism (and is independent of $x'$); see Proposition 2.8.4. Thus each $T^k$-orbit in $X'$ lifts to an $\mathbb{R}^k$-orbit in $\widetilde{X}$. Since $\mathbb{R}^k$ is contractible, the $\mathbb{R}^k$-action on $X'$ splits into a product $\mathbb{R}^k \times W$ (since $X'$ is paracompact) with the $\mathbb{R}^k$-action just a translation on the first factor. The bundle $X' \to W$ is recaptured by dividing out by the covering transformations $\mathbb{Z}^k \subset \mathbb{R}^k$ on each $\mathbb{R}^k$-orbit. Therefore, $(T^k, X') = (T^k, T^k \times W)$, where $T^k$ just acts as left translations on the first factor. $\square$

We remark that the group $T^k \times Q$ is acting properly on $T^k \times W$, and $W$ is simply connected. The $T^k$-action does not lift to the universal covering $\widetilde{X}$ of $X$ but the induced ineffective $\mathbb{R}^k$-action on $T^k \times W$ lifts to an effective $\mathbb{R}^k$-action on $\mathbb{R}^k \times W$ and commutes with the group $\pi_1(X, x)$ of covering transformations on $\mathbb{R}^k \times W$.

COROLLARY 3.5.3. *Let $(T^k, X)$ be an injective action and suppose the exact sequence*

$$0 \longrightarrow \pi_1(T^k, 1) \xrightarrow{\mathrm{ev}_*^x} \pi_1(X, x) \longrightarrow Q \longrightarrow 1,$$

*where $Q = \pi_1(X,x)/\mathrm{Im}(\mathrm{ev}_*^x)$, splits. Then the action $(T^k, X)$ splits as $(T^k, T^k \times T^k\backslash X)$.*

PROOF. Since $\mathrm{ev}_*^x(\pi_1(T^k, 1)) \cong \mathbb{Z}^k$ is a central subgroup, $\pi_1(X,x)$ is naturally isomorphic to $\mathbb{Z}^k \times Q$. Therefore, for every subgroup $S^1 \subset T^k$, the generator of $\pi_1(S^1, 1)$ is mapped to a primitive element of $\pi_1(X,x) = \mathbb{Z}^k \times Q$ by $\mathrm{ev}_*^x : \pi_1(S^1, 1) \to \pi_1(X, x)$. By Lemma 3.5.1, the $S^1$-action is free. Consequently, the torus action $(T^k, X)$ must be a free action.

Next, we want to show that this principal $T^k$-bundle $T^k \to X \to T^k\backslash X$ is a *product bundle*. This follows from the Hirsh-Leray-Dold theorem as in [**CR71**], but we prove this directly. The principal $T^k$-bundle $T^k \to X \to T^k\backslash X$ induces a principal $\mathbb{R}^k$-bundle $\mathbb{R}^k \to X_Q \to T^k\backslash X$ so that the diagram

$$\begin{array}{ccccc} \mathbb{R}^k & \longrightarrow & X_Q & \longrightarrow & T^k\backslash X \\ \downarrow {\scriptstyle /\mathbb{Z}^k} & & \downarrow {\scriptstyle /\mathbb{Z}^k} & & \downarrow = \\ T^k & \longrightarrow & X & \longrightarrow & T^k\backslash X \end{array}$$

is commutative. The $\mathbb{R}^k$-bundle is a product bundle since $\mathbb{R}^k$ is contractible. The covering transformations are just the quotient of the $\mathbb{R}^k$-action by the free $\mathbb{Z}^k$-action yielding $(T^k, X)$ as the splitting product action $(T^k, T^k \times (T^k\backslash X))$. □

S. Schwartzman [**Sch81b**] and G. Dula and D. Gottlieb [**DG90**] have found a generalization of this splitting theorem for a general compact Lie group; see also M. Sadowski [**Sad91a**] for another proof of the torus splitting.

EXERCISE 3.5.4. Let $(S^1, M)$ be a semifree $S^1$-action on a connected closed orientable 3-manifold $M$ as in Example 2.6.1. Show that the restriction of the $S^1$-action to $(M - M^{S^1})$ is injective. Show that the lifted action $(S^1, (M - M^{S^1}))$, as before, is equivalent to $(S^1, S^1 \times \mathbb{R}^2)$ with the $S^1$-action translation on the first factor $S^1$.

EXERCISE 3.5.5. If $(T^k, X)$ is an injective action, $k > 0$, and $X$ is a connected finite simplicial complex, show that the Euler characteristic of $X$ is 0. (Hint: Consider the Lefschetz fixed point formula.)

EXERCISE 3.5.6. Let $(G, M)$ be an effective action of a compact connected Lie group on a closed manifold $M$. Suppose that $H = \mathrm{Im}(\mathrm{ev}_*^x)$ contained in $\pi_1(M, x)$ has rank $s$. Show that there is a finite covering $p : G^* \to G$ and action $(G^*, M)$, given by $g^*(x) = p(g^*)(x)$, with a torus subgroup $T^s$, a direct summand of $G^*$, acting injectively on $M$.

3.5.7 (Homologically injective torus actions). There is a concept stronger than injective torus action. Let $(T^k, M)$ be a torus action on a topological space. It is *homologically injective* if the evaluation map induces an injective homomorphism $\mathrm{ev}_* : H_1(T^k, \mathbb{Z}) \to H_1(M; \mathbb{Z})$. For a Riemannian manifold of nonpositive sectional curvature, the existence of a nontrivial center $\mathbb{Z}^k$ of $\pi_1(M)$ guarantees that the manifold has an action of torus $T^k$; and all such actions are homologically injective. This will be the main topic of Section 11.6.

CHAPTER 4

# Definition of Seifert Fibering

Throughout this chapter, $G$ will be a Lie group, even though it is enough to have $G$ be a locally compact topological group until Lemma 4.2.5.

A Seifert fibering is a map $\tau : X \to B$, which is supposed to generalize the notion of a fiber bundle with a homogeneous space as fiber. The inverse images $\tau^{-1}(b)$ are called fibers. There are typical fibers and singular fibers. The typical fibers are all isomorphic to a fixed homogeneous space of $G$, and the singular fibers will be quotients of the typical fiber by a compact group. We need to impose sufficient conditions to get an effectively computable theory and one which will yield interesting geometric applications. Our approach will be global rather than local. We shall begin with the underlying idea which when later refined will yield a precise definition.

### 4.1. Examples

4.1.1. Suppose on a space $P$, there are two group actions $(G, P, \varphi)$ and $(\Pi, P, \psi)$ with $\Pi$ normalizing the action of $G$. That is, for each $f \in \Pi$, there exists a continuous automorphism $\alpha_f$ of $G$ such that $f(au) = \alpha_f(a)f(u)$, for all $a \in G$, $u \in P$. Hence, each $f \in \Pi$ is a weak $G$-self-equivalence; see Subsection 1.1.4. Note that $\Pi \to \text{Aut}(G)$, given by $f \mapsto \alpha_f$, is a homomorphism. The elements of $\Pi$ thus map the $G$-orbits to $G$-orbits. Consequently, there is an action of $\Pi$ induced on the orbit space $G\backslash P = W$. Let $B = \Pi\backslash W$ denote the orbit space of this $\Pi$-action on $W$, and let $\tilde{\tau} : P \longrightarrow W = G\backslash P$ and $\bar{\nu} : W \longrightarrow B = \Pi\backslash W$ be the orbit mappings.

On the other hand, we could first let $\Pi$ act on $P$. Denote this orbit space by $X = \Pi\backslash P$ and the orbit mapping by $\nu$. If $u \in \nu^{-1}(x)$, then $f(u)$ and $f(au) = \alpha_f(a)f(u)$ lie on the same $G$-orbit of $P$. Therefore, there is induced a map $\tau : X \to B$ which makes the diagram

$$\begin{array}{ccc} P & \xrightarrow{\tilde{\tau}}_{G\backslash} & W \\ \nu \downarrow \Pi\backslash & & \bar{\nu} \downarrow \Pi\backslash \\ X & \xrightarrow{\tau} & B \end{array}$$

commutative. If the action of $\Pi$ were to centralize the $G$-action, that is, if $\alpha_f = \text{id}_G$ for each $f \in \Pi$, then the actions commute and there would be *a $G$-action induced on $X$ whose orbit space is $B$*.

This setup is the *prototype* of a Seifert fibering $\tau : X \to B$ modeled on the $G$-action on $P$. Without further restrictions, the map $\tau$ may have undesirable properties. For example, in Example 1.9.5, $X$ is a torus and $\tau^{-1}(b)$ is a line injectively immersed in the torus. $\mathbb{R}$ acts freely on this immersed line, but the action is not proper. While $\tau$ is a foliation of the torus by lines, the base space $B$ is not Hausdorff and the induced $\mathbb{Z} \times \mathbb{Z}$-action on $W = \mathbb{R}$ is also not proper.

*What is $\tau^{-1}(b)$ in our prototype?* This is $\nu(\tilde{\tau}^{-1}(\bar{\nu}^{-1}(b)))$. Suppose $\tau(x) = b$ and $\nu(u) = x$, then $\nu(G(u)) = \tau^{-1}(b)$. If $\tilde{\tau}(u) = w$, then $\Pi_w$ sends $G(u)$ onto itself since $\Pi_w$ fixes $w$. Therefore $\tau^{-1}(b)$ is the $\nu$-image of the action of $\Pi_w$ on the orbit $G(u)$. If $G$ is a Lie group acting properly on $P$, then the orbit $G(u)$ is the homogeneous space $G_u \backslash G$ and $\Pi_w \backslash (G_u \backslash G)$ is mapped continuously, one-to-one and onto $\tau^{-1}(b)$.

$$\begin{array}{ccc} u & \xrightarrow{\tilde{\tau}} & w \\ \nu \downarrow & & \bar{\nu} \downarrow \\ x & \xrightarrow{\tau} & b \end{array} \qquad \begin{array}{ccc} Gu & \xrightarrow{\tilde{\tau}} & w \\ \nu \downarrow & & \bar{\nu} \downarrow \\ \Pi_w \backslash Gu & \xrightarrow{\tau} & b \end{array}$$

The example above shows that $\Pi_w \backslash Gu \to \tau^{-1}(b)$ still may fail to be a homeomorphism. We shall impose additional restrictions to remedy this situation and to discern and control the nature of the $\Pi_w$-action on $G_u \backslash G$. For each fixed $G$-action on $P$, there are obviously many $\Pi$-actions that normalize it. We shall study the normalizer in TOP($P$), the group of homeomorphisms of $P$, of the $G$-actions on $P$. After this study, we will then be able to determine more precisely how $\Pi_w$ acts on $G_u \backslash G$. But first we shall treat a few elementary examples.

4.1.2. The map from Example 1.6.3

$$\tau : G \times_H S \to H \backslash S = B$$

is a prototypical fibering. The fibers are the $G$-orbits of the $G$-action on $G \times_H S$.

EXERCISE 4.1.3. Let $P = S^1 \times I$. Describe an $S^1 \times \mathbb{Z}_2$-action on $P$ so that $\mathbb{Z}_2 \backslash P$ is the Möbius band and the orbit mapping of the induced $S^1$-action $\mathbb{Z}_2 \backslash P \to \mathbb{Z}_2 \backslash I$ is a prototype Seifert fibering such that each principal orbit (see Remark 1.7.9) is a free orbit and there is exactly one singular orbit.

EXAMPLE 4.1.4. Let $P = S^1 \times T^2$ and define an action of $S^1 \times \mathbb{Z}_2$ on $P$ by

$$z(z_1, (z_2, z_3)) = (zz_1, (z_2, z_3)),$$
$$\alpha(z_1, (z_2, z_3)) = (-z_1, (\bar{z}_2, \bar{z}_3))$$

for $z, z_1 \in S^1$, $(z_2, z_3) \in T^2$, and $\alpha$ is the generator of $\mathbb{Z}_2$. We obtain the following diagram of orbit mappings.

$$\begin{array}{ccc} (S^1 \times \mathbb{Z}_2, S^1 \times T^2) & \xrightarrow{S^1 \backslash}_{\tilde{\tau}} & (\mathbb{Z}_2, T^2) \\ \mathbb{Z}_2 \backslash \downarrow \nu & & \mathbb{Z}_2 \backslash \downarrow \bar{\nu} \\ (S^1, S^1 \times_{\mathbb{Z}_2} T^2) & \xrightarrow{S^1 \backslash}_{\tau} & \mathbb{Z}_2 \backslash T^2 \end{array}$$

The action of $\mathbb{Z}_2$ on $P = T^3$ is by isometries in the usual flat metric, and $X = S^1 \times_{\mathbb{Z}_2} T^2$ is a flat 3-dimensional Riemannian manifold. (In [**Wol77**, Chapter 3], $S^1 \times_{\mathbb{Z}_2} T^2$ is called $\mathfrak{G}_2$ in the Hantzsche-Wendt classification of flat 3-dimensional manifolds.) The action of $\mathbb{Z}_2$ on $T^2$ has four fixed points $(\pm 1, \pm 1)$ and is free otherwise. The quotient $\mathbb{Z}_2 \backslash T^2$ is homeomorphic to the 2-sphere. The induced $S^1$-action on $X$ has four singular orbits (fibers) with isotropy $\mathbb{Z}_2$ over the image in $S^2$ of the four fixed points in $S^2$. The *typical* fibers are the free $S^1$ orbits.

Let us take the same $\mathbb{Z}_2$-action on $S^1 \times T^2$ but use a translational $T^2$-action on the $T^2$-factor. The $\mathbb{Z}_2$-action normalizes (in the homeomorphisms of $T^3$) the

$T^2$-action but does not centralize it. By projecting to the $S^1$-factor, we induce the nonprincipal fiber bundle
$$T^2 \to S^1 \times_{\mathbb{Z}_2} T^2 \to S^1/\mathbb{Z}_2.$$
The typical fiber is now $T^2$. There are no singular fibers.

One can also think of $T^3 = P$ acting by left translations and $\Pi = \mathbb{Z}_2$ acting on $T^3$ as before. This induces the map $\mathbb{Z}_2 \backslash T^3 \to$ point. Here the typical fiber is $T^3$ and the fiber is $\mathfrak{G}_2$ itself.

4.1.5. The base space $\mathbb{Z}_2 \backslash T^2$ of the first Seifert fibering in the previous example is a Seifert fibering itself again, over an arc with $S^1$ as a *typical* fiber over the interior points and arcs over the end points. We may think of this as the projection of a "square pillow" (a topological 2-sphere) onto an arc as shown in Figure 1.

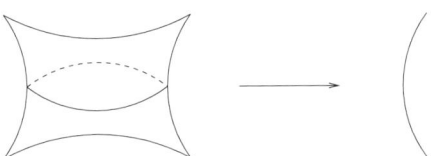

FIGURE 1. Square pillow

Our setup is
$$P = T^2 = S^1 \times S^1, \qquad G = S^1, \qquad \Pi = \mathbb{Z}_2,$$
where $G$ acts on the first factor of $P$ as multiplication.

Take the torus parametrized by $(z_1, z_2)$ and define an action of $S^1 \rtimes \mathbb{Z}_2 = O(2, \mathbb{R})$ on the torus by
$$(z, \epsilon)(z_1, z_2) = (z\bar{z}_1, \bar{z}_2),$$
where $\epsilon$ is the generator of $\mathbb{Z}_2$. Then $S^1$ is normalized but not centralized by $\mathbb{Z}_2$. Dividing out by the normal $S^1$ gives the circle as the orbit space $(z_1, z_2) \mapsto z_2$. This, followed by the induced $\mathbb{Z}_2$-action $(z_2 \mapsto \bar{z}_2)$, gives us an arc for orbit space. On the other hand, if we divide out first by the diagonal $\mathbb{Z}_2$ isometric action, we get $S^1 \times_{\mathbb{Z}_2} S^1$.

This is a topological 2-sphere (the surface of a square pillow). There is then a map induced from $S^1 \times_{\mathbb{Z}_2} S^1 \xrightarrow{\tau} S^1/\mathbb{Z}_2$ which is our fibering and $\tau$ is not an orbit mapping.

$$\begin{array}{ccc} (S^1 \rtimes \mathbb{Z}_2, S^1 \times S^1) & \xrightarrow{S^1 \backslash} & (\mathbb{Z}_2, S^1) \\ \mathbb{Z}_2 \backslash \downarrow \nu & & \mathbb{Z}_2 \backslash \downarrow \bar{\nu} \\ S^1 \times_{\mathbb{Z}_2} S^1 & \xrightarrow{\tau} & \mathbb{Z}_2 \backslash S^1 \end{array}$$

The two singular fibers are quotients of $S^1$ by $\mathbb{Z}_2$ where the $\mathbb{Z}_2$ does not act as translations but as automorphisms of $S^1$.

4.1.6. Note the composite $S^1 \times_{\mathbb{Z}_2} T^2 \to \mathbb{Z}_2 \backslash T^2 \to \mathbb{Z}_2 \backslash S^1$ can be viewed as a Seifert fibering (2-dimensional fiber, 1-dimensional base space)
$$T^2 \longrightarrow T^2 \times_{\mathbb{Z}_2} S^1 \longrightarrow \mathbb{Z}_2 \backslash S^1 (=\text{arc})$$

modeled on $T^2 \times S^1$ with typical fiber $T^2$ and singular fibers homeomorphic to a Klein bottle over the end points of the arc. (Just think of what lies above each horizontal section of the square pillow in Figure 1.) Now, by examining what lies above the top half of the square pillow, we see that $\mathfrak{G}_2$ is also homeomorphic to the double mapping cylinder of the double covering of the Klein bottle by the torus. Using this identification, one sees that $\mathfrak{G}_2$ fibers without singularities over the Klein bottle with fiber $S^1$. This $S^1$-bundle is nontrivial over each orientation reversing curve of the Klein bottle. Thus $\mathfrak{G}_2$ is the total space of two different Seifert fiberings with typical fiber $S^1$ and also the total space of two distinct Seifert fiberings with typical fiber $T^2$. Note all the bases are distinct. In Example 4.5.10, we analyze this manifold in great detail. We can view it as a Seifert manifold modeled on principal bundles $\mathbb{R}^i \to \mathbb{R}^3 \to \mathbb{R}^j$ in different ways ($i + j = 3$), yielding different Seifert structures for the same space.

EXERCISE 4.1.7. Define an action of $\mathbb{Z}_2 \times \mathbb{Z}_2$ on $T^3 = S^1 \times T^2$ so that we get a Seifert fibering $S^1 \times_{\mathbb{Z}_2 \times \mathbb{Z}_2} T^2 \to K$ over the Klein bottle with typical fiber $S^1$ and two singular fibers doubly covered by a typical fiber. (Hint: Use the first $\mathbb{Z}_2$-action in defining $\mathfrak{G}_2$ and a new involution $(z, z_1, z_2) \mapsto (\bar{z}, -z_1, \bar{z}_2)$. Note that $\mathfrak{G}_2$ freely doubly covers this orbit space.)

4.1.8. Let us consider the important example where $G$ is a closed subgroup of a Lie group $P$ and $\Pi$ is another closed subgroup of $P$. We let $G = \ell(G)$ act on the left as translations and $\Pi$ act on $P$ by

$$\psi(\alpha)(x) = x\alpha^{-1}, \ \alpha \in \Pi, \ x \in P.$$

This yields a left $\Pi$-action which commutes with $\ell(G)$ on $P$. We have

$$\psi(\Pi) \cap \ell(G) = (G \cap \Pi) \cap \mathcal{Z}(P) = \Gamma,$$

a closed central subgroup of $G$. Our fibering diagram becomes

$$\begin{array}{ccc} P & \xrightarrow{G\backslash}_{\tilde{\tau}} & G\backslash P \\ {\scriptstyle \Pi\backslash}\downarrow\nu & & {\scriptstyle (\Pi/\Gamma)\backslash}\downarrow\bar{\nu} \\ P/\Pi & \xrightarrow[\tau]{(G/\Gamma)\backslash} & G\backslash P/\Pi. \end{array}$$

A very interesting case occurs when $P$ has a finite number of connected components and $G$ is a *maximal compact subgroup* of $P$. (For example, take $P = \mathrm{GL}(n, \mathbb{R})$ with $G = \mathrm{O}(n) \subset \mathrm{GL}(n, \mathbb{R})$.) Then $G\backslash P$ is diffeomorphic to $\mathbb{R}^n$ for some $n \geq 0$. If $\Pi$ is a discrete and torsion-free subgroup, then $G \cap \Pi = 1$ so $\Gamma = 1$, and $Q = \Pi/\Gamma = \Pi$ acts freely on $\mathbb{R}^n = G\backslash P$. The $G \times \Pi$-action is proper and $G\backslash P/\Pi$ is a $K(\Pi, 1)$-manifold, and $P/\Pi$ is a principal $G$-bundle over $G\backslash P/\Pi$. However, usually a discrete $\Pi$ in $P$ will not be torsion free. In this case, $\tau$ is Seifert fibering where $\tau^{-1}(b) = G/\Pi_w = \Gamma\backslash G/Q_w$, $w \in \bar{\nu}^{-1}(b)$, $1 \to \Gamma \to \Pi_w \to Q_w \to 1$ is exact and the finite $Q_w$ is the isotropy of $Q$ at $w$. The typical fiber is $\Gamma\backslash G$.

4.1.9. We must exercise caution in the above example. For $G$ and $\Pi$ could both act properly and commute, but the subgroup $G \times \Pi \subset P \times P$, acting on $P$, may not act properly. For example, in $P = \mathbb{R}$, let $G \cong \mathbb{Z}$ be generated by 1 and $\Pi$ be generated by $\sqrt{2}$. Then $G$ and $\Pi$ act freely, properly and commute with each other, but the free action of $G \times \Pi$ is not proper.

## 4.2. $\mathrm{TOP}_G(P)$, the group of weak $G$-equivalences

As mentioned in the previous section, the prototype of a Seifert fibering is too general and we cannot avoid pathological situations. Furthermore, if we want to get as much control as possible on the singular fibers, we need to impose additional conditions. We take the actions of $G$ and $\Pi$ to be effective on $P$; i.e., $G$ and $\Pi$ are subgroups of $\mathrm{TOP}(P)$, the group of homeomorphisms of $P$. Furthermore, as $\Pi$ normalizes $G$, $\Pi$ is in the normalizer of $G$ in $\mathrm{TOP}(P)$. This normalizer turns out to be the group of all weak $G$-equivalences of $P$. There is a natural homomorphism from this normalizer into $\mathrm{Aut}(G) \times \mathrm{TOP}(W)$. To study this homomorphism, we shall examine its kernel. The kernel becomes effectively computable if we assume that the action of $G$ on $P$ is free. Then to remove any pathology on these actions, we assume that $G$ acts locally properly. If we also assume that this $G$-action has local cross sections, then $P$ becomes a principal $G$-bundle and the $G$-action is just given by left translations of $G$ along the fibers. (If $G$ is a Lie group acting locally properly, then the $G$-action has local cross section because of the existence of a slice.) Under the assumptions, the normalizer, which is the group of weak $G$-equivalences of $P$, is the same as the group of weak bundle automorphisms of the principal $G$-bundle $P$. Thus for a group $\Pi$ in $\mathrm{TOP}(P)$ to normalize the free $G$-action on the bundle $P$, it must be a subgroup of the weak bundle automorphisms of $P$.

Recall that (Subsection 1.1.4) a homeomorphism $f : P \to P$ is a weak $G$-equivalence if there exists a continuous automorphism $\alpha_f$ of $G$ such that

$$f(xu) = \alpha_f(x)f(u),$$

for all $x \in G$, $u \in P$. Note that $\alpha_f$ is *unique* because the $G$-action on $P$ is effective.

NOTATION 4.2.1. $\mathrm{TOP}_G(P)$ denotes the group of all weak $G$-equivalences of $P$.

4.2.2 (Topology of $\mathrm{TOP}_G(P)$). $\mathrm{TOP}_G(P)$ has the compact-open topology. The groups $\Pi$ and $G$ act locally properly on $P$, and consequently are embedded as closed subsets of $\mathrm{TOP}_G(P)$ in the point-open topology and hence also as closed subsets in the compact-open topology. Our Lie groups $\Pi$ and $\ell(G)$ (the principal left $G$-action) are acting properly on $P$ and so they are embedded as closed subsets in the point-open topology and hence also as closed subsets in the compact-open topology. $\mathrm{TOP}_G(P)$ is not in general a topological group. However, if $P$ is locally compact and either connected or locally connected, then $\mathrm{TOP}_G(P)$ is a topological group under the compact-open topology; see Remark 1.2.6. We will not use this fact unless we specifically mention it.

LEMMA 4.2.3 ([**LR89**]). *$\mathrm{TOP}_G(P)$ is the normalizer of $G$ in $\mathrm{TOP}(P)$, and there exists a natural homomorphism $\mathrm{TOP}_G(P) \to \mathrm{Aut}(G) \times \mathrm{TOP}(G\backslash P)$.*

PROOF. For each $f \in \mathrm{TOP}_G(P)$, we have $f \circ a = \alpha_f(a) \circ f$ for each $a \in G$. Or, in other words, $f \circ a \circ f^{-1} = \alpha_f(a)$. So $f$ normalizes $G$.

Conversely, suppose $f \in \mathrm{TOP}(P)$ normalizes $G$; that is, $f \circ G \circ f^{-1} = G$. Then $f \circ a \circ f^{-1} = a_f$, for some $a_f \in G$. Then

$$(ab)_f = f \circ (ab) \circ f^{-1} = (f \circ (a) \circ f^{-1})(f \circ (b) \circ f^{-1}) = a_f \circ b_f$$

shows that $f$ induces, by conjugation, an automorphism of $G$, by sending $a \mapsto a_f$. If we denote this automorphism by $\alpha_f$, then we have $f(xu) = \alpha_f(x)f(u)$ so that $f \in \mathrm{TOP}_G(P)$.

Since each $G$-orbit is mapped homeomorphically onto another $G$-orbit for $f \in \mathrm{TOP}_G(P)$, there is induced a homeomorphism $\bar{f} : G\backslash P \to G\backslash P$. Furthermore, associated with $f$, we have the unique automorphism $\alpha_f \in \mathrm{Aut}(G)$. It is easy to check that $f \mapsto (\alpha_f, \bar{f})$ is a homomorphism with the group operation in $\mathrm{TOP}_G(P)$ being composition. Also, if we use the compact-open topology (when $P$ is locally compact), this is a continuous homomorphism of topological groups but we will not need that now. $\square$

The homomorphism in the lemma, as we shall see later, may not be onto even if $P$ is a principal $G$-bundle; see Subsections 4.2.13 and 4.2.14. On the other hand, if $P$ is a product $G \times W$, the homomorphism will be onto and split.

To get a handle on the kernel of this homomorphism, we now assume the action of $G$ on $P$ is free. On $G$ itself, $G$ also acts by conjugation, giving a homomorphism $G \to \mathrm{Inn}(G)$.

NOTATION 4.2.4. Let $\mathrm{M}_G(P,G)$ be the (continuous) $G$-equivariant maps from $P$ to $G$, where the action of $G$ on $G$ is given by conjugation. Therefore, for a map $\eta : P \to G$, $\eta \in \mathrm{M}_G(P,G)$ if and only if $\eta(au) = a\eta(u)a^{-1}$ for all $a \in G$ and $u \in P$.

These maps form a group by $(\eta_1 \cdot \eta_2)(u) = \eta_1(u)\eta_2(u)$, for $\eta_1, \eta_2 \in \mathrm{M}_G(P,G)$ and $u \in P$. Note $\eta^{-1}(u) = (\eta(u))^{-1}$.

LEMMA 4.2.5. *The kernel of the homomorphism* $\mathrm{TOP}_G(P) \to \mathrm{Aut}(G) \times \mathrm{TOP}(G\backslash P)$ *is isomorphic to* $\mathrm{M}_G(P,G)$.

PROOF. We can view $\mathrm{M}_G(P,G)$ as a subgroup of $\mathrm{TOP}_G(P)$; namely, we define a homomorphism
$$\psi : \mathrm{M}_G(P,G) \hookrightarrow \mathrm{TOP}_G(P)$$
by $\psi(\eta)(u) = \eta(u)^{-1}u$, for $\eta \in \mathrm{M}_G(P,G)$, $u \in P$. Now,
$$\psi(\eta)(au) = \eta(au)^{-1}au = a\eta(u)^{-1}a^{-1}(au) = a\eta(u)^{-1}u = a\psi(\eta)(u).$$
So $\psi(\eta) \in \mathrm{TOP}_G(P)$, and is in the kernel. Further, $\psi$ is a homomorphism because
$$\psi(\eta_1\eta_2)(u) = (\eta_1\eta_2)(u)^{-1}u = (\eta_1(u)\eta_2(u))^{-1}u$$
$$= (\eta_2(u))^{-1}(\eta_1(u))^{-1}u = (\eta_2(u))^{-1}(\psi(\eta_1)(u))$$
$$= \psi(\eta_1)((\eta_2(u))^{-1}u) = \psi(\eta_1)\psi(\eta_2)(u).$$

On the other hand, if $f \in \mathrm{Kernel}$, then $\bar{f} = 1_{G\backslash P}$ and $\alpha_f = 1_G$. Therefore, because of freeness, $f(u) = (\eta_f(u))^{-1}u$, for some well-defined function $\eta_f : P \to G$. Define $\theta(f) = \eta_f$. We shall show $\theta : \mathrm{Kernel} \to \mathrm{M}_G(P,G)$ is the inverse of $\psi$. Since $af(u) = f(au)$, then $\eta_f(au)^{-1}(au) = a(\eta_f(u))^{-1}u$. Because the $G$-action is free, we have $a\eta_f(u)a^{-1} = \eta_f(au)$. Thus, $\eta_f \in \mathrm{M}_G(P,G)$. Furthermore, $\theta(g \circ f) = \theta(g)\theta(f)$ because $g(f(u)) = (g \circ f)(u) = (\eta_{g \circ f}(u))^{-1}(u) = g(\eta_f(u)^{-1})(u) = (\eta_f(u))^{-1}g(u) = (\eta_f(u))^{-1}(\eta_g(u))^{-1}u = ((\eta_g\eta_f)(u))^{-1}u$.

If $\eta \in \mathrm{M}_G(P,G)$, then define $f_\eta$ by $f_\eta(u) = \psi(\eta)(u) = \eta(u)^{-1}u$. The formula $(\eta_{f_\eta})^{-1}(u) = f_\eta(u) = (\eta(u))^{-1}(u)$ shows that $\theta$ and $\psi$ are inverses to each other, and so the inclusion $\psi : \mathrm{M}_G(P,G) \to \mathrm{TOP}_G(P)$ is the kernel of $\mathrm{TOP}_G(P) \to \mathrm{Aut}(G) \times \mathrm{TOP}(G\backslash P)$. $\square$

REMARK 4.2.6. Since the locally proper $G$-action on $P$ is free, the orbit mapping $P \to G\backslash P$ is a principal $G$-bundle provided we also assume that the $G$-action on $P$ has local cross sections. For the remainder of the section, we assume that $G$

and $\Pi$ are Lie groups acting locally properly with the action of $G$ also being free. (The result will also be valid if we do not assume the locally compact $G$ to be Lie but require that the free locally proper $G$ action has local cross sections on $P$. In other words, that $P$ is a principal $G$-bundle.) In the parlance of principal $G$-bundles, the group $\mathrm{TOP}_G(P)$ is the group of *weakly $G$-equivariant bundle automorphisms*.

NOTATION 4.2.7. Let $G$ be a group. We always denote *conjugation* by $\mu$, so that
$$\mu(a)(x) = axa^{-1}$$
for $a, x \in G$; $\mathrm{Inn}(G)$ denotes the group of inner automorphisms $\mu(a)$ of $G$; $\mathcal{Z}(G)$ denotes the center of $G$. Therefore, $\mathrm{Inn}(G) \cong G/\mathcal{Z}(G)$. For a subset $K$ of $G$, we denote the set of centralizers of $K$ in $G$ by $C_G(K)$.

Since $G$ acts on the left, we shall denote this by $\ell(G)$. Then every $\ell_a = \ell(a) \in \ell(G)$ induces the identity on $W = G\backslash P$. However,
$$\ell_a(xu) = axu = axa^{-1}au = \mu(a)(x) \cdot \ell_a(u)$$
shows that $\ell_a$ induces $\mu(a)$ under $\mathrm{TOP}_G(P) \to \mathrm{Aut}(G)$. If we put $\mathrm{Out}(G) = \mathrm{Aut}(G)/\mathrm{Inn}(G)$, the group of (continuous) outer automorphisms of $G$, the kernel of
$$\mathrm{TOP}_G(P) \to \mathrm{Out}(G) \times \mathrm{TOP}(W)$$
is generated by $\ell(G)$ and $\mathrm{M}_G(P, G)$. If $\ell_a = \psi(\eta) = f_\eta \in \ell(G) \cap \mathrm{M}_G(P, G)$, then
$$\ell_a(bu) = f_\eta(bu) = (\eta(bu))^{-1}bu = b\eta(u)^{-1}(b^{-1}b)u = b\ell_a u = \ell_b \ell_a(u).$$
Therefore, $\ell(G) \cap \psi(\mathrm{M}_G(P, G)) = \mathcal{Z}(G)$. Thus, $\ell(G) \times_{\mathcal{Z}(G)} \psi(\mathrm{M}_G(P, G))$, the quotient of $\ell(G) \times \psi(\mathrm{M}_G(P, G))$ by the normal subgroup $\{(z, z) | z \in \ell(G) \cap \mathrm{M}_G(P, G)\} = \{(z, z^{-1}) | z \in \ell(G) \cap \psi(\mathrm{M}_G(P, G))\}$ is the kernel of $\mathrm{TOP}_G(P) \to \mathrm{Out}(G) \times \mathrm{TOP}(W)$. We suppress $\psi : \mathrm{M}_G(P, G) \hookrightarrow \mathrm{TOP}_G(P)$ and simply consider $\mathrm{M}_G(P, G)$ as a subgroup of $\mathrm{TOP}_G(P)$.

PROPOSITION 4.2.8. *For a principal $G$-bundle $P$ with $W = G\backslash P$, we have the exact sequence*
$$1 \to \ell(G) \times_{\mathcal{Z}(G)} \mathrm{M}_G(P, G) \to \mathrm{TOP}_G(P) \to \mathrm{Out}(G) \times \mathrm{TOP}(W). \quad \square$$

COROLLARY 4.2.9. *For the principal fibration $\tilde\tau : P \to W = G\backslash P$, the subgroup of $\mathrm{TOP}_G(P)$ which leaves the fiber $\tilde\tau^{-1}(w)$ invariant lies in $(\ell(G) \times_{\mathcal{Z}(G)} \mathrm{M}_G(\tilde\tau^{-1}(w), G)) \cdot \mathrm{Aut}(G) = \mathrm{M}_G(\tilde\tau^{-1}(w), G) \rtimes \mathrm{Aut}(G) = \ell(G) \rtimes \mathrm{Aut}(G)$.*

The corollary holds because $r(G) \subset \mathrm{M}_G(\tilde\tau^{-1}(w), G)$ implies $\ell(G) \cdot r(G) \subset \mathrm{Aut}(G)$. There are two important situations where we can show that $\mathrm{M}_G(P, G)$ only depends upon $W$: they are $P = G \times W$ (i.e., trivial bundle) and when $G$ is Abelian.

COROLLARY 4.2.10 ($P = G \times W$). *The exact sequence in Proposition 4.2.8 becomes*
$$1 \to \mathrm{M}(W, G) \rtimes \mathrm{Inn}(G) \to \mathrm{TOP}_G(G \times W) \to \mathrm{Out}(G) \times \mathrm{TOP}(W) \to 1.$$
*In fact*, $\mathrm{M}(W, G) \times_{\mathcal{Z}(G)} \ell(G) = \mathrm{M}(W, G) \rtimes \mathrm{Inn}(G)$, *and*
$$\mathrm{TOP}_G(G \times W) = \mathrm{M}(W, G) \rtimes (\mathrm{Aut}(G) \times \mathrm{TOP}(W)),$$

whose group law is given by
$$\begin{aligned}(\lambda_1,\alpha_1,h_1)\cdot(\lambda_2,\alpha_2,h_2) &= (\lambda_1\cdot{}^{(\alpha_1,h_1)}\lambda_2,\alpha_1\circ\alpha_2,h_1\circ h_2)\\ &= (\lambda_1\cdot(\alpha_1\circ\lambda_2\circ h_1^{-1}),\alpha_1\circ\alpha_2,h_1\circ h_2).\end{aligned}$$

*The action of* $\mathrm{TOP}_G(G\times W)$ *on* $G\times W$ *is given by*
$$\begin{aligned}(\lambda,\alpha,h)\cdot(x,w) &= ((\lambda,1,1)\circ(1,\alpha,h))(x,w)\\ &= (\lambda,1,1)(\alpha(x),h(w))\\ &= (\alpha(x)\cdot(\lambda(h(w)))^{-1},h(w)).\end{aligned}$$

PROOF. Choose a global cross section $W\to P$ once and for all. For $\alpha\times h\in \mathrm{Aut}(G)\times\mathrm{TOP}(W)$, define
$$(\alpha,h)(x,w)=(\alpha(x),h(w)),\text{ for }(x,w)\in G\times W.$$
Then $(\alpha,h)\in\mathrm{TOP}_G(G\times W)$ because $(\alpha,h)(ax,w)=(\alpha(ax),h(w))=\alpha(a)(\alpha(x),h(w))$, for all $a\in G$. So, $\mathrm{TOP}_G(G\times W)\to\mathrm{Aut}(G)\times\mathrm{TOP}(W)$ *is onto and splits.*

We now examine the kernel $\mathrm{M}_G(G\times W,G)$. For $u=(x,w)$, $\eta\in\mathrm{M}_G(G\times W,G)$, we have $\eta(x,w)=x\eta(1,w)x^{-1}$. Thus $\eta(x,w)$ is determined by $\eta(1,w)$ and $x$. Put $\lambda(w)=\eta(1,w)$. The assignment $\eta\mapsto\lambda$ is a homomorphism from $\mathrm{M}_G(G\times W,G)$ to $\mathrm{M}(W,G)$. Conversely, given $\lambda:W\to G$, define $\eta\in\mathrm{M}_G(G\times W,G)$ by $\eta(x,w)=x\lambda(w)x^{-1}$. This gives an isomorphism between $\mathrm{M}_G(G\times W,G)$ and $\mathrm{M}(W,G)$. Thus the isomorphisms are given by

$$\mathrm{M}(W,G)\longrightarrow\mathrm{M}_G(G\times W,G),\qquad \mathrm{M}_G(G\times W,G)\longrightarrow\mathrm{M}(W,G),$$
$$\lambda\longrightarrow\mu\times\lambda,\qquad\qquad\qquad \eta\longrightarrow\eta(1,-).$$

We have that $\psi:\mathrm{M}(W,G)\hookrightarrow\mathrm{TOP}_G(G\times W)$ injects and $\lambda$ acts, via $\psi(\lambda)$, on $G\times W$ by $\psi(\lambda)(x,w)=(x(\lambda(w))^{-1},w)$. We shall suppress the $\psi$.

We claim the action of $\mathrm{Aut}(G)\times\mathrm{TOP}(W)$ on $\mathrm{M}(W,G)$ is given by
$$(\alpha,h)\times\lambda\mapsto\alpha\circ\lambda\circ h^{-1}.$$
For,
$$\begin{aligned}((\alpha,h)\circ\lambda\circ(\alpha,h)^{-1})(x,w) &= ((\alpha,h)\circ\lambda)(\alpha^{-1}(x),h^{-1}(w))\\ &= (\alpha,h)(\alpha^{-1}(x)\lambda^{-1}(h^{-1}(w)),h^{-1}(w))\\ &= (x\alpha(\lambda^{-1}(h^{-1}(w))),w)\\ &= (x\alpha(\lambda(h^{-1}(w)))^{-1},w),\\ &= (\alpha\circ\lambda\circ h^{-1})(x,w),\end{aligned}$$

which yields the desired formula.

For $a\in G$, the constant map $W\to G$ sending $W$ to $a$ is denoted by $r(a)=r_a$. That is,
$$r_a=(a,1,1)\in\mathrm{M}(W,G)\rtimes(\mathrm{Aut}(G)\times\mathrm{TOP}(W)),$$
and $r_a(x,w)=(x\cdot a^{-1},w)$. We may therefore write $\ell_a$ as
$$\ell_a=r_{a^{-1}}\circ\mu(a)=(a^{-1},\mu(a),1)\in\mathrm{M}(W,G)\rtimes(\mathrm{Aut}(G)\times\mathrm{TOP}(W)).$$

Finally, the kernel $\mathrm{M}(W,G)\times_{\mathcal{Z}(G)}\ell(G)$ of $\mathrm{TOP}_G(P)\to\mathrm{Out}(G)\times\mathrm{TOP}(W)$ can be written as $\mathrm{M}(W,G)\rtimes\mathrm{Inn}(G)$ with the specific isomorphism given by
$$\lambda\cdot\ell_a\mapsto(\lambda,1)(r_{a^{-1}},\mu(a))=(\lambda r_{a^{-1}},\mu(a)).\qquad\square$$

EXERCISE 4.2.11. Show that $\ell(G)$ and $r(G)$ are subgroups of $\mathrm{M}(W,G) \rtimes \mathrm{Aut}(G)$; and $\ell(G) \rtimes \mathrm{Aut}(G)$ and $r(G) \rtimes \mathrm{Aut}(G)$ are identical subgroups of $\mathrm{M}(W,G) \rtimes \mathrm{Aut}(G)$ with the correspondence given by

$$\ell_a = (a^{-1}, \mu(a), 1), \ r_a = (a, 1, 1) \in \mathrm{M}(W,G) \rtimes (\mathrm{Aut}(G) \times \mathrm{TOP}(W)).$$

COROLLARY 4.2.12 ($G$ is Abelian). *The exact sequence in Proposition* 4.2.8 *becomes*

$$1 \to \mathrm{M}(W,G) \to \mathrm{TOP}_G(P) \to \mathrm{Aut}(G) \times \mathrm{TOP}(W).$$

Let $\eta \in \mathrm{M}_G(P,G)$. Then $\eta(au) = a\eta(u)a^{-1} = \eta(u)$. Therefore, $\eta$ factors through $\mathrm{M}(W,G)$ and we get the exact sequence. Since $G$ is Abelian, $\mathrm{M}(W,G)$ is Abelian so that the conjugation homomorphism $\mathrm{TOP}_G(P) \longrightarrow \mathrm{M}(W,G)$ factors through $\mathrm{Aut}(G) \times \mathrm{TOP}(W)$. In other words, if $f \in \mathrm{TOP}_G(P)$ maps to $(\alpha, h) \in \mathrm{Aut}(G) \times \mathrm{TOP}(W)$, then, for $\eta \in \mathrm{M}(W,G)$,

$$f \circ \psi(\eta) \circ f^{-1} = \alpha \circ \psi(\eta) \circ h^{-1},$$

just as in the product case.

4.2.13. *The map* $\mathrm{TOP}_G(P) \to \mathrm{Aut}(G) \times \mathrm{TOP}(W)$ *is not necessarily onto* even for $G$ Abelian in general. Take the Hopf bundle, that is, the principal $S^1$-bundle over the 2-sphere whose total space is $S^3$. Let $h: S^2 \to S^2$ be the antipodal map, and take $(1, h) \in \mathrm{Aut}(S^1) \times \mathrm{TOP}(S^2)$. If there exists $f \in \mathrm{TOP}_{S^1}(S^3)$ such that $f \mapsto (1, h)$, then $f$ would have to be orientation reversing. Therefore, the Lefschetz number of $f$ is $1 + (-1)^3(-1) = 2$, implying that $f$ has a fixed point on $S^3$. But this is impossible since $h$ is fixed point free.

EXERCISE 4.2.14. Use the argument above to show $(1_{S^1}, h) \in \mathrm{Aut}(G) \times \mathrm{TOP}(W)$ does not lift to any element in $\mathrm{TOP}_G(P)$, where $P$ is any principal nontrivial $S^1$-bundle over the 2-sphere. Also it can be shown [**NR78, KLR86**] that the total space of each nontrivial principal $S^1$-bundle over an orientable closed surface, different from $S^2$, admits no orientation reversing self-homeomorphism. Use this fact to show that the map $\mathrm{TOP}_G(P) \to \mathrm{Aut}(G) \times \mathrm{TOP}(W)$ is not onto.

LEMMA 4.2.15. *The subgroup* $\mathrm{id} \times \mathrm{TOP}_0(W)$ *is in the image of* $\mathrm{TOP}_G(P)$.

PROOF. Here $\mathrm{TOP}_0(W)$ denotes the subgroup of $\mathrm{TOP}(W)$ which consists of maps isotopic to the identity. Let $h$ be a homeomorphism of $W$, and $H: W \times I \to W$ be a homotopy so that $H|_{W \times 0} = \mathrm{id}$ and $H|_{W \times 1}$ is $h$. Then by the covering homotopy theorem for bundles (e.g., Steenrod [**Ste51**]), there exists a homotopy $\tilde{H}: P \times I \to P$ so that $\tilde{H}|_{P \times 0} = \mathrm{id}$ and $\tilde{H}$ covers $H$. Thus $f = \tilde{H}|_{P \times 1}$ covers the homeomorphism $h$. Since $f$ is a bundle map and covers a homeomorphism, it too is a homeomorphism and is homotopic, through bundle maps, to the identity. In particular, if $h$ itself was isotopic to the identity, then $f$ is also isotopic to the identity. □

REMARK 4.2.16. We have actually shown that if $h$ is homotopic to the identity, then there is a $G$-equivariant map, i.e., a bundle isomorphism $f \in \mathrm{TOP}_G(P)$ which maps onto $\mathrm{id} \times h$. We assumed that our spaces are paracompact Hausdorff spaces to employ the covering homotopy theorem. Assume now that every open subspace of $W$ is paracompact. From the covering homotopy theorem for weakly $G$-equivariant maps between principal $G$-bundles (cf. [**Par91**]), we can characterize the entire image of $\mathrm{TOP}_G(P)$ in $\mathrm{Aut}(G) \times \mathrm{TOP}(W)$: Let $f \in \mathrm{TOP}_G(P)$ with $f(x \cdot u) = \alpha_f(x) \cdot$

$f(u)$. Let $\mathrm{TOP}(W)_h$ be the isotopy classes (i.e., path components) of $\mathrm{TOP}(W)$ whose elements are homotopic to $h$, where $h : W \to W$ is induced by $f$. (Let $[\alpha_f]$ denote the image of $\alpha_f$ in $\mathrm{Aut}(G)$.) Then $[\alpha_f] \times \mathrm{TOP}(W)_h$ is in the image of $\mathrm{TOP}_G(P)$. The union of all such is the complete image of $\mathrm{TOP}_G(P)$ in $\mathrm{Aut}(G) \times \mathrm{TOP}(W)$.

REMARK 4.2.17. The projection $p : \mathrm{TOP}_G(P) \to \mathrm{Aut}(G) \times \mathrm{TOP}(W)$ induces the projections

$$p_1 : \mathrm{TOP}_G(P) \to \mathrm{Aut}(G),$$
$$p_2 : \mathrm{TOP}_G(P) \to \mathrm{TOP}(W),$$
$$\bar{p} : \mathrm{TOP}_G(P) \to \mathrm{Out}(G) \times \mathrm{TOP}(W).$$

If $P$ is locally compact, these maps are continuous homomorphisms of topological groups, in the compact-open topology. While this statement should seem plausible to the reader, the details involve careful computations with the compact-open topology and can be found in [**Par89**].

From Subsection 4.2.5 through Proposition 4.2.8, we described $\psi(\mathrm{M}_G(P,G))$ as the kernel of $p$, and $\ell(G) \times_{Z(G)} \psi(\mathrm{M}_G(P,G))$ as the kernel of $\bar{p}$. If $f \in \mathrm{TOP}_G(P)$ is in the kernel of $p_2$, Park defines $\eta_f \in M_G^w(P,G)$ to be an element $\eta \in M(P,G)$ satisfying

$$\eta(au) = a\eta(u)(p_1(f)(a))^{-1} = a\eta(u)(\alpha_f(a))^{-1},$$

for $a \in G$, $u \in P$, $p_1(f) = \alpha_f$. This map is $G$-equivariant where the $G$-action on $G$ is given by $(a,x) \mapsto ax\,\alpha_f(a)^{-1}$. (Note, if $\eta_f \in M_G(P,G)$, then $\alpha_f = \mathrm{id}$.) The group operation is $(\eta_f \cdot \eta_g)(u) = \eta_f(u)\alpha_{\eta_f}(\eta_g(u))$ making $M_G^w(P,G)$ into a (topological–if $P$ is locally compact) group. Park defines $M_G^s(P,G) \subset M_G^w(P,G)$ if $\alpha_f \in \mathrm{Inn}(G)$. So we have

$$\mathrm{M}_G(P,G) \subset M_G^s(P,G) \subset M_G^w(P,G).$$

Define $\psi : M_G^w(P,G) \to \mathrm{TOP}_G(P)$ by $\psi(\eta_f)(u) = \eta_f(u)^{-1}u$. Then $\psi$ is an isomorphism onto the kernel of $p_2$. By restricting $\psi$ to $M_G^s(P,G)$ and $\mathrm{M}_G(P,G)$, we obtain an isomorphism onto the kernels of $\bar{p}$ and $p$.

## 4.3. Seifert fiberings modeled on a principal $G$-bundle

We will now define a Seifert fibering by taking the prototype of Subsection 4.1.1 and imposing appropriate restrictions that eliminate pathology on $\Pi \backslash P$ and $B$. With these restrictions in place, a useful theory incorporating a conventional geometric interpretation of the fibers can be developed.

4.3.1. Let $G$ be a connected Lie group and $P$ a principal $G$-bundle over the space $W$. Assume that $P$ is connected, completely regular, and admits covering space theory. This implies that $W$ is completely regular and also admits covering space theory. Let $\Pi \subset \mathrm{TOP}_G(P)$ be a Lie group ($\Pi$ could be discrete), act effectively and properly on $P$.

Since $\Pi$ normalizes $\ell(G)$, it acts on $W$. If in addition, we assume that the effective part of the $\Pi$-action on $W$ (i.e., the image of $\Pi \hookrightarrow \mathrm{TOP}_G(P) \to \mathrm{TOP}(W)$) is proper, then the orbit space $\Pi \backslash P$ is called a *Seifert fibered space modeled on the principal $G$-bundle $P$*. The map

$$\tau : \Pi \backslash P \longrightarrow \Pi \backslash W$$

## 4.3. SEIFERT FIBERINGS MODELED ON A PRINCIPAL $G$-BUNDLE

is called a *Seifert fibering* and the space $\Pi \backslash W = B$ is called its *base*. The principal $G$-bundle $P$ is called the *model space* for the Seifert fibering.

Since $\Pi$ and $\ell(G)$ act properly on $P$, they are closed subsets of $\text{TOP}_G(P)$ in the compact-open topology; see Subsection 4.2.2. Let

$$\Gamma_\ell = \Pi \cap \ell(G).$$

This is a closed normal subgroup of $\Pi$ which acts properly and freely as left translations on each principal $G$-fiber of $P$. Thus on $W$, there is an induced action of

$$Q_\ell = \Pi / \Gamma_\ell$$

on $W$. (Notice that the action $(Q_\ell, W)$ may not be effective because it may have an ineffective part, but the image $Q_\ell \to \text{TOP}(W)$ must act properly on $W$ because the image of $\Pi \hookrightarrow \text{TOP}_G(P) \to \text{TOP}(W)$ acts properly on $W$.) More precisely, we get the following commutative diagram.

(4.3.1)
$$\begin{array}{ccc} (\ell(G) \cdot \Pi, P) & \xrightarrow{\ell(G)\backslash}_{\widetilde{\tau}} & (Q_\ell, W) \\ \Pi \backslash \downarrow \nu & & Q_\ell \backslash \downarrow \bar\nu \\ X = \Pi \backslash P & \xrightarrow{\tau} & Q_\ell \backslash W = B \end{array}$$

Then $\tau : \Pi \backslash P \longrightarrow Q_\ell \backslash W$ is a *Seifert fibering modeled on the principal $G$-bundle $P$* with typical fiber $\Gamma_\ell \backslash G$.

**4.3.2** (The typical and regular fibers). Let $\widehat{\Gamma}$ be the kernel of

$$\Pi \longrightarrow \text{TOP}_G(P) \xrightarrow{p_2} \text{TOP}(W).$$

Note that $\Gamma_\ell \subset \widehat{\Gamma}$ and both are closed normal subgroups of $\Pi$ which leave each principal $G$-fiber of $P$ invariant. The action $(\Pi, P)$ induces an action $(\Pi, W)$. The quotient group $\Pi / \widehat{\Gamma} = \widehat{Q}$ is the effective part of the induced $\Pi$-action on $W$. Then $\widehat{Q}$, by definition, acts properly on $W$ with $B = \widehat{Q} \backslash W = Q_\ell \backslash W = \Pi \backslash W$.

We have the following commutative diagram of exact sequences of groups.

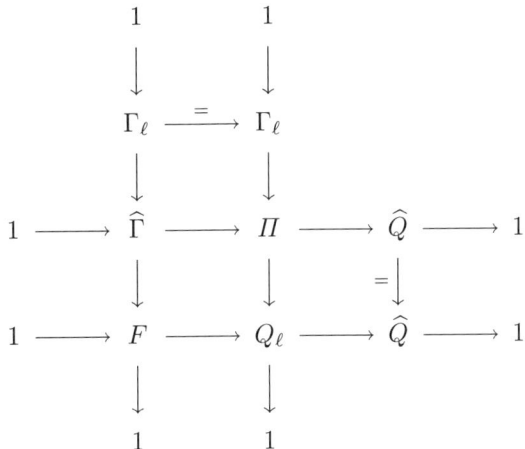

This yields

$$\begin{array}{ccc}
(\Gamma_\ell, G) \longrightarrow & (\Pi, P) \longrightarrow & (Q_\ell, W) \\
\downarrow & \downarrow & \downarrow \\
\Gamma_\ell\backslash G \longrightarrow & \Pi\backslash P \longrightarrow & Q_\ell\backslash W
\end{array}$$

and

$$\begin{array}{ccc}
(\widehat{\Gamma}, G) \longrightarrow & (\Pi, P) \longrightarrow & (\widehat{Q}, W) \\
\downarrow & \downarrow & \downarrow \\
\widehat{\Gamma}\backslash G \longrightarrow & \Pi\backslash P \longrightarrow & \widehat{Q}\backslash W.
\end{array}$$

Note that $F$ is ineffective on $W$ and $Q_\ell$ acts properly on $W$ if and only if $F$ is compact.

Let $\Pi_w$ be the isotropy of the action $(\Pi, W)$ at $w \in W$. Recall that $\Pi_w$ leaves the $G$-fiber over $w$ invariant and $\tau^{-1}(b) = \Pi_w\backslash G$, where $\bar{\nu}(w) = b$; see Subsection 4.1.1. Recall the following notations

$$\Gamma_\ell = \Pi \cap \ell(G)$$
$$\widehat{\Gamma} = \text{the kernel of } \Pi \to \text{TOP}(W)$$
$$\Pi_w = \text{the isotropy of the (ineffective) action } (\Pi, W) \text{ at } w \in W$$

in increasing order by inclusion.

DEFINITION 4.3.3. $\Gamma_\ell\backslash G$ is called the *typical fiber* and $\widehat{\Gamma}\backslash G$ is called the *regular fiber*. If $\Pi_w$ is strictly bigger than $\widehat{\Gamma}$, then $\Pi_w\backslash G$ is a *singular fiber*.

Since $\Gamma_\ell$ is a subgroup of $\ell(G)$, the typical fiber $\Gamma_\ell\backslash G$ is a homogeneous space; see Subsection 4.4.3. $\widehat{\Gamma}$ is the ineffective part of the $\Pi$-action on $W = G\backslash P$. By Corollary 4.2.9, $\Pi_w$ lies in $\ell(G) \rtimes \text{Aut}(G)$. Therefore, the singular fiber (also the regular fiber) is an infra-homogeneous space. See Subsection 4.4.3 for a definition. Notice that typical fibers and regular fibers are independent of the point on the base space.

This fiber nomenclature is motivated by a consequence of Smith theory: if $Q$ is discrete and acts effectively and properly on a connected manifold $W$, then there is an open and dense subset of $W$ for which $Q_w = 1$. We have adopted this notation because most of our applications fit these conditions. In general, observe that $\widehat{Q}_w = 1$ if and only if $\Pi_w = \widehat{\Gamma}$, and $Q_w = 1$ if and only if $\Pi_w = \Gamma_\ell$.

EXAMPLE 4.3.4. In Exercise 4.1.3, $\mathbb{Z}_2 = \Pi$, $G = S^1$, $P = S^1 \times I$, and $W$ is an arc. Double $P$ along its boundary to form $P' = S^1 \times I \bigcup_{\partial(S^1 \times I)} S^1 \times I = S^1 \times S^1$. Extend the $(S^1 \times \mathbb{Z}_2)$-action to an action on $S^1 \times S^1$. Then $S^1 \times_{\mathbb{Z}_2} S^1$ is the Klein bottle with an induced $S^1$-action. The orbit mapping of the $S^1$-action is a Seifert fibering over an arc. Each interior point of the arc corresponds to principal orbits which are both regular and typical fibers while the orbits over the end points of the arc are singular fibers.

EXAMPLE 4.3.5. In Subsection 4.1.5, $\mathbb{Z}_2 = \Pi$, $G = S^1$, $P = S^1 \times S^1$, and $W$ is an arc. Each fiber over an interior point of the arc is both a regular and a typical fiber. The fibers over the end points are singular fibers.

In Subsection 4.1.8, $P$ is not necessarily connected. (Connectedness of $P$ and $G$ is only convenience and not essential to the notion of Seifert fibering.) $W = G\backslash P$

and $\psi(\Pi) \cap \ell(G) = (G \cap \Pi) \cap \mathcal{Z}(P) = \Gamma$ is the kernel of $\Pi \longrightarrow \mathrm{TOP}(W)$. Hence $\Gamma \backslash G$ is both the regular and typical fiber and $\tau$ is the orbit mapping of a $\Gamma \backslash G$-action. For the special case where $G$ is a maximal compact subgroup of $P$, $\Pi$ is discrete and torsion free, all fibers are both regular and typical and isomorphic to $G$. If $\Pi$ is discrete but not torsion free, then the principal $\Gamma \backslash G$-orbits will also be typical and regular fibers, and the nonprincipal orbits of the $\Gamma \backslash G$-action on $\Pi \backslash P$ will be the singular fibers.

REMARK 4.3.6. As mentioned above, connectedness of $P$ and $G$ is a convenience and is not essential to the notion of Seifert fiberings. Most of the concepts and definitions still make sense without these restrictions. In some of our prototype examples, we did not require that $G \cdot \Pi$ act properly (e.g., Subsection 4.1.9). Without the restriction of properness, orbit spaces fail to be Hausdorff in the usual quotient topology. The injection $\theta$ will still exist on the algebraic level but $\theta$ will not necessarily be a topological isomorphism onto its image in $\mathrm{TOP}_G(P)$. The fibers then are more likely to resemble the leaves of a foliation as in descent to the torus of the plane field of parallel lines with irrational slope; see Example 1.1.9. Such considerations have much dynamical significance, but exploration of these matters entails different methods.

4.3.7. Let $\theta : \Pi \to \mathrm{TOP}_G(P)$ be an injective homomorphism so that $\Pi \backslash P \to \Pi \backslash W$ is a Seifert fibering. Let $\Gamma$ be a closed normal subgroup of $\Pi$ so that $\theta(\Gamma) \subset \theta(\Pi) \cap \big(\ell(G) \times_{\mathcal{Z}(G)} \mathrm{M}_G(P, G)\big)$. Denote this restriction by $i$. The following diagram commutes.

$$(4.3.2) \quad \begin{array}{ccccccccc} 1 & \longrightarrow & \Gamma & \longrightarrow & \Pi & \longrightarrow & Q & \longrightarrow & 1 \\ & & \downarrow i & & \downarrow \theta & & \downarrow \varphi \times \rho & & \\ 1 & \longrightarrow & \ell(G) \times_{\mathcal{Z}(G)} \mathrm{M}_G(P,G) & \longrightarrow & \mathrm{TOP}_G(P) & \longrightarrow & \mathrm{Out}(G) \times \mathrm{TOP}(W) & & \end{array}$$

Here $\rho : Q \to \mathrm{TOP}(W)$ is the homomorphism induced by the $\Pi$-action on $W$. The effective part of this action is proper on $W$. The horizontal rows are exact. If $\Gamma_\ell \subset i(\Gamma)$, then $i(\Gamma)$ lies in between $\Gamma_\ell = \theta(\Pi) \cap \ell(G)$ and $\widehat{\Gamma} = \ker\{\Pi \longrightarrow \mathrm{TOP}(W)\}$.

## 4.4. The topology and geometry of the fibers

4.4.1. Let $1 \to \Gamma \to \Pi \to Q \to 1$ be as in Subsection 4.3.7, and the exact sequence $1 \to \Gamma \to \Pi_w \to Q_w \to 1$ be the pullback induced by the inclusion $Q_w \hookrightarrow Q$. The group $\Pi_w$ acts properly by Proposition 1.2.4(6(c)) on the fiber $\tilde{\tau}^{-1}(w)$ as a subgroup of $\mathrm{TOP}_G(\tilde{\tau}^{-1}(w))$. By Corollary 4.2.9, we have

$$\mathrm{M}_G(\tilde{\tau}^{-1}(w), G) \times_{\mathcal{Z}(G)} \ell(G) = r(G) \times_{\mathcal{Z}(G)} \ell(G) = \ell(G) \rtimes \mathrm{Inn}(G).$$

The diagram (4.3.2) becomes

$$\begin{array}{ccccccccc} 1 & \longrightarrow & \Gamma & \longrightarrow & \Pi_w & \longrightarrow & Q_w & \longrightarrow & 1 \\ & & \downarrow i & & \downarrow \theta & & \downarrow \varphi & & \\ 1 & \longrightarrow & \ell(G) \rtimes \mathrm{Inn}(G) & \longrightarrow & \ell(G) \rtimes \mathrm{Aut}(G) & \longrightarrow & \mathrm{Out}(G) & \longrightarrow & 1. \end{array}$$

Now, $\tau^{-1}(b) = \Pi_w \backslash \tilde{\tau}^{-1}(w) = Q_w \backslash (\Gamma \backslash G)$, where $\Pi_w$ lies in $\ell(G) \rtimes \mathrm{Aut}(G)$; the fiber $\tau^{-1}(b)$ is obtained as the quotient of $\Gamma \backslash G$ by the group $Q_w$. If $i(\Gamma) \subset \ell(G)$, we can interpolate the exact sequence $1 \to \ell(G) \to \ell(G) \rtimes \mathrm{Aut}(G) \to \mathrm{Aut}(G) \to 1$

between the top and bottom sequence. The homomorphism $Q_w \xrightarrow{\tilde{\varphi}} \operatorname{Aut}(G)$ is a lift of the homomorphism $Q_w \xrightarrow{\varphi} \operatorname{Out}(G)$.

4.4.2. To make more precise the geometry carried by the singular fibers, we endow $G$ with the linear connection defined by the left invariant vector fields. Since the parallel transport is the effect of the left translations on the tangent vectors of $G$, and hence clearly independent of paths, the connection is flat. A geodesic through the identity element $e \in G$ is a 1-parameter subgroup of $G$ and thus defined for any real value of the affine parameter. All geodesics are translates of geodesics through $e$ and thus the connection is complete. One easily checks that the torsion tensor has vanishing covariant derivative. According to [**KT68**, Proposition 2.1],

$$\operatorname{Aff}(G) = \ell(G) \rtimes \operatorname{Aut}(G)$$

is the group of *affine diffeomorphisms* (connection-preserving diffeomorphisms) of $G$, and $(a, \alpha) \in G \rtimes \operatorname{Aut}(G)$ acts on $G$ by $(a, \alpha)(x) = a \cdot \alpha(x)$ for all $x \in G$. For example, if $G = \mathbb{R}^n$, $\operatorname{Aff}(\mathbb{R}^n) = \mathbb{R}^n \rtimes \operatorname{GL}(n, \mathbb{R})$, the ordinary affine group of $\mathbb{R}^n$.

The theorem of Kamber and Tondeur is stated for a simply connected Lie group $G$. It also hold for any connected Lie group $G$: If $\tilde{G}$ is the universal covering group of $G$, then the lifting sequence of $\ell(G) \rtimes \operatorname{Aut}(G)$ becomes the exact sequence

$$1 \longrightarrow \pi_1(G) \longrightarrow \ell(\tilde{G}) \rtimes \operatorname{Aut}(\tilde{G}, \pi_1(G)) \longrightarrow \ell(G) \rtimes \operatorname{Aut}(G) \longrightarrow 1.$$

$\pi_1(G)$, the group of covering transformations, is a central subgroup of $\tilde{G}$. An automorphism of $G$ lifts uniquely to an automorphism of $\tilde{G}$ (the other lifts are not automorphisms). Thus, the group $\operatorname{Aut}(\tilde{G}, \pi_1(G))$, the automorphisms of $\tilde{G}$ that leave $\pi_1(G)$ invariant, is naturally isomorphic to $\operatorname{Aut}(G)$.

4.4.3. If $H$ is a closed subgroup of a Lie group $G$, the quotient space $G/H$ is called a *homogeneous space*. It is a smooth manifold on which $G$ acts transitively, (the principal isotropy group is $H$). Now suppose $\Pi$ is a closed subgroup of $\operatorname{Aff}(G)$ acting properly on $G$. Now suppose $\Pi$ is a closed subgroup of $\operatorname{Aff}(G) = \ell(G) \rtimes \operatorname{Aut}(G)$ acting properly on $G$. If $\Gamma = \Pi \cap \ell(G)$, then $\Gamma$ is normal in $\Pi$ and $Q = \Pi/\Gamma$ acts properly on the homogeneous space $\Gamma \backslash G$. The quotient $Q \backslash (\Gamma \backslash G) = \Pi \backslash G$ is called an *infra-homogeneous space*. It may not be a manifold. However, if $\Pi$ acts freely, then $\Pi \backslash G$ is called an *infra-homogeneous manifold*. Anyways, the map $\Gamma \backslash G \to \Pi \backslash G$ is extremely nice because the action of $Q = \Pi/\Gamma$ comes from $\operatorname{Aff}(G)$. If $\Pi$ is discrete, then $\Pi \backslash G$ is an orbifold (i.e., a $V$-manifold) with the map $\Gamma \backslash G \to \Pi \backslash G$ being a regular (possibly branched) covering.

For a Seifert fibering modeled on $P$, take $w \in W$ on the orbit $b \in B$. Then $\tau^{-1}(b)$ is homeomorphic to $\Pi_w \backslash G$, where $G$ is a principal fiber over $w$ and $1 \to \Gamma \to \Pi_w \to Q_w \to 1$ is exact. The group $\Pi_w$ acts as affine transformations on $G$. The infra-homogeneous space $\tau^{-1}(b)$ is the quotient of the homogeneous space $\Gamma \backslash G$ by the compact group $Q_w$ of affine transformations on $\Gamma \backslash G$. If $\Pi_w$ acts freely on $G$, then $\Pi_w \backslash G$ is an *infra-homogeneous manifold*.

4.4.4. It is of great interest to decide when $\Pi$ acts freely on $P$. We have seen that this reduces to showing that $\Pi_w$ acts freely on each fiber $\tilde{\tau}^{-1}(w)$, $w \in W$. Assume $\Gamma$ is a discrete subgroup of $G$ via $i : \Gamma \to \ell(G)$, but not necessarily that $\theta(\Pi) \cap \ell(G) = i(\Gamma)$. We have the following

PROPOSITION 4.4.5. *With $\Pi$, $G$ and $P$ as in Subsection 4.4.1, $\Pi$ acts freely on $P$ if and only if $\Pi_w$ acts freely on each $\tilde{\tau}^{-1}(w)$ as elements of $\operatorname{Aff}(G)$. In particular,* if $G$ is simply connected and solvable (so that $G$ is diffeomorphic to $\mathbb{R}^n$), then $\Pi$ acts freely on $P$ if and only if each $\Pi_w$ is torsion free. Furthermore, if $W$ is a finite dimensional contractible space, then $\Pi$ acts freely on $P \approx G \times W$ if and only if $\Pi$ is torsion free.

PROOF. If $G$ is simply connected and solvable, it is diffeomorphic to $\mathbb{R}^n$. The group $\Pi_w$ acts properly on $G \times w$. If $\Pi_w$ has torsion, then it contains a nontrivial finite $p$-subgroup $H$ for some prime $p$. By Lemma 3.1.6, $(G \times w)^H \neq \emptyset$. So $\Pi_w$ cannot act freely. Of course, if $\Pi_w$ is torsion free and some element other than $1 \in H$ fixes a point in $G \times w$, this would contradict the properness of the action of $\Pi$. □

## 4.5. Examples with $\Pi$ discrete

The theory of Seifert fiberings is developed most thoroughly for discrete $\Pi$. Theorems 1.9.2 and 1.9.3 offer sufficient easily checked conditions that a discrete $\Pi$-action on $P$ will lead to a Seifert fibering. If $P$ is locally compact, then the assumptions of Theorem 1.9.2 frequently hold. If $P$ is not locally compact, one may be able to use Theorem 1.9.3 to check if the $\Pi$-action leads to a Seifert fibering by putting the $E$ of this theorem equal to the subgroup in $\mathrm{TOP}_G(P)$, generated by $\Pi$ and $\ell(G)$. In this case, note that the $Q$ in Theorem 1.9.3 is

$$Q_\ell = \ell(G) \cdot \Pi / \ell(G) = \Pi / (\ell(G) \cap \Pi) = \Pi / \Gamma_\ell.$$

The example below shows that these sufficient conditions may fail to be necessary conditions.

4.5.1. Let $\Delta$ be a lattice of $G$, and let $P = G \times W$. Suppose $\Pi = \Delta \times \widehat{Q}$, where $\Delta \subset r(G)$ and $\widehat{Q}$ acts trivially on the $G$-factor and properly and effectively on the $W$-factor. Note $\widehat{Q}$ is the effective part of the $\Pi$-action on $W$. The regular fiber is $G/\Delta$ and the typical fiber $\Gamma_\ell \backslash G$ is $(\Delta \cap \ell(G)) \backslash G = (\mathcal{Z}(G) \cap \Delta) \backslash G$. If the $(\widehat{Q}, W)$-action is free, each $\tau^{-1}(b)$ is a regular fiber but not a typical fiber unless $\mathcal{Z}(G) \cap \Delta = \Delta$.

4.5.2 (The orbit mapping of a locally injective $T^k$-action as a Seifert fibering). The notion of a generalization to the classical 3-dimensional Seifert fiberings was inspired, in part, by the analysis of injective and locally injective toral actions as described in Section 2.8. Recall from there that $(T^k, X)$ is locally injective if and only if the lifted action $(T^k, X_{\mathrm{Im}(\mathrm{ev}_*^x)})$ is free. The lifted action commutes with the covering transformations $Q = \pi_1(X, x)/\mathrm{Im}(\mathrm{ev}_*^x)$. Thus, assuming in addition that $X$ is completely regular, the orbit mapping $X_{\mathrm{Im}(\mathrm{ev}_*^x)} = P \xrightarrow{\tilde{\tau}} T^k \backslash P = W$ is a principal $T^k$-bundle mapping and $\tilde{\tau}$ is $(T^k \times Q)$-equivariant, where the $T^k$-action on $W$ is trivial. We have a diagram, as in Subsection 4.1.1.

$$\begin{array}{ccc} (T^k \times Q, P) & \xrightarrow{\tilde{\tau}} & (Q, T^k \backslash P = W) \\ \nu \downarrow Q\backslash & & \bar{\nu} \downarrow Q\backslash \\ (T^k, X) & \xrightarrow{\tau} & B = Q \backslash W \end{array}$$

The map $\tau$ is our Seifert fibering. It is modeled on the principal $T^k$-bundle $P$. In this situation, $W$ is simply connected and the homomorphism $\partial : \pi_2(W) \longrightarrow \pi_1(T^k)$, from the exact homotopy sequence of the principal bundle $P$ (see Theorem

1.3.3), is an element of $\mathrm{Hom}(H_2(W;\mathbb{Z}), \pi_1(T^k)) \subset H^2(W;\mathbb{Z}^k)$. It can be easily seen that $[\partial] \in H^2(W;\mathbb{Z}^k)$ is the characteristic class of the bundle $P$ and classifies the bundle $P$ over $W$ (when $W$ is assumed to be locally semisimply connected and paracompact). Here the role of $\Pi$ in Subsection 4.3.1 is played by the covering $Q$-action on $P$.

Let $K$ be the kernel of the $T^k$-action (i.e., the finite subgroup of $T^k$ that leaves $X$ fixed). This coincides with $\widehat{\Gamma}$, the kernel of the induced $Q$-action on $T^k \backslash P = W$. Thus $K \backslash T^k$ is the regular fiber. If some $T^k$-orbit on $X$ has trivial isotropy group, for example if $X$ is a manifold and the action of $T^k$ is effective, then $\widehat{\Gamma} = \Gamma = K = 1$. These principal orbits coincide with the regular fibers.

4.5.3. If $(T^k, X)$ is injective (see Definition 3.1.10), the lifted action $(T^k, X_H)$ splits to $(T^k, T^k \times W)$, with $T^k$ acting just as translations on the first factor. Replace the $(T^k, X_H)$ by the obvious (ineffective) $(\mathbb{R}^k, X_H)$. The $\mathbb{R}^k$-action lifts to $(\mathbb{R}^k, \mathbb{R}^k \times W)$ on the universal covering $\widetilde{X_H} = \mathbb{R}^k \times W$, with the $\mathbb{R}^k$-action being the translations along the first factor. The group $\pi_1(X, x)$ of covering transformations centralizes this $\mathbb{R}^k$-action. The group $\pi_1(X, x)$ acts on $W$ with $\pi_1(T^k)$ acting trivially and the quotient $Q$ acting properly. Now, the orbit mapping $\tau : (T^k, X) \to B = Q \backslash W$, is a Seifert fibering modeled on the trivial principal $\mathbb{R}^k$-bundle $\mathbb{R}^k \times W$. The group $\Pi = \pi_1(X, x)$ is a central extension of $\mathbb{Z}^k$ by $Q$ and commutes with the $\mathbb{R}^k$-action on $\mathbb{R}^k \times W$. The $Q$-action is effective on $W$ if and only if the $T^k$-action on $X$ is effective. If $(T^k, X)$ is effective, $\pi_1(T^k) = \mathbb{Z}^k = \Gamma = \Pi \cap \mathbb{R}^k$ and $\widehat{\Gamma} = \Gamma$. Then, both the regular and typical fibers are the principal orbits of $X$. In terms of Subsection 4.3.1, we have

$$\begin{array}{ccccccccc}
1 & \to & \pi_1(T^k, e) & \to & \pi_1(X, x) & \to & Q & \to & 1 \\
& & \downarrow & & \downarrow \theta & & \downarrow 1 \times \rho & & \\
1 & \to & \mathrm{M}(W, \mathbb{R}^k) & \to & \mathrm{M}(W, \mathbb{R}^k) \rtimes (\mathrm{Aut}(\mathbb{R}^k) \times \mathrm{TOP}(W)) & \to & \mathrm{Aut}(\mathbb{R}^k) \times \mathrm{TOP}(W) & \to & 1
\end{array}$$

in the injective case. Suppose the isomorphism of $\mathbb{Z}^k$ into $\mathbb{R}^k$ is the standard inclusion. Then we may write

$$\begin{aligned}
\theta(n, \alpha)(x, w) &= \theta(n)\theta(1, \alpha)(x, w) \\
&= \ell(n)(\lambda_\alpha, 1, \rho(\alpha))(x, w) \\
&= (-n, \mu(n), 1)(\lambda_\alpha, 1, \rho(\alpha))(x, w) \\
&= (-n + \lambda_\alpha, 1, \rho(\alpha))(x, w) \\
&= (x - (-n + \lambda(\rho(\alpha)(w))), \rho(\alpha)(w)) \\
&= (x + n - \lambda(\rho(\alpha)(w)), \rho(\alpha)(w)),
\end{aligned}$$

where $(n, \alpha) \in \pi_1(X, x)$ and $(x, w) \in \mathbb{R}^k \times W$. The homomorphism $1 \times \rho$ maps $Q$ trivially into the first factor and, with kernel, the ineffective part of $(T^k, X)$ into the second factor.

4.5.4. Analysis of a locally injective $(T^k, X)$ (see Definition 2.8.1) is more complicated. We have seen that $(T^k, X)$ lifts to $(T^k, X_H)$ where $X_H$ is a principal $T^k$-bundle $P$ over the simply connected $W = T^k \backslash P$. The group $H = \mathrm{ev}_*^x(\pi_1(T^k))$ is isomorphic to a group $C \oplus F$, $C$ is a free Abelian group of rank $s$, with $s \leq k$, and $F$ is finite Abelian. For convenience of notation, we shall assume that $(T^k, X)$-action is effective. We have the following

LEMMA 4.5.5. *There is a splitting of $T^k$ into $T^s \times T^{k-s}$ so that $\mathrm{ev}^x_*|_{\pi_1(T^s)}$ is injective, $\mathrm{ev}^x_*|_{\pi_1(T^{k-s})}$ has finite image, and $\mathrm{ev}^x_*(\pi_1(T^k)) \cong C \oplus F$.*

PROOF. Let $K$ be the kernel of $\mathrm{ev}^x_* : \pi_1(T^k, 1) \longrightarrow \pi_1(X, x)$. This is a free Abelian group of rank $k - s$ and is contained in a summand $B$, of $\pi_1(T^k)$ of rank $k - s$. Let $A$ be any summand of $\pi_1(T^k)$ so that $A \times B = \pi_1(T^k)$. Observe, first, that $\mathrm{Im}(B) = F$. For if $b \in B$, then $nb \in K$, for some $n$. Thus, $\bar{b} = \mathrm{Im}(b)$ has order a divisor of $n$. Consequently, $\mathrm{Im}(B) \subset F$. If $(a \times b) \subset A \times B$ has image $\bar{a} + \bar{b} = x \in F$, then $n\bar{a} + n\bar{b} = 0$, if $nx = 0$. This implies $n\bar{a} \in F$ and so $ma \in K$, for some $m$. Therefore, $a$ must be 0, $\mathrm{Im}(B) = F$, and $A \to C \oplus F$ must be injective. Because $\mathrm{Im}(B) = F$, $\mathrm{Im}(A) \cap \mathrm{Im}(B) = 0$, and $\mathrm{Im}(A)$ and $\mathrm{Im}(B)$ generates $C \oplus F$, $\mathrm{Im}(A)$ must be a free summand of $C \oplus F$. We replace the splitting of $C \oplus F$ with another splitting $\mathrm{Im}(A) \oplus \mathrm{Im}(B)$. Thus the lemma now follows by factoring the torus $T^k$ into $T^s \times T^{k-s}$, where $\pi_1(T^s) = A$ and $\pi_1(T^{k-s}) = B$. □

4.5.6. By examining the homotopy exact sequence for $P$, as we have $\pi_2(W) \xrightarrow{\partial} A \times B \longrightarrow \mathrm{Im}(A) \times \mathrm{Im}(B) \to 1$, where $\mathrm{Im}(\partial) = K \subset B$. Therefore, $[\partial] \in \mathrm{Hom}(\pi_2(W), A \times B) = \mathrm{Hom}(\pi_2(W), A) \times \mathrm{Hom}(\pi_2(W), B)$. The coordinate of $[\partial]$ on the first factor $\mathrm{Hom}(\pi_2(W), A) = H^2(W; \mathbb{Z}^s)$, is trivial. Consequently, the $T^s$-subbundle of $P$ is trivial and the $T^s$-action is a product action. Therefore, the free $(T^k, P)$-action factors in two ways by

$$\begin{array}{ccc}
(T^s \times T^{k-s}, P) & \xrightarrow{T^{k-s}\backslash} & (T^s, T^{k-s}\backslash P) = (T^s, T^s \times W) \\
{\scriptstyle T^s\backslash} \downarrow & & \downarrow {\scriptstyle T^s\backslash} \\
(T^{k-s}, T^s\backslash P) & \xrightarrow{T^{k-s}\backslash} & W.
\end{array}$$

The horizontal actions are locally injective with no injective part. The vertical actions are injective. Summarizing we have

THEOREM 4.5.7. *If $(T^k, X)$ is a locally injective action, there is a splitting of $T^k = T^s \times T^{k-s}$ so that the action is equivalent to $(T^s \times T^{k-s}, X)$, where the $T^s$-action is injective and the $(T^{k-s}, X)$-action is locally injective with the property that no nontrivial torus subgroup of $T^{k-s}$ acts injectively.*

4.5.8. We should mention one technical caveat. We have used $[\partial]$ to classify the principal $T^k$-bundle $P$ over $W$. This may not work for an arbitrary and somewhat pathological space $W$. What is involved is an identification of the Čech cohomology $\check{H}^2(W; \mathbb{Z})$ as a subgroup of the singular cohomology $H^2(W; \mathbb{Z})$. This is no problem if, for each $x \in X$ and neighborhood $U$ of $x$, there is a neighborhood $V$ with $x \in V \subset U$, such that $H_i(V; \mathbb{Z}) \to H_i(U; \mathbb{Z})$ is trivial for singular homology, $i \leq 1$. Without this assumption, a more complicated cohomology argument is needed to ensure that the theorem holds.

4.5.9. The case where $G$ is a simply connected Abelian or nilpotent Lie group is especially important for us. For $G = \mathbb{R}^n$, $\mathrm{O}(n)$ is a maximal compact subgroup of $\mathrm{GL}(n, \mathbb{R})$. A *uniform* discrete subgroup (i.e., cocompact discrete subgroup) $\Pi$ of $\mathbb{R}^n \rtimes \mathrm{O}(n)$ is called a *crystallographic group*. By a theorem of Bieberbach, $\Pi \cap \mathbb{R}^n$ is isomorphic to $\mathbb{Z}^n$, and is a lattice of $\mathbb{R}^n$. If $\Pi$ is torsion free, we call $\Pi$ a *Bieberbach group* (torsion-free crystallographic group). Flat manifolds are the orbit spaces $\Pi\backslash\mathbb{R}^n$, where $\Pi$ is a Bieberbach group. Note that each flat manifold is finitely covered by a flat torus $\mathbb{Z}^n\backslash\mathbb{R}^n$; see Theorem 8.1.2 for more details.

EXAMPLE 4.5.10. $E(3) = \mathbb{R}^3 \rtimes O(3)$ is the group of isometries of $\mathbb{R}^3$ (3-dimensional Euclidean space). As a set, it is the Cartesian product $\mathbb{R}^3 \times O(3)$, where $O(3)$ is the orthogonal group. The group operation is given by
$$(a, A)(b, B) = (a + Ab, AB).$$
This acts on $\mathbb{R}^3$ by
$$(a, A) \cdot x = a + Ax$$
for $x \in \mathbb{R}^3$. The matrix $A$ is called the *rotational part*, and $a$ is called the *translational part* of $(a, A)$. It is easy to check that this is actually an action:
$$\begin{aligned}(a, A)((b, B) \cdot x) &= (a, A) \cdot (b + Bx)\\ &= a + A(b + Bx)\\ &= (a + Ab) + ABx\\ &= ((a, A)(b, B)) \cdot x.\end{aligned}$$

Consider the subgroup $\Pi \subset E(3)$ generated by
$$\Pi = \langle t_1 = (e_1, I),\ t_2 = (e_2, I),\ t_3 = (e_3, I),\ \alpha = (a, A)\rangle,$$
where
$$e_1 = \begin{bmatrix}1\\0\\0\end{bmatrix},\ e_2 = \begin{bmatrix}0\\1\\0\end{bmatrix},\ e_3 = \begin{bmatrix}0\\0\\1\end{bmatrix},\ a = \frac{1}{2}e_1,\ A = \begin{bmatrix}1 & 0 & 0\\0 & -1 & 0\\0 & 0 & -1\end{bmatrix}.$$
Since $t_i \cdot x = x + e_i$, each $t_i$ is a translation by the $i$th unit vector. On the other hand,
$$\alpha \cdot x = \left(\begin{bmatrix}1/2\\0\\0\end{bmatrix}, \begin{bmatrix}1 & 0 & 0\\0 & -1 & 0\\0 & 0 & -1\end{bmatrix}\right) \cdot \begin{bmatrix}x_1\\x_2\\x_3\end{bmatrix} = \begin{bmatrix}x_1 + 1/2\\-x_2\\-x_3\end{bmatrix}$$
shows that $\alpha$ rotates the $x_2 x_3$-plane by $180°$, while it advances the $x_1$ direction by a half unit. Moreover,
$$\alpha^2 = t_1.$$
Since
$$\alpha t_1 \alpha^{-1} = t_1,\ \alpha t_2 \alpha^{-1} = t_2^{-1},\ \alpha t_3 \alpha^{-1} = t_3^{-1},$$
the center of $\Pi$ is generated by $t_1$, and is isomorphic to $\mathbb{Z}$. The quotient is
$$Q = \Pi/\mathbb{Z} = \langle \bar{t}_2,\ \bar{t}_3, \bar{\alpha}\rangle \cong \mathbb{Z}^2 \rtimes \mathbb{Z}_2.$$
Clearly, $\Pi \cap \mathbb{R}^3 = \mathbb{Z}^3$ (the translation part), and $\Pi/\mathbb{Z}^3 = \mathbb{Z}_2$ generated by $\alpha$. Thus our space is
$$M = \Pi\backslash\mathbb{R}^3 = \mathbb{Z}_2\backslash(\mathbb{Z}^3\backslash\mathbb{R}^3) = \mathbb{Z}_2\backslash T^3.$$
It is exactly the same as the space $X = \mathfrak{G}_2$ of Example 4.1.4. It is $S^1 \times_{\mathbb{Z}_2} (S^1 \times S^1)$, where $\mathbb{Z}_2$ acts on $S^1 \times S^1$ by $(z_1, z_2) \mapsto (\bar{z}_1, \bar{z}_2)$ and on $S^1$ by $z \mapsto -z$. Then $S^1\backslash X$ is the surface of the square pillow space; see Subsection 4.1.5. There are five different ways of looking at the same space $M$ as Seifert manifolds modeled on various different principal bundles.
(1) Split $\mathbb{R}^3$ as $\mathbb{R} = \mathbb{R}\{e_1\}$, $\mathbb{R}^2 = \mathbb{R}\{e_2, e_3\}$.

(1a) Model space with the principal fibering $\mathbb{R}\{e_1\} \to \mathbb{R}^3 \to \mathbb{R}\{e_2, e_3\}$. Using the subgroup $\ell(\mathbb{R}^1) \rtimes \big(\mathrm{Aut}(\mathbb{R}^1) \times \mathrm{Isom}(\mathbb{R}^2)\big) = \ell(\mathbb{R}^1) \rtimes \big(\mathrm{Aut}(\mathbb{R}^1) \times (\mathbb{R}^2 \rtimes O(2))\big)$, the nontrivial element $\alpha$ is represented by

$$\alpha = \left( \tfrac{1}{2}, 1, \left( \begin{bmatrix} 0 \\ 0 \end{bmatrix}, \begin{bmatrix} -1 & 0 \\ 0 & -1 \end{bmatrix} \right) \right).$$

This $\alpha$ rotates $\mathbb{R}^2$ by 180 degrees while it advances by half a unit on the fiber $\mathbb{R}$. Clearly, $\mathbb{Z}\backslash\mathbb{R}^1 = S^1$, and $S^1 \to M \to \Pi\backslash\mathbb{R}^2$ is a Seifert fibering modeled on the principal $\mathbb{R}^1$-bundle over $\mathbb{R}^2$ and with typical fiber $S^1$. Since $\Pi/\mathbb{Z}$ acts effectively on $\mathbb{R}^2$, the regular fibers are equal to the typical fiber. There are four singular fibers, all of which are $\mathbb{Z}_2\backslash S^1$.

(1b) Model space with the principal fibering $\mathbb{R}\{e_2, e_3\} \to \mathbb{R}^3 \to \mathbb{R}\{e_1\}$. Using the subgroup $\ell(\mathbb{R}^2) \rtimes \big(\mathrm{Aut}(\mathbb{R}^2) \times \mathrm{Isom}(\mathbb{R}^1)\big) = \ell(\mathbb{R}^2) \rtimes \big(\mathrm{Aut}(\mathbb{R}^2) \times (\mathbb{R}^1 \rtimes O(1))\big)$, the nontrivial element $\alpha$ is represented by

$$\alpha = \left( \begin{bmatrix} 0 \\ 0 \end{bmatrix}, \begin{bmatrix} -1 & 0 \\ 0 & -1 \end{bmatrix}, (\tfrac{1}{2}, 1) \right).$$

This $\alpha$ acts on the base $\mathbb{R}^1$ as a translation by $\tfrac{1}{2}$ freely yielding a circle $\mathbb{Z}\backslash\mathbb{R} = \mathbb{Z}_2\backslash S^1$ again while it rotates the fiber $\mathbb{R}^2$ by 180 degrees. Thus our space is a $T^2$-bundle over the circle with structure group $\pm I$. This resulting bundle is not a principal bundle. Every fiber is a typical (so regular) fiber $T^2$.

(2) Split $\mathbb{R}^3$ as $\mathbb{R} = \mathbb{R}\{e_3\}$, $\mathbb{R}^2 = \mathbb{R}\{e_1, e_2\}$.

(2a) Model space with the principal fibering $\mathbb{R}\{e_3\} \to \mathbb{R}^3 \to \mathbb{R}\{e_1, e_2\}$. Using the subgroup $\ell(\mathbb{R}^1) \rtimes \big(\mathrm{Aut}(\mathbb{R}^1) \times \mathrm{Isom}(\mathbb{R}^2)\big) = \ell(\mathbb{R}^1) \rtimes \big(\mathrm{Aut}(\mathbb{R}^1) \times (\mathbb{R}^2 \rtimes O(2))\big)$, the nontrivial element $\alpha$ is represented by

$$\alpha = \left( 0, -1, \left( \begin{bmatrix} \tfrac{1}{2} \\ 0 \end{bmatrix}, \begin{bmatrix} 1 & 0 \\ 0 & -1 \end{bmatrix} \right) \right).$$

This $\alpha$ acts on $T^2$ freely yielding the Klein bottle while it also flips the fiber. This is a $S^1$-bundle over the Klein bottle with structure group $\mathbb{Z}_2 = \langle -1 \rangle$. This resulting bundle is not a principal bundle. Since the induced action of $\Pi$ on the base $\mathbb{R}^2$ is free, every fiber is a typical (so regular) fiber $S^1$.

(2b) Model space with the principal fibering $\mathbb{R}\{e_1, e_2\} \to \mathbb{R}^3 \to \mathbb{R}\{e_3\}$. Using the subgroup $\ell(\mathbb{R}^2) \rtimes \big(\mathrm{Aut}(\mathbb{R}^2) \times \mathrm{Isom}(\mathbb{R}^1)\big) = \ell(\mathbb{R}^2) \rtimes \big(\mathrm{Aut}(\mathbb{R}^2) \times (\mathbb{R}^1 \rtimes O(1))\big)$, the nontrivial element $\alpha$ is represented by

$$\alpha = \left( \begin{bmatrix} \tfrac{1}{2} \\ 0 \end{bmatrix}, \begin{bmatrix} 1 & 0 \\ 0 & -1 \end{bmatrix}, (0, -1) \right).$$

The base space is an arc with two end points singular. Regular fibers (and singular fibers) are $T^2$, and the two singular fibers are Klein bottles.

(3) On the other hand, $X$ can also be regarded as a Seifert fibering $X \to X \to$ point, modeled on the principal $\mathbb{R}^3$-bundle over a point. There is just one fiber, $X = \Pi\backslash\mathbb{R}^3$, itself. It is a regular fiber. The typical fiber is $(\Pi \cap \mathbb{R}^3)\backslash\mathbb{R}^3 = \mathbb{Z}^3\backslash\mathbb{R}^3$, the 3-torus. Therefore the typical fiber does not appear here as an actual fiber in this Seifert fibering.

4.5.11. Let $G$ be a connected, simply connected nilpotent Lie group; see Subsection 6.1.2. Choose a maximal compact subgroup $C$ of $\mathrm{Aut}(G)$. A uniform (i.e., cocompact) discrete subgroup $\Pi$ of $G \rtimes C$ is called an *almost crystallographic group*. If it is torsion free, it is called an *almost Bieberbach group*. An almost Bieberbach group $\Pi$ yields an *infra-nilmanifold* $\Pi\backslash G$. Note here again that any infra-nilmanifold is finitely covered by the nilmanifold $\Gamma\backslash G$, where $\Gamma = \Pi \cap G$.

Thus, nilmanifolds are a generalization of tori, and infra-nilmanifolds are generalizations of flat manifolds. It is also known that a manifold is diffeomorphic to an infra-nilmanifold if and only if it is *almost flat*. This term is due to Gromov. See [**FH83**] for a proof of the above fact. See Chapter 8 for an extensive discussion.

EXAMPLE 4.5.12. We shall now formulate the 3-dimensional nilmanifolds as homogeneous spaces in terms of their lattices and also as classical 3-dimensional Seifert manifolds with typical fiber $S^1$ and base the 2-torus.

Consider the *Heisenberg group*

$$N = \left\{ \begin{bmatrix} 1 & x & z \\ 0 & 1 & y \\ 0 & 0 & 1 \end{bmatrix} : x, y, z \in \mathbb{R} \right\},$$

which is connected, simply connected, and two-step nilpotent. We denote such a matrix by $(z, x, y)$ so that

$$\begin{bmatrix} 1 & x & z \\ 0 & 1 & y \\ 0 & 0 & 1 \end{bmatrix} \longleftrightarrow (z, x, y).$$

Then the group operation is

$$(z', x', y') \cdot (z, x, y) = (z' + z + x'y,\ x' + x,\ y' + y),$$

and the center of $N$ is 1-dimensional $\mathcal{Z} = \mathbb{R}$, consisting of all matrices with $x = y = 0$. The quotient $W = N/\mathcal{Z}$ is isomorphic to $\mathbb{R}^2$ so that

$$1 \to \mathbb{R} \to N \to \mathbb{R}^2 = W \to 1$$

is an exact sequence of Lie groups. As spaces, this is a smooth fibration which is also a product $N = \mathbb{R} \times W$. Let

$$\alpha = (0, 1, 0),\ \beta = (0, 0, 1),\ \text{and } \gamma = (1/p, 0, 0) \in N,\ \text{with } p \text{ an integer} \neq 0.$$

These three elements generate a discrete group $\Pi$ in $N$. (The coordinates of the group generated are of the form $(n+s/p, q, r)$ where $n$, $s$, $q$, $r$ are all integers and so $\Pi$ is a discrete subset of $\mathbb{R}^3$ which is diffeomorphic to $N$.) Moreover $\Pi$ is uniform (i.e., cocompact) because $\Pi\backslash N$ is compact as we shall now show. Observe that the left translational action of $\Pi$ on $N$ commutes with the free central $\mathbb{R}$ translational action on $N \cong \mathbb{R} \times \mathbb{R}^2$ and $\mathbb{R} \cap \Pi$ is the center of the subgroup $\Pi$ generated by $\gamma = (1/p, 0, 0)$. The $\mathbb{R}$-action descends to a free $S^1 = \langle\gamma\rangle\backslash\mathbb{R}$-action on $\Pi\backslash N$. The map $\Pi\backslash(\mathbb{R} \times \mathbb{R}^2) \to \mathbb{Z}^2\backslash\mathbb{R}^2 = T^2$ is the orbit mapping of the $S^1$-action on $\Pi\backslash N$, and as $\Pi/\langle\gamma\rangle \cong \mathbb{Z}^2$ acts freely on $\mathbb{R}^2$, $\mathbb{Z}^2\backslash\mathbb{R}^2$ has a 2-torus as orbit space.

It is not hard to see that the Euler class of this bundle is $-p$; see Subsection 15.4.1(1). Since $\Pi\backslash N$ is compact, $\Pi$ is a uniform (i.e., cocompact) discrete subgroup of $N$, and so $\Pi\backslash N$ is a nilmanifold.

Let us look at $\Pi\backslash N$ from a different point of view. We take $W = \mathbb{R}^2$. A group $Q = \mathbb{Z}^2$ acts on $W$ as translations. Let

$$1 \to \mathbb{Z} \to \Pi \to Q \to 1$$

be a central extension of $\mathbb{Z}$ by $Q$. Then $\Pi$ has a presentation
$$\Pi = \langle \alpha, \beta, \gamma \mid [\alpha, \beta] = \gamma^p, [\alpha, \gamma] = [\beta, \gamma] = 1\rangle,$$
where $\gamma$ is a generator of the center $\mathbb{Z}$ and the images of $\alpha, \beta$ in $Q$ are generators of $Q$. Suppose $p \neq 0$. Using $\mathbb{Z} \subset \mathbb{R}$, one can obtain an effective action of $\Pi$ on the product $\mathbb{R} \times W$ as follows: for $(z, x, y) \in \mathbb{R} \times W$,

(4.5.1)
$$\begin{array}{rcl}
\alpha(z, x, y) & = & (z + y, \quad x + 1, \quad y), \\
\beta(z, x, y) & = & (z, \quad\quad x, \quad\quad y + 1), \\
\gamma(z, x, y) & = & (z + \frac{1}{p}, \quad x, \quad\quad y).
\end{array}$$

The elements $\alpha, \beta, \gamma$ in the description of the Heisenberg group also satisfy the same relations as those satisfied by $\Pi$. Furthermore, the actions of $\Pi$ on $N$ by left translations are the same as the actions of $\Pi$ on $\mathbb{R} \times W$. Notice that these maps are of the form
$$(z, x, y) \mapsto (\phi(z) - \lambda(h(x, y)), h(x, y)),$$
where $\phi$ is an automorphism of $\mathbb{R}$ and $h$ is an action of $Q$ on $W$, and $\lambda$ is a map $W \to \mathbb{R}$. Consequently, the group $\Pi$ lies in $\mathrm{TOP}_{\mathbb{R}}(\mathbb{R} \times W)$ as
$$(\lambda, \phi, h) \in \mathrm{M}(W, \mathbb{R}) \rtimes (\mathrm{GL}(1, \mathbb{R}) \times \mathrm{TOP}(W)).$$

This represents $\Pi \backslash N = \Pi \backslash (\mathbb{R} \times W) \longrightarrow \mathbb{Z}^2 \backslash W$ as a Seifert fibering modeled on the trivial $\mathbb{R}$-bundle $\mathbb{R} \times \mathbb{R}^2$ with typical fiber $\mathbb{Z} \backslash \mathbb{R} = S^1$ and base the 2-torus. We shall discuss both classical 3-dimensional Seifert fiberings and nilmanifolds in great detail in Chapters 14 and 15. As in Example 4.5.10, the same space $\Pi \backslash N$ also fibers over a point when $G$ is taken to be $N$ itself instead of $\mathbb{R}$ in $N$, and so here $\Gamma_\ell = \widehat{\Gamma} = \Pi$.

EXERCISE 4.5.13. Let $\alpha' = (0, p, 0)$, $\beta' = (0, 0, 1)$, $\gamma' = (p, 0, 0) \in N$, with $p$ an integer $\neq 0$. Show that the group generated by $\{\alpha', \beta', \gamma'\}$ is isomorphic to the group generated by $\alpha = (0, 1, 0)$, $\beta = (0, 0, 1)$, $\gamma = (1, 0, 0) \in N$. Also show that the nilmanifolds can cover themselves nontrivially.

4.5.14. Many of the examples that we used for illustrations have featured $\Pi$ as a *discrete* group and a *trivial* principal bundle $G \times W \to W$. In analogy to injective torus actions (see Definition 3.1.10), we call such a Seifert fibering an *injective Seifert fibering*. If $\Pi$ is *discrete* and the bundle is *not necessarily trivial*, we call this a *locally injective Seifert fibering*; cf. Definition 2.8.1.

EXAMPLE 4.5.15. Let us consider $\mathrm{SO}(3)$, the group of orientation preserving linear isometries of $\mathbb{R}^3$ fixing the origin. Let $G = \mathrm{SO}(2)$ be the subgroup of rotations about the $z$-axis, and let $H = \mathbb{Z}_2 \times \mathbb{Z}_2$ be rotations of $180°$ about $x$, $y$ and $z$-axes together with the identity. Thus $G$ and $H$ are subgroups of $\mathrm{SO}(3)$. Let $G$ act freely on $\mathrm{SO}(3)$ on the right, and let $H$ act on the left as multiplications. Since these actions commute, the $\mathrm{SO}(2)$-action descends to $(\mathbb{Z}_2 \times \mathbb{Z}_2) \backslash \mathrm{SO}(3)$. To examine the descended $\mathrm{SO}(2)$-action, we instead look at the induced $\mathbb{Z}_2 \times \mathbb{Z}_2$-action on $S^2 = \mathrm{SO}(3)/\mathrm{SO}(2)$ of the principal $\mathrm{SO}(2)$-bundle $\mathrm{SO}(3) \longrightarrow \mathrm{SO}(3)/\mathrm{SO}(2)$. Recall $\mathrm{SO}(3)$ is diffeomorphic to $\mathbb{R}P^3$. This $(\mathbb{Z}_2 \times \mathbb{Z}_2)$-action on $S^2$ is equivalent to the restriction of the $(\mathbb{Z}_2 \times \mathbb{Z}_2)$-action on the unit sphere. We see that we have three pairs of poles corresponding to the intersections of the axes with the unit sphere which are fixed by different elements of order 2 in $\mathbb{Z}_2 \times \mathbb{Z}_2$. Otherwise, the actions

are free. Thus the orbit space of $(\mathbb{Z}_2 \times \mathbb{Z}_2)\backslash S^2$ is again a 2-sphere and there are exactly three singular orbits with isotropy $\mathbb{Z}_2$. Therefore, the $S^1$-action on
$$M = (\mathbb{Z}_2 \times \mathbb{Z}_2)\backslash \mathrm{SO}(3)$$
is free off of three singular orbits where the isotropy is $\mathbb{Z}_2 \subset \mathrm{SO}(2)$. The free $\mathrm{SO}(2)$-action on $\mathrm{SO}(3)$ generates the fundamental group of $\mathrm{SO}(3)$ and so does not lift to the universal covering group $S^3 = \mathrm{Spin}(3)$, the group of unit quaternions. If we take the extended lifting sequence of the free $(\mathbb{Z}_2 \times \mathbb{Z}_2)$-action, we get a central extension of $\mathbb{Z}_2$ by $\mathbb{Z}_2 \times \mathbb{Z}_2$; this is the quaternion group $H$ generated by $\{1, i, j, k\}$. The lifting sequence for $\mathrm{SO}(2)$ gives a connected double covering group $\widehat{\mathrm{SO}}(2)$ of $\mathrm{SO}(2)$ and is the maximal torus of $\mathrm{Spin}(3)$. Therefore, $M = (\mathbb{Z}_2 \times \mathbb{Z}_2)\backslash \mathrm{SO}(3)$ has as its fundamental group the quaternion group of order 8. $M$ is a locally injective Seifert fibering over the 2-sphere with three singular fibers and modeled on the $\mathrm{SO}(2)$-bundle $\mathrm{SO}(3) \to S^2 = \mathrm{SO}(3)/\mathrm{SO}(2)$. We have the following commutative diagram of orbit mappings.

$$
\begin{array}{ccc}
(\mathbb{Z}_2 \times \mathbb{Z}_2, \mathrm{Spin}(3), \widehat{\mathrm{SO}}(2)) & \xrightarrow{/\widehat{\mathrm{SO}}(2)} & S^2 \\
{\scriptstyle /\mathbb{Z}_2}\Big\downarrow & & =\Big\downarrow \\
(\mathbb{Z}_2 \times \mathbb{Z}_2, \mathrm{SO}(3), \mathrm{SO}(2)) & \xrightarrow{/\mathrm{SO}(2)} & (\mathbb{Z}_2 \times \mathbb{Z}_2, S^2) \\
{\scriptstyle (\mathbb{Z}_2 \times \mathbb{Z}_2)\backslash}\Big\downarrow & & \Big\downarrow \\
((\mathbb{Z}_2 \times \mathbb{Z}_2)\backslash \mathrm{SO}(3), \mathrm{SO}(2)) & \xrightarrow{/\mathrm{SO}(2)} & (\mathbb{Z}_2 \times \mathbb{Z}_2)\backslash S^2 \approx S^2
\end{array}
$$

The $\mathrm{SO}(2)$-action on $(\mathbb{Z}_2 \times \mathbb{Z}_2)\backslash \mathrm{SO}(3)$ is locally injective, with three singular orbits of multiplicity 2, by Proposition 2.8.7.

In addition, if we take $\mathbb{Z}_p \subset \mathrm{SO}(2)$, $p$ odd, the $\mathrm{SO}(2)$-action descends to an effective $(\mathrm{SO}(2)/\mathbb{Z}_p) \approx S^1$-action on the 3-manifold $M/\mathbb{Z}_p$ covered by $M$. The quotient by $\mathrm{SO}(2)/\mathbb{Z}_p$ is still $(\mathbb{Z}_2 \times \mathbb{Z}_2)\backslash S^2 \approx S^2$, with the same multiplicities, and $\mathrm{SO}(2)/\mathbb{Z}_p$ is locally injective on $M/\mathbb{Z}_p$.

$$
\begin{array}{ccc}
(M, \mathrm{SO}(2)) & \xrightarrow{/\mathrm{SO}(2)} & M/\mathrm{SO}(2) \approx S^2 \\
{\scriptstyle \mathbb{Z}_p}\Big\downarrow & & =\Big\downarrow \\
(M/\mathbb{Z}_p, \mathrm{SO}(2)/\mathbb{Z}_p) & \xrightarrow{/(\mathrm{SO}(2)/\mathbb{Z}_p)} & (M/\mathbb{Z}_p)/(\mathrm{SO}(2)/\mathbb{Z}_p)
\end{array}
$$

The manifolds just constructed are examples of 3-dimensional spherical space forms.

4.5.16. A *spherical space form* is the quotient of a sphere by a finite group of freely acting isometries. That is, if one takes the unit sphere $S^n \subset \mathbb{R}^{n+1}$, then the group of isometries of $S^n$, with the metric induced from $\mathbb{R}^{n+1}$, is $\mathrm{O}(n+1, \mathbb{R})$ acting on $S^n$ in the usual fashion. If the finite group $\Pi \subset \mathrm{O}(n+1, \mathbb{R})$ acts freely, then the orbit space $\Pi\backslash S^n$ is called a spherical space form. In differential geometric terms, the spherical space forms are the smooth manifolds with a Riemannian metric whose sectional curvature is constant and positive; see [**Wol77**] for details.

EXERCISE 4.5.17. Show that in dimension 2, $S^2$ and $\mathbb{R}P_2$ are the only spherical space forms. Similarly, in dimension $2n$, only $S^{2n}$ and $\mathbb{R}P_{2n}$ appear as spherical space forms. (Use the Lefschetz fixed point formula to show that $\Pi$ must be $\mathbb{Z}_2$.) See [**Wol77**].

4.5.18. In each odd dimensions $\geq 3$, there are infinitely many topologically distinct spherical space forms with distinct non-Abelian fundamental groups. Those with Abelian fundamental groups are the lens spaces.

A remarkable characterization was given by H. Zassenhaus for a finite solvable group to admit a free linear action on a sphere. In general it states: *For $\Pi$ to admit a free faithful representation into $U(n)$, (i.e., $\theta : \Pi \to U(n)$ as above) so that $\Pi$ acts freely on $S^{2n-1}$ (and hence $\Pi\backslash S^{2n-1}$ is a spherical space form), every subgroup of $\Pi$ of order $pq$, $p$ and $q$ primes, is a cyclic group. For solvable groups, this necessary condition is also sufficient. To achieve sufficiency for nonsolvable $\Pi$, one must also require that the only noncyclic composition factor allowed is the simple group $I \cong A_5 \cong \widetilde{PSL}(2,5)$.* We discuss more aspects of spherical space forms in the next section as well as in Section 11.8, Chapters 14 and 15. In Section 15.3, we analyze the 3-dimensional space forms in the spirit of Example 4.5.15 and 4.6.3(5) when $n = 1$. Then the space forms are characterized by their Seifert invariants and their fundamental groups; see also [**Orl72**, §6.2] for another complete discussion of 3-dimensional spherical space forms. For a general discussion with emphasis on the topological classification of finite groups acting freely and not necessarily linearly on spheres, see [**AD02**]

We claim that each $(2n-1)$-dimensional spherical space form has the structure of a Seifert fibering modeled on a principal $S^1$-bundle over $\mathbb{C}P_{n-1}$. Let $\Delta_n = \{\lambda I_n : \lambda \in \mathbb{C}, |\lambda| = 1\}$ be the subgroup of $U(n)$. Then $\Delta_n$ is isomorphic to $S^1 \cong U(1)$ and is the center of $U(n)$. The quotient group $PU(n) = U(n)/\Delta_n$ is called the *projective unitary group*. It is a simple Lie group in adjoint form. In the linear action of $U(n)$ on $S^{2n-1}$, the group $\Delta_n$ acts freely, and the orbit mapping $S^{2n-1} \to \Delta_n\backslash S^{2n-1} \cong \mathbb{C}P_{n-1}$ is the Hopf fibering. The action of $U(n)$ on $S^{2n-1}$ projects to an effective action of $PU(n)$ on $\mathbb{C}P_{n-1}$. It is known that this induced action is as isometries with respect to the natural Kähler metric on $\mathbb{C}P_{n-1}$.

Let $\varphi : \Pi \to U(n)$ be a free representation and put $\Gamma = \Delta_n \cap \varphi(\Pi)$ where $\Pi$ is a finite group. The induced action of $\Delta_n$ on $\varphi(\Pi)\backslash S^{2n-1} = M$ has no fixed points, for otherwise its lift back to $\Delta_n$ on $S^{2n-1}$ would have fixed points. The orbit mapping, $(\Delta_n, \Pi\backslash S^{2n-1}) \to \Pi\backslash \mathbb{C}P_{n-1}$ is a locally injective Seifert fibering modeled on the principal $\Delta_n/\Gamma$-bundle $(\Delta_n/\Gamma, M) \to \mathbb{C}P_{n-1}$ with typical and regular fiber $\Delta_n/\Gamma \cong S^1$. The bundle has the first Chern class (or Euler class) $-|\Gamma|$. We may also describe this Seifert fibering as being modeled over the principal $\Delta_n$-bundle $S^{2n-1} \to \mathbb{C}P_{n-1}$ with typical fiber $\Delta_n$ and regular fiber $\Delta_n/\Gamma$. The spherical space form $M$ may admit different Seifert fiberings than the one described. However, in the case where $2n-1$ is the minimal dimension for which $\Pi$ has a free orthogonal representation, the Seifert fibering just described is essentially unique; see Proposition 11.8.7.

## 4.6. The Seifert Construction

As before, $P$ is a principal $G$-bundle, $W = G\backslash P$. Recall that $\text{TOP}_G(P)$ is the group of all weakly $G$-equivariant homeomorphisms of $P$.

Let $\mathcal{U}$ be a closed subgroup of $\text{TOP}_G(P)$. A Seifert Construction for $\Pi$ into $\mathcal{U}$ is simply a homomorphism $\theta : \Pi \to \mathcal{U} \hookrightarrow \text{TOP}_G(P)$. This $\mathcal{U}$ is called the *uniformizing group* for the Seifert construction. The smaller $\mathcal{U}$ is, the more restrictive the fiber space structure will be. Therefore, the more the fiber structure is restricted, the more likely the geometric structure is enhanced.

DEFINITION 4.6.1. A *Seifert Construction* for
(1) a group extension $1 \longrightarrow \Gamma \longrightarrow \Pi \xrightarrow{p} Q \longrightarrow 1$, $\Pi$ discrete,
(2) a homomorphism $i : \Gamma \to \ell(G) \times_{\mathcal{Z}(G)} \mathrm{M}_G(P, G)$,
(3) a proper action $\rho : Q \to \mathrm{TOP}(W)$,

with the *uniformizing group* $\mathcal{U} \subset \mathrm{TOP}_G(P)$ (closed subgroup), is a homomorphism

$$\theta : \Pi \longrightarrow \mathcal{U}$$

such that

$$\theta|_\Gamma = i,$$

and the diagram

$$\begin{array}{ccccc}
\Pi & \xrightarrow{\theta} & \mathcal{U} & \xhookrightarrow{\subset} & \mathrm{TOP}_G(P) \\
{=}\downarrow & & & & \downarrow \\
\Pi & \xrightarrow{p} & Q & \xrightarrow{\rho} & \mathrm{TOP}(W)
\end{array}$$

is commutative. (That is, the $\Pi$-action on $W$ via $\Pi \xrightarrow{\theta} \mathcal{U} \subset \mathrm{TOP}_G(P) \longrightarrow \mathrm{TOP}(W)$ is the same as $\Pi \xrightarrow{p} Q \xrightarrow{\rho} \mathrm{TOP}(W)$.)

4.6.2. We may require a Seifert construction to satisfy additional natural conditions as more information on $\Pi$ is given. For example, when $P = G \times W$, the trivial bundle, we may require $i = \ell : \Gamma \hookrightarrow \ell(G)$ to be a cocompact discrete subgroup such that every automorphism of $\Gamma$ extends to a unique automorphism of $G$ (i.e., $(G, \Gamma)$ has the Unique Automorphism Extension Property (UAEP), see Definition 5.3.3) and $\mathcal{U} = \mathrm{TOP}_G(P)$. In this case, the Seifert Construction $\theta : \Pi \to \mathrm{TOP}_G(G \times W)$ must make the diagram

$$\begin{array}{ccccccccc}
1 & \longrightarrow & \Gamma & \longrightarrow & \Pi & \longrightarrow & Q & \longrightarrow & 1 \\
& & \downarrow{\ell} & & \downarrow{\theta} & & \downarrow{\varphi \times \rho} & & \\
1 & \longrightarrow & \mathrm{M}(W, G) \rtimes \mathrm{Inn}(G) & \longrightarrow & \mathrm{TOP}_G(G \times W) & \longrightarrow & \mathrm{Out}(G) \rtimes \mathrm{TOP}(W) & \longrightarrow & 1
\end{array}$$

commutative, where $\rho : Q \to \mathrm{Out}(G)$ is the homomorphism induced from the abstract kernel $Q \to \mathrm{Out}(\Gamma)$ and $\mathrm{Out}(\Gamma) \to \mathrm{Out}(G)$ from the UAEP.

In Chapters 7 and 9, we show for various classes of discrete groups that every extension, $1 \to \Gamma \to \Pi \to Q \to 1$, admits a Seifert Construction. To what extent such a construction is unique is also examined. The existence and uniqueness of the Seifert Construction are essential for many of our applications.

4.6.3 (Examples of uniformizing groups). (1) Let $P = \widetilde{\mathrm{PSL}}(2, \mathbb{R})$, the universal covering group of $\mathrm{PSL}(2, \mathbb{R})$. It is a principal $G$-bundle ($G = \mathbb{R}$) over $\mathbf{H} = \mathbb{R} \backslash \widetilde{\mathrm{PSL}}(2, \mathbb{R}) = \mathrm{SO}(2) \backslash \mathrm{PSL}(2, \mathbb{R})$, the 2-dimensional real hyperbolic plane. Take $\mathcal{U} = \mathrm{Isom}_0(P)$, the connected component of the identity of the group of isometries of $P$. Let $\rho : Q \subset \mathrm{PSL}(2, \mathbb{R})$ be a Fuchsian group. Then a Seifert construction for $1 \to \mathbb{Z} \to \Pi \to Q \to 1$ with the uniformizing group $\mathrm{Isom}_0(\widetilde{\mathrm{PSL}}(2, \mathbb{R}))$ is a homomorphism $\theta : \Pi \to \mathrm{Isom}_0(\widetilde{\mathrm{PSL}}(2, \mathbb{R}))$ such that

$$\begin{array}{ccccccccc}
1 & \longrightarrow & \mathbb{Z} & \longrightarrow & \Pi & \longrightarrow & Q & \longrightarrow & 1 \\
& & \downarrow{\ell} & & \downarrow{\theta} & & \downarrow{\varphi \times \rho} & & \\
1 & \longrightarrow & \mathbb{R} & \longrightarrow & \mathrm{Isom}_0(\widetilde{\mathrm{PSL}}(2, \mathbb{R})) & \longrightarrow & \mathrm{PSL}(2, \mathbb{R}) & \longrightarrow & 1
\end{array}$$

is commutative; see Chapter 13.

(2) $G = \mathbb{R}^k$ and $P = \mathbb{R}^k \times \mathbb{R}^n$. The group $\mathrm{GL}(k, \mathbb{R}) \times \mathrm{Aff}(\mathbb{R}^n)$ acts on $L(\mathbb{R}^n, \mathbb{R}^k)$, the group of all maps of the form $x \mapsto Ax + a$, where $A$ is a $(k \times n)$-matrix and $a \in \mathbb{R}^k$. Then $L(\mathbb{R}^n, \mathbb{R}^k)$ is not invariant under $\mathrm{GL}(k, \mathbb{R}) \times \mathrm{TOP}(\mathbb{R}^n)$. One cannot form a semidirect product $L(\mathbb{R}^n, \mathbb{R}^k) \rtimes (\mathrm{GL}(k, \mathbb{R}) \times \mathrm{TOP}(\mathbb{R}^n))$, but can form a semidirect product $\mathcal{U} = L(\mathbb{R}^n, \mathbb{R}^k) \rtimes (\mathrm{GL}(k, \mathbb{R}) \times \mathrm{Aff}(\mathbb{R}^n))$. Clearly, this is a subgroup of $\mathrm{Aff}(k+n)$, and is a closed subgroup of $\mathrm{TOP}_{\mathbb{R}^k}(\mathbb{R}^k \times \mathbb{R}^n)$; see [**Lee83**] and Section 11.4.

(3) If $W$ is a smooth manifold, one can take $\mathcal{U} = \mathrm{Diff}_G(G \times W) = \mathcal{C}(W, G) \rtimes (\mathrm{Aut}(G) \times \mathrm{Diff}(W))$, where $\mathcal{C}(W, G)$ is the group of all smooth maps from $W$ to $G$.

(4) One can even take $\mathcal{U} = r(G) \rtimes (\mathrm{Aut}(G) \times \mathrm{TOP}(W)) = \ell(G) \rtimes (\mathrm{Aut}(G) \times \mathrm{TOP}(W))$ for $G \times W$.

(5) To create a spherical space form in dimension $2n-1$, begin with a representation $\varphi : \Pi \to \mathrm{O}(2n)$ such that no $\varphi(g)$ with $g \neq 1$ has an eigenvalue equal to 1. This means that $\Pi$ acts on $\mathbb{R}^n$ as isometries fixing the origin and freely elsewhere. The unit sphere $S^{2n-1} \subset \mathbb{R}^{2n}$ is left invariant, and the orbit space $\varphi(\Pi) \backslash S^{2n-1}$ is a spherical space form. The representation is called a free representation. Suppose $\varphi(\Pi) \backslash S^{2n-1}$ and $\psi(\Pi) \backslash S^{2n-1}$ are spherical space forms. They will be isometric (respectively, diffeomorphic, homeomorphic) if and only if there exists an automorphism $a : \Pi \to \Pi$ and an equivariant homeomorphism

$$h : (\varphi(\Pi), S^{2n-1}) \longrightarrow (\psi(a\Pi), S^{2n-1}),$$

where $h \in \mathrm{O}(2n)$ (respectively, $\mathrm{Diff}(S^{2n-1}), \mathrm{TOP}(S^{2n-1})$). In this case, the two representations are said to be orthogonally (respectively, differentiably, topologically) equivalent.

To classify spherical space forms up to isometry (respectively, diffeomorphism, homeomorphism) in dimension $2n - 1$, one must:

1. Find all finite groups $\Pi$ which admit free orthogonal representations into $\mathrm{O}(n)$;

2. Classify the equivalence classes of these representations of $\Pi$ into $\mathrm{O}(n)$ (respectively, $\mathrm{Diff}(S^{2n-1}), \mathrm{TOP}(S^{2n-1})$).

The classification of spherical space forms, up to isometry, is the work of many group theorists and geometers: Burnside, Zassenhaus, Killing, Vincent, Wolf, and others; see [**Wol77**, Chapters 4, 5, 6]. The answers for the most part are algorithmic. The classification up to diffeomorphism agrees with the isometric classification and is due to Franz, DeRham, and others. The topological classification of spherical space forms agrees with the diffeomorphic classification because of the topological invariance of Whitehead torsion. However, the topological classification of free finite group actions on spheres is only partially solved.

The first part is solved by finding all $\Pi$ which have free unitary representations $\Pi \to U(n)$. Now, $U(n)$ embeds into $\mathrm{SO}(2n) \subset \mathrm{O}(2n)$ by the realification homomorphism

$$j(z_{r,s}) = \begin{bmatrix} x_{2r-1, 2s-1} & -y_{2r-1, 2s} \\ y_{2r, 2s-1} & x_{2r, 2s} \end{bmatrix},$$

a $(2 \times 2)$-block in the $(2n \times 2n)$-matrix of $\mathrm{O}(2n)$, for each $r, s$ entry with $z = x + iy$. If $\theta : \Pi \to U(n)$ is free (respectively, free and irreducible), $j \circ \theta$ is free (respectively, free and irreducible). Moreover, if $\varphi : \Pi \to \mathrm{O}(2n)$ is free, then there is a free $\theta : \Pi \to U(n)$ such that $j \circ \theta = \hat{\theta}$. $\hat{\theta}$ and $\varphi$ are orthogonally equivalent. Unitarily equivalent $\theta$'s yield orthogonally equivalent $\hat{\theta}$'s. Therefore, no generality is lost in considering only free unitary representations when considering the topological types of odd dimensional spherical space forms. Metrically, however, distinct unitary representations may represent the same spherical space form.

For a $(2n-1)$-dimensional spherical space form with a free representation $\varphi : \Pi \to U(n)$, and the Hopf fibering $\Delta_n \to S^{2n-1} \to \mathbb{C}P_{n-1}$, we get the following embeddings.

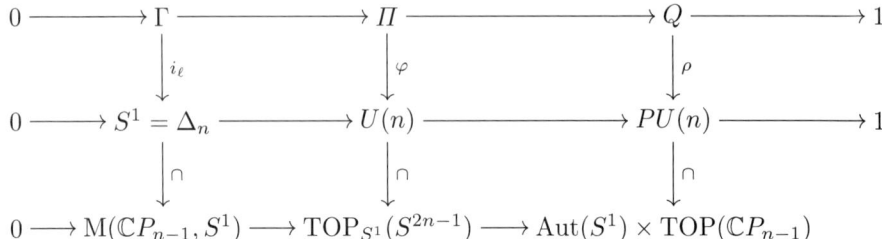

This yields a Seifert construction for $1 \to \Gamma \to \Pi \to Q \to 1$ with uniformizing group $U(n)$. For an embedding of the top exact sequence into the bottom exact sequence, we get a Seifert construction into the universal uniformizing group $\text{TOP}_{S^1}(S^{2n-1})$. If the image of this $\Pi$ can be conjugated in $\text{TOP}_{S^1}(S^{2n-1})$ to lie in $U(n)$ and act freely on $S^{2n-1}$, then $\Pi \backslash S^{2n-1}$ is topologically conjugate to a spherical space form.

EXERCISE 4.6.4. Let $\text{SU}(n) = U(n) \cap \text{SL}(n, \mathbb{C})$. Then $\text{SU}(n)$ is a maximal compact subgroup of $\text{SL}(n, \mathbb{C})$. Show the center of $\text{SU}(n)$ is $\mathbb{Z}_n$ generated by the diagonal matrix $e^{\frac{2\pi i}{n}} I_n$. Prove $\pi_1(\text{PU}(n)) = \pi_1(\text{PSU}(n)) = \mathbb{Z}_n$.

For example, $\text{PU}(2)$ is $\text{SO}(3)$. $\text{PU}(n)$ is the maximal compact subgroup of the group of holomorphic automorphisms of $\mathbb{C}P_{n-1}$ and is the group of isometries of $\mathbb{C}P_{n-1}$ with respect to the natural Kähler metric on $\mathbb{C}P_{n-1}$. Thus $U(n)$ is a group of unitary isometries of $S^{2n-1}$ which centralizes the principal $S^1$-action, and so $U(n) \subset \text{TOP}_{S^1}(S^{2n-1})$.

CHAPTER 5

# Group Cohomology

## 5.1. Introduction

The locally injective Seifert Construction (see Subsection 4.5.14) entails embedding a discrete group $\Pi$ into the universal group $\mathrm{TOP}_G(P)$ of a principal $G$-bundle $P$ in such a way that the following diagram of short exact sequences of groups commutes.

$$
\begin{array}{ccccccccc}
1 & \longrightarrow & \Gamma & \longrightarrow & \Pi & \longrightarrow & Q & \longrightarrow & 1 \\
& & \downarrow \ell & & \downarrow \theta & & \downarrow \varphi \times \rho & & \\
1 & \longrightarrow & \ell(G) \times_{\mathcal{Z}(G)} \mathrm{M}_G(P,G) & \longrightarrow & \mathrm{TOP}_G(P) & \longrightarrow & \mathrm{Out}(G) \times \mathrm{TOP}(W) & &
\end{array}
$$

Therefore, in order to understand when such constructions are possible and to classify them when they do exist, we need to investigate the general procedure for mapping one short exact sequence of groups into another. These results will be formulated in terms of the first and second cohomology of groups with not necessarily Abelian coefficients. We will need the full generality of these results later as we build in more geometry to our Seifert Constructions by "reducing" our universal groups $\mathrm{TOP}_G(P)$.

First, we shall recall some definitions and elementary properties of group extensions. Then we explain the correspondence between congruence classes of extensions of Abelian group $A$ by a group $Q$ in terms of the second cohomology of $Q$ with coefficients in $A$, $H^2(Q; A)$. If $A$ is replaced by a non-Abelian group $G$, we show that the congruence classes of extensions of $G$ by $Q$, $\mathrm{Opext}(G, Q, \varphi)$ is in one-to-one correspondence with $H^2(Q; \mathcal{Z}(G))$, where $\mathcal{Z}(G)$ is the center of $G$.

Next, we study the problem of mapping homomorphically one short exact sequence of groups into another. Given one such homomorphism, we show that all other possible homomorphisms are measured by the first cohomology group. Finally, if we change the extension of the initial exact sequence (measured by a second cohomology group) we determine when it too maps into the target exact sequence.

## 5.2. Group extensions

A group extension is a short exact sequence

$$ 1 \longrightarrow G \longrightarrow E \xrightarrow{p} Q \longrightarrow 1 $$

of not necessarily Abelian groups. We call $E$ an extension of $G$ by $Q$. There is a naturally associated homomorphism $\varphi : Q \to \mathrm{Out}(G)$, called the *abstract kernel* of the extension. This comes about as follows. Pick a "section" $s : Q \to E$ so that the composite $Q \xrightarrow{s} E \xrightarrow{p} Q$ is the identity map ($s$ is not necessarily a

homomorphism). We pick $s(1) = 1$. This defines a map $\widetilde{\varphi} : Q \to \mathrm{Aut}(G)$ (which is a lift of the abstract kernel $\varphi$) given by
$$\widetilde{\varphi}(\alpha) = \mu(s(\alpha)),$$
where $\mu$ is the conjugation map. Even if $\widetilde{\varphi}$ is not a homomorphism, it induces a homomorphism $\varphi : Q \to \mathrm{Out}(G)$, which is our abstract kernel. Of course, $\varphi$ does not depend on the choice of the section $s$. We say *the group $E$ is an extension associated with the abstract kernel $\varphi : Q \to \mathrm{Out}(G)$*.

Since $p(s(\alpha) \cdot s(\beta)) = p(s(\alpha \cdot \beta))$, the element $(s(\alpha) \cdot s(\beta)) \cdot s(\alpha \cdot \beta)^{-1}$ lies in the kernel $G$. Define $f : Q \times Q \to G$ by
$$s(\alpha) \cdot s(\beta) = f(\alpha, \beta) \cdot s(\alpha\beta).$$

Then one can easily verify that

(5.2.1) $$\widetilde{\varphi}(\alpha) \circ \widetilde{\varphi}(\beta) = \mu(f(\alpha, \beta)) \circ \widetilde{\varphi}(\alpha\beta),$$

(5.2.2) $$f(\alpha, 1) = 1 = f(1, \beta),$$

(5.2.3) $$f(\alpha, \beta) \cdot f(\alpha\beta, \gamma) = \widetilde{\varphi}(\alpha)(f(\beta, \gamma)) \cdot f(\alpha, \beta\gamma)$$

for every $\alpha, \beta, \gamma \in Q$. In fact, (5.2.1) follows from the definition of the map $f$ above and (5.2.2) follows from $s(1) = 1$. The associative law of $G$ ensures the right-hand sides of the following two equalities are the same:

$$\begin{aligned}
(s(\alpha)s(\beta)) \cdot s(\gamma) &= f(\alpha,\beta)s(\alpha\beta) \cdot s(\gamma) \\
&= f(\alpha,\beta)f(\alpha\beta,\gamma)s(\alpha\beta\gamma), \\
s(\alpha) \cdot (s(\beta)s(\gamma)) &= s(\alpha) \cdot f(\beta,\gamma)s(\beta\gamma) \\
&= s(\alpha)f(\beta,\gamma)s(\alpha)^{-1} \cdot s(\alpha)s(\beta\gamma) \\
&= \widetilde{\varphi}(\alpha)(f(\beta,\gamma)) \cdot f(\alpha,\beta\gamma)s(\alpha\beta\gamma),
\end{aligned}$$

and this results in (5.2.3).

Conversely, suppose we have maps $\widetilde{\varphi} : Q \to \mathrm{Aut}(G)$ (not necessarily a homomorphism) lifting a homomorphism $\varphi : Q \to \mathrm{Out}(G)$ and $f : Q \times Q \to G$ satisfying (5.2.1), (5.2.2), and (5.2.3). Then there exists an extension $1 \to G \to E \to Q \to 1$ with the given abstract kernel $\varphi$. In fact, $E = G \times Q$ as a set and has group operation
$$(a, \alpha) \cdot (b, \beta) = (a \cdot \widetilde{\varphi}(\alpha)(b) \cdot f(\alpha, \beta),\ \alpha\beta).$$

Thus we have

PROPOSITION 5.2.1. *For a given abstract kernel $\varphi : Q \to \mathrm{Out}(G)$, pick a lift $\widetilde{\varphi} : Q \to \mathrm{Aut}(G)$. Then the set of all extensions with abstract kernel $\varphi$ is in one-to-one correspondence with the set of all maps $f : Q \times Q \to G$ satisfying (5.2.1), (5.2.2), and (5.2.3).*

DEFINITION 5.2.2. Let $\varphi : Q \to \mathrm{Out}(G)$ be a homomorphism, $\widetilde{\varphi} : Q \to \mathrm{Aut}(G)$ be any lift, and $f : Q \times Q \to G$ a map satisfying (5.2.1), (5.2.2), and (5.2.3). Then the group $E = G \times Q$ with group operation
$$(a, \alpha) \cdot (b, \beta) = (a \cdot \widetilde{\varphi}(\alpha)(b) \cdot f(\alpha, \beta),\ \alpha\beta)$$
is denoted by $G \times_{(f, \widetilde{\varphi})} Q$.

Note that some abstract kernel $\varphi : Q \to \text{Out}(G)$ may not have any extension at all. When there is a lift $\widetilde{\varphi}$ which is a homomorphism (for example, when $G$ is Abelian), there exists at least one extension (by taking $f = 1$, the constant map), the *semidirect product* $G \times_{(1,\widetilde{\varphi})} Q$. As a set, this is $G \times Q$, and the group operation is

$$(a, \alpha) \cdot (b, \beta) = (a \cdot \widetilde{\varphi}(\alpha)(b), \alpha\beta).$$

This semidirect product is usually written by $G \rtimes_{\widetilde{\varphi}} Q$.

DEFINITION 5.2.3. Two extensions

$$1 \to G \to E \to Q \to 1 \quad \text{and} \quad 1 \to G \to E' \to Q \to 1$$

with the same abstract kernel are *congruent* if there is a homomorphism $\theta : E \to E'$ which restricts to the identity map on $G$, and induces the identity on $Q$. That is, the diagram

$$\begin{array}{ccccccccc}
1 & \to & G & \to & E & \to & Q & \to & 1 \\
& & \downarrow = & & \downarrow \theta & & \downarrow = & & \\
1 & \to & G & \to & E' & \to & Q & \to & 1
\end{array}$$

is commutative.

DEFINITION 5.2.4. Let $\varphi : Q \to \text{Out}(G)$ be a homomorphism, and $\widetilde{\varphi} : Q \to \text{Aut}(G)$ be a lift of $\varphi$. Then $\text{Opext}(Q, G, \widetilde{\varphi})$ denotes the set of all congruence classes of extensions of the group $G$ by $Q$ with operator $\widetilde{\varphi}$; see [**ML75**] for the notation. Therefore, an element $[f] \in \text{Opext}(Q, G, \widetilde{\varphi})$ is represented by an extension $1 \to G \to E \to Q \to 1$ with $E = G \times_{(f,\widetilde{\varphi})} Q$.

## 5.3. Pullback and pushout of short exact sequences

5.3.1 (Pullback). Let

$$1 \longrightarrow G \longrightarrow \widetilde{P} \longrightarrow P \longrightarrow 1$$

be exact, and let $\rho : Q \to P$ be a homomorphism. Then there is a group $\widetilde{Q}$ fitting into the commuting diagram with exact rows

$$\begin{array}{ccccccccc}
1 & \longrightarrow & G & \longrightarrow & \widetilde{Q} & \longrightarrow & Q & \longrightarrow & 1 \\
& & \downarrow = & & \downarrow & & \downarrow \rho & & \\
1 & \longrightarrow & G & \longrightarrow & \widetilde{P} & \longrightarrow & P & \longrightarrow & 1
\end{array} \qquad \textit{pullback by } \rho : Q \to P.$$

Such a group $\widetilde{Q}$ can be constructed as follows. Suppose $[f] \in \text{Opext}(P, G, \widetilde{\varphi})$ is represented by the extension $\widetilde{P} = G \times_{(f,\widetilde{\varphi})} P$, that is, $\widetilde{P} = G \times P$ as a set, and it has group operation

$$(a, \alpha) \cdot (b, \beta) = (a \cdot \widetilde{\varphi}(\alpha)(b) \cdot f(\alpha, \beta), \; \alpha\beta),$$

where $\widetilde{\varphi} : P \to \text{Aut}(G)$ and $f : P \times P \to G$. Now, $\widetilde{\varphi} : P \to \text{Aut}(G)$ gives rise to

$$\widetilde{\varphi}' : Q \xrightarrow{\rho} P \xrightarrow{\widetilde{\varphi}} \text{Aut}(G),$$

and $f : P \times P \to G$ gives rise to

$$f' : Q \times Q \xrightarrow{\rho \times \rho} P \times P \xrightarrow{f} G.$$

With these new interpretations of $\widetilde{\varphi}$ and $f$, the equalities (5.2.1), (5.2.2), and (5.2.3) still hold. Now we can define $\widetilde{Q}$ by, $\widetilde{Q} = G \times_{(f', \widetilde{\varphi}')} Q$, that is, $\widetilde{Q} = G \times Q$ as a set, and it has group operation

$$(a, \alpha) \cdot (b, \beta) = (a \cdot \widetilde{\varphi}'(\alpha)(b) \cdot f'(\alpha, \beta), \ \alpha\beta)$$

for $a, b \in G$ and $\alpha, \beta \in Q$.

EXERCISE 5.3.2. Under the same hypothesis as above, define $Q'$ to be the set of pairs $(u, \alpha) \in \widetilde{P} \times Q$ with $\rho(\alpha) = \eta(u)$, where $\eta : \widetilde{P} \to P$. Then using group multiplication in $\widetilde{P} \times Q$, show $Q'$ is a group congruent to $\widetilde{Q}$.

DEFINITION 5.3.3. Let $\Gamma$ be a subgroup of a Lie group $G$. We say $(G, \Gamma)$ has a Unique Automorphism Extension Property (UAEP) if every automorphism of $\Gamma$ extends to an automorphism of $G$ uniquely. For example, if $G$ is a connected, simply connected nilpotent Lie group, and $\Gamma$ is any lattice, then $(G, \Gamma)$ has UAEP; see Corollary 6.2.6.

5.3.4 (Pushout). Let $i : \Gamma \hookrightarrow G$ be an injective homomorphism. Suppose $(G, i(\Gamma))$ has UAEP, and

$$1 \longrightarrow \Gamma \longrightarrow \Pi \longrightarrow Q \longrightarrow 1$$

is exact. Then there is a group $E$ fitting into the commuting diagram

$$\begin{array}{ccccccccc}
1 & \longrightarrow & \Gamma & \longrightarrow & \Pi & \longrightarrow & Q & \longrightarrow & 1 \\
& & \downarrow i & & \downarrow & & \downarrow = & & \\
1 & \longrightarrow & G & \longrightarrow & E & \longrightarrow & Q & \longrightarrow & 1
\end{array} \qquad \textit{pushout by } i : \Gamma \to G.$$

Such a group $E$ can be constructed as follows. Suppose $[f] \in \mathrm{Opext}(Q, \Gamma, \widetilde{\varphi})$ is represented by the extension $\Pi = \Gamma \times_{(f, \widetilde{\varphi})} Q$. That is, $\Pi = \Gamma \times Q$ as a set, and has group operation

$$(a, \alpha) \cdot (b, \beta) = (a \cdot \widetilde{\varphi}(\alpha)(b) \cdot f(\alpha, \beta), \ \alpha\beta)$$

for $a, b \in \Gamma$ and $\alpha, \beta \in Q$. Recall that $\widetilde{\varphi}$ and $f$ satisfy the equalities (5.2.1), (5.2.2), and (5.2.3) in Section 5.2. Since $(G, \Gamma)$ has UAEP, $\widetilde{\varphi} : Q \to \mathrm{Aut}(\Gamma)$ can be viewed as

$$\widetilde{\varphi}' : Q \to \mathrm{Aut}(\Gamma) \to \mathrm{Aut}(G),$$

and $f : Q \times Q \to \Gamma$ as

$$f' : Q \times Q \to \Gamma \xrightarrow{i} G.$$

With these new interpretations of $\widetilde{\varphi}$ and $f$, the equalities (5.2.1), (5.2.2) and (5.2.3) still hold. Now we can define $E$ by $E = G \times_{(f', \widetilde{\varphi}')} Q$, that is, $E = G \times Q$ as a set, and it has group operation

$$(a, \alpha) \cdot (b, \beta) = (a \cdot \widetilde{\varphi}'(\alpha)(b) \cdot f'(\alpha, \beta), \ \alpha\beta)$$

for $a, b \in G$ and $\alpha, \beta \in Q$.

When $i$ is an obvious inclusion $i : \Gamma \hookrightarrow G$, we denote such an extension $E$ simply by $G\Pi$.

In the above construction, UAEP is not really necessary. We can do more generally.

PROPOSITION 5.3.5 (General pushout). *Let $i: G \hookrightarrow H$ be an injective homomorphism, and let*
$$1 \longrightarrow G \longrightarrow E \longrightarrow Q \longrightarrow 1$$
*be exact. Let $E = G \times_{(f, \widetilde{\varphi})} Q$. Suppose there exists a map $\widetilde{\psi}: Q \to \mathrm{Aut}(H)$ such that*

(1) $\widetilde{\psi}(\alpha) \circ i = i \circ \widetilde{\varphi}(\alpha)$, *for all $\alpha \in Q$,*
(2) $\widetilde{\psi}(\alpha) \circ \widetilde{\psi}(\beta) = \mu'(if(\alpha, \beta)) \circ \widetilde{\psi}(\alpha\beta)$ *in $\mathrm{Aut}(H)$, for all $\alpha, \beta \in Q$.*

*Then there is a* pushout, *a group $\mathcal{E}$ fitting into the commuting diagram*

(5.3.1)
$$\begin{array}{ccccccccc} 1 & \longrightarrow & G & \longrightarrow & E & \longrightarrow & Q & \longrightarrow & 1 \\ & & \downarrow i & & \downarrow & & \downarrow = & & \\ 1 & \longrightarrow & H & \longrightarrow & \mathcal{E} & \longrightarrow & Q & \longrightarrow & 1. \end{array}$$

PROOF. The homomorphism $\mu'$ in the second condition is conjugation in the group $H$. Let $h = i \circ f : Q \times Q \to G \to H$. We examine the pair $(h, \widetilde{\psi})$:
$$\widetilde{\psi}: Q \to \mathrm{Aut}(H), \qquad h: Q \times Q \to H.$$
The second condition is the requirement (5.2.1). We have $f(\alpha, \beta) \cdot f(\alpha\beta, \gamma) = \widetilde{\varphi}(\alpha)(f(\beta, \gamma)) \cdot f(\alpha, \beta\gamma)$. Applying $i$ to both sides and using $h = i \circ f$ and the first condition, we get $h(\alpha, \beta) \cdot h(\alpha\beta, \gamma) = \widetilde{\psi}(\alpha)(h(\beta, \gamma)) \cdot h(\alpha, \beta\gamma)$, which is the requirement (5.2.3). Finally, condition (5.2.2) is trivial. Consequently, we can form $\mathcal{E}$ by $\mathcal{E} = H \times_{(h, \widetilde{\psi})} Q$.

We claim that there is a homomorphism $\theta: E \to \mathcal{E}$ making the diagram commutative. Obviously, we define $\theta$ by
$$\theta(a, \alpha) = (i(a), \alpha).$$
The following two equalities show that $\theta$ is indeed a homomorphism:
$$\begin{aligned} \theta(a, \alpha) \cdot \theta(b, \beta) &= (i(a), \alpha) \cdot (i(b), \beta) \\ &= (i(a) \cdot \widetilde{\psi}(\alpha)(i(b)) \cdot h(\alpha, \beta), \alpha\beta) \\ &= (i(a) \cdot i(\widetilde{\varphi}(\alpha)(b)), \cdot i(f(\alpha, \beta)), \alpha\beta) \\ &= (i(a \cdot \widetilde{\varphi}(\alpha)(b) \cdot f(\alpha, \beta)), \alpha\beta), \\ \theta((a, \alpha) \cdot (b, \beta)) &= \theta(a \cdot \widetilde{\varphi}(\alpha)(b) \cdot f(\alpha, \beta), \alpha\beta) \\ &= (i(a \cdot \widetilde{\varphi}(\alpha)(b) \cdot f(\alpha, \beta)), \alpha\beta). \end{aligned}$$

Note that the second condition ensures that the composite $Q \xrightarrow{\widetilde{\psi}} \mathrm{Aut}(H) \to \mathrm{Out}(H)$ is a homomorphism, which is the abstract kernel for our new extension $\mathcal{E}$. $\square$

COROLLARY 5.3.6 (General pushout with centralizer). *Let $i: G \hookrightarrow H$ be an injective homomorphism, and let*
$$1 \longrightarrow G \longrightarrow E \longrightarrow Q \longrightarrow 1$$
*be exact. Let $E = G \times_{(f, \widetilde{\varphi})} Q$. Let $C = C_H(i(G))$ be the centralizer of $i(G)$ in $H$. Suppose there exists a map $\widetilde{\psi}: Q \to \mathrm{Aut}(H)$ such that*

(1) $\widetilde{\psi}(\alpha) \circ i = i \circ \widetilde{\varphi}(\alpha)$, *for all $\alpha \in Q$,*
(2) *$H$ is generated by the image $i(G)$ and $C$,*
(3) $\widetilde{\psi}: Q \to \mathrm{Aut}(H)$ *restricts to a homomorphism $\widetilde{\psi}_C: Q \to \mathrm{Aut}(C)$.*

*Then there is a* pushout, *a group $\mathcal{E}$ fitting into the commuting diagram*

(5.3.2)
$$\begin{array}{ccccccccc} 1 & \longrightarrow & G & \longrightarrow & E & \longrightarrow & Q & \longrightarrow & 1 \\ & & \downarrow i & & \downarrow & & \downarrow = & & \\ 1 & \longrightarrow & H & \longrightarrow & \mathcal{E} & \longrightarrow & Q & \longrightarrow & 1. \end{array}$$

PROOF. Since $\widetilde{\varphi}(\alpha) \circ \widetilde{\varphi}(\beta) = \mu(f(\alpha,\beta)) \circ \widetilde{\varphi}(\alpha\beta)$ in $\mathrm{Aut}(G)$, $\widetilde{\psi}(\alpha) \circ \widetilde{\psi}(\beta) = \mu'(if(\alpha,\beta)) \circ \widetilde{\psi}(\alpha\beta)$ in $\mathrm{Aut}(i(G))$ already. On $\mathrm{Aut}(C)$, this equality holds also by the third condition because $\mu'(if(\alpha,\beta))$ is trivial in $\mathrm{Aut}(C)$. Since $H$ is generated by $i(G)$ and $C$, this equality holds on whole $\mathrm{Aut}(H)$. Now apply Proposition 5.3.5. $\square$

5.3.7 (Converse of general pushout). The *general pushout* construction is very general. In fact, any commutative diagram of exact sequences as in (5.3.1) with $i$ injective comes from a general pushout construction. Consider a commutative diagram

$$\begin{array}{ccccccccc} 1 & \longrightarrow & G & \longrightarrow & E & \longrightarrow & Q & \longrightarrow & 1 \\ & & \downarrow i & & \downarrow \theta & & \downarrow = & & \\ 1 & \longrightarrow & H & \longrightarrow & \mathcal{E} & \longrightarrow & Q & \longrightarrow & 1 \end{array}$$

where $E = G \times_{(f,\widetilde{\varphi})} Q$. Recall that

$$f(\alpha,\beta) = (1,\alpha)(1,\beta)(1,\alpha\beta)^{-1}, \qquad \widetilde{\varphi}(\alpha) = \mu(1,\alpha)$$

for all $\alpha, \beta \in Q$. Define

$$h(\alpha,\beta) = \theta(1,\alpha)\theta(1,\beta)\theta(1,\alpha\beta)^{-1}, \qquad \widetilde{\psi}(\alpha) = \mu'(\theta(1,\alpha))$$

for all $\alpha, \beta \in Q$. Then clearly $\mathcal{E} = H \times_{(h,\widetilde{\psi})} Q$. Furthermore,

$$\begin{aligned}(i \circ \widetilde{\varphi}(\alpha))(a) &= i(\mu(1,\alpha)(a,1)) \\ &= \theta(1,\alpha)(i(a),1)\theta(1,\alpha)^{-1} \\ &= \mu'(\theta(1,\alpha))(i(a)) \\ &= (\widetilde{\psi}(\alpha) \circ i)(a)\end{aligned}$$

so that $i \circ \widetilde{\varphi}(\alpha) = \widetilde{\psi}(\alpha) \circ i$. Also, by applying $\mu' \circ \theta$ to the equality $(1,\alpha)(1,\beta) = (1,\alpha\beta)f(\alpha,\beta)$, we get

$$\widetilde{\psi}(\alpha) \circ \widetilde{\psi}(\beta) = \mu'(if(\alpha,\beta)) \circ \widetilde{\psi}(\alpha\beta) \quad \text{in } \mathrm{Aut}(H)$$

for all $\alpha, \beta \in Q$ (note that $\theta = i$ on $G$). Thus we get both equalities in Subsection 5.3.5.

## 5.4. Extensions with Abelian kernel $A$ and $H^2(Q;A)$

5.4.1. Consider the case where $G = A$ is Abelian in the previous sections. We try to classify all group extensions of an Abelian group $A$ by $Q$ with abstract kernel $\varphi : Q \to \mathrm{Out}(A)$. Since $A$ is Abelian, $\mathrm{Aut}(A) = \mathrm{Out}(A)$, and there is no need of having a lift $\widetilde{\varphi}$. Moreover (5.2.1) holds automatically. A map $f : Q \times Q \to A$ satisfying the two equalities (5.2.2) and (5.2.3) is called a *factor set* (or *2-cocycle*) for the abstract kernel $\varphi$. The set of all factor sets is denoted by $Z^2_\varphi(Q;A)$ or simply, by $Z^2(Q;A)$. For $f_1, f_2 \in Z^2_\varphi(Q;A)$, we define

$$(f_1 + f_2)(\alpha,\beta) = f_1(\alpha,\beta) + f_2(\alpha,\beta),$$

which makes $Z^2_\varphi(Q;A)$ into an Abelian group.

Let $\lambda : Q \to A$ be any map. Define $\delta\lambda : Q \times Q \to A$ by
(5.4.1) $$\delta\lambda(\alpha, \beta) = \lambda(\alpha) + \varphi(\alpha)(\lambda(\beta)) - \lambda(\alpha\beta).$$
It turns out that such a $\delta\lambda$ is a 2-cocycle and is called a *2-coboundary*. The set of all 2-coboundaries is denoted by $B^2_\varphi(Q; A)$. Clearly, $B^2_\varphi(Q; A)$ is a subgroup of $Z^2_\varphi(Q; A)$. Let $f_1, f_2 : Q \times Q \to A$ be 2-cocycles. We say $f_1$ is *cohomologous* to $f_2$ if $f_1 - f_2 \in B^2_\varphi(Q; A)$; that is, if there is a map $\lambda : Q \to A$ such that $f_2 - f_1 = \delta\lambda$.

We define the second cohomology group as the quotient group
$$H^2_\varphi(Q; A) = Z^2_\varphi(Q; A)/B^2_\varphi(Q; A).$$

5.4.2 (Extensions). Since the abstract kernel $\varphi : Q \to \operatorname{Out}(A)$ lifts to a homomorphism $\varphi : Q \to \operatorname{Out}(A) = \operatorname{Aut}(A)$, there is always an extension; namely, the semidirect product $A \rtimes_\varphi Q = A \times_{(1,\varphi)} Q$. Then every 2-cocycle $f \in Z^2_\varphi(Q; A)$ gives rise to an extension of $A$ by $Q$, which can be denoted by $A \times_{(f,\varphi)} Q$.

PROPOSITION 5.4.3. *Two cocycles $f_1$ and $f_2$ yield congruent extensions $A \times_{(f_1,\varphi)} Q$ and $A \times_{(f_2,\varphi)} Q$ if and only if $f_1$ and $f_2$ are cohomologous. Consequently, we have*
$$\operatorname{Opext}(A, Q, \varphi) \approx H^2_\varphi(Q; A)$$
*as sets, and they classify the extensions of $A$ by $Q$ with abstract kernel $\varphi$, up to congruence.*

PROOF. Suppose two cocycles $f_1$ and $f_2$ yield congruent extensions $A \times_{(f_1,\varphi)} Q$ and $A \times_{(f_2,\varphi)} Q$. Let $\theta$ be a homomorphism fitting into the commutative diagram

$$\begin{array}{ccccccccc} 1 & \to & A & \to & A \times_{(f_1,\varphi)} Q & \to & Q & \to & 1 \\ & & \downarrow = & & \downarrow \theta & & \downarrow = & & \\ 1 & \to & A & \to & A \times_{(f_2,\varphi)} Q & \to & Q & \to & 1. \end{array}$$

Then $\theta$ is of the form
$$\theta(1, \alpha) = (-\lambda(\alpha), \alpha)$$
for some $\lambda : Q \to A$. We claim that $f_2 - f_1 = \delta\lambda$. Since $\theta$ is a homomorphism, the two left-hand sides of the following
$$\begin{aligned}
\theta(0,\alpha)\theta(0,\beta) &= (-\lambda(\alpha), \alpha)(-\lambda(\beta), \beta) \\
&= (-\lambda(\alpha) - \varphi(\alpha)(\lambda(\beta)) + f_2(\alpha, \beta), \alpha\beta), \\
\theta((0,\alpha)(0,\beta)) &= \theta(f_1(\alpha, \beta), \alpha\beta) \\
&= \theta((f_1(\alpha,\beta), 1)(0, \alpha\beta)) \\
&= (f_1(\alpha,\beta), 1) \cdot \theta(0, \alpha\beta)) \\
&= (f_1(\alpha,\beta), 1)(-\lambda(\alpha\beta), \alpha\beta) \\
&= (f_1(\alpha,\beta) - \lambda(\alpha\beta), \alpha\beta)
\end{aligned}$$
are equal. This shows that $f_2 - f_1 = \delta\lambda$.

Conversely, suppose $f_2 - f_1 = \delta\lambda$. We simply define $\theta : A \times_{(f_1,\varphi)} Q \to A \times_{(f_2,\varphi)} Q$ by
$$\theta(a, \alpha) = (a - \lambda(\alpha), \alpha).$$
Then $\theta$ is a homomorphism fitting into the commutative diagram above, showing the two extensions $A \times_{(f_1,\varphi)} Q$ and $A \times_{(f_2,\varphi)} Q$ are congruent. $\square$

It is customary to identify the zero element of $H^2_\varphi(Q; A)$ with the extension class of the semidirect product $A \rtimes_\varphi Q$. Then addition of the 2-cocycles in $Z^2_\varphi(Q; A)$ induces the group operations in $H^2_\varphi(Q; A)$.

## 5.5. Central extensions

In this section we shall prove that a torsion-free central extension of a group by a finite group is Abelian; see Corollary 5.5.3 below. This section is from [**GLÖ85**, Section 1].

LEMMA 5.5.1. *Let $C$ be an Abelian group, and let $0 \to C \to E \to G \to 1$ be a central extension with $G$ a finite group of order $n$. Then $\varphi : E \to C$, defined by $\varphi(x) = x^n$, is a homomorphism.*

PROOF. Let $f : G \times G \to C$ be a 2-cocycle giving rise to the extension $E$. In other words, for all $\alpha, \beta, \gamma \in G$,
$$(\delta f)(\alpha, \beta, \gamma) = f(\beta\gamma) - f(\alpha\beta, \gamma) + f(\alpha, \beta\gamma) - f(\alpha, \beta) = 0.$$
$E$ is $C \times G$ with the group operation
$$(a, \alpha)(b, \beta) = (a + b + f(\alpha, \beta), \alpha\beta)$$
for all $a, b \in C$ and all $\alpha, \beta \in G$. We may choose $f$ so that $f(\alpha, 1) = f(1, \alpha) = 0$. Define $g : G \to C$ by
$$g(\alpha) = \sum_{t \in G} f(\alpha, t).$$
It is easily checked that, for all $\alpha, \beta \in G$,
$$(\delta g)(\alpha, \beta) = g(\beta) - g(\alpha\beta) + g(\alpha) = nf(\alpha, \beta).$$
If we define $\varphi : E \to C$ by
$$\varphi(a, \alpha) = na + g(\alpha),$$
$\varphi$ is seen to be a homomorphism. To show that $\varphi(a, \alpha) = (a, \alpha)^n$, we see that, by computation,
$$(a, \alpha)^n = \left( na + \sum_{i=1}^{n} f(\alpha, \alpha^i), 1 \right).$$
Hence we will be done when we demonstrate $g(\alpha) = \sum_{i=1}^{n} f(\alpha, \alpha^i)$.

Putting $\alpha^i = \beta$ in $\delta f \equiv 0$ gives, for all $\gamma \in G$,
$$f(\alpha, \alpha^i) - f(\alpha, \alpha^i \gamma) = f(\alpha^i, \gamma) - f(a^{i+1}+, \gamma).$$
Summing over $1 \leq i \leq m = $ order of $\alpha$, the right-hand side disappears, i.e., for all $\gamma \in G$,
$$\sum_{i=1}^{m} f(\alpha, \alpha^i) = \sum_{i=1}^{m} f(\alpha, \alpha^i \gamma).$$
When $\gamma$ ranges over coset representatives for the subgroup generated by $\alpha$ in $G$, any $t \in G$ can be expressed uniquely as $t = a^i \gamma (1 \leq i \leq m)$. Therefore
$$\sum_{t \in G} f(\alpha, t) = \sum^{\gamma} \sum_{i=1}^{m} f(\alpha, \alpha^i \gamma) = \sum_{i=1}^{n} f(\alpha, \alpha^i). \quad \square$$

The next corollary was shown, by a geometric argument, in [**LR81**, Fact 2] when $C$ is finitely generated and torsion free (i.e., $C \cong \mathbb{Z}^k$).

COROLLARY 5.5.2. *Let $C$ be an Abelian group, and let $0 \to C \to E \to G \to 1$ be a central extension with $G$ finite. Then $t(E)$, the elements of finite order in $E$, form a fully invariant subgroup. Moreover, $t(E)$ is finite when $t(C)$ is finite.*

PROOF. Any endomorphism of $E$ maps $t(E)$ into itself, so $t(E)$ is fully invariant. Since $t(E) = \varphi^{-1}(t(C))$ ($\varphi$ is in Lemma 5.5.1) and $t(C)$ is a subgroup, $g(E)$ is a subgroup. The quotient $t(E)/t(C)$ is finite. Therefore, $t(E)$ is finite when $t(C)$ is. $\square$

The following corollary will be crucial in our future arguments. It is a trivial consequence of Lemma 5.5.1. It was already known [**LR81**] when $C \cong \mathbb{Z}^k$.

COROLLARY 5.5.3. *Let $0 \to C \to E \to G \to 1$ be a central extension with $E$ torsion free and $G$ finite. Then $E$ and hence $G$ is Abelian.*

PROOF. When $E$ is torsion free, $\varphi(x) = x^n$ of Lemma 5.5.1 is injective. Hence $E$ is isomorphic to a subgroup of the Abelian group $C$. $\square$

REMARK 5.5.4. If $0 \to C \to E \to G \to 1$ is a central extension, the Lyndon spectral sequence (with $C$ coefficients) gives the exact sequence [**ML75**, p.354, (10.6)]

$$0 \longrightarrow \mathrm{Hom}(G,C) \longrightarrow \mathrm{Hom}(E,C) \longrightarrow \mathrm{Hom}(C,C) \xrightarrow{\delta} H^2(G;C).$$

The boundary map $\delta$ takes the identity map $1_C \in \mathrm{Hom}(C,C)$ to $[E]$, the class representing the extension $E$. If $m[E] = 0$, then there is a $\varphi \in \mathrm{Hom}(E,C)$ such that $\varphi|_C = m$. When $G$ is torsion and $C$ is torsion free, $\mathrm{Hom}(G,C) = 0$; hence $\varphi$ is unique. It is not true in general that $\varphi(x) = x^m$ for all $x \in E$. (If $E \cong C \times G$, we can take $m = 1$ and clearly that does not work. Even when $x^m \in C$ for all $x \in E$, $x \to x^m$ may not be a homomorphism. Take $E = \{\pm 1, \pm i, \pm j, \pm k\}$, the quaternions, $C = \{\pm 1\}$, and $m = 2$.) When $G$ is finite of order $n$, $n[E] = 0$. If $C$ is torsion free, $\varphi(x)^n = \varphi(x^n) = (x^n)^n$ ($x^n$ and $\varphi(x)$ both lie in the torsion-free Abelian group $C$) implies $\varphi(x) = x^n$. This gives an alternate proof of Corollary 5.5.3.

PROPOSITION 5.5.5. *If all the finite subgroups of a compact Lie group $G$ are Abelian, then $G$ is Abelian.*

PROOF. First we show that $G_0$, the connected component of $G$, is a torus. Let $T^r$ be a maximal torus (of rank $r$), $N(T^r)$ its normalizer in $G_0$, and $W$ the Weyl group with $n = |W|$. For any natural number $k$, $(\mathbb{Z}_{kn})^r = \{t \in T^r : t^{kn} = 1\}$ is a characteristic subgroup of $T^r$. Hence the standard action of $W$ on $T^r$ restricts to an action on $(\mathbb{Z}_{kn})^r$, giving the exact sequence of $W$-modules

$$0 \longrightarrow (\mathbb{Z}_{kn})^r \xrightarrow{i} T^r \xrightarrow{kn} T^r \longrightarrow 0.$$

In cohomology, we have the exact sequence

$$H^2(W;(\mathbb{Z}_{kn})^r) \xrightarrow{i_*} H^2(W;T^r) \xrightarrow{kn} H^2(W;T^r).$$

Now $H^2(W;-)$ is annihilated by $n = |W|$, so $i_*$ is onto. Let $i_*[W_k] = [N(T^r)]$, where $[W_k]$ is in $H^2(W;(\mathbb{Z}_{kn})^r)$. Then

$$\begin{array}{ccccccccc}
0 & \longrightarrow & (\mathbb{Z}_{kn})^r & \longrightarrow & W_k & \longrightarrow & W & \longrightarrow & 1 \\
& & \downarrow & & \downarrow & & \downarrow = & & \\
1 & \longrightarrow & T^r & \longrightarrow & N(T^r) & \longrightarrow & W & \longrightarrow & 1
\end{array}$$

is a commutative diagram. Since $W_k$ is isomorphic to a finite subgroup of $G$, it is Abelian. Then $W$ acts trivially on $(\mathbb{Z}_{kn})^r$ (the action is induced by conjugation in

$W_k$). But $\bigcup_k (\mathbb{Z}_{kn})^r$ is dense in $T^r$, hence $W$ acts trivially on $T^r$. The action of the Weyl group is always effective, so $W$ must be trivial. Therefore $G_0 = T^r$; see, e.g., [**Bor72**, Proposition 20.2].

Next consider the exact sequence
$$1 \to G_0 = T^r \to G \to F \to 1,$$
where $F$ is finite, say of order $m$. Choose $[F_k]$ in $H^2(F; (\mathbb{Z}_{km})^r)$ as above, i.e., $[F_k]$ is the preimage of $[G] \in H^2(F; T^r)$. Again $F_k$ is Abelian since it is isomorphic to a finite subgroup of $G$, and $G$ is Abelian because $\bigcup_k F_k$ is dense in $G$. □

Note that a compact Abelian Lie group $G$ is isomorphic to a direct product of a torus (the connected component of the identity) with a finite Abelian group. That is, the exact sequence $1 \to T \to G \to F \to 1$ splits. To get the isomorphism $F$ into $G$, one modifies a preimage of a generator $\alpha$ of $F$ using the fact that $T$ is a divisible group.

## 5.6. Extensions with non-Abelian kernel $G$ and $H^2(Q; \mathcal{Z}(G))$

For a homomorphism $\varphi : Q \to \mathrm{Out}(G)$, take any lift $\widetilde{\varphi} : Q \to \mathrm{Aut}(G)$. This, in turn, induces a unique *homomorphism* $\widetilde{\varphi} : Q \to \mathrm{Aut}(\mathcal{Z}(G))$ by restriction (regardless which lift of $\varphi$ is chosen, and even if $\widetilde{\varphi}$ was not a homomorphism). We shall define an action of $H^2_{\widetilde{\varphi}}(Q; \mathcal{Z}(G))$ on the set $\mathrm{Opext}(G, Q; \widetilde{\varphi})$ which will turn out to be simply transitive (i.e., free and transitive).

Let $g \in Z^2_{\widetilde{\varphi}}(Q; \mathcal{Z}(G))$; that is, $g : Q \times Q \to \mathcal{Z}(G)$ satisfying the equalities (5.2.2) and (5.2.3) of Section 5.2. For any $[f_0] \in \mathrm{Opext}(G, Q; \widetilde{\varphi})$, define
$$(g \cdot f_0)(\alpha, \beta) = g(\alpha, \beta) \cdot f_0(\alpha, \beta).$$
Then, clearly, $g \cdot f_0$ satisfies the equalities (5.2.1), (5.2.2), and (5.2.3). Moreover, if two extensions $G \times_{(f_0, \widetilde{\varphi})} Q$ and $G \times_{(f_1, \widetilde{\varphi})} Q$ are congruent, so are $G \times_{(g \cdot f_0, \widetilde{\varphi})} Q$ and $G \times_{(g \cdot f_1, \widetilde{\varphi})} Q$. Therefore
$$(g, f_0) \mapsto g \cdot f_0$$
is an action of $Z^2(Q; \mathcal{Z}(G))$ on $\mathrm{Opext}(G, Q; \widetilde{\varphi})$.

Suppose $[g \cdot f] = [f]$ for some $[f] \in \mathrm{Opext}(G, Q; \widetilde{\varphi})$. Then one easily sees that $g \in B^2_{\widetilde{\varphi}}(Q; \mathcal{Z}(G))$. This gives us a free action of $H^2_{\widetilde{\varphi}}(Q; \mathcal{Z}(G))$ on $\mathrm{Opext}(G, Q; \widetilde{\varphi})$.

Now we show that this action is transitive. Suppose $f, f' \in \mathrm{Opext}(G, Q; \widetilde{\varphi})$. Let
$$g(\alpha, \beta) = f(\alpha, \beta) \cdot f'(\alpha, \beta)^{-1}$$
for all $\alpha, \beta \in Q$. Then $g(\alpha, \beta)$ lies in $\mathcal{Z}(G)$, and it is easy to check that
$$g : Q \times Q \to \mathcal{Z}(G)$$
satisfies the equalities (5.2.2) and (5.2.3). Thus, $g \in Z^2_{\widetilde{\varphi}}(Q; \mathcal{Z}(G))$. Consequently, we have shown the action of $H^2_{\widetilde{\varphi}}(Q; \mathcal{Z}(G))$ on $\mathrm{Opext}(G, Q; \widetilde{\varphi})$ is simply transitive. Thus,
$$\mathrm{Opext}(G, Q, \widetilde{\varphi}) \approx H^2_{\widetilde{\varphi}}(Q; \mathcal{Z}(G))$$
as sets, and they classify the extensions of $G$ by $Q$ with abstract kernel $\varphi$, up to congruence. In general, however, there may not exist any extension of $G$ by $Q$ with the given abstract kernel. Moreover, even when there exists one, there is no *a priori* special element, like the semidirect product $G \rtimes Q$, because the semidirect product can be formed when and only when the abstract kernel lifts to an homomorphism into $\mathrm{Aut}(G)$.

## 5.7. $H^1(Q;C)$ with a non-Abelian $C$

**5.7.1.** The first cohomology is useful in classifying homomorphisms from one group to another. Let $C$ be a group (not necessarily Abelian), and let $\psi : Q \to \mathrm{Aut}(C)$ be a *homomorphism*. We shall define the *first cohomology* set $H^1_\psi(Q;C)$. A map $\eta : Q \to C$ is called a *1-cocycle* or a *crossed homomorphism* if it satisfies

$$\eta(\alpha\beta) = \eta(\alpha) \cdot \psi(\alpha)(\beta)$$

for all $\alpha, \beta \in Q$. The set of all 1-cocycles is denoted by $Z^1_\psi(Q;C)$. Two 1-cocycles $\eta, \eta' : Q \to C$ are *cohomologous* if there exists $c \in C$ such that

$$\eta'(\alpha) = c \cdot \eta(\alpha) \cdot \psi(\alpha)(c^{-1})$$

for all $\alpha \in Q$. The set of cohomologous classes of $Z^1_\psi(Q;C)$ is denoted by $H^1_\psi(Q;C)$. When $C$ is Abelian, $H^1_\psi(Q;C)$ is the ordinary first cohomology group.

THEOREM 5.7.2. *Consider a commutative diagram*

$$\begin{array}{ccccccccc}
1 & \to & G & \to & E & \to & Q & \to & 1 \\
 & & \downarrow i & & \downarrow \theta_0 & & \downarrow = & & \\
1 & \to & H & \to & \mathcal{E} & \to & Q & \to & 1
\end{array}$$

*with injective $i$. Then all homomorphisms $E \to \mathcal{E}$ fitting into the diagram, up to conjugation by elements of $H$, are in one-to-one correspondence with $H^1(Q;C)$, where $C = C_H(i(G))$, the centralizer of $i(G)$ in $H$.*

PROOF. For $x \in E$, $\widetilde{\psi}(x) = \mu'(\theta_0(x))$ leaves $i(G)$ invariant so it leaves $C$ invariant, where $\mu'$ means conjugation in $\mathcal{E}$. Therefore,

$$\widetilde{\psi} = \mu' \circ \theta_0 : E \to \mathcal{E} \to \mathrm{Aut}(C)$$

is a homomorphism. For $a \in G$ and $x \in E$,

$$\begin{aligned}
\widetilde{\psi}(xa) &= \mu'(\theta_0(x))\mu'(i(a)) \\
&= \mu'(\theta_0(x)), \quad \text{on } C \text{ since } i(a) \text{ commutes with } C \\
&= \widetilde{\psi}(x)
\end{aligned}$$

so that $\widetilde{\psi}$ factors through $Q$. Consequently, we have a homomorphism

$$\widetilde{\psi} : Q \to \mathrm{Aut}(C).$$

With this homomorphism $\widetilde{\psi}$, we shall talk about the first cohomology $H^1(Q;C)$.

Let $\theta : E \to \mathcal{E}$ be a homomorphism which fits into the diagram. For any $a \in G$, $x \in E$,

$$\theta_0(x)i(a)\theta_0(x)^{-1} = \theta_0(xax^{-1}) = i(xax^{-1}) = \theta(xax^{-1}) = \theta(x)i(a)\theta(x)^{-1}.$$

Therefore, the difference $\lambda(x) = \theta(x)\theta_0(x)^{-1}$ lies in $C = C_H(i(G))$. For any $a \in G$, $\lambda(xa) = \lambda(x)$. So $\lambda$ factors through $Q$ and we consider $\lambda$ as a map

$$\lambda : Q \to C.$$

It is easy to see, for $x, y \in E$, that

$$\begin{aligned}
\lambda(xy) &= \lambda(x) \cdot \theta_0(x)\lambda(y)\theta_0(x)^{-1} \\
&= \lambda(x) \cdot \widetilde{\psi}(x)(\lambda(y)).
\end{aligned}$$

Recall that $\widetilde{\psi}$ and $\lambda$ factors through $Q$. Thus we get $\lambda \in Z^1(Q;C)$. Conversely, let $\lambda : Q \to C$ lie in $Z^1(Q;C)$. Then, interpreting $\lambda$ as $\lambda : E \to Q \to C$, one can define

$\theta : E \to \mathcal{E}$ by $\theta(x) = \lambda(x)\theta_0(x)$. Then one easily sees that $\theta$ is a homomorphism fitting into the diagram.

Let $\lambda, \lambda' \in Z^1(Q; C)$ be two maps coming from two homomorphisms $\theta, \theta' : E \to \mathcal{E}$ fitting into the diagram. Suppose there is $c \in H$ such that
$$\theta'(x) = c \cdot \theta(x) \cdot c^{-1}$$
for all $x \in E$. Then $c$ must be in $C$. Moreover,
$$\begin{aligned}\lambda'(x) &= \theta'(x)\theta_0(x)^{-1}\\&= (c \cdot \theta(x) \cdot c^{-1}) \cdot \theta_0(x)^{-1}\\&= c \cdot (\lambda(x)\theta_0(x))c^{-1}\theta_0(x)^{-1}\\&= c \cdot \lambda(x) \cdot \theta_0(x)c^{-1}\theta_0(x)^{-1}\\&= c \cdot \lambda(x) \cdot \widetilde{\psi}(x)(c^{-1})\end{aligned}$$
so that $\lambda'$ is cohomologous to $\lambda$. Conversely, if $\lambda'$ is cohomologous to $\lambda$, it is easy to see that the corresponding $\theta'$ is $\theta$ followed by a conjugation. Therefore, the set of all homomorphisms $\theta$ fitting into the diagram up to conjugation by elements in $H$ is in one-to-one correspondence with $H^1(Q; C)$. □

COROLLARY 5.7.3. *Consider a commutative diagram*

$$\begin{array}{ccccccccc}1 & \longrightarrow & G & \longrightarrow & E & \longrightarrow & Q & \longrightarrow & 1 \\ & & \downarrow i & & \downarrow \theta_0 & & \downarrow j & & \\ 1 & \longrightarrow & H & \longrightarrow & \mathcal{E} & \longrightarrow & L & \longrightarrow & 1\end{array}$$

*with injective $i$. Then all homomorphisms $E \to \mathcal{E}$ fitting into the diagram, up to conjugation by elements of $H$, is in one-to-one correspondence with $H^1(Q; C)$, where $C = C_H(i(G))$, the centralizer of $i(G)$ in $H$.*

PROOF. First get the pullback $1 \longrightarrow H \longrightarrow \mathcal{E}_1 \longrightarrow Q \longrightarrow 1$ of $1 \longrightarrow H \longrightarrow \mathcal{E} \longrightarrow L \longrightarrow 1$ via $j : Q \to L$. Then the homomorphism $\theta_0$ is really a map $E \to \mathcal{E}_1$. Thus we have a commutative diagram

$$\begin{array}{ccccccccc}1 & \longrightarrow & G & \longrightarrow & E & \longrightarrow & Q & \longrightarrow & 1 \\ & & \downarrow i & & \downarrow \theta_0 & & \downarrow = & & \\ 1 & \longrightarrow & H & \longrightarrow & \mathcal{E}_1 & \longrightarrow & Q & \longrightarrow & 1,\end{array}$$

and we can apply the theorem. □

COROLLARY 5.7.4. *Consider a commutative diagram*

$$\begin{array}{ccccccccc}1 & \longrightarrow & G & \longrightarrow & E & \longrightarrow & Q & \longrightarrow & 1 \\ & & \parallel & & \downarrow \theta_0 & & \parallel & & \\ 1 & \longrightarrow & G & \longrightarrow & E & \longrightarrow & Q & \longrightarrow & 1.\end{array}$$

*Then all automorphisms $E \to E$ fitting into the diagram, up to conjugation by elements of $G$, is in one-to-one correspondence with $H^1(Q; \mathcal{Z}(G))$, where $\mathcal{Z}(G)$ is the center of $G$.*

REMARK 5.7.5. The first cohomology is used for counting certain types of algebraic and geometric constructions. This idea is exploited in many parts of this book. For an interesting application that is not explored here, we refer the reader to [**CR72d**, A.10], where the first cohomology is used to count the connected components of the fixed points of finite $p$-group actions on aspherical manifolds and spherical space forms.

CHAPTER 6

# Lie Groups

## 6.1. Introduction

There are many good places to find information about Lie groups. To find comprehensive texts on discrete subgroups of Lie groups is more difficult. A well-known text is the book by M.S. Raghunathan, *Discrete Subgroups of Lie Groups* (Ergebnisse der Mathematik und ihrer Grenzgebiete (vol. 68), Springer, 1972). Of course much has happened since then (1972) especially with regards to semisimple groups, such as Mostow-Prasad-Margulis Rigidity and the extensive work in algebraic groups. We will discuss a few general comments about Lie groups and Lie algebras.

6.1.1. Let $G$ be a Lie group. Let $G_0$ denote the *connected component* containing the identity. Then $G_0$ is an *open and closed normal* subgroup of $G$ and $G/G_0$ is countable (since $G$ is a manifold; smooth—actually real analytic—and we usually assume that it has a countable basis).

*Closed* subgroups of a Lie group are again Lie groups and, for a closed subgroup $H$, $G \to G/H$ is a principal fibering. For the following, see [**Hoc65**, p.186]:

For each Lie group $G$ such that $G/G_0$ is finite, there exists $K$ a compact subgroup of $G$ such that

(1) $G/K$ is diffeomorphic to $\mathbb{R}^n$ for some $n$.
(2) $K$ is *maximal* in the sense that if $K \subset K' \subset G$ with $K'$ compact, then $K = K'$.
(3) If $C$ is any compact subgroup of $G$, there exists $x \in G$ such that $xCx^{-1} \subset K$.

Note that (3) implies (2). $K$ is called a *maximal compact subgroup* of $G$.

6.1.2. Let $G$ be a connected Lie group. For subgroups $H$ and $K$, the *commutator* $[H,K]$ is the subgroup generated by $xyx^{-1}y^{-1}$, for $x \in H$ and $y \in K$. Define inductively $C_k(G)$ and $D_k(G)$ as the groups

$$C_k(G) = [G, C_{k-1}(G)], \quad D_k(G) = [D_{k-1}(G), D_{k-1}(G)]$$

with $C_0(G) = D_0(G) = G$. A group $G$ is called *nilpotent* if $C_k(G) = 1$ for some finite $k$. The smallest such $k$ is called the *nilpotency* of $G$, and $G$ is called $k$-step nilpotent. A group $G$ is called *solvable* if $D_k(G) = 1$ for some finite $k$.

A finite system of normal subgroups

$$G = G_1 \supset G_2 \supset G_3 \supset \cdots \supset G_s \supset 1$$

of $G$ is called a *decreasing central series* of $G$ if the commutator subgroups satisfy the relations

$$[G, G_i] \subset G_{i+1}$$

for each $i = 1, 2, \ldots, s$. A group $G$ is nilpotent if it contains a decreasing central series terminating with the identity.

LEMMA 6.1.3 ([**Gar66**, 3.1]). *Let $G$ be a Lie group, and let $\Gamma$ be a cocompact discrete subgroup of $G$. For any closed normal subgroup $H$ of $G$, the following are equivalent:*
  (1) *$\Gamma \cap H$ is uniform in $H$ (that is, $H/(\Gamma \cap H)$ is compact).*
  (2) *$\Gamma/(\Gamma \cap H)$ is discrete in $G/H$.*
  (3) *$\Gamma/(\Gamma \cap H)$ is uniform in $G/H$.*

DEFINITION 6.1.4. A Lie algebra is *simple* if it has no proper ideals. A Lie algebra is *semisimple* if it is a direct sum of simple ideals. A Lie group $G$ is *simple* (respectively, *semisimple*) if its Lie algebra is simple (respectively, semisimple). A *radical* of a Lie algebra (respectively, Lie group) is its maximal solvable ideal (respectively, maximal connected normal solvable subgroup). A solvable, connected, simply connected Lie group is diffeomorphic to $\mathbb{R}^n$ for some $n$.

Let $G$ be a connected Lie group, and let $S$ be its radical. If $G$ is simply connected, then $G = S \rtimes H$ where $H$ is semisimple, and both $S$ and $H$ are simply connected. If $G$ is not simply connected, then $G = S \cdot H$, where $S$ is the radical and $H$ is connected, semisimple, and $H \cap S$ is discrete. This is called the *Levi decomposition*. Each connected solvable Lie group contains a unique maximal normal connected closed nilpotent subgroup. It is called the (connected) *nilradical*.

6.1.5. Matrix groups supply many nice types of Lie groups. Many of them have close relations with geometry. Let

$$J = \begin{bmatrix} -I & 0 \\ 0 & I \end{bmatrix} \in \mathrm{GL}(p, \mathbb{R}) \times \mathrm{GL}(q, \mathbb{R}).$$

$\mathrm{O}(p, q)$ is defined as the subgroup of $\mathrm{GL}(p+q, \mathbb{R})$,

$$\mathrm{O}(p, q) = \{ A \in \mathrm{GL}(p+q, \mathbb{R}) : AJA^t = J \}.$$

The subgroup

$$\mathrm{SO}(p, q) = \{ A \in \mathrm{O}(p, q) : \det(A) = 1 \}$$

has index 2 in $\mathrm{O}(p, q)$.

If $p, q > 0$, $\mathrm{SO}(p, q)$ has two connected components. The connected component containing the identity matrix is denoted by $\mathrm{SO}_0(p, q)$. Clearly, $\mathrm{O}(0, q) \cong \mathrm{O}(q)$ and $\mathrm{O}(p, 0) \cong \mathrm{O}(p)$.

EXAMPLE 6.1.6. The group of $n \times n$ upper triangular matrices of nonzero determinant is solvable. The group of $n \times n$ upper triangular unipotent matrices (all diagonal entries are 1) is nilpotent; $\mathrm{SO}(n)$ ($n \neq 1, 2, 4$), $\mathrm{PSL}(2, \mathbb{R})$, $\mathrm{O}(1, n)$ ($n \geq 2$) are simple.

6.1.7. Every locally compact topological group has a left (respectively, right) invariant measure $m$, called a *Haar measure*. If $\Gamma$ is a discrete group and $p : G \to \Gamma \backslash G$ is the projection, then $m$ induces on the space of right cosets a measure $\bar{m}$. $\Gamma$ is called a *lattice* if this measure is *finite* (i.e., $\bar{m}(\Gamma \backslash G) < \infty$, or in other words *finite volume*). It follows if $\bar{m}(\Gamma \backslash G) < \infty$, that $m$ is 2-sided and invariant under action of $G$.

If $\Gamma\backslash G$ is compact, then $\Gamma$ is called a *uniform lattice* or a *cocompact lattice*. In general, a subgroup $H$ of $G$ is *uniform* in $G$ if $G/\bar{H}$ is compact. For a simply connected solvable $G$, a lattice is always uniform. For $G = \mathbb{R}^n$, in order that a subgroup $H$ be uniform in $G$, it is necessary and sufficient that $H$ contains $n$ linearly independent elements ($\mathbb{R}^n$ as a vector space).

6.1.8. We shall use an important fact proved by A. Selberg [**Sel60**, Lemma 8] quite often.

**Selberg's Lemma.** *Let $H$ be a finitely generated group of $(n \times n)$ matrices (they need not be real, nor does $H$ need to be discrete). Then $H$ possesses a normal subgroup of finite index, which contains no element of finite order other than the identity element.*

## 6.2. Nilpotent Lie groups

6.2.1. A compact manifold which admits a transitive action of a connected, simply connected nilpotent Lie group is called a *nilmanifold*. Every nilmanifold is diffeomorphic to the space of cosets of a connected simply connected nilpotent Lie group by a discrete subgroup. Nilmanifolds are uniquely determined by their fundamental groups. Moreover, an abstract group $\Gamma$ is the fundamental group of some nilmanifold if and only if $\Gamma$ is a torsion-free, finitely generated nilpotent group.

Not every nilpotent Lie group $G$ can act transitively on some compact manifold; it is necessary and sufficient that $G$ contains a cocompact discrete subgroup, which is equivalent to saying that the Lie algebra of the Lie group $G$ has rational *structural constants* with respect to an appropriate basis [**Rag72**, Theorem 2.12]. (This means that the Lie algebra has a linear basis $\{X_1, \ldots, X_n\}$ such that

$$[X_i, X_j] = \sum_k c_k X_k$$

with all $c_k$ rational, for every $i, j$ and $k$.)

6.2.2. Each nilpotent Lie group has a nontrivial center. If $G$ is a simply connected nilpotent Lie group, it is diffeomorphic to $\mathbb{R}^n$ for some $n$. Thus $G$ contains no elements of finite order, and its center $\mathcal{Z}(G)$ is a connected closed nontrivial subgroup. Thus, for a simply connected nilpotent Lie group $G$,

$$1 \to \mathcal{Z}(G) \to G \to G/\mathcal{Z}(G) \to 1$$

is *exact* and $G/\mathcal{Z}(G)$ is simply connected and nilpotent again.

LEMMA 6.2.3 ([**Mal51**]). *Suppose the Lie group $G$ has a system of 1-parameter subgroups $x_1(t), x_2(t), \ldots, x_r(t)$ satisfying the following properties:*

(1) *Each element of $G$ can be represented in the form*

$$x_1(t_1)x_2(t_2)\cdots x_r(t_r);$$

(2) *The collection of elements of the form $x_i(t_i)x_{i+1}(t_{i+1})\cdots x_r(t_r)$ forms a normal subgroup $G_i$ of $G$; and*

(3) *The quotient groups $G_i/G_{i+1}$ are 1-parameter vector groups.*

*Then if a subgroup $H$ contains the elements $x_1(1), x_2(1), \ldots, x_r(1)$, then $H$ is uniform in $G$; i.e., $G/\bar{H}$ is compact.*

The idea is to construct an analytic diffeomorphism $G \longrightarrow \mathbb{R}^r$ by

$$(x_1(t_1), x_2(t_2), \ldots, x_r(t_r)) \longleftrightarrow (t_1, t_2, \ldots, t_r).$$

THEOREM 6.2.4. *Let $N$ and $V$ be two simply connected nilpotent groups, and let $H$ be a uniform subgroup of $N$. Then any continuous homomorphism $\rho : H \to V$ can be extended uniquely to a continuous homomorphism $\widetilde{\rho} : N \to V$.*

PROOF. Let $H' \subset N \times V$ be the subgroup
$$H' = \{(x, \rho(x)) | x \in H\}.$$
Let $U$ be the Zariski closure of $H'$ in $N \times V$. Let
$$N \xleftarrow{\pi_1} N \times V \xrightarrow{\pi_2} V$$
be projections. Since $\pi_1(U)$ is connected and contains $H$, $\pi_1(U) = N$. Thus $\pi_1|_U$ is onto. One can show that $\dim(U) = \dim(N)$. Since $U$ and $N$ are nilpotent simply connected groups, we conclude that $\pi_1|_U$ is one-to-one.

The graph of any extension of $\rho$ in $N \times V$ is a connected closed subgroup containing $H'$, and of dimension equal to $N$, hence it must be equal to the group $U$. On the other hand, since $\pi_1|_U$ is one-to-one, $U$ is the graph of a map $\widetilde{\rho} = \pi_2 \circ (\pi_1|_U)^{-1}$. Clearly, $\widetilde{\rho}$ is the unique extension of $\rho$. □

COROLLARY 6.2.5 (Rigidity). *If $G_1$ and $G_2$ are simply connected nilpotent Lie groups, $\Gamma_1, \Gamma_2$ closed discrete subgroups such that $G_i/\Gamma_i$ is compact, and $\varphi : \Gamma_1 \to \Gamma_2$ is an isomorphism, then there exists a unique isomorphism $\widetilde{\varphi} : G_1 \to G_2$ such that $\widetilde{\varphi}|_{\Gamma_1} = \varphi$.*

COROLLARY 6.2.6. *Let $N$ be a simply connected nilpotent Lie group and $H$ a uniform subgroup. Then any automorphism of $H$ extends to an automorphism of $N$ uniquely. That is, $(N, H)$ has UAEP (see Definition 5.3.3).*

THEOREM 6.2.7. *Let $\mathfrak{N}$ be a nilpotent Lie algebra with a basis with respect to which the structural constants are rational. Let $\mathfrak{N}_0$ be the vector space over $\mathbb{Q}$ spanned by this basis. If $\mathcal{L}$ is any lattice of maximal rank in $\mathfrak{N}$ contained in $\mathfrak{N}_0$ and $\exp : \mathfrak{N} \to N$ is the exponential map, the group generated by $\exp(\mathcal{L})$ is a lattice in $N$. Conversely, if $\Gamma$ is a lattice in $N$, then the $\mathbb{Z}$-span of $\exp^{-1}(\Gamma)$ is a lattice (of maximal rank) $\mathcal{L}$ in the vector space $\mathfrak{N}$ such that the structural constants of $\mathfrak{N}$ with respect to any basis contained in $\mathcal{L}$ belong to $\mathbb{Q}$.*

6.2.8. Consider the subgroup of $\mathrm{GL}(3, \mathbb{R})$
$$N = \left\{ \begin{bmatrix} 1 & x & z \\ 0 & 1 & y \\ 0 & 0 & 1 \end{bmatrix} : x, y, z \in \mathbb{R} \right\}.$$

The group operations are by matrix multiplication. For simplicity, we denote this matrix by $(x, y, z)$. Then
$$(x, y, z) \circ (x', y', z') = (x + x', y + y', z + z' + xy').$$

Therefore we see that $(0, 0, z)$ is the center isomorphic to $\mathbb{R}$. Note that $N/\langle z \rangle \cong \mathbb{R} \oplus \mathbb{R}$ generated by the images of $x$ and $y$, and $[N, N] = \{(0, 0, z)\} \cong \mathbb{R}$, the center. So by Corollary 6.2.11 or Example 6.4.1(2), any lattice of $N$ intersects $\mathcal{Z}(N)$, the center of $N$ at a lattice. Also in general, for a connected, simply connected nilpotent Lie group, the center is nontrivial and simply connected. Hence $N/\mathcal{Z}(N)$ is again a simply connected nilpotent Lie group of strictly smaller dimension.

The following subgroup $\Gamma$ generated by $a = (1,0,0), b = (0,1,0)$, and $c = (0,0,\frac{1}{k})$, $k \neq 0$ integer, is a lattice. We have that

$$1 \longrightarrow \mathbb{Z} \longrightarrow \Gamma \xrightarrow{j} \mathbb{Z} \oplus \mathbb{Z} \longrightarrow 1$$

is exact where $\mathbb{Z}$ is generated by $c$.

Let $\bar{a} = j(a)$, $\bar{b} = j(b)$ denote the images under $j$. Since $aba^{-1}b^{-1} \in \mathbb{Z} \subset \langle c \rangle$, in $\mathbb{Z} \oplus \mathbb{Z}$, we have $[\bar{a},\bar{b}] = 0 \in \mathbb{Z} \oplus \mathbb{Z}$.

In fact a computation shows that $aba^{-1}b^{-1} = c^k$ since $aba^{-1}b^{-1} = (0,0,1)$. $\Gamma$ has a presentation then of

$$\{a,b,c \mid aba^{-1}b^{-1} = c^k\}.$$

By Section 14.7, the Euler number of $\Pi$ is $-k$. Abelianizing, we get

$$(6.2.1) \qquad \begin{aligned} H_1(\Gamma, \mathbb{Z}) &= \{a,b,c \mid kc = 0\} \\ &\cong \mathbb{Z} \oplus \mathbb{Z} \oplus \mathbb{Z}_k. \end{aligned}$$

Note that there is a left (and right) $S^1$-action on $\Gamma \backslash N$ given by $\mathbb{Z} \backslash \mathbb{R}$, where $\mathbb{Z}$ is generated by $c$. Thus

$$\pi_1(\mathbb{Z}\backslash\mathbb{R} = S^1) \xrightarrow{ev_*} \pi_1(\Gamma) = \Gamma$$

is injective (see Section 3.1.10), just sending the generator of $\pi_1(S^1)$ to $c$. On $H_1$ though, $H_1(S^1; \mathbb{Z})$ maps onto $\mathbb{Z}_k \subset H_1(\Gamma, \mathbb{Z})$ so it does not inject on homology.

6.2.9. A faithful representation $\tau : N \to \mathrm{GL}(n, \mathbb{R})$ of a Lie group is *unipotent* if $\tau(x)$ is unipotent for all $x \in N$ (all diagonal entries are 1). If $N$ is nilpotent and simply connected, $N$ admits a faithful representation $\rho : N \to \mathrm{GL}(n, \mathbb{R})$. If $\mathbf{N}$ denotes the Zariski closure of $\rho(N)$ in $\mathrm{GL}(n, \mathbb{C})$, $\rho(N) = \mathbf{N} \cap \mathrm{GL}(n, \mathbb{R})$.

THEOREM 6.2.10 ([**Rag72**]). *Let $N$ be a simply connected nilpotent Lie group and $\Gamma \subset N$ a closed subgroup. The following conditions on $\Gamma$ are equivalent:*

(1) *For some faithful unipotent representation $\rho : N \to \mathrm{GL}(n, \mathbb{R})$, $\rho(N)$ and $\rho(\Gamma)$ have the same Zariski closure in $\mathrm{GL}(n, \mathbb{C})$.*
(2) *$N/\Gamma$ is compact.*
(3) *$N/\Gamma$ carries a finite invariant measure.*
(4) *There are no proper connected closed subgroups of $N$ containing $\Gamma$.*
(5) *For any unipotent representation $\rho : N \to \mathrm{GL}(n, \mathbb{R})$, $\rho(N)$ and $\rho(\Gamma)$ have the same Zariski closure in $\mathrm{GL}(n, \mathbb{C})$.*

COROLLARY 6.2.11. (1) *Let $N$ be a simply connected nilpotent Lie group, and $H$ be a uniform subgroup of $N$. Then $H \cap D_k(N)$ and $H \cap C_k(N)$ (see Subsection 6.1.2 for definitions) are uniform subgroups of $D_k(N)$ and $C_k(N)$, respectively.*

(2) *Let $N$ be a simply connected nilpotent Lie group and $H$ be a closed uniform subgroup. A connected Lie subgroup $U$ of $N$ is normal in $N$ if and only if it is normalized by $H$.*

(3) *Let $H$ be a closed uniform subgroup of a nilpotent Lie group $N$, and $H_0$ the connected component of $e$ in $H$. Then $H_0$ is a normal subgroup of $N$.*

PROPOSITION 6.2.12 ([**Rag72**, Theorem 2.20]). *Let $N$ be a simply connected nilpotent Lie group and $H$ any subgroup. Then $H$ is contained in a unique minimal connected closed subgroup $H'$ of $N$. If $H$ is closed, then $H'/H$ is compact.*

THEOREM 6.2.13 ([**Rag72**, Corollary 3.9]). *Let $\Gamma$ be a finitely generated nilpotent group. Then every subgroup of $\Gamma$ is finitely generated.*

## 6.3. Solvable Lie groups

The main references for the following are [**Gor71**] and [**Gor73**].

DEFINITION 6.3.1. A Lie algebra $\mathfrak{g}$ is said to be of *type* (R) if for each $x \in \mathfrak{g}$ the eigenvalues of $\mathrm{ad}(x)$ are real. A Lie group $G$ is of type (R) if its Lie algebra is of type (R). This is also called *completely solvable*. Note that this definition of type (R) is different from that in [**Aus73**]. A Lie group $G$ is of *type* (E) if $\exp : \mathfrak{g} \to G$ is surjective, where $\mathfrak{g}$ is the Lie algebra of $G$. Clearly,

$$\text{Abelian} \implies \text{nilpotent} \implies \text{type (R)} \implies \text{type (E)}.$$

6.3.2 (Caution). The analogous rigidity result for solvable groups vis-a-vis nilpotent groups does not hold in general. It is true that if $\Gamma_1$ and $\Gamma_2$ are lattices in $G_1$ and $G_2$, $G_i$ simply connected solvable Lie groups and $\Gamma_1$ is isomorphic to $\Gamma_2$, then $G/\Gamma_1$ is *diffeomorphic* to $G/\Gamma_2$ (Mostow). But, $G_1$ and $G_2$ may not be isomorphic. For example, $\mathbb{Z}^3$ is a lattice in $\mathbb{R}^3$ and $\mathbb{Z}^3$ is also a lattice in the universal covering group of $E_0(2) \cong \mathbb{R}^2 \rtimes \mathrm{SO}(2)$, (so $\widetilde{E_0}(2) = \mathbb{R}^2 \rtimes \mathbb{R}$, not a product).

If $G$ is semisimple, then the map $\mathrm{ad} : G \to \mathrm{Endo}(\mathfrak{g})$ is injective on the maximal compact subgroup (so on a nontrivial torus). This shows that a semisimple Lie algebra cannot be of type (R), so that by the Levi decomposition, every Lie algebra of type (R) is solvable.

PROPOSITION 6.3.3. *Let $S$ be a connected, simply connected solvable Lie group and $L(S)$ its Lie algebra. The following are equivalent:*
 (1) $\exp : L(S) \to S$ *is surjective (that is, $S$ is of type* (E)*)*.
 (2) $\exp$ *is bijective*.
 (3) $\exp$ *is a diffeomorphism of real analytic manifolds*.
 (4) *For every $X \in L(S)$, $\mathrm{ad}(X)$ has no pure imaginary eigenvalues*.

PROOF. (3) $\implies$ (2) $\implies$ (1) is obvious. (1) $\implies$ (3) is proved in [**Dix57**] and (1) $\iff$ (4) is proved in [**Sai51**]. □

PROPOSITION 6.3.4 (Properties of a group $S$ of type (E)).
 (1) *Every connected subgroup and every quotient group of a group of type* (E) *is of type* (E).
 (2) *For every $g \in S$, $g \neq 1$, there is a unique 1-parameter subgroup $x(t)$ such that $x(1) = g$.*
 (3) *If $S$ is simply connected, then for every $g \in S$, the centralizer of $g$ in $S$ is connected.*
 (4) *The center $\mathcal{Z}(S)$ of $S$ is connected.*
 (5) *$S$ contains a unique maximal compact subgroup, which is a central, maximal torus.*
 (6) *For any $a \in S$ and $p > 0$ integer, there exists only one solution to $x^p = a$ in $S$.*

THEOREM 6.3.5 ([**Gor73**, Proposition 2.1]). *Let $S$ and $S'$ be two simply connected Lie groups of type* (R)*, and let $\Gamma$ and $\Gamma'$ be lattices in $S$ and $S'$, respectively. Then every isomorphism $\rho : \Gamma \to \Gamma'$ can be extended uniquely to a continuous isomorphism $\widetilde{\rho} : S \to S'$.*

Let $\Gamma$ be a subgroup of $G$. Recall that $(G, \Gamma)$ has UAEP (see Definition 5.3.3) if every automorphism of $\Gamma$ extends to an automorphism of $G$ uniquely. Here is a stronger concept.

DEFINITION 6.3.6. A simply connected Lie group $G$ has Unique Lattice Isomorphism Extension Property (ULIEP) if any isomorphism between such lattices of $G$ extends uniquely to an automorphism of $G$.

COROLLARY 6.3.7. *A simply connected Lie group of type* (R) *has* ULIEP.

EXAMPLE 6.3.8. Theorem 6.3.5 is not true for groups of type (E) in general. Let

$$A = \begin{pmatrix} e^k & 0 & 0 & 0 \\ 0 & e^k & 0 & 0 \\ 0 & 0 & e^{-k} & 0 \\ 0 & 0 & 0 & e^{-k} \end{pmatrix}.$$

Let $z = e^k + e^{-k}$. Consider the matrix $B = \begin{bmatrix} 1 & 1 \\ z-2 & z-1 \end{bmatrix}$. This matrix belongs to $\mathrm{GL}(2, \mathbf{Z})$. The characteristic polynomial of this matrix is $\lambda^2 - z\lambda + 1 = 0$. So the eigenvalues of $B$, $\lambda$ and $\frac{1}{\lambda}$, satisfy $\lambda - z + \frac{1}{\lambda} = 0$ (dividing the characteristic polynomial by $\lambda$). It is now clear that $e^k$ and $e^{-k}$ must be the two solutions and hence be the eigenvalues of $B$. It now follows that the $(4 \times 4)$-matrix $\begin{bmatrix} B & 0 \\ 0 & B \end{bmatrix} \in \mathrm{GL}(4, \mathbf{Z})$ is diagonalizable and hence must be equivalent to the original $A$. That is, there exists $P \in \mathrm{GL}(4, \mathbf{R})$ such that $P^{-1} \begin{bmatrix} B & 0 \\ 0 & B \end{bmatrix} P = A$. Let

$$\varphi_1(t) = P \begin{pmatrix} e^{kt} \cos 2\pi t & e^{kt} \sin 2\pi t & 0 & 0 \\ -e^{kt} \sin 2\pi t & e^{kt} \cos 2\pi t & 0 & 0 \\ 0 & 0 & e^{-kt} & 0 \\ 0 & 0 & 0 & e^{-kt} \end{pmatrix} P^{-1},$$

$$\varphi_2(t) = P \begin{pmatrix} e^{kt} & 0 & 0 & 0 \\ 0 & e^{kt} & 0 & 0 \\ 0 & 0 & e^{-kt} & 0 \\ 0 & 0 & 0 & e^{-kt} \end{pmatrix} P^{-1}.$$

Form

$$S_1 = \mathbf{R}^4 \rtimes_{\varphi_1} \mathbf{R}, \qquad S_2 = \mathbf{R}^4 \rtimes_{\varphi_2} \mathbf{R}.$$

Clearly, both $S_1$ and $S_2$ are solvable Lie groups of type (E) as one sees from the eigenvalues of $\varphi_i(t)$. Also form

$$D_i = \mathbf{Z}^4 \rtimes_{\varphi_i|_\mathbf{Z}} \mathbf{Z}, \qquad i = 1, 2.$$

Clearly, $\varphi_1(1) = \varphi_2(1) \in \mathrm{GL}(4, \mathbf{Z})$, and hence $D_i$ is a lattice in $S_i$. We have $D_1 = D_2$ and $S_1$ and $S_2$ nonisomorphic. Thus, the identity map $\rho : D_1 \to D_2$ does not extend to any isomorphism $S_1 \to S_2$.

## 6.4. Semisimple Lie groups

*Lattices* of connected Lie groups are *finitely presentable*. Let $\Gamma$ be a lattice in $G$, and let $H$ be a subgroup of $G$. It is important to know when the intersection $\Gamma \cap H$ is a lattice in $H$.

6.4.1. If $L$ is a normal subgroup of $G$, we say that $L$ has the *Bieberbach Property* in $G$ or *Property* (B) if, for each lattice $\Gamma$ of $G$, the subgroup $\Gamma \cdot L$ is *closed* in $G$. Note that Property (B) implies $\Gamma \cap L$ is a lattice in $L$ and $L$ is closed. Here are some examples:

(1) If $G = \mathbb{R}^n \rtimes O(n)$, then $\mathbb{R}^n$ has Property (B). (This is the so-called *First Bieberbach Theorem*; see Theorem 8.1.2.)
(2) If $G$ is a simply connected nilpotent Lie group, then $[G, G] \subset G$ has Property (B) [**Rag72**, Proposition 2.17].
(3) If $G$ is a simply connected solvable Lie group, then the nil radical (i.e., maximal connected nilpotent normal subgroup of $G$) has Property (B) [**Rag72**, Corollary 3.5].
(4) Let $G$ be a connected Lie group, $R$ its radical (not necessarily simply connected), $S$ a maximal semisimple subgroup of $G$, and $C$ the *maximal connected compact normal* subgroup of $S$. Then $CR$ has Property (B) in $G$. Moreover, each lattice of $G$ intersects $CR$ in a uniform lattice of $CR$. Note the special case when $C = 1$ (Auslander).

Statement (4) implies that if $G$ is a connected Lie group, $L$ a connected closed normal subgroup of $G$, and $\Gamma$ a discrete subgroup of $G$, then *if $\Gamma \cap L$ is a lattice of $L$, then $\Gamma L$ is closed (in $G$)*.

The Borel density theorem (see Chapter V of [**Rag72**]) is used to deduce many properties of lattices in Lie groups. We need the following corollary of the Borel density theorem:

PROPOSITION 6.4.2 ([**Rag72**, 5.17]). *Let $\Gamma$ be a lattice of a connected, semisimple Lie group without compact factor, and let $N = N_G(\Gamma)$ be the normalizer of $\Gamma$ in $G$. Then $N/\Gamma$ is finite. Also $\mathcal{Z}(G) \cdot \Gamma$ is discrete.*

Let $\Gamma$ be a lattice as above. Then $\Gamma$ is contained in only a finite number of discrete subgroups of $G$.

6.4.3 (Structure of Lie groups). Let $G$ be a compact connected Lie group, then $G$ has a finite central covering $\widetilde{G}$ such that $\widetilde{G} = T^k \times G_1 \times \cdots \times G_p$, where $T^k$ is a $k$-torus and $G_i$ are simply connected, compact, simple Lie groups. This fact yields the following propositions.

PROPOSITION 6.4.4. *Let $G$ be a compact connected Lie group and $i : T \hookrightarrow G$ its maximal torus. If $i_* : \pi_1(T) \to \pi_1(G)$ is injective, then $G = T$.*

PROOF. Let $\widetilde{G} = T^k \times G_1 \times \cdots \times G_p$ be a finite central covering of $G$, where $T^k$ is a $k$-torus and $G_i$ are simply connected, compact, simple Lie groups. Suppose $p > 0$. Let $\widetilde{i} : T_1 \hookrightarrow G_1$ be a maximal torus of $G_1$. Then $\dim(T_1) > 0$. By conjugating $T$ if necessary, we may assume that the covering map $\pi : \widetilde{G} \to G$ maps $T_1$ into $T$. We have the following commuting diagram.

$$\begin{array}{ccc} \pi_1(T_1) & \xrightarrow{\widetilde{i}_*} & \pi_1(\widetilde{G}) \\ \downarrow \pi_* & & \downarrow \pi_* \\ \pi_1(T) & \xrightarrow{i_*} & \pi_1(G) \end{array}$$

Since $\widetilde{i}_*$ factors through $\pi_1(G_1) = 1$, it is trivial. Since $\pi$ is a covering map, $\pi_*$'s are injective. We get a contradiction because $\pi_* \circ \widetilde{i}_*$ is trivial, while $i_* \circ \pi_*$ is injective. Consequently, $p = 0$, and $G = T$. □

PROPOSITION 6.4.5. *Let $G$ be a compact connected Lie group. If every finite subgroup of $G$ is Abelian, then $G$ is Abelian.*

PROOF. Let $\widetilde{G} = T^k \times G_1 \times \cdots \times G_p$ be a finite central covering, where $T^k$ is a $k$-torus, and $G_i$ are simply connected, compact, simple Lie groups. Suppose $p > 0$. Since any compact simple Lie group contains a finite nonsolvable subgroup (or at least a solvable group $F$ such that $F/\text{Center}(F)$ is not Abelian), the quotient $G = \widetilde{G}/C$ contains a finite non-Abelian subgroup. (See Proposition 5.5.5 for a different proof.) □

When $G$ is not compact, the structure, as expected, is more complicated. We already have seen the Levi decomposition; see Definition 6.1.4.

There is a decomposition of a *semisimple Lie group* into a *compact, Abelian* and *nilpotent* subgroup called the Iwasawa decomposition. For $G$, there exists $K$ a *maximal compact*, $A$ an *Abelian*, and $N$ a *nilpotent* subgroup of $G$ such that the space $K \times A \times N$ is mapped *diffeomorphically onto* $G$. Every element of $G$ is written uniquely as $k \cdot a \cdot n \in KAN$. This is sometimes called the *Iwasawa decomposition* or *KAN decomposition*.

EXAMPLE 6.4.6.
$$\text{SL}(2,\mathbb{R}) = \text{SO}(2) \times \begin{pmatrix} \lambda & 0 \\ 0 & \lambda^{-1} \end{pmatrix} \times \begin{pmatrix} 1 & r \\ 0 & 1 \end{pmatrix} \approx \text{SO}(2) \times \mathbb{R}^+ \times \mathbb{R}$$

is actually given by the Gram-Schmidt process in this case.

EXAMPLE 6.4.7.
$$\text{PSL}(2,\mathbb{C}) = \text{SO}(3) \times \begin{pmatrix} \lambda & 0 \\ 0 & \lambda^{-1} \end{pmatrix} \times \begin{pmatrix} 1 & z \\ 0 & 1 \end{pmatrix} \approx \text{SO}(3) \times \mathbb{R}^+ \times \mathbb{C}$$

is again given by unitary Gram Schmidt for $\text{SL}(2,\mathbb{C}) \subset \text{GL}(2,\mathbb{C})$ so $\text{SO}(3) = \text{SU}(2)/\pm I$, and $\lambda$ would be the length of the first column vector in $\text{PSL}(2,\mathbb{C})$ and $\lambda^{-1}$ is the length of the second column vector.

If $G$ is semisimple and connected and simply connected without normal compact factors, then $G = G_1 \times \cdots \times G_p$ is a direct product with simple factors. Moreover, if it is in adjoint form (i.e., trivial center, center always being discrete), then $G^* = G_1^* \times \cdots \times G_p^*$, where $G_i^*$ is the adjoint form of $G_i$.

THEOREM 6.4.8 (Mostow-Prasad-Margulis rigidity theorem [**Mos73**, **Pra73**]). *Let $\Gamma_1$ and $\Gamma_2$ be lattices in $G_1$ and $G_2$. Assume that*
  (1) *$G_i$ are semisimple without normal compact factors and in adjoint form (i.e., $\mathcal{Z}(G_i) = 1$).*
  (2) *If $\text{PSL}(2,\mathbb{R})$ is a factor of $G_i$, then the projection of $\Gamma$ onto that factor is not* discrete.
  (3) *$\theta : \Gamma_1 \to \Gamma_2$ is an isomorphism.*
*Then $\theta$ extends to a unique isomorphism $\Theta : G_1 \to G_2$ such that $\Theta|_{\Gamma_1} = \theta$.*

CHAPTER 7

# Seifert Fiber Space Construction for $G \times W$

## 7.1. Introduction

In this chapter, the model space will be $P = G \times W$. We assume that the space $W$ is path-connected, locally path-connected, semilocally simply connected paracompact Hausdorff. Recall, from Corollary 4.2.10, that the group of all weakly $G$-equivariant self-homeomorphisms of $G \times W$ is

$$\text{TOP}_G(G \times W) = \text{M}(W, G) \rtimes (\text{Aut}(G) \times \text{TOP}(W)).$$

We shall take $\text{TOP}_G(G \times W)$ as the *uniformizing group* and prove the existence and uniqueness of the Seifert fiber space construction.

Let $\Gamma \subset G$ be a discrete subgroup, $\rho : Q \to \text{TOP}(W)$ a proper action of a discrete group $Q$, and $1 \to \Gamma \to \Pi \to Q \to 1$ a group extension (with abstract kernel $\varphi : Q \to \text{Out}(\Gamma)$).

If the pair $(\Gamma, G)$ has UAEP (see Definition 5.3.3), $\varphi$ induces a homomorphism $Q \to \text{Out}(G)$, which is denoted by $\varphi$ again. Our goal, the *Seifert construction*, is to find a homomorphism $\theta : \Pi \to \text{TOP}_G(G \times W)$ which makes the diagram

$$
\begin{array}{ccccccccc}
1 & \longrightarrow & \Gamma & \longrightarrow & \Pi & \longrightarrow & Q & \longrightarrow & 1 \\
& & {\scriptstyle \ell}\downarrow & & {\scriptstyle \theta}\downarrow & & {\scriptstyle \varphi \times \rho}\downarrow & & \\
1 & \longrightarrow & \text{M}(W, G) \rtimes \text{Inn}(G) & \longrightarrow & \text{TOP}_G(G \times W) & \longrightarrow & \text{Out}(G) \rtimes \text{TOP}(W) & \longrightarrow & 1
\end{array}
$$

commutative.

The construction of a homomorphism $\theta : \Pi \to \text{TOP}_G(G \times W)$ will be achieved in two steps. First, we map $\Pi$ into a group $E = G\Pi$ by the pushout construction (see Subsection 5.3.4) and then $G\Pi$ into $\text{TOP}_G(G \times W)$. For the first step, we need UAEP. Therefore, we require, in addition, that $\Gamma$ be a lattice in a simply connected completely solvable Lie group or a semisimple Lie group for which the Mostow-Prasad-Margulis rigidity theorem holds. Then we stack the two steps together to get our desired homomorphisms. The case when $G$ is Abelian is a basis for all other cases; the other cases rely heavily on the Abelian case.

As it turns out, the proofs of the existence also yields, simultaneously, uniqueness, and rigidity theorems for these homomorphisms. The meaning of existence, uniqueness and rigidity are then explored, for it is these properties on which many of the applications of the Seifert Construction are based. Finally, we observe that the main theorem, Theorem 7.3.2 of this chapter, remains valid in the smooth category. We treat the second step first.

## 7.2. Cohomological criteria

**7.2.1.** Suppose we are given a proper action of a discrete group

(7.2.1) $$\rho : Q \longrightarrow \mathrm{TOP}(W)$$

and a short exact sequence

(7.2.2) $$1 \longrightarrow G \longrightarrow E \longrightarrow Q \longrightarrow 1$$

with abstract kernel $\varphi : Q \to \mathrm{Out}(G)$. Our goal is to embed $E$ into $\mathrm{TOP}_G(G \times W)$, carrying $G$ to $\ell(G)$ and mapping $Q$ by $\varphi \times \rho$.

**7.2.2.** Recall that $\mathrm{Aut}(G) \times \mathrm{TOP}(W)$ acts on $\mathrm{M}(W, G)$ by

(7.2.3) $$(\alpha, h)(\eta) = \alpha \circ \eta \circ h^{-1}.$$

We can lift the abstract kernel $\varphi : Q \to \mathrm{Out}(G)$ to a map (not necessarily a homomorphism) $\widetilde{\varphi} : Q \to \mathrm{Aut}(G)$. Two distinct lifts $\widetilde{\varphi}(\alpha)$ and $\widetilde{\varphi}'(\alpha)$ will differ by an inner automorphism. But the center $\mathcal{Z}(G)$ is fixed by any inner automorphism, so the restrictions of these two automorphisms agree on $\mathcal{Z}(G)$. Furthermore, since $\widetilde{\varphi}(\alpha)\widetilde{\varphi}(\beta) = \mu(a)\widetilde{\varphi}(\alpha\beta)$ for some $a \in G$, the restriction of $\widetilde{\varphi}$ to $\mathcal{Z}(G)$ induces a homomorphism $\widetilde{\varphi} : Q \to \mathrm{Aut}(\mathcal{Z}(G))$. Therefore, by the equality (7.2.3), we get an action of $Q$ on $\mathrm{M}(W, \mathcal{Z}(G))$:

(7.2.4) $$\widetilde{\varphi} \times \rho : Q \longrightarrow \mathrm{Aut}(\mathrm{M}(W, \mathcal{Z}(G))),$$

which is given by $\alpha \cdot \eta = \widetilde{\varphi}(\alpha) \circ \eta \circ \rho(\alpha)^{-1}$. Also, the homomorphism $\mathrm{Aut}(G) \times \mathrm{TOP}(W) \longrightarrow \mathrm{Aut}(\mathrm{M}(W, G))$, given by the equality (7.2.3), induces a commuting diagram

$$\begin{array}{ccccccccc} 1 & \longrightarrow & \mathrm{Inn}(G) & \longrightarrow & \mathrm{Aut}(G) \times \mathrm{TOP}(W) & \longrightarrow & \mathrm{Out}(G) \times \mathrm{TOP}(W) & \longrightarrow & 1 \\ & & \downarrow & & \downarrow & & \downarrow & & \\ 1 & \longrightarrow & \mathrm{Inn}(\mathrm{M}(W,G)) & \longrightarrow & \mathrm{Aut}(\mathrm{M}(W,G)) & \longrightarrow & \mathrm{Out}(\mathrm{M}(W,G)) & \longrightarrow & 1. \end{array}$$

LEMMA 7.2.3. $\mathcal{Z}(\mathrm{M}(W, G) \rtimes \mathrm{Inn}(G)) = \mathrm{M}(W, \mathcal{Z}(G))$.

PROOF. $\mathcal{Z}(\mathrm{M}(W,G) \rtimes \mathrm{Inn}(G)) = \mathcal{Z}(\mathrm{M}(W,G) \times_{\mathcal{Z}(G)} \ell(G)) = \mathcal{Z}(\mathrm{M}(W,G) \times_{\mathcal{Z}(G)} \mathcal{Z}(\ell(G))) = \mathcal{Z}(\mathrm{M}(W,G)) = \mathrm{M}(W, \mathcal{Z}(G))$. □

Let $\varphi : Q \to \mathrm{Out}(G)$ be a homomorphism, and let $\rho : Q \to \mathrm{TOP}(W)$ be an action. Pick any lift $\widetilde{\varphi} : Q \to \mathrm{Aut}(G)$, and define $\widetilde{\psi}_1 : Q \to \mathrm{Aut}(\mathrm{M}(W,G) \rtimes \mathrm{Inn}(G))$ by

(7.2.5) $$\begin{aligned} \widetilde{\psi}_1(\alpha)(\eta, \mu(a)) &= \mu(1, \widetilde{\varphi}(\alpha), \rho(\alpha))(\eta, \mu(a), 1) \\ &= (\widetilde{\varphi}(\alpha) \circ \eta \circ \rho(\alpha)^{-1}, \mu(\widetilde{\varphi}(\alpha)(a)), 1) \\ &= (\widetilde{\varphi}(\alpha) \circ \eta \circ \rho(\alpha)^{-1}, \mu(\widetilde{\varphi}(\alpha)(a))) \end{aligned}$$

for $\alpha \in Q$ and $(\eta, \mu(a)) \in \mathrm{M}(W,G) \rtimes \mathrm{Inn}(G)$. The assignment $\alpha \mapsto \widetilde{\psi}_1(\alpha)$ is not a homomorphism of $Q$ into $\mathrm{Aut}(\mathrm{M}(W,G))$ but induces a homomorphism of $Q$ into $\mathrm{Out}(\mathrm{M}(W,G))$.

THEOREM 7.2.4. *Let $\rho : Q \to \mathrm{TOP}(W)$ be a proper action of a discrete group $Q$, and $1 \to G \longrightarrow E \longrightarrow Q \to 1$ an extension with abstract kernel $\varphi : Q \to \mathrm{Out}(G)$.*

## 7.2. COHOMOLOGICAL CRITERIA

*Suppose*

(1) $H^2(Q; \mathrm{M}(W, \mathcal{Z}(G))) = 0$ *with the action of $Q$ on $\mathrm{M}(W, \mathcal{Z}(G))$ as in (7.2.4), and*

(2) *the homomorphism $Q \xrightarrow{\varphi \times \rho} \mathrm{Out}(G) \times \mathrm{TOP}(W) \longrightarrow \mathrm{Out}(\mathrm{M}(W, G))$ lifts to a homomorphism $Q \longrightarrow \mathrm{Aut}(\mathrm{M}(W, G))$.*

*Then there is a homomorphism $\theta : E \to \mathrm{TOP}_G(G \times W)$ so that*

$$\begin{array}{ccccccccc}
1 & \longrightarrow & G & \longrightarrow & E & \longrightarrow & Q & \longrightarrow & 1 \\
& & \downarrow \ell & & \downarrow \theta & & \downarrow \varphi \times \rho & & \\
1 & \longrightarrow & \mathrm{M}(W, G) \rtimes \mathrm{Inn}(G) & \longrightarrow & \mathrm{TOP}_G(G \times W) & \longrightarrow & \mathrm{Out}(G) \rtimes \mathrm{TOP}(W) & \longrightarrow & 1
\end{array}$$

*is commutative.*

PROOF. First observe that $\mathrm{M}(W, G) \rtimes \mathrm{Inn}(G) = \ell(G) \cdot \mathrm{M}(W, G)$ is generated by $\ell(G)$ and $\mathrm{M}(W, G)$; see Corollary 4.2.10. Moreover, $\mathrm{M}(W, G)$ is the centralizer of $\ell(G)$. Therefore, one can apply Corollary 5.3.6. Let $E = G \times_{(f, \widetilde{\varphi})} Q$.

Any homomorphism $Q \longrightarrow \mathrm{Aut}(\mathrm{M}(W, G))$ lifting $\psi : Q \xrightarrow{\varphi \times \rho} \mathrm{Out}(G) \times \mathrm{TOP}(W) \to \mathrm{Out}(\mathrm{M}(W, G))$ is obtained by adjusting $\widetilde{\psi}_1(\alpha)(\eta) = \widetilde{\varphi}(\alpha) \circ \eta \circ \rho(\alpha)^{-1}$ by an inner automorphism of $\mathrm{M}(W, G)$. Therefore, the adjusted $\widetilde{\psi}_1$ must be of the form

$$\begin{aligned}
\widetilde{\psi}(\alpha)(\eta) &= (\mu(\lambda) \circ \widetilde{\psi}_1)(\alpha)(\eta) \\
&= \mu(\lambda(\alpha))(\widetilde{\varphi}(\alpha) \circ \eta \circ \rho(\alpha)^{-1}) \\
&= \lambda(\alpha) \cdot (\widetilde{\varphi}(\alpha) \circ \eta \circ \rho(\alpha)^{-1}) \cdot \lambda(\alpha)^{-1},
\end{aligned}$$

where $\lambda : Q \to \mathrm{M}(W, G)$ is an adjustment to make $\widetilde{\psi} = \mu(\lambda) \circ \widetilde{\psi}_1 : Q \to \mathrm{Aut}(\mathrm{M}(W, G))$ a homomorphism, lifting $\psi : Q \to \mathrm{Out}(\mathrm{M}(W, G))$. Let $\lambda : Q \to \mathrm{M}(W, G)$ be such a map. From equality (7.2.5),

$$\widetilde{\psi}_1(\ell(a)) = \widetilde{\psi}_1(a^{-1}, \mu(a)) = \ell(\widetilde{\varphi}(\alpha)(a)).$$

Therefore, we have

$$\begin{aligned}
(\widetilde{\psi}(\alpha) \circ \ell)(a) &= \mu(\lambda)(\widetilde{\psi}_1(\ell(a))) \\
&= \mu(\lambda)(\ell(\widetilde{\varphi}(\alpha)(a))) \\
&= \ell(\widetilde{\varphi}(\alpha(a))) \quad \text{(since } \mathrm{M}(W, G) \text{ and } \ell(G) \text{ commutes)} \\
&= (\ell \circ \widetilde{\varphi}(\alpha))(a),
\end{aligned}$$

so that $\widetilde{\psi}(\alpha) \circ \ell = \ell \circ \widetilde{\varphi}(\alpha)$. Thus, the three conditions in Corollary 5.3.6 are satisfied. Consequently, we obtain a commuting diagram of exact sequences:

$$\begin{array}{ccccccccc}
1 & \longrightarrow & G & \longrightarrow & E & \longrightarrow & Q & \longrightarrow & 1 \\
& & \downarrow \ell & & \downarrow \theta' & & \parallel & & \\
1 & \longrightarrow & \mathrm{M}(W, G) \rtimes \mathrm{Inn}(G) & \longrightarrow & \mathcal{E}' & \longrightarrow & Q & \longrightarrow & 1.
\end{array}$$

Let $\mathcal{E}$ be the pullback of the bottom sequence by the homomorphism $\varphi \times \rho : Q \to \mathrm{Out}(G) \times \mathrm{TOP}(W)$. We have a commuting diagram

$$\begin{array}{ccccccccc}
1 & \longrightarrow & \mathrm{M}(W, G) \rtimes \mathrm{Inn}(G) & \longrightarrow & \mathcal{E} & \longrightarrow & Q & \longrightarrow & 1 \\
& & \downarrow \ell & & \downarrow & & \downarrow \varphi \times \rho & & \\
1 & \longrightarrow & \mathrm{M}(W, G) \rtimes \mathrm{Inn}(G) & \longrightarrow & \mathrm{TOP}_G(G \times W) & \longrightarrow & \mathrm{Out}(G) \rtimes \mathrm{TOP}(W) & \longrightarrow & 1.
\end{array}$$

We have two extensions $\mathcal{E}$ and $\mathcal{E}'$ of $\mathrm{M}(W,G) \rtimes \mathrm{Inn}(G)$ by $Q$ with the same $\widetilde{\psi}: Q \to \mathrm{Aut}(\mathrm{M}(W,G) \rtimes \mathrm{Inn}(G))$. Now $\mathcal{Z}(\mathrm{M}(W,G) \rtimes \mathrm{Inn}(G)) = \mathrm{M}(W, \mathcal{Z}(G))$ by Lemma 7.2.3. Since $H^2(Q; \mathrm{M}(W, \mathcal{Z}(G))) = 0$, by the first condition of the hypothesis, the two extensions $\mathcal{E}$ and $\mathcal{E}'$ are congruent. Therefore, we have found a desired homomorphism $E \to \mathcal{E}' \stackrel{\cong}{\to} \mathcal{E}$ into $\mathrm{TOP}_G(G \times W)$. This concludes the proof of theorem. □

## 7.3. Main Construction Theorem for $\mathrm{TOP}_G(G \times W)$

7.3.1. We shall say a discrete group $\Gamma$ is *special* if $\Gamma$ is isomorphic to a lattice in any one of the following Lie groups $G$:

- (S1) $\mathbb{R}^k$ for some $k > 0$.
- (S2) A simply connected nilpotent Lie group.
- (S3) A simply connected, completely solvable Lie group (see Definition 6.3.1); that is, for each $X \in \mathcal{G}$, the Lie algebra of $G$, $\mathrm{ad}(X): \mathcal{G} \to \mathcal{G}$ has only real eigenvalues.
- (S4) A semisimple centerless Lie group without any normal compact factors and if $G$ contains any 3-dimensional factors (i.e., $\mathrm{PSL}(2, \mathbb{R})$), then the projection of the lattice to each of these factors is dense.

We shall also call the Lie group $G$ *special*. Such groups possess ULIEP. That is, any isomorphism between such lattices extends uniquely to an isomorphism of $G$. Even more generally, if $\Gamma_i$ is a lattice in a special $G_i$ ($i = 1, 2$), then any isomorphism $\varphi: \Gamma_1 \to \Gamma_2$ extends uniquely to an isomorphism $\Psi: G_1 \to G_2$. Notice that

$$\text{type (S1)} \Longrightarrow \text{type (S2)} \Longrightarrow \text{type (S3)}.$$

The following is the main construction. Its proof is deferred until Section 7.6.

THEOREM 7.3.2 (Seifert Construction Theorem [**CR69**], [**KLR83**], [**RW77**], [**LLR96**], [**LR89**]). *Let $\Gamma \subset G$ be a special lattice. Let $\rho: Q \to \mathrm{TOP}(W)$ be a proper action of a discrete group with $Q \backslash W$ paracompact. Then for any extension $1 \to \Gamma \to \Pi \to Q \to 1$ (with abstract kernel $\varphi: Q \to \mathrm{Out}(G)$), the following are true:*

*(1) Existence: There exists $\theta: \Pi \to \mathrm{TOP}_G(G \times W)$ making the diagram (7.3.1)*

$$\begin{array}{ccccccccc}
1 & \longrightarrow & \Gamma & \longrightarrow & \Pi & \longrightarrow & Q & \longrightarrow & 1 \\
& & \downarrow \ell & & \downarrow \theta & & \downarrow \varphi \times \rho & & \\
1 & \longrightarrow & \mathrm{M}(W,G) \rtimes \mathrm{Inn}(G) & \longrightarrow & \mathrm{TOP}_G(G \times W) & \longrightarrow & \mathrm{Out}(G) \rtimes \mathrm{TOP}(W) & \longrightarrow & 1
\end{array}$$

*commutative, provided that, in the case of (S4), $\mathrm{Aut}(G)$ splits as $\mathrm{Inn}(G) \rtimes \mathrm{Out}(G)$.*

*If $G$ is special of type (S3), then*

*(2) Uniqueness: Congruent extensions are conjugate in $\mathrm{M}(W,G) \subset \mathrm{TOP}_G(G \times W)$. More precisely, suppose $\theta_1, \theta_2: \Pi \to \mathrm{TOP}_G(G \times W)$ are two homomorphisms which fit into diagram (7.3.1) with fixed $\ell$ and $\varphi \times \rho$. Then there exists $\lambda \in \mathrm{M}(W,G)$ such that $\theta_2 = \mu(\lambda) \circ \theta_1$.*

*(3) Rigidity: Suppose $\theta_1, \theta_2: \Pi \to \mathrm{TOP}_G(G \times W)$ are two homomorphisms which fit into the diagram (7.3.1) (possibly with distinct $\ell$ and $\rho$). Then there exists $(\lambda, a, h) \in \mathrm{TOP}_G(G \times W)$ such that $\theta_2 = \mu(\lambda, a, h) \circ \theta_1$, provided that $\rho_2 = \mu(h) \circ \rho_1$.*

Uniqueness for lattices of type S(4) does not hold in general; see Example 7.4.7, Subsection 7.4.8 and Exercise 7.4.9. However, as we observe in 7.6.9, by replacing $G \times W$ by $G/K \times W$, where $K$ is a maximal compact subgroup of $G$, a very strong version of the theorem for lattices in $G$ of type S(4) will be proved in Chapter 9; see Theorem 9.4.5.

## 7.4. The meaning of existence, uniqueness and rigidity

7.4.1 (Existence). Suppose there exists $\theta : \Pi \to \mathrm{TOP}_G(G \times W)$ making the diagram (7.3.1) commutative. Then $\theta(\Pi)$ acts properly on $G \times W$. This yields an injective Seifert fibering

$$\theta(\Pi)\backslash(G \times W) \xrightarrow{p} Q\backslash W$$

with the *typical fiber* $\Gamma\backslash G$, where $\Gamma = \theta(\Pi) \cap \ell(G)$, and base $Q\backslash W$. We have explained the properties of such a fibering and its singular fibers in Sections 4.3 and 4.4.

7.4.2 (Uniqueness). Let $\theta_0, \theta_1$ be homomorphisms of $\Pi$ into $\mathrm{TOP}_G(G \times W)$ such that

$$\begin{array}{ccccccccc}
1 & \longrightarrow & \Gamma & \longrightarrow & \Pi & \longrightarrow & Q & \longrightarrow & 1 \\
& & \downarrow{\ell} & & \downarrow{\theta_i} & & \downarrow{\varphi \times \rho} & & \\
1 & \longrightarrow & \mathrm{M}(W,G) \rtimes \mathrm{Inn}(G) & \longrightarrow & \mathrm{TOP}_G(G \times W) & \longrightarrow & \mathrm{Out}(G) \rtimes \mathrm{TOP}(W) & \longrightarrow & 1
\end{array}$$

is commutative. If there exists $\lambda \in \mathrm{M}(W,G)$ such that $\theta_1 = \mu(\lambda) \circ \theta_0$, the map $\lambda : W \to G$ induces a homeomorphism of $G \times W$ by

$$\lambda(x, w) = (x \cdot (\lambda(w))^{-1}, w) = (\lambda, 1, I)(x, w);$$

see Corollary 4.2.10. This, in turn, yields a homeomorphism $[\lambda] : \theta_0(\Pi)\backslash(G \times W) \longrightarrow \theta_1(\Pi)\backslash(G \times W)$. As the commuting diagram

$$\begin{array}{ccc}
M_0 = \theta_0(\Pi)\backslash(G \times W) & \xrightarrow{[\lambda]} & \theta_1(\Pi)\backslash(G \times W) = M_1 \\
\downarrow & & \downarrow \\
\rho(Q)\backslash W & \xrightarrow{=} & \rho(Q)\backslash W
\end{array}$$

shows, the map $[\lambda]$ induces the identity map on the base space. In fact, $\lambda$ is $G$-equivariant (that is, it commutes with the left principal $G$-action). Such spaces $M_0$ and $M_1$ are said to be *strictly equivalent*. Also, the actions $\theta_0$ and $\theta_1$ are called strictly equivalent actions on $G \times W$.

When $G$ is special of type (S1), (S2) or (S3), or $W$ is contractible of type (S4) (see Subsection 7.3.1), $\lambda$ is homotopic to the constant map $e : W \to G$ ($e$ is the identity element of $G$). Then the path $\{\lambda_t : 0 \leq t \leq 1\}$ gives rise to a continuous family of homomorphisms $\theta_t : \Pi \to \mathrm{TOP}_G(G \times W)$ by

$$\theta_t(\widetilde{\alpha}) = \lambda_t \cdot \theta_0(\widetilde{\alpha}) \cdot \lambda_t^{-1}$$

for $\widetilde{\alpha} \in \Pi$, and consequently a continuous family of homeomorphisms

$$[\theta_t] : \theta_0(\Pi)\backslash(G \times W) \longrightarrow \theta_t(\Pi)\backslash(G \times W).$$

Thus $\theta_0(\Pi)\backslash(G \times W)$ can be deformed isotopically to $\theta_1(\Pi)\backslash(G \times W)$ by moving just along the fibers. In fact, the family $\lambda_t : G \times W \to G \times W$ is $G$-equivariant. If $\theta_0(\Pi)$ commutes with $\ell(G)$, then the deformation $[\theta_t]$ is $G$-equivariant.

When $M_0 = M_1$, a Seifert isomorphism is called a *Seifert automorphism*. If the conjugation $\mu = (\lambda, a, h)$ induces an equivariant isotopy, then $M_1$ can be isotoped so that it becomes strictly equivalent to $M_0$.

7.4.3 (Rigidity). Let $\theta_0, \theta_1$ be homomorphisms of $\Pi$ into $\mathrm{TOP}_G(G \times W)$ such that

$$
\begin{array}{ccccccccc}
1 & \longrightarrow & \Gamma & \longrightarrow & \Pi & \longrightarrow & Q & \longrightarrow & 1 \\
& & \downarrow \ell_i & & \downarrow \theta_i & & \downarrow \phi_i \times \rho_i & & \\
1 & \longrightarrow & \mathrm{M}(W,G) \rtimes \mathrm{Inn}(G) & \longrightarrow & \mathrm{TOP}_G(G \times W) & \longrightarrow & \mathrm{Out}(G) \times \mathrm{TOP}(W) & \longrightarrow & 1
\end{array}
$$

is commutative. If there exists $(\lambda, a, h) \in \mathrm{TOP}_G(G \times W)$ such that $\theta_1 = \mu(\lambda, a, h) \circ \theta_0$, the map $(\lambda, a, h) : G \times W \to G \times W$ induces a homeomorphism $[\lambda, a, h]$

$$
\begin{array}{ccc}
M_0 = \theta_0(\Pi)\backslash(G \times W) & \xrightarrow{[\lambda,a,h]} & \theta_1(\Pi)\backslash(G \times W) = M_1 \\
\downarrow & & \downarrow \\
\rho_0(Q)\backslash W & \xrightarrow{\overline{h}} & \rho_1(Q)\backslash W
\end{array}
$$

which preserves the fibers. More precisely, $[\lambda, a, h]$ is defined by

$$[\lambda, a, h]\,([x,w]) = [(\lambda, a, h)(x,w)].$$

Since $(\lambda, a, h) \circ \theta_0(\alpha) = \theta_1(\alpha) \circ (\lambda, a, h)$ for all $\alpha \in \Pi$, the map $[\lambda, a, h]$ is well defined. Further, $(\lambda, a, h)(x, w) = \big(a(x) \cdot (\lambda h(w))^{-1}, h(w)\big)$ shows that $[\lambda, a, h]$ is a map from $M_0$ to $M_1$. That is, $[\lambda, a, h]$ is the descent of the weakly equivariant fiber preserving map $(\lambda, a, h)$ sending $G$-fibers to $G$-fibers. The conjugation of $\theta_0$ by $(\lambda, a, h)$ is called a *Seifert isomorphism* of $G \times W$. The induced homeomorphism $[\lambda, a, h] : M_0 \to M_1$ is called a *Seifert isomorphism*.

REMARK 7.4.4. If $\Gamma$ is characteristic in $\Pi$, any automorphism of $\Pi$ induces an automorphism of $\Gamma$ and $Q$. Consequently, if $M_0$ and $M_1$ are Seifert fiber spaces modeled on $G \times W$ which have the same fundamental group (or *orbifold fundamental group*), they are equivalent as Seifert fiber spaces, provided that the base spaces are *rigidly related* (i.e., there exists $h \in \mathrm{TOP}(W)$ for which $\rho_1 = \mu(h) \circ \rho_0$).

EXAMPLE 7.4.5 (When $W$ is a point). Suppose $W = \{p\}$, a point. Then
$$
\begin{aligned}
\mathrm{TOP}_G(G \times W) &= \mathrm{M}(p, G) \rtimes (\mathrm{Aut}(G) \times \mathrm{TOP}(p)) \\
&= r(G) \rtimes \mathrm{Aut}(G) \\
&= \ell(G) \rtimes \mathrm{Aut}(G) \\
&= \mathrm{Aff}(G).
\end{aligned}
$$

Since $\Gamma$ maps into the translational part $\ell(G)$, the image $\Phi$ of $\Pi$ in $\mathrm{Aut}(G)$ is finite. This means that there exists a Riemannian metric on $G$ for which $\Phi \subset \mathrm{Isom}(G)$. Thus an injective Seifert fiber space modeled on $\mathbb{R}^n \times$ point is a flat manifold if $\Pi$ is torsion free, and a flat orbifold otherwise. Similarly, if $G$ is a connected, simply connected nilpotent Lie group, an injective Seifert fiber space modeled on $G \times$ point is an infra-nilmanifold if $\Pi$ is torsion free, and an infra-nilorbifold otherwise. Since $W$ is a point, the rigidity on $W$ always holds, and by the ULIEP and the rigidity theorem, any two infra-$G$-manifolds modeled on special $G$ with isomorphic fundamental groups are affinely diffeomorphic. (Recall that $\text{TOP}_G(G \times W)$ reduces to $G \rtimes \text{Aut}(G)$ in this case.) This generalizes the classical result of Bieberbach's from crystallographic groups to almost crystallographic groups in the nilpotent case; see Chapter 8 for more details.

EXAMPLE 7.4.6. Let
$$G = \mathbb{R}^2, \quad W = \mathbb{R}, \quad \Gamma = \mathbb{Z}^2 \subset G, \quad Q = \mathbb{Z},$$
with $Q$ acting on $W$ as translations. Consider the two Seifert constructions of $\Pi = \Gamma \times Q$ into $\text{TOP}_G(G \times W)$. Let $\alpha \in Q$ be a generator. Using the notation $\text{TOP}_G(G \times W) = \text{M}(W, G) \rtimes (\text{Aut}(G) \times \text{TOP}(W))$, $\theta_0$ and $\theta_1$ are defined for $(0, \alpha) \in \Gamma \times Q$ as follows:
$$\theta_0(0, \alpha) = (\mathbf{0}, I, \alpha),$$
$$\theta_1(0, \alpha) = (\mathbf{a}, I, \alpha),$$
where $\mathbf{a} \in \mathbb{R}^2$ is a fixed element. Thus,
$$\theta_0(0, \alpha)(x, w) = (x, w + \alpha),$$
$$\theta_1(0, \alpha)(x, w) = (x - \mathbf{a}, w + \alpha).$$

Since $H^1(Q; \text{M}(W, G)) = 0$, these two constructions must be equivalent. In fact, let $\lambda \in \text{M}(W, G)$ be given by
$$\lambda(w) = w \cdot \mathbf{a}.$$
Then
$$(\lambda, I, 1)(\mathbf{0}, I, \alpha)(\lambda, I, 1)^{-1} = (\lambda \cdot (\lambda^{-1} \circ \alpha^{-1}), I, \alpha).$$
Now
$$\begin{aligned}(\lambda \cdot (\lambda^{-1} \circ \alpha^{-1}))(w) &= \lambda(w) - \lambda(\alpha^{-1}(w)) \\ &= \lambda(w) - \lambda(w - 1) \\ &= w \cdot \mathbf{a} - (w - 1) \cdot \mathbf{a} \\ &= \mathbf{a}.\end{aligned}$$
This shows that $\mu(\lambda, I, 1) \circ \theta_0 = \theta_1$, and the two constructions are equivalent.

EXAMPLE 7.4.7 (Nonuniqueness). Let $G$ be a Lie group with trivial center and having both a nontrivial compact subgroup $K$ and a torsion-free lattice $\Gamma$. Suppose $Q$ is a closed subgroup of $K$, so
$$Q \subset K \subset G,$$
and $\rho : Q \to \text{TOP}(W)$ is a semifree action (that is, $Q$ acts so that all orbits are free or fixed). We assume that both free orbits and fixed points occur. The next examples show that we cannot have uniqueness theorems in these situations.

For $G \times W$, there are two Seifert Constructions $\theta$ and $\theta'$ of $\Gamma \times Q$ into $\text{TOP}_G(G \times W)$ having the same fixed $\ell$ and $\rho$ which are not strictly equivalent. Consequently, if $Q$ is finite, the set $H^1(Q; \text{M}(W, G))$ is not a singleton. In fact, there is no element of $\text{TOP}_G(G \times W)$ that will conjugate $\theta$ to $\theta'$. Thus, there is no Seifert isomorphism from $\theta$ to $\theta'$.

Two actions $\theta, \theta' : \Gamma \times Q \to \text{M}(W, G) \rtimes (\text{Aut}(G) \times \text{TOP}(W))$ are defined by
$$\theta(z, \alpha) = (\alpha z^{-1}, \mu(z), \rho(\alpha)),$$
$$\theta'(z, \alpha) = (z^{-1}, \mu(z), \rho(\alpha))$$

for $(z, \alpha) \in \Gamma \times Q$. (The $\alpha$ in $\alpha z^{-1}$ for $\theta(z, \alpha)$ is as $\alpha \in Q \subset K \subset G$.) Therefore,
$$\theta(z, \alpha)(x, w) = (zx\alpha^{-1}, \rho(\alpha)w),$$
$$\theta'(z, \alpha)(x, w) = (zx, \rho(\alpha)w)$$
for $(x, w) \in G \times W$. Note both $\theta$ and $\theta'$ map $\Gamma \times Q$ injectively into $\ell(G) \times r(G) \times \text{TOP}(W) \subset \text{TOP}_G(G \times W)$. We obtain two Seifert fiberings

$$X = \theta(\Gamma \times Q)\backslash(G \times W) = (\Gamma\backslash G) \times_Q W \xrightarrow{\tau} Q\backslash W$$
$$= \downarrow$$
$$X' = \theta'(\Gamma \times Q)\backslash(G \times W) = (\Gamma\backslash G) \times (Q\backslash W) \xrightarrow{\tau'} Q\backslash W$$

with $\tau'$ being just projection onto the second factor. Observe that

$$\tau^{-1}(b) = \begin{cases} \Gamma\backslash G, & \text{if } b \text{ is a free } Q\text{- orbit,} \\ \Gamma\backslash G/Q, & \text{if } b \text{ is a fixed orbit,} \end{cases}$$
$$\tau'^{-1}(b) = \Gamma\backslash G, \text{ for each } b \in Q\backslash W,$$

where $Q$ acts freely on $\Gamma\backslash G$. To verify that $Q$ acts freely, we shall show that if $\Gamma x\alpha = \Gamma x$ for some $x \in G$ and $\alpha \in Q$, then $\alpha$ must be 1. Clearly $x\alpha x^{-1} \in \Gamma$ if the equality is to hold. If $\alpha$ is of finite order, then $x\alpha x^{-1}$ is of finite order in a torsion-free group $\Gamma$. Thus, $\alpha = 1$. If $\alpha$ is of infinite order, the cyclic group generated by $x\alpha x^{-1}$ cannot be discrete because $xQx^{-1}$ is compact. But the cyclic group is also in the discrete group $\Gamma$. Therefore $\alpha = 1$. Hence, the action of $Q$ on $\Gamma\backslash G$ must be free. Now $\theta(Q)$ and $\theta'(Q)$ are not strictly equivalent since they have different fibers over each $b$ representing a fixed point in the $Q$-action on $W$.

Moreover, no element of $\text{TOP}_G(G \times W)$ conjugates $\theta$ to $\theta'$ since $\theta$ will always have some fibers, $\tau^{-1}(b)$, homeomorphic to $\Gamma\backslash G/Q$ whereas each of $\theta'$'s fibers are $\Gamma\backslash G$.

7.4.8. In Theorem 5.7.2, we have seen that there is a one-to-one correspondence between the strict equivalence classes of homomorphisms of $\Gamma \times Q$ into $\text{TOP}_G(G \times W)$ keeping $\ell$ and $\rho$ fixed and the elements of the set $H^1(Q; C_{\ell(G)}(\Gamma) \cdot M(W, G))$. (The centralizer of $\ell(\Gamma)$ in $\ell(G) \cdot M(W, G)$ is $C_{\ell(G)}(\Gamma) \cdot M(W, G))$, where $C_{\ell(G)}(\Gamma)$ is the centralizer of $\Gamma$ in $\ell(G)$.) Therefore, $H^1(Q; C_{\ell(G)}(\Gamma) \cdot M(W, G))$ is not a singleton. In general, it seems that showing directly that this cohomology set is not a singleton is a formidable problem.

As an illustration, we will take

$$G = \text{SO}_0(3, 1) \cong \text{PSL}(2, \mathbb{C}),$$

a group of type (S4). Even though the existence theorem always holds for (S4), the uniqueness will, in general, fail. Take

$$Q = \mathbb{Z}_2 \subset \text{SO}(3) \subset \text{SO}_0(3, 1),$$

and let $Q$ act on $W = S^1$ by $\epsilon \times z \mapsto z^\epsilon$, where $\epsilon = \pm 1$, $z \in S^1$. If $\Gamma$ is any torsion-free lattice in $\text{SO}(3, 1)$, the conditions for constructing an example are now satisfied. Also, note that $C_{\ell(G)}(\Gamma) = 1$ for any lattice $\Gamma$ in a group of type (S4), because $C_{\ell(G)}(\Gamma) \subset \mathcal{Z}(G) = 1$. Then form the constructions $\theta$ and $\theta'$ for $\Gamma \times \mathbb{Z}_2$. The space $X$ is the smooth manifold $(\Gamma\backslash G) \times_{\mathbb{Z}_2} S^1$ with empty boundary. In fact it is the *double mapping cylinder* of the double covering $\Gamma\backslash G \to (\Gamma\backslash G)/\mathbb{Z}_2$. $X'$ is the manifold $(\Gamma\backslash G)\times I$ with boundary $(\Gamma\backslash G)\times \partial I$. Therefore, $H^1(\mathbb{Z}_2;\mathrm{M}(S^1,\mathrm{SO}(3,1)))$ is not a singleton.

The examples in this section all map $\Gamma\times Q$ into $\mathcal{U}=(\ell(G)\times r(G))\times \mathrm{TOP}(W)$, a proper subgroup of $\mathrm{TOP}_G(G\times W)$. Assume now that $G$ is of type (S4). Then, using Theorem 5.7.2, the strict equivalence classes of homomorphisms of $\Gamma\times Q$ into $\mathcal{U}$ keeping $\ell$ and $\rho$ fixed are in one-to-one correspondence with the set of conjugacy classes of homomorphisms of $Q$ into $r(G)=C_{\ell(G)\times r(G)}(\ell(\Gamma))$. For finite $Q$, this can be written as the non-Abelian cohomology set $H^1(Q;r(G))$.

For $K$ a maximal compact simple subgroup of $G$, any continuous homomorphism is either an isomorphism or trivial up to conjugation in $r(G)$ (since all maximal compact subgroups are conjugate in $G$). Thus, in this case, $\theta$ and $\theta'$ as described in Example 7.4.7 are the only possible strict equivalence classes with fixed $\ell$ and $\rho$.

EXERCISE 7.4.9. Let $W=S^3$, $G=\mathrm{PSL}(2,\mathbb{C})$, $Q=\mathrm{SO}(3)$ acting on $W$ by suspending the transitive action of $\mathrm{SO}(3)$ on $S^2$. For $\Gamma\times Q$, define $\theta$ and $\theta'$ of $\Gamma\times Q$ into $\ell(G)\times r(G)\times \mathrm{TOP}(W)$ with the same $\ell$ and $\rho$ and show that they are not strictly equivalent in $\mathrm{TOP}_G(G\times S^3)$.

## 7.5. $H^p(Q;\mathrm{M}(W,\mathbb{R}^k))=0$, $p>0$

This section, in preparation for the proof of Theorem 7.3.2, is devoted to some vanishing theorems. In particular, if $W$ and $Q\backslash W$ are paracompact, we shall show the following.

THEOREM 7.5.1 ([**CR69**, Lemma 8.4] and [**CR72b**, Theorem 7.1 and 12.1]). *Let $\rho:Q\to \mathrm{TOP}(W)$ be a proper action of a discrete group $Q$. For any homomorphism $\varphi:Q\to \mathrm{GL}(k,\mathbb{R})$ and an action of $Q$ on $\mathrm{M}(W,\mathbb{R}^k)$ by $\alpha\cdot\lambda=\varphi(\alpha)\circ\lambda\circ\rho(\alpha)^{-1}$, $H^p(Q;\mathrm{M}(W,\mathbb{R}^k))=0$ for all $p>0$.*

The argument appears in full generality in [**CR72b**], with a less comprehensive version in [**Con68**] and [**CR69**]. Let

$$\pi: W \longrightarrow Q\backslash W$$

denote the natural projection. We shall offer, in increasing generality, three arguments. The easiest and most straightforward occurs for the case where (a) $Q\backslash W$ *is compact*. The second requires (b) $Q\backslash W$ *to be finite dimensional* while the third is (c) *the most general* and requires some sheaf-theoretic arguments. We put the third proof off to Chapter 10 where it will fit into a much larger context. The reader interested in only the compact and or finite dimensional cases will find the proofs in this section complete and adequate.

We begin with some important algebraic steps which are useful for all three arguments.

**7.5.2. Shapiro's Lemma.** (see [**Bro82**, III 6.2, p.73]) *If $H\subset Q$ is a subgroup and $M$ is an $H$-module, then $H^*(H;M)\cong H^*(Q;\mathrm{Hom}_{\mathbb{Z}H}(\mathbb{Z}Q,M))$.*

Note that $\mathbb{Z}Q$ is a free $\mathbb{Z}H$-module with a basis given by choosing coset representatives. Moreover, it is easy to see that $\mathrm{Hom}_{\mathbb{Z}H}(\mathbb{Z}Q,M)=\prod_{i\in I}M_i$, where $I$ indexes the cosets $M_i\cong M$.

Now let $x\in W$ and $U_x$ be a $Q_x$-slice at $x$. If we let $\{x_i:i\in I\}$ denote the orbit $Qx$, then there exists $q_i\in Q$ so that $q_ix=x_i$. We may take $U_{x_i}=q_iU_x$ to be $Q_{x_i}$-slice at $x_i$. Then $Q_{x_i}=q_iQ_xq_i^{-1}$. We can choose $U_x$ so small that

$$U_{x_i}\cap U_{x_j}=\emptyset, \text{ if } i\neq j.$$

Then $QU_x$ is a tubular neighborhood of $Qx$ where each of the $U_{x_i}$ are pairwise disjoint and $QU_x = \bigcup_{i \in I} U_{x_i}$. Moreover, the index set $I$ is in one-to-one correspondence with the quotient set $Q/Q_x$. Thus we have

$$\operatorname{Hom}_{\mathbb{Z}Q_x}(\mathbb{Z}Q, \mathrm{M}(U_x, \mathbb{R}^k)) \cong \mathrm{M}(QU_x, \mathbb{R}^k).$$

Similarly, if $\mathfrak{M}(U_x, \mathbb{R}^k)$ and $\mathfrak{M}(QU_x, \mathbb{R}^k)$ denote the subspace of all maps of $W$ into $\mathbb{R}^k$ with support in $U_x$ and in $QU_x$, respectively, then

$$\operatorname{Hom}_{\mathbb{Z}Q_x}(\mathbb{Z}Q, \mathfrak{M}(U_x, \mathbb{R}^k)) \cong \mathfrak{M}(QU_x, \mathbb{R}^k).$$

Consequently, we have the following by Shapiro's Lemma.

7.5.3. $H^*(Q_x; \mathrm{M}(U_x, \mathbb{R}^k)) \cong H^*(Q; \mathrm{M}(QU_x, \mathbb{R}^k))$ and $H^*(Q_x; \mathfrak{M}(U_x, \mathbb{R}^k)) \cong H^*(Q; \mathfrak{M}(QU_x, \mathbb{R}^k))$.

Now since $\mathbb{R}^k$ is a real vector space and $Q_x$ is finite (recall that the action $(Q, W)$ is proper), we have that $H^p(Q_x; \mathrm{M}(U_x, \mathbb{R}^k)) = 0$ and $H^p(Q_x; \mathfrak{M}(U_x, \mathbb{R}^k)) = 0$. Consequently,

PROPOSITION 7.5.4. $H^p(Q; M(QU_x, \mathbb{R}^k)) = 0$ and $H^p(Q; \mathfrak{M}(QU_x, \mathbb{R}^k)) = 0$ for all $p > 0$.

7.5.5 ((a) when $Q\backslash W$ is compact). There is a finite set of points $x_1, \ldots, x_n$ with slice neighborhoods $U_{x_1}, \ldots, U_{x_n}$ as above whose images $\pi(U_{x_1}), \ldots, \pi(U_{x_n})$ in $Q\backslash W$ cover $Q\backslash W$. (The notation $\{x_i\}$ is different from the above. Each $x_i$ here corresponds to $x$ above.) Choose a partition of unity $\widehat{\epsilon}_1, \ldots, \widehat{\epsilon}_n$ subordinate to this open covering. Compose each partition function with the quotient map, $\epsilon_j = \widehat{\epsilon}_j \circ \pi$, to get a partition of unity $\epsilon_1, \ldots, \epsilon_n$ subordinate to the covering $QU_{x_1}, QU_{x_2}, \ldots, QU_{x_n}$ of $W$. We define

$$\epsilon_j : \mathrm{M}(W, \mathbb{R}^k) \to \mathfrak{M}(QU_{x_j}, \mathbb{R}^k)$$

by multiplication by $\epsilon_j$. More precisely, $\epsilon_j(\lambda) = \epsilon_j \cdot \lambda$, where

$$(\epsilon_j \cdot \lambda)(w) = \epsilon_j(w) \cdot \lambda(w),$$

where $\epsilon_j(w) \in \mathbb{R}$ is a scalar, and $\lambda(w) \in \mathbb{R}^k$ is a vector. We claim that $\epsilon_j$ is a $Q$-module homomorphism. We need to show that the following diagram

$$\begin{array}{ccc} \mathrm{M}(W, \mathbb{R}^k) & \xrightarrow{\epsilon_j} & \mathfrak{M}(QU_{x_j}, \mathbb{R}^k) \\ \downarrow \alpha & & \downarrow \alpha \\ \mathrm{M}(W, \mathbb{R}^k) & \xrightarrow{\epsilon_j} & \mathfrak{M}(QU_{x_j}, \mathbb{R}^k) \end{array}$$

is commutative for every $\alpha \in Q$. Some calculations show, for every $\lambda \in \mathrm{M}(W, \mathbb{R}^k)$ and $\alpha \in Q$, that

$$\begin{aligned}
(\epsilon_j(\alpha \cdot \lambda))(w) &= \epsilon_j(w) \cdot (\alpha \cdot \lambda)(w) \\
&= \epsilon_j(w) \cdot [\varphi(\alpha) \circ \lambda \circ \rho(\alpha)^{-1}(w)], \\
(\alpha \cdot (\epsilon_j \cdot \lambda))(w) &= \varphi(\alpha)(\epsilon_j \cdot \lambda(\rho(\alpha)^{-1})(w)) \\
&= \varphi(\alpha)[\epsilon_j(\rho(\alpha)^{-1}(w)) \cdot \lambda(\rho(\alpha)^{-1})(w)] \\
&= \epsilon_j(\rho(\alpha)^{-1}(w)) \cdot \varphi(\alpha)(\lambda(\rho(\alpha)^{-1})(w)) \\
&\quad (\text{since } \epsilon_j(\rho(\alpha)^{-1}(w)) \text{ is a scalar}) \\
&= \epsilon_j(w) \cdot [\varphi(\alpha) \circ \lambda \circ \rho(\alpha)^{-1}(w)].
\end{aligned}$$

Thus $\epsilon_j$ is a $Q$-module homomorphism.

## 7.5. $H^P(Q; M(W, \mathbb{R}^K)) = 0$, $P > 0$

Clearly the diagram

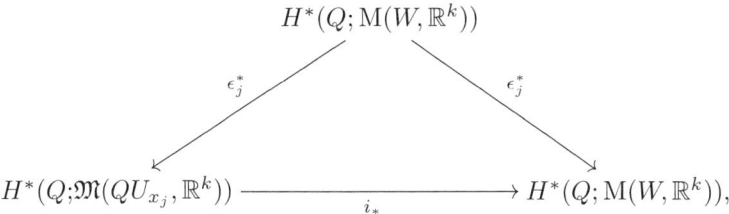

where $i_*$ induced by inclusion $\mathfrak{M}(QU_j, \mathbb{R}^k) \hookrightarrow M(W, \mathbb{R}^k)$, is commutative. Recall $\epsilon_1^* + \epsilon_2^* + \cdots + \epsilon_n^* = \mathrm{id}$. Thus, the identity homomorphism $\sum_j \epsilon_j^* = \mathrm{id}$ : $H^p(Q; M(W, \mathbb{R}^k)) \to H^p(Q; M(W, \mathbb{R}^k))$ factors through $\bigoplus_j H^p(Q; \mathfrak{M}(QU_{x_j}, \mathbb{R}^k))$. However, $\bigoplus_j H^p(Q; \mathfrak{M}(QU_{x_j}, \mathbb{R}^k)) = 0$ by Proposition 7.5.4. This completes the proof of Theorem 7.5.1 when $Q\backslash W$ is compact. $\square$

7.5.6 ((b) when $Q\backslash W$ is finite dimensional). Let us recall that the *order of an open covering* $\{U_i : i \in I\}$ is $n$ if $n$ is the largest integer so that there are $n+1$ elements of the covering $U_{i_0}, \ldots, U_{i_n}$ with $U_{i_0} \cap \cdots \cap U_{i_n} \neq \emptyset$. If no such integer exists, we say the order of the covering is infinite. A paracompact $T_1$ space $X$ is said to have *dimension* less than or equal to $n$ if every open covering has a locally finite open refinement of order less than or equal to $n$. It is of dimension $n$ if it is of dimension less than or equal to $n$ and is not of dimension less than or equal to $n-1$. If $X$ is nonempty, and is not of dimension less than or equal to $n$ for any $n$, then $X$ is said to have infinite dimension. The empty set is said to have dimension $-1$. We have defined the so-called *covering dimension* of $X$, $\dim X$.

Assume that $Q\backslash W$ has dimension $n$ and is paracompact. For each orbit in $Q\backslash W$, choose an $x$ in the orbit and take a slice neighborhood $U_x$. Let $\pi : W \to Q\backslash W$ be the orbit mapping, and denote the image $\pi(U_x)$ by $U_x^*$. Then $\{U_x^*\}$ is a covering of $Q\backslash W$. Take a locally finite refinement $\{U_\alpha^*\}$ of $\{U_x^*\}$ such that $\{U_\alpha^*\}$ has order $n$. For each $\alpha$, choose $\tau(\alpha) = x$, for some $x$ for which $U_\alpha^* \subset U_x^*$. We use the following

LEMMA 7.5.7 ([**Bor60**, 3.7, p.111]). *There is an open covering* $\{G_{i,\beta}^* : \beta \in B_i\}$, $i = 0, 1, \ldots, n$ *such that* $\{G_{i,\beta}^*\}$ *refines* $\{U_\alpha^*\}$ *and* $G_{i,\beta}^* \cap G_{i,\beta'}^* = \emptyset$ *if* $\beta \neq \beta'$.

Here $B_i$ is the set of unordered $i+1$ tuples from the indexing set of $\{U_\alpha^*\}$. So if $\beta = (\alpha_0, \alpha_1, \ldots, \alpha_i)$, then $G_{i,\beta}^* \subset U_{\alpha_0}^* \cap U_{\alpha_1}^* \cap \cdots \cap U_{\alpha_i}^*$. For each $i, \beta$, choose one of the $\alpha_0, \ldots, \alpha_i$, say $\alpha_j$ and we have $G_{i,\beta}^* \subset U_{\alpha_j}^* \subset U_{\tau(\alpha_j)=x_j}^*$. For convenience, set $x = x_j$ and $\pi_x = \pi|_{U_x}$. Then $\pi_x^{-1}(G_{i,\beta}^*)$ will be a $Q_x$-invariant open set in $U_x$, where $\tau(\alpha_j) = x_j = x$. As in Proposition 7.5.4, we have $H^*(Q; \mathfrak{M}(\pi^{-1}(G_{i,\beta}^*), \mathbb{R}^k)) = 0$. This follows as above, i.e., $H^*(Q_x; \mathfrak{M}(\pi_x^{-1}(G_{i,\beta}^*), \mathbb{R}^k)) = 0$ implies

$$H^*(Q; \mathfrak{M}(Q\pi_x^{-1}(G_{i,\beta}^*), \mathbb{R}^k)) = H^*(Q; \mathfrak{M}(\pi^{-1}(G_{i,\beta}^*), \mathbb{R}^k)) = 0.$$

Now $G_{i,\beta}^* \cap G_{i,\beta'}^* = \emptyset$ if $\beta \neq \beta'$. So $\pi^{-1}(G_{i,\beta}^*) \cap \pi^{-1}(G_{i,\beta'}^*) = \emptyset$. Define

$$\widetilde{S}_i = \bigcup_{\beta \in B_i} \pi^{-1}(G_{i,\beta}^*).$$

Therefore,

$$H^*(Q; \mathfrak{M}(\widetilde{S}_i, \mathbb{R}^k)) = H^*(Q; \Pi_{\beta \in B_i} \mathfrak{M}(\pi^{-1}(G_{i,\beta}^*), \mathbb{R}^k))$$

because $\mathfrak{M}(\widetilde{S}_i, \mathbb{R}^k) = \prod_{\beta \in B_i} \mathfrak{M}(\pi^{-1}(G^*_{i,\beta}), \mathbb{R}^k)$.

Let us denote $\pi^{-1}(G^*_{i,\beta})$ by $\widetilde{G}_{i,\beta}$. Now $i$ is fixed between 0 and $n$, and $\widetilde{G}_{i,\beta} \cap \widetilde{G}_{i,\beta'} = \emptyset$ if $\beta \neq \beta'$. So, using [**CE56**, Proposition 91 and 92, pp.97–98], we may pull the product sign out of the coefficients and we get

$$H^p(Q; \mathfrak{M}(\widetilde{S}_i, \mathbb{R}^k)) = \prod_{\beta \in B_i} H^p(Q; \mathfrak{M}(\pi^{-1}(G^*_{i,\beta}), \mathbb{R}^k)).$$

But as we have seen, each of the terms of the product is 0 when $p > 0$. Therefore, $H^p(Q; \mathfrak{M}(\widetilde{S}_i, \mathbb{R}^k)) = 0$. The union of the $\widetilde{S}_i$ covers $W$. Again choose as in the compact case (Subsection 7.5.5), a partition of unity subordinate to $\{S^*_i : i = 0, 1, \ldots, n\}$, where $S^*_i = \bigcup_{\beta \in B_i} G^*_{i,\beta}$. Then if we compose each partition function with the quotient map, we get a partition of unity $\epsilon_0, \epsilon_1, \ldots, \epsilon_n$ subordinate to the covering $\{\widetilde{S}_0, \widetilde{S}_1, \ldots, \widetilde{S}_n\}$. Since $\epsilon_j(qx) = \epsilon_j(x)$, we get that each $\epsilon_j$ is a $Q$-module homomorphism of $\mathrm{M}(W, \mathbb{R}^k)$ into $\mathfrak{M}(\widetilde{S}_j, \mathbb{R}^k)$. Then as in Subsection 7.5.5, we have $H^*(Q; \mathrm{M}(W, \mathbb{R}^k)) = 0$.

7.5.8. To finish the proof in the finite dimensional case, we need to make enough assumptions on $W$ to guarantee that $Q\backslash W$ is paracompact and finite dimensional. It is sufficient to assume that $W$ is Lindelöf; i.e., every open covering has a countable subcovering. This is guaranteed, for example, if $W$ is separable. Then the quotient $Q\backslash W$ is Lindelöf, and since it is regular, $Q\backslash W$ is paracompact. (K. Morita has shown that for Lindelöf (Hausdorff) spaces, regularity is equivalent to paracompactness.) Moreover, in paracompact spaces, the covering dimension is a local property. Since $Q\backslash W$ is locally just the quotient of a slice by a finite group, the local dimension of $W$ is preserved by the quotient map. Therefore, assuming $W$ is Lindelöf and finite dimensional guarantees that $Q\backslash W$ is paracompact and finite dimensional. This completes the proof of the theorem when $Q\backslash W$ is finite dimensional. □

7.5.9 ((c) The general case). For the general case, we do not assume $W$ is finite dimensional. The technique of the proof is different since we cannot find a finite partition of unity. Instead we employ sheaf-theoretic methods which are used for $T^k$-bundles $P$ over $W$ in the holomorphic, smooth, and topological cases. Since this entails additional background, we put off the vanishing proof for the general case until Corollary 10.3.9.

REMARK 7.5.10 (Extending vanishing theorems to compact Lie groups). The vanishing theorems play a crucial role in the existence and uniqueness theorems for Seifert Constructions. They also play an important role in the classification of the possible liftings of $Q$-actions on $W$ to principal $G$-bundles $P$ over $W$.

If we replace the discrete $Q$ by a compact Lie group, we may address similar problems. Hattori and Yoshida [**HY76**] have done this by taking $\mathrm{M}(W, \mathbb{R}^k)$ as a topological $Q$-module and cohomology as the continuous cohomology of Mostow [**Mos61**]. In the continuous case, the role of partition of unity is replaced by integration over the group $Q$ for the vanishing of $H^p(Q; \mathrm{M}(W, \mathbb{R}^k))$. We shall not prove this vanishing theorem as it will take us too far afield. However, we shall use it when we discuss the Hattori and Yoshida theorem of lifting compact group actions to principal $T^k$-bundles; see Chapter 12.

## 7.6. Proof of the Construction Theorem

We prove the Construction Theorem 7.3.2. There are two steps. First we map $\Pi$ into an extension $E$ of $G$ by $Q$, and then we map $E$ into $\text{TOP}_G(G \times W)$.

7.6.1. Every special Lie group $G$ has the ULIEP (Subsection 7.3.1). For every $G$ with ULIEP, let $1 \to \Gamma \to \Pi \to Q \to 1$ be an extension which represents $f \in \text{Opext}(\Gamma, Q, \widetilde{\varphi})$ satisfying (5.2.1), (5.2.2), and (5.2.3) of Section 5.2. One can form a pushout by the inclusion $i : \Gamma \hookrightarrow G$ to obtain $E = G\Pi$; see Subsection 5.3.4. Then $f \in \text{Opext}(G, Q, \widetilde{\varphi})$ and the diagram

$$\begin{array}{ccccccccc} 1 & \longrightarrow & \Gamma & \longrightarrow & \Pi & \longrightarrow & Q & \longrightarrow & 1 \\ & & \downarrow & & \downarrow & & \downarrow & & \\ 1 & \longrightarrow & G & \longrightarrow & E = G\Pi & \longrightarrow & Q & \longrightarrow & 1 \end{array}$$

is commutative.

7.6.2. Now in order to map $E$ into $\text{TOP}_G(G \times W)$, we apply Theorem 7.2.4. We need to prove the two conditions. First note that $H^i(Q; \text{M}(W, \mathcal{Z}(G))) = 0$ for $i = 1, 2$ for a special Lie group $G$, because $\mathcal{Z}(G)$ is isomorphic to some $\mathbb{R}^k$, $k \geq 0$.

We want to verify the second condition of Theorem 7.2.4: the homomorphism $\psi : Q \xrightarrow{\varphi \times \rho} \text{Out}(G) \times \text{TOP}(W) \to \text{Out}(\text{M}(W, G))$ lifts to a homomorphism $\widetilde{\psi} : Q \longrightarrow \text{Aut}(\text{M}(W, G))$. The following two lemmas will suffice.

LEMMA 7.6.3. *Let $A$ be a closed normal subgroup of a connected Lie group $G$. Suppose the quotient homomorphism $p : G \to G/A$ has a continuous cross section $s : G/A \to G$. Then*

$$1 \longrightarrow \text{M}(W, A) \longrightarrow \text{M}(W, G) \longrightarrow \text{M}(W, G/A) \longrightarrow 1$$

*is exact and splits.*

PROOF. For $\lambda \in \text{M}(W, G)$, $\overline{\lambda} \in \text{M}(W, G/A)$ is given by the composite $p \circ \lambda$. This defines a continuous map $\text{M}(W, G) \longrightarrow \text{M}(W, G/A)$. For $\overline{\lambda} \in \text{M}(W, G/A)$, define $\lambda = s \circ \overline{\lambda}$. Then this defines a splitting $\text{M}(W, G/A) \longrightarrow \text{M}(W, G)$. □

LEMMA 7.6.4. *Let $G$ be a special group, and let*

$$1 \to \text{M}(W, G) \to E_1 \to Q \to 1$$

*be any short exact sequence whose abstract kernel is induced from a homomorphism $\varphi \times \rho : Q \to \text{Out}(G) \times \text{TOP}(W)$. Then it splits. Therefore, $Q \to \text{Out}(\text{M}(W, G))$ lifts to a homomorphism $Q \to \text{Aut}(\text{M}(W, G))$. Consequently, the exact sequence*

$$1 \to \text{Inn}(\text{M}(W, G)) \to E_2 \to Q \to 1$$

*splits also.*

PROOF. (Solvable case). Suppose $G$ is a connected, simply connected completely solvable Lie group (in fact, type (E) is sufficient for our statement). We use induction on the dimension of $G$. If dimension of $G$ is 1, $G$ is Abelian and we are done because $H^2(Q; \text{M}(W, G)) = 0$ (Theorem 7.5.1) which implies that the sequence splits. Suppose the statement is true for groups of dimension less than $n$. Let $G$ be a connected, simply connected solvable Lie group of dimension $n$, and let $N$ be the nilradical (the maximal connected normal nilpotent subgroup) of $G$. Then the center of $N$, $\mathcal{Z}(N) \cong \mathbb{R}^k$ for some $k > 0$, is a nontrivial, connected Abelian Lie group is characteristic in $N$. Therefore $\mathcal{Z}(N)$ is characteristic in $G$. Furthermore, every connected normal subgroup of $G$ and the quotient group $G_1 = G/\mathcal{Z}(N)$ are again of the same type (this is also true for Lie groups of type (E)).

$\mathrm{M}(W, \mathcal{Z}(N))$ is normal in $E_1$ since it is characteristic in $\mathrm{M}(W, G)$. By Lemma 7.6.3, $\mathrm{M}(W, G)/\mathrm{M}(W, \mathcal{Z}(N)) = \mathrm{M}(W, G_1)$. Thus we get the following commutative diagram of exact sequences.

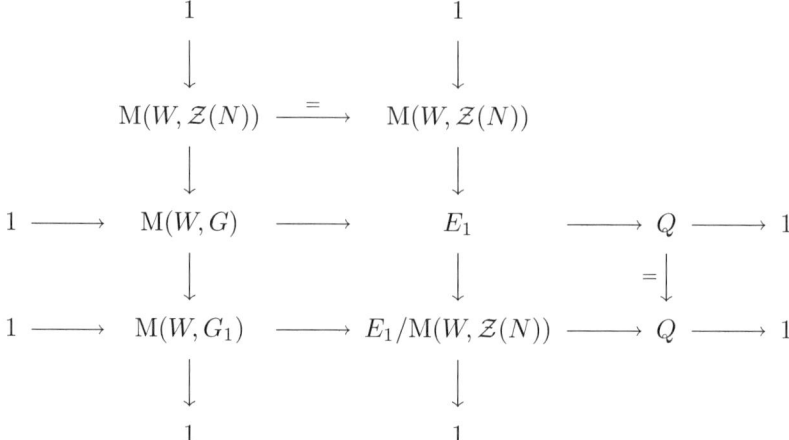

Since $G_1$ is simply connected solvable again, the bottom sequence splits by the induction assumption. Pick a section (homomorphism) $s : Q \to E_1/\mathrm{M}(W, \mathcal{Z}(N))$, and form a pullback of the second vertical short exact sequence via $s$ to get

$$1 \longrightarrow \mathrm{M}(W, \mathcal{Z}(N)) \longrightarrow E' \longrightarrow s(Q) \longrightarrow 1$$
$$\downarrow \qquad\qquad \downarrow \qquad\qquad \downarrow$$
$$1 \longrightarrow \mathrm{M}(W, \mathcal{Z}(N)) \longrightarrow E_1 \longrightarrow E_1/\mathrm{M}(W, \mathcal{Z}(N)) \longrightarrow 1.$$

The abstract kernel of $E'$ is still induced by the original homomorphism $\varphi \times \rho : Q \to \mathrm{Out}(G) \times \mathrm{TOP}(W)$. In fact, $\varphi : Q \to \mathrm{Out}(G)$ induces a homomorphism $Q \to \mathrm{Aut}(\mathcal{Z}(G))$. With this abstract kernel, the top sequence splits by Lemma 7.5.1. This yields a splitting of $Q$ into $E_1$.

(Semi-simple case). Assume for the semisimple case of S(4) that $\mathrm{Aut}(G)$ splits as $\mathrm{Inn}(G) \rtimes \mathrm{Out}(G)$. Since G is in adjoint form, $\mathcal{Z}(G) = 0$. Recall that our abstract kernel for the exact sequence $1 \to \mathrm{M}(W, G) \to E_1 \to Q \to 1$ came from

$$Q \to \mathrm{Out}(G) \times \mathrm{TOP}(W) \to \mathrm{Out}(\mathrm{M}(W, G)).$$

Since $\mathrm{Aut}(G)$ splits as $\mathrm{Inn}(G) \rtimes \mathrm{Out}(G)$, $Q \to \mathrm{Out}(G)$ lifts to $Q \to \mathrm{Out}(G) \to \mathrm{Inn}(G) \rtimes \mathrm{Out}(G) = \mathrm{Aut}(G)$, and our abstract kernel lifts to a homomorphism

$$Q \to \mathrm{Aut}(G) \times \mathrm{TOP}(W) \to \mathrm{Aut}(\mathrm{M}(W, G)).$$

Therefore, one can form a semidirect product $\mathrm{M}(W, G) \rtimes Q$. However, since $\mathcal{Z}(G) = 0$, this semidirect product is the only such an extension so that our original $E_1$ is congruent to $\mathrm{M}(W, G) \rtimes Q$.

The last statement is obvious now because

$$\mathrm{Inn}(\mathrm{M}(W, G)) = \mathrm{M}(W, G)/\mathcal{Z}(\mathrm{M}(W, G)) = \mathrm{M}(W, G)/\mathrm{M}(W, \mathcal{Z}(G)) = \mathrm{M}(W, G). \quad \square$$

7.6.5. The previous lemma and Subsection 7.6.2 ensure the two conditions of Theorem 7.2.4 are satisfied. Thus there is a homomorphism $\theta : G \to \mathrm{TOP}_G(G \times W)$

## 7.6. PROOF OF THE CONSTRUCTION THEOREM

which fits into the diagram

$$
\begin{array}{ccccccccc}
1 & \longrightarrow & G & \longrightarrow & E & \longrightarrow & Q & \longrightarrow & 1 \\
& & \ell \downarrow & & \theta \downarrow & & \downarrow & & \\
1 & \longrightarrow & \mathrm{M}(W,G) \rtimes \mathrm{Inn}(G) & \longrightarrow & \mathrm{TOP}_G(G \times W) & \longrightarrow & \mathrm{Out}(G) \times \mathrm{TOP}(W) & \longrightarrow & 1.
\end{array}
$$

Now we stack the above diagram with the one in Subsection 7.6.1 to obtain a homomorphism $\Pi \to \mathrm{TOP}_G(G \times W)$:

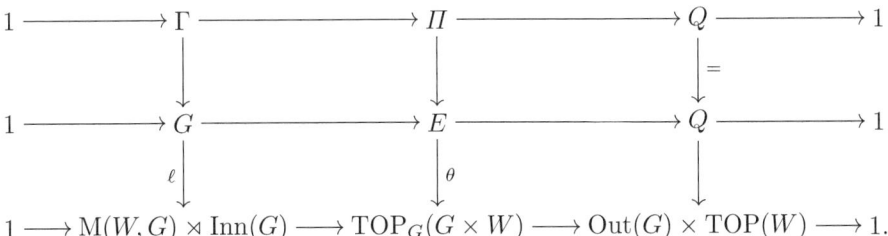

This completes the proof of the existence part (1) of our Theorem 7.3.2.

7.6.6 (Proof of uniqueness). For the uniqueness, we appeal to Theorem 5.7.2. Since the centralizer of $\ell(G)$ in $\mathrm{M}(W,G) \rtimes \mathrm{Inn}(G)$ is $\mathrm{M}(W,G)$, we need to show that $H^1(Q; \mathrm{M}(W,G))$ is a singleton set.

Let $\eta : Q \to \mathrm{M}(W,G)$ be a crossed homomorphism. We want to find $d \in \mathrm{M}(W,G)$ such that
$$\eta(\alpha) = d \cdot \psi(\alpha)(d^{-1})$$
for all $\alpha \in Q$.

If $G$ is Abelian, then Lemma 7.5.1 shows that $H^1(Q; \mathrm{M}(W,G)) = 0$. We use induction for the cases (2) and (3). Suppose $G$ is completely solvable. Let $G_1 = G/\mathcal{Z}(N)$, where $N$ is the nilradical (the maximal connected normal nilpotent subgroup) of $G$. As is well known, $\mathcal{Z}(N)$ is nontrivial so that $\dim(G_1) < \dim(G)$. Since $\mathcal{Z}(N)$ is a characteristic subgroup of $G$, the homomorphism $\psi : Q \to \mathrm{Aut}(\mathrm{M}(W,G))$ induces homomorphisms

$$\hat{\psi} : Q \longrightarrow \mathrm{Aut}(\mathrm{M}(W, \mathcal{Z}(N))) \quad \text{and} \quad \bar{\psi} : Q \longrightarrow \mathrm{Aut}(\mathrm{M}(W, G_1)).$$

By the induction hypothesis, $H^1_{\bar{\psi}}(Q; \mathrm{M}(W, G_1))$ is trivial. Therefore, there exists $c \in \mathrm{M}(W,G)$ such that its image $\bar{c} \in \mathrm{M}(W, G_1)$ satisfies

$$\bar{\eta}(\alpha) = \bar{c} \cdot \bar{\psi}(\alpha)(\bar{c}^{-1})$$

for all $\alpha \in Q$. This implies that $c^{-1} \cdot \eta(\alpha) \cdot \psi(\alpha)(c)$ lies in $\mathrm{M}(W, \mathcal{Z}(N))$. Let

$$\lambda : Q \to \mathrm{M}(W, \mathcal{Z}(N))$$

be defined by

(7.6.1) $$\lambda(\alpha) = c^{-1} \cdot \eta(\alpha) \cdot \psi(\alpha)(c).$$

Some calculation yields

$$\lambda(\alpha\beta) = \lambda(\alpha) \cdot \psi(\alpha)(\lambda(\beta)),$$

which shows that $\lambda \in Z^1(Q; \mathrm{M}(W, \mathcal{Z}(N)))$ with the action of $Q$ on $\mathcal{Z}(N)$ via $\hat{\psi}$. Since $\mathcal{Z}(N) = \mathbb{R}^k$ for some $k$, we have $H^1(Q; \mathrm{M}(W, \mathcal{Z}(N))) = 0$. This implies that there exists an element $\hat{c} \in \mathrm{M}(W, \mathcal{Z}(N))$ such that

$$\lambda(\alpha) = \hat{c} \cdot \hat{\psi}(\alpha)(\hat{c}^{-1}) \tag{7.6.2}$$

for all $\alpha \in Q$.

Let $d = \hat{c} \cdot c \in \mathrm{M}(W, G)$. Then clearly we have

$$\begin{aligned}
\eta(\alpha) &= c \cdot \lambda(\alpha) \cdot \psi(\alpha)(c^{-1}) \quad \text{(from (7.6.1))} \\
&= c \cdot (\hat{c} \cdot \hat{\psi}(\alpha)(\hat{c}^{-1})) \cdot \psi(\alpha)(c^{-1}) \quad \text{(from (7.6.2))} \\
&= (c \cdot \hat{c}) \cdot \psi(\alpha)(c \cdot \hat{c})^{-1} \\
&= d \cdot \psi(\alpha)(d^{-1})
\end{aligned}$$

as we desired. This completes the proof of the uniqueness part (2) of Theorem 7.3.2 for types (S1), (S2), and (S3).

The uniqueness (and the rigidity) for the semisimple case is not true. See Example 7.4.7.

7.6.7 (Proof of rigidity). Let $\theta_1, \theta_2 : \Pi \to \mathrm{TOP}_G(G \times W)$ be two homomorphisms which fit into the diagram (7.3.1), and suppose there exists $h \in \mathrm{TOP}(W)$ such that $\rho_2 = \mu(h) \circ \rho_1$. By the ULIEP, there exists $a \in \mathrm{Aut}(G)$ such that $\ell_2 = a \circ \ell_1$. Let $\theta_1' = \mu(a, h) \circ \theta_1$. Then $\theta_1'$ and $\theta_2$ have the same $\ell$ and $\rho$. Now one can apply the uniqueness to find $\lambda \in \mathrm{M}(W, G)$ to have

$$\theta_2 = \mu(\lambda) \circ \theta_1' = \mu(\lambda, a, h) \circ \theta_1.$$

This concludes the proof of Theorem 7.3.2. □

7.6.8. When G is of type S(4) and where $\mathrm{Aut}(G)$ splits as $\mathrm{Inn}(G) \rtimes \mathrm{Out}(G)$, we can prove a stronger existence theorem. Replace $\mathrm{M}(W, G)$ by the subgroup $r(G) \rtimes (\mathrm{Aut}(G) \times \mathrm{TOP}(W)) \subset \mathrm{TOP}_G(G \times W)$. Then there is a homomorphism $\theta : E \to r(G) \rtimes (\mathrm{Aut}(G) \times \mathrm{TOP}(W))$, with $\theta|_G = \ell$ on $G$. The reason is that (1) $H^2(Q; \mathcal{Z}(G)) = 0$ trivially, and (2) the homomorphism $Q \xrightarrow{\varphi \times \rho} \mathrm{Out}(G) \times \mathrm{TOP}(W) \to \mathrm{Out}(r(G))$ lifts to a homomorphism $Q \longrightarrow \mathrm{Aut}(r(G))$. Thus $E$ maps into

$$r(G) \rtimes (\mathrm{Aut}(G) \times \mathrm{TOP}(W)) = (\ell(G) \rtimes \mathrm{Aut}(G)) \times \mathrm{TOP}(W).$$

This induces two homomorphisms $E \to \ell(G) \rtimes \mathrm{Aut}(G)$ and $E \to \mathrm{TOP}(W)$. Consequently, the action of $E$ on $G \times W$ is diagonal.

For many $G$ of type S(4), $\mathrm{Aut}(G)$ actually splits as $\mathrm{Inn}(G) \rtimes \mathrm{Out}(G)$. As far as we know, there is no general statement of the splitting of $\mathrm{Aut}(G)$.

7.6.9. In [**RW77**], a slightly different construction, valid for any $G$ of type S(4), is given. Instead of $G \times W$ being the uniformizing space, $G/K \times W$ is chosen, where $K$ is a maximal compact subgroup of $G$. Consequently, $G/K$ is a simply connected symmetric space. The regular fiber (Section 4.3) becomes the locally symmetric space $F = \Gamma \backslash G/K$ and singular fiber is the quotient of $F$ by a *finite group of isometries*. The space $E$ has a finite covering $E'$ (perhaps branched), where $E'$ is $F \times (Q' \backslash W)$ and $Q'$ is the kernel of $Q \to \mathrm{Out}(\Gamma)$. Since $\mathrm{Out}(\Gamma)$ is finite, $Q'$ has finite index in $Q$ and the finite group $Q/Q'$ acts diagonally on $F \times (Q' \backslash W)$ (cf. also [**LR96**]). We also obtain uniqueness and rigidity in this context. We shall study this in more detail in Chapter 9.

Theorem 7.3.2 states the main existence, uniqueness, and rigidity theorems for an extension $1 \to \Gamma \to \Pi \to Q \to 1$ where $\Gamma$ is a special lattice in $G$. For some applications we need the following variations.

THEOREM 7.6.10 (Strong rigidity, rigidity for distinct extensions). *Suppose there exists the commutative diagram*

$$\begin{array}{ccccccccc} 1 & \to & \Gamma & \to & \Pi & \to & Q & \to & 1 \\ & & \downarrow \hat{\eta} & & \downarrow \eta & & \downarrow \overline{\eta} & & \\ 1 & \to & \Gamma' & \to & \Pi' & \to & Q' & \to & 1 \end{array} \qquad \begin{array}{ccc} Q & \xrightarrow{\rho} & \text{TOP}(W) \\ \downarrow \overline{\eta} & & \downarrow \mu(h) \\ Q' & \xrightarrow{\rho'} & \text{TOP}(W), \end{array}$$

*where $\Gamma, \Gamma'$ are special lattices in $G$ and $\eta$ is an isomorphism $\Pi \to \Pi'$ inducing the isomorphisms $\hat{\eta}$ and $\overline{\eta}$, and $h \in \text{TOP}(W)$. Suppose $\theta$ is a Seifert Construction for $1 \to \Gamma \to \Pi \to Q \to 1$ with $\epsilon : \Gamma \to G$ and $\rho : Q \to \text{TOP}(W)$ and that $\theta'$ is a Seifert Construction for $1 \to \Gamma' \to \Pi' \to Q' \to 1$ with $\epsilon' : \Gamma' \to G$ and $\rho' : Q \to \text{TOP}(W)$. Then there exists $\hat{h} \in \text{TOP}_G(G \times W)$ such that $\theta' \circ \eta = \mu(\hat{h}) \circ \theta$.*

PROOF. Suppose $\Pi = \Gamma \times_{(\tilde{\varphi},f)} Q$. The map $\tilde{\varphi} : Q \to \text{Aut}(\Gamma)$ arises from conjugation by a lift of $Q$ to $\Pi$. Then $\eta$ induces a compatible lift of $\overline{\eta}(Q) = Q'$ to $\Pi'$, and $\Pi' = \Gamma' \times_{(\tilde{\varphi}',f')} Q'$, where $f'_{(\overline{\eta}(q_1),\overline{\eta}(q_2))} = \hat{\eta}(f(q_1,q_2))$.

From the definition of $\theta'$, the following is commutative:

$$\begin{array}{ccccccccc} 1 & \to & \Gamma' & \to & \Pi' & \to & Q' & \to & 1 \\ & & \downarrow {}_\rho|\epsilon' & & \downarrow \theta' & & \downarrow \varphi' \times \rho' & & \\ 1 & \to & \ell(G) \times \text{M}(W,G) & \to & \text{TOP}_G(G \times W) & \to & \text{Out}(G) \rtimes \text{TOP}(W) & \to & 1, \end{array}$$

where $\tilde{\varphi}'_{\epsilon'} : Q' \xrightarrow{\tilde{\varphi}'} \text{Aut}(\Gamma') \to \text{Aut}(G)$. Define $\theta'' = \mu(1,1,h^{-1}) \circ \theta' \circ \eta$. Then $\theta''$ is another construction for $1 \to \Gamma \to \Pi \to Q \to 1$ with $\epsilon'' = \epsilon' \circ \hat{\eta}$ and $\rho$.

Note that $\rho = \mu(1,1,h^{-1}) \circ \rho' \circ \overline{\eta}$ and $\tilde{\varphi}_{\epsilon''} = \tilde{\varphi}'_{\epsilon'} \circ \hat{\eta}$. Therefore, $\theta$ and $\theta''$ have the same $\rho$ but different $\epsilon$ and $\epsilon''$. There exists an automorphism $a \in \text{Aut}(G)$ so that $\epsilon = a \circ \epsilon''$. Then, there is, by uniqueness, $\lambda \in \text{M}(W,G)$ such that $\theta = \mu(\lambda,1,1) \circ \theta'' = \mu(\lambda,1,1) \circ \mu(1,a,1) \circ \mu(1,1,h^{-1}) \circ \theta' \circ \eta = \mu(\lambda,a,h^{-1}) \circ \theta' \circ \eta$. □

Note that under the hypothesis above, there is induced a homeomorphism $\theta(\Pi) \backslash (G \times W) \to \theta'(\eta(\Pi)) \backslash (G \times W) = \theta'(\Pi') \backslash (G \times W)$ which maps the fibers of the first onto those of the second. In Subsection 7.4.3, we called such a map a *Seifert isomorphism*.

## 7.7. When is $\theta$ injective?

7.7.1. We assume that there exists a construction and study when it will be injective. So let $\rho : Q \to \text{TOP}(W)$ be a proper action of a discrete group $Q$, and $1 \to \Gamma \to \Pi \to Q \to 1$ an extension with abstract kernel $\varphi : Q \to \text{Out}(G)$. Suppose there is a homomorphism $\theta : \Pi \to \text{TOP}_G(G \times W)$ so that

$$\begin{array}{ccccccccc} 1 & \to & \Gamma & \to & \Pi & \to & Q & \to & 1 \\ & & \downarrow \ell & & \downarrow \theta & & \downarrow \varphi \times \rho & & \\ 1 & \to & \text{M}(W,G) \rtimes \text{Inn}(G) & \to & \text{TOP}_G(G \times W) & \to & \text{Out}(G) \rtimes \text{TOP}(W) & \to & 1 \end{array}$$

is commutative. If $\varphi \times \rho$ is injective, then obviously $\theta$ is also injective. Let
$$K = \ker(\varphi \times \rho).$$
Then $K$ is finite since the $Q$-action is proper. Let
$$1 \longrightarrow \Gamma \longrightarrow \Pi_K \longrightarrow K \longrightarrow 1$$
be the pullback sequence from $1 \to \Gamma \to \Pi \to Q \to 1$ induced by the inclusion $i : K \to Q$. Then, $\theta$ is injective if and only if $\theta|_{\Pi_K}$ is injective.

LEMMA 7.7.2. *For a solvable Lie group of type* (E), $\mathrm{M}(W,G)/\mathcal{Z}(G)$ *has no element of finite order.*

PROOF. Let $\lambda \in \mathrm{M}(W,G)$ be an element which has finite order in the quotient. That is, $\lambda^p \in \mathcal{Z}(G)$. This means that
$$(\lambda(w))^p = a \in \mathcal{Z}(G).$$
Since $G$ is of type (E), there exists only one solution to $x^p = a$; see Proposition 6.3.4 for properties of type (E). Since $\lambda(w) \in G$, $\lambda(w)$ must be that solution. The above equality holds for every $w \in W$. Thus, $\lambda$ is a constant function. That is, $\lambda(w) = b \in G$ with $b^p = a$. Therefore, $\lambda = b$, a constant map $W \to \mathcal{Z}(G)$ so that $\lambda$ represents the identity element of $\mathrm{M}(W,G)/\mathcal{Z}(G)$. □

PROPOSITION 7.7.3. *For a solvable Lie group of type* (E),
(1) $\theta(\Pi_K) \subset \ell(G)$,
(2) $\ker(\theta) = \mathrm{Torsion}(\Pi_K)$.

PROOF. (1) Consider the commutative diagram of exact rows

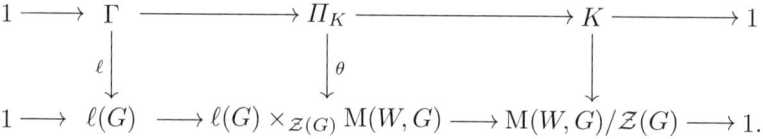

Since $\mathrm{M}(W,G)/\mathcal{Z}(G)$ is torsion free by Lemma 7.7.2 and $K$ is a finite group, $K$ maps trivially into $\mathrm{M}(W,G)/\mathcal{Z}(G)$. Thus, $\Pi_K$ maps into $\ell(G)$.

(2) Let $\tilde{\alpha} \in \Pi_K$ with $\tilde{\alpha}^p = 1$. Then $(\theta(\tilde{\alpha}))^p = \theta(\tilde{\alpha}^p) = 1$ with $\theta(\tilde{\alpha}) \in \ell(G)$ by (1). But $\ell(G)$ has no torsion so that $\theta(\tilde{\alpha}) = 1$.

Conversely, suppose $\theta(\tilde{\alpha}) = 1$. Then $\tilde{\alpha} \in \ker(\varphi \times \rho) = \Pi_K$. Let $q$ be the order of the finite group $K$. Then $\tilde{\alpha}^q \in \Gamma$ and $\Gamma$ is torsion free. Thus, $\tilde{\alpha}^q = 1$. □

COROLLARY 7.7.4. *For a solvable Lie group of type* (E),
(1) $\theta$ *is injective if and only if* $\Pi_K$ *is torsion free.*
(2) $\theta(\Pi) \cap \ell(G) = \theta(\Pi_K)$.
(3) *If* $\theta$ *is injective,* $\theta(\Pi) \cap \ell(G) = \theta(\Gamma)$ *if and only if* $\varphi \times \rho$ *is injective.*

PROOF. (1) If $\Pi_K$ is torsion free, then $\ker(\theta) = \mathrm{Torsion}(\Pi_K)$ is trivial. Conversely, suppose $\Pi_K$ has a torsion element $\tilde{\alpha}$. Then $\theta(\hat{\alpha}) \in \ell(G)$ by Proposition 7.7.3(1). Since $\ell(G)$ is torsion free, $\theta(\hat{\alpha})$ must be trivial.

(2) $\varphi(\ell(G))$ is trivial, and also $\ell(G)$ acts trivially on the base space $W$. That is, $(\varphi \times \rho)(\ell(G)) = 1$. $\theta(\Pi) \cap \ell(G) \subset \ker(\varphi \times \rho) = \theta(\Pi_K)$.

(3) This is obvious. □

## 7.8. Smooth case

If $W$ is a smooth manifold and $Q$ acts on it smoothly so that $\rho : Q \to \mathrm{Diff}(W)$, then the construction can be done smoothly. The universal group is now $\mathcal{C}(W, G) \rtimes (\mathrm{Aut}(G) \times \mathrm{Diff}(W))$, which is certainly the subgroup of weakly $G$-equivariant diffeomorphisms of $G \times W$; see Subsection 4.6.3. In the statement of the main theorem, Theorem 7.3.2, one can replace $\mathrm{TOP}(W)$, $\mathrm{M}(W, G)$, $\mathrm{TOP}_G(P)$ by $\mathrm{Diff}(W)$, $\mathcal{C}(W, G)$ and $\mathrm{Diff}_G(P)$, respectively. All the facts such as existence, uniqueness, and rigidity in the smooth category rely on the vanishing of $H^i(Q; \mathcal{C}(W, \mathbb{R}^k)) = 0$ for $i > 0$. The proof of the vanishing in the smooth case is the same the proof in the continuous case because of the existence of the smooth partition of unity. In the smooth category, we refer to the construction as the smooth Seifert fiber space construction.

If $W$ is a complex manifold and $Q$ acts holomorphically, then one can also ask whether the construction can be done holomorphically on $\mathbb{C}^k \times W$. There are two types of obstructions. It could happen that the necessary $\mathbb{C}^k$-bundle over $W$, while trivial as a smooth bundle, may not be trivial as a holomorphic bundle. A more serious matter is that $H^2\left(Q; \mathcal{H}(W, \mathbb{C}^k)\right)$, where $\mathcal{H}(W, \mathbb{C}^k)$ denotes holomorphic maps, does not necessarily vanish and so not every smooth realization has a holomorphic one. Even when the first difficulty does not arise, the latter one may still persist. The solution to these problems and the corresponding theory is carefully worked out in [**CR72b**]. See Section 14.10 for illustrations of the holomorphic Seifert construction. One particular feature of the extended theory is that the uniformizing space need no longer be a product $G \times W$ but may be any principal $G$-bundle over $W$. See [**LR89**] for Seifert fibered spaces modeled on principal bundles and Section 14.10 for an illustration of the holomorphic theory.

Another general approach to Seifert fiber spaces is due to A. Holmann [**Hol64**]. His procedure extends the classical fiber bundle methods.

When $P$ is not a trivial bundle, as in Chapter 4 and in the aforementioned holomorphic case, one must proceed from the local situation over which the bundle restricts to a product and then piece things together over the overlaps. This was introduced by [**Hol64**], [**CR72b**], and [**LR89**]. See Chapter 10 for more details.

CHAPTER 8

# Generalization of Bieberbach's Theorems

## 8.1. Bieberbach's theorems

8.1.1. We recall some notation first. Let $G$ be a Lie group, and let $\mathrm{Aut}(G)$ be the group of continuous automorphisms of $G$. The group $\mathrm{Aff}(G)$ is the semidirect product $\mathrm{Aff}(G) = G \rtimes \mathrm{Aut}(G)$ with multiplication

$$(a, A) \cdot (b, B) = (a \cdot A(b), AB).$$

It has a Lie group structure and acts on $G$ by

$$(a, A) \cdot x = a \cdot A(x)$$

for all $x \in G$. With the linear connection on $G$ defined by the left invariant vector fields, it is known that $\mathrm{Aff}(G)$ is the group of connection-preserving diffeomorphisms of $G$; see Subsection 4.4.2.

For $G = \mathbb{R}^n$, $\mathrm{Aut}(\mathbb{R}^n) = \mathrm{GL}(n, \mathbb{R})$ and $\mathrm{Aff}(\mathbb{R}^n) = \mathbb{R}^n \rtimes \mathrm{GL}(n, \mathbb{R})$; and $\mathrm{Aff}(\mathbb{R}^n)$ contains the Euclidean group $E(n) = \mathbb{R}^n \rtimes \mathrm{O}(n)$, the group of isometries of $\mathbb{R}^n$. By a *crystallographic group*, we shall mean a discrete cocompact subgroup of $E(n)$. A torsion-free crystallographic group is called a *Bieberbach group*. The following three theorems have been proven by Bieberbach.

THEOREM 8.1.2 (Bieberbach).
(A) *Let $\Pi \subset \mathbb{R}^n \rtimes \mathrm{O}(n)$ be a crystallographic group. Then $\Gamma = \Pi \cap \mathbb{R}^n$ is a lattice of $\mathbb{R}^n$, and $\Gamma$ has finite index in $\Pi$.*
(B) *Let $\Pi, \Pi' \subset \mathbb{R}^n \rtimes \mathrm{O}(n)$ be crystallographic groups. Then every isomorphism $\theta : \Pi \to \Pi'$ is a conjugation by an element of $\mathbb{R}^n \rtimes \mathrm{GL}(n, \mathbb{R})$.*
(C) *In each dimension $n$, there are only finitely many flat manifolds up to affine diffeomorphism.*

COROLLARY 8.1.3. (A) *Every flat Riemannian manifold is finitely covered by a flat torus.*
(B) *Homotopy equivalent flat manifolds are affinely diffeomorphic.*

8.1.4. The Bieberbach theorems have been generalized to nilpotent Lie groups, or even to some solvable Lie groups by Auslander ([**Aus60**], [**Aus61a**]), Lee, Lee-Raymond, and Dekimpe-Lee-Raymond ([**Lee88**], [**Lee95a**], [**LR85a**], [**DLR96**]), etc. In this chapter, we shall discuss such generalizations.

We shall use the term lattice of a Lie group $G$ in this section to denote a discrete cocompact subgroup of $G$.

## 8.2. Proof of Bieberbach's theorems

The following proof is based on an argument of F. Bonahon and L. Siebenmann [**BS83**, Y-13].

THEOREM 8.2.1 (The First Bieberbach Theorem). ; *Let $\Pi \subset E(n) = \mathbb{R}^n \rtimes O(n)$ be a crystallographic group. Then $\Pi \cap \mathbb{R}^n \cong \mathbb{Z}^n$ and $\mathbb{Z}^n$ is discrete in $\mathbb{R}^n$; $\Pi \cap \mathbb{R}^n$ is the unique maximal normal Abelian subgroup of $\Pi$, and $\Pi/(\Pi \cap \mathbb{R}^n)$ is a finite subgroup of $O(n)$ and is called the* holonomy group.

PROOF. We make some initial observations. Let $h : \mathbb{R}^n \rtimes O(n) \to O(n)$ be the natural projection. Conjugation by an element $\gamma \in E(n)$ is given by

$$\gamma v \gamma^{-1} = h(\gamma)(v),$$

for $v \in \mathbb{R}^n \subset \mathbb{R}^n \rtimes O(n)$. Let

$$\Gamma = \Pi \cap \mathbb{R}^n.$$

Then $\Pi/\Gamma$ maps injectively into $O(n)$.

The group $\Gamma = \Pi \cap \mathbb{R}^n$ is a discrete subgroup of $\mathbb{R}^n$ because $\mathbb{R}^n$ is a closed (normal) subgroup of $E(n)$. Thus $\Gamma \cong \mathbb{Z}^k$ for some $k \geq 0$, and it spans a subvector space of dimension $k$ in $\mathbb{R}^n$ denoted by $V_0$. Recall $h : E(n) \to O(n)$ is the canonical projection, and put $h(\Pi) = \Pi_\infty = \Pi/\Gamma \subset O(n)$.

Observe that $\Pi$ leaves $V_0$ invariant because $\Gamma$ is normal in $\Pi$. Form the pushout $V_0 \cdot \Pi$. Then $V_0$ is normal in $V_0 \cdot \Pi$. Also let $\overline{V_0 \cdot \Pi}$ denote its closure in $E(n)$. We claim $\overline{V_0 \cdot \Pi}$ leaves $V_0$ invariant. For if $\gamma \in \overline{V_0 \cdot \Pi}$ and $v \in V_0$, then if $\gamma v \gamma^{-1} \notin V_0$, a closed subgroup of $E(n)$, then there exists $V$ a neighborhood of $\gamma$ in $\overline{V_0 \cdot \Pi}$ such that $\gamma' v {\gamma'}^{-1} \notin \overline{V_0 \cdot \Pi}$ for $\gamma' \in V$, since $\overline{V_0 \cdot \Pi}$ is closed. This contradicts that there exists a sequence $\{\gamma_i\}$ converging to $\gamma$ such that $\gamma_i v \gamma_i^{-1} \in \overline{V_0 \cdot \Pi}$ for $i$ sufficiently large. Observe that

$$(V_0 \cdot \Pi)/V_0 = \Pi/(\Pi \cap V_0) = \Pi/\Gamma = \Pi_\infty.$$

Let $\overline{\Pi}_\infty$ be the closure of $\Pi_\infty$ in $O(n)$.

Consider the following commutative diagram of short exact sequences.

$$\begin{array}{ccccccccc}
1 & \longrightarrow & \Gamma & \longrightarrow & \Pi & \longrightarrow & \Pi_\infty & \longrightarrow & 1 \\
& & \downarrow & & \downarrow & & \downarrow = & & \\
1 & \longrightarrow & V_0 & \longrightarrow & V_0 \cdot \Pi & \longrightarrow & \Pi_\infty & \longrightarrow & 1 \\
& & \downarrow = & & \downarrow \cap & & \downarrow \cap & & \\
1 & \longrightarrow & V_0 & \longrightarrow & \overline{V_0 \cdot \Pi} & \longrightarrow & \overline{V_0 \cdot \Pi}/V_0 & \longrightarrow & 1
\end{array}$$

Then $\overline{\Pi}_\infty \subset \overline{V_0 \cdot \Pi}/V_0$. If not equal, then the inverse image of $\overline{\Pi}_\infty$ in $\overline{V_0 \cdot \Pi}$ would give a closed subset of $\overline{V_0 \cdot \Pi}$ between $V_0 \cdot \Pi$ and $\overline{V_0 \cdot \Pi}$. So $\overline{\Pi}_\infty = \overline{V_0 \cdot \Pi}/V_0$.

We want to prove now that

(1) $V_0 = \mathbb{R}^n$ (therefore $\Gamma = \mathbb{R}^n \cap \Pi \cong \mathbb{Z}^n$), and
(2) $\Pi_\infty$ is a finite group.

Let

$$\Pi_1 = (\overline{\Pi}_\infty)_0$$

be the connected component of the identity of $\overline{\Pi}_\infty$. Take its inverse image in $\overline{V_0 \cdot \Pi}$. The inverse image will have kernel $V_0$ which is connected, and so the inverse image will be connected. Also it will be the connected component of the identity of $\overline{V_0 \cdot \Pi}$. Denote this connected component identity by $\overline{(V_0 \cdot \Pi)}_0$.

Also observe that
$$\overline{V_0 \cdot \Pi}/\overline{(V_0 \cdot \Pi)}_0 = (\overline{V_0 \cdot \Pi}/V_0)/((\overline{V_0 \cdot \Pi})_0/V_0)$$
$$= \overline{\Pi}_\infty/(\overline{\Pi}_\infty)_0$$
$$= \text{finite group}.$$

As mentioned earlier, $\overline{V_0 \cdot \Pi}$ leaves $V_0$ invariant. Also $V_0$ fixes $\Gamma$, because $\Gamma \subset V_0 \subset \mathbb{R}^n$. Therefore, $\overline{V_0 \cdot \Pi}$ leaves $\Gamma$ invariant. Then $\overline{(V_0 \cdot \Pi)}_0$, which is connected, must also fix $\Gamma$, which is discrete. Consequently $\overline{(V \cdot \Pi)}_0$ fixes $V_0$. This implies the representation $\Pi_1 \to \text{Aut}(\mathbb{R}^n)$ is trivial on $V_0$ and faithful on $V_0^\perp \subset \mathbb{R}^n$.

Now, the orbit space of the action of $\overline{(V \cdot \Pi)}_0$ on $E(n)/O(n) \cong \mathbb{R}^n$ is given by
$$\mathbb{R}^n/\overline{(V \cdot \Pi)}_0 = (V_0 \oplus V_0^\perp)/\overline{(V_0 \cdot \Pi)}_0 = V_0^\perp/\Pi_1.$$

If $V_0^\perp \neq 0$, then it is a Euclidean subspace. Since $\Pi_1$ is compact and connected, $V_0^\perp/\Pi_1$ *is not compact*. Until now we have not used the hypothesis that $\mathbb{R}^n/\Pi$ is compact. So now assume that and we shall show that $V_0^\perp \neq 0$ leads to a contradiction which forces $V_0$ to be $\mathbb{R}^n$.

Thus, assume $\mathbb{R}^n/\Pi$ is compact. In the diagram
$$\mathbb{R}^n/\Pi \xrightarrow{q_1} \mathbb{R}^n/\overline{V_0 \cdot \Pi} \xleftarrow{q_2} \mathbb{R}^n/\overline{(V_0 \cdot \Pi)}_0,$$

$q_1$ has compact image $\mathbb{R}^n/\overline{V_0 \cdot \Pi}$. But we have seen that $\overline{V_0 \cdot \Pi}/\overline{(V_0 \cdot \Pi)}_0$ is a finite group. So, this implies that $\mathbb{R}^n/\overline{(V_0 \cdot \Pi)}_0$ is compact, and above we saw that $V_0^\perp/\Pi_1$ is compact only if $V_0^\perp = 0$ and $\Pi_1$ is a point. Therefore, $\overline{\Pi}_0 = \overline{\Pi}_\infty$ is a finite group and $\Gamma = \mathbb{Z}^n$, a full lattice in $\mathbb{R}^n$.

Next we shall show that $\Gamma = \mathbb{Z}^n$ is the *unique maximal normal Abelian subgroup*. Suppose there existed $A$, an Abelian subgroup of $\Pi$, that contains $\Gamma$. Then

$$\begin{array}{ccccccccc} 1 & \longrightarrow & \Gamma & \longrightarrow & A & \longrightarrow & A/\Gamma & \longrightarrow & 1 \\ & & \downarrow & & \downarrow & & \downarrow & & \\ 1 & \longrightarrow & \Gamma & \longrightarrow & \Pi & \longrightarrow & \Pi_\infty & \longrightarrow & 1. \end{array}$$

Consequently, $A/\Gamma$, if not zero, acts faithfully on $\mathbb{R}^n$ but $\Pi$, as we have seen, leaves $\Gamma$ and $\mathbb{R}^n$ invariant, so $A/\Gamma$ acts faithfully on $\Gamma$.

So let $a \in A$ be such that $h(a) \in A/\Gamma$ and for some $v \in \Gamma$, $h(a)v \neq v$. Then $ava^{-1} = h(a) \cdot v$. This implies that $A$ cannot be Abelian unless $A = \Gamma$. Actually more is true: if $A \subset \Pi$ is Abelian and normal, then $A \subset \Gamma$. $\square$

Note in $E(1) = \mathbb{R} \rtimes \mathbb{Z}_2$, the maximal normal Abelian subgroup in any cocompact subgroup is $\mathbb{Z}$. Observe that $\mathbb{Z}_2$ is also a maximal Abelian subgroup, but it is not normal. Also a group $\Pi$ can be discrete, but $\overline{\Pi}_\infty$ can be nondiscrete if $\Pi$ is not cocompact.

THEOREM 8.2.2 (The Second Bieberbach Theorem). *Let $\Pi$ and $\Pi'$ be crystallographic groups in $E(n)$, and let $\psi : \Pi \to \Pi'$ be an isomorphism. Then there exists $\alpha \in \text{Aff}(\mathbb{R}^n) = \mathbb{R}^n \rtimes \text{GL}(n, \mathbb{R})$ such that $\psi(\beta) = \alpha\beta\alpha^{-1}$, for all $\beta \in \Pi$. That is, an isomorphism of crystallographic groups can be realized by affine change of coordinate. More precisely, let*
$$\theta : \Pi \hookrightarrow E(n) \quad \text{and} \quad \theta' : \Pi' \hookrightarrow E(n)$$
*be the original inclusions. Then there exists $\alpha \in \text{Aff}(\mathbb{R}^n)$ such that*
$$\theta' \circ \psi = \mu(\alpha) \circ \theta.$$

Actually there is a *"fourth" Bieberbach Theorem* which was proved by H. Zassenhaus in 1948. It states that *if $\Pi$ is an abstract group such that $\Pi$ contains a normal free Abelian subgroup $\mathbb{Z}^n$ and $Q = \Pi/\mathbb{Z}^n$ is finite and $Q$ acts faithfully on $\mathbb{Z}^n$, there exists a crystallographic group in $E(n)$ isomorphic to $\Pi$*. (Before Zassenhaus, everything in the literature was proved *assuming* that already things were embedded in $E(n)$. Here it is not assumed that $\Pi \subset E(n)$.)

However we have an easy proof of the Fourth Bieberbach Theorem. Our existence theorem for our Seifert Construction yields this result. Let's see how. First we have the injective homomorphism $\varphi : Q \to \operatorname{Aut}(\mathbb{Z}^n)$ which is in $\operatorname{GL}(n,\mathbb{Z})$. Actually we want it to be in $\operatorname{GL}(n,\mathbb{Z}) \cap \operatorname{O}(n)$. Our Seifert Construction sends $\theta : \Pi \to \operatorname{Aff}(\mathbb{R}^n)$ using $\varphi$. Note that $\theta$ is injective since the kernel of $\varphi \times \rho$ is trivial; see Subsection 7.7.1. The idea is now to take $\theta(\Pi)$ and conjugate $\operatorname{TOP}_{\mathbb{R}^n}(\mathbb{R}^n \times w) = \operatorname{Aff}(n)$ by an element $\alpha \in \operatorname{GL}(n,\mathbb{R})$ so that $\bar{\theta}(Q) \subset \operatorname{GL}(n,\mathbb{R})$ is sent into $\operatorname{O}(n)$. ($\varphi(Q)$ is compact and contained in some maximal compact. All maximal compact subgroups are conjugate, so conjugate into standard $\operatorname{O}(n)$.) So $\mu(\alpha) \circ \theta$ is our desired embedding in $E(n)$. It also follows that $\theta(\Pi) \cap \ell(\mathbb{R}^n) = \theta(\mathbb{Z}^n)$ (see Corollary 7.7.4), and it is maximal normal Abelian. So we can always think of putting an abstract group into $E(n) \subset \operatorname{Aff}(n)$.

How do we obtain the Second Bieberbach Theorem? Let $\Pi_1$ and $\Pi_2$ be subgroups of $E(n)$, and suppose they are isomorphic. Note, under the isomorphism, the maximal normal Abelian subgroups are mapped isomorphically since they are characteristic:

$$\begin{array}{ccccccccc} 1 & \longrightarrow & \Pi_1 \cap \mathbb{R}^n & \longrightarrow & \Pi_1 & \longrightarrow & \Gamma_\infty^1 & \longrightarrow & 1 \\ & & A \downarrow & & \psi \downarrow & & \bar{\psi} \downarrow & & \\ 1 & \longrightarrow & \Pi_2 \cap \mathbb{R}^n & \longrightarrow & \Pi_2 & \longrightarrow & \Gamma_\infty^2 & \longrightarrow & 1. \end{array}$$

Therefore, there exists an abstract group $\Pi$ fitting into the exact sequence $1 \to \mathbb{Z}^n \to \Pi \to Q \to 1$. We can identify $\Pi$ with $\Pi_1 \subset E(n)$, and we call this embedding $\theta_1$. Then the other group is given by $\psi \circ \theta_1 = \theta_2$. Note that the map $A$ in the above diagram can be thought as a matrix in $\operatorname{GL}(n,\mathbb{R})$. The homomorphism $\theta_1$ followed by conjugation by $(\mathbf{0}, A) \in \operatorname{Aff}(n)$ maps $\mathbb{Z}^n$ exactly the same as $\theta_2$ does; that is,

$$\mu(\mathbf{0}, A) \circ \theta_1|_{\mathbb{Z}^n} = \theta_2|_{\mathbb{Z}^n}.$$

Let $\theta_3 = \mu(\mathbf{0}, A) \circ \theta_1$. The two homomorphisms $\theta_2$ and $\theta_3$ are considered as Seifert Constructions for the same exact sequence $1 \to \mathbb{Z}^n \to \Pi \to Q \to 1$ with the same embedding $\mathbb{Z}^n \to \mathbb{R}^n$ and $Q \to \operatorname{TOP}(W)$ with $W$ a point.

Now by the rigidity of the construction, there exists $a \in \mathbb{R}^n$ such that $\mu(a, I) \circ \theta_3 = \theta_2$. Consequently, we have

$$\mu(a, A) \circ \theta_1 = \theta_2.$$

Actually we have a stronger result. If $\Pi$ is isomorphic to a crystallographic group and $\theta_1$ and $\theta_2$ are two isomorphisms into $\operatorname{Aff}(n)$, *not* necessarily into $E(n)$ but carrying the maximal normal Abelian subgroups into $\mathbb{R}^n$, then they are still conjugate in $\operatorname{Aff}(n)$.

THEOREM 8.2.3 (The Third Bieberbach Theorem). *For each dimension, there exists at most a finite number of isomorphic crystallographic groups.*

8.2.4. How would we prove the Third Bieberbach Theorem? We sketch the stronger result: we will look at *all* the possible abstract crystallographic groups

$$1 \to \mathbb{Z}^n \to \Pi \to Q \to 1.$$

We want to find all the isomorphism classes of the extensions. If finite, then this will imply the theorem. Note that any lattice in a connected Lie group is finitely presented (the lattice here is not necessarily uniform); see [**Wan72**].

By the Second Theorem, isomorphic crystallographic groups are affinely conjugate. Therefore, we only need to count isomorphism classes of such $\Pi$'s.

(1) We can see, by Selberg's Lemma 6.1.8, that $\mathrm{GL}(n,\mathbb{Z})$ (since it is a lattice of $\mathrm{GL}(n,\mathbb{R})$, it is finitely presented) has a normal torsion-free subgroup $D$ such that $\mathrm{GL}(n,\mathbb{Z})/D$ is finite. Then as $\varphi : Q \to \mathrm{GL}(n,\mathbb{Z})$ must be faithful and $Q$ is finite, such a group $Q$ must be a subgroup of $\mathrm{GL}(n,\mathbb{Z})/D$. Thus there are only finitely many finite groups that can be represented into $\mathrm{GL}(n,\mathbb{Z})$.

(2) According to well-known integral representation theory, for each finite group $Q$, there are only a finite number of distinct conjugacy classes of representations into $\mathrm{GL}(n,\mathbb{Z})$; see Charlap [**Cha86**] or Wolf [**Wol77**] for proof.

(3) Given a finite group $Q$ and a fixed representation $\varphi : Q \to \mathrm{GL}(n,\mathbb{Z})$, there are only finitely many equivalence classes, (hence, isomorphism classes) of extensions of $\mathbb{Z}^n$ by $Q$ because $H^2_\varphi(Q,\mathbb{Z}^n)$ is finite.

8.2.5. According to the book [**BBN+78**], we have the following list.

| dimension | 1 | 2 | 3 | 4 |
|---|---|---|---|---|
| number of crystallographic groups | 2 | 17 | 219 (11) | 4783 (12) |
| number of Bieberbach groups | 1 | 2 | 10 (3) | 74 (1) |

The numbers in parenthesis mean extra *enantiomorphic pairs*. For example, in dimension 3, there are ten isomorphism classes of Bieberbach groups. Three of them have the following property: Fix a representation of one group, into $\mathrm{Aff}(\mathbb{R}^3)$, say $\Pi$. Then the conjugacy class of $\Pi$ by $\mathrm{GL}^+(3,\mathbb{R})$ (determinant $+1$) and the conjugacy class of $\Pi$ by $\mathrm{GL}^-(3,\mathbb{R})$ (determinant $-1$) are disjoint. If we use this finer equivalence relation, then there are thirteen classes of Bieberbach groups in dimension 3.

## 8.3. The First Bieberbach Theorem

The purpose of this section is to generalize the First Bieberbach Theorem to more general class of Lie groups. In other words, we investigate the validity of the following statement:

8.3.1. *Let $G$ be a connected, simply connected Lie group, and let $C$ be a compact subgroup of $\mathrm{Aut}(G)$. If $\Pi \subset G \rtimes C$ is a lattice, then $\Gamma = \Pi \cap G$ is a lattice of $G$, and $\Gamma$ has finite index in $\Pi$.*

The statement is true for nilpotent Lie groups as proven by Auslander.

THEOREM 8.3.2. *Let $G$ be a connected, simply connected nilpotent Lie group, and let $C$ be a compact subgroup of $\mathrm{Aut}(G)$. If $\Pi \subset G \rtimes C$ is a lattice, then $\Gamma = \Pi \cap G$ is a lattice of $G$, and $\Gamma$ has finite index in $\Pi$.*

We shall prove this theorem in Subsection 8.3.10 after Lemma 8.3.9.

Certain Lie groups of type (E) allow a generalization of the First Bieberbach Theorem. For this, we need the following definition.

DEFINITION 8.3.3 ([**DLR96**]). Let $G$ be a connected, simply connected solvable Lie group with nilradical $N$. We say that $G$ has the *strong lattice property* if for any torus $T \subseteq \mathrm{Aut}(G)$ and for any lattice $\Gamma$ of $G \rtimes T$, there exists a lattice $\tilde{\Gamma}$ of $G$ such that $N \cap \Gamma = N \cap \tilde{\Gamma}$.

The rest of this section is drawn from [**DLR96**]; see also [**DLR01**].

THEOREM 8.3.4. *Let $G$ be a connected, simply connected solvable Lie group of type* (E), *and let $C$ be a compact subgroup of* $\mathrm{Aut}(G)$. *Suppose $G$ has the strong lattice property. If $\Pi \subset G \rtimes C$ is a lattice, then $\Gamma = \Pi \cap G$ is a lattice of $G$ and $\Gamma$ has finite index in $\Pi$.*

REMARK 8.3.5. Obviously, Theorem 8.3.4 states that the First Bieberbach Theorem generalizes to a given Lie group $G$ of type (E) *if and only if $G$* has the strong lattice property.

The following is an important theorem of Auslander [**Aus61a**, Theorem 1], and it is crucial for generalizations of Bieberbach's theorems. If one is familiar with Theorem 8.24 in [**Rag72**], then the following theorem is easily deduced from it.

THEOREM 8.3.6 ([**Aus61a**, Theorem 1]). *Let $S$ be a connected, simply connected solvable Lie group, and let $C$ be a compact subgroup of* $\mathrm{Aut}(S)$. *Further, let $\Gamma$ be a discrete (not necessarily uniform) subgroup of $S \cdot C$, and let $\overline{\Gamma S}$ be the closure of the group generated by the elements of $\Gamma$ and $S$. Then the identity component of $\overline{\Gamma S}$ is solvable.*

Before we can actually start the proof of Theorem 8.3.4, we need to establish three technical lemmas which will be used later. The third lemma will use the setup of Auslander [**Aus61a**, Theorem 2].

LEMMA 8.3.7. *Let $G$ be a connected, simply connected solvable Lie group of type* (E) *with nilradical $N$. Then $N$ is maximal nilpotent in $G$. Consequently, the elements $g \in G$ acting unipotently on $N$ (via conjugation in $G$) are precisely all the elements of $N$.*

PROOF. Let $h$ be an element of $G$. The group generated by $N$ and $h$ is nilpotent if and only if $h$ acts unipotently on $N$. As $G/N$ is Abelian this is equivalent to $\mathrm{Ad}(h)$ being unipotent (i.e., only has 1 as an eigenvalue). As $\mathrm{Ad}(h) = \exp \mathrm{ad}_{\log(h)}$ (note that the exponential mapping is a bijection in our case, so log is well defined), this is in turn equivalent to $\mathrm{ad}_{\log(h)}$ being nilpotent. From this it now follows that for any element $h'$ in the one parameter subgroup $H$ passing through $h$, $\mathrm{ad}_{\log(h')}$ is nilpotent. This implies that the group $NH$ is a connected, nilpotent normal subgroup of $G$, which must be a part of $N$ and so $h \in N$. □

LEMMA 8.3.8. *Let $T$ be a torus and suppose that $A_1$ is a lattice of $\mathbb{R}^n$ and $A_2$ is a finitely generated subgroup of $\mathbb{R}^n \oplus T$.*

*Let $f : \mathbb{R}^n \oplus T \to \mathrm{GL}(k, \mathbb{R})$ be a faithful continuous representation, such that $A_1$ and $A_2$ leave the same uniform lattice of $\mathbb{R}^k$ invariant, then $A_2 \cap \mathbb{R}^n$ is of finite index in $A_2$.*

PROOF. Without loss of generality, we can assume that $A_1$ and $A_2$ leave the standard lattice $\mathbb{Z}^k$ invariant (i.e., $f(A_i) \subset \mathrm{GL}(k, \mathbb{Z})$, for $i = 1, 2$).

We consider the usual metric $d$ on $\mathrm{GL}(k, \mathbb{R})$ obtained by regarding $\mathrm{GL}(k, \mathbb{R})$ as being a subspace of $\mathbb{R}^{k^2}$, with the usual Euclidean metric.

Pick any element $a_2 \in A_2$, and consider the group $A$ generated by $A_1$ and $a_2$. Then $f(A) \subset \mathrm{GL}(k, \mathbb{Z})$. Since $\mathrm{GL}(k, \mathbb{Z})$ is a lattice of $\mathrm{GL}(k, \mathbb{R})$, $f(A)$ is discrete. However, $f$ is continuous and faithful so that the preimage of a discrete subset is discrete. Therefore, $A$ is discrete in $\mathbb{R}^n \oplus T$. Since $A_1$ is cocompact in $\mathbb{R}^n \oplus T$ already, $A_1$ has finite index in $A$. In particular, the $T$-component of $a_2 \in \mathbb{R}^n \oplus T$ has finite order. Since $A_2$ is finitely generated, this shows that $A_2 \cap \mathbb{R}^n$ is of finite index in $A_2$. $\square$

LEMMA 8.3.9 ([**Aus61a**]). *Let $G$ be a connected, simply connected solvable Lie group of type* (E) *with nilradical $N$, and let $C$ be a compact subgroup of* $\mathrm{Aut}(G)$. *Let $\Pi$ be a lattice of $G \rtimes C$. Denote by $G^*$ the identity component of the closure of $\Pi G$ in $G \rtimes C$, and let $\Pi^* = \Pi \cap G^*$. Then*

(1) *$\Pi^*$ has finite index in $\Pi$,*
(2) *$G^*$ is solvable,*
(3) *The nilradical $N^*$ of $G^*$ coincides with $N$, and*
(4) *$\Pi^* \cap N = \Pi \cap N$ is a lattice of $N$.*

PROOF. Closure of $\Pi G$ contains $G$, so $G \rtimes C/G^*$ is compact. The projection $G \rtimes C \to G \rtimes C/G^*$ carries $\Pi$ to $\Pi/\Pi^*$. The first claim is obvious. Theorem 8.3.6 ensures that $G^*$ is solvable, our second statement follows. Since $G^*/G$ is a compact solvable Lie group, $G^*/G = T$ is Abelian, and the commutator subgroup of $G^*$, $[G^*, G^*]$ is contained in $G$. Since $N$ is a characteristic subgroup of $G$, it is normal in $G^*$ and hence $N \subset N^*$. Moreover, as $N$ is maximal nilpotent in $G$ (see Lemma 8.3.7), $G \cap N^* = N$. It follows that the canonical map $N^*/N \to G^*/G = T$ is faithful. Therefore, $N^*/N$ is compact, and so we can write $N^* = N \cdot T^*$ for some torus $T^* \subset G^*$. (We note here that in the proof of Theorem 2 in [**Aus61a**], Auslander seems to claim that $N^*/N$ is compact, also in case $G$ is not of type (E). However, as Example 8.3.14 will show, this need not be true.) By Proposition 6.3.4(5) of the properties of a solvable Lie group of type (E), $T^*$ is in the center of $N^*$ so that the action of $T^*$ on $N$ is trivial, and $N^*$ splits as a direct product $N^* = N \times T^*$.

We claim that $T^*$ is trivial. For any $t \in T^*$, $x \in N$ and $y \in G$,

$$\begin{aligned}[t,y] \cdot x \cdot [t,y]^{-1} &= tyt^{-1}(y^{-1} \cdot x \cdot y)ty^{-1}t^{-1} \\ &= ty(y^{-1} \cdot x \cdot y)y^{-1}t^{-1} \\ &\text{(since } y^{-1} \cdot x \cdot y \in N \text{ and } t \text{ acts trivially on } N) \\ &= t \cdot x \cdot t^{-1} \\ &= x.\end{aligned}$$

This shows that $[t, y]$ lies in the centralizer of $N$. However, $[t, y] \in [N^*, G] \subset N^* \cap G \subset N$. Consequently, $[t, y]$ lies in the center of $N$. We have

$$[t^p, y] = [t, y]^p \in N$$

for all $p$. Thus, for any $t \in T^*$ of finite order, we must have $[t, y] = 1$. However, the set of torsion elements is dense in $T^*$. Therefore, $[t, G] = 1$ for all $t \in T^*$, i.e. $T^*$ acts trivially on $G$. As $T^*$ is a compact subgroup, it is conjugated inside $G^*$ to a subgroup of (the maximal compact group) $T$. Let $g_* \in G^*$ be such that

$g_*T^*g_*^{-1} \subset T$. It is easy to check that $g_*T^*g_*^{-1}$ also acts trivially on $G$. But as $g_*T^*g_*^{-1} \subset T \subset \text{Aut}(G)$, the group $g_*T^*g_*^{-1}$, and so also $T^*$, has to be trivial.

This shows that $N = N^*$ is the maximal normal nilpotent analytic subgroup of $G^*$. Then, $[G^*, G^*] \subset N$, and $G^*/N$ is Abelian and isomorphic to the direct sum $T \oplus (G/N)$. By the theorem in §5 of Mostow [**Mos54**] $\Pi \cdot N$ is closed in $G^*$. Therefore,
$$\Pi \cap N = \Pi \cap N^*$$
is a lattice in $N$. □

8.3.10 (Proof of Theorem 8.3.4). We can now prove the theorem using the notation and the results of the lemmas above.

As $G$ has the strong lattice property, there is a lattice $\Lambda$ of $G$ with the property that
$$\Lambda \cap N = \Pi^* \cap N = \Delta = \text{ a lattice of the nilradical } N \text{ of } G.$$
Denote $A_1 = \Lambda/(\Lambda \cap N) \subset G/N$ and $A_2 = \Pi^*/(\Pi^* \cap N) \subset (G \rtimes T)/N$. There is an induced action of $(G \rtimes T)/N$ on each factor group $\gamma_i(N)/\gamma_{i+1}(N) \cong \mathbb{R}^{k_i}$ inducing a continuous representation $f_i : (G \rtimes T)/N \to \text{GL}(k_i, \mathbb{R})$.

Moreover, the representation $f_i$ is such that both $A_1$ and $A_2$ leave the lattice $(\Delta \cap \gamma_i(N))/(\Delta \cap \gamma_{i+1}(N))$ of $\gamma_i(N)/\gamma_{i+1}(N)$ fixed.

Now, the representation (where $c$ denotes the nilpotency class of $N$)
$$f = f_1 \oplus f_2 \oplus \cdots \oplus f_c : (G \rtimes T)/N \to \text{GL}(k_1, \mathbb{R}) \oplus \cdots \oplus \text{GL}(k_c, \mathbb{R}) \subset \text{GL}(k_1 + \cdots + k_c, \mathbb{R})$$
is a representation of $(G \rtimes T)/N$. We claim that this representation is faithful. We postpone, however, the proof of this claim until the end.

As $A_1$ and $A_2$ leave the same lattice of $\mathbb{R}^{k_1 + \cdots + k_c}$ invariant, Lemma 8.3.8 implies that $G \cap \Pi^*$ is of finite index in $\Pi^*$ which finishes the proof, provided we can show that $f$ is a faithful representation of $(G \rtimes T)/N = G/N \oplus T$.

For any element $g \in G$, we will use $\bar{g}$ to denote its image in $G/N$. Now, let $g$ and $t$ be such that $f(\bar{g}t) = 1$. As $\bar{g}$ and $t$ commute, we have that $f(\bar{g}t) = f(\bar{g})f(t) = f(t)f(\bar{g})$. The eigenvalues of $f(t)$ are all of modulus 1, and those of $f(\bar{g})$ are equal to 1 or of modulus different from 1 (property (3)(b) of groups of type (E) mentioned above). But as the eigenvalues of $f(\bar{g}t)$ (which are all supposed to be 1) are obtained by multiplying the eigenvalues of $f(\bar{g})$ and $f(t)$ (as they commute), we must have that the eigenvalues of $f(\bar{g})$ are all equal to one. Now, Lemma 8.3.7 implies that $g \in N$ or that $\bar{g} = 1$.

So, assume that $f(t) = 1$. This means that the automorphism $t \in \text{Aut}(G)$ restricts to a unipotent automorphism of $N$. As $t$ also acts trivially on $G/N$, the action of $t$ on $G$ is unipotent. However, the only element of $T$ acting unipotently on $G$ is the identity element, showing that $f$ is faithful. □

THEOREM 8.3.11. *Let $G$ be a connected, simply connected solvable Lie group of type (E) with nilradical $N$, and let $G/N = \mathbb{R}^n$. Let $\rho : \mathbb{R}^n \to \text{Out}(N)$ be the canonical representation. Assume:*

*The centralizer of $\rho(\mathbb{R}^n)$ in $\text{Out}(N)$ has trivial maximal torus.*

*Then Statement 8.3.1 holds for this $G$ and hence $G$ has the strong lattice property.*

PROOF. In Lemma 8.3.9, $N = N^*$. Since $G^*/N$ is connected Abelian, $G^*/N = \mathbb{R}^n \oplus T$, where $T$ is a torus. On the other hand, $G/N = \mathbb{R}^n$. Consider the natural representation $\rho^* : G^*/N = \mathbb{R}^n \oplus T \to \text{Out}(N)$. This restricts to $\rho : G/N = \mathbb{R}^n \to \text{Out}(N)$. Then $\rho(\mathbb{R}^n)$ commutes with $\rho^*(T)$. Since $\rho^*|_T$ is injective, $T$ must be trivial by the hypothesis. This shows that $G^* = G$. By Lemma 8.3.9 (1), $\Gamma = \Pi^* = \Pi \cap G$ has finite index in $\Pi$. By Remark 1, $G$ has the strong lattice property. □

REMARK 8.3.12. Suppose there is a counterexample to Statement 8.3.1. Then one can find a subgroup $\Pi^*$ of $\Pi$, of finite index, such that the image of $\Pi^*$ in $C$ has positive dimensional closure which is a torus. This implies that the centralizer of $\rho(\mathbb{R}^n)$ in $\mathrm{Out}(N)$ has nontrivial maximal torus, violating the condition of Theorem 8.3.11.

EXAMPLE 8.3.13. Let $G = \mathbb{R}^2 \rtimes_\varphi \mathbb{R}$, with
$$\varphi(t) = \begin{pmatrix} e^t & 0 \\ 0 & e^{-t} \end{pmatrix} \in \mathrm{SL}(2,\mathbb{R}).$$
The group $G$ is called Sol, is completely solvable, and is a model space for a 3-dimensional Sol-*geometry*. Its automorphism group is $G \rtimes (\mathbb{Z}_2 \times \mathbb{Z}_2)$. It trivially satisfies Statement 8.3.1. Also, note that it satisfies the condition of Theorem 8.3.11.

Below, we will list two examples of simply connected, connected solvable Lie groups $G$, for which the generalization of the First Bieberbach Theorem (Theorem 8.3.4) does not hold. The first example is easy but consists of a Lie group which is not of type (E), while the second, slightly more sophisticated example consists of a Lie group of type (R).

EXAMPLE 8.3.14. There is an example in [**Aus60**, pp.589–590] of dimension 5, where the First Bieberbach Theorem does not hold. Here is a 3-dimensional example. Let $G = \mathbb{R}^2 \rtimes \mathbb{R}$ be the universal covering group of $E_0(2) = \mathbb{R}^2 \rtimes \mathrm{SO}(2)$. Therefore $t \in \mathbb{R}$ acts on $\mathbb{R}^2$ by
$$\psi_t = \begin{pmatrix} \cos 2\pi t & \sin 2\pi t \\ -\sin 2\pi t & \cos 2\pi t \end{pmatrix}.$$
This solvable Lie group $G$ is not of type (E). Let $C = \mathrm{SO}(2)$, acting on $\mathbb{R}^2$ in the standard way; that is, $e^{2\pi i t} \in C$ acts via the matrix $\psi_t$. These two actions of $\mathbb{R}$ and $C$ on $\mathbb{R}^2$ commute with each other, so, there is an action of $C$ on $G$. Let $t_0 \in \mathbb{R}$. Clearly, the element
$$\gamma = ((\mathbf{0}, t_0), e^{-2\pi i t_0}) \in G \rtimes C$$
acts on $\mathbb{R}^2$ trivially. The $\psi_{t_0}$ part acts as the element of $G$, and the $e^{-2\pi i t_0}$ part from $C$ undoes it. Take a lattice $\mathbb{Z}^2 = \langle z_1, z_2 \rangle$ of $\mathbb{R}^2$, and let it act as left translations. Consider the subgroup generated by $\mathbb{Z}^2$ and $\gamma$:
$$\Pi = \langle z_1, z_2, \gamma \rangle \subset G \rtimes C.$$
Suppose $t_0$ is irrational. Then $\Pi$ is isomorphic to $\mathbb{Z}^3$ and acts properly discontinuously on our $G$. It is obvious that $\Pi$ does not contain any lattice of $G$ of finite index. Note that the image of $\Pi$ in $C$ is dense.

In this example $N^* = \mathbb{R}^3$, namely, it consists of all elements of the form $((x,y,z), e^{-2\pi i z})$, with $x,y,z \in \mathbb{R}$, while $N = \mathbb{R}^2$. It follows that $N^*/N \cong \mathbb{R}$. This shows that Lemma 8.3.9 is not valid for solvable Lie groups not being of type (E).

Look at the representation $\mathbb{R} \to \mathrm{Aut}(\mathbb{R}^2) = \mathrm{GL}(2,\mathbb{R})$. It has image $\mathrm{SO}(2)$, and its centralizer contains $\mathrm{SO}(2)$ itself. Thus the requirement in Theorem 8.3.11 is trivially not satisfied.

EXAMPLE 8.3.15. Now, we give an example of a Lie group of type (R) for which the First Bieberbach Theorem is not valid.

Consider the following matrix in $\operatorname{GL}(4,\mathbb{Z})$:

$$A = \begin{pmatrix} 0 & 1 & 0 & 0 \\ 0 & 0 & 1 & 0 \\ 0 & 0 & 0 & 1 \\ -1 & 1 & 1 & 1 \end{pmatrix}.$$

This matrix has as characteristic polynomial $x^4 - x^3 - x^2 - x + 1$, having two real roots

$$r_1, r_2 = \frac{1}{4}\left(1 + \sqrt{13} \pm \sqrt{2\sqrt{13} - 2}\right)$$

and two complex roots

$$r_3, r_4 = \frac{1}{4}\left(1 - \sqrt{13} \pm i\sqrt{2\sqrt{13} + 2}\right),$$

for which the following hold:

(1) $r_1 r_2 = 1$, and $r_3 = \overline{r_4}$, so $r_3$ is of modulus 1;
(2) $r_3$ is not a root of unity.

There is a matrix $P \in \operatorname{GL}(4, \mathbb{R})$ such that

$$P.A.P^{-1} = \begin{pmatrix} r_1 & 0 & 0 & 0 \\ 0 & r_2 & 0 & 0 \\ 0 & 0 & b_1 & b_2 \\ 0 & 0 & b_3 & b_4 \end{pmatrix} = \begin{pmatrix} r_1 & 0 & 0 & 0 \\ 0 & r_2 & 0 & 0 \\ 0 & 0 & 1 & 0 \\ 0 & 0 & 0 & 1 \end{pmatrix} \begin{pmatrix} 1 & 0 & 0 & 0 \\ 0 & 1 & 0 & 0 \\ 0 & 0 & b_1 & b_2 \\ 0 & 0 & b_3 & b_4 \end{pmatrix} = M_1 M_2,$$

where the eigenvalues of $\begin{bmatrix} b_1 & b_2 \\ b_3 & b_4 \end{bmatrix}$ are $r_3$ and $r_4$. Note that $M_1$ and $M_2$ commute. We can define both $M_1^t$ and $M_2^t$ for a real parameter $t$. Remark that $M_2^{t+2k\pi} = M_2^t$. Define

$$\varphi : \mathbb{R} \to \operatorname{GL}(4, \mathbb{R}) : t \mapsto P^{-1} M_1^t P$$

and

$$\psi : S^1 = \mathbb{R}/(2\pi\mathbb{Z}) \to \operatorname{GL}(4, \mathbb{R}) : t \mapsto P^{-1} M_2^t P.$$

Let $G = \mathbb{R}^4 \rtimes_\varphi \mathbb{R}$, where $\mathbb{R}$ acts on $\mathbb{R}^4$ via $\varphi$. Then $G$ is a solvable Lie group of type (R). The action of $\mathbb{R}/2\pi\mathbb{Z}$ on $\mathbb{R}^4$ by $\psi$ commutes with the action of $\mathbb{R}$ by $\psi$, and hence it induces an action of $\mathbb{R}/2\pi\mathbb{Z}$ on $G$. Let $S^1 = \mathbb{R}/(2\pi\mathbb{Z})$ be the compact group of automorphisms of $G$ which is defined by taking $t(x, y) = (\psi(t)(x), y)$, $\forall t \in \mathbb{R}/(2\pi\mathbb{Z})$, $\forall x \in \mathbb{R}^4, \forall y \in \mathbb{R}$.

Now, the group $\Gamma = \mathbb{Z}^4 \rtimes \mathbb{Z}$, where $1 \in \mathbb{Z}$ acts on $\mathbb{Z}^4$ as $A$, can be embedded into $G \rtimes S^1$ by mapping $(z_1, z_2) \mapsto ((z_1, z_2), z_2)$, $\forall z_1 \in \mathbb{Z}^4$ and $z_2 \in \mathbb{Z}$. In this way $\Gamma$ is realized as a lattice of $G \rtimes S^1$ without $\Gamma \cap G = \mathbb{Z}^4$ being a lattice of $G$. So, the First Bieberbach Theorem does not hold for $G$.

Note, for the representation $\mathbb{R} \to \operatorname{Aut}(\mathbb{R}^4) = \operatorname{GL}(4, \mathbb{R})$, the condition of Theorem 8.3.11 does not hold. For, the image of $\mathbb{R}$ (which is generated by $M_1$) has centralizer which contains $\operatorname{SO}(2)$ rotating the 2-dimensional factors of $\mathbb{R}^4$.

8.3.16. Although Theorems 8.3.4 and 8.3.11 give us a necessary and sufficient condition for a Lie group of type (R) to admit a Bieberbach theory, they do not provide concrete criteria to check whether or not a given Lie group allows a generalization of the First Bieberbach Theorem. Therefore, in this section we describe a substantial class of Lie groups of type (R), satisfying the conditions of Theorem 8.3.4.

DEFINITION 8.3.17. Let $G$ be a connected, simply connected solvable Lie group of type (R), with nilradical $N$. Then $G$ is said to be NICE if and only if there exists a filtration of closed analytic, characteristic subgroups $N_i$ of $N$:
$$1 = N_0 \subseteq N_1 \subseteq N_2 \subseteq N_3 \subseteq \cdots \subseteq N_c = N$$
with $N_i/N_{i-1} \cong \mathbb{R}^{k_i}$ for all $i \in \{1, 2, \ldots, c\}$ and such that:
(1) If $\Gamma$ is a lattice of $N$, then $(\Gamma \cap N_i)/(\Gamma \cap N_{i-1})$ is a lattice in $N_i/N_{i-1}$ for all $i \in \{1, 2, \ldots, c\}$;
(2) The matrix representing the action of any element of $G/N$ on a factor $N_i/N_{i-1}$ is unipotent or has only positive real eigenvalues, each of multiplicity one.

REMARK 8.3.18. If $N$ is a connected and simply connected nilpotent Lie group, then both the upper central series and the lower central series of $N$ satisfy property (1) of the definition above. As property (2) is trivially satisfied in this case, $N$ is NICE.

EXAMPLE 8.3.19. Let $H$ be the 3-dimensional Heisenberg group consisting of matrices of the form $\begin{pmatrix} 1 & x & z \\ 0 & 1 & y \\ 0 & 0 & 1 \end{pmatrix}$, with $x, y, z \in \mathbb{R}$. There is an automorphism $\varphi_1$ of $H$ which is given by

$$\varphi_1 \begin{pmatrix} 1 & 1 & 0 \\ 0 & 1 & 0 \\ 0 & 0 & 1 \end{pmatrix} = \begin{pmatrix} 1 & 2 & 0 \\ 0 & 1 & 1 \\ 0 & 0 & 1 \end{pmatrix}, \; \varphi_1 \begin{pmatrix} 1 & 0 & 0 \\ 0 & 1 & 1 \\ 0 & 0 & 1 \end{pmatrix} = \begin{pmatrix} 1 & 1 & 0 \\ 0 & 1 & 1 \\ 0 & 0 & 1 \end{pmatrix} \text{ and } \varphi_1 \begin{pmatrix} 1 & 0 & 1 \\ 0 & 1 & 0 \\ 0 & 0 & 1 \end{pmatrix} = \begin{pmatrix} 1 & 0 & 1 \\ 0 & 1 & 0 \\ 0 & 0 & 1 \end{pmatrix}.$$

This automorphism induces a commutative diagram

$$\begin{array}{ccccccccc}
1 & \longrightarrow & \mathcal{Z}(H) \cong \mathbb{R} & \longrightarrow & H & \longrightarrow & H/\mathcal{Z}(H) \cong \mathbb{R}^2 & \longrightarrow & 1 \\
& & \downarrow = & & \downarrow \varphi_1 & & \downarrow \left(\begin{smallmatrix} 2 & 1 \\ 1 & 1 \end{smallmatrix}\right) & & \\
1 & \longrightarrow & \mathcal{Z}(H) \cong \mathbb{R} & \longrightarrow & H & \longrightarrow & H/\mathcal{Z}(H) \cong \mathbb{R}^2 & \longrightarrow & 1.
\end{array}$$

It follows that there is a morphism $\varphi : \mathbb{R} \to \mathrm{Aut}(H)$, with $\varphi(1) = \varphi_1$. Moreover, the resulting group $H \rtimes_\varphi \mathbb{R}$ is a Lie group of type (R). Now, we define for all positive integers $k$, the following Lie group of type (R):
$$G_k = \mathbb{R}^k \times (H \rtimes_\varphi \mathbb{R}).$$

As both the upper and the lower central series of the nilradical of a Lie group of type (R) satisfy the first criterion of the definition of a NICE group, we focus on these filtrations of the nilradical $N = \mathbb{R}^k \times H$ of $G_k$.

The *lower central series* of $N$ is $1 \subseteq \mathbb{R} \; (= [N, N] = [H, H]) \subseteq N$, with $N/\mathbb{R} \cong \mathbb{R}^{k+2}$. Note that the action of an element of $G_k/N$ ($\cong \mathbb{R}$) on $\mathbb{R}^{k+2}$ is not unipotent and has an eigenvalue 1 which is of multiplicity $k$. So, if $k > 1$, this filtration of $N$ does not satisfy the second property of the definition of a NICE group.

The *upper central series* of $N$ is $1 \subseteq \mathbb{R}^{k+1} \; (= \mathcal{Z}(N) = \mathbb{R}^k \times \mathcal{Z}(H)) \subseteq N$, with $N/\mathbb{R}^{k+1} \cong \mathbb{R}^2$. Now, the action of $G_k/N$ on $\mathbb{R}^{k+1}$ is unipotent, while each nonidentity element of $G/N$ acts on $\mathbb{R}^2$ via an automorphism having two different eigenvalues. This implies that the upper central series of $G_k$ satisfies the second criterion of the definition of a NICE group, which allows us to conclude that all groups $G_k$ are NICE.

THEOREM 8.3.20. *Let G be NICE, then G has the strong lattice property.*

Before we are able to prove this theorem, we need to establish one more lemma.

LEMMA 8.3.21. *Let $\mathbb{Z}^n \subset \mathbb{R}^n \oplus T$, where $T$ is a torus, be a lattice such that the image $\mathbb{Z}^n \subset \mathbb{R}^n \oplus T \to T$ is dense in $T$. Let $f : \mathbb{R}^n \oplus T \to \mathrm{GL}(k, \mathbb{R})$ be a representation such that $f(x, 1)$ is unipotent or has only positive real eigenvalues, all being of multiplicity one, for all $x \in \mathbb{R}^n$. If $\mathbb{Z}^n$ leaves a lattice of $\mathbb{R}^k$ invariant, then the $T$-action on $\mathbb{R}^k$ is trivial.*

PROOF. The elements $f(1, t)$ ($t \in T$) all have eigenvalues which are of modulus one. As $T$ commutes with $\mathbb{R}^n$ it is known that the eigenvalues of $f(x, t)$ can be obtained as the product of the eigenvalues of $f(x, 1)$ and $f(1, t)$ (in some order).

Now, consider any element $\gamma = (x_\gamma, t_\gamma) \in \mathbb{Z}^n$. We claim that $f(1, t_\gamma)$ has only eigenvalues which are roots of unity. Indeed, suppose first that the action of $f(x, 1)$ only has positive real eigenvalues of multiplicity one. If $z$ is a complex eigenvalue of $f(x_\gamma, t_\gamma)$, then also $\bar{z}$, the complex conjugate of $z$, is an eigenvalue of $f(x_\gamma, t_\gamma)$. However, as the eigenvalues of $f(x_\gamma, t_\gamma)$ are obtained by multiplying the eigenvalues of $f(x_\gamma, 1)$ with the eigenvalues of $f(1, t_\gamma)$, all these eigenvalues have a different modulus. This implies that $z = \bar{z} \in \mathbb{R}$. But this implies in turn that $f(1, t_\gamma)$ has only 1 or $-1$ as eigenvalues.

Now, suppose that the action of $f(x_\gamma, 1)$ is unipotent. Then all eigenvalues of $f(x_\gamma, t_\gamma)$ are of modulus 1. As $f(x_\gamma, t_\gamma)$ fixes a lattice of $\mathbb{R}^k$, the matrix $f(x_\gamma, t_\gamma)$ is conjugated, inside $\mathrm{GL}(k, \mathbb{R})$, to a matrix in $\mathrm{GL}(k, \mathbb{Z})$, which has of course the same eigenvalues. Now [**ST87**, Lemma 11.6] implies that all eigenvalues of $f(x_\gamma, t_\gamma)$, which are the same as the eigenvalues of $f(1, t_\gamma)$, are roots of unity.

As $\mathbb{Z}^n$ is finitely generated, there is a subgroup of finite index in $\mathbb{Z}^n$, for which the elements $(x_\gamma, t_\gamma)$ are such that the eigenvalues of $f(1, t_\gamma)$ are all 1. As this set of $t_\gamma$'s is still dense in $T$, it follows that the eigenvalues of $f(1, t)$ for all elements of $t \in T$ must be 1. This means that $T$ acts unipotently on $\mathbb{R}^k$. However, the only unipotent action of $T$ on $\mathbb{R}^k$ is the trivial action, which finishes the proof of the lemma. □

8.3.22 (Proof of Theorem 8.3.20). Let $T$ be a torus in $\mathrm{Aut}(G)$ and choose a lattice $\Pi$ in $G \rtimes T$. As before, we let $G^*$ be the identity component of the closure of $\Pi G$ in $G \rtimes T$ and we let $\Pi^* = G^* \cap \Pi$. We know that $G^* = G \rtimes T'$ for some torus $T' \subset \mathrm{Aut}(G)$ and by Lemma 3, we have that

(1) $\Pi^*$ is of finite index in $\Pi$,
(2) the nilradical of $G^*$ coincides with the nilradical $N$ of $G$, and
(3) $\Pi^* \cap N = \Pi \cap N$ is a lattice of $N$.

We will show that $T'$ is the trivial group. Suppose that
$$1 = N_0 \subseteq N_1 \subseteq N_2 \subseteq N_3 \subseteq \cdots \subseteq N_c = N$$
is the filtration satisfying the properties of the definition of a NICE Lie group of type (R). As $\Pi^*/(\Pi^* \cap N)$ is a lattice of $G/N \times T'$, it follows from Lemma 8.3.21 that $T'$ acts trivially on any quotient $N_i/N_{i-1}$ ($1 \leq i \leq c$). Therefore, $T'$ acts unipotently on $N$. However, as said before, this implies that the action of $T'$ on $N$ is trivial. This combined with the fact that $T'$ acts trivially on $G/N = \mathbb{R}^n$ implies that $T'$ acts trivially on $G$. But as $T' \subseteq \mathrm{Aut}(G)$, this implies that $T'$ is trivial.

Hence $G = G^*$ and $\Pi^* = \Gamma = \Pi \cap G$ itself is a lattice of $G$, with $\Gamma \cap N = \Pi \cap N$. This completes the proof of the theorem. □

REMARK 8.3.23. If $G$ is a nilpotent connected and simply connected Lie group, the conditions of the above theorem are automatically satisfied.

## 8.4. The Second Bieberbach Theorem

The main references for this section are [**Lee88**], [**Lee95a**], and [**Lee95b**]. Let $G$ be a connected, simply connected solvable Lie group of type (R); $\Gamma \in G$ a *lattice* of $G$; and $\Pi \in \mathrm{Aff}(G)$ a torsion-free finite extension of $\Gamma$. Then $\Pi$ acts freely on $G$, and the manifold $\Pi \backslash G$ is called an *infra-solvmanifold of type* (R). (Note that a solvmanifold is usually defined as a quotient of a solvable Lie group by a closed subgroup.)

We show in this section that homotopy equivalent infra-solvmanifolds of type (R) are affinely diffeomorphic. Note that the class of infra-solvmanifolds of type (R) includes all flat manifolds, and more generally, all infra-nilmanifolds.

LEMMA 8.4.1. *Let $G$ be a connected, simply connected solvable Lie group, and let $\Psi$ be a finite group. Then every extension $1 \to G \to E \to \Psi \to 1$ splits. Therefore, if $\mathcal{Z}(G)$ is connected, then every homomorphism $\Psi \to \mathrm{Out}(G)$ lifts to a homomorphism $\Psi \to \mathrm{Aut}(G)$.*

PROOF. We use induction on the dimension of $G$. If the dimension of $G$ is 1, $G$ is Abelian and clearly any extension of $\mathbb{R}^1$ by a finite group splits. Suppose the statement is true for groups of dimension less than $n$. Let $G$ be a connected, simply connected solvable Lie group of dimension $n$, and let $N$ be the nilradical (the maximal connected normal nilpotent subgroup) of $G$. Then the center of $N$, $\mathcal{Z}(N)$, is a nontrivial, connected Abelian Lie group, isomorphic to $\mathbb{R}^k$ for some $k$, is characteristic in $N$. Therefore, $\mathcal{Z}(N)$ is characteristic in $G$. Consequently, $\mathcal{Z}(N)$ is normal in $E$. By dividing out by $\mathcal{Z}(N)$, we obtain $1 \to G/\mathcal{Z}(N) \to E/\mathcal{Z}(N) \to \Psi \to 1$. Since $\mathcal{Z}(N)$ is connected, $G/\mathcal{Z}(N)$ is again a connected, simply connected solvable Lie group. By the induction hypothesis, this sequence splits. Let $\Psi' \subset E/\mathcal{Z}(N)$ be the image of a splitting. This induces an exact sequence $1 \to \mathcal{Z}(N) \to E' \to \Psi' \to 1$, where $E'$ is a subgroup of $E$. Such a sequence splits since $H^2(\Psi'; \mathcal{Z}(N)) = 0$. Consequently, we have a splitting of $\Psi$ into $E$.

Let $\Psi \to \mathrm{Out}(G)$ be a homomorphism. Consider the pullback of the exact sequence $1 \to \mathrm{Inn}(G) \to \mathrm{Aut}(G) \to \mathrm{Out}(G) \to 1$ by $\Psi \to \mathrm{Out}(G)$, say $1 \to \mathrm{Inn}(G) \to \tilde{\Psi} \to \Psi \to 1$. Then $\Psi \to \mathrm{Out}(G)$ lifts to a homomorphism $\Psi \to \mathrm{Aut}(G)$ if and only if the short exact sequence $1 \to \mathrm{Inn}(G) \to \tilde{\Psi} \to \Psi \to 1$ splits. But $\mathrm{Inn}(G)$ is simply connected since $\mathcal{Z}(G)$ is connected by hypothesis. Therefore, one can apply the first statement for $\mathrm{Inn}(G)$. □

THEOREM 8.4.2 (Embedding problem; cf. Example 7.4.5 and [**LR81**, 1.2]). *Let $G$ be a connected, simply connected solvable Lie group, and $\Gamma$ a lattice of $G$ such that $(\Gamma, G)$ has UAEP (see Definition 5.3.3). Let $\Psi$ be a finite group. Then for every extension $\Pi$, $1 \to \Gamma \to \Pi \to \Psi \to 1$, there is a homomorphism $\theta : \Pi \to \mathrm{Aff}(G)$ such that $\theta|_\Gamma = \mathrm{id}$. The set of all torsion elements of $C_\Pi(\Gamma)$ forms a characteristic subgroup and is the kernel of $\theta$. Therefore, $\theta$ is injective if and only if $C_\Pi(\Gamma)$ is torsion free.*

PROOF. Since $(\Gamma, G)$ has UAEP, there is a natural inclusion $\mathrm{Aut}(\Gamma) \hookrightarrow \mathrm{Aut}(G)$, and a homomorphism $\mathrm{Out}(\Gamma) \to \mathrm{Out}(G)$. From $\Gamma \hookrightarrow G$ and the short exact sequence given, one can form a pushout $E$; see Subsection 5.3.4. Thus $E$ fits into the following commutative diagram with exact rows:

$$\begin{array}{ccccccccc} 1 & \longrightarrow & \Gamma & \longrightarrow & \Pi & \longrightarrow & \Psi & \longrightarrow & 1 \\ & & \downarrow i & & \downarrow \theta & & \downarrow = & & \\ 1 & \longrightarrow & G & \longrightarrow & E & \longrightarrow & \Psi & \longrightarrow & 1. \end{array}$$

By Lemma 8.4.1, the abstract kernel for the extension $\Pi$ lifts to a homomorphism $\Psi \to \mathrm{Aut}(G)$ and $E$ splits as $G \rtimes \Psi$, and we get $\theta : \Pi \hookrightarrow G \rtimes \Psi \to \mathrm{Aff}(G)$.

To compute the kernel $K$ of $\theta : \Pi \to \mathrm{Aff}(G)$, let $\alpha \in K$. For every $z \in \Gamma$,

$$\theta(\alpha z \alpha^{-1}) = \theta(z).$$

But $\alpha z \alpha^{-1} \in \Gamma$ and $\theta$ is injective on $\Gamma$. Therefore, $\alpha z \alpha^{-1} = z$. This shows that $\alpha$ commutes with every element of $\Gamma$; that is, $\alpha \in C_\Pi(\Gamma)$. Since $\Pi/\Gamma$ is finite, there is $p$ such that $\alpha^p \in \Gamma$. Then $\theta(\alpha^p) = 1$ and $\theta|_\Gamma$ injective imply that $\alpha^p = 1$. Thus, $\alpha$ must be a torsion element.

Conversely, let $\alpha$ be a torsion element of $C_\Pi(\Gamma)$. Let $\theta(\alpha) = (a, A) \in \mathrm{Aff}(G)$. For any $z \in \Gamma$,

$$(z, I) = (a, A)(z, I)(a, A)^{-1} = (a \cdot A(z) \cdot a^{-1}, I).$$

So, $A(z) = a^{-1} \cdot z \cdot a = \mu(a^{-1})(z)$ for all $z \in \Gamma$. By UAEP, this is true for all $z \in G$. Therefore $A = \mu(a^{-1}) \in \mathrm{Inn}(G)$ and $\theta(\alpha) = (a, \mu(a^{-1}))$. For this to be a torsion, $a$ must be 1 so that $\theta(\alpha) = \mathrm{id}$. □

THEOREM 8.4.3 (Rigidity of infra-solvmanifolds). *Let $G$ be a connected, simply connected solvable Lie group of type* (R). *Let $\Pi, \Pi' \in \mathrm{Aff}(G)$ be finite extensions of lattices in $G$. Then every isomorphism $\theta : \Pi \to \Pi'$ is a conjugation by an element of $\mathrm{Aff}(G)$.*

PROOF. The theorem is a direct consequence of Example 7.4.5 where $W$ is a point. However we shall give full details following the proof of rigidity in Proof 7.6.7.

Let $\Gamma = G \cap \Pi$, $\Gamma' = G \cap \Pi'$ be the pure translations. Let $\Lambda$ be the normal subgroup of $\Pi$ generated by $\{x^n : x \in \Pi\}$, where $n$ is the product of the orders of groups $\Pi/\Gamma$ and $\Pi'/\Gamma'$. Then $\Lambda$ is a characteristic subgroup of $\Pi$ contained in $\Gamma \subset G$. Then $\theta|_\Lambda : \Lambda \to \theta(\Lambda)$ is an isomorphism of lattices of $G$. By Corollary 6.3.7, $G$ has the ULIEP, and $\theta|_\Lambda$ extends uniquely to an automorphism $C : G \to G$. Thus,

$$\theta|_\Lambda = C|_\Lambda,$$

and hence, $\theta(z, I) = (Cz, I)$ for all $(z, I) \in \Lambda$.

Take $\widehat{\Pi} = \Pi$ as an abstract group, and take the identity map $\Pi \to \Pi$ as $\theta_1$:

$$\theta_1 = \mathrm{id} : \Pi \to \Pi \subset \mathrm{Aff}(G).$$

Also, let

$$\theta_2 = \theta \circ \theta_1 : \Pi \to \Pi' \subset \mathrm{Aff}(G).$$

Define a new homomorphism $\theta_3$ by

$$\theta_3 = \mu(1, C)^{-1} \circ \theta_2 : \Pi \to \Pi' \subset \mathrm{Aff}(G).$$

Then the two homomorphisms $\theta_1$ and $\theta_3$ coincide on $\Lambda$. That is,
$$\theta_1|_\Lambda = \ell = \theta_3|_\Lambda.$$
Now we use the Seifert fiber space construction to complete the proof. We take the singleton as $W$ and the trivial action of the finite group on $W$, $\rho : \Psi = \Pi/\Lambda \to$ TOP$(W)$. Thus, $\theta_1$ and $\theta_3$ are two Seifert Constructions for the exact sequence
$$1 \to \Lambda \to \Pi \to \Psi \to 1$$
with the same $\ell : \Lambda \to \ell(G)$ and $\varphi \times \rho : \Psi \to \mathrm{Out}(G) \times \mathrm{TOP}(W)$. Now apply the uniqueness part of Theorem 7.3.2 to find a conjugation from $\theta_3$ to $\theta_1$. If one replaces $W$ by a point $\{p\}$ in the proof of Theorem 7.3.2, one gets a complete proof. It is basically the statement $H^1(\Pi/\Lambda; G) = 0$. □

COROLLARY 8.4.4. *Homotopy equivalent infra-solvmanifolds of type* (R) *are affinely diffeomorphic. Therefore, homotopy equivalent infra-nilmanifolds are affinely diffeomorphic.*

PROOF. Let $M = \Pi \backslash G$, $M' = \Pi' \backslash G'$ be infra-solvmanifolds of type (R). A homotopy equivalence $M \to M'$ induces an isomorphism $\theta : \Pi \to \Pi'$. Since $\theta$ maps some subgroup of $G \cap \Pi$ into a lattice of $G'$, there is an isomorphism of $G$ onto $G'$. Using this isomorphism, we identify $G$ with $G'$. Now apply Theorem 8.4.3 to $\theta : \Pi \to \Pi'$ to find an element $h = (d, D) \in \mathrm{Aff}(G)$ which conjugates $\Pi$ onto $\Pi'$. This gives a weakly equivariant map
$$(h, \mu(h)) : (G, \Pi) \to (G, \Pi')$$
and $h$ gives rise to an affine diffeomorphism $M \to M'$, which is homotopic to the original map. □

We recall the notion of the Fitting subgroup of a polycyclic-by-finite group (e.g., see [**Seg83**]).

DEFINITION 8.4.5. Let $\Gamma$ be a polycyclic-by-finite group. Then the *Fitting subgroup* of $\Gamma$, denoted by $\mathrm{Fitt}(\Gamma)$ is the unique maximal normal nilpotent subgroup of $\Gamma$.

8.4.6. For a finitely generated group $\Pi$, the maximal normal, nilpotent subgroup of $\Pi$ is called the *discrete nilradical* [**Aus63**]. For a polycyclic-by-finite $\Pi$, a discrete nilradical exists and it is the same as the Fitting subgroup of $\Pi$.

Let $G$ be a connected, simply connected solvable Lie group of type (R), and let $\Pi \in \mathrm{Aff}(G)$ be a finite extension of a lattice in $G$. Let $\Gamma = G \cap \Pi$. It is known that if $G$ is Abelian, $\Gamma$ is the maximal normal Abelian subgroup of $\Pi$. If $G$ is nilpotent, $\Gamma$ is the maximal normal, nilpotent subgroup of $\Pi$. A solvable analog of this is not true.

EXAMPLE 8.4.7. Let $G = \mathbb{R}^3$, and let $\Pi$ be an orientable Bieberbach group with holonomy group $\mathbb{Z}_2$. (This is $\mathfrak{G}_2$ in [**Wol77**].) Clearly, $\Pi \cap G$ is not the maximal normal solvable subgroup of $\Pi$ since $\Pi$ itself is solvable.

PROPOSITION 8.4.8. *Let $G$ be a connected, simply connected solvable Lie group of type* (R), *$N$ its nilradical. Let $\Pi \in \mathrm{Aff}(G)$ be a finite extension of a lattice in $G$, and let $\Gamma = G \cap \Pi$, $\Delta = N \cap \Pi$. Then $\Delta$ is the discrete nilradical of $\Pi$, $\Delta$ and $\Gamma$ are characteristic in $\Pi$.*

PROOF. Let $\Delta'$ be the discrete nilradical of $\Pi$. We shall show $\Delta' = \Delta$. If the rank of $\Delta'$ is greater than that of $\Delta$, $(\Delta' \cap \Gamma) \ne \Delta$. This would imply then $\Delta' \cap \Gamma$ is a nilpotent normal subgroup of $\Gamma$. But [**Gor71**, 4.2] implies that $\Delta$ is the discrete nilradical of $\Gamma$, which is a contradiction. Therefore, $[\Delta : \Delta']$ is finite.

We claim that such a $\Delta'$ must lie in $N$. Let $\Phi = \Pi/\Gamma$ and $\mathbb{R}^s = G/N$. Consider the map $\lambda : \Psi \to \mathbb{R}^s$ defined as follows. If $\alpha = (a, A) \in \mathrm{Aff}(G)$, let $\lambda(\alpha)$ be the image of $a \in G$ in $\mathbb{R}^s$ under the projection map $G \to \mathbb{R}^s$. It is easy to see that $\lambda$ is well defined and satisfies the cocycle condition

$$\lambda(\alpha\beta) = \lambda(\alpha) \, A(\lambda(\beta)).$$

However, $H^1(\Psi, \mathbb{R}^s) = 0$ since $\Psi$ is a finite group. This implies that there is an element $\bar{d}$ of $\mathbb{R}^s$ such that $\lambda(\alpha) = \bar{d} \, A(\bar{d}^{-1})$. Let $d \in G$ be a preimage of $\bar{d}$. Then the conjugation of the group $\Delta'$ by $d$ sends $\Delta'$ into $\mathrm{Aff}(N)$. Therefore, we may think that $\Delta'$ lies in $\mathrm{Aff}(N)$. However, $\Delta = N \cap \Delta'$ is the maximal normal nilpotent subgroup of $\Delta'$. Since $\Delta'$ itself is nilpotent, $\Delta'$ must equal to $\Delta$.

It is easy to see that $\Gamma$ is characteristic in $\Pi$: Every automorphism of $\Pi$ is conjugation by an element of $\mathrm{Aff}(G)$ (Theorem 8.4.3), and all such conjugations map $\Gamma$ onto itself. $\square$

8.4.9. With the same notation as in Proposition 8.4.8, $\Gamma$ need not be the maximal normal solvable subgroup of $\Pi$ such that $\Gamma/\Delta$ is free Abelian; see [**Dek97**, pp.518–519] for a counterexample.

8.4.10 (Nilpotent case). Since any simply connected nilpotent Lie group $N$ is completely solvable, the theorems stated in this section also hold for nilpotent Lie groups. However, as we have shown, the First Bieberbach Theorem always holds for lattices in $N \rtimes C$; the hypothesis that our lattice $\Gamma$ in $N \rtimes C$ intersects $N$ in a lattice (with finite index) is redundant.

8.4.11 (Counterexamples). Many statements which we proved are not true for general solvable Lie groups. Consider the 3-dimensional orientable Bieberbach group $\mathfrak{G}_6$ with holonomy $\mathbb{Z}_2 \oplus \mathbb{Z}_2$; see [**Wol77**, §3.5.5]. Then $\mathfrak{G}_6$ has a presentation

$$\begin{aligned}
\mathfrak{G}_6 = \langle t_1, t_2, t_3, \alpha, \beta, \gamma \mid \ & [t_i, t_j] = 1 \, (i, j = 1, 2, 3), \\
& \alpha t_i \alpha^{-1} = t_i^{-1} \, (i = 2, 3), \\
& \beta t_i \beta^{-1} = t_i^{-1} \, (i = 1, 3), \\
& \gamma t_i \gamma^{-1} = t_i^{-1} \, (i = 1, 2), \\
& \alpha^2 = t_1, \ \beta^2 = t_2, \ \gamma^2 = t_3 \rangle.
\end{aligned}$$

An explicit representation of $\mathfrak{G}_6$ into $E(3) = \mathbb{R}^3 \rtimes \mathrm{SO}(3)$ is given as follows:

$t_i = (e_i, I)$, where $e_i$ is the unit vector, $(i = 1, 2, 3)$,

$$\alpha = \left( \begin{bmatrix} \frac{1}{2} \\ 0 \\ 0 \end{bmatrix}, \begin{bmatrix} 1 & 0 & 0 \\ 0 & -1 & 0 \\ 0 & 0 & -1 \end{bmatrix} \right),$$

$$\beta = \left( \begin{bmatrix} 0 \\ \frac{1}{2} \\ \frac{1}{2} \end{bmatrix}, \begin{bmatrix} -1 & 0 & 0 \\ 0 & 1 & 0 \\ 0 & 0 & -1 \end{bmatrix} \right),$$

$$\gamma = \left( \begin{bmatrix} \frac{1}{2} \\ \frac{1}{2} \\ \frac{1}{2} \end{bmatrix}, \begin{bmatrix} -1 & 0 & 0 \\ 0 & -1 & 0 \\ 0 & 0 & 1 \end{bmatrix} \right).$$

Let
$$H = \mathbb{R}^2 \rtimes \mathbb{R},$$
where $t \in \mathbb{R}$ acts on $\mathbb{R}^2$ by $R(t) = \begin{bmatrix} \cos 2\pi t & \sin 2\pi t \\ -\sin 2\pi t & \cos 2\pi t \end{bmatrix}$. This is the universal covering group $\widetilde{E_0(2)}$ of the connected component of the Euclidean group $E(2)$. Note that $H$ is not of type (R). We claim that the group $\mathfrak{G}_6$ embeds into $H \rtimes \mathrm{Aut}(H)$ so that $\mathfrak{G}_6 \backslash H$ is an infra-solvmanifold, diffeomorphic to the original flat manifold $\mathfrak{G}_6 \backslash \mathbb{R}^3$.

The group operation of $H$ is, for $(x, s), (y, t) \in \mathbb{R}^2 \rtimes \mathbb{R}$,
$$(x, s)(y, t) = (x + R(s)y, s + t).$$

Let $\Omega: H \to H$ be an automorphism defined by
$$\Omega(x, s) = (Ux, -s),$$
where $U = \begin{bmatrix} 1 & 0 \\ 0 & -1 \end{bmatrix}$. In fact, the following equalities show that $\Omega$ is an automorphism of $H$:
$$\begin{aligned} \Omega(x,s)\Omega(y,t) &= (Ux, -s)(Uy, -t) \\ &= (Ux + R(-s)Uy, -s - t) \\ &= (Ux + UR(s)y, -s - t) \\ &= \Omega(x + R(s)y, s + t) \\ &= \Omega((x,s)(y,t)). \end{aligned}$$

One can check that the following is a representation of $\mathfrak{G}_6$ into $H \rtimes \mathrm{Aut}(H)$:

$$t_1 = \left( \left( \begin{bmatrix} 0 \\ 0 \end{bmatrix}, 1 \right), I \right), \quad t_2 = \left( \left( \begin{bmatrix} 1 \\ 0 \end{bmatrix}, 0 \right), I \right), \quad t_3 = \left( \left( \begin{bmatrix} 0 \\ 1 \end{bmatrix}, 0 \right), I \right),$$

$$\alpha = \left( \left( \begin{bmatrix} 0 \\ 0 \end{bmatrix}, \frac{1}{2} \right), I \right), \quad \beta = \left( \left( \begin{bmatrix} \frac{1}{2} \\ \frac{1}{2} \end{bmatrix}, 0 \right), \Omega \right), \quad \gamma = \left( \left( \begin{bmatrix} \frac{1}{2} \\ \frac{1}{2} \end{bmatrix}, \frac{1}{2} \right), \Omega \right).$$

The discrete nilradical of $\mathfrak{G}_6$ is $\Gamma \cong \mathbb{Z}^3$; the nilradical of $H$ is $\mathbb{R}^2$. Therefore, $N \cap \mathfrak{G}_6$ is not the discrete nilradical of $\mathfrak{G}_6$. Thus Proposition 8.4.8 fails. Since not every automorphism of $\Gamma$ extends to an automorphism of the Lie group $H$, the rigidity theorem, Theorem 8.4.3, fails for this $H$.

EXERCISE 8.4.12. Prove the "Fourth" Bieberbach theorem for nilpotent groups. That is, show that the following is a characterization of almost crystallographic groups. *Let $\Pi$ be a group such that its discrete nilradical (maximal normal nilpotent subgroup) $\Delta$ is torsion free and of finite index in $\Pi$ and such that the induced homomorphism $\Pi/\Delta \to \mathrm{Out}(\Delta)$ is faithful. Then, there exists simply connected* nilpotent Lie group $N$ and an embedding $\theta$ of $\Pi$ into $\mathrm{Aff}(N)$ so that $\theta(\Delta) = \theta(\Pi) \cap N$; $\Pi/\Delta$ is called the holonomy group *of this almost crystallographic group.* If $\Pi$ is torsion free, $\theta(\Pi)\backslash N$ is an infra-nilmanifold.

### 8.5. The Third Bieberbach Theorem

Bieberbach proved that there are only finitely many flat Riemannian manifolds in each dimension. This can be stated as: under each tori, there are only finitely many "infra-tori". There has been a question if the analogous statement for nilmanifolds is true. It turns out that this is not true for nilmanifolds or solvmanifolds in general.

First of all, there is only one connected, simply connected Abelian Lie group in each dimension. Secondly, with a fixed Lie group $\mathbb{R}^n$, there is only one flat torus modulo conjugation in $\mathrm{Aff}(\mathbb{R}^n)$. That is, all lattices of $\mathbb{R}^n$ give rise to affinely diffeomorphic tori. When a non-Abelian connected, simply connected, nilpotent Lie group $N$ has a lattice, $N$ has infinitely many nonisomorphic lattices. This can be seen as follows: Let $\Delta$ be a lattice of $N$. The center $\mathcal{Z}(N)$ is nontrivial and is a proper subgroup of $N$. The torus $\mathcal{Z}(N)/\mathcal{Z}(\Delta)$ acts on the nilmanifold $\Delta\backslash N$ as isometries (with any left invariant metric on $N$) freely. The liftings of any finite group of this torus will yield a lattice of $N$. If one picks a finite subgroup from $(\mathcal{Z}(N) \cap [N,N])/(\mathcal{Z}(N) \cap [N,N]) \cap \Delta$, the resulting lattices are not isomorphic to each other. They are even homologically distinct. Consequently, there are infinitely many distinct nilmanifolds covered by the nilmanifold $\Delta\backslash N$. This reveals a big difference between lattices in Abelian Lie groups and nilpotent Lie groups: All lattices in $\mathbb{R}^n$ are isomorphic to each other, while there are infinitely many nonisomorphic lattices in a non-Abelian nilpotent Lie group. It turns out that the above example is the only way to get an infinite family of infra-nilmanifolds covered by a given nilmanifold. We need a definition.

A covering of aspherical manifolds $M \to M'$ is called an *essential covering* if no element of the deck transformation group is homotopic to the identity. Let $\Psi$ be the deck transformation group. The map sending a free homotopy class into $\mathrm{Out}(\pi_1(M)) = \mathrm{Aut}(\pi_1(M))/\mathrm{Inn}(\pi_1(M))$ defines a natural homomorphism $\rho: \Psi \to \mathrm{Out}(\pi_1(M))$. The covering is essential if and only if $\rho$ is injective.

Since they are aspherical spaces, the group of free homotopy classes is isomorphic to $\mathrm{Out}(\pi_1(M))$. Observe that this is weaker than the definition given in [**Lee88**] as the following example shows.

EXAMPLE 8.5.1. Let
$$X = \begin{bmatrix} 1 & 1 & 0 \\ 0 & 1 & 0 \\ 0 & 0 & 1 \end{bmatrix}, Y = \begin{bmatrix} 1 & 0 & 0 \\ 0 & 1 & 1 \\ 0 & 0 & 1 \end{bmatrix}, Z = \begin{bmatrix} 1 & 0 & 1 \\ 0 & 1 & 0 \\ 0 & 0 & 1 \end{bmatrix}$$
be elements of the Heisenberg group $N$. We denote the $k$th root of $X$ by $\frac{1}{k}X$. Consider the lattice $\Delta = \langle X, Y, \frac{1}{6}Z \rangle$. Let $\Pi = \langle \frac{1}{3}X, Y, \frac{1}{6}Z \rangle$. Then $1 \to \Delta \to \Pi \to \mathbb{Z}_3 \to 1$ is exact, and the abstract kernel $\mathbb{Z}_3 \to \mathrm{Out}(\Delta)$ is injective. Of course, $\mathbb{Z}_3 \to \mathrm{Out}(N)$ is trivial. This means that $\Delta\backslash N \to \Pi\backslash N$ is an essential covering in our sense. In [**Lee88**], for a nilmanifold $M = \Delta\backslash G$, $M \to M'$ is an essential covering if the composite $\pi_1(M')/\Delta \to \mathrm{Out}(\Delta) \to \mathrm{Out}(G)$ is nontrivial. Therefore the above example is not essential in that sense. Our current definition is weaker

than the one in [**Lee88**]. Thus the following statement is more general than the one there, even for the nilpotent case.

THEOREM 8.5.2. *Under each solvmanifold of type* (R), *there are only finitely many infra-solvmanifolds which are essentially covered by the solvmanifold.*

PROOF. Let $M = \Gamma\backslash G$, where $G$ is a connected, simply connected solvable Lie group of type (R), and $\Gamma$ is a lattice in $G$. Let $M' = \Pi\backslash G$ be an infra-solvmanifold which is essentially covered by $M$. So $\Gamma \subset \Pi \subset \mathrm{Aff}(G)$ and the finite deck transformation group $\Psi = \Pi/\Gamma$ injects into $\mathrm{Out}(\Gamma)$. We shall show first that there are only finitely many such finite groups $\Psi$ in $\mathrm{Out}(\Gamma)$, up to conjugacy. Let $N$ be the nilradical of $G$. By Proposition 8.4.8, $\Delta = N \cap \Gamma$ is the discrete nilradical of $\Gamma$. Since $\Gamma$ is poly-cyclic and $\Gamma/\Delta$ is free Abelian, $\mathrm{Aut}(\Gamma)$ is arithmetic; see [**Seg83**, p.172, Exercise 10]. Since $\mathrm{Inn}(\Gamma)$ is a torsion-free solvable subgroup of the arithmetic group $\mathrm{Aut}(\Gamma)$, we can apply [**Seg83**, p.176, Proposition 10] to see that there are only a finite number of conjugacy classes of finite extensions of $\mathrm{Inn}(\Gamma)$ in $\mathrm{Aut}(\Gamma)$. Consequently, there are only finitely many conjugacy classes of finite subgroups $\Psi$ in $\mathrm{Out}(\Gamma)$. (This step was missing in the proof of [**Lee88**]. See also [**DIM94**].)

For a fixed abstract kernel $\Psi \subset \mathrm{Out}(\Gamma)$, consider how many extensions of $\Gamma$ by $\Psi$ there are, realizing the abstract kernel. We have $H^2(\Psi; \mathcal{Z}(\Gamma) \otimes \mathbb{Q}) = 0$ since $\Psi$ is finite. This implies $H^2(\Psi; \mathcal{Z}(\Gamma))$ is finite, and hence, there are only finitely many extensions of $\Gamma$ by $\Psi$, up to equivalence; see, for example, [**Mos54**, Chapter IV]. Since conjugate abstract kernels give rise to the same (up to isomorphism) sets of extensions, we conclude that there are only finitely many isomorphism classes of extensions of $\Gamma$ by finite groups. The above step may be argued differently. The following can be found in [**Seg83**, p.176, Theorem 6]. Let $\Gamma$ be polycyclic-by-finite, and let $d$ be a positive integer. Then the isomorphism classes of $\{\Pi : \Gamma$ is a normal subgroup of $\Pi, |\Pi/\Gamma| \leq d\}$ is finite. Since there are only finitely many possible $d$'s by the previous paragraph, we get the same conclusion.

To complete the proof, we appeal to the rigidity theorem, Theorem 8.4.3: if $\Pi \cong \Pi'$, then they are conjugate to each other by an element of $\mathrm{Aff}(G)$. This means that the infra-solvmanifolds $\Pi\backslash G$ and $\Pi'\backslash G$ are affinely diffeomorphic. Consequently, up to affine diffeomorphisms, there are only finitely many infra-solvmanifolds essentially covered by $M$. □

As a special case, we have

COROLLARY 8.5.3 ([**Lee88**]). *Under each nilmanifold, there are only finitely many infra-nilmanifolds which are essentially covered by the nilmanifold.*

CHAPTER 9

# Seifert Manifolds with $\Gamma\backslash G/K$-Fiber

## 9.1. Introduction

In Chapter 7, we worked out the injective Seifert fiber space construction with model space $G \times W$ and uniformizing group $\text{TOP}_G(G \times W)$.

Let $G$ be a connected Lie group and $K$ a closed subgroup. Consider the product space $(G/K) \times W$ of the space of left cosets $\{xK\}$ with $W$, and let $\text{TOP}_{G,K}(G/K \times W)$ be the group of self-homeomorphisms of $G/K \times W$ induced from weakly $G$-equivariant self-homeomorphisms of $G \times W$.

The aim of this chapter is to study Seifert fiber spaces modeled on

$$(G/K \times W, \ \text{TOP}_{G,K}(G/K \times W)).$$

Given a lattice $\Gamma \subset G$, a proper action $\rho : Q \to W$ of a discrete group, and an extension $1 \to \Gamma \to \Pi \to Q \to 1$, we construct a proper action of $\Pi$ on $G/K \times W$. Such spaces will have Seifert fiber structures

$$\Gamma\backslash G/K \to \Pi\backslash(G/K \times W) \to Q\backslash W,$$

where the double coset space $\Gamma\backslash G/K$ is a typical fiber. Of course, when $K = 1$, we have the ordinary model space $(G \times W, \ \text{TOP}_G(G \times W))$. We shall pay special attention to the case where $G$ is a semisimple Lie group in its adjoint form or $G$ is a special kind of solvable Lie group. In both cases we take $K$ to be a maximal compact subgroup so that the typical fibers are $K(\Gamma, 1)$-spaces. The material in this chapter is taken from [**LR96**].

9.1.1. One of the important geometric problems that has motivated the development of Seifert fiberings has been the construction of closed aspherical manifolds realizing Poincaré duality groups $\Pi$ of the form $1 \to \Gamma \to \Pi \to Q \to 1$, where $\Gamma$ is a cocompact torsion-free lattice in a noncompact Lie group. In Chapter 7, closed aspherical manifolds were found for commutative, nilpotent, and solvable $G$ provided that $Q$ could be made to act properly on some contractible manifold with compact quotient. There, $G$ was a simply connected Lie group diffeomorphic to a Euclidean space.

Hence it seems advantageous to enlarge the concept of Seifert Constructions to include Seifert fiber spaces modeled on $G/K \times W$ as well, where $G$ is an arbitrary Lie group and $K$ is a closed subgroup. Then a Seifert Construction would yield a *bundle with singularities*, $\Gamma\backslash G/K \to \Pi(\backslash G/K \times W) \to Q\backslash W$, where the double coset space is a typical fiber. In particular, if $G$ is an arbitrary noncompact Lie group with a finite number of connected components and $K$ is a maximal compact subgroup of $G$, then $G/K$ is diffeomorphic to an Euclidean space. An earlier paper of Raymond and Wigner[**RW77**] dealt with the construction of closed aspherical manifolds $M = K(\Pi, 1)$, where $\Gamma$ was a lattice in a noncompact semisimple Lie group $G$ in adjoint form. The method which was rather ad hoc yielded the desired manifolds. The groups $\Gamma$ and $G$ had to be replaced by the isomorphic groups $\operatorname{Inn}(\Gamma)$ and $\operatorname{Inn}(G)$, respectively. However, it has not been clear just how that construction fit into the general theory of Seifert Constructions. Also, the geometry of the spaces obtained could not be easily explained in terms of geometries of $G$ and $W$. In this chapter, we explain how one goes about creating a general theory for Seifert Constructions on $G/K \times W$ where $G$ is connected Lie group and $K$ is a closed subgroup. As a consequence, the main result of [**RW77**] is recaptured in Theorem 9.4.5. An advantage of the present approach over the earlier one is Corollary 9.4.6 where it is easily shown that the constructed manifold $M(\Pi)$ inherits the product geometry from $G/K$ and $W$.

NOTATION 9.1.2. For a subgroup $K$ of $G$, $\operatorname{Aut}^0(G,K) = \{\alpha \in \operatorname{Aut}(G) : x^{-1}\alpha(x) \in K \text{ for all } x \in G\}$.

We show that $\operatorname{Aut}^0(G,K)$ is a subgroup of $\operatorname{Aut}(K)$. Let $\alpha \in \operatorname{Aut}^0(G,K)$. First, we show that $\alpha \in \operatorname{Aut}(K)$. Clearly, $\alpha(K) \subset K$. Conversely, let $k \in K$ and $k = \alpha(x)$ for some $x \in G$ since $\alpha$ is an automorphism of $G$. Then

$$x^{-1} = x^{-1}(kk^{-1}) = (x^{-1}k)k^{-1} = (x^{-1}\alpha(x))k^{-1} \in K \cdot K = K$$

shows that $K \subset \alpha(K)$, proving $\alpha(K) = K$. It is not difficult to show that $\operatorname{Aut}^0(G,K)$ forms a subgroup.

9.1.3. Section 9.2 describes the general setup culminating in a complete description of the uniformizing group $\operatorname{TOP}_{G,K}(G/K \times W)$ into which discrete groups $\Pi$ need to be mapped to create Seifert fiberings modeled on $G/K \times W$. Section 9.3 calculates $\operatorname{TOP}_{G,K}(G/K \times W)$ when $N_G(K) = K$ and $\operatorname{Aut}^0(G,K) = 1$. The main result is Theorem 9.3.2 which is crucial for the later sections. Section 9.4 then solves the embedding and uniqueness problem for $1 \to \Gamma \to \Pi \to Q \to 1$ into $\operatorname{TOP}_{G,K}(G/K \times W)$ when $(G,K)$ is a Riemannian symmetric pair of noncompact type.

Section 9.5 is concerned with solvable $G$. If $G$ is solvable, not all the nice features of the theory of Seifert Constructions that work so well in the nilpotent or semisimple cases remain valid. With the aid of the earlier sections, we show that under certain conditions, the usual embedding and uniqueness theorems are valid if we enlarge $G$ to $G'$ by extending it by a compact Abelian group $K$. Then, using the pair $(G', K)$ and the technique of the earlier sections, we obtain the desired solution. This explains the geometry of the infra-solvmanifolds constructed by Auslander and Johnson in [**AJ76**] as Seifert manifolds.

Applications and specific examples are given to illustrate how the theory works in practice.

## 9.2. The group $\operatorname{TOP}_{G,K}(G/K \times W)$

9.2.1. Let $\operatorname{TOP}_{G,K}(G \times W)$ be the subgroup of $\operatorname{TOP}_G(G \times W)$ consisting of elements which induce maps on $G/K \times W$ (where $G/K$ is the space of left $K$-cosets), and let $\operatorname{TOP}_{G,K}(G/K \times W)$ be the image of $\operatorname{TOP}_{G,K}(G \times W)$ in $\operatorname{TOP}(G/K \times W)$. Therefore, $\operatorname{TOP}_{G,K}(G/K \times W)$ is the group of homeomorphisms on $G/K \times W$

which are induced from the weakly $G$-equivariant homeomorphisms on $G \times W$.

$$\mathrm{TOP}_{G,K}(G \times W) \subset \mathrm{TOP}_G(G \times W) \subset \mathrm{TOP}(G \times W)$$
$$\downarrow$$
$$\mathrm{TOP}_{G,K}(G/K \times W) \subset \mathrm{TOP}(G/K \times W)$$

We need to study the groups $\mathrm{TOP}_{G,K}(G \times W)$ and $\mathrm{TOP}_{G,K}(G/K \times W)$ in detail.

PROPOSITION 9.2.2.
$$\mathrm{TOP}_{G,K}(G \times W) = \ell(G) \cdot \bigl(\mathrm{M}(W, N_G(K)) \rtimes \mathrm{Aut}(G, K)\bigr) \rtimes \mathrm{TOP}(W).$$

PROOF. We need to prove that $f \in \mathrm{TOP}_G(G \times W)$ belongs to $\mathrm{TOP}_{G,K}(G \times W)$ if and only if $f$ is of the form $\ell(a) \cdot (\lambda, \alpha, h) = (\lambda a^{-1}, \mu(a)\alpha, h)$, where $a \in G$ and

(1) $\alpha \in \mathrm{Aut}(G, K)$, and
(2) $\lambda \in \mathrm{M}(W, N_G(K))$.

Suppose $f = (\lambda_1, \alpha_1, h) \in \mathrm{TOP}_{G,K}(G \times W)$. Let $(\lambda_1, \alpha_1, h)(xK, w) = (x'K, h(w))$ for some $x' \in G$. Then

$$x'^{-1}\alpha_1(x)\alpha_1(K)\lambda_1(hw)^{-1} = K.$$

In particular, we must have

$$x'^{-1}\alpha_1(x)\lambda_1(hw)^{-1} = x'^{-1}\alpha_1(x)\alpha_1(1)\lambda_1(hw)^{-1} \in K.$$

The above two equalities yield

$$\alpha_1(K) = \lambda_1(hw)^{-1} K \lambda_1(hw).$$

Notice that the left-hand side is independent of $w$. Fix $w_0 \in W$, and let $a^{-1} = \lambda_1(hw_0)$; $\lambda = \lambda_1 a$ and $\alpha = \mu(a^{-1}) \circ \alpha_1$. Since $\alpha(K) = a^{-1}\alpha_1(K)a = K$, we have

$$\alpha \in \mathrm{Aut}(G, K).$$

Also $\lambda(hw) \cdot K \cdot \lambda(hw)^{-1} = \lambda(hw) \cdot \alpha(K) \cdot \lambda(hw)^{-1} = \lambda_1(hw) \cdot \alpha_1(K) \cdot \lambda_1(hw)^{-1} = K$ for all $w \in W$ shows that $\lambda(w) \in N_G(K)$ for all $w \in W$. Thus,

$$\lambda \in \mathrm{M}(W, N_G(K)).$$

Consequently, $f = (\lambda a^{-1}, \mu(a)\alpha, h) = \ell(a) \cdot (\lambda, \alpha, h)$ where $a \in G$, $\alpha \in \mathrm{Aut}(G, K)$ and $\lambda \in \mathrm{M}(W, N_G(K))$.

Conversely, let $\alpha \in \mathrm{Aut}(G, K)$ and $\lambda \in \mathrm{M}(W, N_G(K))$. Then it is easy to see that $(\lambda, \alpha, h)(xK, w) = (\alpha(x)\lambda(hw)^{-1}K, hw)$ so that $(\lambda, \alpha, h)$ maps $K$-cosets to $K$-cosets. It is clear that $\ell(G)$ preserves $K$-cosets also. This completes the proof. $\square$

LEMMA 9.2.3 (Ineffective part of $\mathrm{TOP}_{G,K}(G \times W)$). *An element* $(\lambda a^{-1}, \mu(a)\alpha, h) \in \mathrm{TOP}_{G,K}(G \times W)$ *acts trivially on* $G/K \times W$ *if and only if*

(1) $h = \mathrm{id}$ *on* $W$,
(2) $\lambda a^{-1} \in \mathrm{M}(W, K)$, *and*
(3) $x^{-1} \cdot (a\alpha(x)a^{-1}) \in K$ *for all* $x \in G$.

*Consequently, the kernel of* $\mathrm{TOP}_{G,K}(G \times W) \to \mathrm{TOP}_{G,K}(G/K \times W)$ *is exactly* $\mathrm{M}(W, K) \rtimes \mathrm{Aut}^0(G, K)$.

PROOF. Suppose $(\lambda a^{-1}, \mu(a)\alpha, h) \in \text{TOP}_{G,K}(G \times W)$ acts trivially on $G/K \times W$. Two points $(\lambda a^{-1}, \mu(a)\alpha, h)(x, w) = (a\alpha(x)\lambda(hw)^{-1}, hw)$ and $(x, w)$ represent the same point in $G/K \times W$ if and only if $hw = w$ and $x^{-1}a\alpha(x)\lambda(w)^{-1} \in K$. These should hold for all $x \in G$ and all $w \in W$. For $x = 1$, the latter reduces to $a\lambda(w)^{-1} \in K$ so that $a\lambda^{-1} \in \text{M}(W, K)$. Now, $x^{-1}a\alpha(x)a^{-1}(a\lambda(w)^{-1}) = x^{-1}a\alpha(x)\lambda(w)^{-1} \in K$ and $a\lambda(w)^{-1} \in K$ yield $x^{-1}(\mu(a)\alpha)(x) = x^{-1}a\alpha(x)a^{-1} \in K$ for all $x \in G$ so that $\mu(a)\alpha \in \text{Aut}^0(G, K)$. Thus $(\lambda a^{-1}, \mu(a)\alpha, 1) \in \text{M}(W, K) \rtimes \text{Aut}^0(G, K)$. Conversely, for $(\lambda_1, \alpha_1) \in \text{M}(W, K) \rtimes \text{Aut}^0(G, K)$, we have $(\lambda_1, \alpha_1)(xK, w) = (xK, w)$. □

COROLLARY 9.2.4 (Ineffective part of $(\ell(G) \cdot \text{M}(W, N_G(K))) \rtimes \text{Inn}(G, K)$). *The part of the kernel of* $\text{TOP}_{G,K}(G \times W) \to \text{TOP}_{G,K}(G/K \times W)$ *in* $\text{M}(W, G) \rtimes \text{Inn}(G)$ *is exactly* $\text{M}(W, K) \rtimes \text{Inn}^0(G, K)$. □

COROLLARY 9.2.5.
$$\text{TOP}_{G,K}(G/K \times W) = \frac{\ell(G) \cdot (\text{M}(W, N_G(K)) \rtimes \text{Aut}(G, K))}{\text{M}(W, K) \rtimes \text{Aut}^0(G, K)} \rtimes \text{TOP}(W). \qquad \square$$

Also, because $\mu(a) = \ell(a)r(a)$, $(\ell(G) \cdot \text{M}(W, N_G(K))) \rtimes \text{Inn}(G, K) = \ell(G) \cdot \text{M}(W, N_G(K))$. Therefore,
$$1 \to \ell(G) \cdot \text{M}(W, N_G(K)) \to \text{TOP}_{G,K}(G \times W) \to \text{Out}(G, K) \times \text{TOP}(W) \to 1$$
is exact. Thus,

COROLLARY 9.2.6. *There exists a commuting "9-diagram" (with exact rows and columns):*

$$\begin{array}{ccccccccc}
& & 1 & & 1 & & 1 & & \\
& & \downarrow & & \downarrow & & \downarrow & & \\
1 & \to & \text{M}(W,K) \rtimes \text{Inn}^0(G,K) & \to & \text{M}(W,K) \rtimes \text{Aut}^0(G,K) & \to & \text{Out}^0(G,K) & \to & 1 \\
& & \downarrow & & \downarrow & & \downarrow & & \\
1 & \to & \ell(G) \cdot \text{M}(W, N_G(K)) & \to & \text{TOP}_{G,K}(G \times W) & \to & \text{Out}(G,K) \times \text{TOP}(W) & \to & 1 \\
& & \downarrow & & \downarrow & & \downarrow & & \\
1 & \to & \frac{\ell(G) \cdot \text{M}(W, N_G(K))}{\text{M}(W,K) \rtimes \text{Inn}^0(G,K)} & \to & \text{TOP}_{G,K}(G/K \times W) & \to & \overline{\text{Out}(G,K)} \times \text{TOP}(W) & \to & 1 \\
& & \downarrow & & \downarrow & & \downarrow & & \\
& & 1 & & 1 & & 1 & &
\end{array}$$

PROPOSITION 9.2.7. *Suppose $H$ is a closed subgroup of $K$ and is normal in $G$. Let $G/H = \overline{G}$ and $K/H = \overline{K}$. Then* $\text{TOP}_{G,K}(G/K \times W) = \text{TOP}_{\overline{G},\overline{K}}(\overline{G}/\overline{K} \times W)$.

PROOF. Since $H$ is normal in $G$, $[G, H] \subset H \subset K$. This implies that $\mu(H) \subset \text{Aut}^0(G, K)$. Therefore,
$$\ell(H) = \{(h^{-1}, \mu(h)) : h \in H\} \subset \text{M}(W, H) \rtimes \mu(H) \subset \text{M}(W, K) \rtimes \text{Aut}^0(G, K).$$
Also, $N_G(K)/H = N_{\overline{G}}(\overline{K})$. Finally, $\text{Aut}(G, K)/(\text{Aut}(H) \cap \text{Aut}(G, K)) = \text{Aut}(\overline{G}, \overline{K})$ so that
$$\frac{\text{Aut}(G, K)}{\text{Aut}^0(G, K)} = \frac{\text{Aut}(G, K)/(\text{Aut}(H) \cap \text{Aut}(G, K))}{\text{Aut}^0(G, K)/(\text{Aut}(H) \cap \text{Aut}^0(G, K))} = \frac{\text{Aut}(\overline{G}, \overline{K})}{\text{Aut}^0(\overline{G}, \overline{K})}.$$

Consequently, we get $\text{TOP}_{G,K}(G/K \times W) = \text{TOP}_{\overline{G},\overline{K}}(\overline{G}/\overline{K} \times W)$. □

In particular, suppose $K$ itself is normal in $G$ (i.e., $H = K$ in the above proposition). With $\overline{G} = G/K$ (and $\overline{K} = K/K = 1$), we have
$$\text{TOP}_{G,K}(G/K \times W) = \text{M}(W, \overline{G}) \rtimes (\text{Aut}(\overline{G}) \rtimes \text{TOP}(W))$$
which is exactly the same as $\text{TOP}_{\overline{G}}(\overline{G} \times W)$.

Let $H$ be a subgroup of $G$ and let $\alpha \in \text{Aut}(G)$. Even though $\alpha(H) \subset H$, $\alpha|_H : H \to H$ may not be an automorphism of $H$ in general. For example, any integral matrix $A \in \text{GL}(n, \mathbb{R})$ of determinant $> 1$ maps $\mathbb{Z}^n$ into itself, but it is not an isomorphism of $\mathbb{Z}^n$. The following provides a sufficient condition, which will be used in the subsequent sections.

LEMMA 9.2.8. *Let $H$ be a closed subgroup of $G$ with finitely many connected components, and let $\alpha \in \text{Aut}(G)$ such that $\alpha(H) \subset H$. Then $\alpha(H) = H$.*

PROOF. Let $H_0$ be the connected component of the identity of $H$. Then $H_0$ is closed in $G$. Since $\alpha$ is a global homeomorphism, $\alpha(H_0)$ is closed in $G$ and, hence, closed in $H_0$. Because $\alpha(H_0)$ is a connected manifold having the same dimension as $H_0$ and is embedded in $H_0$ as a closed subset, invariance of domain implies that $\alpha(H_0) = H_0$.

Now, $H_0$ is normal in $H$, and $\alpha$ induces a homomorphism $\overline{\alpha}$ of $H/H_0$ into itself. It is enough to show that $\overline{\alpha}$ is onto. But it is true because $H/H_0$ is a finite group. □

9.2.9. If we wish to find all Seifert fiberings over the space $Q\backslash W$ with typical fiber $\Gamma\backslash G/K$, we need to do the following:

Step A. Find a proper action of $Q$ on $W$; that is, find a representation $\rho : Q \to \text{TOP}(W)$ so that $Q$ acts properly on $W$. This ensures that $Q\backslash W$ is Hausdorff and inherits some of the geometry of $W$.

Step B. For each group extension
$$1 \to \Gamma \to \Pi \to Q \to 1,$$
find all homomorphisms $\theta : \Pi \to \text{TOP}_{G,K}(G/K \times W)$ such that $\theta|_\Gamma : \Gamma \to \ell(G)$ restricts to an injective homomorphism into a lattice of $G = \ell(G)$ and such that the diagram

$$\begin{array}{ccccccccc}
1 & \to & \Gamma & \to & \Pi & \to & Q & \to & 1 \\
& & \downarrow & & \theta\downarrow & & \varphi\times\rho\downarrow & & \\
1 & \to & \frac{\ell(G)\cdot \text{M}(W, N_G(K))}{\text{M}(W,K)\rtimes\text{Inn}^0(G,K)} & \to & \text{TOP}_{G,K}(G/K \times W) & \to & \overline{\text{Out}(G,K)} \times \text{TOP}(W) & \to & 1
\end{array}$$

is commutative. Since $\Pi$ acts on $\Gamma$ by conjugation, the action of $\theta(\Pi)$ extends to an action on the kernel of $\text{TOP}_{G,K}(G/K \times W) \to \overline{\text{Out}(G,K)} \times \text{TOP}(W)$. Induced will be the homomorphism $\varphi : Q \to \overline{\text{Out}(G,K)}$. Because $\theta(\Gamma)$ is a lattice and $Q$ acts properly on $W$, it can be seen that $\Pi$ acts properly on $G/K \times W$ via $\theta(\Pi)$.

9.2.10. We call the homomorphism $\theta$ a *Seifert Construction*. If $\theta$ is injective, we call $\theta$ an *embedding* (into $\text{TOP}_{G,K}(G/K \times W)$).

The space $X = \theta(\Pi)\backslash(G/K \times W)$ is a *Seifert fiber space* and the induced mapping $\theta(\Pi)\backslash(G/K \times W) \to Q\backslash W$ is called a *Seifert fibering* (or Seifert bundle) with *typical fiber* $\Gamma\backslash G/K$ over the base $Q\backslash W$.

We also say that the fibering is *modeled on* $G/K \times W$. If $G/K \times W$ is a manifold and $\theta(\Pi)$ acts freely, we call $X$ a *Seifert manifold*. Sometimes we abuse the technical meaning of "orbifold" and call $X$ an orbifold. The constructions are done *smoothly* if we replace $\mathrm{TOP}(W)$ by diffeomorphisms of $W$ when $W$ is a smooth manifold.

9.2.11. For a fixed $\theta_0$, a conjugation of $\theta_0$ by an element of $\mathrm{TOP}_{G,K}(G/K \times W)$ is called an automorphism of the Seifert Construction $\theta_0$. Running through all conjugacy classes (conjugating by elements of $\mathrm{TOP}_{G,K}(G/K \times W)$ and varying $i$ and $\rho$) yields all the constructions modeled on $G/K \times W$ for the group $\Pi$. The group $\mathrm{TOP}_{G,K}(G/K \times W)$ is called the *uniformizing group*. A construction $\theta$ into a subgroup of $\mathrm{TOP}_{G,K}(G/K \times W)$ is called a *reduction* of the uniformizing group. Since $\mathrm{TOP}_{G,K}(G/K \times W)$ contains many geometrically interesting subgroups, a reduction to one of these groups will induce the interesting geometric structure on $X$. In practice, we shall begin with a fixed $\rho : Q \to \mathrm{TOP}(W)$ and a fixed $i : \Gamma \to \ell(G)$. Then we find all $\theta$'s so that $\theta$ induces $i$ and $\rho$.

If $K$ is a trivial group, then $\mathrm{TOP}_{G,K}(G/K \times W)$ becomes $\mathrm{TOP}_G(G \times W) = \mathrm{M}(W, G) \rtimes (\mathrm{Aut}(G) \times \mathrm{TOP}(W))$.

## 9.3. When $N_G(K) = K$ and $\mathrm{Aut}^0(G, K) = 1$

Most of the Lie groups treated in the later sections (symmetric spaces of non-compact type and solv-manifolds) satisfy the two conditions of this section's title. Consequently, as we shall see, the kernel of $\mathrm{TOP}_{G,K}(G/K \times W) \to \mathrm{TOP}(W)$ becomes a Lie group. This has the tendency to make Seifert Constructions, if they exist, more rigid. On the other hand, being able to work with kernels which are Lie groups instead of much larger kernels, enables us to prove the strong existence and uniqueness theorems for the Seifert Constructions investigated in the later sections.

We begin with a sufficient condition for $\mathrm{Aut}^0(G, K) = 1$.

LEMMA 9.3.1. *Let $G$ be a Lie group, and let $K$ be a closed subgroup of $G$. Suppose the largest normal subgroup of $G$ contained in $K$ is trivial. Then $\mathrm{Aut}^0(G, K)$ is trivial.*

PROOF. Suppose $\alpha \in \mathrm{Aut}^0(G, K)$. By the definition of $\mathrm{Aut}^0(G, K)$, $x^{-1}\alpha(x) \in K$ for every $x \in G$. Let

$K' = $ the subgroup of $K$ generated by $\{x^{-1}\alpha(x) : x \in G, \alpha \in \mathrm{Aut}^0(G, K)\}$.

For any $b \in G$,
$$b^{-1} \cdot (x^{-1}\alpha(x)) \cdot b = (xb)^{-1}\alpha(xb) \cdot (b^{-1}\alpha(b))^{-1},$$
which is an element of $K'$. Thus $K'$ is a subgroup of $K$ which is normal in $G$. However, the largest normal subgroup of $G$ contained in $K$ is trivial. Therefore $K'$ is trivial. This implies $\alpha = \mathrm{id}$. Consequently, $\mathrm{Aut}^0(G, K)$ is a trivial group. □

For simplicity, we fix some notation:

$$\mathrm{Aff}(G, K) = \ell(G) \rtimes \mathrm{Aut}(G, K).$$

Since $r(K)$ is normal in $\mathrm{Aff}(G, K)$, we define

$$\overline{\mathrm{Aff}}(G, K) = \frac{\ell(G) \rtimes \mathrm{Aut}(G, K)}{r(K)},$$

$$\overline{\mathrm{Aff}}_0(G, K) = \frac{\ell(G) \rtimes \mathrm{Inn}(G, K)}{r(K)}.$$

THEOREM 9.3.2. *Let $G$ be a connected Lie group and $K$ be a closed subgroup of $G$. Suppose*

(1) $N_G(K) = K$, *and*
(2) $\mathrm{Aut}^0(G, K)$ *is trivial.*

*Then,* $\mathrm{TOP}_{G,K}(G/K \times W) = \overline{\mathrm{Aff}}(G, K) \times \mathrm{TOP}(W)$.

PROOF. From Corollary 9.2.5, we have

$$\mathrm{TOP}_{G,K}(G/K \times W) = \frac{\ell(G) \cdot (\mathrm{M}(W, N_G(K)) \rtimes \mathrm{Aut}(G,K))}{\mathrm{M}(W,K) \rtimes \mathrm{Aut}^0(G,K)} \rtimes \mathrm{TOP}(W)$$

$$= \frac{\ell(G) \cdot (\mathrm{M}(W,K) \rtimes \mathrm{Aut}(G,K))}{\mathrm{M}(W,K)} \rtimes \mathrm{TOP}(W),$$

from the two conditions given in the statement. Now the factors $\mathrm{M}(W, K)$ drop out. However, notice that $(\ell(G) \rtimes \mathrm{Aut}(G,K)) \cap \mathrm{M}(W,K) = r(K)$. Therefore,

$$\mathrm{TOP}_{G,K}(G/K \times W) = \frac{\ell(G) \rtimes \mathrm{Aut}(G,K)}{r(K)} \times \mathrm{TOP}(W).$$

This is a direct product rather than a semidirect product, since $(1, 1, h) \in \mathrm{TOP}(W)$ commutes with $(a, \alpha, 1) \in \mathrm{Aff}(G, K)$, because $a$ is a constant map. Of course, the group $\ell(G) \rtimes \mathrm{Aut}(G, K)$ acts on $G/K$ by $(a, \alpha) \cdot xK = a\alpha(x)K$. □

PROPOSITION 9.3.3. *Let $G$ be a connected Lie group and $K$ a closed subgroup of $G$. Suppose $N_G(K) = K$ and every closed subgroup of $G$ isomorphic to $K$ is a conjugate of $K$. Then there exists an isomorphism $\Psi : \overline{\mathrm{Aff}}(G, K) \to \mathrm{Aut}(G)$ making the square*

$$\begin{array}{ccc} G & \xrightarrow{\mu} & \mathrm{Aut}(G) \\ \cong \downarrow \ell & & \cong \uparrow \Psi \\ \ell(G) & \longrightarrow & \overline{\mathrm{Aff}}(G, K) \end{array}$$

*commutative.*

PROOF. Since $G$ is normal in $G \rtimes \mathrm{Aut}(G, K)$, conjugation by elements $(a, \alpha) \in G \rtimes \mathrm{Aut}(G, K)$ on $(x, 1) \in G$,

$$(a, \alpha)(x, 1)(a, \alpha)^{-1} = (a \cdot \alpha(x) \cdot a^{-1}, 1),$$

yields a homomorphism $\Psi : G \rtimes \mathrm{Aut}(G, K) \to \mathrm{Aut}(G)$ given by

$$\Psi(a, \alpha) = \mu(a)\alpha.$$

For $k \in K$, $(k, \mu(k^{-1})) \in G \rtimes \mathrm{Aut}(G, K)$ and $\Psi(k, \mu(k^{-1})) = 1$. Conversely, suppose $\Psi(a, \alpha) = 1$. Then $\alpha = \mu(a^{-1})$ so that $\mu(a^{-1}) \in \mathrm{Aut}(G, K)$. Therefore, $a \in N_G(K)$. However, $N_G(K) = K$ so that $a \in K$. We have shown that the kernel of $\Psi$ is exactly

$$K \cong \{(k, \mu(k^{-1})) : k \in K\}.$$

To show $\Psi$ is surjective, let $\beta \in \mathrm{Aut}(G)$. Since $\beta(K)$ is isomorphic to $K$, there exists $a \in G$ for which $\beta(K) = aKa^{-1}$. Then $\mu(a)^{-1}\beta \in \mathrm{Aut}(G, K)$. This shows $(a, \mu(a)^{-1}\beta) \in G \rtimes \mathrm{Aut}(G, K)$ maps to $\alpha$ by $\Psi$.

Since $K$ is closed and $\Psi$ is continuous and surjective, it induces an isomorphism of groups

$$\frac{G \rtimes \mathrm{Aut}(G, K)}{K} \longrightarrow \mathrm{Aut}(G)$$

which is a diffeomorphism and so they are isomorphic as Lie groups.

The group $G \rtimes \mathrm{Aut}(G,K)$ acts on $G$ by $(a,\alpha) \cdot x = a\alpha(x)$. In other words, $G \rtimes \mathrm{Aut}(G,K)$ is naturally identified with $\ell(G) \rtimes \mathrm{Aut}(G,K)$. Under this identification, the kernel of $\Psi$ is exactly $r(K)$ since $(k, \mu(k^{-1}))x = k \cdot k^{-1}xk = xk$. □

COROLLARY 9.3.4. *With the same conditions as in* Proposition 9.3.3, *the following diagram is commutative.*

$$\begin{array}{ccccccccc}
1 & \longrightarrow & \mathrm{Inn}(G) & \longrightarrow & \mathrm{Aut}(G) & \longrightarrow & \mathrm{Out}(G) & \longrightarrow & 1 \\
& & \ell \downarrow \cong & & \Psi^{-1} \downarrow \cong & & \downarrow = & & \\
1 & \longrightarrow & \overline{\mathrm{Aff}}_0(G,K) & \longrightarrow & \overline{\mathrm{Aff}}(G,K) & \longrightarrow & \mathrm{Out}(G) & \longrightarrow & 1
\end{array}$$

## 9.4. Symmetric spaces of noncompact type

A *symmetric space* is a triple $(G, K, \sigma)$ consisting of a connected Lie group $G$, a closed subgroup $K$ of $G$, and an involutive (i.e., order 2) automorphism $\sigma$ of $G$ such that $(G^\sigma)_0 \subset K \subset G^\sigma$, where $G^\sigma$ is the fixed-point set of $\sigma$. $(G, K, \sigma)$ is (*almost*, respectively) *effective* if the largest normal subgroup $N$ of $G$ contained in $K$ is trivial (discrete, respectively). If $(G, K, \sigma)$ is a symmetric space, then $(G/N, K/N, \sigma^*)$ is an effective symmetric space, where $\sigma^*$ is the automorphism of $G/N$ induced from $\sigma$. If, in addition, the group $\mathrm{Ad}_G(K)$ is compact, $(G, K, \sigma)$ is said to be a Riemannian symmetric space.

Throughout this section, $(G, K, \sigma)$ will be an effective Riemannian symmetric space of noncompact type. Therefore *$G$ is a connected, semisimple Lie group in its adjoint form with no compact normal factors, and $K$ is a closed maximal compact subgroup of $G$*. We collect some facts for such groups:

PROPOSITION 9.4.1. *Let $G$ be a connected, centerless, semisimple Lie group without any normal compact factors. Let $K$ be a maximal compact subgroup of $G$. Then*

(1) $N_G(K) = K$, *and $K$ is connected.*
(2) *Every closed subgroup of $G$ isomorphic to $K$ is a conjugate of $K$.*
(3) $\mathrm{Out}(G)$ *is finite.*
(4) $\mathrm{TOP}_{G,K}(G/K \times W) = \overline{\mathrm{Aff}}(G,K) \times \mathrm{TOP}(W)$. □

For (1), see [**Hel62**, p.275; A3(i)]. For (2), see [**Hel62**, VI Theorem 2.1]. (3) is well known. Now Theorem 9.3.2 characterizes our uniformizing group completely. (4) is a consequence of Theorem 9.3.2 by (1) and (2).

LEMMA 9.4.2. *Let $\overline{G} = \mathrm{Aut}(G)$, and let $\overline{K}$ be its maximal compact subgroup containing $\mu(K) = \mathrm{Ad}_G(K)$. Then $\overline{K} = \mathrm{Aut}(G,K)$ and $\overline{K}/K \cong \mathrm{Out}(G)$.*

PROOF. Since $\mathrm{Out}(G)$ is finite, $\overline{K}/K$ is discrete. Hence $\overline{K}_0 = \mu(K) \subset \overline{G}$ is the connected component of $\overline{K}$ containing the identity element. Thus, $\mu(K)$ is normal in $\overline{K}$. Let $\alpha \in \overline{K}$ be any element. We claim that $\alpha(K) = K$. Pick any $k \in K$. We shall show $\alpha(k) \in K$. Since $\mu(K)$ is normal in $\overline{K}$, $\alpha \cdot \mu(k) \cdot \alpha^{-1} = \mu(k')$ for some $k' \in K$. Note that these are equal as elements of $\overline{K}$ (and hence, as elements of $\overline{G} = \mathrm{Aut}(G)$). Since $\alpha \cdot \mu(k) \cdot \alpha^{-1} = \mu(\alpha(k))$, we have

$$\mu(k') = \mu(\alpha(k)).$$

That is, conjugations by $k'$ and $\alpha(k)$ produce the same automorphisms of $G$. Consequently, $\alpha(k) \cdot k'^{-1} \in \mathcal{Z}(G)$, the center of $G$, which is trivial. We have shown that $\alpha(k) \cdot k'^{-1} = 1$, so, $\alpha(k) = k' \in K$. Hence $\alpha(K) \subset K$. Since $K$ is compact, one can apply Lemma 9.2.8: $\alpha(K) \subset K$ implies that $\alpha$ induces an automorphism of $K$ so that $\alpha \in \mathrm{Aut}(G, K)$. Thus $\overline{K} \subset \mathrm{Aut}(G, K)$.

Since $N_G(K) = K$, we have $\mathrm{Inn}(G, K) = \mathrm{Ad}_G(K) \cong K$. Thus we have $\mathrm{Aut}(G,K)/\mathrm{Inn}(G,K) \subset \mathrm{Out}(G)$, a finite group. However, all maximal compact subgroups of $G$ are conjugate to each other. Therefore, for every $\beta \in \mathrm{Aut}(G)$, there exists $a \in G$ for which $\beta(K) = aKa^{-1}$. Then $\mu(a)^{-1}\beta \in \mathrm{Aut}(G, K)$. This implies $\mathrm{Aut}(G,K)/\mathrm{Inn}(G,K) \to \mathrm{Out}(G)$ is surjective. Thus we have a short exact sequence
$$1 \to K \to \mathrm{Aut}(G, K) \to \mathrm{Out}(G) \to 1.$$
Since $K$ and $\mathrm{Out}(G)$ are compact, so is $\mathrm{Aut}(G, K)$. By maximality of $\overline{K}$, we have $\overline{K} = \mathrm{Aut}(G, K)$. $\square$

The action of $\mathrm{Aut}(G)$ on $G/K$ above can also be interpreted as follows: By Lemma 9.4.2, the conjugation map induces an identification
$$\overline{\mu} : G/K \xrightarrow{\approx} \overline{G}/\overline{K}.$$
In fact, $\overline{\mu}$ is bijective because $N_G(K) = K$ and $\overline{K} = \mathrm{Aut}(G, K)$. Recall that $\Psi : \overline{\mathrm{Aff}}(G,K) \to \mathrm{Aff}(G)$ sending $(a, \alpha)$ to $\mu(a) \circ \alpha$ is an isomorphism.

PROPOSITION 9.4.3. *The diffeomorphism $\overline{\mu} : G/K \to \overline{G}/\overline{K}$ is weakly $\mathrm{Aut}(G)$-equivariant. More precisely, the following diagram is commutative.*

$$\begin{array}{ccc} \overline{\mathrm{Aff}}(G,K) \times G/K & \longrightarrow & G/K \\ {\scriptstyle \Psi \times \overline{\mu}} \downarrow & & \downarrow {\scriptstyle \overline{\mu}} \approx \\ \overline{G} \times \overline{G}/\overline{K} & \longrightarrow & \overline{G}/\overline{K} \end{array}$$

PROOF. Let $(a, \alpha) \in \overline{\mathrm{Aff}}(G, K)$, $xK \in G/K$. Then

$$\begin{array}{ccc} (a, \alpha) \cdot xK & \longrightarrow & a\alpha(x)K \\ {\scriptstyle \Psi \times \overline{\mu}} \downarrow & & \downarrow {\scriptstyle \overline{\mu}} \approx \\ (\mu(a) \circ \alpha, \mu(x)\overline{K}) & \longrightarrow & \mu(a\alpha(x))\overline{K} \end{array}$$

Note that $\mu(a\alpha(x))\overline{K} = (\mu(a) \circ \alpha \circ \mu(x))\overline{K}$. $\square$

PROPOSITION 9.4.4. *With the $G$-invariant Riemannian metric on $G/K$ induced by the Killing-Cartan form of $\mathfrak{g}$, $\overline{\mathrm{Aff}}(G, K) = \mathrm{Isom}(G/K)$.*

PROOF. Recall the Killing-Cartan form is defined by
$$B(X, Y) = \mathrm{Trace}(\mathrm{ad}\, X \cdot \mathrm{ad}\, Y).$$
Any left invariant metric on $G$ which is also right invariant on $K$ is a constant times of this metric. Let $\alpha \in \mathrm{Aut}(G, K)$, and $\alpha_* = d\alpha \in \mathrm{Aut}(\mathfrak{g})$. Then $\mathrm{ad}(\alpha_* X) = \alpha_* \circ \mathrm{ad} X \circ \alpha_*^{-1}$. A calculation shows that
$$B(\alpha_* X, \alpha_* Y) = B(X, Y).$$
Therefore, $\alpha_*$ leaves the quadratic form $B$ on $\mathfrak{g}$ invariant. Let $\mathfrak{g} = \mathfrak{k} \oplus \mathfrak{p}$ be the orthogonal decomposition, where $\mathfrak{k}$ is the Lie algebra of $K$. Since $\alpha_*$ maps $\mathfrak{k}$ onto itself and $B$ is invariant under $\alpha_*$, $\alpha_*$ leaves the orthogonal complement $\mathfrak{p}$ invariant. Thus, $\alpha_*$ maps $\mathfrak{p}$ to itself, and preserves the quadratic form $B$ on $\mathfrak{p}$. The metric on $G/K$ is just a scalar multiple of the restriction of the Killing-Cartan form $B$ on $\mathfrak{p}$.

We have shown that $\alpha_*$ is an isometry on $G/K$. Since $\ell(G) \subset \mathrm{Isom}(G/K)$ clearly, we have
$$\ell(G) \rtimes \mathrm{Aut}(G,K) \to \mathrm{Isom}(G/K).$$

We show this homomorphism is surjective. It is well known that $\ell(G) = \mathrm{Isom}_0(G/K)$. Suppose $\overline{f} \in \mathrm{Isom}(G/K)$. By transitivity of the action of $\ell(G)$ on $G/K$, we may assume that $\overline{f}$ fixes the point $K = eK \in G/K$. Since $\ell(G)$ is normal in $\mathrm{Isom}(G/K)$, conjugation by $\overline{f}$ defines an automorphism of $\ell(G)$. Let $f = \ell^{-1} \circ \mu(\overline{f}) \circ \ell \in \mathrm{Aut}(G)$. Then

$$\begin{array}{ccc} G & \xrightarrow{f} & G \\ \ell \downarrow & & \downarrow \ell \\ \ell(G) & \xrightarrow{\mu(\overline{f})} & \ell(G) \end{array}$$

commutes and
$$\ell(f(a)) = \overline{f} \circ \ell(a) \circ \overline{f}^{\,-1}.$$

With $\overline{f}(eK) = eK$, an easy calculation shows that
$$\overline{f}(aK) = \overline{f} \circ \ell(a)(eK) = \ell(f(a)) \circ \overline{f}(eK) = \ell(f(a))(eK) = f(a)K$$

for all $a \in G$. Thus, for $k \in K$, $f(k)K = \overline{f}(kK) = \overline{f}(K) = K$ so that $f(k) \in K$. We have shown that the automorphism $f$ of $G$ maps $K$ to itself. This proves the surjectivity of $\ell(G) \rtimes \mathrm{Aut}(G,K) \to \mathrm{Isom}(G/K)$. Since the kernel of this homomorphism is $r(K)$, we have completed the proof. $\square$

THEOREM 9.4.5 (cf. [**RW77**, Theorem 2]). *Let $G$ be a connected, centerless, semisimple Lie group without any normal compact factors. Let $K$ be a maximal compact subgroup of $G$. Let $\Gamma$ be a lattice of $G$. If there is a 3-dimensional factor, assume that the image of the lattice $\Gamma$ by the projection onto that factor is dense. Let $\rho : Q \to \mathrm{TOP}(W)$ be a proper action of a discrete group $Q$, and let $1 \to \Gamma \to \Pi \to Q \to 1$ be an exact sequence. Then there exists a homomorphism $\theta : \Pi \to \mathrm{TOP}_{G,K}(G/K \times W) = \overline{\mathrm{Aff}}(G,K) \times \mathrm{TOP}(W)$ so that the diagram with exact rows*

$$\begin{array}{ccccccccc} 1 & \to & \Gamma & \to & \Pi & \to & Q & \to & 1 \\ & & i \downarrow & & \theta \downarrow & & \downarrow \varphi \times \rho & & \\ 1 & \to & \overline{\mathrm{Aff}}_0(G,K) & \to & \mathrm{TOP}_{G,K}(G/K \times W) & \to & \mathrm{Out}(G) \times \mathrm{TOP}(W) & \to & 1 \end{array}$$

*is commutative, yielding a Seifert fiber space with typical fiber the double coset space $\Gamma\backslash G/K$. Such a homomorphism $\theta$ with fixed $i$ and $\varphi \times \rho$ is unique. The action is free if and only if the preimage of each isotropy $Q_w$ in $\Pi$ is torsion free.*

PROOF. (Existence). Since $\Gamma$ is normal in $\Pi$, there is a natural homomorphism $\mu : \Pi \to \mathrm{Aut}(\Gamma)$. Under the conditions on $G$ stated, Mostow's rigidity theorem ensures that the pair $(\Gamma, G)$ has UAEP. UAEP gives rise to a homomorphism $\mathrm{Aut}(\Gamma) \to \mathrm{Aut}(G)$. Consequently, we have a homomorphism $\Pi \to \mathrm{Aut}(G)$ so that the following diagram with exact rows is commutative.

$$\begin{array}{ccccccccc} 1 & \to & \Gamma & \to & \Pi & \to & Q & \to & 1 \\ & & \downarrow & & \downarrow & & \downarrow & & \\ 1 & \to & \mathrm{Inn}(G) & \to & \mathrm{Aut}(G) & \to & \mathrm{Out}(G) & \to & 1 \end{array}$$

Composing $\Pi \to \mathrm{Aut}(G)$ with the isomorphism $\Psi^{-1} : \mathrm{Aut}(G) \to \overline{\mathrm{Aff}}(G, K)$ in Proposition 9.3.3, we get a homomorphism $\Pi \to \mathrm{Aut}(G) \to \overline{\mathrm{Aff}}(G, K)$, under which $\Gamma$ is mapped into $\overline{\mathrm{Aff}}_0(G, K)$. This together with the action $\Pi \to Q \to \mathrm{TOP}(W)$ gives rise to a homomorphism $\Pi \to \overline{\mathrm{Aff}}(G, K) \times \mathrm{TOP}(W) = \mathrm{TOP}_{G,K}(G/K \times W)$.

(Uniqueness). We apply Corollary 5.7.3 to the commuting diagram in the statement of the theorem, with $i : \Gamma \to \overline{\mathrm{Aff}}_0(G, K)$ for $i : G \to H$. To this end, we need to calculate the centralizer of $\Gamma$ in $\overline{\mathrm{Aff}}_0(G, K) = \frac{\ell(G) \rtimes \mathrm{Inn}(G,K)}{r(K)}$. An element $(a, \mu(b)) \in \ell(G) \rtimes \mathrm{Inn}(G, K)$ represents an element in the centralizer if and only if

$$[(a, \mu(b)), (x, 1)] = (abxb^{-1}a^{-1}x^{-1}, 1) = (\mu(ab)(x) \cdot x^{-1}, 1) \in r(K)$$

for every $x \in \Gamma$. Recall that elements of $r(K)$ in $\ell(G) \rtimes \mathrm{Aut}(G, K)$ are of the form $(k^{-1}, \mu(k))$ with $k \in K$. Therefore, it happens if and only if $\mu(ab)(x) \cdot x^{-1} = 1$ for for every $x \in \Gamma$. By UAEP, this should happen for every $x \in G$. Then $\mu(ab) \in \mathrm{Inn}^0(G, K)$. But $\mathrm{Inn}^0(G, K)$ is trivial by Lemma 9.3.1. Therefore, $\mu(ab) = 1$ so that $ab = 1$. Since $\mu(b) \in \mathrm{Inn}(G, K)$, $(a, \mu(b)) = (b^{-1}, \mu(b)) \in r(K)$, which represents the identity element of $\overline{\mathrm{Aff}}_0(G, K)$. We have shown $C_{\overline{\mathrm{Aff}}_0(G,K)}(\Gamma)$ is trivial so that $Z^1(Q; C_{\overline{\mathrm{Aff}}_0(G,K)}(\Gamma)) = 0$. (Note that we do not need conjugation for the uniqueness here; i.e., there is only one homomorphism $\theta$ for a fixed $\ell$ and $\rho$). Finally, since $G/K$ is diffeomorphic to $\mathbb{R}^n$, the action of $\Pi$ is free if and only if the preimage of each isotropy $Q_w$ in $\Pi$ is torsion free. □

In [**LLR96**], we have discussed a situation where $G$ is semisimple and $K$ is trivial. Lemma 2.2 in [**LLR96**] describes a necessary and sufficient condition for an extension $\Pi$ to be mapped into $\mathrm{TOP}_G(G \times W)$. In the applications there, $G$ was of compact type and the lattice $\Gamma$ was assumed to go into $\ell(G) \times r(G)$. Section 5 of [**LLR96**] discusses uniqueness for that situation.

When $W$ is a Riemannian manifold, the space $G/K \times W$ acquires the natural product metric. Then,

$$\overline{\mathrm{Aff}}(G, K) \times \mathrm{Isom}(W) = \mathrm{Isom}(G/K) \times \mathrm{Isom}(W) \subset \mathrm{Isom}(G/K \times W).$$

COROLLARY 9.4.6. *Suppose $W$ is a Riemannian manifold, and $Q$ acts on $W$ as isometries (i.e., $\rho$ maps $Q$ into $\mathrm{Isom}(W)$). Then the construction yields a representation*

$$\Pi \to \mathrm{Isom}(G/K) \times \mathrm{Isom}(W) \subset \mathrm{Isom}(G/K \times W),$$

*yielding a Riemannian orbifold $\Pi \backslash (G/K \times W)$.* □

The space $\Pi \backslash (G \times W)$ has a Seifert fiber structure

$$\Gamma \backslash G/K \to \Pi \backslash (G/K \times W) \to Q \backslash W,$$

where $\Gamma \backslash G/K$, the typical fiber, is a Riemannian symmetric space. Singular fibers are finite quotients of the typical fiber, where the finite actions are via isometries of $G/K$.

Here is a more precise account. Let $Q_0$ be the kernel of $\phi : Q \to \mathrm{Out}(\Gamma)$. Then the preimage of $Q_0$ in $\Pi$ splits as a direct product $\Gamma \times Q_0$ because $\Gamma$ has trivial center. Now $Q$ in $\Pi$ acts trivially on $\Gamma$ by conjugation so each trivial automorphism extends uniquely to the trivial automorphism of $G$. But $G$ maps by conjugation injectively into $\mathrm{Aut}(G)$ which injects into $\overline{\mathrm{Aff}}(G, K)$ by the isomorphism $\Psi^{-1}$. Thus the image of $Q_0$ into $\overline{\mathrm{Aff}}(G, K)$ is trivial. Consequently, the normal subgroup $\Gamma \times Q_0$ of $\Pi$ acts on $G/K \times W$ in such a way that $\Gamma$ acts only on the $G/K$-factor as left translations, $Q_0$ acts only on the $W$-factor via $\rho$, yielding $(\Gamma\backslash G/K) \times (Q_0\backslash W)$. Because Out$(\Gamma)$ is finite, $Q/Q_0$ is finite. The finite quotient group $F = \Pi/(\Gamma \times Q_0)$ acts diagonally on $(\Gamma\backslash G/K) \times (Q_0\backslash W)$.

$$\begin{array}{ccc}
G/K \times W & \xrightarrow{Q_0\backslash} & G/K \times Q_0\backslash W \\
{\scriptstyle \Gamma\backslash}\downarrow & & {\scriptstyle \Gamma\backslash}\downarrow \\
\Gamma\backslash G/K \times W & \xrightarrow{Q_0\backslash} & (\Gamma\backslash G/K) \times (Q_0\backslash W) \\
& & {\scriptstyle F\backslash}\downarrow \\
& & (\Gamma\backslash G/K) \times_F (Q_0\backslash W) \xrightarrow{=} \Pi\backslash(G/K \times W)
\end{array}$$

EXAMPLE 9.4.7. Let $G = \mathrm{SO}_0(1,3)$, $W = \mathbb{R}^n$. Let $\Gamma$ be a (respectively, torsion-free) lattice of $G$, and $Q \subset E(n)$ be a crystallographic group. Then $K = \mathrm{SO}(3)$ and $\mathrm{SO}_0(1,3)/\mathrm{SO}(3) = \mathbf{H}^3$, the 3-dimensional hyperbolic space. Thus we have

$$\overline{\mathrm{Aff}}(G,K) \times \mathrm{Isom}(\mathbb{R}^n) = \mathrm{Isom}(G/K) \times \mathrm{Isom}(\mathbb{R}^n)$$
$$= \mathrm{Isom}(\mathbf{H}^3) \times E(n)$$
$$\subset \mathrm{Isom}(\mathbf{H}^3 \times \mathbb{R}^n).$$

For any extension $\Pi$ of $\Gamma$ by $Q$, there exists a Seifert fibering (respectively, aspherical manifold) $\Pi \to \mathrm{Isom}(\mathbf{H}^3) \times E(n)$,

$$\Gamma\backslash\mathbf{H}^3 \to \Pi\backslash(\mathbf{H}^3 \times \mathbb{R}^n) \to Q\backslash\mathbb{R}^n$$

with typical fiber the hyperbolic space form $\Gamma\backslash\mathbf{H}^3$ and base orbifold $Q\backslash\mathbb{R}^n$.

EXAMPLE 9.4.8. Seifert Constructions for $G = \mathbb{R}^n$ are more numerous and more twisted than constructions with $G$ semisimple or $G/K$ as above. For example, for $Q$ take the Fuchsian group whose orbit space $Q\backslash\mathbf{H}^2$ is the 2-sphere with multiplicities 2, 3, and 7. For Seifert fiberings modeled on $\mathbb{R}^3 \times \mathbf{H}^2$, $\Gamma = \mathbb{Z}^3$, we obtain an infinite number of different fiberings parametrized by $\mathbb{Z}^3$ when $Q \to \mathrm{Aut}(\mathbb{Z}^3)$ is trivial. If we take just those constructions that yield $K(\Pi, 1)$'s, we get an infinite number of distinct 5-manifolds that fiber over the 2-torus with finite Abelian structure group; see [**CR72c**, 2.2]. The infinite number of distinct extensions can be seen from the fact $H^2(Q; \mathbb{Z}^3) = \mathbb{Z}^3$. These aspherical 5-manifolds all exhibit $\mathbb{R}^2 \times \widetilde{\mathrm{PSL}(2,\mathbb{R})}$ geometry. (From [**CR72c**, 2.2], we can view $\Pi$ as having a finite indexed normal subgroup $\mathbb{Z}^2 \times$(the fundamental group of a closed Seifert 3-manifold $M$) with a cyclic quotient. The aspherical 5-manifold is then *diagonally* covered by $T^2 \times M$.) In other words, the uniformizing group can be reduced to $\mathrm{Isom}(\mathbb{R}^2) \times \mathrm{Isom}_0(\widetilde{\mathrm{PSL}(2,\mathbb{R})}) = E(2) \times (\mathbb{R} \times_{\mathbb{Z}} \widetilde{\mathrm{PSL}(2,\mathbb{R})})$.

On the other hand, take $G/K \times \mathbf{H}^2$, where $G = \mathrm{SO}_0(1,3) \cong \mathrm{PSL}(2,\mathbb{C}) \cong \mathrm{Isom}_0(\mathbf{H}^3)$ and $K = \mathrm{SO}(3)$. For any lattice $\Gamma \subset G$ and homomorphism $\phi : Q \to \mathrm{Aut}(\Gamma)$, there exists just one extension $1 \to \Gamma \to \Pi \to Q \to 1$, because the center of $\Gamma$ is trivial. Furthermore, $\Pi$ contains a finite indexed normal subgroup $\Gamma \times Q_0$, where $Q_0$ is the kernel of $Q \to \mathrm{Out}(\Gamma)$ (Out$(\Gamma)$ is finite). The Seifert Construction $M(\Pi) = \theta(\Pi)\backslash(G/K \times \mathbf{H}^2)$ is regularly covered (possibly branched) by the product space $(\Gamma\backslash G/K) \times (Q_0\backslash W)$, where $Q/Q_0$ acts diagonally and isometrically as in Theorem 9.4.5. This means that $M(\Pi)$ has $\mathbf{H}^3 \times \mathbf{H}^2$-geometry as an orbifold. If

$Q \to \mathrm{Out}(\Gamma)$ is trivial, then $\Pi$ cannot be torsion free even if $\Gamma$ is torsion free. Consequently, $M(\Pi)$ is not aspherical

If we want $M(\Pi)$ to be a closed aspherical 5-manifold, then $\Pi$ must be torsion free. This means that $\Pi_w$ in $1 \to \Gamma \to \Pi_w \to Q_w \to 1$ must be torsion free for each $Q_w \cong \mathbb{Z}_2, \mathbb{Z}_3$ and $\mathbb{Z}_7$. Moreover, the image of $Q$ in $\mathrm{Out}(\Gamma)$, which is isomorphic to $Q/Q_0$ can have no Abelian quotient other than 1 because $Q$ is perfect. In addition, $\Gamma$ must be normally contained in other torsion-free lattices in $\mathrm{PSL}(2,\mathbb{C}) \rtimes \mathbb{Z}_2$ with quotients $\mathbb{Z}_2, \mathbb{Z}_3$ or $\mathbb{Z}_7$. This is of course difficult to achieve in general (we may easily find $\Gamma$ so that $\mathrm{Out}(\Gamma)$ does not contain each of $Q_w \cong \mathbb{Z}_2, \mathbb{Z}_3$ and $\mathbb{Z}_7$ as subgroups). The point here is that the possible group extensions $\Pi$ of $\Gamma$ by $Q$ and the structure of $\Pi$ severely restrict the possible Seifert Constructions which yield torsion-free $\Pi$, and hence, aspherical manifolds when compared with extensions of $\mathbb{Z}^3$ by $Q$. One caveat, though, is that there are far more nonisomorphic lattices in $\mathrm{SO}_0(1,3)$ than in $\mathbb{R}^3$ and, consequently, leads to a rich supply of Seifert fiberings despite the finiteness of each $\mathrm{Out}(\Gamma)$.

REMARK 9.4.9. Let us now explain more carefully the connection with [**RW77**]. Let $Q_0$ be the kernel of $Q \to \mathrm{Out}(\Gamma)$. Then, $\Gamma \times Q_0$ is a normal subgroup of $\Pi$ with quotient isomorphic to the finite group $Q/Q_0$. The exact sequence $1 \to \Gamma \to \Pi/Q_0 \to Q/Q_0 \to 1$ injects into $1 \to \mathrm{Inn}(\Gamma) \to \mathrm{Aut}(\Gamma) \to \mathrm{Out}(\Gamma) \to 1$. Since every automorphism of $i(\Gamma) \subset G = \ell(G)$ extends uniquely to an automorphism of $G$, $\Pi/Q_0$ is mapped into $\mathrm{Aut}(G) = \overline{G}$ carrying $\Gamma$ to $\mu(i(\Gamma)) = \overline{\Gamma}$ in $\mathrm{Inn}(\Gamma) \subset \mathrm{Inn}(G)$. The action, in [**RW77**], of $\Pi$ on $\overline{G}/\overline{K} \times W$ is given by the composite

$$\overline{G}/\overline{K} \times W \xrightarrow{Q_0\backslash} \overline{G}/\overline{K} \times (Q_0\backslash W) \xrightarrow{\overline{\Gamma}\backslash} (\overline{\Gamma}\backslash\overline{G}/\overline{K}) \times (Q_0\backslash W) \xrightarrow{Q/Q_0} \Pi\backslash(\overline{G}/\overline{K} \times W).$$

Specifically, the action of $\beta \in \Pi$ on $\overline{\alpha K} \times w$ is given by $\overline{\alpha K} \times w \mapsto \overline{\beta \alpha K} \times w'$, where $\overline{\beta}$ is the automorphism of $G$ induced by conjugation by $\beta \in \Pi$, and $w' = \rho(j(\beta))(w)$, where $j : \Pi \to Q$.

Let $\theta : \Pi \to \overline{\mathrm{Aff}}(G,K) \times \mathrm{TOP}(W)$ be a Seifert Construction. Then Remark 9.4.3 tells us that the action of $\theta(\Pi) \subset \overline{\mathrm{Aff}}(G,K) \times \mathrm{TOP}(W)$ on $G/K \times W$ is equivalent to the action of $\Pi$ on $\overline{G}/\overline{K} \times W$ via the isomorphism $\Psi : \overline{\mathrm{Aff}}(G,K) \to \overline{G}$, and the diffeomorphism $\overline{\mu} : G/K \to \overline{G}/\overline{K}$.

If $\rho : Q \to \mathrm{TOP}(W)$, where $W$ is a contractible manifold, and $\Pi$ is torsion free, then the space $M(\Pi) = \theta(\Pi)\backslash(G/K \times W)$ is the $K(\Pi,1)$ manifold constructed in [**RW77**] as mentioned in Subsection 9.1.1, and $M(\Pi)$ will be smooth if $\rho$ is smooth.

Mostow's rigidity theorem does not apply to $G = \mathrm{PSL}(2,\mathbb{R})$. However, by changing the embedding of $\Gamma$, one can still embed the group $\Pi$ into $\mathrm{TOP}_{G,K}(G \times W)$ provided that the image of the abstract kernel in $\mathrm{Out}(\Gamma)$ of the given extension is finite.

THEOREM 9.4.10. *Let $G = \mathrm{PSL}(2,\mathbb{R})$, $K = S^1 \subset \mathrm{PSL}(2,\mathbb{R})$ be a maximal compact subgroup. Let $\Gamma$ be a lattice of $G$, let $\rho : Q \to \mathrm{TOP}(W)$ be a proper action of a discrete group, and let $1 \to \Gamma \to \Pi \to Q \to 1$ be an exact sequence. Assume that the abstract kernel $\varphi : Q \to \mathrm{Out}(\Gamma)$ associated with this extension has finite image. Then there exists a homomorphism $\theta : \Pi \to \mathrm{TOP}_{G,K}(G/K \times W)$ so that* *the diagram with exact rows*

$$\begin{array}{ccccccccc} 1 \longrightarrow & \Gamma & \longrightarrow & \Pi & \longrightarrow & Q & \longrightarrow 1 \\ & i\downarrow & & \theta\downarrow & & \varphi\times\rho\downarrow & \\ 1 \longrightarrow & \mathrm{PSL}(2,\mathbb{R}) \longrightarrow & \mathrm{TOP}_{G,K}(G/K\times W) & \longrightarrow & \mathrm{Out}(G)\times\mathrm{TOP}(W) & \longrightarrow 1 \end{array}$$

is commutative where $i : \Gamma \to \ell(G)$ may be different from the original $\Gamma \subset \ell(G)$. This yields a Seifert fiber space with the surface orbifold $\Gamma\backslash G/K = \Gamma\backslash \mathbf{H}^2$ as typical fiber. The action is free if and only if the preimage of each isotropy $Q_w$ in $\Pi$ is torsion free and, in particular, $\Gamma$ is a surface group.

PROOF. First we need to calculate $\mathrm{TOP}_{G,K}(G/K\times W)$. Since $N_G(K) = K$, the general case still applies. We have

$$\mathrm{TOP}_{G,K}(G/K\times W) = \frac{\ell(G) \rtimes \mathrm{Aut}(G,K)}{r(K) \rtimes \mathrm{Aut}^0(G,K)} \times \mathrm{TOP}(W).$$

Let $Q_0$ be the kernel of $\varphi : Q \to \mathrm{Out}(\Gamma)$. Then $\overline{Q} = Q/Q_0$ is finite. Consider the extension

$$1 \to \Gamma \to \overline{\Pi} \to \overline{Q} \to 1,$$

where $\overline{\Pi} = \Pi/Q_0$. By Nielsen's theorem, as completed by S. Kerckhoff, there exists a homomorphism $\overline{\Pi} \to \mathrm{PSL}(2,\mathbb{R}) \rtimes \mathbb{Z}_2$ realizing this extension as a group action. This together with the action $Q \to \mathrm{TOP}(W)$ gives rise to a desired homomorphism $\theta$. If the abstract kernel of $Q$ into $\mathrm{Out}(\Gamma)$ is not finite, then we cannot apply Nielsen's theorem and a Seifert Construction is not possible by this method. □

9.4.11. Theorem 9.4.5 still holds if $G$ contains 3-dimensional factors (i.e., $\mathrm{PSL}(2,\mathbb{R})$-factors) provided that the projection of $\Gamma$ to each of these factors is dense, because the lattice will still satisfy UAEP condition; see [**Mos73**], [**Pra73**].

The other extreme case will be generalization of Theorem 9.4.10. Suppose $G = \mathrm{PSL}(2,\mathbb{R}) \times \mathrm{PSL}(2,\mathbb{R}) \times \cdots \times \mathrm{PSL}(2,\mathbb{R})$, and assume $\Gamma$ is a lattice in $G$ such that *none* of the images of $\Gamma$ by the projection onto each factor is dense. Then $\Gamma$ lies in a group of the form $\Delta = \Gamma_1 \times \Gamma_2 \times \cdots \times \Gamma_s$ (simply take the images of projections). The argument of the above theorem goes through, and the statement holds true in this more general setting.

EXAMPLE 9.4.12. Let $\Gamma \subset \mathrm{PSL}(2,\mathbb{R})$ be a compact surface group of genus 9, and $\widehat{Q} = \mathbb{Z}^2 \subset E(2)$. A finite group $F = \mathbb{Z}_2$ acts on the surface as a covering transformation yielding a surface of genus 5. It also acts on $\widehat{Q}$ by sending the generators $t_1 \mapsto t_1^{-1}$ and $t_2 \mapsto t_2^{-1}$ so that it has four fixed points on the 2-torus. Let $\Pi$ be an extension of $\Gamma \times \mathbb{Z}^2$ by $F = \mathbb{Z}_2$ so that $1 \to \Gamma \times \mathbb{Z}^2 \to \Pi \to \mathbb{Z}_2 \to 1$ is exact. We view $\Pi$ as $1 \to \Gamma \to \Pi \to Q \to 1$, where $Q = \widehat{Q} \rtimes \mathbb{Z}_2$. Then, $\Pi$ acts freely on $(\mathrm{PSL}(2,\mathbb{R})/S^1) \times \mathbb{R}^2 = \mathbf{H} \times \mathbb{R}^2$. The resulting space $\Pi\backslash(\mathbf{H}\times\mathbb{R}^2)$ is an aspherical Seifert manifold over a flat orbifold (topologically the 2-sphere) with typical fiber the surface of genus 9. There are four singular fibers, all of which are surfaces of genus 5.

## 9.5. Solvmanifolds

A solvmanifold $X$ is a space on which a solvable Lie group acts transitively. This is equivalent to saying $X = G/H$, where $G$ is a solvable Lie group and $H$ is a closed subgroup. An *infra-solvmanifold* is a quotient space of a solvable Lie group $G$ by a closed subgroup $H'$ of $G \times \text{Aut}(G)$ which is finitely covered by a solvmanifold. Therefore, $H' \cap G$ must have finite index in $H'$.

We shall work with a special kind of solvable Lie group. They are the split Lie hull of a predivisible group; see below for the definitions. This is not very restrictive, because every poly-{cyclic or finite} group contains a characteristic predivisible group. The following definition can be found in [**AJ76**].

DEFINITION 9.5.1. A *Mostow-Wang group* $\Gamma$ is one that occurs in an exact sequence $1 \to \Delta \to \Gamma \to \mathbb{Z}^s \to 1$, where $\Delta$ is a torsion-free, finitely generated nilpotent group.

A torsion-free group $\Gamma$ is called a *predivisible group* if it fits into the short exact sequence $1 \to \Delta \to \Gamma \to \mathbb{Z}^k \to 1$, and satisfies

(1) $\Delta$ is nilpotent.
(2) For $\gamma \in \Gamma$, let $\mu(\gamma)$ be the automorphism of $\Delta_\mathbb{R}$ (the Mal'cev completion of $\Delta$). Then for each eigenvalue $\theta$ of $\mu(\gamma)_*$, the automorphism of the Lie algebra of $\Delta_\mathbb{R}$ which is the derivative of $\mu(\gamma)$,

$$\theta \cdot |\theta|^{-1} = \cos 2\pi\rho + i \sin 2\pi\rho$$

with $\rho = 0$ or irrational.

It is known that a torsion-free poly-{cyclic or finite} group $\Pi$ contains a characteristic Mostow-Wang subgroup of finite index, *A Mostow-Wang group contains a characteristic predivisible polycyclic group of finite index*; see [**AJ76**] and the references there.

For a predivisible group $\Gamma$, there exists a connected solvable Lie group $G = S \rtimes K$, called the *split Lie hull of* $\Gamma$, satisfying

⟨P1⟩ $\Gamma$ is a lattice of $S$,
⟨P2⟩ $(\Gamma, G)$ has UAEP,
⟨P3⟩ $S$ is a closed normal subgroup of $G$, and
⟨P4⟩ $K$ is a maximal compact subgroup of $G$ which is a torus.

LEMMA 9.5.2. *With $G$ as above, let $N$ be the largest normal subgroup of $G$ contained in $K$. Then $N$ is fully invariant in $G$.*

PROOF. Since $S$ is normal in $G$, $[S, N] \subset [S, G] \subset S$. Similarly, since $N$ is normal in $G$, $[S, N] \subset N$. Consequently, $[S, N] \subset S \cap N \subset S \cap K = \{1\}$. Since $K$ is Abelian, this implies that $N \subset \mathcal{Z}(G)$. In fact, $N = \mathcal{Z}(G) \cap K$.

We claim that $N$ is fully invariant. Since $N$ is closed, it is compact. Therefore, either $N$ is finite or the set of elements of finite order is dense in $N$. Let $f : G \to G$ be an automorphism. Assume $f(N) \not\subset N$. If $f(n) \in N$ for every element $n$ of finite order, then $f(N) \subset N$. Therefore, there exists $n \in N$ of finite order, say of order $p$, such that $f(1, n) = (a, \alpha)$ with $a \neq 1$. Since $(a, \alpha) \in \mathcal{Z}(G)$, $\alpha = \mu(a^{-1})$ so that $\alpha(a) = a$. Thus,

$$(1, 1) = f((1, n)^p) = (f(1, n))^p = (a, \alpha)^p = (a^p, \alpha^p).$$

However, since $S$ is torsion free, $a^p = 1$ is not possible. This proves $f(1, n)$ is of the form $(1, \alpha)$, which implies $f(N) \subset K$. However, $\mathcal{Z}(G)$ is fully invariant so that a central element maps to a central element. Thus $\alpha \in N$ again. Since $N$ is compact, one can apply Lemma 9.2.8 so that $\alpha(N) \subset N$ implies that $\alpha$ induces an automorphism of $N$. We have proved that $N$ is fully invariant. □

When we divide out $G$ by this group $N$, all the properties $\langle P1\rangle$ through $\langle P4\rangle$ are preserved (with $K$ replaced by $K/N$). We lose nothing by dividing out by $N$, since $G/K$ is diffeomorphic to $(G/N)/(K/N)$ and the uniformizing groups are the same; see Proposition 9.2.7. Therefore, we may assume that $N$ is trivial from the beginning. So we add one more property to the list:

$\langle P5\rangle$ The largest normal subgroup of $G$ contained in $K$ is trivial.

*For the rest of this section, $\Gamma$ is a lattice of $G = S \rtimes K$ satisfying the conditions $\langle P1\rangle$ through $\langle P5\rangle$ above.*

LEMMA 9.5.3. *Every compact subgroup of $G$ isomorphic to $K$ is conjugate to $K$.*

PROOF. Let $K' \subset G$ be a torus. It acts on the coset space $G/K \cong S$ smoothly, as left translations. It is well known that a torus action on a Euclidean space has a fixed point, say $pK$. Now $K' \cdot pK = pK$ implies $p^{-1}K'p \subset K$. If $K' \cong K$, then $p^{-1}K'p = K$ clearly. In fact, the statement is true in more generality: Let $G$ be a Lie group with a finite number of components and $K$ be a maximal compact subgroup. Then every compact subgroup of $G$ can be conjugated into $K$. □

Let
$$S^K = \{s \in S : [k,s] = 1 \quad \text{for all } k \in K\}$$
be the fixed-point set of the $K$-action on $S$. Since $[(1,k),(a,\alpha)] = (k(a) \cdot a^{-1}, 1)$ and $K$ is Abelian, we have

LEMMA 9.5.4. $N_G(K) = C_G(K) = S^K \times K$. □

COROLLARY 9.5.5. *Suppose $S^K$ is trivial. Then*
  (1) *There exists an isomorphism $\Psi : \overline{\mathrm{Aff}}(G,K) \to \mathrm{Aut}(G)$.*
  (2) $\mathrm{TOP}_{G,K}(G/K \times W) = \overline{\mathrm{Aff}}(G,K) \times \mathrm{TOP}(W)$, *where $\overline{\mathrm{Aff}}(G,K) = \frac{\ell(G) \rtimes \mathrm{Aut}(G,K)}{r(K)}$.*

PROOF. Lemma 9.5.4 implies $N_G(K) = K$ in our case. (1) Proposition 9.3.3 applies because of Lemma 9.5.3. (2) Lemma 9.3.1 with the condition $\langle P5\rangle$ implies $\mathrm{Aut}^0(G,K)$ is trivial. Now one applies Theorem 9.3.2. □

THEOREM 9.5.6. *Let $\Gamma$ be a predivisible group, and let $G = S \rtimes K$ be a solvable Lie group satisfying $\langle P1\rangle$ through $\langle P5\rangle$. Also assume that $S^K$ is trivial. Let $\rho : Q \to \mathrm{TOP}(W)$ be a proper action of a discrete group, and let $1 \to \Gamma \to \Pi \to Q \to 1$ be an exact sequence. Then there exists a homomorphism $\theta : \Pi \to \mathrm{TOP}_{G,K}(G/K \times W) = \overline{\mathrm{Aff}}(G,K) \times \mathrm{TOP}(W)$ so that the diagram with exact rows*

$$\begin{array}{ccccccccc}
1 \longrightarrow & \Gamma & \longrightarrow & \Pi & \longrightarrow & Q & \longrightarrow 1 \\
& {\scriptstyle i}\downarrow & & {\scriptstyle \theta}\downarrow & & {\scriptstyle \varphi \times \rho}\downarrow & \\
1 \longrightarrow & \overline{\mathrm{Aff}}(G,K) & \longrightarrow & \mathrm{TOP}_{G,K}(G/K \times W) & \longrightarrow & \mathrm{Out}(G) \times \mathrm{TOP}(W) & \longrightarrow 1
\end{array}$$

*is commutative, yielding a Seifert fiber space with typical fiber the double coset space $\Gamma\backslash G/K$, a solvmanifold. Such homomorphism $\theta$ with fixed $i$ and $\varphi \times \rho$ is unique. The action is free if and only if the preimage of each isotropy $Q_w$ in $\Pi$ is torsion free.*

PROOF. (Existence). Since $\Gamma$ is normal in $\Pi$, there is a natural homomorphism $\mu : \Pi \to \operatorname{Aut}(\Gamma)$. UAEP, by $\langle P2 \rangle$, gives rise to a homomorphism $\operatorname{Aut}(\Gamma) \to \operatorname{Aut}(G)$. Consequently, we have a homomorphism $\Pi \to \operatorname{Aut}(G)$. Composing $\Pi \to \operatorname{Aut}(G)$ with $\Psi^{-1} : \operatorname{Aut}(G) \to \overline{\operatorname{Aff}}(G, K)$ in Proposition 5.5, we get a homomorphism $\Pi \to \operatorname{Aut}(G) \to \overline{\operatorname{Aff}}(G, K)$. Under this homomorphism, $\Gamma$ is mapped into $\overline{\operatorname{Aff}}_0(G, K)$. This together with the action $\Pi \to Q \to \operatorname{TOP}(W)$ gives rise to a homomorphism $\Pi \to \overline{\operatorname{Aff}}(G, K) \times \operatorname{TOP}(W) = \operatorname{TOP}_{G, K}(G/K \times W)$.

(Uniqueness). Same as the proof of Theorem 9.4.5. $\square$

In the previous theorem, we assumed that $S^K$ is trivial. When $S^K$ is nontrivial, the uniformizing group is pretty big and is not so easy to handle. However, when $Q$ is finite, (as in [**AJ76**]), a Seifert Construction can be made.

THEOREM 9.5.7 (when $Q$ is finite; cf.[**AJ76**]). *Let $\Gamma$ be a predivisible group, and $G = S \rtimes K$ be a solvable Lie group satisfying $\langle P1 \rangle$ through $\langle P5 \rangle$, and let $1 \to \Gamma \to \Pi \to Q \to 1$ be an exact sequence with $Q$ finite. Then there exists a homomorphism $\theta : \Pi \to \overline{\operatorname{Aff}}(G, K)$ so that the diagram with exact rows*

$$\begin{array}{ccccccccc} 1 & \longrightarrow & \Gamma & \longrightarrow & \Pi & \longrightarrow & Q & \longrightarrow & 1 \\ & & {\scriptstyle i}\downarrow & & {\scriptstyle \theta}\downarrow & & {\scriptstyle \varphi}\downarrow & & \\ 1 & \longrightarrow & \overline{\operatorname{Aff}}_0(G, K) & \longrightarrow & \overline{\operatorname{Aff}}(G, K) & \longrightarrow & \operatorname{Out}(G) & \longrightarrow & 1 \end{array}$$

*is commutative. Such homomorphism $\theta$ with fixed $i$ and $\varphi$ is unique. The action is free if and only $\Pi$ is torsion free, in which case, $\theta(\Pi) \backslash G/K$ will be an infra-solvmanifold.*

PROOF. Choose $W$ to be a point. Since $Q$ is finite, the trivial action of $Q$ on $W$ is proper. By Corollary 9.2.5, the uniformizing group $\operatorname{TOP}_{G,K}(G/K \times W)$ is then

$$\operatorname{TOP}_{G,K}(G/K) = \frac{\ell(G) \cdot [r(N_G(K)) \rtimes \operatorname{Aut}(G, K)]}{r(K) \rtimes \operatorname{Aut}^0(G, K)}.$$

We can still apply Lemma 9.3.1 to have $\operatorname{Aut}^0(G, K) = 1$ since we have $\langle P5 \rangle$. Thus,

$$\operatorname{TOP}_{G,K}(G/K) = \frac{\ell(G) \cdot [r(N_G(K)) \rtimes \operatorname{Aut}(G, K)]}{r(K)}.$$

Since $r(N_G(K)) \subset \ell(G) \rtimes \operatorname{Aut}(G, K)$, this is equal to $\dfrac{\ell(G) \rtimes \operatorname{Aut}(G, K)}{r(K)} = \overline{\operatorname{Aff}}(G, K)$.

(Existence). We shall first map $\Pi$ into $\operatorname{Aff}(G, K) = \ell(G) \rtimes \operatorname{Aut}(G, K)$. UAEP, by $\langle P2 \rangle$, gives rise to an extension $1 \to G \to \Pi \cdot G \to Q \to 1$. However, it is known that every finite extension of $G$ splits; see [**Aus73**, p.251]. Therefore the group $\Pi \cdot G \cong G \rtimes Q = (S \rtimes K) \rtimes Q$. Let $K'$ be a maximal compact subgroup of $(S \rtimes K) \rtimes Q$ containing $K$. Then, clearly, $(S \rtimes K) \rtimes Q = S \rtimes K'$, where $K' \cong K \rtimes Q$. In other words, $\Pi \cdot G \cong G \rtimes Q \cong S \rtimes (K \rtimes Q)$. Then the conjugation map sends $Q$ into $\operatorname{Aut}(G, K)$. Consequently, we have mapped $\Pi \cdot G$ into $\ell(G) \rtimes \operatorname{Aut}(G, K)$ via

$$\Pi \longrightarrow \Pi \cdot G \longrightarrow G \rtimes Q \longrightarrow S \rtimes (K \rtimes Q) \longrightarrow \ell(G) \rtimes \operatorname{Aut}(G, K).$$

This together with the projection $\ell(G) \rtimes \operatorname{Aut}(G, K) \to \overline{\operatorname{Aff}}(G, K)$ gives a desired homomorphism $\Pi \to \overline{\operatorname{Aff}}(G, K)$.

(Uniqueness). We apply Corollary 5.7.3 to the commuting diagram in the statement of the theorem, with $i : \Gamma \to \overline{\operatorname{Aff}}_0(G, K)$ for $i : G \to H$. We need to calculate the centralizer of $\Gamma$ in $\overline{\mathrm{Aff}}_0(G,K) = \frac{\ell(G) \rtimes \mathrm{Inn}(G,K)}{r(K)}$. An element $(a, \mu(b)) \in \ell(G) \rtimes \mathrm{Inn}(G,K)$ represents an element in the centralizer if and only if

$$[(a,\mu(b)),(x,1)] = (abxb^{-1}ax^{-1}, 1) = (\mu(ab)(x) \cdot x^{-1}, 1) \in r(K)$$

for every $x \in \Gamma$. Recall that elements of $r(K)$ in $\ell(G) \rtimes \mathrm{Aut}(G,K)$ are of the form $(k^{-1}, \mu(k))$ with $k \in K$. Therefore, it happens if and only if $\mu(ab)(x) \cdot x^{-1} \in \mathcal{Z}(G) \cap K$. However, $\mathcal{Z}(G) \cap K$ is trivial so that $\mu(ab)(x) \cdot x^{-1} = 1$ for every $x \in \Gamma$. By UAEP, this should happen for every $x \in G$. Then $\mu(ab) \in \mathrm{Inn}^0(G,K)$. But $\mathrm{Inn}^0(G,K)$ is trivial by Lemma 9.3.1. Therefore, $\mu(ab) = 1$ so that $ab \in \mathcal{Z}(G)$. Let $a = zb^{-1}$ for some $z \in \mathcal{Z}(G)$. Since $\mu(b) \in \mathrm{Inn}(G,K)$,

$$(a, \mu(b)) = (zb^{-1}, \mu(b)) = (z, 1)(b^{-1}, \mu(b)),$$

which represents the element $(z,1) \in \ell(\mathcal{Z}(G))$ of $\overline{\mathrm{Aff}}_0(G,K)$. We have shown $C_{\overline{\mathrm{Aff}}_0(G,K)}(\Gamma) = \ell(\mathcal{Z}(G))$. Since $Q$ is a finite group and $\mathcal{Z}(G)$ is isomorphic to $\mathbb{R}^k$ for some $k$,

$$H^1(Q; C_{\overline{\mathrm{Aff}}_0(G,K)}(\Gamma)) = H^1(Q; \mathbb{R}^k) = 0.$$

Such a homomorphism $\theta$ with fixed $i$ and $\varphi$ is unique, up to conjugation by elements of $\overline{\mathrm{Aff}}(G,K)$. Finally, since $G/K$ is diffeomorphic to $\mathbb{R}^n$, the action is free if and only if $\Pi$ is torsion free. $\square$

The Seifert Construction in Theorem 9.5.7 for $\Pi$, when $Q$ is finite, gives us detailed knowledge of the geometric structure of the spaces constructed earlier by Auslander and Johnson in [**AJ76**]. For example, if $G$ has a left invariant metric which is also right $K$-invariant, then the resulting space $G/K$ will inherit the metric.

More generally, we shall take the case where $\varphi : Q \to \mathrm{Out}(G)$ has a finite image.

COROLLARY 9.5.8 ($\mathcal{Z}(\Gamma)$ trivial and $\varphi : Q \to \mathrm{Out}(G)$ has a finite image). *Let $\Gamma$ be a predivisible group without center, and let $G = S \rtimes K$ be a solvable Lie group satisfying $\langle P1 \rangle$ through $\langle P5 \rangle$. Let $\rho : Q \to \mathrm{TOP}(W)$ be a proper action of a discrete group, and let $1 \to \Gamma \to \Pi \to Q \to 1$ be an exact sequence. Assume the abstract kernel $\varphi : Q \to \mathrm{Out}(G)$, associated to this extension, has a finite image. Then there exists a homomorphism $\theta : \Pi \to \mathrm{TOP}_{G,K}(G/K \times W) = \overline{\mathrm{Aff}}(G,K) \times \mathrm{TOP}(W)$ so that the diagram with exact rows*

$$\begin{array}{ccccccccc}
1 & \longrightarrow & \Gamma & \longrightarrow & \Pi & \longrightarrow & Q & \longrightarrow & 1 \\
& & \downarrow i & & \downarrow \theta & & \downarrow \varphi \times \rho & & \\
1 & \longrightarrow & \overline{\mathrm{Aff}}(G,K) & \longrightarrow & \mathrm{TOP}_{G,K}(G/K \times W) & \longrightarrow & \mathrm{Out}(G) \times \mathrm{TOP}(W) & \longrightarrow & 1
\end{array}$$

*is commutative, yielding a Seifert fiber space with typical fiber the solvmanifold $\Gamma \backslash G/K$. Such homomorphism $\theta$ with fixed $i$ and $\varphi \times \rho$ is unique. The action is free if and only if the preimage of each isotropy $Q_w$ in $\Pi$ is torsion free.*

PROOF. Let $Q_0$ be the kernel of $\varphi : Q \to \mathrm{Out}(G)$. Then, since $\Gamma$ is centerless, $Q_0$ lifts to a normal subgroup of $\Pi$. Now consider the exact sequence $1 \to \Gamma \to \Pi/Q_0 \to Q/Q_0 \to 1$. Since $Q/Q_0$ is finite, Theorem 9.5.7 applies to obtain a homomorphism $\Pi/Q_0 \to \overline{\mathrm{Aff}}(G,K)$. Now the two homomorphisms $\Pi \to \Pi/Q_0 \to \overline{\mathrm{Aff}}(G,K)$ and $\Pi \to Q \to \mathrm{TOP}(W)$ give a desired homomorphism. $\square$

## 9.5. SOLVMANIFOLDS

The structure of the space $\theta(\Pi)\backslash(G/K \times W)$ is similar to the symmetric space case. That is, it has a Seifert fiber structure

$$\Gamma\backslash G/K \to \theta(\Pi)\backslash(G/K \times W) \to Q\backslash W,$$

where the typical fiber $\Gamma\backslash G/K$ is a solvmanifold. Singular fibers are finite quotients of the typical fiber, where the finite actions are via elements of $\overline{\mathrm{Aff}}(G,K)$. Let $Q_0$ be the kernel of $\phi: Q \to \mathrm{Out}(\Gamma)$. Then the preimage of $Q_0$ in $\Pi$ splits as a direct product $\Gamma \times Q_0$ because $\Gamma$ has trivial center. Since $\mathrm{Aut}(G) = \mathrm{Inn}(G) \rtimes \mathrm{Out}(G)$, $Q_0$ maps trivially into $\overline{\mathrm{Aff}}(G,K)$. Consequently, the normal subgroup $\Gamma \times Q_0$ of $\Pi$ acts on $G/K \times W$ in such a way that $\Gamma$ acts only on the $G/K$-factor as left translations, $Q_0$ acts only on the $W$-factor via $\rho$, yielding $(\Gamma\backslash G/K) \times (Q\backslash W)$. Because $\varphi: Q \to \mathrm{Out}(\Gamma)$ has a finite image, $Q/Q_0$ is finite. The finite quotient group $F = \Pi/(\Gamma \times Q_0)$ acts diagonally on $(\Gamma\backslash G/K) \times (Q\backslash W)$.

CHAPTER 10

# Locally Injective Seifert Fiberings with Torus Fibers

## 10.1. Introduction

Chapter 10 is an exploration of Seifert fiberings where the principal $G$-bundle is a principal $T^k$-bundle. The emphasis is on the nonproduct case and the material is independent of Chapter 7. It can also be treated as an introduction to Chapter 12 where the discrete group $Q$ is replaced by a connected compact Lie group.

Let $P$ be a principal $T^k$-bundle over a paracompact space $W$, and let $Q$ be a discrete group acting properly on $W$. We wish to lift the action $(Q,W)$ to an action $(Q,P)$ of weak bundle automorphisms of $P$. Then the lifted $Q$ will normalize the left translational $T^k$-action $P$. The $T^k$-fibers will descend to fibers, perhaps singular, on $Q\backslash P$ and the natural map $Q\backslash P \to Q\backslash W$ will be a Seifert fibering.

Of course, in general, we cannot always expect to accomplish this lifting of $Q$ to $P$. There are several obstructions. The first is that $\varphi \times \rho : Q \to \mathrm{Aut}(T^k) \times \mathrm{TOP}(W)$ may fail to be in the image of the weak bundle automorphisms $\mathrm{TOP}_{T^k}(P)$ as in Subsection 4.2.13. Here $\varphi$ is a homomorphism from $Q$ to $\mathrm{Aut}(T^k)$ and $\rho : Q \to \mathrm{TOP}(W)$ is the homomorphism describing the $Q$-action on $W$. Secondly, even if $(\varphi \times \rho)(Q)$ is contained in the image of $p : \mathrm{TOP}_{T^k}(P) \to \mathrm{Aut}(T^k) \times \mathrm{TOP}(W)$, the action of $Q$ may still not lift. The kernel of $p$ is $\mathrm{M}(W, T^k)$ and is a $Q$-module as in Corollary 4.2.12.

If $(\varphi \times \rho)(Q) \subset \mathrm{image}(p)$, then we may form the pullback:

$$
\begin{array}{ccccccccc}
1 & \longrightarrow & \mathrm{M}(W, T^k) & \longrightarrow & E(P, Q) & \longrightarrow & Q & \longrightarrow & 1 \\
& & \Big\downarrow = & & \Big\downarrow \epsilon & & \Big\downarrow \varphi \times \rho & & \\
1 & \longrightarrow & \mathrm{M}(W, T^k) & \longrightarrow & \mathrm{TOP}_{T^k}(P) & \stackrel{p}{\longrightarrow} & \mathrm{Aut}(T^k) \times \mathrm{TOP}(W). & &
\end{array}
$$

The top row is an extension and can be represented by $[E(P,Q)] \in H^2(Q; \mathrm{M}(W, T^k))$ where "[ ]" means the cohomology class that yields the extension $E(P,Q)$ up to congruence. Therefore, $Q$ lifts to a group of weak bundle automorphisms if the extension $E(P,Q)$ splits. That is, if $[E(P,Q)] = 0$.

In the most general set-up for a Seifert fibering, we would expect to find an extension

$$1 \to F \to \Pi \to Q \to 1$$

with $F$ a discrete lattice in $T^k$ and $\Pi$ an extension of $F$ by $Q$. There would be a homomorphism $\theta$ of $\Pi$ into $E(P,Q)$. This factors through the pushout $T^k \cdot \Pi$. So one will want to investigate extensions

$$1 \to T^k \to E \to Q \to 1$$

which map into $E(P,Q)$.

We will show in Proposition 10.1.4 that the pullback extension $E(P,Q)$ exists if and only if $[P] \in H^2(W; \mathbb{Z}^k)^Q$. For most $Q$, a homomorphism $\theta : E \to E(P,Q)$ for some $E$ will exist if and only if $[E(P,Q)]$ has finite order; see Subsection 10.1.2.

The existence questions are explored in Section 10.1. In Section 10.2, we characterize the bundles $P$ over $W$ for which $E(P,Q)$ splits. The answer is intimately connected with the Borel space of $(Q, W)$. Section 10.2 can also be thought of as a preparation for the case of continuous $Q$ investigated in Chapter 12. The same characterization problem is independently investigated in Section 10.3 but from a local (sheaf theoretic) point of view. This leads to two spectral sequences which enable us to independently recapture all of Chapter 7 for product $T^k$-bundles as well as a complete analysis when the bundle is not a product bundle (Sections 10.3, 10.4, and 10.5). The results and methods of this chapter are illustrated in Chapter 15 by making specific calculations in the special case of the classical 3-dimensional Seifert manifolds. These methods and applications provide a different perspective to the presentation in Chapter 14 obtained by more traditional methods.

10.1.1. In this section, we formulate the existence and uniqueness of Seifert fiberings modeled on principal $T^k$-bundles over $W$ in cohomological terms. As usual $\rho : Q \to \text{TOP}(W)$ is a proper action of a discrete group $Q$ on a paracompact space $W$. If we are not already given a Seifert fibering, but instead are trying to construct one, we will also need a homomorphism $\varphi : Q \to \text{Aut}(T^k)$.

10.1.2. Since we will be considering Seifert fiberings modeled on principal $T^k$-bundles, let us recall that a real torus $T^k$ is determined by a real basis in $\mathbb{R}^k$ such that each torus $T^k$ is associated with a short exact sequence

$$1 \to \Lambda^k \to \mathbb{R}^k \to T^k \to 1.$$

In this sequence, $\Lambda^k \cong \mathbb{Z}^k$ is the integral lattice in $\mathbb{R}^k$ generated by the basis vectors. By $\text{Aut}(T^k) \subset \text{GL}(k, \mathbb{R})$, we mean the subgroup of elements which preserve the images of $\Lambda^k$ in $\mathbb{R}^k$. An element of $\text{Aut}(T^k)$ may be regarded as an automorphism of any one of three groups $\Lambda^k$, $\mathbb{R}^k$, or $T^k$. It will be convenient for us to think of the standard orthonormal basis in $\mathbb{R}^k$ to define $T^k$. Then $\text{Aut}(\mathbb{Z}^k) = \text{GL}(k, \mathbb{Z}) = \text{Aut}(T^k) \subset \text{GL}(k, \mathbb{R})$.

For a principal $G$-bundle $P$ where $G$ is connected Abelian group, we have, by Corollary 4.2.12, the exact sequence

$$1 \to \text{M}(W, G) \xrightarrow{\psi} \text{TOP}_G(P) \to \text{Aut}(G) \times \text{TOP}(W).$$

The last homomorphism may not be onto if $P$ is not a product bundle (see Subsection 4.2.13, Exercise 4.2.14, and Lemma 4.2.15). If $f \in \text{TOP}_G(P)$, then conjugation of $\text{M}(W,G)$ by $f$ is given by $f \circ \psi(\eta) \circ f^{-1} = \alpha \circ \psi(\eta) \circ h^{-1}$, where $f \mapsto (\alpha, h) \in \text{Aut}(G) \times \text{TOP}(W)$.

Suppose $F \xhookrightarrow{i} T^k$ is a discrete subgroup. If we are to map $1 \to F \to \Pi \to Q \to 1$ homomorphically into $\text{TOP}_{T^k}(P)$, then we are looking for a homomorphism $\theta$ as indicated below:

(10.1.1)
$$\begin{array}{ccccccccc}
0 & \longrightarrow & F & \longrightarrow & \Pi & \longrightarrow & Q & \longrightarrow & 1 \\
& & \downarrow{\iota} & & \downarrow{\bar{\iota}} & & \downarrow{=} & & \\
0 & \longrightarrow & T^k & \longrightarrow & T^k \cdot \Pi & \longrightarrow & Q & \longrightarrow & 1 \\
& & \downarrow{\ell} & & \downarrow{\theta} & & \downarrow{=} & & \\
0 & \longrightarrow & \mathrm{M}(W, T^k) & \longrightarrow & E(P, Q) & \longrightarrow & Q & \longrightarrow & 1 \\
& & \downarrow{=} & & \downarrow & & \downarrow{\varphi \times \rho} & & \\
0 & \longrightarrow & \mathrm{M}(W, T^k) & \longrightarrow & \mathrm{TOP}_{T^k}(P) & \longrightarrow & \mathrm{Aut}(T^k) \times \mathrm{TOP}(W) & &
\end{array}$$

where the third sequence is the pullback via $\varphi \times \rho : Q \to \mathrm{Aut}(T^k) \times \mathrm{TOP}(W)$. Of course, this will make sense only if $(\varphi \times \rho)(Q)$ is contained in the image of $\mathrm{TOP}_{T^k}(P)$. We are also using some homomorphism $\varphi : \Pi \to \mathrm{Aut}(T^k)$ which is compatible with $\Pi \to \mathrm{Aut}(F)$, to form a pushout $T^k \cdot \Pi$. Note that $F \subset T^k \subset \mathrm{M}(W, T^k)$ are $Q$-modules. As usual, $\alpha \cdot \lambda = \varphi(\alpha) \circ \lambda \circ \rho(\alpha)^{-1}$.

If we represent these extensions in terms of their representatives in second cohomology as in Section 5.4, we see that our problem reduces to understanding the homomorphisms

$$[\Pi] \xrightarrow{\iota_*} [T^k \cdot \Pi] \xrightarrow{\ell_*} [E(P, Q)]$$

$$\cap \qquad\qquad \cap \qquad\qquad \cap$$

$$H^2(Q; F) \xrightarrow{\iota_*} H^2(Q; T^k) \xrightarrow{\ell_*} H^2(Q; \mathrm{M}(W, T^k)).$$

If there is such a $\theta$, then $\ell_*([T^k \cdot \Pi]) = [E(P, Q)]$. If $F$ is a finite subgroup of $T^k$ and if $\bar{\iota}$ exists, $[T^k \cdot \Pi]$ is a torsion class in $H^2(Q; T^k)$. Therefore, if $\theta \circ \bar{\iota}$ exists, as in the diagram, $[E(P, Q)]$ *must be a torsion class in $H^2(Q; M(W, T^k))$*.

10.1.3. As we mentioned, $(\varphi \times \rho)(Q) \in \mathrm{Aut}(T^k) \times \mathrm{TOP}(W)$ may not be in the image of $\mathrm{TOP}_{T^k}(P)$ for some nonproduct bundles $P$ over $W$. In this situation, no extension $E(P, Q)$ will exist (i.e., we cannot pullback $Q$ via $\varphi \times \rho$), and so no such $\theta$ exists. The principal $T^k$-bundles, up to bundle equivalence, are determined by their characteristic class $c_1(P) \in H^2(W, \mathbb{Z}^k)$. (Recall that $H^2(W, \mathbb{Z}^k)$ is naturally equivalent to $[W, B_{T^k}]$, the homotopy classes of maps of $W$ into the classifying space $B_{T^k}$ for principal $T^k$-bundles, see Section 1.4, and $B_{T^k} \simeq \mathbb{C}P_\infty \times \cdots \times \mathbb{C}P_\infty$ ($k$ copies) is a $K(\mathbb{Z}^k, 2)$-space.) In this section, we will write $[P]$ for $c_1(P)$. That is,

$$[P] = c_1(P) \in H^2(W; \mathbb{Z}^k).$$

Now $\mathrm{Aut}(T^k) \times \mathrm{TOP}(W)$ acts on the set of bundles $[W, BT^k]$ by

$$P \mapsto (\omega, h)(P) = \omega_*(h^*(P)),$$

for $(\omega, h) \in \mathrm{Aut}(T^k) \times \mathrm{TOP}(W)$, and $P \in [W, BT^k]$. The bundle $\omega_*(h^*(P))$ is

$$\begin{array}{ccccc}
\omega_*(h^*(P)) & \xrightarrow{\omega^{-1}} & h^*(P) & \longrightarrow & P \\
\downarrow & & \downarrow & & \downarrow \\
W & \xrightarrow{=} & W & \xrightarrow{h} & W.
\end{array}$$

PROPOSITION 10.1.4. *An extension $E(P,Q)$ exists for a given bundle $P$ if and only if $[P] \in H^2(W; \mathbb{Z}^k)^Q$.*

PROOF. Suppose $E(P,Q)$ exists for a given bundle $P$. Then, for each $\alpha \in Q$, pick a preimage of $\alpha$, $f \in E(P,Q) \subset \mathrm{TOP}_{T^k}(P)$ so that $f(au) = \varphi(\alpha)(a)f(u)$, and $f$ induces $\rho(\alpha)$ on $W$. That is, $\alpha$ gives rise to $\varphi(\alpha) \in \mathrm{Aut}(T^k)$ and $\rho(\alpha) \in \mathrm{TOP}(W)$ so that the diagram

$$\begin{array}{ccc} P & \xrightarrow{f} & P \\ \downarrow & & \downarrow \\ W & \xrightarrow{\rho(\alpha)} & W \end{array}$$

is commutative. Thus, there is a weak bundle automorphism $f$ between $P$ and $(\varphi(\alpha), \rho(\alpha))(P) = \varphi(\alpha)_* \rho(\alpha)^*(P) = P$. Thus $\alpha([P]) = [P]$ for each $\alpha \in Q$. This means that $[P] \in H^2(W; \mathbb{Z}^k)$ will be left fixed by the action of $Q$ on $H^2(W; \mathbb{Z}^k)$. Consequently, $E(P,Q)$ exists only if $[P] \in H^2(W; \mathbb{Z}^k)^Q$.

On the other hand, if $[P] \in H^2(W; \mathbb{Z}^k)^Q$, then the composition $\varphi(\alpha)_* \circ \rho(\alpha)^*$ sends the bundle $P$ to $P$ as a weak bundle automorphism. If $[P] \notin H^2(W; \mathbb{Z}^k)^Q$, then there is no extended lift of the $Q$-action to a group of weak bundle automorphisms of $P$. □

EXAMPLE 10.1.5. Suppose $Q = \mathbb{Z}_2$ generated by the anti-podal map $A$ on $S^2$. For a principal bundle $P$ over $S^2$, we cannot lift $Q$ to a group of bundle automorphisms of $P$ unless $P$ is the trivial bundle. For, if $A$ lifts to bundle automorphisms of $P$, then $\varphi_*(A)(P) = P$. Therefore,

$$(\varphi(A), \rho(A))[P] = \varphi_*(A)(\rho(A)^*[P]) = \varphi_*(A)(-[P]) = -[P]$$

and consequently, $Q$ lifts only if $P$ is the product bundle. In cohomological terms, the action of $\mathbb{Z}_2$ on $H^2(S^2; \mathbb{Z})$ is nontrivial and $H^2(S^2; \mathbb{Z})^Q = 0$.

On the other hand, if we allow $Q$ to lift to weak bundle automorphisms, then $\varphi_*(A)(P) = -[P]$ and therefore,

$$(\varphi(A), \rho(A))[P] = \varphi_*(A)(\rho(A)^*[P]) = \varphi_*(A)(-[P]) = -(-[P]) = [P].$$

Consequently, $A$ lifts to a weak bundle automorphism of $P$. In cohomological terms, this says that $H^2(S^2; \mathbb{Z})^Q = \mathbb{Z}$.

Observe also, that if $A$ is an orientation reversing homeomorphism of a compact closed oriented surface $W$, then $A$ lifts to a weak bundle automorphism of any principal $S^1$-bundle $P$ over $W$. But $A$ will not lift to a bundle automorphism of $P$ unless $P$ is the product bundle.

Let $\beta : H^2(W; \mathbb{Z}^k)^Q \to H^2(Q; \mathrm{M}(W, T^k))$ be the assignment to each bundle $P$, $[P] \in H^2(W; \mathbb{Z}^k)^Q$, its extension class $[E(P,Q)] \in H^2(Q; \mathrm{M}(W, T^k))$. Some of the bundles in $H^2(W, \mathbb{Z}^k)^Q$ map to the 0-class in $H^2(Q; \mathrm{M}(W, T^k))$. That is, the particular extension $E(P,Q)$ splits. Let

$$\mathcal{E}(P,Q) = \mathrm{M}(W, T^k) \rtimes Q$$

be a particular splitting for the split extension $[E(P,Q)]$. There are many such splittings and they are given by the elements of $H^1(Q; \mathrm{M}(W, T^k))$. We take the totality of all the $\mathcal{E}(P,Q)$'s for each of the bundles $P$ for which $[E(P,Q)]$ is trivial.

In Section 10.2, we shall show that this set forms a group isomorphic to $H^2(EQ \times_Q W; \mathbb{Z}^k)$, the second cohomology group of the Borel space for the action $(Q, W)$ with coefficients in $\mathbb{Z}^k$, which is to be regarded as a $Q$-module. The natural map of the fiber $j : W \to EQ \times_Q W$ induces a homomorphism

$$j : H^2(EQ \times_Q W; \mathbb{Z}^k) \longrightarrow H^2(W, \mathbb{Z}^k),$$

whose image lies in $H^2(W, \mathbb{Z}^k)^Q$. In Section 10.2, we shall see that if $\mathcal{E}(P, Q) \in H^2(EQ \times_Q W; \mathbb{Z}^k)$, then $j(\mathcal{E}(P, Q))$ is exactly what we would expect it to be, namely, the characteristic class of the bundle $P$. The kernel of $j$ is $H^1(Q; \mathrm{M}(W, T^k))$. See Theorem 10.3.12. $[H^i(Q; \mathrm{M}(W, T^k))$ is naturally isomorphic to $H^{i+1}(Q; \mathbb{Z}^k)$, $i \geq 1$, if $W$ is simply connected.]

In Subsection 10.2.3 as well as in Theorem 10.3.12, this takes the form of the following proposition when $W$ is simply connected.

PROPOSITION 10.1.6. [**CR72b**, 3.8]. *The sequence*

$$0 \to H^2(Q; \mathbb{Z}^k) \longrightarrow H^2(EQ \times_Q W; \mathbb{Z}^k) \xrightarrow{j}$$
$$H^2(W; \mathbb{Z}^k)^Q \xrightarrow{\beta} H^3(Q; \mathbb{Z}^k) \longrightarrow H^3(EQ \times_Q W; \mathbb{Z}^k)$$

*is exact. It coincides with the exact sequence derived from the terms of low degree of the spectral sequence associated to the fibering of the Borel space $EQ \times_Q W$ over the classifying space $BQ$ for the group $Q$ and with fiber the $Q$-space $W$.*

10.1.7. In the following three subsections, we will examine some of the cohomological criteria that must be satisfied if either $\theta : T^k \cdot \Pi \to E(P, Q)$ or $\theta \circ \bar{\iota} : E \to E(P, Q)$ in the diagram (10.1.1) exists.

LEMMA 10.1.8.

$$\begin{array}{ccccccccc}
0 & \longrightarrow & \mathbb{Z}^k & \xrightarrow{i} & \mathbb{R}^k & \xrightarrow{\exp} & T^k & \longrightarrow & 0 \\
& & \downarrow \ell & & \downarrow \ell & & \downarrow \ell & & \\
0 & \longrightarrow & \mathrm{M}(W, \mathbb{Z}^k) & \xrightarrow{i} & \mathrm{M}(W, \mathbb{R}^k) & \xrightarrow{\exp} & \mathrm{M}(W, T^k) & \xrightarrow{d} & H^1(W; \mathbb{Z}^k) \longrightarrow 0
\end{array}$$

*is a commutative diagram of $Q$-modules with exact rows. (The maps $\ell$ from $\mathbb{Z}^k$, $\mathbb{R}^k$, and $T^k$ are inclusions of constant maps of $W$ into $\mathbb{Z}^k$, $\mathbb{R}^k$, and $T^k$, respectively.)*

PROOF. If $W$ is connected, which we are assuming, $\ell$ is an isomorphism on the first vertical map and injections on the other two. $H^1(W; \mathbb{Z}^k)$ is identified with the homotopy classes of maps from $W$ into $T^k$, because $[W, T^k] = [W, K(\mathbb{Z}^k, 1)] = H^1(W; \mathbb{Z}^k)$ as a $Q$-module. We get two exact sequences from the bottom sequence by setting the image of exp as $\mathrm{M}_0(W, T^k)$:

$$0 \longrightarrow \mathrm{M}(W, \mathbb{Z}^k) \xrightarrow{i} \mathrm{M}(W, \mathbb{R}^k) \xrightarrow{\exp} \mathrm{M}_0(W, T^k) \longrightarrow 0$$
$$0 \longrightarrow \mathrm{M}_0(W, T^k) \longrightarrow \mathrm{M}(W, T^k) \longrightarrow H^1(W, \mathbb{Z}^k) \longrightarrow 0.$$

If $H^1(W, \mathbb{Z}^k) = 0$, then $\mathrm{M}_0(W, T^k) = \mathrm{M}(W, T^k)$. In any case, $\mathrm{M}_0(W, T^k)$ is the subgroup of elements of $\mathrm{M}(W, T^k)$ which are homotopically trivial and hence liftable to $\mathrm{M}(W, \mathbb{R}^k)$. □

PROPOSITION 10.1.9. *If $H^1(W,\mathbb{Z}^k) = 0$, $H^2(Q,\mathbb{Z}^k)$ and $H^3(Q,\mathbb{Z}^k)$ are finitely generated as Q-modules and $[P] \in H^2(W,\mathbb{Z}^k)^Q$, then*

$$\ell_* : H^2(Q, T^k) \to H^2(Q; \mathrm{M}(W, T^k))$$

*is an isomorphism on the nondivisible torsion subgroup and trivial on the divisible elements.*

PROOF. Recall from Theorem 7.5.1 that $H^i(Q; \mathrm{M}(W, \mathbb{R}^k)) = 0$, $i > 0$, with $Q\backslash W$ compact or finite dimensional. (In Corollary 10.3.10, we reprove this independently of Theorem 7.5.1 and show, in addition, that these restrictions on $Q\backslash W$ can be dropped.) We have $H^i(Q; \mathrm{M}_0(W, T^k)) \xrightarrow{\delta} H^{i+1}(Q; \mathbb{Z}^k)$ is an isomorphism for all $i > 0$, and

$$H^{i+1}(Q; \mathbb{Z}^k) = H^i(Q; \mathrm{M}_0(W, T^k)) \to H^i(Q; \mathrm{M}(W, T^k)) \to H^i(Q; H^1(W; \mathbb{Z}^k))$$

is exact. If $H^3(Q; \mathbb{Z}^k)$ and $H^2(Q; H^1(W; \mathbb{Z}^k))$ contain no nontrivial divisible elements, then $H^2(Q; \mathrm{M}(W, T^k))$ also contains no nontrivial divisible elements. In particular, if $H^1(W; \mathbb{Z}^k) = 0$ and $H^3(Q; \mathbb{Z}^k)$ is finitely generated as a Q-module, then the diagram

(10.1.2)
$$\begin{array}{ccc} H^2(Q; T^k) & \xrightarrow{\delta_1} & H^3(Q; \mathbb{Z}^k) \\ \ell_{2*}\downarrow & & =\downarrow \\ H^2(Q; \mathrm{M}(W, T^k)) & \xrightarrow[=]{\delta} & H^3(Q; \mathbb{Z}^k) \end{array}$$

commutes and $\delta_1$, $\ell_{2*}$ are isomorphisms on the nondivisible torsion subgroup of $H^2(Q; T^k)$. □

COROLLARY 10.1.10. *If P, Q, and W are as above and $E(P,Q)$ exists, then there exists $\theta : E \to E(P,Q)$ fitting into the commutative diagram*

$$\begin{array}{ccccccccc} 1 & \longrightarrow & T^k & \longrightarrow & E & \longrightarrow & Q & \longrightarrow & 1 \\ & & \downarrow & & \downarrow \theta & & \downarrow = & & \\ 1 & \longrightarrow & \mathrm{M}(W, T^k) & \longrightarrow & E(P,Q) & \longrightarrow & Q & \longrightarrow & 1 \end{array}$$

*if and only if $[E(P,Q)] \in H^2(Q; \mathrm{M}(W, T^k)) = H^3(Q; \mathbb{Z}^k)$ is a torsion class (cf. Subsection 10.1.2). In particular, if Q is a finite group, then each of the groups in the diagram (10.1.2) is finite and each map is an isomorphism.*

Of course, if $[E]$ is a torsion class, then $\ell_{2*}[E] = \ell_{2*}([E]+(\text{a divisible element}))$. So, $E$ will usually not be unique. However, if we want $\theta \circ \bar{\iota}(\Pi)$ to exist (see Subsection 10.1.2), $\bar{\iota}(\Pi) = E$ is unique up to congruence. Note, if $E(P,Q)$ is not a torsion class, no $\theta : E \to E(P,Q)$ can exist; see Proposition 10.4.9 for such examples.

COROLLARY 10.1.11. *If P, Q and $E(P,Q)$ are as above, then $E(P,Q)$ splits if and only if there is a split E with $\theta(E) = E(P,Q)$ for some $\theta$.*

10.1.12. We now recall some information from Section 5.3. Let $G$ and $H$ be Abelian groups. Consider the commutative diagram of group extensions

$$\begin{array}{ccccccccc} 1 & \longrightarrow & G & \longrightarrow & E & \longrightarrow & Q & \longrightarrow & 1 \\ & & \downarrow \ell & & \downarrow \theta_0 & & \downarrow = & & \\ 1 & \longrightarrow & H & \longrightarrow & \mathcal{E} & \longrightarrow & Q & \longrightarrow & 1 \end{array}$$

where $\ell$ is injective, $\theta_0$ is a homomorphism, $E = G \times_{(f,\varphi)} Q$ and $\mathcal{E} = H \times_{(k,\psi)} Q$ are as in the converse of the general push-out constructions of Subsection 5.3.7. Put $\theta_0(0, q) = (\lambda_0(q), q) \in \mathcal{E}$ for some $\lambda_0 : Q \to H$. Then

$$\theta_0((0, q), (0, q')) = \theta_0(f(q, q'), qq') = (\ell \circ f(q, q') + \lambda_0(qq'), qq').$$

Now,

$$\theta_0(0, q)\theta_0(0, q') = (\lambda_0(q), q)(\lambda_0(q'), q') = (\lambda_0(q) +^q \lambda_0(q') + k(q, q'), qq').$$

Consequently, $\ell \circ f(q, q') = k(q, q') + \lambda_0(q) +^q \lambda_0(q') - \lambda_0(qq') = k(q, q') + (\delta\lambda_0)(q, q')$. That is, $\ell \circ f$ and $k$ are homologous. Then we have a congruence

$$\omega : \mathcal{E} = H \times_{(k,\psi)} Q \longrightarrow H \times_{(\ell \circ f,\psi)} Q = \mathcal{E}'$$

given by $\omega(h, q) = (h - \lambda_0(q), q)$.

If $\theta$ is another map making the diagram commute, then the difference is

$$\theta(0, q) \cdot \theta_0(0, q)^{-1} = (\lambda(q) - \lambda_0(q), 1) = (\gamma(q), 1),$$

where $\gamma \in Z^1(Q; H)$. That is,

$$\gamma(qq') = \gamma(q) +^q \gamma(q').$$

In fact, there is a one-to-one correspondence between the injections $\theta$, keeping $\ell$ and $Q$ fixed and $Z^1(Q; H)$; see Theorem 5.7.2.

Let us make this correspondence more specific. For a fixed $\theta_0$, let us change $\mathcal{E}$ by the congruence $\omega$ so that now $\mathcal{E}' = H \times_{(\ell \circ f,\psi)} Q$. Then $\omega \circ \theta_0(0, q) = (0, q)$ and $\omega \circ \theta(0, q) = (\gamma(q), q)$ in $\mathcal{E}'$, where $\gamma \in Z^1(Q; H)$ is as above and the cocycle associated to the injection $\theta_0$ is the zero 1-cocycle.

Another interpretation of $Z^1(Q; H)$ assigns to each 1-cocycle $\gamma$ an automorphism $\widetilde{\gamma} : \mathcal{E}' \to \mathcal{E}'$ keeping $H$ and $\mathcal{E}'/H = Q$ element-wise fixed, i.e., $\widetilde{\gamma} \in \mathrm{Aut}(\mathcal{E}', H, Q)$; see [**Gru70**, p.45]. The automorphism $\widetilde{\gamma}$ is given by $\widetilde{\gamma}(h, q) = (h + \gamma(q), q)$ yielding an isomorphism of $Z^1(Q; H)$ with $\mathrm{Aut}(\mathcal{E}', H, Q)$, where the addition in $Z^1(Q; H)$ corresponds to the composition of automorphisms in $\mathrm{Aut}(\mathcal{E}')$. For, if $\theta_1(g, q) = (g + \gamma_1(q), q)$ and $\theta_2(g, q) = (g + \gamma_2(q), q)$ in $\mathcal{E}'$, then define $\theta_3$ by $\theta_3(g, q) = (g + \gamma_1(q) + \gamma_2(q), q) = (g + \gamma_3(q), q)$. The automorphism $\widetilde{\gamma}_1$ carries $\theta_0(E)$ to $\theta_1(E)$ while it sends $\theta_2(E)$ to $\theta_3(E)$. So $\widetilde{\gamma}_3$ is the composition of $\widetilde{\gamma}_1$ with $\widetilde{\gamma}_2$. Thus we have

PROPOSITION 10.1.13. *After fixing the choices as above, there is an isomorphism from $Z^1(Q; H)$ to $\mathrm{Aut}(\mathcal{E}', H, Q)$. Each injection $\theta$ corresponds to a cocycle $\gamma$ which in turn, corresponds to an automorphism $\widetilde{\gamma} : \mathcal{E}' \to \mathcal{E}'$ such that $\theta(E) = \widetilde{\gamma} \circ \theta_0(E)$. The addition of cocycles corresponds to composition of automorphisms.*

10.1.14. Two cocycles $\gamma_1$ and $\gamma_2$ are cohomologous if and only if $\widetilde{\gamma}_2 = \mu(c) \circ \widetilde{\gamma}_1$, where $\mu(c)$ is conjugation of $\mathcal{E}'$ by an element $c \in H$; see Theorem 5.7.2 and Corollary 5.7.3. Therefore we have

PROPOSITION 10.1.15. $H^1(Q;H)$ is isomorphic to $\mathrm{Aut}(\mathcal{E}',H,Q)/\mathrm{Inn}(H)$.

10.1.16. Let us apply the observations made in Subsections 10.1.7–10.1.14 to the problems posed in Subsection 10.1.2. Let $P$ be a principal $T^k$-bundle over $W$, and let $\varphi \times \rho : Q \to \mathrm{Aut}(T^k) \times \mathrm{TOP}(W)$ be fixed, where $\rho$ is a proper action of a discrete group $Q$. We assume that $(\varphi \times \rho)(Q)$ is contained in the $\mathrm{Im}(\mathrm{TOP}_{T^k}(P) \to \mathrm{Aut}(T^k) \times \mathrm{TOP}(W))$ so that $E(P,Q)$ is well defined. We wish to find:
  (i) For what extensions $1 \to T^k \to E \to Q \to 1$, does $\theta : E \to E(P,Q)$ exist as in the diagram (10.1.1)?
  (ii) If there exists a $\theta_0 : E \to E(P,Q)$, how are the different $\theta$'s fitting into the diagram classified?
  (iii) Which $E$ corresponds to a push-out $T^k \cdot \Pi$?
  (iv) What happens when the extension $E(P,Q)$ splits?

PROPOSITION 10.1.17. (i) $\theta : E \to E(P,Q)$ exists if and only if $\ell_*[E] = [E(P,Q)] \in H^2(Q;\mathrm{M}(W,T^k))$. If $[E(P,Q)]$ is of infinite order, no $\theta$ exists for most reasonable $Q$. If $[E(P,Q)]$ splits as $\mathrm{M}(W,T^k) \rtimes_\varphi Q$, then $\theta : T^k \rtimes_{\varphi|_{\mathrm{Aut}(T^k)}} Q \to E(P,Q)$ exists.

PROOF. Recall that the action $\widetilde{\varphi} : Q \to \mathrm{Aut}(\mathrm{M}(W,T^k))$, in Corollary 4.2.12, is given by
$$\widetilde{\varphi}(\psi)(\eta) =^q \eta = (\varphi(q) \times \rho(q))(\eta) = \varphi(q) \circ \eta \circ \rho(q)^{-1}.$$
If $\eta : W \to T^k$ is a constant map, then $\eta \circ \rho(q)^{-1} = \eta$ and $\varphi(q)(\eta) \in T^k$ because $E(P,Q)$ normalizes $\ell(T^k)$. Therefore $\varphi : Q \to \mathrm{Aut}(\mathrm{M}(W,T^k))$ when restricted to the constant maps $\ell(T^k)$ is an automorphism of $T^k$. Therefore, if we take an extension $1 \to T^k \to E \to Q \to 1$ with $Q \to \mathrm{Aut}(T^k)$ given by $\varphi(q)$, $q \in Q$, then $\ell_* : H^2_\varphi(Q;T^k) \to H^2(Q;\mathrm{M}(W,T^k))$ is defined. Furthermore, $\theta : E \to E(P,Q)$ exists if and only if $\ell_*([E]) = [E(P,Q)]$.

Given $\varphi$, there is an extension $T^k \rtimes_{\varphi|_{\mathrm{Aut}(T^k)}} Q$. Consequently, if $E(P,Q)$ splits as $\mathrm{M}(W,T^k) \rtimes_\varphi Q$, then $\theta : T^k \rtimes_{\varphi|_{\mathrm{Aut}(T^k)}} Q \to E(P,Q)$ exists and is induced by $\ell_*$. (Of course, for the same split $E(P,Q)$, there may be other $(E,\theta')$ which do not split and $\theta' : E \to E(P,Q)$).

As seen in the proof of Proposition 10.1.9,
$$H^3(Q;\mathbb{Z}^k) \to H^2(Q;\mathrm{M}(W,T^k)) \to H^2(Q;H^1(W;\mathbb{Z}^k))$$
is exact. For our reasonable $Q$, $H^2(Q;T^k) \cong$ Divisible $\oplus$ Torsion, and the two extreme groups of the exact sequence contain no divisible elements. Therefore, the image of $\ell_*[E]$ must be a torsion class. See Proposition 10.4.9 for examples of $P$ where $[E(P,Q)]$ has infinite order and no $\theta$ exists. □

REMARK 10.1.18. If $P$ is a product $T^k$-bundle, then by Corollary 4.2.10,
$$\mathrm{TOP}_{T^k}(P) = \mathrm{M}(W,T^k) \rtimes_{(\varphi \times \rho)} (\mathrm{Aut}(T^k) \times \mathrm{TOP}(W)).$$
Therefore, $E(P,Q)$ exists and splits.

10.1.19 (iii). If $E$ is to be the push-out of $1 \to F \to \Pi \to Q \to 1$ as in the diagram (10.1.1), then the torsion class $[\Pi] \in H^2(Q;F)$ must map by $\bar{\iota}_*$ into a torsion class $[E] \in H^2(Q;T^k)$. Furthermore, on the principal $F\backslash T^k$-bundle $F\backslash P$, the group $E(F\backslash P, Q)$ splits. One can reconstruct the $\Pi$-action on $P$ from the $Q$-action on $F\backslash P$. In principle, knowing all the $Q$-actions on a $T^k$-bundle $P'$, one should be able to construct all the various $\Pi$-actions on the various $T^k$-bundles covering $P'$. In Section 10.3, we determine all the possible liftings of the $Q$-action on $W$ to weak-bundle automorphisms of the $T^k$-bundles over $W$.

PROPOSITION 10.1.20 (ii). *If $\theta_0 : E \to E(P,Q)$ exists for some extension $E$ and fits into the third row of the diagram (10.1.1), then $H^1(Q; \mathrm{M}(W, T^k)) \cong \mathrm{Aut}(E(P,Q), \mathrm{M}(W, T^k), Q)/\mathrm{Inn}(\mathrm{M}(W, T^k))$ and is isomorphic to the strict equivalence classes of all the homomorphisms $\theta$ fitting into that part of the diagram.*

Two homomorphisms fitting into the diagram are *strictly equivalent* if and only if their images differ in $\mathrm{TOP}_{T^k}(P)$ by conjugation by an element of $\mathrm{M}(W, T^k)$, (compare with Subsection 7.4.2).

PROOF. We may identify, as in Subsection 10.1.12, each homomorphism $\theta$ fitting into the diagram with a 1-cocycle $\gamma \in Z^1(Q; \mathrm{M}(W, T^k))$. Then $\gamma$ in turn induces an automorphism $\widetilde{\gamma} \in \mathrm{Aut}(E(P,Q), \mathrm{M}(W, T^k), Q)$.

For, if $\theta_0 : E \to E(P,Q)$ is a fixed homomorphism, we, as in Subsection 10.1.12, associate the identity isomorphism and the trivial 1-cocycle to $E = T^k \times_{(f,\varphi)} Q \xrightarrow{\theta_0} E(P,Q) = \mathrm{M}(W, T^k) \times_{(f,\psi)} Q$. If $\theta_1 = \gamma \circ \theta_0$, then $\theta_1(t,q) = (t + \gamma(q), q)$. The induced automorphism $\widetilde{\gamma} : E(P,Q) \to E(P,Q)$, where $\widetilde{\gamma}(k,q) = (k + \gamma(q), q)$, sends $\theta_0(E)$ to $\theta_1(E)$. Addition of cocycles corresponds to composition of automorphisms of $E(P,Q)$. If $\gamma_1$ and $\gamma_2$ are cohomologous, then $\widetilde{\gamma}_2 - \widetilde{\gamma}_1$ is an automorphism given by conjugation by an element $c \in \mathrm{M}(W, T^k)$. This means that the two images $\theta_2(E)$ and $\theta_1(E)$ differ by $\mu(c)$ in $\mathrm{Aut}(E(P,Q), \mathrm{M}(W, T^k), Q)$, and therefore $\theta_2$ and $\theta_1$ are strictly equivalent in $\mathrm{TOP}_{T^k}(P)$. Of course, it must be remembered that this specific identification depends upon the choices of $\theta_0$ and the map $s : Q \to E$ which determines the cocycle $f$. Different choices will still yield a one-to-one correspondence between strict equivalence classes of homomorphisms fitting into the diagram and the elements of $H^1(Q; \mathrm{M}(W, T^k))$. $\square$

REMARK 10.1.21. If $\theta_0 \circ \overline{\iota} : \Pi \to \mathrm{TOP}_{T^k}(P)$ exists, then $H^1(Q; \mathrm{M}(W, T^k))$ also measures all the homomorphisms $\theta \circ \overline{\iota} : \Pi \to \mathrm{TOP}_{T^k}(P)$ up to strict equivalence.

## 10.2. When does $E(P,Q)$ split?

We investigate (iv) of Subsection 10.1.16 in this and the next section. Each $\theta : E \to E(P,Q)$ injects an *extended lifting* of a $Q$-action on $W$ into the group of weak bundle automorphisms of $P$. These extended liftings to $P$ can be determined if one understands the liftings of $Q$ to a specific principal $T^k$-bundle covered by $P$, as mentioned in Remark 10.1.18. Therefore we shall study those bundles for which $E(P,Q)$ splits.

If $P$ is a product bundle, then $E(P,Q)$ does split and we can measure the distinct liftings, i.e., distinct $Q$-actions, to $P$, up to strict equivalence, by $H^1(Q; \mathrm{M}(W, T^k))$.

For a possibly nonproduct bundle $P$, the group $E(P,Q)$ splits if and only if the principal $T^k$-bundle is the pullback of a $T^k$-bundle (not necessarily principal) over the Borel-space $W_Q = EQ \times_Q W$. This will be demonstrated by two different and independent methods in this and the next sections, respectively. In the first, as treated in Section 10.2, the plan is to show that this condition is equivalent to the vanishing of $[E(P,Q)]$ in $H^2(Q; \mathrm{M}(W, T^k))$. In Section 10.3, we adopt a more comprehensive hands-on approach and construct all the possible liftings $Q$ to the principal $T^k$-bundles over $W$.

In Chapter 12 we treat the same problem as in Sections 10.2 and 10.3 except that, in Chapter 12, $Q$ is assumed to be a compact connected Lie group instead of a discrete group and the lifting of $Q$ is to bundle automorphisms. Section 10.2 is a good introduction to Chapter 12. See Section 15.7 for a detailed illustration of the methods of Sections 10.1 and 10.2.

THEOREM 10.2.1. *Let $P$ be a principal $T^k$-bundle over $W$. Let $Q$ be discrete and act properly on $W$. Then $Q$ lifts to a group of weak bundle automorphisms of $P$ if and only if $P$ is the pullback of a $T^k$-bundle on $W_Q = EQ \times_Q W$, the Borel-space of $(Q, W)$, and whose structure group is in $T^k \rtimes \mathrm{Aut}(T^k)$.*

PROOF. We will also show that the $T^k$-bundle over $W_Q$ is principal if and only if $\varphi : Q \to \mathrm{Aut}(T^k)$ is trivial. The "only if" part of the theorem is easily demonstrated in Subsection 12.2.5 for any Lie group but with the restriction that $Q$ lifts to bundle automorphisms. To obtain the stronger liftings to weak-bundle automorphisms, we follow the argument and notation in Subsection 12.2.5 except that for the principal $T^k$-bundle $\pi_2^*(P)$ on $EQ \times W$, we induce instead a lift of the $Q$-action as a group of weak bundle automorphisms if $\varphi : Q \to \mathrm{Aut}(T^k)$ is nontrivial. (This is achieved by citing Lemma 12.2.2 which is valid for any group.) By dividing out the free $Q$-action on the principal $T^k$-bundle over $EQ \times W$, we obtain a $T^k$-bundle $\widetilde{P}$ (not necessarily principal), over $W_Q$ and $i^* \circ \pi^*(\widetilde{P}) = P$. Here $\pi$ is the orbit mapping $EQ \times W \to EQ \times_Q W$ and $i$ is the inclusion $W \hookrightarrow EQ \times W$ given by $i(w) = e_0 \times w$, for some fixed $e_0$ in $EQ$.

For the "if" part, let us assume there is a $T^k$-bundle $\widetilde{P}$ over $W_Q$ such that $P \cong (\pi \circ i)^* \widetilde{P}$ is a principal $T^k$-bundle over $W$ via the maps $\pi \circ i : W \to EQ \times W \to EQ \times_Q W$. The $T^k$-bundle $\pi^*(\widetilde{P})$ over $EQ \times W$ is a principal $T^k$-bundle because $i : W \to EQ \times W$ is a homotopy equivalence and $i^*(\pi^* \widetilde{P}) = P$ is a principal $T^k$-bundle. Furthermore, the action $Q$ on $EQ \times W$ is free and $EQ \times W \to W_Q$ is a covering map. This implies also that $\pi^*(\widetilde{P}) \to \widetilde{P}$ is a covering map. In fact, we have a locally injective Seifert fibering

$$\begin{array}{ccc} \pi^*(\widetilde{P}) & \xrightarrow{T^k \backslash} & EQ \times W \\ Q \backslash \downarrow & & \downarrow Q \backslash \\ \widetilde{P} & \longrightarrow & EQ \times_Q W. \end{array}$$

The $Q$-action on $\pi^*(\widetilde{P})$ is the same as that induced on $\pi^*(\widetilde{P})$ via the map $\widetilde{P} \to EQ \times_Q W$ and the $Q$-action on $EQ \times W$. The map $i$ is not $Q$-equivariant so we need another reason why $E(P,Q)$ splits.

Since $\pi^*(\widetilde{P})$ has a $T^k \rtimes Q$-action, we know $\pi^*(\widetilde{P}) \in H^2(EQ \times W, \mathbb{Z}^k)^Q$. This implies that $[i^* \circ \pi^*(\widetilde{P})] = [P] \in H^2(Q; \mathbb{Z}^k)^Q$ since $i$ is a homotopy equivalence. Therefore, we have an extension

$$1 \to \mathrm{M}(W, T^k) \to E(P, Q) \to Q \to 1$$

as an element of $H^2(Q; \mathrm{M}(W, T^k))$. We can find a section $s$ of $Q$ to $E(P,Q)$ which maps injectively into $\mathrm{TOP}_{T^k}(P)$. Thus, we get a cocycle

$$f(q, q') = s(q) \cdot s(q') \cdot s(qq')^{-1}$$

whose cohomology class characterizes the extension $E(P,Q)$. Therefore, the cohomology class $[f(q,q')] = [E(P,Q)]$ and $[f(q,q')]$ can be regarded as the obstruction to splitting $E(P,Q)$. On the other hand, with the projection $\pi_2 : EQ \times W \to W$, $E(\pi_2^* \circ i^* \circ \pi^*(\widetilde{P}), Q) = E(\pi_2^*(P), Q)$ does split. An element in $\pi_2^*(P)$ can be written as $(e, w) \times u$, where $(e, w) \in EQ \times W$ and $u \in P$ such that $\pi_2(e, w) = w$ and $u$ is an element of $P$ that projects to $w$. Then we have a section

$$\widetilde{s} : Q \to E(\pi_2^*(P), Q) \subset \text{TOP}_{T^k}(EQ \times W),$$

where $\widetilde{s}(q)((e,w) \times u) = ((qe, qw) \times s(q)(u))$. This defines the cocycle

$$\widetilde{f}(q, q') = \widetilde{s}(q) \cdot \widetilde{s}(q') \cdot \widetilde{s}(qq')^{-1}.$$

Now $\pi_2^* : f(q, q') \mapsto \widetilde{f}(q, q')$. Since $\widetilde{f}(q, q')$ is cohomologous to 0, $[\pi_2^*(f(q, q'))] = 0$.

It remains to show that

$$\pi_2^* : H^2(Q; \mathrm{M}(W, T^k)) \to H^2(Q; \mathrm{M}(EQ \times W, T^k))$$

is injective. If so, $[f(q, q')] = 0$ and $E(P, Q)$ splits. To see this, we have the exact sequences and vertical homomorphisms induced by $\pi_2$:

$$\begin{array}{ccccccccc}
0 & \longrightarrow & \mathrm{M}_0(W, T^k) & \longrightarrow & \mathrm{M}(W, T^k) & \longrightarrow & H^1(W; \mathbb{Z}^k) & \longrightarrow & 0 \\
& & \downarrow {\scriptstyle (\pi_2^*)^0} & & \downarrow {\scriptstyle \pi_2^*} & & \downarrow {\scriptstyle (\pi_2^*)^1} & & \\
0 & \longrightarrow & \mathrm{M}_0(EQ \times W, T^k) & \longrightarrow & \mathrm{M}(EQ \times W, T^k) & \longrightarrow & H^1(EQ \times W; \mathbb{Z}^k) & \longrightarrow & 0.
\end{array}$$

We treat these groups as coefficients for cohomology of $Q$. Note $(\pi_2^*)^1$ is an isomorphism of discrete groups, and so

$$H^2(Q; H^1(W; \mathbb{Z}^k)) \to H^2(Q; H^1(EQ \times W; \mathbb{Z}^k))$$

is an isomorphism.

By Subsection 7.5.6(b), $H^p(Q; \mathrm{M}(W, \mathbb{R}^k))$ and $H^p(Q; \mathrm{M}((EQ)^n \times W, \mathbb{R}^k))$ vanishes for $p > 0$, where $(EQ)^n$ is a finite dimensional approximation of $EQ$ (assuming that $W$ is finite dimensional). Then we have

$$\begin{array}{ccc}
0 & & 0 \\
\downarrow & & \downarrow \\
H^2(Q; \mathrm{M}_0(W, T^k)) & \xrightarrow{(\pi_2^*)^0} & H^2(Q; \mathrm{M}_0((EQ)^n \times W, T^k)) \\
\downarrow {\scriptstyle \delta} & & \downarrow {\scriptstyle \delta} \\
H^3(Q; \mathbb{Z}^k) & \xrightarrow{(\pi_2^*)^0} & H^3(Q; \mathbb{Z}^k) \\
\downarrow & & \downarrow \\
0 & & 0
\end{array}$$

where both $\delta$'s are isomorphisms. So $(\pi_2^*)^0$ is an isomorphism. Thus the maps $(\pi_2^*)^0$ and $(\pi_2^*)^1$ both induce isomorphisms on cohomology. The 5-lemma completes the argument. □

REMARK 10.2.2. As mentioned earlier, in Theorem 10.3.8, it will be shown independently of Subsection 7.5.6(b), that the cohomology vanishes without the need of finite dimensional assumptions nor the use of finite dimensional approximations.

10.2.3. The Borel space $EQ \times_Q W = W_Q$ fibers over $BQ$ with projection map $\pi$ and fiber $W$. We obtain a spectral sequence from this fibering whose $E_2$-term is
$$E_2^{p,q} = H^p(BQ; H^q(W; \mathbb{Z}^k)).$$
The group $H^q(W; \mathbb{Z}^k)$, in the $E_2^{p,q}$-term, is local coefficient system if $\varphi : Q \to \text{Aut}(\mathbb{Z}^k)$ is nontrivial. The spectral sequence simplifies when $W$ is simply connected and $\varphi : Q \to \text{Aut}(\mathbb{Z}^k)$ is trivial. Then we get the following exact sequence from the terms of low degree:

$$0 \to H^2(BQ; \mathbb{Z}^k) \xrightarrow{\pi^*} H^2(W_Q; \mathbb{Z}^k) \xrightarrow{i^*} H^2(W; \mathbb{Z}^k)^Q$$
$$\xrightarrow{d} H^3(BQ; \mathbb{Z}^k) \xrightarrow{\pi^*} H^3(W_Q; \mathbb{Z}^k)$$

and $H^1(BQ; \mathbb{Z}^k) \longrightarrow H^1(W_Q; \mathbb{Z}^k)$ is an isomorphism.

A second spectral sequence arises from the map $p : W_Q \to Q \backslash W$. If $v \in Q \backslash W$ and $w$ is on an orbit over $v$, then $p^{-1}(v) \cong Q_w \backslash EQ = BQ_w$. The Leray spectral sequence of the map $p$ has for $E_2$ term $H^2(Q \backslash W; h^q)$, where $h^q$ is the Leray sheaf associated to the cohomology of the fibers.

In the next two sections, we take a more local approach to the problems. This approach is independent of Section 10.2. However, it will turn out that one of the main quantities studied, $H^n(Q; \mathcal{Z}^k)$, is naturally isomorphic to $H^n(EQ \times_Q W; \mathbb{Z}^k)$. Two spectral sequences derived to analyze $H^n(Q; \mathcal{Z}^k)$ are similar (in fact, equivalent) to the ones mentioned above. However, in their new guise they are easier to interpret when $W$ is not simply connected.

## 10.3. From local to global: $H^*(Q; \mathcal{T}^k)$ and $H^*(Q; \mathcal{Z}^k)$

This section is taken from [**CR72b**]; compare also [**Suw79**] and [**Gro57**]. We know that $E(P, Q)$ splits if the principal $T^k$-bundle is a product bundle. The converse is not true. However, by taking a covering $\mathcal{V} = \{V_i : i \in I\}$ of the base $Q \backslash W$, we can find a covering of $W$, $\mathcal{W} = \{W_i = \bar{\nu}^{-1}(V_i) : i \in I\}$ of $Q$-invariant open subsets, where $\bar{\nu} : W \to Q \backslash W$ is the orbit mapping. Over $W_i$, the bundle can be chosen trivial, and so the $Q$-action on $W_i$ lifts to those parts of the bundle over $W_i$. The transition functions $g_{ij}(W_i \cap W_j) \to T^k$ tell us how to patch the product structures over $W_i$ and $W_j$ together so that we have a bundle over $W_i \cup W_j$. The collection of transition functions $\{g_{ij} : (i, j) \in I \times I\}$ forms a 1-cocycle on the nerve of the covering with values in the presheaf of germs of continuous functions from $W$ into $T^k$. If we can do the patching so that it is compatible with the $Q$-action on each $W_i$, we will get a 1-cocycle $\gamma_{ji}$ dependent upon both $q \in Q$ and $w \in W_i \cap W_j$. This 1-cocycle encodes the bundle $P$ as well as the lifting of the $Q$-action to $P$. This, as we shall see, is part of a Čech version of the cohomology of $Q$ with coefficients in a certain $Q$-sheaf, $\mathcal{T}^k$. This procedure works in the holomorphic, smooth, and continuous categories and enormous amount of information can be gleaned from studying these cohomology groups. We focus here on the continuous category but invite the reader to pursue the holomorphic Seifert fibering category in [**CR72b**] or Section 14.10 for an illustration of this technique. In cross checking details in [**CR72b**], the reader needs to make allowances for [**CR72b**]'s use of right actions of $Q$ on $W$. On the other hand, in the summary of relevant parts of [**CR72b**] in [**Suw79**] left actions are used as used here.

## 10.3.1.
Let $P$, a principal $T^k$-bundle over $W$, be equivalent to a product bundle. Let $s: W \to P$ be a cross section and identify $T^k \times W$ with $P$ by

$$h: T^k \times W \longrightarrow P,$$
$$(x, w) \longmapsto x \cdot s(w)$$

for $x \in T^k$.

As we have seen, $E(P, Q)$ exists and is split. We take the obvious lifting of $Q$ to $P$ given by

$$q \cdot (x, w) = (0, \varphi(q), \rho(q))(x, w) = \big(\varphi(q)(x), \rho(q)(w)\big) = \varphi(q)(x) \cdot s(\rho(q))(w).$$

This is associated to the trivial cocycle $\gamma_0: Q \to Z^1(Q; \mathrm{M}(W, T^k))$. For ease of notation, we suppress the symbols $\varphi \times \rho$ and $s$ whenever it is not confusing to do so. We then can think of our $\theta_0(T^k \rtimes Q)$-action on $P = T^k \times W$ to be given by

$$\theta_0(t, q)(x, w) = (t \cdot \varphi(q)(x), qw).$$

Recall from Corollary 4.2.10 that $(\lambda, \alpha, h) \in \mathrm{M}(W, T^k) \rtimes (\mathrm{Aut}(T^k) \times \mathrm{TOP}(W))$ acted on $T^k \times W$ as follows

$$(\lambda, \alpha, h)(x, w) = (\alpha(x) \cdot (\lambda(h(w)))^{-1}, h(w)).$$

Let $\theta: T^k \rtimes Q \to E(P, Q)$ be another injection, making the diagram (10.1.1) commute, with $\gamma: Q \to Z^1(Q; \mathrm{M}(W, T^k))$ its associated 1-cocycle. Then $\theta(t, q)$, in terms of $E(P, Q) \subset \mathrm{TOP}_{T^k}(T^k \times W)$, can be written as

$$\ell(t)(\gamma(q), \varphi(q), \rho(q)) = (t^{-1}, \mu(t), 1)(\gamma(q), \varphi(q), \rho(q)) = (t^{-1} \cdot \gamma(q), \varphi(q), q)$$

since $\mu(t)$ is trivial. Therefore,

$$\theta(t, q)(x, w) = (t^{-1} \cdot \gamma(q), \varphi(q), q)(x, w)$$
$$= (\varphi(q)(x) \cdot (t^{-1} \cdot \gamma(q)(qw))^{-1}, qw)$$
$$= (t\varphi(q)(x) \cdot (\gamma(q)(qw))^{-1}, qw),$$

where, for the convenience, we write $qw$ instead of $\rho(q)(w)$. We sometimes denote $\gamma(q)(qw)$ by $\gamma(q; qw)$ to more closely resemble the notation in [**CR72b**]. Then

$$\gamma(q)(w) = \gamma(q)(q(q^{-1}w)) = \gamma(q; w).$$

The cocycle condition $\gamma(q_1 q_2) = \gamma(q_1) \cdot^{q_1} \gamma(q_2)$ becomes

$$\gamma(q_1 q_2; w) = \gamma(q_1; w) \cdot \varphi(q_1)(\gamma(q_2; q_1^{-1} w)).$$

## 10.3.2.
We no longer assume that $P$ is a product bundle. We shall show how one constructs a covering of $Q$-invariant open subsets $\{O_i : i \in I\}$ of $P$ where the restriction of $P$ to each subset is a product.

Let $P \xrightarrow{\bar{\tau}} W$, $W \xrightarrow{\bar{\nu}} Q \backslash W$ be the orbit mappings of the $T^k$-action and the $Q$-action, respectively.

$$\begin{array}{c} P \\ \bar{\tau} \downarrow /T^k \\ W \xrightarrow[\bar{\nu}]{/Q} Q \backslash W \end{array}$$

For each $v \in Q \backslash W$, pick a $w \in W$ such that $\bar{\nu}(w) = v$. Choose a connected $Q_w$-slice, $S_w$, at $w$ such that $\bar{\tau}^{-1}(S_w)$ is a product $T^k \times S_w$. Then $S_w \times_{Q_w} Q$ is a disjoint union of open sets, namely $Q(S_w)$. Over each of the components of

$Q(S_w)$, the bundle is trivial and therefore the bundle is trivial over $Q(S_w)$. The image $\bar{\nu}(S_w) = Q_w \backslash S_w = V_v$ is open, $\bar{\nu}^{-1}(V_v) = W_w = Q(S_w)$ is $Q$-invariant in $W$ and $\bar{\tau}^{-1}(W_w)$ is a product $T^k \times W_w$. In this way, we find a locally finite covering $\{V_i : i \in I\}$ of $Q \backslash W$ and a locally finite open covering $\{\bar{\nu}^{-1}(V_i) = W_i : i \in I\}$ of $W$ such that $\{\bar{\tau}^{-1}(W_i) = O_i : i \in I\}$ is a locally finite open covering of $P$ with $P$ restricting to a product bundle over each $W_i$. We may now construct liftings of the $Q$-action on the $Q$-invariant $W_i$ to $O_i$. Our plan is to determine the conditions that guarantee that certain lifts agree on the overlaps $O_i \cap O_j$, for all $i, j$, to obtain a global lift of the $Q$-action to $P$.

10.3.3. Let $\{(W_i, h_i) : i \in I\}$ be a coordinate bundle for $P$ as described in the preceding subsection. Here $h_i : (x, w) \mapsto xs_i(w)$ is a homeomorphism $T^k \times W_i \to O_i$, where $s_i : W_i \to O_i$ is a section given by $s_i(w) = h_i(1, w)$. If $w \in W_i \cap W_j$, then $h_j^{-1} \circ h_i : (x, w) \mapsto (g_{ji} w \cdot x, w)$ where $g_{ji}$ is a transition function for the bundle $P$. Put

$$\gamma_{ji}(1; w) = g_{ji}(w).$$

Clearly, if $w \in W_i \cap W_j \cap W_k$, then

$$\gamma_{ki}(1; w) = \gamma_{kj}(1; w) \cdot \gamma_{ji}(1; w),$$

and is a 1-cocycle for the presheaf of maps of the $W_i$ into $T^k$. Suppose there is a $Q$-action on $O_i$ and $O_j$ given by

$$h_i^{-1}(q \cdot xs_i(w)) = (\varphi(q)(x) \cdot \gamma_{ii}(q; qw), qw),$$
$$h_j^{-1}(q \cdot xs_j(w)) = (\varphi(q)(x) \cdot \gamma_{jj}(q; qw), qw).$$

If the actions agree on $O_i \cap O_j$, then

$$\gamma_{ii}(q; qw) \cdot \gamma_{ji}(1; qw) = \varphi(q) \gamma_{ji}(1; w) \cdot \gamma_{jj}(q; qw)$$

must be satisfied. Conversely, if the condition holds for all $q, w$, then the $Q$-actions agree on $O_i \cap O_j$.

Let us denote this common value by $\gamma_{ji}(q; qw)$. Then if there is also a $Q$-action on $O_k$ such that the actions agree on $O_i \cap O_j \cap O_k$, then, on the intersection $w \in W_i \cap W_j \cap W_k$, the cocycle condition

$$\gamma_{ki}(q'q; q'qw) = \varphi(q') \gamma_{ji}(q; qw) \cdot \gamma_{kj}(q'; q'qw)$$

must be satisfied. The condition can also be written as

$$\gamma_{ki}(q'q; w) = \varphi(q') \gamma_{ji}(q; {q'}^{-1} w) \cdot \gamma_{kj}(q'; w)$$

or

$$\gamma_{ki}(q'q) = {}^{q'}\gamma_{ji}(q) \cdot \gamma_{kj}(q').$$

Conversely, if this cocycle condition is satisfied for all $w, q, q'$ and all triples $\{i, j, k\}$, then a global $Q$-action covering $(Q, W)$ is defined on all of $P$.

Note this last cocycle condition implies all the previous conditions because $i, j, k$ are not necessarily distinct. Also, cohomologous cocycles (equivalent to choosing different cross sections for $W_i \to O_i$) yield actions equivalent up to conjugation by elements of $\mathrm{M}(W, T^k)$.

**10.3.4** (Digression for sheaf cohomology). Take an exact sequence of sheaves
$$0 \to \mathcal{S}' \to \mathcal{S} \to \mathcal{S}'' \to 0$$
over $X$. The functor $\Gamma$ taking sections from sheaves to Abelian groups is a covariant left exact functor so that
$$0 \to \Gamma(X, \mathcal{S}') \to \Gamma(X, \mathcal{S}) \to \Gamma(X, \mathcal{S}'')$$
is exact. Now take an exact injective resolution of $\mathcal{S}$
$$0 \to \mathcal{C}^0(X; \mathcal{S}) \to \mathcal{C}^1(X; \mathcal{S}) \to \cdots .$$
The cohomology is defined as
$$H^q(X;\mathcal{S}) = \frac{\ker\{\Gamma(C^q(X;\mathcal{S})) \xrightarrow{d^q} \Gamma(C^{q+1}(X;\mathcal{S}))\}}{\operatorname{Im}\{\Gamma(C^{q-1}(X;\mathcal{S})) \xrightarrow{d^{q-1}} \Gamma(C^q(X;\mathcal{S}))\}}.$$

**10.3.5.** Let $\mathcal{T}^k$ be the sheaf of germs of local functions from $W$ into $T^k$. $\mathcal{T}^k$ is a $Q$-sheaf because the sections $\Gamma(W, \mathcal{T}^k)$ of the sheaf $\mathcal{T}^k$ have a $Q$-module structure given by
$$({}^q\Gamma)(w) = \varphi(q)\Gamma(q^{-1}w).$$
The $Q$-invariant sections $\Gamma(W; \mathcal{T}^k)^Q$ are left-exact covariant functor and its right derived functor (the cohomology of an injective resolution of $\Gamma(W; \mathcal{T}^k)^Q$) is defined to be $H^*(Q; \mathcal{T}^k)$. $\Gamma(W; \mathcal{T}^k)^Q$ is a composite of two covariant functors $\Gamma^Q \circ \Gamma(W; \mathcal{T}^k)$, where $\Gamma^Q$ assigns to a $Q$-module the $Q$-invariant elements. This leads to a well-known construction of a double complex and a spectral sequence

(10.3.1) $\quad ''E_2^{p,q} = H^p(Q; H^q(W, \mathcal{T}^k)) \Longrightarrow ''E^n = H^n(Q; \mathcal{T}^k).$

We can also write $\Gamma(W; \mathcal{T}^k)^Q$ as a composite $\Gamma_{Q \backslash W} \circ \bar{\nu}_*$, where $\bar{\nu}_*$ assigns to the $Q$-sheaf $\mathcal{T}^k$ over $W$ a sheaf over $Q \backslash W$ constructed from the presheaf $V \to \Gamma(\bar{\nu}^{-1}(V), \mathcal{T}^k)^Q$ and $\Gamma_{Q \backslash W}$ is the resulting sections over $Q \backslash W$. This composite leads to another double complex and spectral sequence

(10.3.2) $\quad 'E_2^{p,q} = H^p(Q \backslash W; \mathfrak{h}^q) \Longrightarrow 'E^n = H^n(Q; \mathcal{T}^k).$

Here $\mathfrak{h}^q$ is the sheaf whose presheaf is given by $V \to H^q(Q; \Gamma(\bar{\nu}^{-1}(V), \mathcal{T}^k))$. The stalk of $\mathfrak{h}^q$ at $\bar{\nu}(w)$ is given by $H^q(Q_w; \mathcal{T}^k{}_w) = H^q_\varphi(Q_w; T^k) \approx H^{q+1}_\varphi(Q_w; \mathbb{Z}^k)$, where $\mathcal{T}^k{}_w$ is the stalk of $\mathcal{T}^k$ at $w$. The last isomorphism is a consequence of $H^j(Q_w; \mathbb{R}^k) = 0$ for all $j > 0$, since $Q_w$ is finite; see [**CR72b**, p.129] or [**Suw79**, p.78].

The idea behind the computation of the stalk of the sheaf $\mathfrak{h}^q$ is to take a $Q$-invariant neighborhood of the $Q$-orbit through $w \in W$, $(Q, Q \times_{Q_w} S)$, with $S$ a slice at $w$ and $Q_w$ the isotropy subgroup at $w$. Then, one shows that the maps of this invariant neighborhood $M(\bar{\nu}^{-1}(\bar{\nu}(S)), T^k)$ are isomorphic, as a $Q$-module, to $\operatorname{Hom}_{\mathbb{Z}Q_w}(\mathbb{Z}Q, M(S, T^k))$. Consequently, $H^q(Q; M(\bar{\nu}^{-1}(\bar{\nu}(S)), T^k))$ is isomorphic to $H^q(Q_w; M(S, T^k))$. Passing to a limit with smaller slices gives the desired isomorphisms.

**10.3.6.** With $\{\gamma_{ji}\}$ we have, in effect, constructed a 1-dimensional Čech cohomology version of $H^1(Q; \mathcal{T}^k)$. The Čech version of $H^p(Q; \mathcal{T}^k)$ is described as follows.

Take a locally finite open covering $\mathcal{V} = \{V_i : i \in I\}$ of $Q \backslash W$ and set $W_i = \bar{\nu}^{-1}(V_i)$, a $Q$-invariant open subset of $W$. We can always choose $V_i$ in $Q \backslash W$ small

enough so that the bundle over the $W_i$ is a product bundle as described in Subsection 10.3.2. For the covering $\mathcal{W} = \{W_i : i \in I\}$, let $N(\mathcal{W})$ be the nerve of this covering, and let $C^q(\mathcal{W}; \mathcal{T}^k)$ be the group of $q$-cochains on the nerve $\mathcal{W}$ with coefficients in the sheaf $\mathcal{T}^k$. This cochain group has a natural $Q$-module structure. Let
$$C^{p,q}_{\mathcal{W}}(Q; \mathcal{T}^k) = C^p(Q; C^q(\mathcal{W}; \mathcal{T}^k)),$$
the group of $p$-cochains of $Q$ with coefficients in $C^q(\mathcal{W}; \mathcal{T}^k)$. Note this leads to a double complex and we are interested in the cohomology of the total complex. It is the direct limit of the cohomology of this total complex that yields $H^*(Q; \mathcal{T}^k)$.

10.3.7. Each element $\sigma \in C^{p,q}_{\mathcal{W}}(Q; \mathcal{T}^k)$ is represented by a $\{\sigma^{i_0,\ldots,i_q}(w; q_1, \ldots, q_p)\}$ where $w \in W_{i_0} \cap \cdots \cap W_{i_q}$. That is, for each $(i_0, \ldots, i_q) \in I^{q+1}$ and each $(q_1, \ldots, q_p) \in Q^p$, $\sigma^{i_0,\ldots,i_q}(w; q_1, \ldots, q_p)$ is a section of $\mathcal{T}^k$ over $W_{i_0} \cap \cdots \cap W_{i_q}$. Put
$$\mathcal{C}^p_{\mathcal{W}}(Q; \mathcal{T}^k) = C^{p,p}_{\mathcal{W}}(Q; \mathcal{T}^k),$$
where $\delta : \mathcal{C}^p_{\mathcal{W}}(Q; \mathcal{T}^k) \longrightarrow \mathcal{C}^{p+1}_{\mathcal{W}}(Q; \mathcal{T}^k)$ is given by
$$(\delta\sigma)^{i_0,\ldots,i_{p+1}}(w; q_1, \ldots, q_{p+1}) = \varphi(q_1) \sigma^{i_1,\ldots,i_{p+1}}(q_1^{-1}w; q_2, \ldots, q_{p+1})$$
$$+ \sum_{j=1}^{p} (-1)^j \sigma^{i_0,\ldots,\hat{i_j},\ldots,i_{p+1}}(w; q_1, \ldots, q_j q_{j+1}, \ldots, q_{p+1})$$
$$+ (-1)^{p+1} \sigma^{i_0,\ldots,i_p}(w; q_1, \ldots, \cdots, q_p).$$

(Note $(\delta\sigma)^{k,j,i}(w; q', q) = 0$ if and only if $\sigma^{j,i}(w; q) = \gamma_{ji}(q; qw)$ is our previously defined 1-cocycle.) With this coboundary operator, we define $H^p_{\mathcal{W}}(Q; \mathcal{T}^k)$ as the $p$th cohomology group of the complex $\mathcal{C}^*_{\mathcal{W}}(Q; \mathcal{T}^k, \delta)$ and
$$\check{H}^p(Q; \mathcal{T}^k) = \text{dir lim}_{\mathcal{V}} H^p_{\mathcal{W}}(Q; \mathcal{T}^k),$$
where $\mathcal{V}$ runs through the open coverings of $Q \backslash W$. $H^*_{\mathcal{W}}(Q; \mathcal{T}^k)$ is also the cohomology of the total complex of the double complex $\{C^{p,q}_{\mathcal{W}}(Q; \mathcal{T}^k)\}$. We get spectral sequences for each of these coverings and it is the direct limit of each of the terms that yields spectral sequences isomorphic to spectral sequences (10.3.1) and (10.3.2); see [**CR72b**] or [**Suw79**] for details.

Instead of using maps into $T^k$, we could just well consider maps into $\mathbb{Z}^k$ and $\mathbb{R}^k$. The same constructions and spectral sequences hold for the corresponding sheaves $\mathcal{Z}^k$ and $\mathcal{R}^k$, respectively. These sheaves form an exact sequence
$$0 \to \mathcal{Z}^k \to \mathcal{R}^k \to \mathcal{T}^k \to 0$$
of sheaves over $W$.

THEOREM 10.3.8. *$H^p(Q; \mathcal{R}^k) = 0$, for all $p > 0$.*

PROOF. We examine the first spectral sequence $'E_2^{p,q} = H^p(Q \backslash W; \mathfrak{h}^q_{\mathbb{R}})$ and the sheaf $\mathfrak{h}^q_{\mathbb{R}}$. Recall that the stalk $\mathfrak{h}^q_{\bar{\nu}(w)} = H^q(Q_w; \mathbb{R}^k)$ is 0 since $Q_w$ is finite. Therefore, $\mathfrak{h}^q_{\mathbb{R}}$ is the trivial sheaf for all $q > 0$ and $'E_2^{p,q}$ collapses so that $H^n(Q; \mathcal{R}^k) = H^n(Q \backslash W; \mathfrak{h}^0_{\mathbb{R}})$. The sheaf $\mathfrak{h}^0_{\mathbb{R}}$ is a module over the fine sheaf of rings of germs of continuous functions of $Q \backslash W \to \mathbb{R}$. Therefore, $\mathfrak{h}^0_{\mathbb{R}}$ is also fine and consequently $H^n(Q; \mathcal{R}^k) = 0$, $n > 0$. □

(We can also be explicit as to why $\mathfrak{h}_\mathbb{R}^0$ is fine. The sections of $\mathfrak{h}_\mathbb{R}^0$ are the maps $f: W \to \mathbb{R}^k$ which satisfy $f(w) = {}^q f(w) = \varphi(q) f(q^{-1}(w))$. If $\mathcal{V} = \{V_i : i \in I\}$ is a locally finite covering of $Q\backslash W$, there is a partition of unity $\{\epsilon_i : i \in I\}$ subordinate to $\mathcal{V}$ such that $\epsilon_i' = \epsilon_i \circ \bar\nu : W \to \mathbb{R}$, and $\epsilon_i'(qw) = \epsilon_i'(w)$. Then $\epsilon_i'(w) \cdot f(w)$ is a global section of $\mathfrak{h}_\mathbb{R}^0 \to Q\backslash W$ with support in $V_i$. We have

$$f(w) = \sum_i \epsilon_i'(w) \cdot f(w) = \sum_i \epsilon_i'(w) \cdot \varphi(q) f(q^{-1}(w)) = \sum_i \varphi(q)(\epsilon_i'(w) \cdot f(q^{-1}(w)))$$
$$= ({}^q f)(w).$$

Thus, $f(w) = {}^q f(w)$.)

COROLLARY 10.3.9. *If $Q\backslash W$ is paracompact, then $H^p(Q; \mathrm{M}(W, \mathbb{R}^k)) = 0$, for all $p > 0$.*

This is a generalization of Subsection 7.5.6(b) where it was assumed $Q\backslash W$ is also finite dimensional or compact. The proof here is independent of Chapter 7.

PROOF. The sheaf $\mathcal{R}^k$ of germs of continuous local functions from a paracompact $W$ into $\mathbb{R}^k$ is a module over the fine sheaf of germs of continuous local functions into $\mathbb{R}$. Therefore, $\mathcal{R}^k$ is fine and $''E_2^{p,q} = H^p(Q; H^q(W; \mathcal{R}^k)) = 0$, for all $q > 0$. So $''E_2^{p,q}$ collapses and $H^p(Q; H^0(W; \mathcal{R}^k)) = H^p(Q; \mathrm{M}(W; \mathbb{R}^k)) = H^p(Q; \mathcal{R}^k) = 0$, if $p > 0$. □

We also observe that if $W$ is a smooth manifold, $P$ is a smooth bundle over $W$ and $\rho : Q \to \mathrm{TOP}(W)$ is a smooth action, then, using germs of smooth functions for our sheaves and smooth partitions of unity, we have the same results in the smooth category.

COROLLARY 10.3.10. *$\delta : H^n(Q; \mathcal{T}^k) \longrightarrow H^{n+1}(Q; \mathcal{Z}^k)$ is an isomorphism for all $n > 0$.*

PROOF. We refer the reader to the elementary proof on page 132–133 of [**CR72b**] that the sequence

$$\to H^n(Q; \mathcal{R}^k) \to H^n(Q; \mathcal{T}^k) \xrightarrow{\delta} H^{n+1}(Q; \mathcal{Z}^k) \to H^{n+1}(Q; \mathcal{R}^k) \to \cdots$$

is exact. Since $H^p(Q; \mathcal{R}^k)$ vanishes for $p > 0$, $\delta$ is an isomorphism. □

For $n = 0$, we have, by the $''E_2$-spectral sequence, the exact sequence:

$$0 \to \mathrm{M}(W, \mathbb{Z}^k)^Q \to \mathrm{M}(W, \mathbb{R}^k)^Q \to \mathrm{M}(W, T^k)^Q \to H^1(Q; \mathcal{Z}^k) \to 0.$$

10.3.11. In contrast to the topological and smooth categories, the homomorphism $\delta$ of Corollary 10.3.10 is generally not an isomorphism in the holomorphic category. We have the following.

Let $T^k$ be a complex torus of real dimension $2k$, and let $\mathfrak{C}^k$, usually denoted by $\mathcal{O}(W)^k$, be the sheaf of germs of holomorphic functions from the complex manifold $W$ into $\mathbb{C}^k$. Then the stalk is

$$(\mathfrak{h}^j)_{\bar\nu(w)} = H^j(Q_w; \mathfrak{C}^k{}_w) \to H^j(Q_w; \mathbb{C}^k) = 0, \quad j > 0.$$

Therefore, the $'E_2^{p,q}$-spectral sequence collapses and we get that $H^p(Q; \mathcal{O}(W)^k) = H^p(Q; \mathfrak{C}^k) \to H^p(Q\backslash W; \mathfrak{h}^0{}_\mathbb{C})$ is an isomorphism for $p > 0$. Hence, in the holomorphic category, the long exact sequence in the proof of Corollary 10.3.10 becomes

$$\to H^n(Q\backslash W; \mathfrak{h}^0{}_\mathbb{C}) \to H^n(Q; \mathcal{T}^k) \xrightarrow{\delta} H^{n+1}(Q; \mathcal{Z}^{2k}) \to H^{n+1}(Q\backslash W; \mathfrak{h}^0{}_\mathbb{C}) \to .$$

Since there is no holomorphic partition of unity, $H^n(Q\backslash W; \mathfrak{h}^0{}_\mathbb{C})$ will not vanish in general. Consequently, $\delta$, unlike in the topological and smooth cases, is not an isomorphism.

See Subsection 14.10.2 for an interpretation where $n = 1$, $k = 1$, and $W$ is the unit disk. We also remark but will not use that if $\varphi|_{Q_w} : Q_w \to \mathrm{Aut}(\mathbb{Z}^{2k}) \subset \mathrm{Aut}(\mathbb{C}^k)$ is trivial, then $(\mathfrak{h}^0)_\mathbb{C} \to Q\backslash W$ is a locally free sheaf of rank $k$ as a module over $\mathcal{O}(Q\backslash W)$; see [**CR72b**, 2.4]. We now return to the topological and smooth categories.

THEOREM 10.3.12. *The following sequence is exact:*
$$0 \to H^1(Q; \mathrm{M}(W, \mathcal{T}^k)) \xrightarrow{e_1} H^1(Q; \mathcal{T}^k) \xrightarrow{e_2} H^2(W; \mathbb{Z}^k)^Q \xrightarrow{d} H^2(Q; \mathrm{M}(W, \mathcal{T}^k)) \xrightarrow{e_3} H^2(Q; \mathcal{T}^k).$$

PROOF. If we take $''E_2^{p,q}$ and write out the exact sequence arising from terms of low degree, we have
$$0 \to {''E_2^{1,0}} \xrightarrow{e_1} H^1(Q; \mathcal{T}^k) \xrightarrow{e_2} {''E_2^{0,1}} \xrightarrow{d} E_2^{2,0} \xrightarrow{e_3} H^2(Q; \mathcal{T}^k),$$
where $e_1$, $e_2$, and $e_3$ are edge homomorphisms.
$${''E_2^{1,0}} = H^1(Q; H^0(W; \mathcal{T}^k)) = H^1(Q; \mathrm{M}(W, \mathcal{T}^k)),$$
for $H^0(W; \mathcal{T}^k)$ is just the sections from $W$ to $\mathcal{T}^k$. Similarly for $''E_2^{2,0}$.
$${''E_2^{0,1}} = H^0(Q; H^1(W; \mathcal{T}^k)) = H^1(W; \mathcal{T}^k)^Q.$$
Since $0 \to \mathcal{Z}^k \to \mathcal{R}^k \to \mathcal{T}^k \to 0$ is exact and $H^n(W; \mathcal{R}^k) = 0$, for all $n > 0$, we have the isomorphism. Therefore, $H^1(W; \mathcal{T}^k)^Q \cong H^2(W; \mathbb{Z}^k)^Q$. (The $\mathbb{Z}^k$-coefficients may be twisted if $\varphi : Q \to \mathrm{Aut}(\mathbb{Z}^k)$ is not trivial.) $\square$

If $W$ is simply connected, the exact sequence of the theorem in terms of $\mathbb{Z}^k$ takes the form:
$$0 \to H^2(Q; \mathbb{Z}^k) \xrightarrow{e_1} H^2(Q; \mathcal{Z}^k) \xrightarrow{e_2} H^2(W; \mathbb{Z}^k)^Q \xrightarrow{d} H^3(Q; \mathbb{Z}^k) \to H^3(Q; \mathcal{Z}^k).$$

One may wish to compare this exact sequence with the exact sequence mentioned in Subsection 10.2.3.

10.3.13. We interpret the exact sequence of Theorem 10.3.12 as follows.

1. Each element $\gamma \in H^1(Q; \mathcal{T}^k)$ represents a strict equivalence class of lifts of the $Q$-action on $W$ to weak bundle automorphisms on some principal $T^k$-bundle $P$ over $W$. Furthermore, $H^1(Q; \mathcal{T}^k)$ is isomorphic to $H^2(Q; \mathcal{Z}^k)$.

2. The homomorphism $e_2$ assigns to each $\gamma$ the characteristic class $e_2(\gamma) = [P] \in H^2(W; \mathbb{Z}^k)^Q$ of the bundle $P$ associated with $\gamma$. The elements of $H^2(W; \mathbb{Z}^k)^Q$ represent all the bundles for which there is an *extended lifting* of $Q$ to a group $E(P, Q)$ of weak bundle automorphism of $P$, as in the third line of the diagram (10.1.1).

3. If $[P] = e_2(\gamma)$, then all $\gamma' \in H^1(Q; \mathcal{T}^k)$ such that $e_2(\gamma') = [P]$ are of the form $\gamma + e_1(\delta)$ for some $\delta \in H^1(Q; \mathrm{M}(W, T^k))$, and conversely.

4. The homomorphism $d$ assigns to the class $[P] \in H^1(W; \mathcal{T}^k)^Q \cong H^2(W; \mathbb{Z}^k)^Q$ the group extension
$$1 \to \mathrm{M}(W, \mathcal{T}^k) \to E(P, Q) \to Q \to 1$$
corresponding to the extended lifting associated to $P$. The elements of $H^3(Q; \mathbb{Z}^k) \cong H^2(Q; \mathrm{M}(W, T^k))$ represent all possible congruence classes of extensions of $\mathrm{M}(W, T^k)$ by $Q$.

5. The edge homomorphism $e_3 : H^2(Q; \mathrm{M}(W, T^k)) \to H^2(Q; \mathcal{T}^k) \cong H^3(Q; \mathcal{Z}^k)$ can be thought as the obstruction for some extension $u \in H^2(Q; \mathrm{M}(W, T^k))$ to be of the form $d[P] = E(P,Q)$, and consequently, the obstruction for $u$ to be mapped by $\epsilon$ into $\mathrm{TOP}_{T^k}(P)$ for some $P$ as in diagram (10.1.1).

6. $H^n(Q; \mathcal{Z}^k) = H^n(W_Q; \mathbb{Z}^k)$, $n > 1$. In particular,

$$H^1(Q; \mathcal{T}^k) \xrightarrow[\cong]{\delta} H^2(Q; \mathcal{Z}^k) \xrightarrow{\cong} H^2(W_Q; \mathbb{Z}^k).$$

PROOF. 1. We discussed in Subsections 10.3.3 and 10.3.7 a Čech version of $H^n(Q; \mathcal{T}^k)$. In the Čech version when $n = 1$, we saw that an element $[\gamma] \in H^1(Q; \mathcal{T}^k)$ is represented by a cocycle $\gamma_{ji}(q; w)$ on some $Q$-invariant covering on some bundle $P$. A cohomologous cocycle represents another lifting of $Q$ to $P$ strictly equivalent to the lifting determined by $\gamma_{ji}$.

2. It is easy to check that $e_2(\gamma_{ji}(q; w)) = \gamma_{ji}(1, w) = g_{ji}$. Therefore, $e_2$ maps $H^1(Q; \mathcal{T}^k)$ into $H^0(Q; H^1(W; \mathcal{T}^k)) \cong H^1(W; \mathcal{T}^k)^Q \cong H^2(W; \mathbb{Z}^k)^Q$. The rest of part 2 is a consequence of Subsections 10.1.3 and 10.1.4.

Parts 3, 4, and 5 follow from exactness.

6. We already have an isomorphism between $H^n(Q; \mathcal{T}^k)$ and $H^{n+1}(Q; \mathcal{Z}^k)$, for all $n > 0$. It is not difficult but it does take some effort, as alluded to in [**CR72b**, 3.9], to establish an isomorphism between $H^{n+1}(Q; \mathcal{Z}^k)$ and $H^{n+1}(W_Q; \mathbb{Z}^k)$ using the methods of Section 10.3.

$H^1(Q; \mathcal{T}^k) \cong H^2(Q; \mathcal{Z}^k)$ are the strict equivalence classes of lifts of $Q$ to bundles over $P$. In Theorem 10.2.1 we showed that a bundle $P$ over $W$ admits a lift of the $Q$-action on $W$ if and only if $P$ is the pullback of a $T^k$-bundle (with structure group $T^k \times \mathrm{GL}(k, \mathbb{Z})$) over $W_Q$ via $(\pi \circ i)^*$. Therefore, there is a one-to-one correspondence between $H^2(Q; \mathcal{Z}^k) \cong H^2(W_Q; \mathbb{Z}^k)$.

Another observation in support of the plausibility that the assertion can be demonstrated solely from the point of view of Section 10.3 is the similarity of the $''E_2$-spectral sequence for $H^*(Q; \mathcal{Z}^k)$ and the spectral sequence associated with the fibering $W_Q \to BQ$. Also, the $'E_2$-spectral sequence and the Leray spectral sequence of the map $W_Q \to Q \backslash W$ both have their $E_2^{p,q}$-terms $H^p(Q \backslash W; \mathfrak{h}^q) \cong H^p(Q; \mathcal{L}^q)$, where $\mathcal{L}^q$ is the Leray sheaf whose stalk over $\nu(w) \in Q \backslash W$ is given by $H^q(Q_w \backslash EQ; \mathbb{Z}^k) \cong \mathfrak{h}^q{}_{\bar{\nu}(w)}$.

Let us describe the coefficient system $\mathbb{Z}^k$ in $H^*(W_Q; \mathbb{Z}^k)$. On $EQ \times W$, we have a simple system since we are considering principal $T^k$-bundles on $EQ \times W$ and $W$. This simple system is the constant sheaf $(EQ \times W) \times \mathbb{Z}^k$. There is a $Q$-action on $EQ \times W$ and $\mathbb{Z}^k$. By taking the diagonal quotient $(EQ \times W) \times_Q \mathbb{Z}^k$ is, over $W_Q$, a locally constant sheaf locally isomorphic to $\mathbb{Z}^k$. This is the local coefficient system for $H^*(W_Q; \mathbb{Z}^k)$.

The isomorphism of part 6 is used in this book only for illustrative purposes. We have given a proof for $n = 2$, the most important case. Also, note if $W$ is contractible, then all bundles over $W$ are products and $H^n(Q; \mathbb{Z}^k) \to H^n(Q; \mathcal{Z}^k)$ is an isomorphism. Moreover, the spectral sequence for $W_Q \to BQ$ collapses since $H^q(W; \mathbb{Z}^k) = 0$, and so $H^p(BQ; \mathbb{Z}^k) \to H^p(W_Q; \mathbb{Z}^k)$ is an isomorphism. But as $H^p(Q; \mathbb{Z}^k)$ is canonically isomorphic to $H^p(BQ; \mathbb{Z}^k)$, we have also established the isomorphism for part 6 when $W$ is contractible. □

## 10.4. The product case, $\mathbb{R}^k \times W$

The product case bbrk x W In this section, we recapture, using the preceding sections, the existence and uniqueness theorems for Seifert fiberings modeled on $\mathbb{R}^k \times W$. This is achieved independently of Chapter 7 but is restricted to $G = \mathbb{R}^k$. We shall also investigate when a lift of a $Q$-action on $W$ to a group of weak bundle automorphism on $T^k \times W$ has an extended lifting to a group $\Pi$ of weak bundle automorphisms on $\mathbb{R}^k \times W$.

10.4.1 (Existence and uniqueness for Seifert fiberings modeled on $\mathbb{R}^k \times W$). Let $i: \mathbb{Z}^k \subset \mathbb{R}^k$ and $\ell: \mathbb{R}^k \to M(W, \mathbb{R}^k)$ send $r \in \mathbb{R}^k$ to the constant map $W \to r$. Let $\varphi: Q \to \text{Aut}(\mathbb{Z}^k) \subset \text{Aut}(\mathbb{R}^k)$ and $1 \to \mathbb{Z}^k \to \Pi \to Q \to 1$ be any extension with $\varphi$ as given. Let $\rho: Q \to \text{TOP}(W)$ be a homomorphism with $Q$ acting on $W$ properly.

THEOREM 10.4.2. *There exists a homomorphism $\theta$ such that the diagram*

$$\begin{array}{ccccccccc} 1 & \to & \mathbb{Z}^k & \to & \Pi & \to & Q & \to & 1 \\ & & \downarrow \scriptstyle{\ell \circ i} & & \downarrow \scriptstyle{\theta} & & \downarrow \scriptstyle{\varphi \times \rho} & & \\ 1 & \to & M(W, \mathbb{R}^k) & \to & \text{TOP}_{\mathbb{R}^k}(\mathbb{R}^k \times W) & \to & \text{Aut}(\mathbb{R}^k) \times \text{TOP}(W) & \to & 1 \end{array}$$

*commutes and the rows are exact. Furthermore, if $\theta'$ is another homomorphism with the same $\ell \circ i$ and $\varphi \times \rho$, then there exists $c \in M(W, \mathbb{R}^k)$ such that $\theta' = \mu(c) \circ \theta$.*

This is the existence and uniqueness part of Theorem 7.3.2 for $G = \mathbb{R}^k$. Our proof here is independent of Chapter 7. It also says that $\theta(\Pi)\backslash \mathbb{R}^k \times W \to Q\backslash W$ is a Seifert fibering with $T^k = \mathbb{Z}^k \backslash \mathbb{R}^k$ as typical fiber. Furthermore, $\theta(\Pi)$ and $\theta'(\Pi)$ are strictly equivalent.

PROOF. We interpret the first exact sequence in the proof of Theorem 10.3.12 but use $\mathcal{R}^k$ instead of $\mathcal{T}^k$. Since

$$H^p(Q; \mathcal{R}^k) = 0, \text{ for } p > 0, k > 0$$

(Theorem 10.3.8), and

$${}''E_2^{p,q} = H^p(Q; H^q(W; \mathcal{R}^k)) = 0, \text{ for } q > 0,$$

the terms of the sequence (using $\mathcal{R}^k$) are all 0. Therefore, $E(P,Q) = M(W, \mathbb{R}^k) \rtimes Q$. Let $1 \to \mathbb{R}^k \to E \to Q \to 1$ be any extension with $\varphi: Q \to \text{Aut}(\mathbb{R}^k)$ as above, then we get

$$\begin{array}{ccccccccc} 1 & \to & \mathbb{R}^k & \to & E & \to & Q & \to & 1 \\ & & \downarrow \scriptstyle{\ell} & & \downarrow \scriptstyle{\epsilon} & & \downarrow \scriptstyle{=} & & \\ 1 & \to & M(W, \mathbb{R}^k) & \to & E(P,Q) & \to & Q & \to & 1. \end{array}$$

The homomorphism $\epsilon$ exists since $\ell_*[E] = [E(P,Q)] = 0$.

To define $\theta$, we first form the push-out $\mathbb{R}^k \cdot \Pi$. We get

$$\begin{array}{ccccccccc} 1 & \to & \mathbb{Z}^k & \to & \Pi & \to & Q & \to & 1 \\ & & \downarrow \scriptstyle{i} & & \downarrow \scriptstyle{\bar{i}} & & \downarrow \scriptstyle{=} & & \\ 1 & \to & \mathbb{R}^k & \to & \mathbb{R}^k \cdot \Pi & \to & Q & \to & 1 \end{array}$$

and then we compose this with $\epsilon$. That is, $\theta = \epsilon \circ \bar{i}$. Since $H^1(Q; M(W, \mathbb{R}^k)) = 0$, each $\theta$ is unique up to conjugation by some element $c \in M(W, \mathbb{R}^k)$. □

Note in Theorem 10.4.2, $(\mathbb{R}^k \cdot \Pi, \mathbb{R}^k \times W)$ factors through $(T^k \rtimes Q, T^k \times W)$ by dividing out by $\mathbb{Z}^k \subset \mathbb{R}^k$. Consequently, $Q\backslash T^k \times W = \theta(\Pi)\backslash \mathbb{R}^k \times W$. However, not all lifts of $Q$ to $T^k \times W$ come from $\Pi$-actions on $\mathbb{R}^k \times W$. We will characterize those lifts $\theta$ that come from $\Pi$-actions. But first we have the following

LEMMA 10.4.3. *A principal $T^k$-bundle $P$ over $W$ is covered by a product $\mathbb{R}^k$-bundle $P^*$ where the $\mathbb{R}^k$-fiber covers the $T^k$-fiber of $P$ if and only if the $T^k$-action on $P$ is injective.*

PROOF. Let $(T^k, P)$ be a principal $T^k$-bundle over the base $W$. Let $H$ be the image of $\mathrm{ev}_*^x : \pi_1(T^k, 1) \to \pi_1(P, x)$. Therefore,
$$0 \to K \to \pi_1(T^k, 1) \to H \to 0$$
is exact. Then $G = \mathbb{R}^k/K$ acts on $P$ via the covering projection $G \to T^k = \mathbb{R}^k/\mathbb{Z}^k$. The action of $G$ lifts to the universal covering $\widetilde{P}$ of $P$ and covers the action of $T^k$ on $P$. We claim the action of $G$ is free on $\widetilde{P}$ and is the largest quotient of $\mathbb{R}^k$ covering $T^k$ with the property. (Therefore $G = \mathbb{R}^k$ if and only if $K = 0$ and $(T^k, P)$ is injective.) Let $\pi : \widetilde{W} \to W$ be the universal covering projection. Form the induced principal $T^k$-bundle $\pi^*(P) = P^*$. Now $\pi_1(P^*) = H$ and the lifted $(T^k, P^*)$ is free. The group $G$ covers $T^k$ and lifts to the universal covering $\widetilde{P}$. On $\widetilde{P}$, the $G$-action is free. If $K' \subset K$, then $\mathbb{R}^k/K'$ covers $G$ but the lift to $\widetilde{P}$ is ineffective. If $K' \supset K$, $\mathbb{R}^k/K'$ will not lift to $P^*$. □

Denote $\mathrm{M}_0(W, T^k) \rtimes (\mathrm{Aut}(T^k) \times \mathrm{TOP}(W))$ by $\mathrm{TOP}_{T^k}(T^k \times W)_0$; see Lemma 10.1.7 for definition of $\mathrm{M}_0(W, T^k)$.

EXERCISE 10.4.4. Show that $\mathrm{TOP}_{T^k}(T^k \times W)_0$ is well defined and $\mathrm{M}(W, \mathbb{R}^k) \to \mathrm{M}_0(W, T^k)$, given by $\lambda \mapsto e^{2\pi i \lambda}$, is a regular covering projection with regular fiber $\mathrm{M}(W, \mathbb{Z}^k) = \mathbb{Z}^k$.

10.4.5. If $[\gamma] \in H^1(Q; \mathcal{T}^k)$ represents a lift of the $Q$-action on $W$ to $P = T^k \times W$ (i.e., $e_2[\gamma] = 0$), then $[\gamma] \in H^1(Q; \mathrm{M}(W, T^k)) \subset H^1(Q; \mathcal{T}^k)$ because $e_1$ is injective. We may represent the action on $T^k \times W$ by
$$q \cdot (x, w) = (\varphi(q)(x) \cdot \gamma(q)(qw)^{-1}, q(w)),$$
where $\gamma \in Z^1(Q; \mathrm{M}(W, T^k))$. $H^1(Q; \mathrm{M}(W, T^k))$ fits into the following exact sequence.
$$\to H^1(W; \mathbb{Z}^k)^Q \to H^1(Q; \mathrm{M}_0(W, T^k)) \xrightarrow{i} H^1(Q; \mathrm{M}(W, T^k)) \xrightarrow{j} H^1(Q; H^1(W; \mathbb{Z}^k)) \to$$
This exact sequence arises from the exact sequence of $Q$-modules
$$0 \to \mathrm{M}_0(W, T^k) \to \mathrm{M}(W, T^k) \to H^1(W; \mathbb{Z}^k) \to 0.$$
If $\gamma : Q \to \mathrm{M}_0(W, T^k)$ is a 1-cocycle, we can form a set-theoretic pullback:

$$
\begin{array}{ccccccccc}
1 & \longrightarrow & \mathbb{Z}^k & \longrightarrow & \mathrm{M}(W, \mathbb{R}^k) & \longrightarrow & \mathrm{M}_0(W, T^k) & \longrightarrow & 1 \\
& & \uparrow = & & \uparrow & & \uparrow & & \\
1 & \longrightarrow & \mathbb{Z}^k & \longrightarrow & \Pi & \longrightarrow & Q & \longrightarrow & 1
\end{array}
$$

We take $\Pi$ as a subgroup of $\mathrm{M}(W, \mathbb{R}^k) \rtimes Q$. The coboundary $\delta \colon H^1(Q; \mathrm{M}_0(W, T^k)) \to H^2(Q; \mathbb{Z}^k)$ assigns the class $[\gamma]$ to the extension class $[\Pi]$. See, for example, [**Bro82**, p.94–95] or [**Lee83**] for an explicit construction.

Now one is able to show

EXERCISE 10.4.6. The $Q$-action on $T^k \times W$ has an extended lifting to $1 \to \mathbb{Z}^k \to \Pi \to Q \to 1$ on $\mathbb{R}^k \times W$ as in Theorem 10.4.2 if and only if $\gamma$ is in the image of $H^1(Q; \mathrm{M}_0(W, T^k))$ in $H^1(Q; \mathrm{M}(W, T^k))$.

COROLLARY 10.4.7. *Let $(T^k, X)$ be a locally injective action and $H = \mathrm{Im}(\mathrm{ev}_*^x \pi_1(T^k, 1)) \subset \pi_1(X, x)$. Then $(T^k, X_H)$ is a lift of $(T^k, X)$ to $X_H$ and the lifted action is free. Let $K = \ker(\pi_1(T^k, 1) \to H)$ and $G = T_K^k$ be the covering group of $T^k$ whose fundamental group is $K$. The covering transformations $T_K^k \to T^k$ are naturally isomorphic to $H$. Lift the action of $G$ to $\widetilde{X}_H = \widetilde{X}$, the universal covering of $X$. Show that the action of $G$ on $\widetilde{X}$ is free and factors*

$$(G, \widetilde{X}) \xrightarrow{H\backslash} (T^k, X_H) \xrightarrow{\pi_1(X,x)/H} (T^k, X).$$

(Actually this is contained in Lemma 10.4.3 already.)

EXAMPLE 10.4.8. Consider the bundle projection

$$(S^1, S^1 \times (S^1 \times \mathbb{R})) \xrightarrow{S^1\backslash} (S^1 \times \mathbb{R}).$$

Let the integers, $J$, act freely on $S^1 \times \mathbb{R}$ by just translation on the second factor. Assume $\varphi: J \to \mathrm{Aut}(S^1)$ is trivial. We shall show that there are $\mathbb{Z}$ strict equivalence classes of liftings of the $J$-action to $S^1 \times (S^1 \times \mathbb{R})$ but only one has an extended lifting of the $J$-action to $(\mathbb{R}, \mathbb{R} \times (S^1 \times \mathbb{R}))$.

Since $H^2(S^1 \times \mathbb{R}; \mathbb{Z}) = 0$ and $H^1(J; \mathrm{M}_0(S^1 \times \mathbb{R}, S^1)) = H^2(J; \mathbb{Z}) = 0$, we have that

$$0 \to H^1(J; \mathrm{M}(W, S^1)) \to H^1(J; H^1(S^1 \times \mathbb{R}^1; \mathbb{Z})) = \mathbb{Z} \to 0$$

is exact. Since $e_1$ injects, we have $H^1(J; \mathcal{T}^1) = \mathbb{Z}$. Each element of $\mathbb{Z}$ is a strict equivalence class of $Q(= J)$-actions on $S^1 \times (S^1 \times \mathbb{R})$. Observe that only the 0-class has an extended lifting to $(\mathbb{R}, \mathbb{R} \times (S^1 \times \mathbb{R}))$. The extended lift of the $J$-action representing the 0-class is a $\mathbb{Z} \times J$-action on $(\mathbb{R}^1, \mathbb{R}^1 \times S^1 \times \mathbb{R}^1)$ and is given by

$$(n_1, n_2) \times (r_1, z, r_2) \mapsto (r_1 + n_1, z, r_2 + n_2)$$

with $(n_1, n_2) \in \mathbb{Z} \times J$ and $\pi \cong \mathbb{Z}^3$.

If $[\gamma] = m \in \mathbb{Z}$ and $m \neq 0$, then $\theta(J)\backslash S^1 \times (S^1 \times \mathbb{R}) \to T^2$ is a principal $S^1$-bundle over $T^2$ with characteristic class (Euler class) $= \pm m$. It is a nilmanifold. (If $m = 0$, the principal bundle is the 3-torus.)

PROPOSITION 10.4.9 (cf. Corollary 10.1.10). *There exists a countably infinite number of $Q$-invariant principal $S^1$-bundles $P$ over each member $W$ of a family of simply connected 3-manifolds for which no central extension of $S^1$ by $Q$ embeds into $E(P, Q)$. That is, no $\theta: E \to E(P, Q)$ exists. Moreover, there are, over each $W$, an uncountable number of principal $S^1$-bundles which are not $Q$-invariant.*

PROOF. Let $Q$ be the fundamental group of any orientable closed aspherical 3-manifold $X$. Delete the interior of a nicely embedded 3-disk. Denote the manifold with 2-sphere boundary by $W^*$. The universal covering $W$ of $W^*$ is just the universal covering $\widetilde{X}$ of $X$ with all the lifts of the open 3-disk deleted. Since $W$ is simply connected, the terms of low degree of the $''E_2$ spectral sequence in Theorem 10.3.12 become

$$0 \to H^2(Q; \mathbb{Z}) \xrightarrow{e_1} H^2(Q; \mathcal{Z}) \xrightarrow{e_2} H^2(W; \mathbb{Z})^Q \xrightarrow{d} H^2(Q; \mathrm{M}(W, S^1)) \xrightarrow{e_3} H^3(Q; \mathcal{Z}).$$

We claim that the last term is 0; $e_1$ and $d$ are isomorphisms. Consequently, $e_2$ is trivial. Therefore only the product bundle over $W$ admits a lifting of $Q$. First,

$$H^n(Q; \mathcal{Z}^k) = H^n(Q\backslash W; \mathbb{Z}^k)$$

whenever $Q$ acts freely on $W$. For,

$$'E_2^{p,q} = H^p(Q\backslash W; \mathfrak{h}^q{}_{\mathbb{Z}}) \Longrightarrow H^n(Q; \mathcal{Z}^k).$$

Since $Q$ acts freely, $\mathfrak{h}^q{}_{\nu(w)} = 0$ for $q > 0$. So $'E_2$ collapses and $H^n(Q; \mathcal{Z}^k) = H^n(Q\backslash W; \mathfrak{h}^q{}_{\mathbb{Z}}) = H^n(Q\backslash W; \mathbb{Z}^k)$. (Alternatively, we can use the identification of $H^n(Q; \mathcal{Z}^k)$ with $H^n(W_Q; \mathbb{Z}^k)$. Note $W_Q$ fibers over $Q\backslash W$, since $Q$ acts freely, with fiber $EQ$. Therefore $H^n(Q\backslash W; \mathbb{Z}^k) = H^n(Q; \mathcal{Z}^k)$.) Since $Q\backslash W$ has the homotopy type of a 2-complex, $H^3(Q\backslash W; \mathbb{Z}) = 0$ and therefore, $e_3$ is trivial. Now,

$$H^2(Q; \mathbb{Z}) = H^2(X; \mathbb{Z}) \to H^2(Q\backslash W; \mathbb{Z}) = H^2(Q; \mathcal{Z})$$

is an isomorphism because $Q\backslash W$ is just $X$ minus the interior of a 3-ball. So, $e_1$ and consequently $d$ are isomorphisms.

The isomorphism assigns $[P] \in H^2(W; \mathbb{Z})^Q$ the element $[E(P,Q)] \in H^2(Q; \mathrm{M}(W, S^1)) \cong H^3(Q; \mathbb{Z}) \cong \mathbb{Z}$. Therefore, if $P$ is a nontrivial bundle and $[P] \in H^2(W; \mathbb{Z})^Q$, $[E(P,Q)]$ must be a nontrivial integer. Observe that $H^2(Q; S^1) = T^s$ for some $s$ equal to the rank of $H^2(Q; \mathbb{Z})$. $T^s$ maps to 0 under $H^2(Q; S^1) \to H^2(Q; \mathrm{M}(W, S^1))$ since the target group has no nontrivial divisible elements. Therefore, every central extension of $S^1$ by $Q$ embeds into the group $\mathrm{M}(W, S^1) \rtimes Q$ associated to the product bundle. However, assuming $s \neq 0$, none of these extensions embeds into any of the $E(P,Q)$ where $[E(P,Q)] \in \mathbb{Z} - \{0\}$; cf. Proposition 10.1.9. Note that since $H^2(W; \mathbb{Z})$ is uncountable, "most" of the $S^1$-bundles over $W$ (the non-$Q$-invariant $S^1$-bundles) admit no $E(P,Q)$'s. (Note $H^2(W; \mathbb{Z}) = \mathrm{Hom}(H_2(W; \mathbb{Z}), \mathbb{Z})$. But $H_2(W) = \pi_2(W)$ is countably infinitely generated (the boundary of each of the deleted 3-cells of $\widetilde{X}$). And $\mathrm{Hom}(\sum_{i=1}^{\infty}(\mathbb{Z})_i, \mathbb{Z})$ is uncountable.) $\square$

EXERCISE 10.4.10. What happens if we do the same construction but using 4- or higher dimensional aspherical manifolds?

## 10.5. Some aspects of the $'E^{p,q}$-spectral sequence

Some aspects of the Ep,q spectral sequence

In this section, we explore special situations where the information provided by the $'E$-spectral sequence is crucial.

10.5.1. Let $Q$ be a discrete group acting properly on $W$. Similar to equation (10.3.2) in Subsection 10.3.5, $\mathfrak{h}^0_{\mathbb{Z}}$ is a sheaf whose presheaf is given by

$$V \to H^0(Q; \mathrm{M}(\bar{\nu}^{-1}(V), \mathbb{Z}^k)) = \mathrm{M}(\bar{\nu}^{-1}(V), \mathbb{Z}^k)^Q$$

for each open $V \subset Q\backslash W$. If $\varphi : Q_w \to \mathrm{Aut}(\mathbb{Z}^k)$ is trivial for all $w \in W$, then the sheaf $\mathfrak{h}^0{}_{\mathbb{Z}}$ over $Q\backslash W$ is locally constant and locally isomorphic to $\mathbb{Z}^k$. Therefore,

$$H^p(Q\backslash W; \mathfrak{h}^0) \cong H^p_\varphi(Q\backslash W; \mathbb{Z}^k).$$

Globally, this local system, $\mathbb{Z}^k$, may be twisted, since $\varphi : Q \to \mathrm{Aut}(\mathbb{Z}^k)$ can still be nontrivial while $\varphi|_{Q_w}$ is trivial for all $w$. In particular, we have $\Gamma(W; \mathcal{Z}^k)^Q = H^0(Q; \mathcal{Z}^k) = H^0(Q\backslash W; \mathfrak{h}^0{}_{\mathbb{Z}})$.

10.5.2. Suppose we have a fibering as in Subsection 10.1.2 with $\phi : Q \to \mathrm{Aut}(T^k)$. Then for $w \in W$, the action of $Q_w$ at $W$ induces an action of $\Pi_w$ on the principle fiber $T_w^k$ over $w$, where $1 \to F \to \Pi_w \to Q_w \to 1$ is exact. From Subsection 4.4.1, we see that the action of $\Pi_w$ on $T_w^k$ is given by an affine action of $\Pi_w$ *contained in* $T^k \rtimes \mathrm{Aut}(T^k)$ on $T_w^k$. This action commutes with the $T^k$-action in an invariant neighborhood of $T_w^k$ if and only if $\phi|_{Q_w} : Q_w \to \mathrm{Aut}(T^k)$ is trivial. We say that the Seifert fibering $\tau : \Pi\backslash P \to Q\backslash W$ is a *local $T^k$-action* if for each $w \in W$, the homomorphism $\phi|_{Q_w}$ is trivial. Since $\mathrm{Aut}(T^k)$ is contained in $\mathrm{Aut}(\mathbb{R}^k)$ and $\mathrm{Aut}(\mathbb{Z}^k)$ is isomorphic to $\mathrm{Aut}(T^k)$, this condition is equivalent to saying that $\phi|_{Q_w} : Q_w \to \mathrm{Aut}(\mathbb{Z}^k)$ is trivial. (The automorphisms of $T^k$ are clearly isomorphic to the isomorphisms of $\pi_1(T^k) \cong \mathbb{Z}^k$.) Thus there is a global $T^k$-action on $\Pi\backslash P$ and $\tau$ is the orbit mapping if $\phi : Q \to \mathrm{Aut}(T^k)$ is trivial.

EXERCISE 10.5.3. [**CR69**, 9.4]. Let the discrete $Q$ act properly on $W$, and assume, for each $w \in W$, that $\varphi|_{Q_w} Q_w \to \mathrm{Aut}(\mathbb{Z}^k)$ is trivial. Show, using the $'E_2^{p,q}$-spectral sequence in Subsection 10.3.5, that the sequence

$$0 \to H^2(Q\backslash W; \mathbb{Z}^k) \xrightarrow{e_1} H^2(Q; \mathcal{Z}^k) \xrightarrow{e_2} H^0(Q\backslash W; \mathfrak{h}^2) \xrightarrow{d} H^3(Q\backslash W; \mathbb{Z}^k) \xrightarrow{e_3} H^3(Q; \mathcal{Z}^k)$$

is exact. (Hint: $H^1(Q_w; \mathbb{Z}^k) = 0$, for all $w \in W$, so $\mathfrak{h}^1$ becomes the 0-sheaf.)

EXERCISE 10.5.4. Let $Q$ act freely on $W$. Then $H^p(Q\backslash W; \mathbb{Z}^k) \xrightarrow{e_1} H^p(Q; \mathcal{Z}^k)$ is an isomorphism for $p \geq 0$. (The coefficients in $H^p(Q\backslash W; \mathbb{Z}^k)$ are simple if $\varphi : Q \to \mathrm{Aut}(\mathbb{Z}^k)$ is trivial.) In particular, $H^p(Q; \mathbb{Z}^k) \to H^p(Q\backslash W; \mathbb{Z}^k)$ is the composite of the edge homomorphisms

$$H^p(Q; \mathbb{Z}^k) \xrightarrow{''e_1} H^p(Q; \mathcal{Z}^k) \xrightarrow{'e_1^{-1}} H^p(Q\backslash W; \mathbb{Z}^k).$$

This is an isomorphism if $p = 1$, and is injective if $p = 2$ and $W$ is simply connected.

10.5.5 ($'e_2 : H^2(Q; \mathcal{Z}^k) \longrightarrow H^0(Q\backslash W; \mathfrak{h}_\mathbb{Z}^2)$). An element in $H^2(Q; \mathcal{Z}^k) = H^1(Q; \mathcal{T}^k)$ is represented by a cocycle $\{\gamma_{ji}\}$ on a coordinate bundle for a bundle $P$. Locally on a coordinate neighborhood, as in Subsection 10.3.3, the cocycle is given by $\gamma_{ii} \in Z^1(Q; \mathrm{M}(\bar{\nu}^{-1}(V_i), T^k))$. The edge homomorphism $'e_2 : H^1(Q; \mathcal{T}^k) \to H^0(Q\backslash W; \mathfrak{h}_T^1)$ sends $\gamma_{ji}$ to a section of $\mathfrak{h}^1$ over $Q\backslash W$. Locally this restricts from the 1-cocycle $\gamma_{ii} : Q \to \mathrm{M}(\bar{\nu}^{-1}(V_i), T^k)$ to $\gamma_{ii_v} : Q \to \mathrm{M}(\bar{\nu}^{-1}(v), T^k)$, or equivalently to the 1-cocycle $Q_w \to \mathrm{M}(w, T^k) = T^k$. On the cohomology, this is an element of

$$(\mathfrak{h}^1_{\bar{\nu}(w)})_T = H^1(Q_w; T^k) \cong H^2(Q_w; \mathbb{Z}^k) \cong (\mathfrak{h}^2_{\bar{\nu}(w)})_\mathbb{Z}.$$

Therefore, this section at $w$ defines the action of $Q_w$ on the fiber $T^k$ over $w$. It normalizes the left translational action of the fiber. Or, alternatively, the cocycle $(\gamma_{ii})_v$, where $\bar{\nu}(w) = v$, describes the action of $Q$ on the $T^k$-fiber over $\bar{\nu}^{-1}(v)$. Note, if $Q_w = 1$, the section at $\bar{\nu}^{-1}(v)$ has value the trivial homomorphism of $Q_w$ into $T^k$.

Recall from above that the 1-cocycle $('e_2[\gamma])_v : Q_w \to T^k$ becomes $d('e_2[\gamma])_v \in H^2(Q_w; \mathbb{Z}^k)$. This last cohomology class represents an extension $1 \to \mathbb{Z}^k \to \Pi_w \to Q_w \to 1$; see [**Bro82**, p.95, Exercise 2]. (The extension is central if and only if $\varphi|_{Q_w} : Q_w \to \mathrm{Aut}(T^k) = \mathrm{Aut}(\mathbb{Z}^k)$ is trivial.) The group $\Pi_w$ acts on the universal covering $\mathbb{R}^k$ of $T^k$ and is the extended lift to $\mathbb{R}^k$ of the $Q_w$-action on $T^k$. The action of $\Pi_w$, and consequently of $Q_w$ on $T^k$, is free if and only if $\Pi_w$ contains no nontrivial torsion. We have the following

PROPOSITION 10.5.6. *The lift of a $Q$-action on $W$ to a bundle $P$, represented by $[\gamma] \in H^2(Q; \mathcal{Z}^k)$ is free, if and only if, the extensions of $\mathbb{Z}^k$ by $Q_w$ determined by $('e_2[\gamma])_{\bar{\nu}(w)=v} \in H^2(Q_w; \mathbb{Z}^k)$ are torsion free, for all $v \in Q\backslash W$ and $w \in \bar{\nu}^{-1}(v)$.*

Note, if $\varphi|_{Q_w} : Q_w \to \mathrm{Aut}(\mathbb{Z}^k)$ is trivial for all $w$, then the lift of $Q$ acts freely on $P$, if and only if, $('e_2[\gamma])_{\bar{\nu}(w)=v} \in \mathrm{Hom}(Q_w, T^k) = H^1(Q_w; T^k)$ is injective, for each $w$.

Also, if $P$ is a product and $W$ is simply connected, then $[\gamma]$ is an element of $H^1(Q; \mathrm{M}(W, T^k)) \xrightarrow[\cong]{\delta} H^2(Q; \mathbb{Z}^k)$. This $Q$-action on $P$ has an extended lifting to $\mathbb{R}^k \times W$ whose extension class is

$$\delta[\gamma] : 1 \to \mathbb{Z}^k \to \Pi \to Q \to 1$$

determined by $\delta[\gamma]$; see Subsection 10.4.5. The $Q$-action is free on $P$ and the extended action is free on $\mathbb{R}^k \times W$, if and only if the pullback from $\delta[\gamma]$ is torsion free for each $Q_w$, $w \in W$. If $\varphi|_{Q_w} : Q_w \to \mathrm{Aut}(T^k)$ is trivial, then $\Pi_w$ torsion free is equivalent to $\Pi_w$ being free Abelian.

EXERCISE 10.5.7. Let $W$ be a point and $Q$ any finite group. Let $\varphi : Q \to \mathrm{Aut}(\mathbb{Z}^k)$. Examine both spectral sequences for this case and show $H^n(Q; \mathcal{Z}^k) = H^n_\varphi(Q; \mathbb{Z}^k)$. When does $Q$ act freely on $T^k$? Compare with Chapter 8.

10.5.8. The term $'E_2^{0,j} = H^0(Q\backslash W; \mathfrak{h}^j)$ is the group of sections over $Q\backslash W$ of the sheaf $(\mathfrak{h}^j)_{\mathcal{Z}}$. The stalk $\mathfrak{h}^j_{\bar{\nu}(w)}$ at $\bar{\nu}(w)$ is $H^j_\varphi(Q_w; \mathbb{Z}^k)$. If $w$ is replaced by $qw$, then $Q_w$ is replaced by $Q_{qw} = qQ_wq^{-1}$. In particular, when $j = 2$, each element $[\gamma] \in H^2(Q; \mathcal{Z}^k)$ determines via $'e_2[\gamma]_{\bar{\nu}(w)} \in H^2(Q_w; \mathbb{Z}^k)$ a group extension, $1 \to \mathbb{Z}^k \to \Pi_w \to Q_w \to 1$, one for each point $w \in W$; see Proposition 10.5.6.

Now assume that $Q$ is acting properly on $W$ such that all but a finite number of orbits are free orbits. On $V = Q\backslash W$, let $\{v_1, \ldots, v_n\}$ be the image of the nonfree orbits. Choose a point $w_\ell$ in $\bar{\nu}^{-1}(v_\ell)$, $\ell = 1, \ldots, n$, and assume $\varphi : Q_w \to \mathrm{Aut}(\mathbb{Z}^k)$ is trivial, for all $w \in W$.

We claim: *All the terms not on the edges of the first quadrant spectral sequence $'E_2^{i,j}$ are 0.* That is, $H^i(Q\backslash W; \mathfrak{h}^j) = 0$, if $i > 0$ and $j > 0$. Note, if $v \notin \{v_1, \ldots, v_n\}$, then $\mathfrak{h}^j_v = 0$, $j > 0$. That is, the stalks of the sheaf $\mathfrak{h}^j$ is 0 everywhere except over $\{v_1, \ldots, v_n\}$. The cohomology of such a sheaf locally concentrated on a discrete set of points is 0. That is, $H^i(Q\backslash W; \mathfrak{h}^j) = 0$, if $i > 0$, $j > 0$. Therefore, the only nonzero terms of $'E_2^{i,j}$ lie on the coordinate axis. Now consider $H^0(Q\backslash W; \mathfrak{h}^j)$. This is the group of sections of the sheaf $\mathfrak{h}^j$ which is $\prod_{\ell=1}^n H^j(Q_{w_\ell}; \mathbb{Z}^k)$. $H^i(Q\backslash W; \mathfrak{h}^0_{\mathbb{Z}^k})$ is $H^i(Q\backslash W; \mathbb{Z}^k)$. The coefficient system is locally constant and is constant if $\varphi : Q \to \mathrm{Aut}(\mathbb{Z}^k)$ is trivial.

10.5.9. With such a spectral sequence as in Subsection 10.5.8, we get the long exact sequence

$$\to H^i(Q; \mathcal{Z}^k) \xrightarrow{'e_2} \prod_{\ell=1}^n H^i(Q_{w_\ell}; \mathbb{Z}^k) \xrightarrow{d} H^{i+1}(Q\backslash W; \mathbb{Z}^k) \xrightarrow{'e_1} H^{i+1}(Q\backslash W; \mathcal{Z}^k) \xrightarrow{'e_2}$$

PROPOSITION 10.5.10. *With $Q$ as in Subsection 10.5.8 and $W$ simply connected, we can combine the two spectral sequences in low degrees to get the following diagram* *of exact sequences.*

$$
\begin{array}{c}
0 \\
\downarrow \\
H^2_\varphi(Q; \mathcal{Z}^k) \\
{\scriptstyle \varphi} \downarrow {\scriptstyle ''e_1} \\
0 \to H^2_\varphi(Q\backslash W; \mathcal{Z}^k) \xrightarrow{'e_1} H^2(Q; \mathcal{Z}^k) \xrightarrow{'e_2} \prod_{\ell=1}^n H^2(Q_{w_\ell}; \mathbb{Z}^k) \xrightarrow{d} H^3_\varphi(Q\backslash W; \mathcal{Z}^k) \xrightarrow{'e_1} H^3(Q; \mathcal{Z}^k) \to \\
\downarrow {\scriptstyle ''e_2} \\
H^2_\varphi(W; \mathbb{Z}^k)^Q \\
\downarrow d \\
H^3_\varphi(Q; \mathcal{Z}^k) \\
\downarrow \\
H^3(Q; \mathcal{Z}^k)
\end{array}
$$

Note, if $W$ is not simply connected, we replace $H^i_\varphi(Q; \mathbb{Z}^k)$ by $H^{i-1}_\varphi(Q; \mathrm{M}(W, T^k))$, $i = 2$ and $3$ and still retain exactness. If we also drop the condition that all but a finite number of orbits are free on $W$, then we must replace $\prod_{\ell=1}^n H^2(Q_{w_\ell}; \mathbb{Z}^k)$ by $H^0(Q\backslash W; \mathfrak{h}^2)$ and terminate the horizontal exact sequence at $H^3(Q; \mathcal{Z}^k)$; see Chapter 15 for further ramifications of the diagram above.

10.5.11. An important special situation occurs when $W$ is contractible and $Q$ discrete acts properly on $W$. Then every principal $T^k$-bundle over $W$ is trivial. Furthermore, $H^j(W; \mathcal{Z}^k) = 0$ and so $H^n_\varphi(Q; \mathbb{Z}^k) \xrightarrow{''e_1} H^n(Q; \mathcal{Z}^k)$ is an isomorphism for all $n > 0$.

CHAPTER 11

# Applications

The injective Seifert Construction, which is a special embedding, $\theta : \Pi \to \text{TOP}_G(G \times W)$, of the group $\Pi$ into $\text{TOP}_G(G \times W)$ such that $\Pi$ acts properly on $G \times W$, preserves some of the properties of both $G$ and $W$ on $\theta(\Pi)\backslash(G \times W)$. Furthermore, the action of $\Pi$ on $G \times W$ *twists* the topology and geometry of $G$ and $W$ to create the orbit space $\theta(\Pi)\backslash(G \times W)$ in the same way that the group structures of $\Gamma$ and $Q$ *twists* to create the group $\Pi$. In other words, this algebraic twisting of $\Pi$ makes the geometric twisting of the *bundle with singularities*

$$\Gamma\backslash G \to \theta(\Pi)\backslash(G \times W) \to Q\backslash W,$$

where the homogeneous space $\Gamma\backslash G$ is a typical fiber. In the several applications, we have included here these features seem especially prominent.

One of the important topological problems that has motivated the development of Seifert fiberings is the construction of closed aspherical manifolds realizing Poincaré duality groups $\Pi$ of the form $1 \to \Gamma \to \Pi \to Q \to 1$. The Seifert Construction enables one to find explicit aspherical manifolds $M(\Pi)$ when $Q$ acts on a contractible manifold $W$ and $\Gamma$ is a torsion-free lattice in a Lie group; see Section 11.1.

The rigidity of the Seifert Construction is important for classification problems. We exhibit, in Section 11.2, several instances where rigidity is useful for classification.

For a topological manifold, the homotopy classes of self-homotopy equivalences can be regarded as algebraic data. We show, in Section 11.3, how the Seifert Construction can often be used to lift finite subgroups of homotopy classes to an action on the manifold. The lifting problem consists of two stages. First, one has an abstract kernel $\psi : G \to \text{Out}(\pi_1(M))$ that must be realized as a group extension. If this fails, no lifting of $G$ is possible. In order for an extension to exist, a certain 3-dimensional cohomology class must be 0. When this fails and $\mathcal{Z}(\pi_1(M))$ is finitely generated free Abelian, we show in Theorem 11.3.30 that there is an *inflation*, $H \xrightarrow{\text{inf}} G \xrightarrow{\psi} \text{Out}(\pi_1(M))$ for which the obstruction for the existence of a group extension realizing the abstract kernel $\text{inf} \circ \psi : H \to \text{Out}(\pi_1(M))$ vanishes. Then the method of Theorem 7.3.2 used for Seifert Constructions can be invoked for completing the second stage for the group $H$.

The interaction between the fundamental group and geometry is especially strong for aspherical manifolds. A representation $\theta : \Gamma \to \text{Aff}(\mathbb{R}^k)$ which yields a proper action with $\theta(\Gamma)\backslash\mathbb{R}^k$ compact is called an *affine structure* on $\Gamma$. Analogously, $\theta : \Gamma \to P(\mathbb{R}^k)$, where $P(\mathbb{R}^k)$ is the group of all polynomial diffeomorphisms of $\mathbb{R}^k$, is called a *polynomial structure* on $\Gamma$. The affine diffeomorphisms are polynomial diffeomorphisms of degree less than or equal to 1. Not all torsion-free polycyclic-by-finite groups admit an affine structure but, as sketched in Section 11.4, they do admit a polynomial structure.

The close connection between the Bieberbach theorems and the existence, uniqueness, and rigidity of the Seifert Construction is explored. The rigidity of homeomorphisms between infra-nilmanifolds is extended to continuous maps. This enables us to recapture some recent results of Nielsen fixed point theory on infra-nilmanifolds and infra-solvmanifolds; see Section 11.5.

As mentioned earlier, a torus action $(T^k, X)$ is *homologically injective* if the evaluation map is injective on the first homology group. This is a much stronger concept than being injective on the fundamental group. Homologically injective actions are characterized by being splittable as $(T^k, X) = (T^k, T^k \times_\Delta Y)$, where $\Delta$ is a finite Abelian group acting diagonally and freely as translations on the first factor. Consequently, $X$, which fibers over $G \backslash X = \Delta \backslash Y$, also fibers without singularities over $T^k/\Delta$ with fiber $Y$. There is also induced a $T^k$-equivariant map $(T^k, X) \to (T^k, T^k/\Delta)$. In Section 11.6, we characterize this splitting and also extend the theorems to Seifert fiberings with cocompact lattices in simply connected nilpotent Lie groups. As one would expect, the strong condition leads to interesting examples in topology and geometry, and some of these are also described in this section. These splitting theorems provide analogue for lattices of type (S1) and (S2) to the splitting theorems in Chapter 9, Corollary 9.4.6.

If $M$ is a closed aspherical manifold and $C = \mathcal{Z}(\pi_1(M))$, it is unknown, in general, if there exists an effective torus action on $M$ for which $\operatorname{Im}(\operatorname{ev}_*^x(\pi_1(T^k))) = C$. Such an action, if it exists, is called a maximal torus action on $M$. In Section 11.7, we construct maximal torus actions for a large class of aspherical Seifert manifolds, including solvmanifolds and double coset spaces. In Section 11.8, we construct the analogue of maximal torus actions on spherical space forms.

This chapter consists of the following sections:
(1) Existence of closed $K(\Pi, 1)$-manifolds
(2) Rigidity for Seifert fiberings
(3) Lifting problem for homotopy classes of self-homotopy equivalences
(4) Polynomial structures for solvmanifolds
(5) Applications to fixed-point theory
(6) Homologically injective torus operations
(7) Maximal torus actions on solvmanifolds and double coset spaces
(8) Toral rank of spherical space forms

## 11.1. Existence of closed $K(\Pi, 1)$-manifolds

11.1.1. There are two difficult problems related to the title:
(1) *Which groups can be the fundamental group of a closed aspherical manifold?*
(2) *If $\Pi$ is the fundamental group of an aspherical manifold, can we give an actual explicit construction of an aspherical manifold for the group $\Pi$?*

There are some general criteria for the first problem, such as $\Pi$ must be finitely presented, have finite cohomological dimension, and satisfy the Poincaré duality in that dimension. The Seifert Construction gives answers to both questions for a large class of groups $\Pi$. The idea is that if $1 \to \Gamma \to \Pi \to Q \to 1$ is a torsion-free extension where $\Gamma$ is the fundamental group of a closed aspherical manifold and

$Q$ is a proper action on a contractible manifold $W$ with compact quotient, then $\Pi$ should be the fundamental group of a closed aspherical manifold. We have the following

THEOREM 11.1.2. *Let $\Gamma$ be a cocompact special lattice in $G$; see Section 7.3.1, and let $\rho : Q \to \mathrm{TOP}(W)$ be a proper action of a discrete group on a contractible manifold $W$ with compact quotient. If $1 \to \Gamma \to \Pi \to Q \to 1$ is a torsion-free extension of $\Gamma$ by $Q$, then for any Seifert Construction $\theta : \Pi \to \mathrm{TOP}_G(G \times W)$,*

(1) *$M(\theta(\Pi)) = \theta(\Pi) \backslash (G \times W)$ is a closed aspherical manifold if $\Gamma$ is of type (S3), and*
(2) *$M(\theta(\Pi)) = \theta(\Pi) \backslash ((G/K) \times W)$, where $K$ is a maximal compact subgroup of $G$, is a closed aspherical manifold if $\Gamma$ is of type (S4).*

PROOF. For each extension $1 \to \Gamma \to \Pi \to Q \to 1$, there exists a homomorphism $\theta$ of $\Pi$ into $\mathrm{TOP}_G(G \times W)$, by Theorem 7.3.2 for $G$ of type (S3), (respectively, $\mathrm{TOP}_{(G,K)}(G/K \times W)$ for $G$ of type (S4)).

Since $\Pi$ acts properly on $G \times W$ (or $G/K \times W$), which is contractible and is torsion free, $\Pi$ must act freely since any isotropy subgroup must be finite. We need only to check that $\theta$ is injective. Suppose $Q_0$ is the kernel of $\psi \times \rho : Q \to \mathrm{Out}(G) \times \mathrm{TOP}(W)$. Then $Q_0$ is finite since the $Q$-action on $W$ is proper. Let $1 \to \Gamma \to \Pi_0 \to Q_0 \to 1$ be the pullback via $Q_0 \subset Q$. By Corollary 7.7.4, $\theta$ is injective if and only if $\Pi_0$ is torsion free. But the group $\Pi_0$ is torsion free since $\Pi$ is assumed to be torsion free. $\square$

REMARK 11.1.3. (1) If $W$ is a smooth contractible manifold and $\rho : Q \to \mathrm{Diff}(W)$, then the construction can be done smoothly and $M(\theta(\Pi))$ is smooth.

(2) If $\rho_1$ and $\rho_2$ are *rigidly related* (i.e., there exists $h \in \mathrm{TOP}(W)$ for which $\rho_2 = \mu(h) \circ \rho_1$) and $\Gamma$ is characteristic in $\Pi$, then $M(\theta_1(\Pi))$ and $M(\theta_2(\Pi))$ are homeomorphic via a Seifert automorphism; see Remark 7.4.4. Moreover, if we fix $\ell$ and $\rho$, then the constructed $M(\theta_i(\Pi))$ are all *strictly equivalent*; see Section 7.4, especially Subsection 7.4.2.

(3) When $W = \{p\}$ is a point (a 0-dimensional contractible manifold), then $Q$ must be finite for $Q$ to act properly, and every $\rho : Q \to \mathrm{TOP}(\{p\})$ is rigidly related. The closed aspherical manifolds constructed are infra-$G$-manifolds; cf. Example 7.4.5.

(4) One important application of these constructions is that they provide model aspherical manifolds with often strong geometric properties. If one wants to study the famous conjecture that two closed aspherical manifolds with isomorphic fundamental groups are homeomorphic via the methods of controlled surgery, then the constructed aspherical Seifert manifolds are excellent model manifolds.

This point of view has been taken by H. Rees in his thesis [**Ree83**]; cf. [**HR83**], [**FH83**], and A. Nicas and C. Stark in [**NS85**]. A consequence of the last reference is that if $M^{n+2}$ is a closed aspherical manifold admitting a codimension-two torus action and $f : N \to M$ is a homotopy equivalence with $N$ a closed manifold, then $f$ is homotopic to a homeomorphism provided that $n \neq 3$ or 4.

(5) In order for $M(\theta(\Pi))$ to be closed aspherical in Theorems 11.1.2 and 11.1.4, $W$ needs only be a contractible manifold factor rather than an actual manifold. By a *manifold factor*, we mean a space $W$ such that $W \times \mathbb{R}^1$ is homeomorphic to a topological manifold.

If $W$ is a noncontractible manifold factor, the Seifert construction still produces a topological manifold $M(\theta(\Pi))$ provided $\Pi$ acts freely on $G \times W$ or $G/K \times W$. The group $\Pi$ acts freely if and only if $\Pi_w$ in the extension

$$1 \to \Gamma \to \Pi_w \to Q_w \to 1$$

is torsion free, for each $w \in W$. This extension is the pullback from $1 \to \Gamma \to \Pi \to Q \to 1$, induced by the inclusion of $Q_w$ into $Q$.

The above procedure can be extended for even more general extensions. As an example,

THEOREM 11.1.4. *Let $\Pi$ be a torsion-free extension of a virtually poly-$\mathbb{Z}$ group $\Gamma$ by $Q$, where $Q$ acts on a contractible manifold $W$ properly with compact quotient. Then there exists a closed $K(\Pi, 1)$-manifold.*

PROOF. A torsion-free virtually poly-$\mathbb{Z}$ group $\Gamma$ has a unique maximal normal nilpotent subgroup $\Delta$, which is called the *discrete nilradical* of $\Gamma$; see Subsection 8.4.6. Then the quotient $\Gamma/\Delta$ is virtually free Abelian of finite rank. Furthermore, since $\Delta$ is a characteristic subgroup of $\Gamma$, it is normal in $\Pi$. Consider the commuting diagram with exact rows and columns:

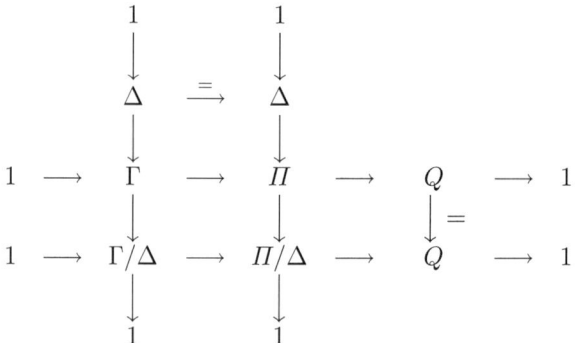

Since $\Gamma/\Delta$ is virtually free Abelian of finite rank (say, of $s$), it contains a characteristic subgroup $\mathbb{Z}^s$. (This can be seen as follows: Let $\mathbb{Z}^s$ be a normal subgroup of $\Gamma/\Delta$ of finite index $q$. Let $\Gamma(q)$ be the subgroup of $\Gamma/\Delta$ generated by $\{x^q : x \in \Gamma/\Delta\}$. Then clearly, $\Gamma(q)$ is a free Abelian group and is a characteristic subgroup of $\Gamma/\Delta$.) Let $Q' = (\Pi/\Delta)/\mathbb{Z}^s$. Then the natural projection $Q' \longrightarrow Q$ has a finite kernel. Therefore, if we let $Q'$ act on $W$ via $Q$, the action will still be proper.

One can do a Seifert fiber space construction with the exact sequence

$$1 \longrightarrow \mathbb{Z}^s \longrightarrow \Pi/\Delta \longrightarrow Q' \longrightarrow 1,$$

which yields a proper action of $\Pi/\Delta$ on $\mathbb{R}^s \times W$ with compact quotient. Using this action of $\Pi/\Delta$ on $\mathbb{R}^s \times W$, one does a Seifert fiber space construction with the exact sequence

$$1 \longrightarrow \Delta \longrightarrow \Pi \longrightarrow \Pi/\Delta \longrightarrow 1.$$

This gives rise to a proper action of $\Pi$ on $N \times (\mathbb{R}^s \times W)$, where $N$ is the unique simply connected nilpotent Lie group containing $\Delta$ as a lattice, with compact quotient.

If the space $W$ is smooth, and the action of $Q$ on $W$ is smooth, both constructions can be done smoothly so that the proper action of $\Pi$ on $N \times (\mathbb{R}^s \times W)$ is smooth.

In any case, since the group $\Pi$ is torsion free, the resulting action of $\Pi$ on $N \times (\mathbb{R}^s \times W)$ is free. Consequently, we get a closed $K(\Pi, 1)$-manifold
$$M = \Pi\backslash(N \times \mathbb{R}^s \times W).$$
It has a Seifert fiber structure
$$F \longrightarrow M \longrightarrow Q\backslash W,$$
where the typical fiber $F$ itself has a Seifert fiber structure
$$\Delta\backslash N \longrightarrow F \longrightarrow T^s = \mathbb{Z}^s\backslash\mathbb{R}^s.$$
In fact, since the action of the characteristic subgroup $\mathbb{Z}^s$ on $\mathbb{R}^s$ is free, $F$ is a genuine fiber bundle, with fiber a nilmanifold $\Delta\backslash N$ over the base torus $T^s$. $\square$

The space $W$ does not have to be aspherical. As long as the action of discrete $Q$ is proper, this iterated construction works. The resulting action of $\Pi$ is free if and only if the preimage of $Q_w$ (the isotropy of the $Q$-action at $w \in W$) in $\Pi$ is torsion free. In this case, the space $\Pi\backslash(G \times W)$ will not be aspherical; see Theorem 7.3.2 and cf. also Theorems 9.5.6 and 9.5.7.

11.1.5. In a slightly different vein, Frank Johnson [**Joh78**] has defined the notion of a poly $\mathbb{L}_+$-group:
• A group is in $\mathbb{L}$ if it is a discrete uniform subgroup in a connected finite covering group of a semisimple Lie group of type (S4).
• A group is in $\mathbb{L}_+$ if either it is in $\mathbb{L}$ or it is virtually poly-$\mathbb{Z}$. A *poly-$\mathbb{L}_+$-group* $\Pi$ is a group having a filtration $1 = \Pi_0 \subset \Pi_1 \subset \Pi_2 \subset \cdots \subset \Pi_k = \Pi$, where $\Pi_i$ is normal in $\Pi_{i+1}$ and $\Pi_{i+1}/\Pi_i \in \mathbb{L}_+$.

Johnson made two other assumptions about the group $\Pi$. In [**LR84**, section 4.6], these assumptions are shown to be redundant, and a proof of the Johnson theorem is given there using the techniques of Seifert fiberings.

THEOREM 11.1.6. *If $\Pi$ is a torsion-free poly-$\mathbb{L}_+$-group, then there exists a closed smooth $K(\Pi, 1)$-manifold.*

EXAMPLE 11.1.7. (Codimension-2 injective Seifert fiberings with the torus $T^k$ as typical fiber). As an illustration of the foregoing sections, let $Q$ act properly and effectively on $\mathbb{R}^2 = W$ with compact quotient. Then $Q$ can be topologically conjugated into the group of Euclidean motions if $Q$ is solvable and to the group of hyperbolic isometries if $Q$ is not solvable. For each extension $1 \to \mathbb{Z}^k \to \Pi \to Q \to 1$, there is a Seifert Construction:

$$\begin{array}{ccccccccc}
1 & \longrightarrow & \mathbb{Z}^k & \longrightarrow & \Pi & \longrightarrow & Q & \longrightarrow & 1 \\
& & \downarrow{\scriptstyle \epsilon} & & \downarrow{\scriptstyle \theta} & & \downarrow{\scriptstyle \varphi \times \rho} & & \\
1 & \longrightarrow & \mathrm{M}(\mathbb{R}^2, \mathbb{R}^k) & \longrightarrow & \mathrm{TOP}_{\mathbb{R}^k}(\mathbb{R}^k \times \mathbb{R}^2) & \longrightarrow & \mathrm{Aut}(\mathbb{R}^k) \times \mathrm{TOP}(\mathbb{R}^2) & \longrightarrow & 1.
\end{array}$$

Let $M(\Pi)$ denote the space $\theta(\Pi)\backslash\mathbb{R}^k \times \mathbb{R}^2$. Since $\rho$ is injective, $\theta$ is injective. The mapping $M(\Pi) \to Q\backslash\mathbb{R}^2$ is a Seifert fibering with typical (and also regular) fiber $T^k = \mathbb{Z}^k\backslash\mathbb{R}^k$. The base of the fibering $B = Q\backslash\mathbb{R}^2$ is a 2-dimensional Euclidean or hyperbolic orbifold. If $\varphi : Q \to \mathrm{Aut}(\mathbb{R}^k)$ is trivial, the fibering is the orbit mapping of an injective $T^k$-action on $M(\Pi)$. Furthermore, $\Pi$ is torsion free if and only if $\Pi$ acts freely on $\mathbb{R}^k \times \mathbb{R}^2$, in which case $M(\Pi)$ is an aspherical manifold. If $\Pi$ is not torsion free, then $M(\Pi)$ is an orbifold.

The group $Q$ contains no nontrivial normal Abelian subgroups when $Q$ is not solvable. Therefore, $\mathbb{Z}^k \subset \Pi$ is the characteristic maximal normal Abelian subgroup of $\Pi$. Since $Q$ is topologically rigid, the Seifert Construction, $\theta$, is unique and rigid in the sense of Theorem 7.3.2(3). Thus, any isomorphism $\theta(\Pi_1) \to \theta(\Pi_2)$ can be realized by a Seifert isomorphism. That is, $M(\Pi_1)$ is homeomorphic to $M(\Pi_2)$ via a Seifert fiber preserving homeomorphism induced by a conjugation of $\theta(\Pi_1)$ to $\theta(\Pi_2)$ by an element of $\mathrm{TOP}_{\mathbb{R}^k}(\mathbb{R}^k \times \mathbb{R}^2)$.

If the fibering is the orbit mapping of a $T^k$-action (i.e., $\varphi : Q \to \mathrm{Aut}(\mathbb{R}^k)$ is trivial), then the Seifert isomorphism is a weak $T^k$-equivalence. Similar results hold for $Q$ solvable and centerless because $\mathrm{ev}^x_*(\pi_1(T^k)) = \mathbb{Z}^k$ is the center and is a characteristic subgroup of $\pi_1(M)$.

11.1.8. In Example 11.1.7, if $k = 1$ and $\Pi$ is torsion free, the Seifert Construction will produce all the possible closed aspherical Seifert 3-manifolds with typical fiber $S^1$. We call all of the Seifert 3-manifolds listed by Seifert in [**Sei33**], (see [**ST80**] for a translation into English), the *classical* Seifert 3-manifolds. For Seifert, each fiber is an $S^1$ with a tubular neighborhood a Seifert fibered solid torus $(S^1, S^1 \times_{\mathbb{Z}_n} D^2)$ where $\mathbb{Z}_n$ acts diagonally by translating freely on the first factor and by rotation on the second factor. The base is a closed surface with a finite number of *cone* singularities. Seifert does not consider *special exceptional* fibers which are $S^1$-fibers with a tubular neighborhood $(S^1, S^1 \times_{\mathbb{Z}_2} D^2)$, which is a solid Klein bottle (that is, a Möbius band $\times I$), where $\mathbb{Z}_2$ acts freely on $S^1$ and by reflection on $D^2$. With special exceptional fibers, the base is a surface with boundary, with possible cone singularities in the interior of the base, where the inverse image of the boundary points are all special exceptional fibers. Any Seifert 3-manifold, not necessarily classical, having a normal $\mathbb{Z}$ in its fundamental group $\Pi$ and quotient group $G = \Pi/\mathbb{Z}$ infinite, will be aspherical.

A thorough investigation of Seifert 3-manifolds can be found in Chapters 14 and 15. There the classification in terms of explicit numerical invariants is exploited to derive significant connections with other lower dimensional phenomena.

11.1.9. In line with the methods of this section, we state two problems where a complete answer is unknown.

(1). If $M$ is a closed aspherical manifold with $1 \to \Gamma \to \pi_1(M) \to Q \to 1$ exact and $\Gamma$ is a cocompact lattice in a group of type (S3) or (S4), does there exist a contractible manifold $W$ on which $Q$ acts properly with compact quotient?

(2) If $Q$ is a discrete group which acts properly and cocompactly on a contractible manifold, does $Q$ have a torsion-free subgroup of finite index?

11.1.10. A large and important class of closed aspherical manifolds has been constructed, using Coxeter groups, by M. Davis [**Dav83**]; see also Remark 3.1.20. They exhibit properties quite different from the aspherical manifolds constructed with the aid of Lie groups. Some of these manifolds are not smoothable nor even admit a PL structure, and their universal coverings are not homeomorphic to Euclidean space. Davis's methods, when combined with the methods of this section, lead to an even larger class of closed aspherical manifolds.

## 11.2. Rigidity of Seifert fibering

11.2.1. The observant reader will have noticed that in Theorem 11.1.2, we did not require that $\rho : Q \to \mathrm{TOP}(W)$ be injective as we did in our illustrations with

## 11.2. RIGIDITY OF SEIFERT FIBERING

$W = \mathbb{R}^2$. For $k = 1$, no torsion-free $\Pi$ occurs that was not already detected when $\rho$ is injective. This is not the case when $k > 1$. In general (see Subsection 7.7.1), for $\theta$ to be injective and $\Pi \cap \mathbb{R}^k = \mathbb{Z}^k$, we can assume, without loss of generality, that $\varphi \times \rho$ is injective.

Consider an example in dimension 4: for the extension $1 \to \mathbb{Z}^2 \to \Pi \to \mathbb{Z}_2 \times Q \to 1$ with $Q$ a surface group, where $\varphi \times \rho : \mathbb{Z}_2 \times Q \to \operatorname{Aut}(\mathbb{Z}^2) \times \operatorname{TOP}(\mathbb{R}^2)$, we assume $\varphi$ is injective on $\mathbb{Z}_2$ and trivial on $Q$, and $\rho$ is trivial on $\mathbb{Z}_2$ and injective on $Q$. (We can choose $\Pi$ so that $\Pi = \pi_1(\text{Klein bottle}) \times Q$ and $\theta(\Pi)\backslash \mathbb{R}^4 =$ Klein bottle $\times Q\backslash \mathbb{R}^2$. The typical fiber is the 2-torus, but each fiber is a regular fiber, the Klein bottle.)

Let $\Gamma$, $\Pi$, $Q$ be as in Theorem 11.1.2. Let $\Pi_L \subset \operatorname{Aff}(G)$ be an extension of a lattice $\Gamma = \Pi_L \cap \ell(G)$ in the completely solvable Lie group $G$ by a finite group $L$. Let $1 \to \Pi_L \to \Pi \xrightarrow{j'} Q' \to 1$ be an extension where $\rho' : Q' \to \operatorname{TOP}(W)$ is an effective proper action.

PROPOSITION 11.2.2. *There exists an injection $\theta : \Pi \to \operatorname{TOP}_G(G \times W)$ such that $\theta|_{\Pi_L} : \Pi_L \to \operatorname{Aff}(G) \subset \operatorname{TOP}_G(G \times W)$ and $\Pi_L\backslash G$ is a regular fiber for the Seifert Construction $\theta$.*

PROOF. The group $\Gamma$ is a characteristic subgroup of $\Pi_L$ and so $1 \to \Gamma \to \Pi \to \Pi/\Gamma = Q \to 1$ is exact. The natural map $Q = \Pi/\Gamma \to \Pi/\Pi_L = Q' \subset \operatorname{TOP}(W)$ has kernel $L$. We have the commutative diagram of exact sequences

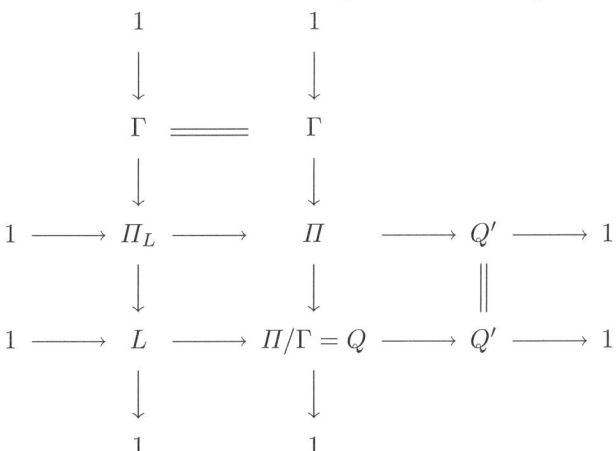

We use the middle vertical sequence to make the Seifert Construction. □

The proposition shows that replacing the lattice $\Gamma$ by the larger $\Pi_L$ does not lead to any new Seifert fiberings.

11.2.3. Let $1 \to \Gamma \to \Pi \to Q \to 1$ be an extension of discrete groups where $\Gamma$ is isomorphic to a special lattice in a connected Lie group $G$ of type (S3); see Subsection 7.3.1. Let $Q$ act properly on a space $W$ via the homomorphism $\rho$, and let $\ell$ be an isomorphism of $\Gamma$ onto a lattice of $G$. Then the existence part of Theorem 7.3.2 asserts that there exists a homomorphism $\theta : \Pi \to \operatorname{TOP}_G(G \times W)$ such that $\theta|_\Gamma = \ell$ and we have the commutative diagram (7.3.1). This yields a Seifert fibering $M = \theta(\Pi)\backslash(G \times W) \to Q\backslash W = B$ with typical fiber $\Gamma\backslash G$. The theorem also says the construction is unique. That is, congruent extensions are conjugate in $\operatorname{M}(W, G) \subset \operatorname{TOP}_G(G \times W)$, Theorem 7.3.2(2); i.e, $M_1$ and $M_2$ are strictly equivalent. Furthermore, for $i = 1, 2$, if there are extensions fitting into the diagram

$$\begin{array}{ccccccccc}
1 & \longrightarrow & \Gamma_i & \longrightarrow & \Pi_i & \longrightarrow & Q_i & \longrightarrow & 1 \\
& & \downarrow \ell_i & & \downarrow \theta_i & & \downarrow \varphi_i \times \rho_i & & \\
1 & \longrightarrow & \ell(G) \rtimes \mathrm{Inn}(G) & \longrightarrow & \mathrm{TOP}_G(G \times W) & \longrightarrow & \mathrm{Out}(G) \times \mathrm{TOP}(W) & \longrightarrow & 1
\end{array}$$

and an isomorphism $\eta : \Pi_1 \to \Pi_2$ so that $\eta|_{\Gamma_1} : \Gamma_1 \to \Gamma_2$ is an isomorphism, inducing an isomorphism $\overline{\eta} : Q_1 \to Q_2$, and a homeomorphism $h \in \mathrm{TOP}(W)$ so that $\mu(h) \circ \rho_1 = \rho_2 \circ \overline{\eta}$, then there exists a conjugation $\mu(\lambda, a, h)$ so that $\theta_2 \circ \eta = \mu(\lambda, a, h) \circ \theta_1$.

This is rigidity of the Seifert Construction and implies that there is induced a Seifert isomorphism $M_1 \to M_2$. Obviously, in classifying Seifert fiberings, rigidity can play an important role and so we shall give some practical conditions that allow us to verify that the hypothesis of rigidity holds.

For our problem it suffices to check when the $\theta_i$ are injective. Corollary 7.7.4 tells how to recognize the kernels of $\theta_i$. So we assume the $\theta_i$ are injective and without loss of generality, $\varphi_i \times \rho_i$ are injective. This implies $\theta_i(\Gamma_i) = \theta(\Pi_i) \cap \ell(G)$.

THEOREM 11.2.4 (cf. [**Ray79**, Section 3]). *Let* $1 \to \Gamma_i \to \Pi_i \to Q_i \to 1$ ($i = 1, 2$), *be extensions of discrete groups where* $\Gamma_i$ *are isomorphic to special lattices in a connected Lie group* $G$ *of type* (S3). *Let* $Q_i$ *act properly on a space* $W$ *via the homomorphisms* $\rho_i$ *with* $\rho_i(Q)\backslash W$ *compact, and* $\ell_i$ *isomorphisms of* $\Gamma_i$ *onto lattices of* $G$. *Let* $\theta_i : \Pi_i \to \mathrm{TOP}_G(G \times W)$ *be Seifert Constructions. Suppose one of the following conditions on* $(Q_i, W)$ *holds:*

(i) $W = \mathbb{R}^2$ *and* $\rho_i(Q_i)$ *is not solvable.*
(ii) $W$ *is a Riemannian symmetric space of noncompact type (Section 9.4) with no compact factors and no 1-dimensional factors,* $\rho_i(Q_i) \subset \mathrm{Isom}(W)$, $Q_i\backslash W$ *compact. If* $W$ *has 2-dimensional factors, then the projection of* $\rho_i(Q_i)$ *to those factors are dense.*

*Then any isomorphism* $\eta : \Pi_1 \to \Pi_2$ *induces a Seifert isomorphism*

$$\theta_1(\Pi_1)\backslash(G \times W) \to \theta_2(\Pi_2)\backslash(G \times W).$$

For the proof of the theorem, we need to show that the isomorphism $\eta$ restricts to an isomorphism of $\Gamma_1$ onto $\Gamma_2$. This will induce an isomorphism of $Q_1$ to $Q_2$. Furthermore, we want this isomorphism to induce an isomorphism of $\rho_1(Q_1)$ to $\rho_2(Q_2)$. We then check that $\rho_1(Q_1)$ is conjugate to $\rho_2(Q_2)$ in $\mathrm{TOP}(W)$. Actually we prove the conjugation to be in $\mathrm{Isom}(W)$.

In (i), if $\rho_i(Q_i)$ is not solvable, then it is topologically hyperbolic. That is, if $\overline{\eta} : \rho_1(Q_1) \to \rho_2(Q_2)$ is an isomorphism, there is a homeomorphism $h : \mathbb{R}^2 \to \mathbb{R}^2$ so that $\rho_2(\overline{\eta}(q)) = h \circ \rho_1(q) \circ h^{-1}$, by a theorem of Macbeath [**Mac67**, Theorem 3]. Furthermore, $\rho_i(Q_i)$ can be conjugated, within $\mathrm{TOP}(\mathbb{R}^2)$, into the subgroup of full isometries of the hyperbolic plane, $\mathrm{PSL}(2, \mathbb{R}) \rtimes \mathbb{Z}_2$. Therefore, according to Lemmas 11.2.5 and 11.2.6, $\rho_i(Q_i)$ contains no normal finite subgroups and no normal solvable subgroups other than the trivial subgroup. This latter fact will be used in showing that $\eta$ restricts to an isomorphism of $\Gamma_1$ to $\Gamma_2$.

A symmetric space as in (ii) is a homogeneous space $G/K$, where $G$ is a semisimple Lie group with no compact factors in adjoint form and $K$ is a maximal compact subgroup. This is exploited to show $\Gamma_1$ is mapped isomorphically onto $\Gamma_2$, and $\rho_1(Q_1)$ is conjugate to $\rho_2(Q_2)$ in $\mathrm{Isom}(G/K)$.

We shall use the following two lemmas.

LEMMA 11.2.5. *Let $Q$ be a discrete group acting effectively and properly on a contractible manifold $W$ so that $Q\backslash W$ is compact. Assume that $Q$ contains a torsion-free subgroup of finite index. Then a normal finite subgroup of $Q$ is the trivial group.*

PROOF. Let $H$ be the intersection of the torsion-free subgroup with all of its conjugates. Then $H$ is torsion free and normal in $Q$. Let $F$ be a normal finite subgroup of $Q$. Then, $F \cap H = \{1\}$. Let $F \cdot H$ be the group generated by $F$ and $H$. Then $(F \cdot H)/H = F/(F \cap H) = F$. Therefore, $F$ splits back to $F \cdot H$. Then as $F$ is normal, and the group $F \cdot H$ is congruent to $F \times H$. $F \times H$ acts effectively on $W$ which is the universal cover of the closed manifold $H\backslash W$. Since $H\backslash W$ is an admissible manifold (Subsection 3.2.1), $F$ must be trivial. □

LEMMA 11.2.6. *Let $\Gamma$ be a cocompact lattice in a semisimple Lie group $G$ without compact factors and in adjoint form. Then any nontrivial normal subgroup $N$ of $\Gamma$ is not solvable. In fact, no subgroup of $N$ of finite index is solvable.*

PROOF. If $N$ is finite, then it is central in $G$. Therefore, $N$ is $\{1\}$ because $G$ is in adjoint form. The lattice $\Gamma$ is dense in the Zariski topology by Borel's density theorem. So the Zariski closure of $N$ is a normal, and hence a semisimple subgroup of $G$. If $N$ were solvable, its Zariski closure would be solvable. This also implies that no subgroup of $N$ of finite index in $N$ can be solvable. □

A symmetric space $W$ of noncompact type with no compact and no 1-dimensional factors is the quotient space $G/K$ of a connected semisimple Lie group $G$ in adjoint form and without compact factors, where $K$ is a maximal compact subgroup. A cocompact lattice $\Gamma$ in $G$ acts on $G/K$ as a group of isometries. The full group of isometries on $G/K$ is $\mathrm{Aut}(G)$ where the connected component of the identity is $G$ itself and $W = G/K$ can be identified with $\mathrm{Aut}(G)/\overline{K} = \overline{G}/\overline{K}$, $\overline{K}$ being the maximal compact subgroup of $\overline{G} = \mathrm{Aut}(G)$. As seen in Section 9.4, there is an exact sequence of Lie groups $1 \to G \to \mathrm{Aut}(G) \to \mathrm{Out}(G) \to 1$, and $\mathrm{Out}(G)$ is finite.

If $\rho(Q)$ is now a discrete, cocompact subgroup of isometries of $W$, then $\rho(Q)$ is a cocompact subgroup of $\overline{G}$. $\overline{G}$ has a faithful representation into $\mathrm{GL}(n,\mathbb{R})$ for some $n$. (Actually we need only that $\rho(Q) \cap G$ has a characteristic torsion free subgroup of finite index to get a torsion-free normal subgroup of finite index in $\rho(Q)$.) Therefore, $\rho(Q)$ has a torsion-free subgroup of finite index by Selberg's Lemma in Subsection 6.1.8. Consequently, Lemma 11.2.5 applies and $\rho(Q)$ contains no nontrivial, finite normal subgroups. For, if $H$ is a normal solvable subgroup of $\rho(Q)$, then $H \cap (\rho(Q) \cap G)$ is solvable in $\rho(Q) \cap G$. By Lemma 11.2.6, $H \cap (\rho(Q) \cap G)$ is trivial, hence $H$ is finite and, consequently by Lemma 11.2.5, it is trivial. Therefore, $\rho(Q)$ contains no normal nontrivial solvable subgroups.

In Remark 9.4.9, we explained the connections between the earlier part of section 9.4 and [**RW77**] where it is observed that the Mostow rigidity theorem in $G$ extends to lattices in the larger group $\mathrm{Aut}(G)$.

Let $\Pi$ be a lattice in $\mathrm{Aut}(G)$, with $G$ of type S(4). Then $\Pi \cap G = \Gamma$ is a lattice in $G$. We may write $\Pi$ as an extension of $\Gamma$ by the finite group $F$. We may assume that $\psi : F \to \mathrm{Out}(\Gamma)$ injects. For, if not, let $L$ be the kernel of $\psi$. Then $\Gamma \times L$ is in $\Pi$. This would lead to a contradiction by choosing a torsion-free sublattice $\Gamma'$ in $\Gamma$ and we would get an effective action of $\Gamma' \times L$ on $\Gamma' \backslash G/K$'s universal covering ($K$ is maximal compact in $G$).

The homomorphism $\psi : F \to \mathrm{Out}(\Gamma)$ with UAEP of $(\Gamma, G)$ induces a homomorphism $\tilde{\psi} : F \to \mathrm{Out}(G)$. We may also assume that $\tilde{\psi}$ is injective. For otherwise, we would get an extension, $\Pi'$, of $\Gamma$ by a subgroup of $F$, with $\Gamma \subset \Pi' \subset G \cap \Pi$.

Now we apply Theorem 9.4.5 with $W$ a point. We have this embedding of $\Pi$ into $\mathrm{Aut}(G) = \overline{\mathrm{Aff}}(G, K)$ which carries $\Gamma$ into $\mathrm{Inn}(G) = \overline{\mathrm{Aff}}_0(G, K)$ (this embedding forces $F \to \mathrm{Aut}(G)$ to be injective).

LEMMA 11.2.7. *Suppose $G$ is centerless and has ULIEP. Then, any finite extension $\bar{G} \subset \mathrm{Aut}(G)$ of $G$ has ULIEP.*

PROOF. Let $Q_1, Q_2 \subset \bar{G}$ be cocompact discrete subgroups, and let $\theta : Q_1 \to Q_2$ be an isomorphism. Since $\bar{G}/G$ is finite, the images $Q_i \to \bar{G} \to \bar{G}/G$ are finite. Let $m$ be the product of the orders of the images of $Q_i$'s in $\bar{G}/G$. Let

$$\Gamma_1 = \langle x^m : x \in Q_1 \rangle,$$
$$\Gamma_2 = \langle y^m : y \in Q_2 \rangle.$$

Then $\Gamma_i$ is a characteristic subgroup of $Q_i$ so that the isomorphism $\theta$ maps $\Gamma_1$ onto $\Gamma_2$ isomorphically. Clearly, $\Gamma_i$'s are lattices of $G$ by the choice of $m$. Then ULIEP of $G$ yields an automorphism $\omega : G \to G$ extending $\theta : \Gamma_1 \to \Gamma_2$. Now $\omega \in \mathrm{Aut}(G)$ can be interpreted in two different ways as an automorphism of $G$. The first is as an automorphism of $G$ as it was. The second way is via conjugation in $\mathrm{Aut}(G)$ as $\mu(\omega)$ is conjugation by $\omega$. With $g \mapsto \mu(g)$ via $G \subset \mathrm{Aut}(G)$, we have

$$\mu(\omega(g)) = \omega \circ \mu(g) \circ \omega^{-1}$$

because

$$(\omega \circ \mu(g) \circ \omega^{-1})(x) = \omega\left(\mu(g)(\omega^{-1}(x))\right)$$
$$= \omega(g \cdot \omega^{-1}(x) \cdot g^{-1})$$
$$= \omega(g) \cdot x \cdot \omega(g)^{-1}$$
$$= \mu(\omega(g))(x).$$

Suppressing all the $\mu$'s, we have

$$\omega(g) = \omega \circ g \circ \omega^{-1}$$

for $\omega \in \mathrm{Aut}(G)$ and $g \in G$. The second interpretation, $\omega$ as $\mu(\omega)$, enables us to look at $\mu(\omega)$ as an automorphism of $\mathrm{Aut}(G)$ (not just as a automorphism of $G$). Let $Q_3 = \mu(\omega^{-1})(Q_2) \subset \mathrm{Aut}(G)$, and let

$$\theta' = \mu(\omega^{-1}) \circ \theta : \ Q_1 \to Q_2 \to Q_3.$$

Notice that $Q_3$ may not be in $\bar{G}$, but it is in $\mathrm{Aut}(G)$. In any case, $\theta' : Q_1 \to Q_3$ is an isomorphism which is the identity map on $Q_1 \cap G$. (In fact, it was the identity map on $\Gamma_1$, but the ULIEP implies that it is the identity on the bigger group $Q_1 \cap G$. Even more is true. Clearly, $\theta'|_{Q_1 \cap G}$ extends to a unique automorphism of $G$, which is the identity map on $G$.)

We have two groups $Q_1, Q_3 \subset \mathrm{Aut}(G)$ and an isomorphism between them such that $Q_1 \cap G = Q_3 \cap G$ and $\theta'|_{Q_1 \cap G}$ is the identity map. We claim that $Q_1 = Q_3$

and $\theta'$ is the identity map. Suppose there is $\alpha \in Q_1$ such that $\theta'(\alpha) \neq \alpha$. Then there exists $g \in G$ such that $\theta'(\alpha)(g) \neq \alpha(g)$. However,

$$\begin{aligned}
\theta'(\alpha)(g) &= \theta'(\alpha) \cdot g \cdot \theta'(\alpha)^{-1} \\
&= \theta'(\alpha) \cdot \theta'(g) \cdot \theta'(\alpha)^{-1} \quad (\text{since } \theta'|_G \text{ is identity}) \\
&= \theta'(\alpha g \alpha^{-1}), \\
\alpha(g) &= \alpha g \alpha^{-1}.
\end{aligned}$$

But, since $\alpha g \alpha^{-1} \in G$, we have $\theta'(\alpha g \alpha^{-1}) = \alpha g \alpha^{-1}$. This implies $\theta'(\alpha)(g) = \alpha(g)$, a contradiction. $\square$

The argument proves also the following

COROLLARY 11.2.8. *Suppose $G$ is centerless, $\Gamma$ a lattice of $G$, and $(\Gamma, G)$ has UAEP. Then, for any finite extension $\bar{G} \subset \mathrm{Aut}(G)$ of $G$ and a lattice $\bar{\Gamma}$ of $\bar{G}$, $(\bar{\Gamma}, \bar{G})$ has UAEP.*

COROLLARY 11.2.9. *If $\rho(Q_1)$ and $\rho(Q_2)$ are groups as in Theorem 11.2.4 and are isomorphic and torsion free, then the locally symmetric Riemannian manifolds $\rho(Q_i)\backslash G/K$ are isometric.*

Thus, if $\bar{\eta} : \rho_1(Q_1) \to \rho_2(Q_2)$ is an isomorphism between cocompact lattices in $\mathrm{Aut}(G)$, then there exists an automorphism $\hat{\eta} : \mathrm{Aut}(G) \to \mathrm{Aut}(G)$ such that $\hat{\eta}|_{Q_1} = \bar{\eta}$. This allows us to conclude that if $\bar{\eta} : \rho_1(Q_1) \to \rho_2(Q_2)$ is an isomorphism of discrete groups of isometries of $W$ with $\rho_i(Q_i)\backslash W$ compact, then there is an isometry $h : G/K \to G/K$ such that $h \circ \rho_1 \circ h^{-1} = \rho_2 \circ \bar{\eta}$ provided $W$ has no 1-dimensional and no compact factors and any projection to a 2-dimensional factor is dense.

All that remains in proving our theorem is to show that $\eta$ restricts to an isomorphism between $\Gamma_1$ and $\Gamma_2$ and induces an isomorphism of $\rho_1(Q_1)$ to $\rho_2(Q_2)$. So, let $L_i = \ker(\rho_i) \subset Q_i$. Then $L_i$ is a finite normal subgroup of $Q_i$. Let $1 \to \Gamma_i \to \Pi_{L_i} \to L_i \to 1$ be the pullback of $1 \to \Gamma_i \to \Pi_i \to Q_i \to 1$ via $L_i \hookrightarrow Q_i$. Since $\rho_2(\eta(\Gamma_1))$ is a normal subgroup of $\rho_2(Q_2)$, and $\rho_2(Q_2)$ has no normal solvable subgroup, $\rho_2(\eta(\Gamma_1))$ must be trivial. That is, $\eta(\Gamma_1) \subset \Pi_{L_2}$. Now $\Gamma_1$ has finite index in $\Pi_{L_1}$. Therefore, $\rho_2(\eta(\Pi_{L_1}))$ is a finite normal subgroup of $\rho_2(Q_2)$, which must be trivial again. Thus, we have $\eta(\Pi_{L_1}) \subset \Pi_{L_2}$. By symmetry,

$$\eta(\Pi_{L_1}) = \Pi_{L_2}.$$

Since $\varphi_i \times \rho_i$ is injective, $\varphi_1 : Q_1 \to \mathrm{Out}(\Gamma_1) \subset \mathrm{Out}(G)$ is injective on $L_1$. Then by Theorem 8.4.3 and its proof (assuming $G$ is of type (S3)), we know that $\Gamma_1$ maps onto $\Gamma_2$ isomorphically by $\varphi$. Consequently, $\eta$ induces an isomorphism $\rho_1(Q_1) \to \rho_2(Q_2)$. Since $\eta$ maps $\Gamma_1$ onto $\Gamma_2$ isomorphically, we can apply the rigidity part of Theorem 7.3.2 to complete the proof of Theorem 11.2.4.

REMARK 11.2.10. (1) The theorem with $\Gamma_1 \cong \Gamma_2 \cong \mathbb{Z}^k$ in (i) and (ii) is a far-reaching extension of Corollary 14.12.5. This classification is not only for manifolds ($\Pi_i$'s are torsion free) but also for all the possible orbifolds ($Q$, of course, being topologically hyperbolic). With $k = 2$ and $\Pi_i$ torsion free, these 4-dimensional Seifert manifolds include all the Kodaira elliptic surfaces with only multiple fibers as singularities and whose fundamental groups are nonsolvable. In [**OR70b**], explicit canonical normal forms of the Seifert fiberings which arise as orbit mappings of $T^2$-actions on 4-manifolds are given. The descriptions in terms of orbit invariants yield an equivariant classification in the spirit of Chapter 14. Presentations of torsion-free $\Pi$ when $k \geq 2$, $\varphi : Q \to \operatorname{Aut}(\mathbb{Z}^k)$, can be found in [**Zie69**]. However, in this case, it is not possible to find an algorithm that decides, in a finite number of steps, whether two presentations of the fundamental groups for these Seifert manifolds are isomorphic if $k > 2$; see [**Zim85**].

(2) For $W = \mathbb{R}^2$, $Q \subset \operatorname{TOP}(\mathbb{R}^2)$, topologically hyperbolic, and $\Gamma = \mathbb{Z}^k$ with $\varphi : Q \to \operatorname{Aut}(\mathbb{R}^k)$ trivial (i.e., admitting an injective $T^k$-action), Nicas and Stark [**NS85**] show that a homotopy equivalence from $M = \theta(\Pi)\backslash(\mathbb{R}^k \times \mathbb{R}^2)$, ($\Pi$ torsion free) to a closed manifold $N$ is homotopic to a homeomorphism provided $k \geq 3$. They use $M$ as a model manifold and then apply surgery procedures to obtain the result. We conjecture that their theorem is also valid when $\varphi : Q \to \operatorname{Out}(G)$ is not necessarily trivial and $\Gamma$ is completely solvable.

(3) Locally symmetric spaces of the noncompact type are Riemannian manifolds whose sectional curvature is less than or equal to 0. Farrell and Jones [**FJ89**] have shown if $N^n$, $n > 4$, is a closed Riemannian manifold whose sectional curvature is less than or equal to 0, then any homotopy equivalence $f : M \to N$ from a closed manifold $M$ can be deformed to a homeomorphism. This important work has many applications; see, for example, [**FJ98**]. A consequence to our Seifert fiberings is the following. Let $\theta$ be a Seifert Construction for the extension $1 \to \Gamma \to \Pi \to Q \to 1$ with $\Gamma$ a special lattice and $Q$ abstractly isomorphic to a cocompact group of isometries $\overline{Q}$ on a nonpositively curved Riemannian manifold diffeomorphic to $\mathbb{R}^n$. Then, $Q$ contains a torsion-free normal subgroup $Q'$ of finite index and $M = \theta(\Pi)\backslash(G \times \mathbb{R}^n)$ is the quotient of $M' = \theta(\Pi')\backslash(G \times \mathbb{R}^n)$ by a finite group of Seifert automorphisms $Q/Q'$. $M'$ is homeomorphic to a fiber bundle over the nonpositively curved manifold $\overline{Q}'\backslash\mathbb{R}^n$ with fiber homeomorphic to $\Gamma\backslash G$. Here we applied the rigidity for the locally symmetric manifolds of negative curvature and the rigidity in $\operatorname{Aut}(\mathbb{R}^n)$ for the flat factors.

## 11.3. Lifting problem for homotopy classes

Let $M$ be a reasonable path-connected space, say a connected ANR, and $\mathcal{E}(M)$ be the $H$-space of homotopy equivalences of $M$ into itself. Any $f \in \mathcal{E}(M)$ induces an isomorphism $f_* : \pi_1(M, x) \to \pi_1(M, f(x))$. By choosing a path $\omega$ from $x$ to $f(x)$, we have an automorphism $f_*^\omega$ of $\pi_1(M,x)$, defined by $f_*^\omega([\tau]) = [\omega^{-1} \cdot (f \circ \tau) \cdot \omega]$. A different choice of $\omega$ alters $f_*^\omega$ only by an inner automorphism. Therefore, we obtain a map

$$\gamma : \mathcal{E}(M) \to \operatorname{Out}(\Pi),$$

where $\Pi = \pi_1(M,x)$. Let $\mathcal{E}_0(M)$ be the space of self-homotopy equivalences which are homotopic to the identity. Then $\gamma$ maps $\mathcal{E}_0(M)$ to the identity in $\operatorname{Out}(\Pi)$. The homotopy class of homotopy equivalences (i.e., $\mathcal{E}(M)$ mod $\mathcal{E}_0(M)$), forms a group, under composition. The map $\gamma$ induces a homomorphism $\pi_0(\mathcal{E}(M)) = \mathcal{E}(M)/\mathcal{E}_0(M) \to \operatorname{Out}(\Pi)$. If $M$ is aspherical, the map $\gamma$ induces an isomorphism. There is a natural map $i : \operatorname{TOP}(M) \to \mathcal{E}(M)$ induced by inclusion (we use the compact-open topology). Composing $i$ with $\gamma$ is a homomorphism. The connected component of the identity maps to the identity in $\pi_0(\mathcal{E}(M))$ and $\operatorname{Out}(\Pi)$. There is induced a sequence of natural homomorphisms $\mathrm{TOP}(M) \to \pi_0(\mathrm{TOP}(M)) \xrightarrow{i_*} \pi_0(\mathcal{E}(M)) \xrightarrow{\gamma} \mathrm{Out}(\Pi)$.

DEFINITION 11.3.1. A homomorphism $\psi : F \to \mathrm{Out}(\Pi)$ is called an *abstract kernel*. A *lifting* of $\psi$ as a group of homeomorphisms is a homomorphism $\hat{\psi} : F \to \mathrm{TOP}(M)$ which makes

$$\begin{array}{ccc} F & \xrightarrow{=} & F \\ \hat{\psi} \downarrow & & \downarrow \psi \\ \mathrm{TOP}(M) \longrightarrow \mathcal{E}(M) & \longrightarrow & \mathrm{Out}(\pi_1(M)) \end{array}$$

commutative. The abstract kernel $F \to \mathrm{Out}(\Pi)$ is *topologically realizable* if it can be realized as an action of $F$ on $M$ (i.e., a lifting as a group of homeomorphisms exists).

J. Nielsen [**Nie43**] had shown that every cyclic group of outer automorphisms on a closed surface could be *topologically* realized. Others had shown, by sometimes different methods, that finite $p$-groups and solvable Lie groups could be topologically realized on compact surfaces ([**Mac62**], [**Zie81**]). In 1983, S. Kerckhoff [**Ker83**] showed that all finite subgroups of $\mathrm{Out}(\pi_1(M))$, where $M$ is a closed surface, can be topologically realized. In 1977, the first examples showing the failure of topological realization on closed aspherical $n$-manifolds $n \geq 3$, were constructed [**RS77**]. These examples were nilmanifolds. Many other examples of failure on closed aspherical manifolds soon followed, e.g., [**ZZ79**], [**LR82**]. The problem of when one can lift finite groups $F$ from $\pi_0(\mathrm{TOP}(M))$, $\pi_0(\mathcal{E}(M))$ or $\mathrm{Out}(\pi_1(M))$ to $\mathrm{TOP}(M)$ or $\mathrm{Diff}(M)$ became known as the *Nielsen realization problem*. Of course, obvious restrictions must be assumed for the problem to have relevance. The problem is certainly relevant for a closed aspherical manifold because homomorphism $\pi_0(\mathcal{E}(M)) \to \mathrm{Out}(\pi_1(M))$ is an isomorphism. In Chapter 3, we saw that admissible manifolds are good generalizations of closed aspherical manifolds and they enjoy some of the same properties as closed aspherical manifolds. The finite groups that may act effectively on admissible manifolds are essentially determined by the finite subgroups of $\mathrm{Aut}(\pi_1(M))$ and $\mathrm{Out}(\pi_1(M))$, see Theorem 3.2.2. We shall produce a simple strong necessary algebraic condition for the existence of a topological realization of an abstract kernel $\psi : F \to \mathrm{Out}(\Pi)$ on an admissible manifold. Unfortunately, on some nonaspherical admissible manifolds, this condition is not sufficient. Whether or not this condition is also sufficient for closed aspherical manifolds is unknown at this time. However, as we shall see, the Seifert Construction permits us to verify that the necessary condition is also sufficient for a topological realization, as a group of Seifert automorphisms, for large classes of aspherical manifolds.

DEFINITION 11.3.2. An extension $1 \to \Pi \to E \to F \to 1$ is called *admissible* [**LR81**] if each torsion element of $C_E(\Pi)$, the centralizer of $\Pi$ in $E$, is an element of $\Pi$ (so, of $\mathcal{Z}(\Pi)$). That is, $\mathcal{Z}(\Pi)$ and $C_E(\Pi)$ have the same torsion elements so that the inclusion $\mathcal{Z}(\Pi) \to C_E(\Pi)$ is an isomorphism when restricted to torsions.

The definition agrees with the definition introduced in [**LR81**] where it is additionally assumed that $\mathcal{Z}(\Pi)$ is torsion free. Recall that (Definition 3.2.1) a closed manifold $M$ is called an *admissible manifold* if the only periodic self-homeomorphisms of $\widetilde{M}$ commuting with the deck transformation group $\pi_1(M)$ are elements of the center of $\pi_1(M)$. This means that, a manifold $M$ is an admissible manifold if and only if, for every finite effective group action $(F, M)$, the lifting exact sequence (see Subsection 2.2.2) of the action is admissible. Theorem 3.2.8 asserts that all closed aspherical, hyper-aspherical, and $K$-manifolds are admissible manifolds.

COROLLARY 11.3.3. *Let $(F, M)$ be an effective action of a finite group on an admissible manifold $M$ with $\pi_1(M) = \Pi$. Then the induced extension $1 \to \Pi \to F^* \to F \to 1$, where $F^*$ denotes the group of all liftings of $F$ to homeomorphisms of $\widetilde{M}$, is admissible.*

Notice that, in the pullback diagram (see Subsection 5.3.1 for pullback),

$$\begin{array}{ccccccccc} 1 & \longrightarrow & \Pi & \longrightarrow & F^* & \longrightarrow & F & \longrightarrow & 1 \\ & & \downarrow & & \downarrow & & \downarrow & & \\ 1 & \longrightarrow & \Pi & \longrightarrow & N_{\mathrm{TOP}(\widetilde{M})}\Pi & \longrightarrow & \mathrm{TOP}(M) & \longrightarrow & 1 \end{array}$$

the bottom sequence is admissible, so is the top one.

REMARK 11.3.4. Let $(F, M)$ be an action (not necessarily effective) of a finite group on an admissible manifold $M$ with $\pi_1(M) = \Pi$. Then there exists an extension (not necessarily admissible) $1 \to \Pi \to E \to F \to 1$ realizing the abstract kernel $\psi : F \xrightarrow{\hat{\psi}} \mathrm{TOP}(M) \xrightarrow{\psi'} \mathrm{Out}(\pi_1(M))$.

PROOF. Since $(\hat{\psi}(F), M)$ is effective, there exists an admissible extension $E'$ of $\Pi$ by $\hat{\psi}(F)$, $1 \to \Pi \to E' \to \hat{\psi}(F) \to 1$. We can pullback this short exact sequence via $F \xrightarrow{\hat{\psi}} \hat{\psi}(F)$ to get

$$\begin{array}{ccccccccc} 1 & \longrightarrow & \Pi & \longrightarrow & E & \longrightarrow & F & \longrightarrow & 1 \\ & & \parallel & & \downarrow & & \downarrow{\hat{\psi}} & & \\ 2 & \longrightarrow & \Pi & \longrightarrow & E' & \longrightarrow & \hat{\psi}(F) & \longrightarrow & 1. \end{array}$$

Certainly the top row is an extension of $\Pi$ by $F$ realizing $(\Pi, F, \psi = \psi' \circ \hat{\psi})$. □

Thus we have a necessary condition for the existence of a lifting of an abstract kernel $\psi : F \to \mathrm{Out}(\Pi)$, as an (effective, respectively,) group action: the existence of an (admissible, respectively,) group extension of $\Pi$ by $F$ realizing the abstract kernel. For finite groups, this necessary condition is also sufficient for some tractable manifolds. However, for some admissible manifolds, as the next examples show, this necessary condition is not always sufficient.

EXAMPLE 11.3.5. The extension $0 \to \mathbb{Z}^4 \to \mathbb{Z}^4 \to \mathbb{Z}_p \to 0$ is admissible when $\psi : \mathbb{Z}_p \to \mathrm{Out}(\mathbb{Z}^4)$ is trivial, and $p$ is prime. However, there is no realization of this abstract kernel as an effective group action on the admissible manifold $M = T^4 \# \mathbb{C}P_2$. We note that the Euler characteristic of $M$ is 1. If $\mathbb{Z}_p$ acts on $M$, $\chi(M^{\mathbb{Z}_p}) \equiv \chi(M)$ mod $p$, by the Smith theorems. Therefore, $M^{\mathbb{Z}_p} \neq \emptyset$. Then $\psi = \theta : \mathbb{Z}_p \to \mathrm{Aut}(\mathbb{Z}^4) = \mathrm{Out}(\mathbb{Z}^4)$ must be injective by Theorem 3.2.2. In this case, the existence of an admissible extension for an abstract kernel on this admissible manifold does not yield a lifting of $\mathbb{Z}_p$ to $\mathrm{TOP}(M)$.

EXERCISE 11.3.6. Show if $F$ acts effectively on $M = T^4 \# \mathbb{C}P^2$, then $\theta : F \to \mathrm{GL}(4, \mathbb{Z})$ is injective; cf. Exercise 3.4.6.

EXAMPLE 11.3.7. Let $M$ be a Seifert 3-manifold that fibers over the 2-sphere with three exceptional fibers of multiplicity $\{p, q, r\}$, see Chapter 14. Assume that the $p, q, r$ are all odd, distinct primes. Therefore, $M$ admits an injective $S^1$-action, is aspherical and admits no orientation reversing self-homotopy equivalence. The sequence $0 \to \mathbb{Z} \to \pi_1(M) \to Q \to 1$ is exact, where $\mathbb{Z}$ is the center of $\pi_1(M)$ and $Q$ is a centerless Fuchsian group normally generated by $\mathbb{Z}_p, \mathbb{Z}_q$ and $\mathbb{Z}_r$. Any finite subgroup of $Q$ is a subgroup of a conjugate of $\mathbb{Z}_p, \mathbb{Z}_q$, or $\mathbb{Z}_r$. It is known that $1 \to \mathrm{Inn}(Q) = Q \to \mathrm{Aut}(Q) \to \mathrm{Out}(Q) = \mathbb{Z}_2 \to 1$ is exact and the $\mathbb{Z}_2$ splits back. ($Q \backslash \mathbb{R}^2$, as an orbifold, is the 2-sphere with three distinct branch points $x, y, z$. $\mathrm{Out}(Q)$ is isomorphic to $\pi_0(\mathrm{TOP}(S^2; x, y, z))$. This latter group is $\pi_0$ of the homeomorphisms of $S^2$ which fix three distinct points and is isomorphic to $\mathbb{Z}_2$.) From [**CR77**, §6, especially Corollary 1 and 2 on page 65] (note in the notation there, that $\Gamma(a) = 1$, $H^1(Q; \mathbb{Z}) = 0$ and $\mathrm{Inn}(Q) = \mathrm{Inn}(\pi_1(M))$), it follows that

$$\mathrm{Aut}(\pi_1(M)) = \mathrm{Inn}(\pi_1(M)) \rtimes \mathrm{Out}(\pi_1(M)) = Q \rtimes \mathrm{Out}(Q) = Q \rtimes \mathbb{Z}_2.$$

LEMMA 11.3.8. *For $M$ as in Example 11.3.7, the finite groups that act effectively on $M$ are the subgroups of the finite dihedral groups.*

PROOF. Clearly any subgroup of $S^1$ acts on $M$. In fact, $O(2)$ acts on $M$. The orbit space of a circle action is $S^2$ with three singular orbits. Arrange these singular orbits along the equator. Then it is easy to see that there is an involution on $M$ which reflects the orbits across the equator. It reverses the orientation of each of the fibers and reverses the orientation of the base. The involution is orientation preserving and compatible with the $S^1$-action giving an $O(2)$-action on $M$. Therefore, any finite dihedral group acts effectively on $M$.

Now, suppose $H$ is an effective action of a finite group on $M$. Let $1 \to \pi_1(M) = \Pi \to E \to H \to 1$ be the lifting exact sequence. Let $\varphi : H \to \mathrm{Out}(\pi_1(M)) \cong \mathbb{Z}_2$ be the abstract kernel and $K$ the kernel of the homomorphism $\varphi$. This extension is admissible and $\mathcal{Z}(\Pi) \cong \mathbb{Z}$. Therefore, $C_E(\Pi)$ is also $\mathbb{Z}$ and $K$ is a finite cyclic group of index at most 2 in $H$. The action of $\mathbb{Z}_2$ in $\mathrm{Out}(\Pi)$ is nontrivial on the center and therefore also on $K$. Then $H$ is dihedral if $K \neq H$, otherwise it is cyclic. □

11.3.9. Suppose now that $A$ is a finite subgroup of $E$ above. Then $A$ maps isomorphically into a subgroup of $H$ since $\Pi$ is torsion free. The map from $E$ to $\mathrm{Aut}(\Pi)$ is injective on torsion, since $C_E(\Pi) \cong \mathbb{Z}$. Therefore, $A$ maps isomorphically into a subgroup of $\mathrm{Aut}(\Pi)$ isomorphic to one of the $\mathbb{Z}_{p_i} \rtimes \mathbb{Z}_2$, $i = 1, 2$ or 3. Furthermore, each $A$ must fix some point of $M$. Since the $p_i$ are prime, a subgroup of $A$ isomorphic to one of the $\mathbb{Z}_{p_i}$ cannot act freely. For if it acted freely, the lifting sequence of the subgroup would have to be torsion free as $M$ is aspherical. Similarly for an element of order 2 which reverses orientation of the base. We have shown the following

COROLLARY 11.3.10. *Any finite subgroup $A$ of the lifting sequence in $E$ acts with fixed points on $M$ and is isomorphic to a subgroup of $\mathbb{Z}_{p_i} \rtimes \mathbb{Z}_2$, for $i = 1, 2$, or 3.*

Let $p_1, q_1, r_1, p_2, q_2, r_2$ be distinct odd primes. Let $N$ be the oriented connected sum $M_1 \# M_2$ where $M_i$ are as above with $p_i, q_i, r_i$ being the orders of the multiplicities of the exceptional fibers of $M_i$, $i = 1, 2$. The manifold $N$ is hyper-aspherical.

THEOREM 11.3.11. *Every abstract kernel $\psi : F \to \mathrm{Out}(\pi_1(N))$ has an algebraic realization as a group extension but only $\mathbb{Z}_2$ can act effectively and smoothly on $N$.*

PROOF. In general, the obstruction to the existence of a group extension realizing this abstract kernel is a cohomology class in $H^3(F; \mathcal{Z}(\pi_1(N)))$. Since $\pi_1(N)$ has trivial center, the obstruction class vanishes and so an extension realizing the abstract kernel always exists.

Meeks and Yau [**MY80**] have shown that any finite group $F$ that acts effectively and smoothly on $M_1 \# M_2$, where $M_i$ are closed, irreducible and aspherical 3-manifolds, has an $F$-invariant 2-sphere $S \subset N$ along which the connected sum is made. Furthermore, the action of $F$ smoothly extends to an action of $F$ on both $M_1$ and $M_2$ by coning over the invariant sphere. Then $F$ must fix the vertices of the cones in $M_1$ and $M_2$. Consequently, $F$ must be a subgroup of $\mathrm{Aut}(\pi_1(M_1))$ and also of $\mathrm{Aut}(\pi_1(M_2))$. Since the odd primes for $\mathrm{Aut}(\pi_1(M_1))$ and $\mathrm{Aut}(\pi_1(M_2))$ are different, $F$ must be $\mathbb{Z}_2$.

By Bloomberg (Subsection 3.4.4), we have that $\mathrm{Out}(\pi_1(N)) = \mathrm{Aut}(\pi_1(M_1)) \times \mathrm{Aut}(\pi_1(M_2))$. A torsion subgroup of $\mathrm{Out}(\pi_1(N))$ is a torsion subgroup of $(A \rtimes \mathbb{Z}_2) \times (B \rtimes \mathbb{Z}_2)$, where $A$ is $\mathbb{Z}_{p_1}$, $\mathbb{Z}_{q_1}$, or $\mathbb{Z}_{r_1}$ and $B$ is $\mathbb{Z}_{p_2}$, $\mathbb{Z}_{q_2}$, or $\mathbb{Z}_{r_2}$. For example, if we choose $\mathbb{Z}_{p_1} \times \mathbb{Z}_{p_2}$ and map it into $\mathrm{Out}(\pi_1(N))$ injectively, we will have an admissible extension. The group $\mathbb{Z}_{p_1} \times \mathbb{Z}_{p_2}$, however, is not realizable as a smooth action on $N$. An action of $\mathbb{Z}_2$ on $N$ can easily be constructed. □

REMARK 11.3.12. Any smooth action of a compact Lie group on $M$ as in Example 11.3.7 can be smoothly conjugated into the O(2)-action described in the theorem. For the connected component, this follows from Chapter 14, and for the dihedral part, it follows from [**MS86**]. Finite topological actions cannot be always be conjugated into the O(2)-action.

11.3.13. While the existence of an admissible extension is a very strong necessary condition for effective topological realization of an abstract kernel on an admissible manifold, the fundamental group does not always capture the homotopy type of the manifold. Moreover, the relationship between homotopy type of an admissible manifold and its homeomorphism type is weaker than it is for aspherical manifolds. Therefore, in seeking to show that the existence of an admissible extension is sufficient for a topological realization of an abstract kernel, it is advisable to confine one's self to the subject of aspherical manifolds. In fact, we have the following

*Unsolved Problem*: Does there exist a closed aspherical manifold $M$ such that there is an extension $1 \to \pi_1(M) \to E \to F \to 1$ with $F$ finite, but $F$ cannot be topologically realized as a group action on $M$?

DEFINITION 11.3.14. Let $Q$ act properly on a space $W$, and let $B$ be the quotient $Q \backslash W$. Suppose for each extension $1 \to Q \to E \to F \to 1$ by a finite group $F$, the action of $Q$ extends to a proper action of $E$ on $W$. Then we say that the $Q$-action on $W$ is *finitely extendable*. In particular, then $F$ acts on $B$ preserving the orbit structure.

If $\Gamma$ is normal in $\Pi$, recall $\mathrm{Aut}(\Pi,\Gamma)$ denotes the automorphisms of $\Pi$ that leave $\Gamma$ invariant. Since $\mathrm{Inn}(\Pi)$ leaves $\Gamma$ invariant, we can put $\mathrm{Aut}(\Pi,\Gamma)/\mathrm{Inn}(\Pi) = \mathrm{Out}(\Pi,\Gamma)$. It is a subgroup of $\mathrm{Out}(\Pi)$.

We are interested in realizing a finite abstract kernel $F \to \mathrm{Out}(\Pi)$ as a group action on a model Seifert fiber space $M(\Pi)$ with a typical fiber $\Gamma\backslash G$. Ideally, we want the $F$-action to be fiber preserving maps; in fact, Seifert automorphisms. This means that, on the group level, the extension must leave the lattice $\Gamma$ invariant. In other words, we consider only those abstract kernels which have images in $\mathrm{Out}(\Pi,\Gamma)$.

THEOREM 11.3.15. *Let $M(\theta(\Pi)) = \theta(\Pi)\backslash(G \times W)$ be a Seifert fiber space with typical fiber $\Gamma\backslash G$, where $G$ is a Lie group of type (S3); see Subsection 7.3.1. Suppose $(Q,W)$ is finitely extendable, where $Q = \Pi/\Gamma$. Then each abstract kernel $\psi : F \to \mathrm{Out}(\Pi,\Gamma)$ of a finite group $F$ can be topologically realized as a group of Seifert automorphisms on $M(\theta(\Pi))$ if and only if the abstract kernel $\psi$ admits some extension.*

REMARK 11.3.16. This theorem proves that if $(Q,W)$ is finitely extendable, then $(\theta(\Pi), G \times W)$ becomes finitely extendable itself, provided that $\Gamma$ is characteristic in $\Pi$. Thus, we can enlarge the class of extendable pairs more and more. Here is a list of finitely extendable pairs:

(1) $W = \{p\}$ a point and $(Q,W)$ any finite group;
(2) hyperbolic space and a cocompact lattice;
(3) $\mathbb{R}^n$ and a crystallographic group;
(4) connected, simply connected nilpotent Lie group and its finitely extended lattice;
(5) connected, simply connected completely solvable Lie group and its almost crystallographic group;
(6) a Riemannian symmetric space with no compact, 1- and 2-dimensional factors and $Q$ a group of isometries with compact quotient.

PROOF. Let $1 \to \Pi \to E \to F \to 1$ be an extension realizing the abstract kernel $\psi$. Since $\psi(F) \subset \mathrm{Out}(\Pi,\Gamma)$, $\Gamma$ is normal in $E$. We have the commutative diagram with exact columns and rows:

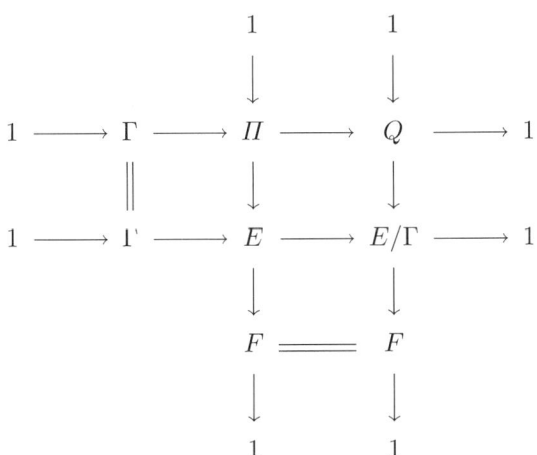

Consider the induced extension
$$1 \longrightarrow Q \longrightarrow E/\Gamma \longrightarrow F \longrightarrow 1.$$
Since $\rho : Q \to \mathrm{TOP}(W)$ is finitely extendable, there exists $\rho' : E/\Gamma \to \mathrm{TOP}(W)$ extending $\rho : Q \to \mathrm{TOP}(W)$. Again by the existence part of Theorem 7.3.2 for special lattices, there exists $\theta' : E \to \mathrm{TOP}_G(G \times W)$, where $\theta'|_\Gamma = \theta|_\Gamma = i : \Gamma \hookrightarrow G$, and $\rho'|_Q = \rho$. Put $\theta'|_\Pi = \theta'$. Of course, $\theta'$ may be different from $\theta$, but as $\theta$ and $\theta'$ agree on $\Gamma$ and $Q$, we can apply Theorem 7.3.2(2) to conjugate $\mathrm{TOP}_G(G \times W)$ by an element of $\mathrm{M}(W,G) \rtimes \mathrm{Inn}(G)$ which carries $\theta'|_\Pi$ to $\theta$ so that the new homomorphism $\theta' : E \to \mathrm{TOP}_G(G \times W)$ is an extension of $\theta : \Pi \to \mathrm{TOP}_G(G \times W)$. This yields an action of $F$ on $\theta(\Pi) \backslash (G \times W)$ as a group of Seifert automorphisms as desired. □

COROLLARY 11.3.17. *Let $M = \Pi \backslash G$ be an infra-$G$-manifold, where $G$ is one of the special Lie groups (see Subsection 7.3.1) and $\Pi \subset G \rtimes \mathrm{Aut}(G)$. Then each abstract kernel $\psi : F \to \mathrm{Out}(\Pi, \Gamma)$ of a finite group $F$ can be topologically realized as an (effective, respectively) group of affine diffeomorphisms on $M$ if and only if the abstract kernel $\psi$ admits an (admissible, respectively) extension.*

PROOF. The infra-$G$-manifold $M(\Pi)$ is modeled on $G \times \{p\}$ ($p$=point), $(G/K \times \{p\}$ for a convenient form of $G$ in the semisimple case), and $\mathrm{TOP}_G(G \times \{p\}) = \mathrm{Aff}(G)$ (respectively, $\overline{\mathrm{Aff}}(G,K)$); see Chapter 9 for this notation. Trivially, every $Q \to \mathrm{TOP}(\{p\})$ extends to $E/\Gamma \to \mathrm{TOP}(\{p\})$. The above theorem then immediately applies, and $F$ acts on $M(\Pi)$ by Seifert automorphisms which are affine diffeomorphisms. □

EXERCISE 11.3.18. *Let $\Pi = \mathbb{Z}^2$ and $M = \Pi \backslash \mathbb{R}^2 = T^2$. Realize the abstract kernel $\psi : \mathbb{Z}_2 \to \mathrm{GL}(2, \mathbb{Z})$. How many distinct actions $(\mathbb{Z}_2, T^2)$ (up to equivalence) do you get?*

EXERCISE 11.3.19. *Verify the claims for (1)–(5) made in Remark 11.3.16.*

REMARK 11.3.20. 1. Since we may introduce a metric structure in the corollary from a left invariant metric on $G$, $M(\Pi)$ has the structure of a flat, almost flat, Riemannian infra-solvmanifold or a locally symmetric spaces. We may also further conjugate $\theta(\Pi)$ in $\mathrm{Aff}(G)$ so that $F$ now acts on the conjugated manifold by isometries preserving the flat, etc., structures.

2. The proper action of $\Pi$ in the theorem is not necessarily free nor effective. Thus $M(\Pi)$ could very well be a Seifert orbifold. The corollary then works for such orbifolds, i.e., infra-$G$-spaces. In the Euclidean case, $M(\Pi)$ would then be a Euclidean *crystal* and $\Pi$ a Euclidean crystallographic group. In Theorem 11.3.15, $F$ sends fibers (which could be $G$-crystals instead of infra-$G$-spaces) to fibers.

For a torsion-free poly-{cyclic or finite} group $\Pi$, we can always find a (characteristic) *predivisible* subgroup $\Gamma$ of finite index in $\Pi$; see Section 9.5. Let $Q$ be the finite quotient $\Pi/\Gamma$, and choose $W = \{\text{point}\}$. Then the Seifert Construction of Theorem 9.5.7 produces an embedding $\theta(\Pi) \subset \overline{\mathrm{Aff}}(G, K)$ and the Seifert manifold $M(\Pi) = \theta(\Pi) \backslash G/K$ is a closed smooth $K(\Pi, 1)$ manifold.

COROLLARY 11.3.21 (Smooth realization of group actions from homotopy data). *Under the conditions of Theorem 9.5.7, let $M(\Pi) = \theta(\Pi) \backslash G/K$ be a Seifert manifold. Suppose now $\psi : F \to \mathrm{Out}(\Pi) = \pi_0 \mathcal{E}(M(\Pi))$ is a homomorphism of a finite* group $F$ into the homotopy classes of self-homotopy equivalences of $M(\Pi)$. Then, $F$ acts on $M(\Pi)$ if and only if there exists an extension,

$$1 \to \Pi \to E \to F \to 1,$$

realizing the abstract kernel $\psi$. Moreover, the action can be chosen to be smooth, induced from smooth Seifert automorphisms contained in $\overline{\mathrm{Aff}}(G,K)$. The action of $F$ is effective if and only if $C_E(\Pi)$ is torsion free.

PROOF. In order to have an action, we must have a lifting sequence and hence an extension, $1 \to \Pi \to E \to F \to 1$, that realizes the abstract kernel $\psi$. Since $\Gamma$ is characteristic in $\Pi$, it is normal in $E$ and and $1 \to Q = \Pi/\Gamma \to E/\Gamma \to F \to 1$ is exact. Because of the commutative diagram

$$\begin{array}{ccccccccc}
1 & \longrightarrow & \Gamma & \longrightarrow & \Pi & \longrightarrow & Q & \longrightarrow & 1 \\
& & \parallel & & \downarrow & & \downarrow & & \\
1 & \longrightarrow & \Gamma & \longrightarrow & E & \longrightarrow & E/\Gamma & \longrightarrow & 1 \\
& & & & \downarrow & & \downarrow & & \\
& & & & F & = & F & &
\end{array}$$

we can find a Seifert construction $\theta' : E \to \overline{\mathrm{Aff}}(G,K)$ which extends $\theta : \Pi \to \overline{\mathrm{Aff}}(G,K)$. Therefore the group $F$ acts on $M(\Pi)$ smoothly as diffeomorphisms preserving the Seifert structure. The action of $F$, since $M$ is aspherical, is effective if and only if, $C_E(\Pi)$ is torsion free. In any case, we have a lift $\widetilde{\psi}$,

$$\begin{array}{ccc}
F & \xrightarrow{\widetilde{\psi}} & \mathrm{Diff}(M(\Pi)) \\
\psi \downarrow & & j \downarrow \\
\mathrm{Out}(\Pi) & = & \mathcal{E}(M(\Pi))
\end{array}$$

where $j$ sends a self-diffeomorphism to its homotopy class. In case there exists one extension realizing the abstract kernel $\psi$, then for each element of $H^2(F, \mathcal{Z}(\Pi))$ there is a congruence class of extensions $E$, realizing the abstract kernel $\psi$. Each of these extensions gives rise to a (not necessarily effective) action of $F$ on $M(\Pi)$. □

If we combine this corollary with surgery results, we can get much stronger statements.

THEOREM 11.3.22. *Let $1 \to \Pi \to E \to F \to 1$ be an extension of a torsion-free poly{cyclic or finite} group $\Pi$ by a finite group $F$ with abstract kernel $\psi : F \to \mathrm{Out}(\Pi)$. Let $M$ be a closed aspherical $n$-manifold, $n > 4$, with $\pi_1(M) = \Pi$. Then there exists an action of $F$ on $M$ which realizes the abstract kernel $\psi : F \to \mathrm{Out}(\Pi) = \pi_0(\mathcal{E}(M))$, the group of homotopy classes of self-homotopy equivalences of $M$. The action is effective if and only if $C_E(\Pi)$ is torsion free, and is free if and only if $E$ is torsion free. In the latter case, any two such actions are weakly equivalent.*

PROOF. Pick $\Gamma$ a characteristic predivisible subgroup in $\Pi$. Then we have a commutative diagram as in the above corollary. With the notation in Definition 9.5.1, we get an action of $E/\Gamma$ on $\Gamma\backslash G/K = M(\Gamma)$ and an action of $F$ on

$M' = \theta(\Pi)\backslash G/K$, realizing $\psi$ on $M'$. Since the torsion-free $\Pi$ is poly-$\mathbb{Z}$ (respectively, poly (cyclic or finite)), a theorem of Wall [**Wal70**] (respectively, Farrell-Jones [**FJ98**]) says that any homotopy equivalence between $M$ and $M'$ is homotopic to a homeomorphism.

Both Wall's and Farrell and Jones's theorems were proven for dimension greater than 4. Their theorems are valid in dimension 4 by surgery results of Freedman and Quinn, and in dimension 3 by the solutions of the Geometric Conjecture. Therefore, we need only to pull back the action of $F$ on $M'$ to obtain the desired action on $M$. By uniqueness, we have that any two free actions on $M'$ will be weakly equivalent. □

THEOREM 11.3.23. *Let $G$ be a semisimple centerless Lie group without any normal compact factors and if $G$ contains any 3-dimensional factors (i.e., $\mathrm{PSL}(2,\mathbb{R})$), then the projection of the lattice to each of these factors is dense. Let $M(\theta(\Pi)) = \theta(\Pi)\backslash(G/K \times W)$ be a Seifert manifold with typical fiber $\Gamma\backslash G/K$. Suppose $(\Pi/\Gamma, W)$ is finitely extendable. Then each abstract kernel $\psi : F \to \mathrm{Out}(\Pi,\Gamma)$ of a finite group $F$ can be topologically realized as an (effective, respectively) group of Seifert automorphisms on $M(\theta(\Pi))$ if and only if the abstract kernel $\psi$ admits an (admissible, respectively) extension.*

PROOF. Same argument as Theorem 11.3.15. □

COROLLARY 11.3.24. *Let $G$ be as in the above theorem, and let $M = \theta(\Pi)\backslash G/K$. Then each abstract kernel $\psi : F \to \mathrm{Out}(\Pi,\Gamma)$ of a finite group $F$ can be topologically realized as an (effective, respectively) group action on $M$ if and only if the abstract kernel $\psi$ admits an (admissible, respectively) extension.*

EXAMPLE 11.3.25 ([**LR82**, (1.1)]). We shall exhibit an abstract kernel on the fundamental group of a flat manifold which has no realization as a group extension. The technique is similar to that employed in [**RS77**].

Let $\mathfrak{G}_2$ (see Example 4.5.10) be the 3-dimensional flat manifold defined by $\mathfrak{G}_2 = (S^1, S^1 \times_{\mathbb{Z}_2} T^2)$. The generator $\omega$ of $\mathbb{Z}_2$ acts on $T^2$ by $\omega(z_1, z_2) = (z_1^{-1}, z_2^{-1})$ and on $S^1$ by $z \mapsto -z$. Denote points of $\mathfrak{G}_2$ by $\langle r, z_1, z_2 \rangle$. Then $\langle r, z_1, z_2 \rangle = \langle r-1, z_1^{-1}, z_2^{-1} \rangle$ for $r \in \mathbb{R}$, $(z_1, z_2) \in T^2$. Define $D : \mathfrak{G}_2 \to \mathfrak{G}_2$ by $D\langle r, z_1, z_2 \rangle = \langle r-1, z_2^{-1}, z_1 \rangle$. Think of $D$ as given by the matrix

$$\begin{bmatrix} -1 & 0 & 0 \\ 0 & 0 & 1 \\ 0 & -1 & 0 \end{bmatrix}.$$

Note that $D$ is well defined, $D\langle 0, 1, 1 \rangle = \langle 0, 1, 1 \rangle$ and $D^2\langle r, z_1, z_2 \rangle = \langle r, z_1^{-1}, z_2^{-1} \rangle$. Define $D_t^2\langle r, z_1, z_2 \rangle = \langle r-t, z_1^{-1}, z_2^{-1} \rangle$. Then $D_0^2 = D^2$, $D_1^2 = \mathrm{id}_{\mathfrak{G}_2}$. Hence $D^2$ is isotopic, through isometries, to the identity. Now $D^2$ restricted to $T^2 = \{\langle 0, z_1, z_2 \rangle\} \subset \mathfrak{G}_2$ is $\omega$, and $\pi_1(T^2, \langle 0, 1, 1 \rangle)$ is a characteristic subgroup of $\pi_1(\mathfrak{G}_2)$ since it is the kernel of $\pi_1(\mathfrak{G}_2) \to H_1(\mathfrak{G}_2) \otimes \mathbb{Q}$.

If there is $H$ homotopic to $D$ such that $H^2 = \mathrm{id}$, then we have the commutative diagram of exact sequences

$$\begin{array}{ccccccccc} 1 & \longrightarrow & \Pi & \longrightarrow & E & \longrightarrow & \langle H \rangle \cong \mathbb{Z}_2 & \longrightarrow & 1 \\ & & \downarrow & & \downarrow & & \downarrow \psi & & \\ 1 & \longrightarrow & \mathrm{Inn}(\pi_1(\mathfrak{G}_2)) & \longrightarrow & \mathrm{Aut}(\pi_1(\mathfrak{G}_2)) & \longrightarrow & \mathrm{Out}(\pi_1(\mathfrak{G}_2)) & \longrightarrow & 1. \end{array}$$

Choose $e \in E$ such that conjugation by $e$, $\mu(e)$ is precisely $D_* \in \operatorname{Aut}(\pi_1(\mathfrak{G}_2))$. $\mathfrak{G}_2$ fibers over $S^1$ by $\langle r, z_1, z_2 \rangle \mapsto e^{2\pi i r}$, and let $\eta$ be the generator of $\pi_1$ of the section $\langle r, 1, 1 \rangle$ in $\pi_1(\mathfrak{G}_2)$. Then $\mu_\eta = \mu_{e^2}$. Therefore $e^2 = c \cdot \eta$ for some $c \in \mathcal{Z}(\pi_1(\mathfrak{G}_2)) \cong \mathbb{Z}$. Note $\mathcal{Z}(\pi_1(\mathfrak{G}_2)) \cong \mathbb{Z}$ is generated by $\eta^2$. Since $D_*(\eta) = \eta^{-1}$, $D_*(e^2) = e^{-2}$. On the other hand, $D_*(e^2) = \mu_e(e^2) = e^2$. Therefore, $e^4 = 1$. But as $e^2 \in \pi_1(\mathfrak{G}_2)$, which is torsion free, we have a contradiction. Therefore, no extension exists and so no involution $H$ topologically realizing the abstract kernel $\psi$ exists.

COROLLARY 11.3.26. *There exists an isometry $D$ on $\mathfrak{G}_2$ such that $D^2$ is isotopic, through isometries, to the identity but $D$ is not homotopic to any involution $H$.*

11.3.27. Observe that $\mathbb{Z}_4 \to \mathbb{Z}_2 \xrightarrow{\psi} \operatorname{Out}(\pi_1(\mathfrak{G}_2))$, in Example 11.3.25, is realized by the group of isometries $(D) \cong \mathbb{Z}_4$. We call $\mathbb{Z}_4$ an *inflation* of the abstract kernel $\psi$. Note, in this case, the new abstract kernel has an admissible extension.

In general, whenever the abstract kernel $\psi : F \to \operatorname{Out}(\Pi)$ fails to have an extension realizing the abstract kernel, we may find a larger finite group $H$ which maps homomorphically onto $F$ so that the composite with $\psi$, has an extension realizing this new abstract kernel, $H \to F \xrightarrow{\psi} \operatorname{Out}(\Pi)$. When we couple this with some of the previous theorems which guarantee topological realization by Seifert automorphisms, we have particular solutions to the Nielsen realization problem.

11.3.28. Let $\Pi$ be a group whose center is $\mathcal{Z}(\Pi) = C$. If $\psi : F \to \operatorname{Out}(\Pi)$ is a homomorphism, we may choose a map $\widetilde{\psi} : F \to \operatorname{Aut}(\Pi)$ such that the composite $F \xrightarrow{\widetilde{\psi}} \operatorname{Aut}(\Pi) \to \operatorname{Out}(\Pi)$ is $\psi$. The map $\widetilde{\psi}$ induces a homomorphism of $F$ to $\operatorname{Aut}(C)$ independent of the choice of lift $\widetilde{\psi}$. If we put $E = \Pi \times F$, we can attempt, using $\psi$, to construct a group structure so that $1 \to \Pi \to E \to F \to 1$ is an extension realizing the abstract kernel. If we follow the procedure of Section 5.2, we find that this is frustrated by the possible failure of the associative law of the group operation. This failure is measured by a cocycle in $Z^3_\psi(F; C)$. The cocycle depends upon choices made in the attempt of constructing a product structure. Varying the suitable choices alters the cocycle by a coboundary. Therefore we obtain an obstruction element $o(\psi) \in H^3_\psi(F; C)$ which vanishes if and only if there is an extension $E$ realizing this abstract kernel. In particular, if $C = 0$, then $o(\psi) = 0$. In this case, there is an extension induced by the pullback from

$$1 \to \operatorname{Inn}(\Pi) = \Pi \to \operatorname{Aut}(\Pi) \to \operatorname{Out}(\Pi) \to 1.$$

For details, see [**ML75**, Chapter 4, §8 and §9].

11.3.29. Now suppose $F$ is finite and $C = \mathbb{Z}^k$, $k > 0$, and $\psi : F \to \operatorname{Out}(\Pi)$, a given abstract kernel. Then the obstruction $o(\psi)$ has finite order, say $n$. Embed $\alpha : \mathbb{Z}^k \to (\frac{1}{n}\mathbb{Z})^k$. Since any automorphism of $\mathbb{Z}^k$ extends uniquely to an automorphism of $(\frac{1}{n}\mathbb{Z})^k$, there is a homomorphism

$$\alpha_* : H^3_\psi(F; \mathbb{Z}^k) \to H^3_\psi\big(F; (\tfrac{1}{n}\mathbb{Z})^k\big)$$

under which $\alpha_*(o(\psi)) = 0$.

We have a commutative diagram of $F$-modules

$$\begin{array}{ccccccccc}
0 & \longrightarrow & \mathbb{Z}^k & \longrightarrow & \mathbb{R}^k & \longrightarrow & T^k & \longrightarrow & 1 \\
& & \downarrow & & \downarrow & & \downarrow & & \\
0 & \longrightarrow & (\frac{1}{n}\mathbb{Z})^k & \longrightarrow & \mathbb{R}^k & \longrightarrow & T^k/(\mathbb{Z}_n)^k & \longrightarrow & 1.
\end{array}$$

Since $H^i(F;\mathbb{R}^k) = 0$, $i > 0$, we have $H^i_\psi(F;T^k) \stackrel{d^i}{\cong} H^{i+1}_\psi(F;\mathbb{Z}^k)$, $i \geq 0$. Let $\gamma = (d^2)^{-1}(o(\psi)) \in H^2(F;T^k)$. Since $\alpha_*(o(\psi)) = 0$,

$$(d^2)^{-1} \circ \alpha_* \circ d^2 = \alpha'_* : H^2(F;T^k) \to H^2(F;T^k/(\mathbb{Z}_n)^k)$$

maps $\gamma$ to 0. Consequently, there is a class $\delta \in H^2(F;(\mathbb{Z}_n)^k)$ for which $i_*(\delta) = \gamma$, where $i : (\mathbb{Z}_n)^k \to (S^1)^k = T^k$ is the natural inclusion. The element $\delta$ determines an extension $1 \to (\mathbb{Z}_n)^k \stackrel{j}{\to} G \stackrel{\eta}{\to} F \to 1$. Consider now the new abstract kernel $\psi \circ \eta : G \to \mathrm{Out}(\Pi)$.

From the commutative diagram

$$\begin{array}{ccccc}
\delta \in & H^2(F;(\mathbb{Z}_n)^k) & \stackrel{\eta^*}{\longrightarrow} & H^2(G;(\mathbb{Z}_n)^k) \\
& i_* \downarrow & & i_* \downarrow \\
(d^2)^{-1} \circ (\alpha\psi) = \gamma \in & H^2(F;T^k) & \stackrel{\eta^*}{\longrightarrow} & H^2(G;T^k) \\
& \alpha'_* \downarrow & & \alpha'_* \downarrow \\
& H^2(F;T^k/(\mathbb{Z}_n)^k) & \stackrel{\eta^*}{\longrightarrow} & H^2(G;T^k/(\mathbb{Z}_n)^k)
\end{array}$$

we see that $\eta^*((d^2)^{-1} \cdot o(\psi)) \in H^2(G;T^k)$ is 0 if $\eta^*\delta = 0$. But $\eta^*\delta$ is the pullback of

$$\begin{array}{ccccccc}
1 & \longrightarrow & (\mathbb{Z}_n)^k & \longrightarrow & \eta^*(G) & \longrightarrow & G \\
& & \| & & \downarrow & & \downarrow \eta \\
1 & \longrightarrow & (\mathbb{Z}_n)^k & \longrightarrow & G & \longrightarrow & F.
\end{array}$$

The top sequence splits, and so $\eta^*\delta = 0$ and $o(\eta \circ \psi) = \eta^*(o(\psi)) \in H^3(G;\mathbb{Z}^3) = 0$. Therefore we have shown

THEOREM 11.3.30 ([**Zim80**]). *If $\mathcal{Z}(\Pi) = C \cong \mathbb{Z}^k$, $k > 0$, and $o(\psi) \in H^3_\psi(F;C)$ has order $n$, then there is an extension $1 \to \Pi \to E \to G \to 1$ realizing the abstract kernel $\eta \circ \psi = \psi'$, where $1 \to (\mathbb{Z}_n)^k \to G \to F \to 1$ is an inflation of the abstract kernel $\psi$ as constructed above.*

There are many extensions for each $\psi \circ \eta : G \to \mathrm{Out}(\Pi)$. The set of congruence classes of extensions realizing the kernel $\psi'$ is in one-to-one correspondence with $H^2_{\psi'}(G;\mathcal{Z}(\Pi))$.

THEOREM 11.3.31 ([**Lee82a**], [**Lee82b**], [**Zim80**]). *Let $M$ be a flat manifold. Given a finite subgroup $F$ of $\pi_0(\mathcal{E}(M))$, there always exists a group $F^*$, together with a surjective homomorphism $F^* \to F$ with a finite Abelian kernel such that it can be realized as a group of affine diffeomorphisms of $M$. Furthermore, the finite Abelian kernels are uniformly bounded by $H^1(M;Z)/\mathrm{Center}(\pi_1(M))$.*

11.3.32. The argument in Subsection 11.3.29 is purely algebraic with no assumption on $\Pi$ other than $\mathcal{Z}(\Pi)$ is $\mathbb{Z}^k$. In any case, for an abstract kernel $\psi : F \to \text{Out}(\Pi)$ with $o(\psi) \neq 0$ and of order $n$, there is always an inflation $\eta : G \to F$ of $F$ with kernel $(\mathbb{Z}_n)^k$ and an extension $1 \to \Pi \to E \to G \to 1$ realizing the abstract kernel $\psi' = \psi \circ \eta$. Furthermore, if $M(G(\Pi))$ is a Seifert manifold as in Subsections 11.3.15–11.3.24, with $(Q, W)$ finitely extendable, we can apply the method of proof of Theorem 11.3.15 to topologically realize the abstract kernel $\psi'$ as a group of Seifert automorphisms. For example, in Corollary 11.3.17, the realization will be by affine diffeomorphisms.

The realization may not be effective. Suppose $\psi : F \to \text{Out}(\Pi)$ is injective and $1 \to (\mathbb{Z}_n)^k \to G \xrightarrow{\eta} F \to 1$ is the inflation of $F$. Let $1 \to \Pi \to E \to G \to 1$ be an extension realizing $\psi'$. Then $\theta : E \to \text{TOP}_G(G \times W)$ (or $\text{TOP}_{G,K}(G \times W)$) is injective (i.e., $G$ will be effective) if and only if $C_E(\Pi)$ is torsion free. If not torsion free, then the torsion $T$ of $C_E(\Pi)$, a subgroup of $(\mathbb{Z}_n)^k$, is precisely the kernel of $\theta$. Then $G/T$ acts effectively on $M(\theta(\Pi))$ as Seifert automorphisms realizing the abstract kernel $G/T \to F \to \text{Out}(\Pi)$ whose admissible extension is $1 \to \Pi \to E/T \to G/T \to 1$. The kernel of $G/T \to F$ is $(\mathbb{Z}_n)^k/T$.

Theorem 11.3.31 formulates Theorem 11.3.30 more sharply for the special case of flat manifolds and uses somewhat different arguments than given in Subsection 11.3.29.

For more about the realizations up to strict equivalences and finding examples where $F$ does not lift because there are no extensions realizing the abstract kernels, the reader is referred to [**KLR83**], [**LR96**], [**LR81**], [**ZZ79**], [**LR82**], [**RS77**], [**Lee82a**], [**Lee82b**], [**Igo84**], [**SY79**], and [**Ray79**].

## 11.4. Polynomial structures for solvmanifolds

11.4.1. John Milnor [**Mil77**] asked if every torsion-free polycyclic-by-finite group $\Gamma$ occurs as the fundamental group of a compact, complete affinely flat manifold. This is equivalent to asking if $\Gamma$ can act on $\mathbb{R}^K$ properly as affine motions with $\Gamma \backslash \mathbb{R}^K$ compact.

However, Benoist ([**Ben92**], [**Ben95**]) constructed an example of a 10-step nilpotent group $\Gamma$ of Hirsch length 11 which does not admit an affine structure. This example was generalized to a family of examples by Burde and Grunewald ([**BG95**]). In [**Bur96**], Burde constructs counterexamples of nilpotency class 9 and Hirsch length 10.

11.4.2. A polynomial diffeomorphism $f$ of $\mathbb{R}^n$ is a bijective polynomial transformation of $\mathbb{R}^n$ for which the inverse mapping is again polynomial. Let us write $\text{P}(\mathbb{R}^n)$ for the group consisting of all polynomial diffeomorphisms. Affine diffeomorphisms clearly are polynomial diffeomorphisms of degree less than or equal to 1; smooth actions could be considered as being *polynomial of infinite degree*.

A representation $\theta : \Gamma \to \text{Aff}(\mathbb{R}^K)$ which yields a proper action with $\theta(\Gamma) \backslash \mathbb{R}^K$ compact is called an *affine structure on* $\Gamma$. It is also common to call $\theta(\Gamma)$ an *affine crystallographic group* (ACG) ([**FG83**], [**GS94**]). Analogously to the affine structure, a representation $\theta : \Gamma \to \text{P}(\mathbb{R}^K)$ which yields a proper action with $\theta(\Gamma) \backslash \mathbb{R}^K$ compact is called a *polynomial structure on* $\Gamma$; $\theta(\Gamma)$ is called a *polynomial crystallographic group*.

THEOREM 11.4.3 ([**DI97**]). *Every polycyclic-by-finite group $\Gamma$ admits a polynomial structure of bounded degree. That is, $\Gamma$ can act on $\mathbb{R}^K$ properly as polynomial diffeomorphisms so that $\Gamma\backslash\mathbb{R}^K$ is compact. Moreover, all polynomials involved consist entirely of a bounded degree.*

The case when $\Gamma$ is nilpotent was proved in [**DIL96**]. The construction of this polynomial structure is a special case of an iterated Seifert fiber space construction, which can be achieved here because of a very strong cohomology vanishing theorem, Theorem 11.4.11.

11.4.4 (Polynomial diffeomorphisms). Write $\mathrm{P}(\mathbb{R}^K, \mathbb{R}^k)$ for the real vector space of polynomial mappings from $\mathbb{R}^K$ to $\mathbb{R}^k$. An element $p(x_1, \ldots, x_K)$ of $\mathrm{P}(\mathbb{R}^K, \mathbb{R}^k)$ consists of $k$ polynomials in $K$ variables:

$$p(x_1, \ldots, x_K) = \begin{pmatrix} p_1(x_1, x_2, \ldots, x_K) \\ p_2(x_1, x_2, \ldots, x_K) \\ \vdots \\ p_k(x_1, x_2, \ldots, x_K) \end{pmatrix}, \text{ with } p_i(x_1, \ldots, x_K) \in \mathrm{P}(\mathbb{R}^K, \mathbb{R}).$$

By the degree of $p$, denoted by $\deg(p)$, we mean the maximum of the degrees of the $p_i$ ($1 \leq i \leq k$). Note in particular, that $\mathrm{P}(\mathbb{R}^K, \mathbb{R}^k)$ contains $\mathbb{R}^k$ as the subgroup of constant mappings (degree-0 mappings).

We denote by $\mathrm{P}(\mathbb{R}^K)$ the group of polynomial diffeomorphisms of $\mathbb{R}^K$. Here, the group-law is composition of mappings (so $\mathrm{P}(\mathbb{R}^K)$ is a subset of $\mathrm{P}(\mathbb{R}^K, \mathbb{R}^K)$, but not a subgroup, because the latter has the addition as the group operation). Elements of $\mathrm{P}(\mathbb{R}^K)$ are polynomial bijections whose inverse mappings are again polynomials.

EXAMPLE 11.4.5. Let $p, q : \mathbb{R}^2 \to \mathbb{R}^2$ be such that

$$p(x, y) = (y + 1, x + y^2) \text{ and } q(x, y) = (y - x^2 + 2x - 1, x - 1).$$

Clearly, they are inverse to each other in $\mathrm{P}(\mathbb{R}^2)$.

11.4.6. The vector space $\mathrm{P}(\mathbb{R}^K, \mathbb{R}^k)$ has $\mathrm{GL}(\mathbb{R}^k) \times \mathrm{P}(\mathbb{R}^K)$-module structure, via

$$\forall (g, h) \in \mathrm{GL}(\mathbb{R}^k) \times \mathrm{P}(\mathbb{R}^K), \forall p \in \mathrm{P}(\mathbb{R}^K, \mathbb{R}^k) : {}^{(g,h)}p = g \circ p \circ h^{-1}.$$

The resulting semidirect product $\mathrm{P}(\mathbb{R}^K, \mathbb{R}^k) \rtimes (\mathrm{GL}(\mathbb{R}^k) \times \mathrm{P}(\mathbb{R}^K))$ embeds into $\mathrm{P}(\mathbb{R}^{k+K})$ as follows: $\forall p \in \mathrm{P}(\mathbb{R}^K, \mathbb{R}^k), \forall g \in \mathrm{GL}(\mathbb{R}^k), \forall h \in \mathrm{P}(\mathbb{R}^K)$ :

$$\forall x \in \mathbb{R}^k, \forall y \in \mathbb{R}^K : (p, g, h)(x, y) = (g(x) - p(h(y)), h(y)).$$

11.4.7. The crux of the construction is the iteration of the following procedure. Let

$$1 \to \mathbb{Z}^k \to \Pi \to Q \to 1$$

be an exact sequence with abstract kernel $\varphi : Q \to \mathrm{GL}(k, \mathbb{R})$. Let

$$\rho : Q \to \mathrm{P}(\mathbb{R}^K)$$

be a representation which yields a proper action of $Q$ on $\mathbb{R}^K$ with $Q\backslash\mathbb{R}^K$ compact. We try to find a homomorphism $\theta : \Pi \to \mathrm{P}(\mathbb{R}^K, \mathbb{R}^k) \rtimes (\mathrm{GL}(k, \mathbb{R}) \times \mathrm{P}(\mathbb{R}^K))$ so that the diagram

$$1 \longrightarrow \mathbb{Z}^k \longrightarrow \Pi \longrightarrow Q \longrightarrow 1$$
$$\downarrow i \qquad \downarrow \theta \qquad \downarrow \varphi \times \rho$$
$$1 \to \mathrm{P}(\mathbb{R}^K, \mathbb{R}^k) \to \mathrm{P}(\mathbb{R}^K, \mathbb{R}^k) \rtimes (\mathrm{GL}(k,\mathbb{R}) \times \mathrm{P}(\mathbb{R}^K)) \to \mathrm{GL}(k,\mathbb{R}) \times \mathrm{P}(\mathbb{R}^K) \to 1,$$

where $i: \mathbb{Z}^k \to \mathbb{R}^k \subset \mathrm{P}(\mathbb{R}^K, \mathbb{R}^k)$ is the standard translations, is commutative. Note that

$$\mathrm{P}(\mathbb{R}^K, \mathbb{R}^k) \subset \mathrm{M}(\mathbb{R}^K, \mathbb{R}^k) \quad \text{and} \quad \mathrm{P}(\mathbb{R}^K) \subset \mathrm{TOP}(\mathbb{R}^K),$$

and therefore,

$$\mathrm{P}(\mathbb{R}^K, \mathbb{R}^k) \rtimes (\mathrm{GL}(k,\mathbb{R}) \times \mathrm{P}(\mathbb{R}^K)) \xrightarrow{\subset} \mathrm{P}(\mathbb{R}^{K+k})$$
$$\cap \downarrow \qquad\qquad \cap \downarrow$$
$$\mathrm{M}(\mathbb{R}^K, \mathbb{R}^k) \rtimes (\mathrm{GL}(k,\mathbb{R}) \times \mathrm{TOP}(\mathbb{R}^K)) \xrightarrow{\subset} \mathrm{TOP}(\mathbb{R}^{K+k});$$

see Corollary 4.2.10.

11.4.8 (Canonical type polynomial representations). It is well known ([**Seg83**, lemma 6, pp.16]) that, if $\Gamma$ is a polycyclic-by-finite group, then there exists an ascending sequence (or filtration) of normal subgroups $\Gamma_i$ ($0 \leq i \leq c+1$) of $\Gamma$

(11.4.1) $\qquad \Gamma_*: \ \Gamma_0 = 1 \subseteq \Gamma_1 \subseteq \Gamma_2 \subseteq \cdots \subseteq \Gamma_{c-1} \subseteq \Gamma_c \subseteq \Gamma_{c+1} = \Gamma$

for which

$$\Gamma_i / \Gamma_{i-1} \cong \mathbb{Z}^{k_i} \text{ for } 1 \leq i \leq c \text{ and some } k_i \in \mathbb{N}_0 \text{ and } \Gamma/\Gamma_c \text{ is finite}.$$

Let us call such a filtration of $\Gamma$ a *torsion-free filtration* (of length $c$). We will also use $K_i = k_i + k_{i+1} + \cdots + k_c$ and $K_{c+1} = 0$. It follows that $h(\Gamma) = K_1$, the *Hirsh number* (or rank) of $\Gamma$.

DEFINITION 11.4.9. For every $i$, write $\varphi_i : \Gamma/\Gamma_i \to \mathrm{Aut}(\mathbb{Z}^{k_i})$ for the morphism induced by the short exact sequence

$$1 \to \mathbb{Z}^{k_i} (\cong \Gamma_i/\Gamma_{i-1}) \to \Gamma/\Gamma_{i-1} \to \Gamma/\Gamma_i \to 1.$$

A polynomial representation $\rho = \rho_0 : \Gamma \to \mathrm{P}(\mathbb{R}^{h(\Gamma)})$ will be called *of canonical type* with respect to $\Gamma_*$ (or simply of canonical type) if and only if it induces a sequence of representations

$$\rho_i : \Gamma/\Gamma_i \to \mathrm{P}(\mathbb{R}^{K_{i+1}}) \ (1 \leq i \leq c)$$

and a sequence of morphisms

$$j_i : \mathbb{Z}^{k_i} \hookrightarrow \mathbb{R}^{k_i} \to \mathrm{P}(\mathbb{R}^{K_{i+1}}, \mathbb{R}^{k_i}) \ (1 \leq i \leq c)$$

such that for all $i$ the following diagram commutes:
(11.4.2)
$$1 \to \mathbb{Z}^{k_i} \approx \Gamma_i/\Gamma_{i-1} \longrightarrow \Gamma/\Gamma_{i-1} \longrightarrow \Gamma/\Gamma_i \longrightarrow 1$$
$$\downarrow j_i \qquad\qquad \downarrow \rho_{i-1} \qquad\qquad \downarrow \psi_i \times \rho_i$$
$$1 \to \mathrm{P}(\mathbb{R}^{K_{i+1}}, \mathbb{R}^{k_i}) \to \mathrm{P}(\mathbb{R}^{K_{i+1}}, \mathbb{R}^{k_i}) \rtimes (\mathrm{GL}(\mathbb{R}^{k_i}) \times \mathrm{P}(\mathbb{R}^{K_{i+1}})) \to \mathrm{GL}(\mathbb{R}^{k_i}) \times \mathrm{P}(\mathbb{R}^{K_{i+1}}) \to 1,$$

where $\psi_i$ is the unique morphism $\psi_i : \Gamma/\Gamma_i \to \mathrm{GL}(\mathbb{R}^{k_i})$ satisfying

$$\forall \bar{\gamma} \in \Gamma/\Gamma_i, \ \forall z \in \mathbb{Z}^{k_i} : \psi_i(\bar{\gamma})(j_i(z)) = j_i(\varphi_i(\bar{\gamma})z);$$

i.e., $\psi_i$ is the abstract kernel for the top extension sequence.

11.4.10. Iterating this procedure, we will have found a desired homomorphism $\Gamma \to P(\mathbb{R}_K)$. The existence of $\rho_{i-1}$ is guaranteed by
$$H^2(\Gamma/\Gamma_i; P(\mathbb{R}^{K_{i+1}}, \mathbb{R}_i^k)) = 0$$
as the proof of the general construction shows; see Theorem 7.3.2. Also,
$$H^1(\Gamma/\Gamma_i; P(\mathbb{R}^{K_{i+1}}, \mathbb{R}_i^k)) = 0$$
guarantees the uniqueness of such $\rho_{i-1}$ (with fixed $j_i$ and $\psi_i \times \rho_i$). These are achieved by the following theorem. In fact, the major work of the paper [**DI97**] is a proof of the following.

THEOREM 11.4.11 (Main cohomology vanishing theorem). *If $\Gamma$ is a polycyclic-by-finite group admitting a canonical type polynomial representation $\rho : \Gamma \to P(\mathbb{R}^m)$, then, for every representation $\varphi : \Gamma \to GL(\mathbb{R}^n)$ and for all $i > 0$, $H^i_{\varphi \times \rho}(\Gamma; P(\mathbb{R}^m, \mathbb{R}^n))$ = 0.*

EXAMPLE 11.4.12. Take $N$ the discrete Heisenberg group
$$N = \langle a, b, c \mid [b, a] = c,\ [c, a] = [c, b] = 1 \rangle$$
equipped with the torsion-free filtration
$$N_* : N_0 = 1 \subseteq N_1 = Z(N) = \langle c \rangle \subseteq N_2 = N \subseteq N_3 = N.$$
In this case $k_1 = 1$ and $k_2 = 2$. Let $q(y)$ be any polynomial over the field of real numbers. Then the morphism $\rho_q : N \to P(\mathbb{R}^3)$ with
$$\rho_q(a)(x, y, z) = (x, y, z+1),\ \rho(b)(x, y, z) = (x + q(y) + z, y + 1, z),$$
$$\rho(c)(x, y, z) = (x + 1, y, z)$$
is a canonical type polynomial representation of $N$ with respect to $N_*$. The upper bound on the degrees of the polynomials involved in $\rho$ can be as big as one wants by choosing a different $q(y)$.

11.4.13 (Rigidity of polynomial structures). Two polynomial actions $\rho_1$ and $\rho_2$ of $\Gamma$ are said to be *polynomially conjugated* if there exists $p$ in $P(\mathbb{R}^n)$ such that, for all $g$ in $\Gamma$, one has $p \circ \rho_1(g) = \rho_2(g) \circ p$.

THEOREM 11.4.14 ([**BD02**, Theorem 1.1]). *Let $\Gamma$ be a polycyclic-by-finite group. Then any two polynomial crystallographic actions of $\Gamma$ of bounded degree on some $\mathbb{R}^n$ are polynomially conjugated.*

Note that this statement does not assume any *canonical type* embeddings. The main tool in the proof is the notion of algebraic hull of a polynomial action of bounded degree. Let $P(\mathbb{R}^n)$ be the group of polynomial bijections of $\mathbb{R}^n$ with polynomial inverse, and let $P^d(\mathbb{R}^n)$ be the subset of polynomial bijections $p$ such that the degrees of $p$ and $p^{-1}$ are bounded by $d$.

A regular map $i : X \to Y$ between two real algebraic varieties is said to be a *closed immersion* if the image $i(X)$ is Zariski closed and if the map $i^* : \mathbb{R}[Y] \to \mathbb{R}[X] : \phi \mapsto \phi \circ i$ is surjective.

Let $\rho : \Gamma \to P(\mathbb{R}^n)$ be a polynomial action of $\Gamma$ on $\mathbb{R}^n$. The Zariski closure $G := A(\rho(\Gamma))$ of $\rho(\Gamma)$ in $P^d(\mathbb{R}^n)$ is a subgroup of $P(\mathbb{R}^n)$ which does not depend on the choice of $d$. The real algebraic group $A(\rho(\Gamma))$ is called the *algebraic hull* of $\rho(\Gamma)$. A subgroup of $P(\mathbb{R}^n)$ is said to be *Zariski closed* if it is of bounded degree and equal to its algebraic hull.

PROPOSITION 11.4.15. *Let $\Gamma \subseteq \mathrm{P}(\mathbb{R}^n)$ be a polycyclic-by-finite crystallographic subgroup of bounded degree. Then the unipotent radical $U(\Gamma)$ of $A(\Gamma)$ acts simply transitively on $\mathbb{R}^n$.*

PROPOSITION 11.4.16. *For $i = 1, 2$, let $G_i$ be a Zariski closed subgroup of $\mathrm{P}(\mathbb{R}^n)$, suppose that the Zariski connected component of $G_i$ is solvable, and that the unipotent radical $U_i$ of $G_i$ acts simply transitively on $\mathbb{R}^n$. Then, for any isomorphism of algebraic groups $F : G_1 \to G_2$, there exists an element $p$ in $\mathrm{P}(\mathbb{R}^n)$ such that, for all $g_1$ in $G_1$, $p \circ g_1 = F(g_1) \circ p$.*

11.4.17 (Proof of Theorem 11.4.14). Let us denote by $\rho_1, \rho_2 : \Gamma \to \mathrm{P}(\mathbb{R}^n)$ these two actions. First of all recall that the kernel of $\rho_i$ ($i = 1, 2$) is the unique maximal finite normal subgroup $F_\Gamma$ of $\Gamma$. Therefore, we can assume, without loss of generality, that $F_\Gamma = 1$ and that $\rho_1$ and $\rho_2$ are injective. Moreover, by [**Rag72**, Lemma 4.41], we know that the isomorphism $\rho_2 \circ \rho_1^{-1} : \rho_1(\Gamma) \to \rho_2(\Gamma)$ extends to an isomorphism of algebraic groups $F : A(\rho_1(\Gamma)) \to A(\rho_2(\Gamma))$. By Propositions 11.4.15 and 11.4.16, there exists an element $p$ in $\mathrm{P}(\mathbb{R}^n)$ such that, for all $a$ in $A(\rho_1(\Gamma))$, one has $F(a) \circ p = p \circ a$. This map $p$ is again the one we are looking for.

## 11.5. Applications to fixed-point theory

We show that Bieberbach's rigidity theorem for flat manifolds still holds true for any continuous maps on infra-nilmanifolds. Namely, every endomorphism of an almost crystallographic group is semiconjugate to an affine endomorphism. Applying this result to fixed-point theory, we obtain a criterion for the Lefschetz number and Nielsen number for a map on infra-nilmanifolds to be equal. Some material is taken from [**Lee95b**].

11.5.1. Let $G$ be a connected Lie group. Consider the semigroup $\mathrm{Endo}(G)$, the set of all endomorphisms of $G$, with the composition as operation. We form the semidirect product $G \rtimes \mathrm{Endo}(G)$ and call it $\mathrm{aff}(G)$. With the binary operation

$$(a, A)(b, B) = (a \cdot Ab, AB),$$

the set $\mathrm{aff}(G)$ forms a semigroup with identity $(e, I)$, where $e \in G$ and $I \in \mathrm{Endo}(G)$ are the identity elements. The semigroup $\mathrm{aff}(G)$ "acts" on $G$ by

$$(a, A) \cdot x = a \cdot Ax.$$

Note that $(a, A)$ is not a homeomorphism unless $A \in \mathrm{Aut}(G)$. Clearly, $\mathrm{aff}(G)$ is a subsemigroup of the semigroup of all continuous maps of $G$ into itself, for $((a, A)(b, B))x = (a, A)((b, B)x)$ for all $x \in G$. We call elements of $\mathrm{aff}(G)$ *affine endomorphisms*.

11.5.2 (Generalization of the Second Bieberbach Theorem). Let $G$ be a connected and simply connected nilpotent Lie group. In Section 8.4, we have seen that, for any isomorphism between two almost crystallographic groups, is a conjugation by an element of $\mathrm{Aff}(G)$. We shall generalize this result to all homomorphisms (not necessarily isomorphisms). Topologically, this implies that every continuous map on an infra-nilmanifold is homotopic to a map induced by an affine endomorphism on the Lie group level. It can be stated as: every endomorphism of an almost crystallographic group is *semiconjugate* to an affine endomorphism.

THEOREM 11.5.3. *Let $\Pi, \Pi' \subset \mathrm{Aff}(G)$ be two almost crystallographic groups. Then for any homomorphism $\theta : \Pi \to \Pi'$, there exists $g = (d, D) \in \mathrm{aff}(G)$ such that $\theta(\alpha) \cdot g = g \cdot \alpha$ for all $\alpha \in \Pi$.*

EXAMPLE 11.5.4. The subgroup $\Gamma = \Pi \cap G$ of an almost crystallographic group $\Pi$ is characteristic, but not fully invariant. The homomorphism $\theta$ in Theorem 11.5.3 may not map the maximal normal nilpotent subgroup $\Gamma$ of $\Pi$ into that of $\Pi'$. This causes a lot of trouble. Let $\Pi$ be an orientable 4-dimensional Bieberbach group with holonomy group $\mathbb{Z}_2$. More precisely, $\Pi \subset \mathbb{R}^4 \rtimes O(4) = E(4) \subset \mathrm{Aff}(\mathbb{R}^4)$ is generated by $(e_1, I), (e_2, I), (e_3, I), (e_4, I)$, and $(a, A)$, where $a = (1/2, 0, 0, 0)^t$, and $A$ is diagonal matrix with diagonal entries $1, -1, -1$, and $1$. Note that $(a, A)^2 = (e_1, I)$. The subgroup generated by $(e_1, I), (e_2, I), (e_3, I)$, and $(a, A)$ forms a 3-dimensional Bieberbach group $\mathcal{G}_2$, and $\Pi = \mathcal{G}_2 \times \mathbb{Z}$. Consider the endomorphism $\theta : \Pi \to \Pi$ which is the composite $\Pi \to \mathbb{Z} \to \Pi$, where the first map is the projection onto $\mathbb{Z} = \langle (e_4, I) \rangle$ and the second map sends $(e_4, I)$ to $(a, A)$. Thus the homomorphism $\theta$ does not map the maximal normal Abelian subgroup $\mathbb{Z}^4$ (generated by the four translations) into itself. Such a $\mathbb{Z}^4$ is characteristic but not fully invariant in $\Pi$. Let

$$d = \begin{bmatrix} x \\ 0 \\ 0 \\ y \end{bmatrix}, \quad D = \begin{bmatrix} 0 & 0 & 0 & 1/2 \\ 0 & 0 & 0 & 0 \\ 0 & 0 & 0 & 0 \\ 0 & 0 & 0 & 0 \end{bmatrix},$$

and let $g = (d, D)$. Then it is easy to see that $\theta(\alpha) \cdot g = g \cdot \alpha$ for all $\alpha \in \Pi$.

It turns out that the element $g = (d, D)$ is the most general form. The matrix $D$ is uniquely determined and the translation part $d$ can vary only in two dimensions.

11.5.5 (Proof of Theorem). Let $\Gamma = \Pi \cap G$, $\Gamma' = \Pi' \cap G$. As the example shows, the characteristic subgroup $\Gamma$ may not go into $\Gamma'$ by the homomorphism $\theta$. Let $p$ be the product of the orders of $\Pi/\Gamma$ and $\Pi'/\Gamma'$. Let $\Lambda, \Lambda'$ be the normal subgroups of $\Pi, \Pi'$ generated by

$$\{x^p : x \in \Pi\} \text{ and } \{y^p : y \in \Pi'\}.$$

Then $\Lambda$ and $\Lambda'$ are fully invariant subgroups of $\Pi$ and $\Pi'$, and they have finite indices, both lying in $G$. Clearly, $\theta$ maps $\Lambda$ into $\Lambda'$. Let $Q = \Pi/\Lambda$.

Consider the homomorphism $\Lambda \xrightarrow{\theta} \Lambda' \hookrightarrow G$. Since $\Lambda$ is a lattice of $G$, by Malćev's work, any such a homomorphism extends uniquely to a continuous homomorphism $C : G \to G$; cf. [**Lee92**, 2.11]. Thus, $\theta|_\Lambda = C|_\Lambda$, where $C \in \mathrm{Endo}(G)$, and hence, $\theta(z, 1) = (Cz, 1)$ for all $z \in \Lambda$ (more precisely, $(z, 1) \in \Lambda$).

Let us denote the composite homomorphism $\Pi \xrightarrow{\theta} \Pi' \hookrightarrow G \rtimes \mathrm{Aut}(G) \to \mathrm{Aut}(G)$ by $\bar{\theta}$ and define a map $f : \Pi \to G$ by

(11.5.1) $$\theta(w, K) = (Cw \cdot f(w, K), \bar{\theta}(w, K)).$$

For any $(z, 1) \in \Lambda$ and $(w, K) \in \Pi$, apply $\theta$ to both sides of $(w, K)(z, 1)(w, K)^{-1} = (w \cdot Kz \cdot w^{-1}, 1)$ to get $Cw \cdot f(w, K) \cdot \bar{\theta}(w, K)(Cz) \cdot f(w, K)^{-1} \cdot (Cw)^{-1} = \theta(w \cdot Kz \cdot w^{-1})$. However, $w \cdot Kz \cdot w^{-1} \in \Lambda$ since $\Lambda$ is normal in $\Pi$, and the latter term equals to $C(w \cdot Kz \cdot w^{-1}) = Cw \cdot CKz \cdot (Cw)^{-1}$ since $C : G \to G$ is a homomorphism. From this we have

(11.5.2) $$\bar{\theta}(w, K)(Cz) = f(w, K)^{-1} \cdot CKz \cdot f(w, K).$$

This is true for all $z \in \Lambda$. Note that $\bar\theta(w,K)$ and $K$ are automorphisms of the Lie group $G$ and $C: G \to G$ is an endomorphism. By the uniqueness of extension of a homomorphism $\Lambda \to G$ to an endomorphism $G \to G$, as mentioned above, the equality (2) holds true for all $z \in G$. It is also easy to see that $f(zw,K) = f(w,K)$ for all $z \in \Lambda$ so that $f: \Pi \to G$ does not depend on $\Lambda$. Thus, $f$ factors through $Q = \Pi/\Lambda$. Moreover, $\bar\theta: \Pi \to \mathrm{Aut}(G)$ also factors through $Q$ since $\Lambda$ maps trivially into $\mathrm{Aut}(G)$. We still use the notation $(w,K)$ to denote elements of $Q$ and $\bar\theta$ to denote the induced map $Q \to \mathrm{Aut}(G)$.

We claim that: *with the $Q$-structure on $G$ via $\bar\theta: Q \to \mathrm{Aut}(G)$, $f \in Z^1(Q; G)$; i.e., $f: Q \to G$ is a crossed homomorphism.*

We shall show $f((w,K) \cdot (w',K')) = f(w,K) \cdot \bar\theta(w,K) f(w',K')$ for all $(w,K)$, $(w',K') \in \Pi$. (Note that we are using the elements of $\Pi$ to denote the elements of $Q$.) Apply $\theta$ to both sides of $(w,K)(w',K') = (w \cdot Kw', KK')$ to get $Cw \cdot f(w,K) \cdot \bar\theta(w,K)[Cw' \cdot f(w',K')] = C(w \cdot Kw') \cdot f((w,K)(w',K'))$. From this it follows that

$$f((w,K)(w',K')) = (CKw')^{-1} \cdot f(w,K) \cdot \bar\theta(w,K)(Cw') \cdot \bar\theta(w,K) f(w',K').$$

From (2) we have $\bar\theta(w,K) Cw' = f(w,K)^{-1} \cdot CKw' \cdot f(w,K)$ so that $f((w,K) \cdot (w',K')) = f(w,K) \cdot \bar\theta(w,K) f(w',K')$.

According to Theorem 8.4.3 (and its proof), it was proved that $H^1(Q;G) = 0$ whenever $Q$ is a finite group and $G$ is a connected and simply connected nilpotent Lie group. The proof uses induction on the nilpotency of $G$ together with the fact that $H^1(Q;G) = 0$ for a finite group $Q$ and a real vector group $G$. This means that any crossed homomorphism is *principal*. In other words, there exists $d \in G$ such that

(11.5.3) $$f(w,K) = d \cdot \bar\theta(w,K)(d^{-1}).$$

Let $D = \mu(d^{-1}) \circ C$ and $g = (d, D) \in \mathrm{aff}(G)$, and we check that $\theta$ is *conjugation* by $g$. Using equalities (11.5.1), (11.5.2), and (11.5.3), one can show $\bar\theta(w,K) \circ \mu(d^{-1}) \circ C = \mu(d^{-1}) \circ C \circ K$. Thus, for any $(w,K) \in \Pi$,

$$\begin{aligned}
\theta(w,K) \cdot (d,D) &= (Cw \cdot f(w,K),\ \bar\theta(w,K)) \cdot (d,\ \mu(d^{-1}) \circ C) \\
&= (Cw \cdot f(w,K) \cdot \bar\theta(w,K)(d),\ \bar\theta(w,K) \circ \mu(d^{-1}) \circ C) \\
&= (Cw \cdot d \cdot \bar\theta(w,K)(d^{-1}) \cdot \bar\theta(w,K)(d),\ \bar\theta(w,K) \circ \mu(d^{-1}) \circ C) \\
&= (Cw \cdot d,\ \mu(d^{-1}) \circ C \circ K) \\
&= (d,D) \cdot (w,K).
\end{aligned}$$

This finishes the proof of theorem. $\square$

COROLLARY 11.5.6. *Let $M = \Pi \backslash G$ be an infra-nilmanifold, and $h: M \to M$ be any map. Then $h$ is homotopic to a map induced from an affine endomorphism $G \to G$.*

PROOF. Since $M$ is a manifold, $h$ can be homotoped to a map with a fixed point, say $x$. We start with the homomorphism $h_*: \pi_1(M,x) \to \pi_1(M,x)$, induced from $h$, as our $\theta$ in Theorem 11.5.3, and obtain $\tilde g = (d,D)$ satisfying

$$h_*(\alpha) \circ \tilde g = \tilde g \circ \alpha.$$

Let $g: M \to M$ be the induced map. Then $h_* = g_*$. Since any two continuous maps on a closed aspherical manifold inducing the same homomorphism on the fundamental group (up to conjugation by an element of the fundamental group) are homotopic to each other, $h$ is homotopic to $g$. This completes the proof of the corollary. $\square$

COROLLARY 11.5.7 (Corollary 8.4.4). *Homotopy equivalent infra-nilmanifolds are affinely diffeomorphic.*

Now we consider the uniqueness problem in Theorem 11.5.3: How many $g$'s are there? Let $\Phi = \Pi/(G \cap \Pi) \subset \mathrm{Aut}(G)$ and $\Phi' = \Pi'/(G \cap \Pi') \subset \mathrm{Aut}(G)$ be the holonomy groups of $\Pi$ and $\Pi'$. Let $\Psi'$ be the image of $\theta(\Pi)$ in $\Phi'$. So $\Psi' \subset \Phi' \subset \mathrm{Aut}(G)$. Let $G^{\Psi'}$ denote the fixed-point set of the action. Recall the notation: For $c \in G$, $\mu(c)$ denotes conjugation by $c$. Therefore, $\mu(c)(x) = cxc^{-1}$ for all $x \in G$.

PROPOSITION 11.5.8 (Uniqueness). *With the same notation as above, suppose $\theta(\alpha) \cdot g = g \cdot \alpha$ for all $\alpha \in \Pi$. Then $\theta(\alpha) \cdot \gamma = \gamma \cdot \alpha$ for all $\alpha \in \Pi$ if and only if $\gamma = \xi \cdot g$, where $\xi = (c, \mu(c^{-1}))$, for $c \in G^{\Psi'}$. Therefore, $D$ is unique up to $\mathrm{Inn}(G)$. If $\theta$ is an isomorphism, then $c \in G^{\Phi'}$. In particular, if $\Pi$ is a Bieberbach group with $H^1(\Pi;\mathbb{R}) = 0$ and $\theta$ is an isomorphism, then such a $g$ is unique.*

PROOF. Let $g = (d, D)$, $\gamma = (c, C)$. Since $\theta(\alpha) \cdot g = g \cdot \alpha$ holds when $\alpha = (z, 1) \in \Lambda$, we have $Dz = d^{-1}z'd$, where $\theta(z, 1) = (z', 1)$. Similarly, $Cz = c^{-1}z'c$. Thus $Cz = \mu(c^{-1}d)Dz$ for all $z \in \Lambda$. Since $\Lambda$ is a lattice, this equality holds on $G$. Consequently, $C = \mu(c^{-1}d)D$. Now $\gamma = (c, C) = (c, \mu(c^{-1}d)D) = (d^{-1}c, \mu(c^{-1}d))(d, D) = (h, \mu(h^{-1}))(d, D)$, if we let $h = d^{-1}c$. Set $\xi = (h, \mu(h^{-1}))$. Then $\gamma = \xi \cdot g$. Now we shall observe that $h \in G^{\Psi'}$. Let $\theta(\alpha) = (b, B)$. Then $\theta(\alpha)\xi g = \theta(\alpha)\gamma = \gamma\alpha = \xi g\alpha = \xi\theta(\alpha)g$ yields $Bh = h$ for all $(b, B) = \theta(\alpha)$. Clearly then $B \in \Psi'$ by definition. For a Bieberbach group $\Pi$, note that rank $H^1(\Pi;\mathbb{Z}) = \dim G^\Phi$. □

11.5.9 (Application to fixed-point theory). Let $M$ be a closed manifold, and let $f : M \to M$ be a continuous map. The *Lefschetz number* $L(f)$ of $f$ is defined by

$$L(f) := \sum_k (-1)^k \mathrm{Trace}\{(f_*)_k : H_k(M;\mathbb{Q}) \to H_k(M;\mathbb{Q})\}$$

To define the *Nielsen number* $N(f)$ of $f$, we define an equivalence relation on $\mathrm{Fix}(f)$ as follows: For $x_0, x_1 \in \mathrm{Fix}(f)$, $x_0 \sim x_1$ if and only if there exists a path $c$ from $x_0$ to $x_1$ such that $c$ is homotopic to $f \circ c$ relative to the end points. An equivalence class of this relation is called a *fixed-point class* (FPC) of $f$. To each FPC $F$, one can assign an integer $\mathrm{ind}(f, F)$. An FPC $F$ is called *essential* if $\mathrm{ind}(f, F) \neq 0$. Now,

$$N(f) := \text{the number of essential fixed-point classes.}$$

These two numbers give information on the existence of fixed-point sets. If $L(f) \neq 0$, every self-map of $M$ homotopic to $f$ has a nonempty fixed-point set. The Nielsen number is a lower bound for the number of components of the fixed-point set of all maps homotopic to $f$. Even though $N(f)$ gives more information than $L(f)$ does, it is harder to calculate. If $M$ is an infra-nilmanifold, and $f$ is homotopically periodic, then it will be shown that $L(f) = N(f)$.

LEMMA 11.5.10. *Let $B \in \mathrm{GL}(n,\mathbb{R})$ with a finite order. Then $\det(I - B) \geq 0$.*

PROOF. Since $B$ has finite order, it can be conjugated into the orthogonal group $O(n)$. Since all eigenvalues are roots of unity, there exists $P \in \mathrm{GL}(n,\mathbb{R})$ such that $PBP^{-1}$ is a block diagonal matrix, with each block being a $(1 \times 1)$- or a $(2 \times 2)$-matrix. All $(1 \times 1)$-blocks must be $D = [\pm 1]$, and hence $\det(I - D) = 0$ or 2. For a $(2 \times 2)$-block, it is of the form $\begin{bmatrix} \cos t & \sin t \\ -\sin t & \cos t \end{bmatrix}$. Consequently, each $(2 \times 2)$-block $D$ has the property that $\det(I - D) = (1 - \cos t)^2 + \sin^2 t = 2(1 - \cos t) \geq 0$. □

THEOREM 11.5.11. *Let $f : M \to M$ be a continuous map on an infra-nilmanifold $M = \Pi\backslash G$. Let $g = (d, D) \in \mathrm{aff}(G)$ be a homotopy lift of $f$ by Corollary 11.5.6. Then $L(f) = N(f)$ (respectively, $L(f) = -N(f)$) if and only if $\det(I - D_*A_*) \geq 0$ (respectively, $\det(I - D_*A_*) \leq 0$) for all $A \in \Phi$, the holonomy group of $M$.*

PROOF. Since $L(f)$ and $N(f)$ are homotopy invariants, we may assume that $f$ is the map induced from $g$. Let $\Gamma = \Pi \cap G$. Then $\Gamma$ is a normal subgroup of $\Pi$, of finite index, say $p$. Let $\Lambda$ be the normal subgroups of $\Pi$ generated by $\{x^p : x \in \Pi\}$. Then $f_* : \Pi \to \Pi$ maps $\Lambda$ into itself. Therefore, $f$ induces a map on the finite-sheeted regular covering space $\Lambda\backslash G$ of $\Pi\backslash G$.

Let $\widetilde{f}$ be a lift of $f$ to $\Lambda\backslash G$. Then

$$L(f) = \frac{1}{[\Pi:\Lambda]} \sum L(\alpha\widetilde{f}) = \frac{1}{|\Phi|} \sum_{A \in \Phi} \frac{\det(A_* - D_*)}{\det A_*},$$

$$N(f) = \frac{1}{[\Pi:\Lambda]} \sum |N(\alpha\widetilde{f})| = \frac{1}{|\Phi|} \sum_{A \in \Phi} |\det(A_* - D_*)|,$$

where the sum ranges over all $\alpha \in \Pi/\Lambda$; see, [**Jia83**, III 2.12] and [**KLL05**, Theorem 3.5] and [**LL06**, Theorem 3.4]. Since $|\det A_*| = 1$ and $\det(A_* - D_*)/\det A_* = \det(I - D_*A_*^{-1})$, it is easy to see from the formula above that the theorem is proved. □

COROLLARY 11.5.12 ([**KL88b**]). *Let $f : M \to M$ be a homotopically periodic map on an infra-nilmanifold. Then $N(f) = L(f)$.*

PROOF. Here is an argument which is completely different from the one in [**KL88b**]. Let $\Gamma = \Pi \cap G$ and $\Phi = \Pi/\Gamma$, the holonomy group. Let $g = (d, D) \in G \rtimes \mathrm{Aut}(G)$ be a homotopy lift of $f$ to $G$. Let $E$ be the lifting group of the action of $\langle g \rangle$ to $G$. That is, $E$ is generated by $\Pi$ and $g$. Then $E/\Gamma$ is a finite group generated by $\Phi$ and $D$. For every $A \in \Phi$, $DA$ lies in $E/\Gamma$, and has a finite order. By Lemma 11.5.10, $\det(I - DA) \geq 0$ for all $A \in \Phi$. By Theorem 11.5.11, $L(f) = N(f)$. □

COROLLARY 11.5.13 ([**McC94**]). *Let $f : M \to M$ be a homotopically periodic map on an infra-solvmanifold. Then $N(f) = L(f)$.*

PROOF. In [**Lee92**], the statement for solvmanifolds was proved. We needed a subgroup invariant under $f_*$. To achieve this, a new model space $M'$ which is homotopy equivalent to $M$, together with a map $f' : M' \to M'$ corresponding to $f$ was constructed. The new space $M'$ is a fiber bundle over a torus with fiber a nilmanifold; and $f'$ is fiber preserving. Moreover, we found a fully invariant subgroup $\Lambda$ of $\Pi$ of finite index (so, it is invariant under $f'_*$). Now we can apply the same argument as in the proof of Theorem 11.5.11. □

EXAMPLE 11.5.14. Let $\Pi$ be an orientable 3-dimensional Bieberbach group with holonomy group $\mathbb{Z}_2$. More precisely, $\Pi \subset \mathbb{R}^3 \rtimes O(3) = E(3)$ is generated by $(e_1, I), (e_2, I), (e_3, I)$, and $(a, A)$, where $a = (1/2, 0, 0)^t$, $A$ is a diagonal matrix with diagonal entries $1, -1$, and $-1$. Note that $(a, A)^2 = (e_1, I)$. Let $M = \mathbb{R}^3/\Pi$ be the flat manifold. Consider the endomorphism $\theta : \Pi \to \Pi$ which is defined by the conjugation by $g = (d, D)$, where

$$d = \begin{bmatrix} 0 \\ 0 \\ 0 \end{bmatrix}, \qquad D = \begin{bmatrix} 3 & 0 & 0 \\ 0 & 0 & 1 \\ 0 & 2 & 0 \end{bmatrix}.$$

Let $f : M \to M$ be the map induced from $g$. There are only two conjugacy classes of $g$; namely, $g$ and $\alpha g$. Fix$(g) = (0,0,0)^t$ and Fix$(\alpha g) = (1/4, 0, 0)^t$. Since $\det(I - D) = \det(I - AD) = +2$, $L(f) = N(f) = 2$.

The Lefschetz number can be calculated from homology groups also.

(1) $H_0(M; \mathbb{R}) = \mathbb{R}$; $f_*$ is the identity map.
(2) $H_1(M; \mathbb{R}) = \mathbb{R}$, which is generated by the element $(e_1, I)$.
    $f_*$ is multiplication by 3 (the $(1,1)$-entry of $D$).
(3) $H_2(M; \mathbb{R}) = \mathbb{R}$; $f_*$ is multiplication by $\det \begin{bmatrix} 0 & 1 \\ 2 & 0 \end{bmatrix} = -2$.
(4) $H_3(M; \mathbb{R}) = \mathbb{R}$; $f_*$ is multiplication by $\det(D) = -6$.

Therefore, $L(f) = \Sigma (-1)^i \text{Trace} f_* = 1 - 3 + (-2) - (-6) = 2$. Note that $f$ has infinite period, and this example is not covered by Corollary 11.5.13.

EXAMPLE 11.5.15. Let $\Pi$ be same as in Example 11.5.14. This time $g = (d, D)$, is given by

$$d = \begin{bmatrix} 0 \\ 0 \\ 0 \end{bmatrix}, \qquad D = \begin{bmatrix} 3 & 0 & 0 \\ 0 & 1 & 1 \\ 0 & 1 & 2 \end{bmatrix}.$$

Let $f : M \to M$ be the map induced from $g$. There are six conjugacy classes of $g$; namely, $g$ and $\alpha g$, $\alpha t_1 g$, $\alpha t_1^2 g$, $\alpha t_1^3 g$, and $\alpha t_1^4 g$. Each class has exactly one fixed point. Clearly, $\det(I - D) = +2$ and $\det(I - AD) = -10$. Therefore, the first fixed point has index $+1$ and the rest have index $-1$. Consequently, $L(f) = -4$, while $N(f) = 6$.

## 11.6. Homologically injective torus operations

11.6.1. There have been many efforts trying to split a manifold as a product of two manifolds. Let $M$ be a flat Riemannian manifold whose fundamental group contains a nontrivial center. Calabi has shown that such an $M$ almost splits. More precisely, there exists a compact flat manifold $N$ and a finite Abelian group $\Phi$ such that $M = T^k \times_\Phi N$, the quotient space of $T^k \times N$ by a free diagonal action of $\Phi$, where $\Phi$ acts freely as translations on the first factor and as isometries on the second factor; see [**Wol77**]. Lawson and Yau [**LY72**] and Eberlein [**Ebe82**] have shown the same fact for closed manifolds $M$ of nonpositive sectional curvature: If $\pi_1(M)$ has nontrivial center $\mathbb{Z}^k$, then $M$ splits as $M = T^k \times_\Phi N$, where $N$ is a closed manifold of nonpositive sectional curvature and $\Phi$ is a finite Abelian group acting diagonally and freely on the $T^k$-factor as translations.

Prior to Lawson and Yau's and Eberlein's work, Conner and Raymond [**CR71**] generalized Calabi's results to homologically injective torus actions. Let $(T^k, M)$ be a torus action on a topological space. For a base point $x_0 \in M$, consider the evaluation map ev $: (T^k, e) \to (M, x_0)$ sending $t \mapsto t x_0$. Recall that the action is *injective* if the evaluation map induces an injective homomorphism ev$_*$ : $\pi_1(T^k, e) \to \pi_1(M, x_0)$. It is *homologically injective* if the evaluation map induces an injective homomorphism $\mathrm{ev}_* : H_1(T^k, \mathbb{Z}) \to H_1(M; \mathbb{Z})$. By Theorem 2.4.2, the injectiveness condition is independent of choice of the base point and the image is a central subgroup of $\pi_1(X)$. For a Riemannian manifold of nonpositive sectional curvature, the existence of a nontrivial center $\mathbb{Z}^k$ of $\pi_1(M)$ guarantees that the manifold has an action of torus $T^k$, and all such actions are homologically injective.

In this section, topological spaces are always assumed to be paracompact, path-connected and locally path-connected, and either (1) locally compact and semisimply connected, or (2) has the homotopy type of the CW-complex. Therefore, our topological spaces admit covering space theory.

THEOREM 11.6.2 (Splitting Theorem) [**CR71**]). *If a topological space $X$ with $H_1(X; \mathbb{Z})$ finitely generated admits a homologically injective (topological) torus action $(T^k, X)$, then $X$ splits as $T^k \times_\Phi N$ for some $N$, where $\Phi$ is a finite Abelian group acting diagonally and acting freely on the $T^k$-factor as translations.*

The *splitting* $X = T^k \times_\Phi N$ as above implies that $X$ has a Seifert fiber space structure with typical fiber $T^k$ and base space $N/\Phi$. All the singular fibers are again tori which are finitely covered by $T^k$. It also gives rise to another genuine fiber structure; namely, $X$ fibers over the torus $T^k/\Phi$ with the fiber $N$ and a finite structure group. The above theorem does not require that the space $X$ be aspherical. On the other hand, the only compact connected Lie group that can act on aspherical manifolds are tori. Therefore, splitting a manifold using a group action for an aspherical manifold forces the group to be a torus. In other words, for aspherical manifolds, there can be no generalization of splitting using compact Lie group actions other than tori.

The condition for a torus action $(T^k, X)$ to be homologically injective is equivalent to the element $[\pi_1(X)]$ in $H^2(Q; \mathbb{Z}^k)$ having finite order [**CR71**]. Keep in mind that the cohomology class $[\pi_1(X)]$ is represented by the extension sequence $1 \to \mathbb{Z}^k \to \pi_1(X) \to Q \to 1$.

DEFINITION 11.6.3. An extension $1 \to \Gamma \to \Pi \to Q \to 1$ is called *inner* if the abstract kernel, $Q \to \mathrm{Out}(\Gamma)$, is trivial. Suppose $\Gamma$ is a subgroup of $G$. The extension $\Pi$ is $G$-*inner* if, for every $\sigma \in \Pi$, $\mu(\sigma) \in \mathrm{Aut}(\Gamma)$ is equal to conjugation by an element of $G$. If $\Pi$ is inner, then it is $G$-inner.

A normal subgroup $A$ of $C$ is said to be *homologically injective* in $C$ if the inclusion induces an injective homomorphism on the first homology, $H_1(A; \mathbb{Z}) \to H_1(C; \mathbb{Z})$, or equivalently, $A \cap [C, C] = \{1\}$.

The Splitting Theorem 11.6.2 can now be generalized to injective Seifert fiberings with typical fiber a compact solvmanifold of type (R).

THEOREM 11.6.4. *Let $X$ be an injective Seifert fibering with typical fiber a compact solvmanifold $\Gamma\backslash G$ of type (R). Assume that $\Pi$, with finitely generated center, acts freely on $\widetilde{X}$. Then the following are equivalent:*

(1) *The abstract kernel, $Q \to \mathrm{Out}(\Gamma)$, of the associated exact sequence $1 \to \Gamma \to \Pi \to Q \to 1$, $\Pi = \pi_1(X)$, has finite image in $\mathrm{Out}(\Gamma)$ and the center of $\Gamma$, $\mathcal{Z}(\Gamma)$, homologically injects into $C_\Pi(\Gamma)$;*
(2) $X = (\Gamma\backslash G) \times_\Phi X'$, *where $\Phi$ is a finite group which acts diagonally, as affine maps on the first factor.*

For a proof, we need the following lemmas. Some part of the first lemma is essentially proved in [**CR71**].

LEMMA 11.6.5. *Let $\Pi$ be a finitely generated group, and let $0 \to \mathbb{Z}^k \to \Pi \to Q \to 1$ be a central extension. Then the following are equivalent:*

(1) *$[\Pi] \in H^2(Q; \mathbb{Z}^k)$ has finite order;*
(2) *$\mathbb{Z}^k$ is homologically injective;*
(3) *$\Pi$ contains a normal subgroup $\widehat{Q}$ such that $\mathbb{Z}^k \cap \widehat{Q} = 1$ and $\Pi/(\mathbb{Z}^k \times \widehat{Q})$ is a finite Abelian group.*

PROOF. ($1 \Rightarrow 2$). Suppose $m[\Pi] = 0 \in H^2(Q; \mathbb{Z}^k)$. Let $\mathbb{Z}^k = A$, $(\frac{1}{m}\mathbb{Z})^k = B$, and $C = B/A$ so that $0 \to A \to B \to C \to 0$ is a short exact sequences of trivial $Q$-modules. Consider the Bockstein cohomology sequence

$$\cdots \to H^1(Q; C) \xrightarrow{\delta^*} H^2(Q; A) \xrightarrow{\iota} H^2(Q; B) \to \cdots.$$

The fact $m[\Pi] = 0 \in H^2(Q; A)$ implies that $\iota[\Pi] = 0 \in H^2(Q; B)$. Thus there exists $[\eta] \in H^1(Q; C) = \mathrm{Homomorphism}(Q, C)$ such that $\delta^*[\eta] = [\Pi]$. Let $\widehat{Q}$ be the kernel of the homomorphism $\eta : Q \to C$. Then the subgroup of $\Pi$ corresponding to $\widehat{Q}$ splits (consider $H^1(\widehat{Q}; C) \to H^2(\widehat{Q}; A)$). Therefore, we have a subgroup $A \times \widehat{Q}$ of $\Pi$ with $\Pi/(A \times \widehat{Q}) = Q/\widehat{Q} \subset C$. Let $\Delta = Q/\widehat{Q}$, a finite Abelian group. Consider the following commutative diagram with exact rows and columns:

$$\begin{array}{ccccccccc}
& & 1 & & 1 & & & & \\
& & \downarrow & & \downarrow & & & & \\
& & \widehat{Q} & = & \widehat{Q} & & & & \\
& & \downarrow & & \downarrow & & & & \\
1 & \to & A \times \widehat{Q} & \to & \Pi & \to & \Delta & \to & 1 \\
& & \downarrow & & \downarrow & & \| & & \\
1 & \to & A & \to & \Pi/\widehat{Q} & \to & \Delta & \to & 1 \\
& & \downarrow & & \downarrow & & & & \\
& & 1 & & 1 & & & &
\end{array}$$

The group $\Pi/\widehat{Q}$ is a central extension of $A$ by a finite Abelian group $\Delta$. Since $A$ is torsion free, $\Pi/\widehat{Q}$ is Abelian; see Corollary 5.5.3. Therefore, $[\Pi, \Pi] \subset \widehat{Q}$, which is disjoint from $A = \mathbb{Z}^k$. This implies $\mathbb{Z}^k$ injects into $\Pi/[\Pi, \Pi] = H_1(\Pi; \mathbb{Z})$.

($2 \Rightarrow 3$). Let $A = H_1(\Pi; \mathbb{Z})/\mathrm{Torsion}$. Then $\mathbb{Z}^k \xrightarrow{i} H_1(\Pi; \mathbb{Z}) \xrightarrow{p} A$ ($p$ projection) is still injective. Now let $A/(p \circ i)(\mathbb{Z}^k) = \Delta \oplus \mathbb{Z}^m$, where $\Delta$ is a finite group. Set $\overline{Q}$ be the preimage of $\Delta$ under the composite

$$\Pi \to H_1(\Pi; \mathbb{Z}) \to H_1(\Pi; \mathbb{Z})/\mathrm{Torsion} = A \to \Delta \oplus \mathbb{Z}^m.$$

Then there is a direct summand to $\overline{Q}$, and hence a homomorphism of $H_1(\Pi; \mathbb{Z}) \to \overline{Q}$. Let $\widehat{Q}$ be the kernel of $\Pi \to H_1(\Pi; \mathbb{Z}) \to \overline{Q}$. Then clearly $\mathbb{Z}^k \cap \widehat{Q} = 1$ and $\Pi/(\mathbb{Z}^k \times \widehat{Q}) = \overline{Q}/\mathbb{Z}^k$ is Abelian.

($3 \Rightarrow 1$). The commutative diagram

$$\begin{array}{ccccccccc}
1 & \longrightarrow & \mathbb{Z}^k & \longrightarrow & \mathbb{Z}^k \times \widehat{Q} & \longrightarrow & \widehat{Q} & \longrightarrow & 1 \\
& & = \downarrow & & \downarrow & & \downarrow \overline{\theta} & & \\
1 & \longrightarrow & \mathbb{Z}^k & \longrightarrow & \Pi & \longrightarrow & Q & \longrightarrow & 1
\end{array}$$

shows that the natural homomorphism $\overline{\theta}^* : H^2(Q;\mathbb{Z}^k) \to H^2(\widehat{Q};\mathbb{Z}^k)$ maps $[\Pi] \in H^2(Q;\mathbb{Z}^k)$ to $[\mathbb{Z}^k \times \widehat{Q}] = 0 \in H^2(\widehat{Q};\mathbb{Z}^k)$. Let $\gamma ; H^2(\widehat{Q};\mathbb{Z}^k) \to H^2(Q;\mathbb{Z}^k)$ be the transfer homomorphism. Then $\gamma \circ \overline{\theta}^* =$ multiplication by $m =$ the order of $Q/\widehat{Q}$. Thus, $m[\Pi] = \gamma \circ \overline{\theta}^*([\Pi]) = \gamma(0) = 0$. $\square$

LEMMA 11.6.6. *Let $\Gamma$ be a group whose center $\mathcal{Z}(\Gamma)$ is a free Abelian group of finite rank. Let $1 \to \Gamma \to \Pi \to Q \to 1$ be an extension whose abstract kernel has finite image. Then the following are equivalent:*
  (1) *$[\Pi]$ has finite order in $H^2(Q; \mathcal{Z}(\Gamma))$;*
  (2) *$\Pi$ contains a normal subgroup $\Gamma \times Q'$ such that $\Phi = \Pi/(\Gamma \times Q')$ is a finite group;*
  (3) *$\mathcal{Z}(\Gamma)$ homologically injects into $C_\Pi(\Gamma)$.*

PROOF. (1) $\iff$ (3). Let $P \subset Q$ be the kernel of $Q \to \mathrm{Out}(\Gamma)$, and let $\Pi' \subset \Pi$ be the preimage of $P$. Since $Q/P$ is finite, the homomorphism $i^* : H^2(Q, \mathcal{Z}(\Gamma)) \to H^2(P, \mathcal{Z}(\Gamma))$, induced by the inclusion $i : P \hookrightarrow Q$, has finite kernel.

To see this, consider the Lyndon spectral sequence for $i : P \hookrightarrow Q$. The $E_2^{p,q}$ terms are, $H^p(Q/P; H^q(P; \mathcal{Z}(\Gamma)))$. We seek kernel of $i^*$. This is given by the exact sequence
$$0 \to E_\infty^{2,0} \to \ker(i^*) \to E_\infty^{1,1} \to 0.$$
As both $E_\infty^{2,0}$ and $E_\infty^{1,1}$ are finite, $\ker(i^*)$ is finite. (As an alternative argument, we can apply the transfer homomorphism provided that homology of $P$ and $Q$ are finitely generated in low dimensions.) Therefore, $[\Pi] \in H^2(Q, \mathcal{Z}(\Gamma))$ has finite order if and only if $[\Pi'] \in H^2(P, \mathcal{Z}(\Gamma))$ has finite order. Also, for the statement (3), note that $C_\Pi(\Gamma) = C_{\Pi'}(\Gamma)$. Therefore, in proving (1) $\iff$ (3), it is enough to work with $\Pi'$ instead of $\Pi$. Hence, we assume that the extension $1 \to \Gamma \to \Pi \to Q \to 1$ has trivial abstract kernel. Then
$$1 \to \mathcal{Z}(\Gamma) \to C_\Pi(\Gamma) \to Q \to 1$$
is a central extension. The extensions $[\Pi]$ and $[C_\Pi(\Gamma)]$ are both classified by the same cohomology group $H^2(Q, \mathcal{Z}(\Gamma))$. Furthermore, since the abstract kernels are trivial, there exist direct products, which correspond to each other naturally. This proves the equivalence of (1) and (3), using Lemma 11.6.5.

(1) $\implies$ (2). The condition (1) implies that $[C_\Pi(\Gamma)]$ has finite order. By Lemma 11.6.5, $C_\Pi(\Gamma)$ contains a normal subgroup $\mathcal{Z}(\Gamma) \times Q'$ such that $C_\Pi(\Gamma)/(\mathcal{Z}(\Gamma) \times Q')$ is a finite group. However, $\mathcal{Z}(\Gamma) \times Q'$ may not be normal in $\Pi$. Let $C'$ be the intersection of all conjugates of $\mathcal{Z}(\Gamma) \times Q'$ by elements of $\Pi$. Since $C_\Pi(\Gamma)$ is normal in $\Pi$, and $\mathcal{Z}(\Gamma) \times Q'$ has finite index in $C_\Pi(\Gamma)$, there are only finitely many conjugacy (by elements of $\Pi$) classes of $\mathcal{Z}(\Gamma) \times Q'$. Therefore $C'$ is normal in $\Pi$ and has finite index in $C_\Pi(\Gamma)$. Moreover $C'$ splits also, which we denote by $\mathcal{Z}(\Gamma) \times Q'$ again. Let $\Pi' = \Gamma \cdot Q'$ so that $1 \to \Gamma \to \Pi' \to Q' \to 1$ is exact. Clearly, this splits as $\Pi' = \Gamma \times Q'$, is normal in $\Pi$, and $[\Pi : \Pi'] = [\Pi : \Gamma \cdot C_\Pi(\Gamma)][\Gamma \cdot C_\Pi(\Gamma) : \Pi']$ is finite.

(2) $\implies$ (3). Since $\Pi/(\Gamma \times Q')$ is finite, $C_\Pi(\Gamma)/(\mathcal{Z}(\Gamma) \times Q')$ is finite. Now apply Lemma 11.6.5. $\square$

11.6.7 (Proof of Theorem 11.6.4). (2)$\Rightarrow$(1). Suppose $X$ is of the form $X = (\Gamma \backslash G) \times_\Phi (X')$. Let $Q' = \pi_1(X')$ and $C = C_\Pi(\Gamma)$. Since $G$ is simply connected, $\pi_1((\Gamma \backslash G) \times X') = \Gamma \times Q'$ and $\Pi/(\Gamma \times Q') = \Phi$, a finite group. Since $C$ contains $Q'$, and the image of $Q'$ in $Q$ has finite index, the abstract kernel $Q \to \mathrm{Out}(\Gamma)$ factors through the finite group $\Phi$. Moreover, $\mathcal{Z}(\Gamma) \times Q'$ has finite index in $C$. By Lemma 11.6.6, this implies that $\mathcal{Z}(\Gamma)$ homologically injects to $C$.

(1)$\Rightarrow$(2). Assume $X$ satisfies (1). Let

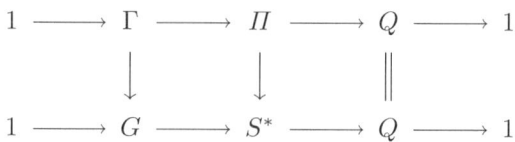

be the diagram associated with $(G, \widetilde{X})$. Here all the vertical maps are inclusions, and $S^*$ is the subgroup of $\mathrm{TOP}(\widetilde{X})$ generated by the subgroups $\Pi$ and $G$.

The abstract kernel $Q \to \mathrm{Out}(\Gamma)$ induces an abstract kernel $Q \to \mathrm{Out}(G)$ since every automorphism of $\Gamma$ uniquely extends to an automorphism of $G$. In fact, this abstract kernel is the abstract kernel for the bottom exact sequence of the diagram above. Since the abstract kernel $\phi : Q \to \mathrm{Out}(G)$ has finite image, it lifts to a homomorphism $Q \to \mathrm{Aut}(G)$ which we denote by $\phi$ again. This is a result of the statement that any finite extension of $\mathrm{Inn}(G)$ splits.

Here is an argument: Let $1 \to G_1 \to H \to F \to 1$ be exact, where $G_1 = \mathrm{Inn}(G)$ above and $F$ is finite. Since $F$ is finite, $[H] \in H^2(F; \mathcal{Z}(G_1))$ has a finite order. If $f : F \times F \to \mathcal{Z}(G_1)$ is a cocycle, $mf = 0$, where $m$ is the order of the cohomology group. If $\lambda : F \to \mathcal{Z}(G_1)$ is such that $\delta \lambda = mf$, since $G$ is simply connected solvable of type (R), so is $G_1$, and hence $G_1$ is *divisible*. We have $f(x, y) = (\delta \frac{1}{m} \lambda)(x, y)$. Thus $f$ is a coboundary.

We claim that $S^* = G \rtimes Q$. Since the abstract kernel lifts to $\phi : Q \to \mathrm{Aut}(G)$, there is a canonical one-to-one correspondence between $\mathrm{Opext}(Q, Z(G), \phi)$ and $\mathrm{Opext}(Q, G, \phi)$ by sending the semidirect product $Z(G) \rtimes Q$ to the semidirect product $G \rtimes Q$; see Definition 5.2.4. If $[E] \in \mathrm{Opext}(Q, Z(G), \phi)$ corresponds to $[S^*] \in \mathrm{Opext}(Q, G, \phi)$, there is a commutative diagram

$$\begin{array}{ccccccccc}
1 & \longrightarrow & \mathcal{Z}(G) & \longrightarrow & E & \longrightarrow & Q & \longrightarrow & 1 \\
& & \downarrow & & \downarrow & & \| & & \\
1 & \longrightarrow & G & \longrightarrow & S^* & \longrightarrow & Q & \longrightarrow & 1,
\end{array}$$

where all the vertical maps are inclusions. To see that $S^*$ splits, it will be enough to show that $E$ splits.

Certainly $C_{S^*}(G) \subset E$ and $C_{S^*}(G)/\mathcal{Z}(G)$ has finite index in $Q$. The hypothesis that $\mathcal{Z}(\Gamma)$ homologically injects into $C = C_\Pi(\Gamma)$ implies that $[C]$ has finite order in $H^2(C/\mathcal{Z}(\Gamma); \mathcal{Z}(\Gamma))$. Therefore $\epsilon_*[C] = 0$ in $H^2(C/\mathcal{Z}(\Gamma); \mathcal{Z}(G))$, where $\epsilon_*$ is induced by $\epsilon : \mathcal{Z}(\Gamma) \to \mathcal{Z}(G)$. Thus we have $[E] = 0$ in $H^2(Q; \mathcal{Z}(G))$ because $[E] = i^*(\epsilon_*[C])$ from the commutative diagram

$$\begin{array}{ccccccccc}
1 & \longrightarrow & \mathcal{Z}(G) & \longrightarrow & C & \longrightarrow & C/\mathcal{Z}(G) & \longrightarrow & 1 \\
& & \| & & \downarrow & & \downarrow i & & \\
1 & \longrightarrow & \mathcal{Z}(G) & \longrightarrow & E & \longrightarrow & Q & \longrightarrow & 1.
\end{array}$$

This proves that $S^*$ splits into $G \rtimes Q$.

## 11.6. HOMOLOGICALLY INJECTIVE OPERATIONS

Clearly, by Lemma 11.6.6, there exists a subgroup $Q'$ of $\Pi$ such that $\Gamma \cap Q' = 1$, $Q' \subset C_\Pi(\Gamma)$, and $\Gamma \times Q'$ is normal in $\Pi$ with $\phi = \Pi/\Gamma \times Q'$ finite. Without loss of generality, we may assume that $Q'$ is in $1 \times Q \subset G \rtimes Q$.

The action $(G, \widetilde{X})$ is free and $(G, \widetilde{X}) = (G, G \times W)$ equivariantly. For the group operation of $\text{TOP}_G(G \times W)$, and the action of $\text{TOP}_G(G \times W)$ on $G \times W$ are as in Corollary 4.2.10:

$$(\lambda_1, \alpha_1, h_1) \cdot (\lambda_2, \alpha_2, h_2) = (\lambda_1 \cdot (\alpha_1 \circ \lambda_2 \circ h_1^{-1}), \alpha_1 \circ \alpha_2, h_1 \circ h_2),$$
$$(\lambda, \alpha, h) \cdot (x, w) = (\alpha(x) \cdot (\lambda(h(w)))^{-1}, h(w)).$$

The group $G \subset G \rtimes Q$ acts on $G \times W$ as left translations on the first factor and trivially on the second. In general, $Q$ factor mixes the second factor into the first. We would like to alter $W$ to $W'$ so that $\widetilde{X} = G \times W'$ on which $Q'$ acts as a subgroup of $\text{TOP}(W) \subset \text{M}(W, G) \rtimes (\text{Aut}(G) \times \text{TOP}(W))$. For $\alpha \in Q \subset G \rtimes Q$, let $(\lambda(\alpha), \tau(\alpha), \alpha)$ denote its representation as an element of $\text{TOP}_G(W \times G)$. Then we check that

$$\lambda(\alpha\beta) = \lambda(\alpha) \cdot \tau(\alpha) \circ \lambda(\beta) \circ \alpha^{-1}.$$

If we introduce the notation $\alpha_* : \text{Maps}(W, G) \to \text{Maps}(W, G)$ by

$$\alpha_*(s) = \tau(\alpha) \circ s \circ \alpha^{-1},$$

the above becomes

$$\lambda(\alpha\beta) = \lambda(\alpha) \cdot \alpha_*(\lambda(\beta)).$$

Let $s \in \text{Maps}(W, G)$, and put $W' = \{(s(w), w) \mid w \in W\} \subset G \times W$. Suppose, under the action of $Q$ on $G \times W$, $W'$ is invariant. That is, $\alpha(s(w), w) = (s(\alpha w), \alpha w)$ for all $\alpha \in Q$. (Notice that, for $w' = (s(w), w)$, this equality means that

$$\alpha(1, w') = (1, (\alpha w)')$$

for $w' \in W'$ so that $Q$ acts on $G \times W'$ as a subgroup of $\text{TOP}(W')$.) Comparing the first slots, we get $s(\alpha w) = \tau(\alpha)(sw) \cdot (\lambda(\alpha)(\alpha w))^{-1}$. Thus,

$$\lambda(\alpha)(\alpha w) = (s(\alpha w))^{-1} \cdot \tau(\alpha)(sw)$$
$$= (s(\alpha w))^{-1} \cdot \tau(\alpha) \ s(\alpha^{-1}\alpha)w$$
$$= (s^{-1} \cdot \tau(\alpha) \circ s \circ \alpha^{-1})(\alpha w)$$
$$= (s^{-1} \cdot \alpha_*(s))(\alpha w).$$

This implies $\lambda(\alpha) = s^{-1} \cdot \alpha_*(s)$ for all $\alpha \in Q$.

Conversely, if there is $s \in \text{M}(W, G)$ such that $\lambda(\alpha) = s^{-1} \cdot \alpha_*(s)$ for all $\alpha \in Q$, then $W'$ defined as above will be $Q$-invariant. Therefore, if we show that the non-Abelian group cohomology $H^1(Q; \text{Maps}(W, G))$ vanishes, the invariant $W'$ will have been shown to exist; see Subsection 5.7.1. We have shown that this cohomology set is trivial in Subsection 7.6.6 and so $W'$ exists and is $Q$-invariant.

It is easy to see that

$$G \times W \xrightarrow{\zeta} G \times W$$
$$(x, w) \longrightarrow (x \cdot s(w), w)$$

is $G$-equivariant and weakly $Q$-equivariant. More precisely, for $a \in G$ and $\alpha \in Q$,

$$\zeta(a(x, w)) = (ax \cdot s(w), w) = a\zeta(x, w),$$
$$\zeta(\alpha(x, w)) = (\tau(\alpha)(x) \cdot (\lambda(a)(\alpha w))^{-1} \cdot s(\alpha w), \alpha w) = \mu(s^{-1})(\alpha)(\zeta(x, w)),$$

where $\mu(s^{-1}) = \mu(s^{-1}, 1, 1)$ for $(s^{-1}, 1, 1) \in M(W, G) \rtimes (\operatorname{Aut}(G) \times \operatorname{TOP}(W))$.
We summarize the facts that we have proved as follows.

(1) $\Pi$ contains a subgroup $Q'$ such that $\Gamma \cap Q' = 1$ and $\Gamma \times Q'$ is normal, $\Phi = \Pi/\Gamma \times Q'$ is finite.
(2) $G \subset G \rtimes Q$ acts only on the first factor of $\widetilde{X} = G \times W'$ as left translations and trivially on $W'$-factor.
(3) $Q \subset G \rtimes Q$ acts on $\widetilde{X} = G \times W'$ as a subgroup of $\operatorname{Aut}(G) \rtimes \operatorname{TOP}(W) \subset \operatorname{Maps}(W, G) \rtimes (\operatorname{Aut}(G) \rtimes \operatorname{TOP}(W))$; i.e., $\alpha \mapsto (1, \tau(\alpha), \alpha)$.
(4) Since $\tau(\alpha)$ is conjugation by $\alpha$, for $\alpha \in Q'$, $\tau(\alpha) = \operatorname{id}$. Therefore, for $\alpha \in Q'$, $\alpha \mapsto (1, 1, \alpha)$.
(5) From (2) and (4), $(a, \alpha) \in G \times Q'$ acts on $G \times W'$ by

$$(a, \alpha)(x, w) = (ax, \alpha w).$$

Consequently,

$$\begin{aligned}
X &= \Pi \backslash (G \times W) \\
&= (\Gamma \times Q' \backslash G \times W') / \Phi \\
&= (\Gamma \backslash G) \times_\Phi (Q' \backslash W').
\end{aligned}$$

Clearly, the $\Phi$-action on $\Gamma \backslash G$ is as a subgroup of $\operatorname{Aff}(\Gamma \backslash G)$ because the action of $G \rtimes Q$ on $G$ is through $G \rtimes \operatorname{Aut}(G)$. This completes the proof of the theorem. $\square$

We may give an explicit Seifert homeomorphism between $\Pi \backslash (G \times W)$ and $(\Gamma \backslash G) \times_\Phi (Q' \backslash W)$ as follows: On $G \times W$, let $\alpha \in Q$ and $s \in M(W, G)$ be such that $\lambda(\alpha) = s^{-1} \cdot \alpha_*(s)$, for all $\alpha \in Q$. We have the following commutative diagram

$$\begin{array}{ccccc}
(x, w) & \longrightarrow & (x \cdot s(w), w) & \longrightarrow & (x, (s(w), w)) \in G \times W' \\
\downarrow \alpha & & \downarrow \alpha & & \downarrow \alpha \\
(\tau(\alpha)(x) \cdot (\lambda(\alpha)(\alpha w))^{-1}, \alpha w) & \to & (\tau(\alpha)(x) \cdot s(\alpha w), \alpha w) & \to & (\tau(\alpha)(x), s\alpha(w), \alpha w).
\end{array}$$

Notice that the bottom middle term is

$$(\tau(\alpha)(x), s\alpha(w), \alpha w) = (\tau(\alpha)(x) \cdot \tau(\alpha)s(w) \cdot (\lambda(\alpha)(\alpha w))^{-1}, \alpha w).$$

In the left-hand square, there is a Seifert homeomorphism between $\Pi \backslash (G \times W)$ and $\Pi \backslash (G \times W)$ achieved by conjugation on $G \times W$ by $s \in M(W, G)$. The second square is a homeomorphism between $G \times W$ and $G \times W'$. The actions of $Q$ are equivariant. We note that in the last column that when $\alpha \in Q'$, $\tau(\alpha)(x) = x$.

COROLLARY 11.6.8. *Let $(\Gamma \backslash G, X)$ be an injective Seifert fiber space where $G$ is solvable of type* (R) *and $\Pi = \pi_1(X)$. Then the following are equivalent when the center of $\Pi$ is finitely generated:*

(1) *The associated exact sequence $1 \to \Gamma \to \Pi \to Q \to 1$ has trivial abstract kernel and $\mathcal{Z}(\Gamma)$ homologically injects to $C_\Pi(\Gamma)$;*
(2) *$X = (\Gamma \backslash G) \times_\Delta X'$, where $\Delta$ is a finite Abelian group which acts diagonally, freely as translations along the torus $\mathcal{Z}(\Gamma) \backslash \mathcal{Z}(G)$.*

The *splitting* $X = (\Gamma \backslash G) \times_\Phi X'$ in the theorem implies that $X$ has a Seifert fiber structure with the typical fiber $\Gamma \backslash G$ and the base space $X'/\Phi$. The singular fibers are *orbifolds* finitely covered by $\Gamma \backslash G$. In the case of $1 \to \Gamma \to \Pi \to Q \to 1$ with trivial abstract kernel, the singular fibers are again homogeneous spaces finitely covered by $\Gamma \backslash G$. In this case, $X$ also has a genuine fiber structure over the homogeneous space $(\Gamma \backslash G)/\Delta$ with fiber $X'$ and finite Abelian structure group $\Delta$.

Let $(\Gamma \backslash G, X)$ be as in the corollary and satisfy condition (1). Let $X^*$ be the covering of $X$ with $\pi_1(X^*) = C$. Then the torus $T = \mathcal{Z}(\Gamma) \backslash \mathcal{Z}(G)$ acts on $X^*$. This torus action is homologically injective. In fact, $T$ acts on the space $(\mathcal{Z}(G) \times W)/C$ homologically injectively. Therefore, $(\mathcal{Z}(G) \times W)/C = T \times_\Delta X'$. Let $G_1 = G/\mathcal{Z}(G)$ and $\Gamma_1 = \Gamma/\mathcal{Z}(\Gamma)$. Then $\Gamma_1$ is a lattice in $G_1$, and $X$ fibers over $\Gamma_1 \backslash G_1$ with fiber $T \times_\Delta X'$. This is explained by the following commutative diagram easily:

$$\begin{array}{ccc} \mathcal{Z}(G) \times W & \xrightarrow{C\backslash} & T^k \times_\Delta X' \\ \downarrow & & \downarrow \\ G \times W & \xrightarrow{\Pi\backslash} & X \\ \downarrow & & \downarrow \\ G_1 & \xrightarrow{\Gamma_1\backslash} & \Gamma_1 \backslash G_1. \end{array}$$

All the horizontal maps are coverings and the vertical rows are fibrations. The homotopy exact sequence of the second fibration is $1 \to C \to \Pi \to \Gamma_1 \to 1$.

REMARK 11.6.9. We cite some instances where conditions (1) or (2) of the theorem or the corollary can be verified.

(1) If $G = \mathbb{R}^n$ and $\Gamma = \mathbb{Z}^n$, then Corollary 11.6.8 becomes a restatement of Theorem 11.6.2.

(2) Consider 3-dimensional manifolds $M$ which are injective Seifert fiberings with $G = \mathbb{R}^1$ and typical fiber $S^1$. They are modeled on $(\mathbb{R}^1, \mathbb{R}^1 \times \mathbb{R}^2)$ or $(\mathbb{R}^1, \mathbb{R}^1 \times S^2)$. If $M$ is noncompact and $\Pi = \pi_1(M)$ is inner (see Definition 11.6.3), then $(S^1, M)$ is homologically injective. If $\Pi$ is not inner, then it has a double cover which is inner and so Theorem 11.6.4 always holds. The same thing holds for $M$ compact with nonempty boundary.

If $M$ is compact with empty boundary, then $M$ satisfies the conditions of Theorem 11.6.4 except for those classical Seifert fiberings where $M$ is of type $\mathfrak{o}_1$ or $n_2$ ($\mathfrak{o}_0$, $\mathfrak{o}_n$ in Seifert's notation, see Subsection 14.13.1) and $e(M)$, the Euler number of $M$, is different from 0. The easiest way to verify this assertion is to examine the diagram in Proposition 10.5.10 from where it can easily be deduced that the conditions of Lemma 11.6.6 is valid. (We also point out that if $\pi_1(M)$ is finite, the Seifert 3-manifold is closed, covered by $S^3$, and is of type $\mathfrak{o}_1$ with Euler number $\neq 0$).

(3) For a specific example, consider the nonprincipal $S^1$-bundles over $\mathbb{R}P_2$. Each bundle is determined by a characteristic class $a \in H^2_\varphi(\mathbb{R}P_2; \mathbb{Z}) \cong \mathbb{Z}$, where $\mathbb{Z}$ is a nontrivial local coefficient system. In these bundles, the orientation of a circle fiber is reversed while traversing an orientation reversing curve in the base $\mathbb{R}P_2$. If $a \neq 0$, the bundle is covered by $S^3$ and $\pi_1(M(a))$ is finite. If $a = 0$, then there is a cross section and $M(a = 0)$ can be identified with $S^1 \times_{\mathbb{Z}_2} S^2$, where $\mathbb{Z}_2$ acts freely on $S^2$ and by reflection on $S^1$. The extension exact sequence for this manifold is $1 \to \mathbb{Z} \to \mathbb{Z} \rtimes \mathbb{Z}_2 \to \mathbb{Z}_2 \to 1$, and it is the noncentral extension associated with this injective Seifert fibering which is modeled on $(\mathbb{R}^1, \mathbb{R}^1 \times S^2)$. There is a projection onto $I = \mathbb{Z}_2 \backslash S^1$ with the inverse image over the interior points being $S^2$ and over the end points being $\mathbb{R}P_2$. Thus, $S^1 \times_{\mathbb{Z}_2} S^2$ is homeomorphic to $\mathbb{R}P_3 \# \mathbb{R}P_3$.

$M(a = 0)$ satisfies Theorem 11.6.4, but not the corollary as the extension exact sequence is not inner.

Similarly, take $M = S^1 \times_{\mathbb{Z}_2} \Sigma_g$, where $\Sigma_g$ is the orientable double cover of the nonorientable surface $\Sigma'_k$ with genera satisfying $g = k - 1$. Again, $M$ satisfies Theorem 11.6.4, but not Corollary 11.6.8. The Euler characteristic class of $M$ is $0 \in H^2_\varphi(\Sigma'_k; \mathbb{Z}) \cong \mathbb{Z}$, where each fiber over each orientation reversing curve is reversed. All the other nonprincipal bundles are injective Seifert fiberings with characteristic class of infinite order. These are classical Seifert manifolds of type $\mathfrak{n}_2$ in the notation of Subsection 14.13.1.

(4) The Seifert 3-manifolds modeled on $(\mathbb{R}^2, \mathbb{R}^2 \times \mathbb{R}^1)$ with typical fiber $T^2$ and holonomy infinite never satisfy Theorem 11.6.4. In fact, these are 3-manifolds with the Sol-geometry.

(5) Consider the closed 4-dimensional manifolds $M$ modeled on $(\mathbb{R}^2, \mathbb{R}^2 \times W)$, with $W = \mathbb{R}^2$, and typical fiber $T^2$. The associated exact sequence to the Seifert structure is
$$1 \to \mathbb{Z}^2 \to \Pi \to Q \to 1$$
with $\Phi : Q \to \operatorname{Aut}(\mathbb{Z}^2)$ and $\Pi$ acting freely. If the image of $\Phi$ is finite, then these 4-manifolds are finitely fiberwise covered by an orientable Seifert fibering with an injective torus action. These coverings all admit a complex structure with a complex 1-torus as typical fiber. These complex manifolds are elliptic surfaces with possible multiple fibers (i.e., singular fibers) but no exceptional fibers. The complex structure is Kähler if and only if $[\Pi] \in H^2(Q; \mathbb{Z}^2)$ has finite order and hence the torus action is homologically injective. In this finite case, the finite central covering can be holomorphically deformed to an algebraic surface $(T^2, T^2 \times_\Delta Y)$. If the order $[\Pi]$ is not finite, the complex structure is never Kähler; see [**CR72b**] for more details. For an extensive discussion as to when holomorphic $T^k$-actions on complex manifolds can holomorphically fibered over the torus, see [**Car72**].

(6) We also have the following:

THEOREM ([**CR72b**, §12.6]). *Let $(T^k, M)$ be an action on a homologically Kählerian manifold. Then all isotropy groups are finite if and only if the action is homologically injective.*

A closed (connected) manifold $M$ is homologically Kählerian [**Bor60**, Chap XII §6] if there exists a class $a \in H^2(M; \mathbb{Q})$ such that $H^s(M; \mathbb{Q}) \xrightarrow{\cup a^{n-s}} H^{2n-s}(M; \mathbb{Q})$ is an isomorphism for $s = 0, 1, \ldots n$ with $2n$ equal to the dimension of $M$. A compact connected complex manifold which admits a Kähler metric is homologically Kählerian.

(7) Let $(T^k, X)$ be an injective torus action whose extension sequence is given by
$$0 \to \mathbb{Z}^k \to \pi_1(X) \to Q \to 1.$$
We have

THEOREM ([**CR72c**, §2.1]). *If $k > \operatorname{rank} H^2(Q; \mathbb{Z})$, then for any $\Pi$ as above, there is an integer $j \geq k - \operatorname{rank} H^2(Q; \mathbb{Z})$ and a direct product decomposition $T^j \times T^{k-j}$ so that $H_1(T^j; \mathbb{Z}) \to H_1(X; \mathbb{Z})$ injects and the image $H_1(T^{k-j}; \mathbb{Z}) \to H_1(X; \mathbb{Z})$ is finite. Consequently, we may write $(T^k, X)$ as $(T^j \times T^{k-j}; T^j \times_\Delta Y)$, where $T^j$ acts only as translations on the first factor, $\Delta$ acts diagonally on $T^j \times Y$ and as translations on the first factor. The group $T^{k-j}$ acts on $T^j \times Y$ and injectively on $Y$.*

For example, in the case of elliptic surfaces treated in (5), which are not homologically injective, we have $(T^2, X) = (S_1^1 \times S_2^1, S_1^1 \times_\Delta Y)$. Therefore, $Y$ is a Seifert 3-manifold with an injective but not homologically injective $S_2^1$-action on $Y$. Thus, any elliptic surface of the kind treated in (5) can be topologically or smoothly analyzed by this method.

(8) Consider all closed $M$ which admit a complete Riemannian metric whose curvature is nonpositive and for which $\pi_1(M)$ contains a nontrivial normal Abelian subgroup. For, the centralizer of the maximal normal Abelian subgroup will be of finite index in $\pi_1(M)$. One then applies the center theorem of [**LY72**] yielding a homologically injective torus action on a finite regular covering. This implies that condition (1) of Theorem 11.6.4 will hold.

EXAMPLE 11.6.10. Let $G$ be a 3-dimensional Heisenberg group, i.e., the group of all upper triangular matrices with diagonal entries 1. Consider
$$x = I + E_{1,2}, \quad y = I + E_{2,3}, \quad z = I + E_{1,3} \in G,$$
where $I$ is the identity matrix, and $E_{i,j}$ a $3 \times 3$ matrix whose $(i,j)$-entry is 1, and 0 elsewhere. Let $\Gamma$ be the lattice generated by $x^2$, $y$, and $z$. Let $N = Q'\backslash \mathbf{H}$ be a hyperbolic surface of genus 2, so $Q' \subset \mathrm{PSL}(2, \mathbb{R})$ is a Fuchsian group. Let $\Phi = \mathbb{Z}_2$ act on $\Gamma\backslash G$ and on $N$ as follows: Let the nontrivial generator $\tau \in \Phi$ act on the universal covering group level as (right) translation by $x$. It also acts on the surface $N$ by a rotation by $180°$ with two fixed points. The quotient $\Phi\backslash N$ is a torus with two singular points. Now the manifold $M = (\Gamma\backslash G) \times_{\mathbb{Z}_2} N$ has associated homotopy exact sequence $1 \to \Gamma \to \Pi \to Q \to 1$, where $Q = Q' \rtimes \mathbb{Z}_2$. Clearly, $\Gamma \times Q'$ is normal in $\Pi$ and has index 2. The only torus action on $M$ is the circle action of $\mathcal{Z}(G)/\mathcal{Z}(\Gamma)$. Clearly,
$$\mathcal{Z}(\Gamma) \cong \mathbb{Z}.$$
The circle action on $M$ is not homologically injective. This is obvious because the center $\mathbb{Z}$ cannot be separated even in $\Gamma$. In other words, $1 \to \mathcal{Z}(\Gamma) \to \Gamma \to \mathbb{Z}^2 \to 1$ represents an element of infinite order in $H^2(\mathbb{Z}^2; \mathbb{Z})$.

Consequently, $[\Pi] \in H^2(Q; \mathcal{Z}(\Gamma))$ has infinite order. This shows that there is no way of splitting off this circle using the action of $\mathcal{Z}(G)/\mathcal{Z}(\Gamma)$. From the construction of the manifold $M$, there is a splitting of $M$ as $(\Gamma\backslash G) \times_{\mathbb{Z}_2} N$. The injective Seifert fiber space with $\Gamma\backslash G$-fiber
$$\Gamma\backslash G \to M \to \Phi\backslash N$$
has two singular points which are the fixed points of the action of $\Phi$ on $N$. The singular fibers are nilmanifolds $(\Gamma\backslash G)/\Phi$. Note that the extension $1 \to \Gamma \to \Pi \to Q \to 1$ is not inner, but just $G$-inner, and hence the $\mathbb{Z}_2$-action on $\Gamma\backslash G$ is not in the $S^1$-action.

Also the action of $\mathbb{Z}_2$ on $\Gamma\backslash G$ lifts to a new lattice $\Gamma' = \langle x, y, z \rangle$, and $M$ has a genuine fibration structure
$$N \to M \to \Gamma\backslash G',$$
where $\Gamma\backslash G'$ is a nilmanifold doubly covered by $\Gamma\backslash G$.

11.6.11 (Nonuniqueness of fibers). This example is due to Tollefson [**Tol69**]. For any closed orientable surface $\Sigma_g$, of genus $g$, there exists a closed oriented surface $\Sigma_m$, $m = k(g-1) + 1$, with a free $\mathbb{Z}_k$-action whose orbit space is $\Sigma_g$. (For example, take $m = 3$, $k = 2$, and $g = 2$.) Form $(S^1, S^1 \times_{\mathbb{Z}_k} \Sigma_m) = (S^1, M)$. $M$ fibers over $S^1 = \mathbb{Z}_k \backslash S^1$, with fiber $\Sigma_m$. At the same time, $(S^1, M) \xrightarrow{S^1} \mathbb{Z}_k \backslash \Sigma_m = \Sigma_g$ is a principal $S^1$-bundle. It has a characteristic class $a \in H^2(\Sigma_g; \mathbb{Z}) = \mathbb{Z}$, which determines the extension exact sequence

$$0 \to \mathbb{Z} \to \pi_1(M) \to \pi_1(\Sigma_g) \to 1.$$

(The class $a$ is the same as $e(M)$, the Euler class of this principal $S^1$-bundle.) Since $M$ has a $\mathbb{Z}_k$ central covering $S^1 \times \Sigma_m$, the order of $a$ is finite. In fact, its order must be 1. Therefore, $(S^1, M) = (S^1, S^1 \times \Sigma_g)$. So the same Seifert fibering $(S^1, M) \to \Sigma_g$ has many different fiberings over $S^1$ (just vary $k$).

The nonuniqueness of fibers can be rather subtle. Charlap [**Cha65**] constructed closed flat manifolds $M_1 = S^1 \times N_1$ and $M_2 = S^1 \times N_2$ such that $M_1$ and $M_2$ are diffeomorphic but $N_1$ and $N_2$ have nonisomorphic fundamental groups.

Somewhat simpler examples can be constructed with algebraic surfaces as discussed in [**CR72a**]. Let $Y$ be a closed orientable 2-manifold with an action of $\mathbb{Z}_n$ such that no nontrivial subgroup of $\mathbb{Z}_n$ acts freely on $Y$. Form $(T^2 \times_{\mathbb{Z}_n \times \mathbb{Z}_n} Y)$. This admits a complex algebraic structure. It can be shown to be homeomorphic to $S^1 \times M_1$ and to $S^1 \times M_2$, where $M_1$ and $M_2$ are Seifert 3-manifolds with a homologically injective $S^1$-action. With a judicious choice of the action of $\mathbb{Z}_n \times \mathbb{Z}_n$ on $T^2 \times Y$, $\pi_1(M_1) \neq \pi_1(M_2)$; see [**CR72a**] for a complete description including the Charlap examples.

11.6.12. From the examples in Subsection 11.6.11, it is clear that a homologically injective Seifert fiber space $(T^k, X)$ can have vastly different splittings $T^k \times_\Delta Y$. It is of interest to classify these splittings. However, to achieve this, one needs to narrow the type of splittings to be considered. Such a classification can be found in [**CR72a**]. Other discussions can also be found in [**Sch81b**] and [**Sad91b**]. Below, we shall treat a special case of this classification.

11.6.13. Let $G$ be a compact connected Lie group acting on a space $X$ and $H$ a closed subgroup. Suppose $f : (G, X) \to (G, G/H)$ is an equivariant map. Then the trivial coset $\{H\}$ in $G/H$ is an $H$-slice in $(G, G/H)$. Put $Y = f^{-1}(\{H\}) \subset X$, then $Y$ is the pullback of the $H$-slice $\{H\}$ and so is a global $H$-slice in $(G, X)$ (Proposition 1.6.4). Hence $(G, X)$ can be written as fiber bundle with fiber $Y$, structure group $H$, and base space $G/H$. In terms of equivariant maps, we have

$$\begin{array}{ccccc} (G \times H, G) & \longleftarrow & (G \times H, G \times Y) & \xrightarrow{G\backslash} & (H, Y) \\ \downarrow {\scriptstyle H\backslash} & & \downarrow {\scriptstyle H\backslash} & & \downarrow {\scriptstyle H\backslash} \\ (G, G/H) & \longleftarrow & (G, G \times_H Y) & \xrightarrow{G\backslash} & H\backslash Y = G\backslash X. \end{array}$$

If $(G, X) = (G, G \times_H Y)$, then $f_0 : (G, X) \to (G, G/H)$ defined by $f_0((g, y)) = gH$, where $((g, y)) = ((gh^{-1}, hy))$ for $h \in H$, is an equivariant map. (Here $((g, y))$ represents the $H$-orbit in $(G, X)$ of the point $(g, y) \in G \times Y$.)

Of course, there may be other $G$-equivariant maps $f : (G, X) \to (G, G/H)$. Then $f^{-1}\{H\} = Y_f$ and we get another splitting of $(G, X)$ as $(G, G \times_H Y_f)$. Here $H \backslash Y_f = G \backslash X = H \backslash Y_{f_0}$, where $Y_{f_0}$ is the original splitting $Y$. To classify these splittings of $(G, X)$ with *fixed* subgroup $H$, we say that two splittings $(G, G \times_H Y_{f_1})$ and $(G, G \times_H Y_{f_2})$ of $(G, X)$ are *strongly equivalent* if there is an $H$-equivariant homeomorphism $\theta : Y_{f_1} \to Y_{f_2}$ such that

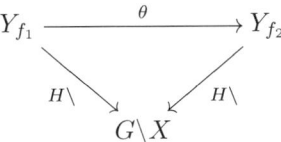

commutes.

11.6.14. As a special case, consider $(G, X)$ of the form $(G, (G \times_H Y)) = (G, G \times_H Y_{f_0})$ where $H$ acts *freely* on $Y$. We look for all the strong equivalence classes $(G, G \times_H, Y_f)$, $G$-equivariantly homeomorphic to $(G, X)$, where $H$ acts freely on $Y_f$. Now, strongly equivalent free actions of $H$ on $Y_f$ must yield equivalent principal $G$-bundles over $G \backslash X$. Since the principal $G$-bundle fibers over $(G, G/H)$ with fiber $Y_f$ and structure group $H$, the structure group of the principal $G$-fibering $(G, X) \to G \backslash X = H \backslash Y_f$ is reducible to the closed subgroup $H$.

Let $a \in H^1(H \backslash Y; \mathfrak{G})$ be a sheaf of germs of continuous functions of $H \backslash Y$ into $G$ representing the principal $G$-bundles $(G, X)$ over $G \backslash X$. Let $b \in H^1(H \backslash Y; \mathfrak{H})$ be the principal $H$-bundles over $H \backslash Y$ representing $(H, Y)$. Since $a$ is represented by $(G, G \times_H Y)$, $j^*(b) = a$ is a reduction of the structure group, where $j^* : H^1(H \backslash Y; \mathfrak{H}) \to H^1(H \backslash Y; \mathfrak{G})$.

THEOREM ([**CR72a**, 2.7]). *The set of strong equivalence classes (relative to the choice $a$), is the set of all bundle reductions of $a$ (to the subgroup $H$); that is, all elements $b \in H^1(H \backslash Y; \mathfrak{H})$ such that $j^*(b) = a$.*

11.6.15. In particular, if $G$ is a torus $T^k$ and $H$ is a closed finite subgroup, then the theorem reduces to the Bockstein sequence

$$0 \longrightarrow H^1(H \backslash Y; \mathbb{Z}^k) \xrightarrow{i} H^1(H \backslash Y; \mathbb{Z}^k) \longrightarrow H^1(H \backslash Y; \mathfrak{H}) \xrightarrow{\beta = j^*} H^2(H \backslash Y; \mathbb{Z}^k).$$

For example, the choice of $a$ is an element of $H^2(H \backslash Y; \mathbb{Z}^k)$. The strong equivalence classes are those elements $b \in H^1(H \backslash Y; H)$ which are carried into $a$ by the Bockstein map $\beta$. The set is identified with the elements of

$$H^1(H \backslash Y; \mathbb{Z}^k)/i(H^1(H \backslash Y; \mathbb{Z}^k)).$$

If we turn to the example in Subsection 11.6.11, we see that the bundle over $\Sigma_g$ is trivial. Then, the set of strongly inequivalent splittings is in one-to-one correspondence with $H^1(\Sigma_g; \mathbb{Z}_k)$ since the Bockstein map is trivial. Note also that many of the inequivalent splittings will have $Y_f$ disconnected.

For another illustration, consider the nonorientable closed 2-manifolds $X$ of nonorientable genus $k$, $k \geq 1$. There are exactly two principal $S^1$-bundles over each $X$, given by an element $a \in H^2(X; \mathbb{Z}) \cong \mathbb{Z}_2$. Let $H = \mathbb{Z}_m$. The homomorphism $m : H^2(X; \mathbb{Z}) \to H^2(X; \mathbb{Z})$ is trivial if $m$ is even and an isomorphism if $m$ is odd. Therefore, the Bockstein map $\beta : H^1(X; \mathbb{Z}_m) \to H^2(X; \mathbb{Z})$ is onto for $m$ even and trivial for $m$ odd.

For example, if $X = \mathbb{R}P_2$, $H^1(\mathbb{R}P_2; \mathbb{Z}_m) = \mathbb{Z}_2$ $m$ even, and $0$ $m$ odd. Consequently, if $m$ is even, then for each of the two principal $S^1$-bundles, there is just one strong equivalence class of splittings for a fixed subgroup, $\mathbb{Z}_m$. For the trivial bundle, the fibers are disconnected and for the nontrivial bundle, the fibers are disconnected for $m > 2$. If $m$ is odd, the nontrivial bundle fibers over $\mathbb{Z}_2 \backslash S^1$ but every $S^1$-equivariant map to $S^1$ must have even degree. In other words, any twisted covering of $\mathbb{R}P_2$, necessary for $X$ to be the nontrivial $S^1$-bundle, must be even order. For the trivial bundle $X = S^1 \times \mathbb{R}P_2$, there is exactly one strong equivalence class of splitting for each odd $m$. The fibers in this case are $m$ disjoint copies of $\mathbb{R}P_2$.

The reader may wish to investigate the principal $S^1$-bundle $X$ over the Klein bottle or surfaces of nonorientable even higher genus. In all these cases, the number of strongly inequivalent splittings is given by the theorem.

## 11.7. Maximal torus actions

DEFINITION 11.7.1. Any compact, connected Lie group which acts effectively on a closed admissible manifold (see Definition 3.2.1) (e.g., aspherical manifolds) is a torus $T^k$ with $k \leq$ rank of $\mathcal{Z}(\pi_1(M))$, the center of $\pi_1(M)$; see Theorem 3.2.2. When $k = $ rank $\mathcal{Z}(\pi_1(M))$, the torus action is called a *maximal torus action*.

LEMMA 11.7.2 ([**CR75**, Lemma 1]). *If $(T^k, M)$ is an effective action of a torus on a closed aspherical manifold and if $H \subset \pi_1(M, x)$ is a central subgroup which contains* $\mathrm{Im}(\mathrm{ev}_*^x)$, *then $H/\mathrm{Im}(\mathrm{ev}_*^x)$ contains no elements of finite order.*

PROOF. Let $M_H$ be the covering space associated to the subgroup $H$ (i.e., with $\pi_1(M_H) = H$). By Theorem 2.5.1, the $T^k$-action lifts to $M_H$. We will show first that $(T^k, M_H)$ is free. Let $b \in M_H$, and suppose $T_b$ be the isotropy of the action $(T^k, M_H)$. Let $p : (M_H, b) \to (M, p(b))$ be the covering projection. If $T_b \neq 1$, then there exists a finite cyclic subgroup $F \subset T_b \subset T_{p(b)}$. Let

$$g : (I, 0, 1) \longrightarrow (T^k, e, f),$$

where $f$ is a generator of $F$. The path $p(g(t) \cdot b)$ is the projection of a loop based at $b \in M_H$ to a loop based at $p(b)$ in $M$. The homotopy class of $p(g(t) \cdot b)$ is an element of $H$ which is in the center of $\pi_1(M)$. Let

$$\alpha : (I, 0, 1) \longrightarrow (M, p(b), p(b))$$

be a loop in $M$ based at $p(b)$. By Lemma 2.7.1,

$$[f \cdot \alpha(s)] = [\overline{p(g(t) \cdot b)} * \alpha(s) * p(g(t) \cdot b)].$$

But as $p(g(t) \cdot b)$ is in the center of $\pi_1(M, p(b))$, we have $f_*(\alpha) = \alpha$. In other words, $F \longrightarrow \mathrm{Aut}(\pi_1(M, p(b)))$ is trivial. This contradicts that $F \longrightarrow \mathrm{Aut}(\pi_1(M, p(b)))$ must be injective since $F$ fixes $p(b)$; see Theorem 3.2.2(2). So $T_b = 1$ for each $b \in M_H$.

Note, $\mathrm{Im}(\mathrm{ev}_*^{p(b)}) = \mathrm{Im}(\mathrm{ev}_*^b) \subset H = \pi_1(M, p(b))$. Put $Q = H/\mathrm{Im}(\mathrm{ev}_*^b)$. Lift the free $(T^k, M_H)$-action to the splitting action $(T^k, T^k \times W)$, where $W$ is contractible. The group $Q$ acts freely on $W$ because the $T^k$-action on $M_H$ is free; see Theorem 3.5.2. The group $Q$ is torsion free, for if it contained some $p$-torsion for some prime $p$, then there is $w \in W$ such that $\mathbb{Z}_p \subset Q_w$, by the Smith theorem, contradicting the freeness of the $Q$-action on $W$.

If $H$ is finitely generated, then $H \cong \mathbb{Z}^n$ for some $n$. $\mathrm{Im}(\mathrm{ev}_*^{p(b)}) \cong \mathbb{Z}^k$ and $H/\mathrm{Im}(\mathrm{ev}_*^{p(b)}) = Q$ torsion free. Therefore, $\mathrm{Im}(\mathrm{ev}_*^{p(b)})$ is a direct summand of $H$. □

COROLLARY 11.7.3. *If $(T^k, M)$ is an effective action of a torus on a closed aspherical manifold and if $H \subset \pi_1(M, x)$ is a finitely generated central subgroup for which* $\mathrm{Im}(\mathrm{ev}_*^x) \subset H$, *then* $\mathrm{Im}(\mathrm{ev}_*^x)$ *is a direct summand of $H$.*

COROLLARY 11.7.4. *Let $(T^k, M)$ is a torus action on a closed aspherical manifold $M$ for which $\mathcal{Z}(\pi_1(M))$ is finitely generated. If it is a maximal torus action, then $\mathrm{Im}(\mathrm{ev}^x_*) = \mathrm{Center}\, \pi_1(M, x)$. Conversely, if $\mathrm{Im}(\mathrm{ev}^x_*) = \mathrm{Center}\, \pi_1(M, x)$, then $(T^k, M)$ is a maximal torus action on $M$.*

REMARK 11.7.5. Let $M$ be a closed aspherical manifold for which $\mathcal{Z}(\pi_1(M)) \cong \mathbb{Z}^k$. Does $M$ admit a maximal torus action? No examples of closed aspherical manifolds that do not admit maximal torus actions are known to us; see Remark 3.1.19.

This question has a negative answer, in general, if we replace "aspherical" by "admissible". Take $M = T^4 \# \mathbb{C}P_2$. This manifold is hyper-aspherical. $\pi_1(M) = \mathbb{Z}^4$ and no $T^k$, $k > 0$, acts effectively on it. Since $\chi(M) = 1$, $M^{T^k} \neq \emptyset$. In fact, $M^{T^k} = M$, $k > 0$. Also no nontrivial finite group can act effectively and homotopically trivially on it. In fact, if a finite $G$ acts on $M$, then $\theta : G \to \mathrm{Aut}(\mathbb{Z}^k)$ is injective; see Exercise 3.4.6.

11.7.6 (Maximal torus action on infra-nilmanifolds). Let $M$ be an infra-nilmanifold. So

$$M = \Pi \backslash G$$

where $G$ is a simply connected nilpotent Lie group,

$$\Pi \subset G \rtimes C \subset G \rtimes \mathrm{Aut}(G) = \mathrm{Aff}(G),$$

$\Pi$ is a lattice of $G \rtimes C$, $C$ a compact subgroup of $\mathrm{Aut}(G)$. Let

$$N_{\mathrm{Aff}(G)}(\Pi)/\Pi = \mathrm{Aff}(M).$$

Then $\mathrm{Aff}(M)$ is a Lie group, the group of *affine* self-diffeomorphisms of $M$.

We claim the following sequence is exact:

$$1 \to \mathrm{Aff}_0(M) \to \mathrm{Aff}(M) \to \mathrm{Out}(\pi_1(M)) \to 1.$$

Every self-homotopy equivalence induces an automorphism of $\Pi$, unique up to an inner automorphism. The homotopy classes of self-homotopy equivalences are in one-to-one correspondence with elements of $\mathrm{Out}(\Pi)$. We think of $M = \theta(\Pi) \backslash (G \times \{\mathrm{pt}\})$ as a Seifert Construction, where $\theta : \Pi = \pi_1(M) \to \mathrm{TOP}_G(G \times \{\mathrm{pt}\}) = \mathrm{Aff}(G \times \{\mathrm{pt}\})$. We use uniqueness and rigidity for this setup. Then every isomorphism $\Pi \to \Pi$ is given by conjugation by a homeomorphism in $\mathrm{Aff}(G)$ and an automorphism of $\Pi$ will have a realization in $N_{\mathrm{Aff}(G)}(\Pi)$. Thus $\mathrm{Aff}(M) = N_{\mathrm{Aff}(G)}(\Pi)/\Pi \to \mathrm{Out}(\Pi)$ is onto. We are interested in the kernel of this homomorphism. Look at the following commutative diagram.

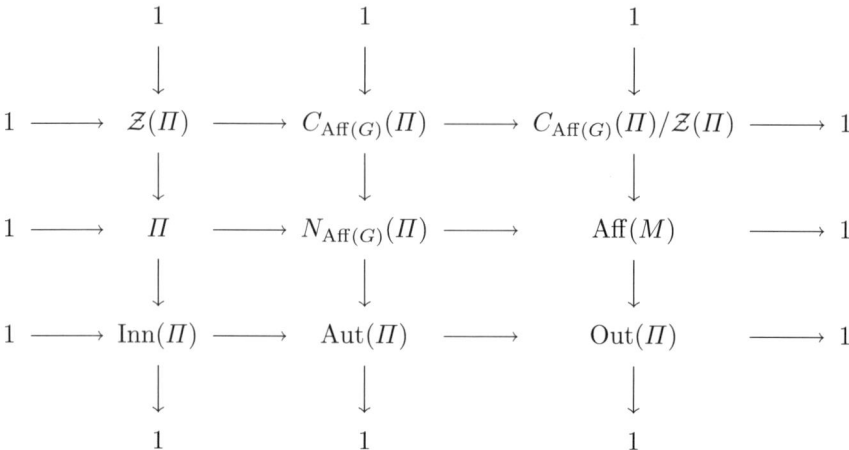

We see the kernel of $\text{Aff}(M) \to \text{Out}(\Pi)$ is $C_{\text{Aff}(G)}(\Pi)/\mathcal{Z}(\Pi)$. There are two different ways of expressing $\text{Aff}(G)$. Namely,

$$\begin{aligned}\text{Aff}(G) &= r(G) \rtimes \text{Aut}(G) \\ &= \ell(G) \rtimes \text{Aut}(G).\end{aligned}$$

The actions on $x \in G$ are as follows:

$$(r(a), \alpha) \cdot x = \alpha(x) \cdot a^{-1}, \qquad (\ell(a), \alpha) \cdot x = a \cdot \alpha(x).$$

Clearly, the correspondence

$$(r(a), \alpha) \longleftrightarrow (\ell(a^{-1}), \mu(a) \circ \alpha)$$

is a bijective map between $r(G) \rtimes \text{Aut}(G)$ and $\ell(G) \rtimes \text{Aut}(G)$. The first expression is more natural with respect to the Seifert Construction because

$$\begin{aligned}\text{TOP}_G(G \times \{w\}) &= \text{M}(\{w\}, G) \rtimes (\text{Aut}(G) \times \text{TOP}(\{w\})) \\ &= r(G) \rtimes \text{Aut}(G).\end{aligned}$$

But, in this section, we shall use $\text{Aff}(G) = \ell(G) \rtimes \text{Aut}(G)$.

THEOREM 11.7.7. *Let $M = \Pi \backslash G$ be an infra-nilmanifold, where $G$ is a simply connected nilpotent Lie group, $\Pi \subset G \rtimes \text{Aut}(G)$ an almost Bieberbach group. Then $\text{Aff}_0(M) = C_{\text{Aff}(G)}(\Pi)/\mathcal{Z}(\Pi) = G^Q/\mathcal{Z}(\Pi)$ ($Q$ the holonomy group), and it contains a maximal torus action $(\mathcal{Z}(G))^Q/\mathcal{Z}(\Pi)$.*

PROOF. We want to show $C_{\text{Aff}(G)}(\Pi)/\mathcal{Z}(\Pi)$ is connected (as a Lie subgroup). Let $(a, \alpha) \in C_{\text{Aff}(G)}(\Pi)$. Then $(a, \alpha)$ must centralize $\Gamma \subset \Pi$ and so also all of $G$. Thus,

$$(a, \alpha)(g, 1) = (g, 1)(a, \alpha),$$

which implies $\alpha(g) = a^{-1}ga$ for all $g \in G$. Thus $\alpha = \mu(a^{-1})$. Therefore each element of $C_{\text{Aff}(G)}(\Pi)$ must be of the form $(a, \mu(a^{-1})) = r(a^{-1})$. Now it must also centralize $\Pi$. Let $\Pi \cap G = \Gamma$ and $\Pi/\Gamma = Q$. Then the holonomy group $Q$ injects into $\text{Aut}(G)$ naturally and we have a commutative diagram

$$\begin{array}{ccccccccc} 1 & \longrightarrow & \Gamma & \longrightarrow & \Pi & \longrightarrow & Q & \longrightarrow & 1 \\ & & \downarrow & & \downarrow & & \downarrow & & \\ 1 & \longrightarrow & G & \longrightarrow & G \rtimes \text{Aut}(G) & \longrightarrow & \text{Aut}(G) & \longrightarrow & 1. \end{array}$$

Let $(y, \beta) \in \Pi$. Then $(a, \mu(a^{-1}))(y, \beta) = (y, \beta)(a, \mu(a^{-1}))$ implies $ya = y\beta(a)$ for all $y \in \Gamma$. Thus $a = \beta(a)$. That is, as we run through all the $\beta \in Q$, $\beta(a) = a$ so that $a \in G^Q$, a closed subgroup of $G$. Clearly $G^Q$ is a simply connected nilpotent subgroup, best seen from the Lie algebra. We have

$$\begin{aligned} \mathrm{Aff}_0(M) &= C_{\mathrm{Aff}(G)}(\Pi)/\mathcal{Z}(\Pi) \\ &= \{(a, \mu(a^{-1})) : a \in G^Q\}/\mathcal{Z}(\Pi) \\ &\cong r(G^Q)/\mathcal{Z}(\Pi). \end{aligned}$$

Since $\mathcal{Z}(\Pi) \subset \mathcal{Z}(G)$, we have $\mathcal{Z}(\Pi) \subset (\mathcal{Z}(G))^Q$. It is easy to see that $\mathcal{Z}(\Pi)$ is a uniform lattice of $(\mathcal{Z}(G))^Q$, and so $(\mathcal{Z}(G))^Q/\mathcal{Z}(\Pi)$ is a torus, acting effectively on $M$. Since $\pi_1((\mathcal{Z}(G))^Q/\mathcal{Z}(\Pi)) = \mathcal{Z}(\Pi)$, this torus action is a maximal torus action by Corollary 11.7.4. Note $G^Q/\mathcal{Z}(\Pi)$ may not be compact. □

EXAMPLE 11.7.8 (Klein bottle). The fundamental group has presentation $\Pi = \{a, b \mid a^2 b^2 = 1\}$. $\mathcal{Z}(\Pi) = \{a^2\}$, the maximal normal Abelian subgroup is

$$\Gamma = \mathbb{R}^2 \cap \Pi = \{a^2, b \mid [a^2, b] = 1\},$$

and the *holonomy group* is $Q = \mathbb{Z}/2$. The universal covering space is the Abelian Lie group $G = \mathbb{R}^2$, and $G^Q = \mathbb{R}^1$ with $G^Q \cap \Pi = \{a^2\} \approx \mathbb{Z}$. Therefore, $\mathrm{Aff}_0(M) \cong G^Q/\mathcal{Z}(\Pi) = \mathbb{R}/\mathbb{Z} = S^1$, a circle. This is the maximal torus action by affine diffeomorphisms (in fact, isometries).

EXERCISE 11.7.9. Find $\mathrm{Aut}(\pi_1(\text{Klein bottle}))$ and $\mathrm{Out}(\pi_1(\text{Klein bottle}))$. (Answer: $\mathbb{Z}_2 \times \mathbb{Z}_2$).

EXAMPLE 11.7.10. Let $G$ be a simply connected nilpotent Lie group, and let $\Pi \subset G \rtimes \mathrm{Aut}(G)$ be an almost Bieberbach group. Suppose $\Gamma = \Pi$. That is, $\Pi$ has trivial holonomy group $Q$. Then $G^Q = G$ since $Q = 1$. Therefore,

$$\mathrm{Aff}_0(M) = G^Q/\mathcal{Z}(\Gamma) = G/\mathcal{Z}(\Gamma).$$

Topologically, this is a product $\mathcal{Z}(G)/\mathcal{Z}(\Gamma) \times G/\mathcal{Z}(G)$, of a torus with a simply connected nilpotent Lie group. This is the covering space of $M$ corresponding to the image of the evaluation homomorphism of the maximal torus action. Note that $\Gamma/\mathcal{Z}(\Gamma)$ is a lattice in $G/\mathcal{Z}(G)$. The maximal torus action on a nilmanifold is free, since $\Gamma/\mathcal{Z}(\Gamma)$ is torsion free. In fact, any effective torus action on a nilmanifold is free. This follows from Corollary 11.7.3.

COROLLARY 11.7.11. *Let $G$ be a simply connected nilpotent Lie group, $\Pi \subset G \rtimes \mathrm{Aut}(G)$ an almost Bieberbach group. Then*

$$\mathrm{Aff}_0(M) = r(G^Q)/\mathcal{Z}(\Pi)$$

*contains a torus subgroup $(\mathcal{Z}(G))^Q/\mathcal{Z}(\Pi)$, with quotient group a simply connected nilpotent Lie group $G^Q/\mathcal{Z}(G)^Q$. Therefore, if $G = \mathbb{R}^n$ (i.e., $\Pi$ is a Bieberbach group), then $\mathrm{Aff}_0(M)$ is a torus $G^Q/\mathcal{Z}(\Pi)$.* □

REMARK 11.7.12. The torus is a maximal torus action and represents also the connected component of the full isometry group.

11.7.13. In [**LR91**], a smooth maximal torus action is constructed on each solvmanifold (quotient of a connected simply connected solvable Lie group by its lattice). Now it is easy to see that every infra-solvmanifold $M = \Pi \backslash G$ of type (R) admits an affine maximal torus action: The center $\mathcal{Z}(\Pi)$ lies in the center of $G$. Since $\mathcal{Z}(G)$ is connected, $\mathcal{Z}(\Pi)\backslash(\mathcal{Z}(G))^Q$ is a torus. The action is via left translations on the universal covering $G$, and hence is affine. If the solvmanifold has a metric coming from a left invariant metric of the universal covering group, this torus lies in the group of isometries of $M$.

11.7.14 (Counterexamples). Most statements which we proved are not true for general solvable Lie groups. Let $H = \mathbb{R}^2 \rtimes \mathbb{R}$, where $t \in \mathbb{R}$ acts on $\mathbb{R}^2$ by $\begin{bmatrix} \cos t & \sin t \\ -\sin t & \cos t \end{bmatrix}$. This is the universal covering group of the Euclidean group $E(2)_0$. Note that $H$ is not of type (R). Take $\Gamma = \mathbb{Z}^2 \times \mathbb{Z} \subset H$, where $\mathbb{Z}^2$ is the standard lattice of $\mathbb{R}^2$ and $\mathbb{Z}$ is the center of $H$. Let $\Pi_1$ be a torsion-free extension of $\Gamma$ by $\Psi = \mathbb{Z}_2 \times \mathbb{Z}_2$, where $\Psi$ acts on $\mathbb{Z}^3$ as diagonal matrices

$$I,\ \alpha = \begin{bmatrix} -1 & 0 & 0 \\ 0 & -1 & 0 \\ 0 & 0 & 1 \end{bmatrix},\ \beta = \begin{bmatrix} 1 & 0 & 0 \\ 0 & -1 & 0 \\ 0 & 0 & -1 \end{bmatrix},\ \gamma = \begin{bmatrix} -1 & 0 & 0 \\ 0 & 1 & 0 \\ 0 & 0 & -1 \end{bmatrix}.$$

This is the 3-dimensional abstract Bieberbach group $\mathfrak{G}_6$ in [**Wol77**]. There is no such a group $\Psi$ in $\mathrm{Aut}(H)$. (The matrices $\beta$ and $\gamma$ map the generator of $\mathbb{Z}$ to its negative. That implies they must act on $\mathbb{R}^2$ as $-I$. But the first $(2 \times 2)$-block of these matrices are not $-I$.) and $\Pi_1$ does not embed into $\mathrm{Aff}(H)$. Thus, Theorem 8.4.2 is not true for this example. The discrete nilradical of $\Pi_1$ is $\Gamma \cong \mathbb{Z}^3$; nilradical of $H$ is $\mathbb{R}^2$. Therefore, $\mathbb{R}^2 \cap \Pi_1$ is not the discrete nilradical of $\Pi_1$. Thus Proposition 8.4.8 also fails. The rigidity theorem, Theorem 8.4.3, holds for this $H$.

It is easy to find a solvable Lie group $G$ on which the rigidity fails. Let $G = H \times \mathbb{R}^3$ be the direct product of the above $H$ with $\mathbb{R}^3$. Let $\Pi$ be a torsion-free extension of $\mathbb{Z}^3$ by $\mathbb{Z}_2$ generated by $\alpha$, an abstract Bieberbach group $\mathfrak{G}_2$ in [**Wol77**]. Then $\Pi$ embeds into $E(3) \subset \mathrm{Aff}(\mathbb{R}^3)$ and also into $H \rtimes \mathrm{Aut}(H)$ so that $H \cap \Pi = \Gamma \cong \mathbb{Z}^3$. Also $\mathbb{Z}^3$ embeds into $H$ as $\Gamma$ above as well as into $\mathbb{R}^3$ as the standard lattice. Take $\Pi' = \mathbb{Z}^3 \times \Pi \subset \mathrm{Aff}(G)$ and $\Pi'' = \Pi \times \mathbb{Z}^3 \subset \mathrm{Aff}(G)$. Then, clearly, $\Pi' \cong \Pi''$ by interchanging the components, but the solvmanifold $\Pi'\backslash G$ is not affinely diffeomorphic to $\Pi''\backslash G$, since there is no automorphism of $G$ which interchanges the two factors of $G$. Therefore the rigidity theorem, Theorem 8.4.3, fails. On $\Pi'\backslash G$, there is an affine maximal torus action $T^4$, but not all of it comes from left translation, since $G$ does not have $\mathcal{Z}(G) \cong \mathbb{R}^4$.

THEOREM 11.7.15. *Suppose $H$ is a compact Lie group of homeomorphisms acting freely and locally smoothly (see Subsection 1.8.4) on an infra-nilmanifold $M$. Suppose $\dim(H\backslash M) \neq 3$. Then the action can be conjugated into $\mathrm{Aff}(M)$ so that the subgroup $H_1 = \{h \in H : h \simeq identity\}$ is contained in the standard maximal torus action on $M$.*

PROOF. The connected component of the identity $H_0$ in $H$ is a torus of dimension, say $s$, (because $M$ is aspherical, see Theorem 3.2.2), and is contained in $H_1$. Let $\pi_1(M) = \Pi$. Then by Corollary 11.7.3, $\pi_1(H_0) \cong \mathbb{Z}^s$ is a direct summand of $\mathcal{Z}(\pi_1(M))$.

Let $E$ be the group of all lifts of the action of $H$ on $M$ so that $1 \to \Pi \to E \to H \to 1$ is exact. Let $Q$, $Q'$ and $F$ be defined by

$$Q = \Pi/\mathbb{Z}^s,\quad Q' = E/\mathbb{R}^s,\quad F = H/H_0$$

so that the following diagram is commutative with exact rows and columns:

## 11.7. MAXIMAL TORUS ACTIONS

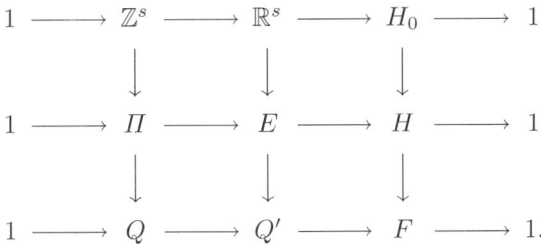

(All vertical lines are also short exact sequences.)

We claim that $Q'$ is torsion free. (Then $Q'$ is a torsion-free, finitely generated, virtually nilpotent group. Thus, it is an *almost Bieberbach group*.) Suppose it has nontrivial torsion. Pick a subgroup $P = \mathbb{Z}_p$ ($p$ prime) of $Q'$. The restriction of the middle vertical exact sequence gives a short exact sequence $1 \to \mathbb{R}^s \to E_P \to P \to 1$. Note that $E_P$ is sitting in $E$. Any such sequence splits so that $E_P = \mathbb{R}^s \rtimes P$. Thus
$$\mathbb{Z}_p = P \subset E_P \subset E.$$
The group $\mathbb{Z}_p$ acts on the universal covering space $\widetilde{M}$ of $M$. Since $\widetilde{M}$ is contractible, this action has a nonempty fixed point; see Subsection 1.8.1. Then this $P \subset E \subset \mathrm{TOP}(\widetilde{M})$ projects to $P \subset H \subset \mathrm{TOP}(M)$ which is acting freely, a contradiction.

We may lift the $T^s$-action on $M$ to $M_{\mathrm{Im}(\mathrm{ev}^x_*)} = M_{\mathbb{Z}^s}$ (the covering space of $M$ with fundamental group $\mathbb{Z}^s$) to get a splitting
$$(T^s, M_{\mathbb{Z}^s}) = (T^s, T^s \times W),$$
and the action of $Q'$ on $M_{\mathbb{Z}^s}$ projects down to $(Q', W)$.

If we lift all of the action of $Q'$ to $\widetilde{M}$, we get an action of $E$ on $\widetilde{M} = \mathbb{R}^s \times W$. As the short exact sequence
$$1 \to \mathbb{R}^s \to E \to Q' \to 1$$
shows, $E$ acts on $\mathbb{R}^s \times W$ as a fiber preserving homeomorphism group. In other words, $E$ is acts as a subgroup of $\mathrm{TOP}_{\mathbb{R}^s}(\mathbb{R}^s \times W) = \mathrm{M}(W, \mathbb{R}^s) \rtimes (\mathrm{GL}(n, \mathbb{R}) \times \mathrm{TOP}(W))$ as follows:

$$\begin{array}{ccccccccc}
1 & \longrightarrow & \mathbb{R}^s & \longrightarrow & E & \longrightarrow & Q' & \longrightarrow & 1 \\
& & \downarrow & & \downarrow{\scriptstyle \Theta_1} & & \downarrow & & \\
1 & \longrightarrow & \mathrm{M}(W, \mathbb{R}^s) & \longrightarrow & \mathrm{TOP}_{\mathbb{R}^s}(\mathbb{R}^s \times W) & \longrightarrow & \mathrm{GL}(n, \mathbb{R}) \times \mathrm{TOP}(W) & \longrightarrow & 1;
\end{array}$$

see Corollary 4.2.10.

Since $H_1 \to \mathrm{Out}(\Pi)$ is trivial, we have induced
$$1 \to \mathcal{Z}(\Pi) \to C_E(\Pi) \to H_1 \to 1$$
with $C_E(\Pi)$, the centralizer of $\Pi$ in $E$, being the kernel of $E \to \mathrm{Aut}(\Pi)$. Thus it is a central extension.

We analyze $C_E(\Pi)$ more carefully. Since $\pi_1(H_0) \cong \mathbb{Z}^s$ is a direct summand of $\mathcal{Z}(\pi_1(M))$, we have $\mathcal{Z}(\Pi) = \mathbb{Z}^s \oplus \mathbb{Z}^{k-s}$. Denote $C_E(\Pi)/\mathbb{R}^s$ by $Z$. Then we have a commutative diagram of exact rows and columns:

$$\begin{CD}
1 @>>> \mathbb{Z}^s @>>> \mathbb{R}^s @>>> H_0 @>>> 1 \\
@. @VVV @VVV @VVV @. \\
1 @>>> \mathbb{Z}^s \oplus \mathbb{Z}^{k-s} @>>> C_E(\Pi) @>>> H_1 @>>> 1 \\
@. @VVV @VVV @VVV @. \\
1 @>>> \mathbb{Z}^{k-s} @>>> Z @>>> F_1 @>>> 1.
\end{CD}$$

All vertical lines are also short exact sequences.

As shown in the last horizontal line, $Z$ is a torsion-free central extension of $\mathbb{Z}^{k-s}$ by a finite group. Such a group is itself free Abelian so that $Z \approx \mathbb{Z}^{k-s}$; see Corollary 5.5.3. Consequently, the middle vertical short exact sequence splits, and we have

$$C_E(\Pi) = \mathbb{R}^s \oplus Z, \quad Z \approx \mathbb{Z}^{k-s}$$

and

$$H_1 = H_0 \oplus F_1, \quad F_1 \text{ finite}$$

is an Abelian group.

On the other hand, $M$ was an infra-nilmanifold. So $\widetilde{M} = G$, a connected, simply connected nilpotent Lie group, and $M = \Pi \backslash G$ with $\Pi \subset \mathrm{Aff}(G)$ as an almost Bieberbach group. There is a canonical maximal torus action $A^k$ on $M$. This comes from the left translations $\ell(G)$. More precisely, the centralizer of $\Pi$ in $\mathrm{Aff}(G)$ is $\mathbb{R}^k$ which lies in the center of $\ell(G)$. We can reparametrize $A^k = A^s \times A^{k-s}$ so that the standard action of $A^s$ has the same evaluation homomorphism as $H_0 = T^s$. That is,

$$\mathrm{ev}_*(\pi_1(T^s)) = \mathrm{ev}_*(\pi_1(A^s)) \subset \mathcal{Z}(\Pi).$$

The action of $A^s$ is free because $\pi_1(M)/\mathrm{ev}_*^x(\pi_1(T^s)) = Q$ is torsion free, for otherwise the $T^s$-action could not be free. We write the centralizer $\mathbb{R}^k$ by

$$\mathbb{R}^k = \widetilde{A}^s \times \widetilde{A}^{k-s}$$

so that $\widetilde{A}^s \approx \mathbb{R}^s$, $\widetilde{A}^{k-s} \approx \mathbb{R}^{k-s}$, and $A^k = (\Pi \cap \mathbb{R}^k)\backslash \mathbb{R}^k$. Clearly $\widetilde{A}^s$ is a normal subgroup of $G$. Let us denote the quotient by $\overline{G}$ so that

$$1 \longrightarrow \widetilde{A}^s \longrightarrow G \longrightarrow \overline{G} \longrightarrow 1$$

is exact. Then $\overline{G}$ is again a connected, simply connected nilpotent Lie group.

Consider the group $Q'$. Its subgroup $Q$ acts on the quotient group $\overline{G}$ as a subgroup of $\mathrm{Aff}(\overline{G})$. Since $Q'$ is a torsion-free extension of $Q$ by a finite group $F$, there is a unique way of embedding $Q'$ into $\mathrm{Aff}(\overline{G})$ so that the original $Q \subset \mathrm{Aff}(\overline{G})$ is untouched (which uses UAEP of the pair $(Q, \overline{G})$).

The extended lift of $Q$ to $G$ is exactly $\Pi$. Therefore, the group $\widetilde{\Pi} = \Pi \cdot \widetilde{A}^s$ naturally lies in $\mathrm{Aff}(G) = G \rtimes \mathrm{Aut}(G)$ already. Our goal is to embed $H$ into $\mathrm{Aff}(G)$.

Consider the commutative diagram with exact rows and columns

$$
\begin{array}{ccccccccc}
& & \widetilde{A}^s & = & \widetilde{A}^s & & & & \\
& & \downarrow & & \downarrow & & & & \\
1 & \longrightarrow & \widetilde{\Pi} = \Pi \cdot \widetilde{A}^s & \longrightarrow & E & \longrightarrow & F & \longrightarrow & 1 \\
& & \downarrow & & \downarrow & & \| & & \\
1 & \longrightarrow & Q & \longrightarrow & Q' & \longrightarrow & F & \longrightarrow & 1.
\end{array}
$$

(All vertical lines are also short exact sequences.)

Since $\Pi$ is normal in $E$, and the lattice $\Pi \cap G$ is characteristic in $\Pi$, the lattice is normalized by $E$. Thus we have a homomorphism

$$\theta : E \to \operatorname{Aut}(\Pi \cap G) \to \operatorname{Aut}(G)$$

by UAEP of the lattice. Define a map

$$\lambda : E \to G$$

in such a way that:

(1) for $\alpha \in \widetilde{\Pi}$, $\alpha \mapsto (\lambda(\alpha), \theta(\alpha))$ is the embedding of $\widetilde{\Pi} \subset \operatorname{Aff}(G)$;
(2) for all $\alpha \in E$, $\alpha \mapsto (\lambda(\alpha), \theta(\alpha)) \mapsto \operatorname{Aff}(\overline{G})$ is a homomorphism, where the second map is the natural $\operatorname{Aff}(G) \to \operatorname{Aff}(\overline{G})$.

Our goal is finding a $\lambda$ such that

$$
\begin{array}{ccc}
E & \longrightarrow & G \rtimes \operatorname{Aut}(G) = \operatorname{Aff}(G) \\
\alpha & \longrightarrow & (\lambda(\alpha), \theta(\alpha))
\end{array}
$$

is a homomorphism extending $\widetilde{\Pi} \longrightarrow \operatorname{Aff}(G)$. It will be a homomorphism if and only if

$$\lambda(\alpha) \cdot \theta(\alpha)(\lambda(\beta)) = \lambda(\alpha\beta)$$

for all $\alpha, \beta \in E$. By the second condition that $\alpha \mapsto (\lambda(\alpha), \theta(\alpha)) \mapsto \operatorname{Aff}(\overline{G})$ is a homomorphism, this equality holds on $\overline{G}$ level. Therefore, $\lambda(\alpha) \cdot \theta(\alpha)(\lambda(\beta)) \cdot \lambda(\alpha\beta)^{-1}$ lies in $\widetilde{A}^s$ for all for all $\alpha, \beta \in E$. Let $f : E \times E \to \widetilde{A}^s$ be defined by

$$f(\alpha, \beta) = \lambda(\alpha) \cdot \theta(\alpha)(\lambda(\beta)) \cdot \lambda(\alpha\beta)^{-1}.$$

We quickly see that $f$ is really a 2-cocycle

$$f : F \times F \to \widetilde{A}^s$$

because $f(\alpha, \beta) = 1$ for $\alpha, \beta \in \widetilde{\Pi}$ by the first condition. As we know very well, $H^2(F; \widetilde{A}^s) = 0$ (recall $\widetilde{A}^s = \mathbb{R}^s$). Consequently, we can find such a desired $\lambda$.

Now we compare the two actions of $E$ on $\widetilde{M} = \mathbb{R}^s \times W$ and $\widetilde{M} = G = \widetilde{A}^s \times \overline{G}$. The first one is topological (even though the part $\Pi$ is affine), and the second one is affine by our construction. The $E$-actions have corresponding $Q'$-actions which we denote by

$$\rho_1 : Q' \to \operatorname{TOP}(W),$$
$$\rho_2 : Q' \to \operatorname{TOP}(\overline{G}).$$

These yield the manifolds $Q'\backslash W$ and $Q'\backslash \overline{G}$. Of course, the latter is an infra-nilmanifold. These two spaces are homeomorphic by virtue of the theorem of Farrell and Hsiang [**FH83**] for dimension not equal to 4, 5 and for dimension 4 by Freedman, Farrell and Jones [**FJ98**]. Therefore, the free $Q'$-actions on $W$ and $\overline{G}$ are weakly equivalent. That is, there is a homeomorphism $h : W \to \overline{G}$ and an automorphism $\omega : Q' \to Q'$ so that

$$\begin{array}{ccc} W & \xrightarrow{h} & \overline{G} \\ \rho_1(\alpha) \downarrow & & \downarrow \rho_2(\omega(\alpha)) \\ W & \xrightarrow{h} & \overline{G} \end{array}$$

for all $\alpha \in Q'$. But, the Second Generalized Bieberbach Theorem 8.4.3 ensures that $\omega$ is a conjugation. More precisely, there exists $(a, A) \in \mathrm{Aff}(\overline{G})$ such that

$$\rho_2(\omega(\alpha)) = (a, A)\rho_2(\alpha)(a, A)^{-1}$$

for all $\alpha \in Q'$. Thus we have a commuting diagram

$$\begin{array}{ccccc} W & \xrightarrow{h} & \overline{G} & \xrightarrow{(a,A)} & \overline{G} \\ \rho_1(\alpha) \downarrow & & \downarrow \rho_2(\omega(\alpha)) & & \downarrow \rho_2(\alpha) \\ W & \xrightarrow{h} & \overline{G} & \xrightarrow{(a,A)} & \overline{G} \end{array}$$

for all $\alpha \in Q'$.

The fibers $\mathbb{R}^s$ and $\widetilde{A}^s$ are isomorphic by extending the natural isomorphism of the lattices $\mathbb{Z}^s$. (Recall we reparametrized the maximal torus action on the infra-nilmanifold so that the evaluation of subtorus $A^s$ is the same as that of the topological torus action $T^s$.) So we identify $\mathbb{R}^s$ with $\widetilde{A}^s$.

Now pick a lift of $W \xrightarrow{h} \overline{G} \xrightarrow{(a,A)} \overline{G}$ to a map

$$\mathfrak{F} : \mathbb{R}^s \times W \longrightarrow \widetilde{A}^s \times \overline{G}$$

so that the fiber map is just the identification. Then the action of $E$ on $\mathbb{R}^s \times W$ given by

$$\alpha \mapsto \mathfrak{F} \circ (\lambda(\alpha), \theta(\alpha)) \circ \mathfrak{F}^{-1}$$

is a new Seifert Construction with the homomorphism $\rho_1 : Q' \to \mathrm{TOP}(W)$, which we denote by

$$\Theta_2 : E \longrightarrow \mathrm{TOP}_{\mathbb{R}^s}(\mathbb{R}^s \times W).$$

Thus we have two Seifert Constructions:

$$\begin{array}{ccccccccc} 1 & \longrightarrow & \mathbb{R}^s & \longrightarrow & E & \longrightarrow & Q' & \longrightarrow & 1 \\ & & \downarrow & & \Theta_1 \downarrow \Theta_2 & & \downarrow \rho_1 & & \\ 1 & \longrightarrow & \mathrm{M}(W, \mathbb{R}^s) & \longrightarrow & \mathrm{TOP}_{\mathbb{R}^s}(\mathbb{R}^s \times W) & \longrightarrow & \mathrm{GL}(n, \mathbb{R}) \times \mathrm{TOP}(W) & \longrightarrow & 1. \end{array}$$

Now the uniqueness part of Theorem 7.3.2 applies. The two constructions are conjugate by an element of $\mathrm{M}(W, \mathbb{R}^s)$. We conclude that the original topological action $(E, \mathbb{R}^s \times W)$ and the constructed affine action $(E, G)$ are related by a topological conjugation. By the affine action $(E, G)$, the centralizer $C_E(\Pi)$ maps into the maximal torus action. $\square$

11.7.16. In our analysis of infra-nilmanifolds, we saw that much of the information about the manifolds is encoded in the Lie group $\mathrm{Aff}(M) = N_{\mathrm{Aff}(M)}(\Phi)/\Phi$. (This is a special case of the geometric information found in $N_{\mathrm{TOP}_G(G \times W)}(\Phi)/\Phi$ for an injective Seifert fibering $M = \Phi \backslash G \times W$.) We have seen that $\pi_0(\mathrm{Aff}(M)) = \mathrm{Aff}(M)/(C_{\mathrm{Aff}(M)}(\Pi)/\mathcal{Z}(\Pi)) = \mathrm{Out}(\Pi)$. $\mathrm{Out}(\Pi)$ can also be identified with $\mathrm{TOP}(M)$ or $\mathrm{Diff}(M)$ modulo those homotopic to the identity in the space of self homotopy equivalences, but $\pi_0(\mathrm{Diff}(M))$ and $\pi_0(\mathrm{TOP}(M))$ may be much larger. In the following commuting diagram, $\mathcal{E}(M)$ is the $H$-space of all self-homotopy equivalences of $M$.

$$\begin{array}{ccccc}
\mathrm{Aff}(M) & \xrightarrow{\mathrm{inc}} & \mathrm{Diff}(M) & \xrightarrow{\mathrm{inc}} & \mathcal{E}(M) \\
\downarrow & & \downarrow & & \downarrow \\
\pi_0(\mathrm{Aff}(M)) & \xrightarrow{\mathrm{inj}} & \pi_0(\mathrm{Diff}(M)) & \xrightarrow{\mathrm{surj}} & \pi_0(\mathcal{E}(M))
\end{array}$$

Note that the composite homomorphism on the bottom $\pi_0(\mathrm{Aff}(M)) \longrightarrow \pi_0(\mathcal{E}(M))$ is an isomorphism.

The kernel, $\pi_0(\mathrm{TOP}(M)) \to \mathrm{Out}(\Phi) = \pi_0(\mathcal{E}(M))$, denoted by $K$ is *exotic* (the indications are that $K$ consists solely of 2-torsions). Farrell and Jones [**FJ93**] have shown that $K$ is isomorphic to $\sum_{i=1}^{\infty}(\mathbb{Z}_2)_i = \mathbb{Z}_2^{\infty}$ when $\dim(M) > 10$ and $M$ is closed and homeomorphic to a nonpositively curved manifold.

PROPOSITION 11.7.17 (cf. [**LR82**, (4.1) Corollary 2]). *If $M$ is a closed manifold homeomorphic to an infra-nilmanifold, then no nontrivial finite subgroup of $K$ (the kernel of $\pi_0(\mathrm{TOP}(M)) \to \mathrm{Out}(\Phi)$) can act freely on $M$.*

PROOF. Let $G$ be a finite subgroup of $K$ and suppose it acts freely on $M$. Each element $g \in G$ is homotopic to the identity but not isotopic to the identity, $g \neq 1$. Since the group $G$ acts freely, it also acts locally smoothly by default. Therefore, $G$ can be embedded in the maximal torus action of $M$. That is, the action can be conjugated in $\mathrm{TOP}(M)$ so that its image is in the standard maximal torus action on $M$. Each such conjugated $g$ is isotopic to the identity inside the maximal torus action. This is a contradiction. $\square$

REMARK 11.7.18. For some admissible manifolds, we can guarantee that any homotopically trivial action is free.

1. Any compact Lie group $G$ that acts effectively and homotopically trivially on an admissible manifold $M$ is Abelian. If the center of $\pi_1(M)$ is finitely generated, then $G$ is isomorphic to a direct product of torus group with a finite Abelian group. Moreover, if $\mathrm{Inn}(\pi_1(M))$ is torsion free, then $G$ acts freely.

2. If $M$ is a nilmanifold, then any homotopically trivial action of a compact Lie group is a free action. If the action is locally smooth, then the action can be conjugated into the standard free torus action. Therefore no nontrivial subgroup of $K$ can act on $M$.

3. ([**LR82**, Corollary 2]) If $\mathcal{Z}(\pi_1(M))$ is a direct product factor of $\pi_1(M)$, where $M$ is a closed infra-nilmanifold, then any locally smooth homotopically trivial action is free and embeds in the maximal torus action. In particular, no nontrivial subgroup of $K$ acts effectively on $M$.

4. If $M$ is a closed flat manifold and the holonomy has odd order, then no nontrivial subgroup of $K$ acts effectively on $M$, assuming $\dim(M) > 10$.

5. If $M$ is an admissible manifold and $\mathcal{Z}(\pi_1(M)) = 1$, then no nontrivial compact Lie group acts homotopically trivially.

6. Suppose $M$ and $M'$ are homotopy equivalent infra-nilmanifolds; $M = \Pi\backslash G$ and $M' = \Pi'\backslash G$. Then by the Generalized Bieberbach Theorem (Theorem 8.4.3), $\Pi'$ is a conjugate of $\Pi$, say by $\alpha \in \text{Aff}(G)$. Then $\alpha$ induces an affine diffeomorphism from $M$ to $M'$. Of course, this induced diffeomorphism sends the standard maximal torus action of $M$ to the standard maximal torus action of $M'$. However, in general, maximal torus actions cannot be compared easily.

PROOF. 1. First we note that a finite subgroup $F$ of $G$ acting on $M$ is Abelian by Theorem 3.2.2(3). Secondly, if the center of $\pi_1(M)$ is finitely generated of rank $k$, say, then $F$ can be embedded in a torus of rank $k$ by Theorem 3.2.2(4). Then by Proposition 5.5.5, $G$ itself is Abelian and $G_0$, the connected component of the identity, is a torus $T^s$ with $0 \leq s \leq k$. Finally, the universal covering $\widetilde{G}$ of $G$ is split by Lemma 8.4.1, and so $G$ is split. Thus, as $G_0$ is a torus and $G$ is Abelian, $G$ is a product of $T^s$ with a finite Abelian group.

Now suppose $\text{Inn}(\pi_1(M))$ is torsion free. Then $G_x \subset G$ is finite and injects into $\text{Aut}(\pi_1(M))$, and hence lies in $\text{Inn}(\pi_1(M)) \subset \text{Aut}(\pi_1(M))$. As $\text{Inn}(\pi_1(M))$ is torsion free, $G_x = 1$.

2. Since $\text{Inn}(\pi_1(M))$ is torsion free, $G$ acts freely, and if it acts locally smoothly, $G$ can be conjugated into the maximal torus action. As each element of the maximal torus action is isotopic to the identity, no subgroup of $K$ can act on $M$.

3. $\pi_1(M) = \mathcal{Z}(\pi_1(M)) \times \pi_1(M)/\mathcal{Z}(\pi_1(M))$. The group $\pi_1(M)/\mathcal{Z}(\pi_1(M)) = \text{Inn}(\pi_1(M))$ and is torsion free. Therefore the previous argument applies.

4. Suppose $G$ is a finite subgroup of $K$ that acts effectively on $M$. Let $E$ be the extended lifting to $\widetilde{M} = \mathbb{R}^n$. Let $e \in E$ be such that its image in $G$ is $g$ where $g$ is not isotopic to the identity on $M$. The image of $e$ in $\text{Aut}(\Pi)$ lies in $\text{Inn}(\Pi)$. Therefore there is $\sigma \in \Pi$ such that $\mu(\sigma) = \text{Im}(e)$. Then $(\mu(\sigma))^2 = \mu(\sigma^2) = 1$ implies $\sigma^2 \in \mathcal{Z}(\Pi)$. Denote $\sigma$ by $(a, A) \in \text{Aff}(\mathbb{R}^n)$. Then $((I + A)a, A^2) = \sigma^2 \in \mathcal{Z}(\Pi)$ implies $A^2 = I$. So $M$ has even order holonomy, a contradiction.

5. If $G$ is a finite group acting on an admissible $M$ and $\mathcal{Z}(\pi_1(M)) = 1$, then $\psi: G \to \text{Out}(\pi_1(M))$ must be faithful. So $G$ cannot act homotopically trivially.

6. A maximal torus action on a closed aspherical manifold is not necessarily unique up to topological equivalence. (Of course it is understood that we do not distinguish between two maximal torus actions if they differ only by an automorphism of the tori.) Without local smoothness assumptions, even a free maximal torus action may have a nonlocally Euclidean orbit space. (For example: $S^1 \times \mathfrak{G}_6$ is a flat manifold with a free $S^1$ maximal torus action. However, $S^1 \times \mathfrak{G}_6$ is homeomorphic to $S^1 \times \mathfrak{G}_6^*$, where $\mathfrak{G}_6^*$ is not locally Euclidean (collapse a badly embedded arc—$\mathfrak{G}_6^*$ cannot have an $S^1$-action on it). Therefore, the maximal torus action with $S^1$ acting as translation on the first factor of $S^1 \times \mathfrak{G}_6^*$ cannot be conjugated in $\text{TOP}(S^1 \times \mathfrak{G}_6^*)$ to the standard maximal torus action.) □

The rest of this section is devoted to the study of maximal torus action on solvmanifolds. In general, a solv-manifold is a quotient of a simply connected solvable Lie group by a closed subgroup. The main reference for this is [**LR91**].

11.7.19. Let $S$ be a connected, simply connected solvable Lie group, and let $H$ be a closed subgroup of $S$. The coset space $H\backslash S$ is called a *solvmanifold*. In this section, we consider only the case when $H = \Gamma$ is discrete so that $\Gamma$ is a uniform lattice of $G$. More generally, let $\Pi$ be a subgroup of $\operatorname{Aff}(S) = S \rtimes \operatorname{Aut}(S)$ acting freely on $S$ such that $\Gamma = \Pi \cap S$ is a lattice of $S$ and $\Pi/\Gamma$ is finite. We call the orbit space $\Pi\backslash S$ an *infra-solvmanifold*. Therefore, a compact infra-solvmanifold is finitely covered by a solvmanifold.

It is a theorem of Mostow that two compact solvmanifolds of the same fundamental group are diffeomorphic. The significance of this statement is seen from the fact, different from the nilpotent theory, that the group $\Gamma$ does not determine the Lie group $S$. In other words, given a group $\Gamma$, there may exist two distinct connected, simply connected solvable Lie groups $S_1, S_2$ both containing a copy of $\Gamma$ as a lattice. Mostow's theorem says that $S_1$ is diffeomorphic to $S_2$, and $\Gamma\backslash S_1$ is diffeomorphic to $\Gamma\backslash S_2$.

For many closed $K(\Pi, 1)$-manifolds such as those with virtually poly-$\mathbb{Z}$ fundamental groups, it is verified in [**LR84**] that $\mathcal{Z}(\Pi)$ is finitely generated and the manifold admits a maximal torus action. In fact, one may find a topological version of Theorem 11.7.20 (see below) in [**LR84**]. It uses surgery results of Wall and does not explain how the torus action arises explicitly from the solvable group $S$. The following theorem does not rely on surgery theory, and the solution is given explicitly in terms of Lie theory.

THEOREM 11.7.20. *Let $S$ be a simply connected solvable Lie group and $\Pi$ be a lattice of $S$. Then the solvmanifold $\Pi\backslash S$ admits a smooth maximal torus action.*

The plan based upon Lie theory is to construct a new connected and simply connected solvable Lie group $S(\Gamma)$ and an embedding of $\Pi$ into $\operatorname{Aff}(S(\Gamma)) = S(\Gamma) \rtimes \operatorname{Aut}(S(\Gamma))$ with the following properties: (a) the infra-solvmanifold $\Pi\backslash S(\Gamma)$ is diffeomorphic to the solvmanifold $\Pi\backslash S$; and (b) $\Pi\backslash S(\Gamma)$ admits a smooth maximal torus action. The torus action constructed on $\Pi\backslash S(\Gamma)$ descends from a central vector group of $S(\Gamma)$ which commutes with the affine action of $\Pi$ on $S(\Gamma)$. Then we can pull back the torus action on $\Pi\backslash S(\Gamma)$ to $\Pi\backslash S$ obtaining a smooth maximal torus action on the solvmanifold $\Pi\backslash S$. The reader should find the elementary Example 11.7.27 instructive. It illustrates, in an explicit fashion, some of steps to be taken in the proof of the theorem. The proof starts in Subsection 11.7.25.

11.7.21. Since $\Pi$ is a lattice of $S$, it is a *strongly torsion-free $\mathcal{S}$ group*; that is, $\Pi$ contains a finitely generated, torsion-free nilpotent normal subgroup $D$ with the quotient $\Pi/D$ free Abelian of finite rank. Such a group $\Phi$ contains a unique *maximal normal nilpotent subgroup* $M$ which automatically contains $[\Pi, \Pi]$. The group $\Pi$ also contains a characteristic subgroup $\Gamma$ of finite index such that $\Gamma$ is *strongly torsion-free $\mathcal{S}$ group of type I (predivisible group)*, and $\Gamma \supset M$. This means that:
 (1) $\Gamma\backslash M$ is torsion free.
 (2) Let $\mu(\gamma)$ be the automorphism of the real nilpotent Lie group $M_\mathbb{R}$ (see below for notation) containing $M$ as a lattice, induced from the conjugation by an element $\gamma \in \Gamma$. If $\theta$ is an eigenvalue of the derivative of $\mu(\gamma)$, then
$$\theta|\theta|^{-1} = \cos 2\pi\rho + i \sin 2\pi\rho,$$
where $\rho$ is either 0 or irrational.

NOTATION 11.7.22. For a finitely generated, torsion-free nilpotent group $D$, the unique connected and simply connected nilpotent Lie group is denoted by $D_\mathbb{R}$. This is the *Mal'cev completion*.

11.7.23. The short exact sequence of groups $1 \to M \to \Gamma \to \mathbb{Z}^k \to 1$ induces an exact sequence $1 \to M_\mathbb{R} \to \Gamma M_\mathbb{R} \to \mathbb{Z}^k \to 1$. One may think of $\Gamma M_\mathbb{R}$ as the pushout of $M \to \Gamma$ with $M \hookrightarrow M_\mathbb{R}$ since $(M, M_\mathbb{R})$ has the unique automorphism extension property; see Definition 5.3.3. In other words, $\Gamma M_\mathbb{R}$ is the unique group fitting into the following commutative diagram:

$$\begin{array}{ccccccccc} 1 & \to & M & \to & \Gamma & \to & \mathbb{Z}^k & \to & 1 \\ & & \downarrow & & \downarrow & & \| & & \\ 1 & \to & M_\mathbb{R} & \to & \Gamma M_\mathbb{R} & \to & \mathbb{Z}^k & \to & 1. \end{array}$$

Does there exist a connected and simply connected solvable Lie group $S(\Gamma)$ containing $\Gamma M_\mathbb{R}$? Using Wang's construction, Auslander constructed such a group $S(\Gamma)$ which fits into the following commutative diagram:

$$\begin{array}{ccccccccc} 1 & \to & M_\mathbb{R} & \to & \Gamma M_\mathbb{R} & \to & \mathbb{Z}^k & \to & 1 \\ & & \| & & \downarrow & & \cap\downarrow & & \\ 1 & \to & M_\mathbb{R} & \to & S(\Gamma) & \to & \mathbb{R}^k & \to & 1, \end{array}$$

where $\mathbb{Z}^k \subset \mathbb{R}^k$ as a lattice; see [**Wan56**] and [**Aus61b**]. Moreover, $S(\Gamma)$ has the property that there exists $\gamma_1, \gamma_2, \ldots, \gamma_k$ whose images form a set of generators for $\Gamma/M$ which lie on 1-parameter groups in $S(\Gamma)$.

11.7.24 (More properties of $S(\Gamma)$). (1) $\Gamma \subset S(\Gamma)$ as a lattice.
(2) There exists a torus subgroup $T^*$ of $\mathrm{Aut}(S(\Gamma))$ such that $S \subset S(\Gamma) \rtimes T^*$. Moreover, the composite $S \hookrightarrow S(\Gamma) \rtimes T^* \to T^*$ is surjective.
(3) Let $N$ be the *nilradical* of $S$; that is, the maximal normal nilpotent connected Lie subgroup of $S$. Then $\Gamma N$ can be naturally identified with $\Gamma M_\mathbb{R}$. With this identification, we have $[S(\Gamma), S(\Gamma)] \subset \Gamma N$, and hence $\Gamma N$ is normal in $S(\Gamma)$.
(4) Any automorphism $\theta$ of $\Gamma M_\mathbb{R}$ which is trivial on $\Gamma M_\mathbb{R}/M_\mathbb{R}$ can be uniquely extended to an automorphism of $S(\Gamma)$.
(5) $N$ is normal in $S(\Gamma) \rtimes T^*$.

We shall study Seifert fiberings of infra-solvmanifolds. Suppose our model space $P$ itself is a connected, simply connected Lie group; $G$ a connected closed normal subgroup and $W = P/G$. We shall consider the short exact sequence of groups $1 \to G \to P \to W \to 1$ as a principal $G$-bundle. The group $\mathrm{Diff}_G(P)$ of all weakly $G$-equivariant smooth diffeomorphisms of $P$ onto itself is exactly the normalizer of $G = \ell(G)$ in $\mathrm{Diff}(P)$, and is equal to $\mathrm{TOP}_G(P) \cap \mathrm{Diff}(P)$. Let $C(W, G)$ be the group of all smooth maps from $W$ to $G$. Suppose $P \to W$ has a *smooth cross section*. Then we have a short exact sequence

$$1 \to C(W, G) \rtimes \mathrm{Inn}(G) \to \mathrm{Diff}_G(P) \to \mathrm{Out}(G) \times \mathrm{Diff}(W) \to 1.$$

The *affine group* $\mathrm{Aff}(P) = P \rtimes \mathrm{Aut}(P)$ acts on $P$ by: $(p, \gamma) \cdot u = p \cdot \gamma(u)$ for $(p, \gamma) \in \mathrm{Aff}(P)$ and $u \in P$. Note that $P$ acts as left translations. For $g \in G$, we have $(p, \gamma)(g, 1)(p, \gamma)^{-1} = (p\gamma(g)p^{-1}, 1)$. Let us denote the subgroup of $\mathrm{Aut}(P)$ which leaves $G$ invariant by $\mathrm{Aut}(P, G)$. An important fact for us is

(11.7.1) $$P \rtimes \mathrm{Aut}(P, G) \subset \mathrm{Diff}_G(P).$$

This is true because $\mathrm{Diff}_G(P)$ is the normalizer of $\ell(G)$ in $\mathrm{Diff}(P)$, and $\ell(G)$ is normal in $P \rtimes \mathrm{Aut}(P,G)$.

11.7.25 (Proof of Theorem 11.7.20). We go back to our solvable Lie groups. Since $N$ is the nilradical of $S$, $S/N$ is commutative, say of dimension $s$. Therefore, we have an exact sequence of groups
$$1 \to N \to S \to S/N = \mathbb{R}^s \to 1$$

On the other hand, since $[S(\Gamma), S(\Gamma)] \subset \Gamma N$ from Property 11.7.24(3) of $S(\Gamma)$, and $[S(\Gamma), S(\Gamma)]$ is connected, we have $[S(\Gamma), S(\Gamma)] \subset N$. Therefore $N$ is normal in $S(\Gamma)$ and $S(\Gamma)/N$ is a commutative Lie group, $\mathbb{R}^s$. Therefore
$$1 \to N \to S(\Gamma) \to S(\Gamma)/N = \mathbb{R}^s \to 1$$
is exact.

Since $N$ is normal in both $S$ and $S(\Gamma)$, the inclusion maps $\Gamma N \hookrightarrow S$ and $\Gamma N \hookrightarrow S(\Gamma)$ induce $\Gamma/(\Gamma \cap N) \hookrightarrow S/N$ and $\Gamma/(\Gamma \cap N) \hookrightarrow S(\Gamma)/N$. By these homomorphisms we identify $S/N = \mathbb{R}^s$ with $S(\Gamma)/N = \mathbb{R}^s$.

The group $\Pi \subset S$ acts on $S$ as left multiplications. Therefore, from the inclusion (11.7.1), we have $\Pi \subset \mathrm{Diff}_N(S)$. In fact, we have the following commutative diagram:

$$\begin{array}{ccccccccc} 1 & \longrightarrow & \Pi \cap N & \longrightarrow & \Pi & \longrightarrow & \Pi/(\Pi \cap N) & \longrightarrow & 1 \\ & & \downarrow & & \downarrow & & \downarrow & & \\ 1 & \longrightarrow & C(\mathbb{R}^s, N) \rtimes \mathrm{Inn}(N) & \longrightarrow & \mathrm{Diff}_N(S) & \longrightarrow & \mathrm{Out}(N) \times \mathrm{Diff}(\mathbb{R}^s) & \longrightarrow & 1. \end{array}$$

Similarly, $\Pi \subset S(\Gamma) \rtimes T^* \subset S(\Gamma) \rtimes \mathrm{Aut}(S(\Gamma), N)$, because $N$ is normal in $S(\Gamma) \rtimes T^*$. $S(\Gamma) \rtimes T^*$ acts on $S(\Gamma)$ as affine maps which implies that $\Pi \subset \mathrm{Diff}_N(S(\Gamma))$ by $(*)$. We have

$$\begin{array}{ccccccccc} 1 & \longrightarrow & \Pi \cap N & \longrightarrow & \Pi & \longrightarrow & \Pi/(\Pi \cap N) & \longrightarrow & 1 \\ & & \downarrow & & \downarrow & & \downarrow & & \\ 1 & \longrightarrow & C(\mathbb{R}^s, N) \rtimes \mathrm{Inn}(N) & \longrightarrow & \mathrm{Diff}_N(S(\Gamma)) & \longrightarrow & \mathrm{Out}(N) \times \mathrm{Diff}(\mathbb{R}^s) & \longrightarrow & 1. \end{array}$$

Let us denote $\Pi/(\Pi \cap N)$ simply by $Q$. Then $Q$ is a free Abelian group of rank $s$, where $s = \dim(S/N)$. Clearly, $\Gamma/(\Gamma \cap N)$ is a subgroup of $Q$ of finite index, because $M \subset \Gamma$ (so, $\Pi \cap N = \Gamma \cap N$). We shall examine the two actions of $Q$ on $S/N$ and $S(\Gamma)/N$.

The action of $Q$ on $S/N$ is induced by the left translation by $\Pi$ on $S$. Therefore, $Q = \mathbb{Z}^s$ acts on $S/N = \mathbb{R}^s$ also as left translations. Moreover, $Q$ is a lattice in $S/N$.

Now the action of $Q$ on $S(\Gamma)/N$ is induced by the affine action of $\Pi$ on $S(\Gamma)$. The projection $S(\Gamma) \to S(\Gamma)/N$ yields a homomorphism $S(\Gamma) \rtimes \mathrm{Aut}(S(\Gamma), N) \to (S(\Gamma)/N) \rtimes \mathrm{Aut}(S(\Gamma)/N)$ naturally. We recall how $S \subset S(\Gamma) \cdot T^*$ of Property 11.7.24(2) was constructed in [**Aus61b**]. $S$ acts on $\Gamma N$ by conjugation, which extends to an automorphism of $\Gamma M_\mathbb{R}$. The latter is trivial on $\Gamma M_\mathbb{R}/M_\mathbb{R}$, and hence it can be extended to an automorphism of $S(\Gamma)$ by Property 11.7.24(4). Since the $S$-action on $\Gamma N/N$ is trivial, and $\Gamma N/N = \mathbb{Z}^s$ sits in $\mathbb{R}^s = S(\Gamma)/N$ as a uniform lattice, the action of $S$ on $S(\Gamma)/N$ must be trivial as well. This implies that $S \subset S(\Gamma) \cdot T^* \subset S(\Gamma) \rtimes \mathrm{Aut}(S(\Gamma), N) \to (S(\Gamma)/N) \rtimes \mathrm{Aut}(S(\Gamma)/N)$ has image in $S(\Gamma)/N \times \{1\}$. Therefore, $Q = \mathbb{Z}^s$ acts on $S(\Gamma)/N = \mathbb{R}^s$ as left translations.

Moreover, $Q$ is a lattice $S(\Gamma)/N$. We conclude that both actions of $Q = \Pi/\Pi \cap N$ on $S/N$ and $S(\Gamma)/N$ are as left translations.

Furthermore, $z \in \Pi \cap N$ goes into $C(S/N, N) \rtimes \mathrm{Inn}(N)$ and $C(S(\Gamma)/N, N) \rtimes \mathrm{Inn}(N)$ as $(z^{-1}, \mu(z))$, as left translations, where $\mu(z)$ is the conjugation by $z$ so that $\mu(z)(a) = zaz^{-1}$. Actually, $\Pi \cap N \subset N$ sits in $C(\mathbb{R}^s, N) \rtimes \mathrm{Inn}(N)$ as constant maps.

Choose an $N$-equivariant diffeomorphism $\tau : S \to S(\Gamma)$. This can be done as follows. Take smooth sections (not homomorphisms) $s_1 : \mathbb{R}^s \to S$ and $s_2 : \mathbb{R}^s \to S(\Gamma)$. With these sections, we define an $N$-bundle equivalence $\tau : S \to S(\Gamma)$ by $\tau(x \cdot s_1(w)) = x \cdot s_2(w)$ for all $x \in N$ and $w \in \mathbb{R}^s$. Let us denote the representations of $\Pi$ into $\mathrm{Diff}_N(S)$ and $\mathrm{Diff}_N(S(\Gamma))$ by $\psi_1, \psi_2$, respectively. More precisely, $\psi_1 : \Pi \to S \subset \mathrm{Diff}_N(S)$, and $\psi_2 : \Pi \to S \subset S(\Gamma) \rtimes T^* \subset S(\Gamma) \rtimes \mathrm{Aut}(S(\Gamma), N) \subset \mathrm{Diff}_N(S(\Gamma))$. Since $\tau$ is $N$-equivariant, $\mu(\tau) \circ \psi_1$ is a representation of $\Pi$ into $\mathrm{Diff}_N(S(\Gamma))$. This bundle map $\tau : S \to S(\Gamma)$ induces an isomorphism $f \mapsto \tau \cdot f \cdot \tau^{-1}$ of $\mathrm{Diff}_N(S)$ onto $\mathrm{Diff}_N(S(\Gamma))$.

Consider the two representations $\mu(\tau) \circ \psi_1, \psi_2 : \Pi \to \mathrm{Diff}_N(S(\Gamma))$. Since they induce the same maps of the kernel $\Pi \cap N$ into $C(\mathbb{R}^s, N) \rtimes \mathrm{Inn}(N)$, and of the quotient $\Pi/(\Pi \cap N)$ into $\mathrm{Out}(N) \times \mathrm{Diff}(\mathbb{R}^s)$, we can now apply the uniqueness of the Seifert Construction. We have a commutative diagram

$$\begin{array}{ccccccccc}
1 & \longrightarrow & \Pi \cap N & \longrightarrow & \Pi & \longrightarrow & \Pi/(\Pi \cap N) & \longrightarrow & 1 \\
& & \downarrow & & \downarrow \psi_2 \,\, \downarrow \mu(\tau)\circ\psi_1 & & \downarrow & & \\
1 & \longrightarrow & C(\mathbb{R}^s, N) \rtimes \mathrm{Inn}(N) & \longrightarrow & \mathrm{Diff}_N(S(\Gamma)) & \longrightarrow & \mathrm{Out}(N) \times \mathrm{Diff}(\mathbb{R}^s) & \longrightarrow & 1.
\end{array}$$

By Theorem 7.3.2, there exists an element $\lambda \in C(\mathbb{R}^s, N)$ which conjugates $\psi_2$ to $\mu(\tau) \circ \psi_1$. Thus

$$\begin{array}{ccc}
\Pi & \xrightarrow{\psi_1} & \mathrm{Diff}_N(S) \\
\psi_2 \downarrow & & \downarrow \mu(\tau) \\
\mathrm{Diff}_N(S(\Gamma)) & \xrightarrow{\mu(\lambda)} & \mathrm{Diff}_N(S(\Gamma))
\end{array}$$

is commutative. The map $\mu(\tau^{-1} \circ \lambda)$ sends $\psi_2(\Pi)$ to $\psi_1(\Pi)$ yielding a diffeomorphism from $\Pi \backslash S(\Gamma)$ onto $\Pi \backslash S$. In this argument, the fact that $N$ is a connected, simply connected nilpotent Lie group is essential.

Now we show the space $\Pi \backslash S(\Gamma)$ admits a smooth maximal torus action. Let $\mathcal{Z}(\Pi) = \mathbb{Z}^k$ be the center of $\Pi$. Since $M$ is the maximal normal nilpotent subgroup of $\Pi$, $\mathbb{Z}^k \subset M$. Let $\mathbb{R}^k$ be the smallest connected subgroup of $M_\mathbb{R}$ containing $\mathbb{Z}^k$. Since $\Pi$ commutes with $\mathbb{Z}^k$ and $\Pi \subset \mathrm{Aff}(S(\Gamma))$, $\Pi$ commutes with $(\mathbb{Z}^k)_\mathbb{R} = \mathbb{R}^k$. This means that $\mathbb{R}^k$ lies in the centralizer of $\Pi$ in $\mathrm{Diff}_N(S(\Gamma))$. Of course, $\mathbb{R}^k \cap \Pi = \mathcal{Z}(\Pi)$. Thus we obtain an action of torus $\mathbb{R}^k/\mathbb{Z}^k$ on the model space $\Pi \backslash S(\Gamma)$. This action is smooth, (actually, it is a group of isometries if we give a left invariant metric on $S(\Gamma)$), and is a maximal torus action on $\Pi \backslash S(\Gamma)$. Now one can pull back this action to a smooth action on $\Pi \backslash S$. This completes the proof of Theorem 11.7.20. □

COROLLARY 11.7.26 (Mostow). *Let $S_1, S_2$ be two connected, simply connected solvable Lie groups. Let $\Gamma_i$ be a lattice in $S_i$, $i = 1, 2$. Suppose $\Gamma_1$ is isomorphic to $\Gamma_2$. Then $S_1/\Gamma_1$ is diffeomorphic to $S_2/\Gamma_2$.*

For Mostow's argument, see [**Rag72**, Theorem 3.6]. We give a different proof. Since $\Gamma_1 \cong \Gamma_2 (= \Pi)$, construct a connected, simply connected solvable Lie group $S(\Gamma)$ on which these groups act. By Theorem 11.7.20, $S_i/\Gamma_i$ is diffeomorphic to $S(\Gamma)/\Pi$, $i = 1, 2$. Therefore, $S_1/\Gamma_1$ is diffeomorphic to $S_2/\Gamma_2$.

The following example illustrates the construction employed in the proof of the theorem. Moreover, the example serves to illustrate why one is compelled to look for a larger group than $S$ if one wishes to construct a maximal torus action from the descent of a vector subgroup.

EXAMPLE 11.7.27. Let $S = \widetilde{E_0(2)} = \mathbb{R}^2 \rtimes \mathbb{R}$ be the universal covering group of the 2-dimensional Euclidean group, where $(0, t)$ acts on $\mathbb{R}^2$ by $x \mapsto e^{2\pi i t} x$, $x$ seen as a complex number. Let $\Pi$ be the lattice generated by

$$t_1 = \left(\begin{bmatrix}1\\0\end{bmatrix}, 0\right), \quad t_2 = \left(\begin{bmatrix}0\\1\end{bmatrix}, 0\right), \quad \alpha = \left(\begin{bmatrix}0\\0\end{bmatrix}, \frac{1}{2}\right).$$

The subgroup $\Gamma$ generated by $t_1, t_2$ and $\alpha^2$ is a characteristic subgroup of $\Pi$, isomorphic to $\mathbb{Z}^3$. Then $S(\Gamma) = \mathbb{R}^3$ and we get an embedding of $S$ into $S(\Gamma) \rtimes S^1 = \mathbb{R}^3 \rtimes \mathrm{SO}(2) \subset \mathbb{R}^3 \rtimes \mathrm{O}(3) = E(3)$. The homomorphism is obvious:

$$\left(\begin{bmatrix}x_1\\x_2\end{bmatrix}, t\right) \mapsto \left(\begin{bmatrix}x_1\\x_2\\t\end{bmatrix}, \begin{bmatrix}\cos 2\pi t & \sin 2\pi t & 0\\ -\sin 2\pi t & \cos 2\pi t & 0\\ 0 & 0 & 1\end{bmatrix}\right).$$

The image of $\Pi$ in $E(3)$ is the orientable Bieberbach group of dimension 3 with holonomy group $\mathbb{Z}_2$. Clearly, the manifold $\Pi \backslash \mathbb{R}^2 \rtimes \mathbb{R}$ is diffeomorphic to the flat manifold $\Pi \backslash \mathbb{R}^3$, i.e., $\mathfrak{G}_2$. On $\Pi \backslash \mathbb{R}^3$, there is a maximal torus action by $S^1$, generated by the left translation by $\mathbb{R} = \{[0\ 0\ s]^t : s \in \mathbb{R}\}$. Note that this subgroup $\mathbb{R}$ of $S(\Gamma)$ is not in the image of $S$. This means that there is no $S^1$-action on $\Pi \backslash S$ coming from the left translation. In fact, it comes from the right translation by the $\mathbb{R}$-factor of $S = \mathbb{R}^2 \rtimes \mathbb{R}$.

If we consider just the subgroup $\Gamma$, it is even clearer what the theorem says. The solvmanifold $\Gamma \backslash \mathbb{R}^2 \rtimes \mathbb{R}$ is diffeomorphic to the torus $\Gamma \backslash \mathbb{R}^3$. On the latter torus, there is a standard $T^3$-action as translations. However, no vector subgroup in $S$ descends to give a maximal torus action on $\Gamma \backslash \mathbb{R}^2 \rtimes \mathbb{R}$.

We now turn to general Lie groups. Little is known for the existence of a maximal torus action on general double coset spaces. Under some strong conditions, we can show an aspherical double coset space of a Lie group admits a maximal torus action.

THEOREM 11.7.28. *Let $G$ be a connected, simply connected Lie group, and let $R$ be its radical. Suppose $S = G/R$ does not contain any normal compact factor. Let $K$ be a maximal compact subgroup of $G$ and $\Gamma$ a torsion-free cocompact lattice in $G$ such that $(\Gamma \cap R, R)$ has the unique automorphism extension property. If $\exp : \mathcal{R} \to R$ is surjective, then the double coset space $\Gamma \backslash G / K$ admits a smooth maximal torus action.*

PROOF. Let $G = R \rtimes S$ be the Levi decomposition of $G$. Let $A = \{a \in R | (a, u) \in \mathcal{Z}(\Gamma) \text{ for some } u \in S\}$. Let $(a, u) \in \mathcal{Z}(\Gamma)$. Then for any $(z, 1) \in \Gamma_R = \Gamma \cap R$, $(z, 1)(a, u) = (a, u)(z, 1)$. This implies that $^u z = a^{-1} z a$. Since $(\Gamma \cap R, R)$ has UAEP, the two automorphisms $u$ and $\mu(a^{-1})$ induce the same automorphisms on $R$. Therefore, $^u x = a^{-1}xa$ for all $x \in R$. Moreover, for any $(b, v) \in \Gamma$, we have $^v a = a$. Now it is easy to see that $A$ is a commutative subgroup of $R$.

Choose generators $(a_i, u_i), i = 1, 2, \ldots, k$ for $\mathcal{Z}(\Gamma)$. We define a homomorphism $\phi_R : \mathbb{R}^k \to R$ as follows: Since $\exp : \mathcal{R} \to R$ is onto, $\log$ is defined on $R$. Let $A_i = \log a_i$. Then $\phi_R$ is the composite $\mathbb{R}^k \to \mathcal{R} \xrightarrow{\exp} R$, where the first map is the linear transformation from $\mathbb{R}^k$ to $\mathcal{R}$ sending the standard basis to $A_1, A_2, \ldots, A_k$. Since $[A_i, A_j] = 0$, the image of $\mathbb{R}^k$ in $\mathcal{R}$ is a commutative Lie subalgebra, and hence the exponential map restricted to this subalgebra is a homomorphism. Consequently, $\phi_R$ is a homomorphism.

Next, we define $\phi_S : \mathbb{R}^k \to S$ as follows: Let $S = S_1 \times S_2 \times \cdots \times S_r$, where each $S_i$ is a simple group. For each $i$, let $S_i^*$ denote the adjoint form of $S_i$, and choose a maximal compact subgroup of $S_i^*$. This maximal compact subgroup is of the form either $S^1 \times H$ or $H$, where $H$ does not have a circle factor, depending on whether $S_i$ has infinite center or not. This determines a subgroup $\mathbb{R}^{\epsilon_i} \times \widetilde{H}_i \subset S_i$, where $\widetilde{H}_i$ is compact, and $\epsilon_i = 1$ or $0$, depending on whether $S_i$ has infinite center or not. In the former case, $\mathbb{R}$ contains the infinite summand of the center of $S_i$. Then $K = \Pi \widetilde{H}_i$ is a maximal compact subgroup of $S$.

Consider the map $\mathcal{Z}(\Gamma) \to \Pi(\mathbb{R}^{\epsilon_i} \times \widetilde{H}_i) \to \Pi \mathbb{R}^{\epsilon_i} \subset \Pi S_i$, where $\Pi(\mathbb{R}^{\epsilon_i} \times \widetilde{H}_i) \to \Pi \mathbb{R}^{\epsilon_i}$ is a projection. We extend this to a homomorphism $\phi_S : \mathbb{R}^k \to \Pi \mathbb{R}^{\epsilon_i} \subset S$. Note that $\phi_S(\mathcal{Z}(\Gamma))$ differs from $p(\mathcal{Z}(\Gamma))$ by elements in $\Pi \widetilde{H}_i \subset K$.

Note that $K$ commutes with $\mathbb{R}^{\epsilon_1} \times \mathbb{R}^{\epsilon_2} \times \cdots \times \mathbb{R}^{\epsilon_r}$. Thus we have an induced action of $\mathbb{R}^k$ on $G/K$. The action of $\mathbb{R}^k$ on $G/K$ will not be effective in general, because $\mathcal{Z}(\Gamma) \to \Pi \mathbb{R}^{\epsilon_i} \subset S$ may have a nontrivial kernel. Even though the actions by $\mathbb{Z}^k \subset \mathbb{R}^k$ and by $\mathcal{Z}(\Gamma)$ are different on $S$, they induce the same one over $S/K$.

A desired action of $\mathbb{R}^k$ on $G/K = R \cdot S/K$ is then given by

$$\phi(t)(x, w) = (x \cdot \phi_R(t), w \cdot \phi_S(t)).$$

Since $\Gamma$ acts on $G$ as left multiplications, it commutes with the $\mathbb{R}^k$-action defined above. Moreover, we have $\mathbb{R}^k \cap \Gamma = \mathcal{Z}(\Gamma)$ on $G/K$. Consequently, we have obtained a smooth action of $T^k = \mathcal{Z}(\Gamma)\backslash \mathbb{R}^k$ on $\Gamma \backslash G/K$. □

THEOREM 11.7.29 ([**LR91**]). *Let $G$ be a connected, simply connected Lie group without any normal compact factors in its semisimple part. Let $\Gamma$ be a torsion-free cocompact lattice and $K$ a maximal compact subgroup of $G$. Then there is a smooth manifold $M$, which is homotopy equivalent to the double coset space $\Gamma \backslash G/K$, admitting a smooth maximal torus action.*

PROOF. We may assume that $\Gamma = \pi_1(\Gamma \backslash G/K)$. Let $R$ be the radical of $G$. Then $G = R \rtimes S$. Let $p : G \to S$ be the projection, and let $\mathcal{Z}(\Gamma)$ denote the center of $\Gamma$. Let $\widetilde{\Gamma} = \Gamma_R \cdot \mathcal{Z}(\Gamma)$, where $\Gamma_R = \Gamma \cap R$. It is poly-{cyclic or finite} since $1 \to \Gamma_R \to \Gamma_R \cdot \mathcal{Z}(\Gamma) \to p(\mathcal{Z}(\Gamma)) \to 1$ is exact, $\Gamma_R$ is a lattice of $R$, and $p(\mathcal{Z}(\Gamma))$ is a finitely generated Abelian group. Such a group $\widetilde{\Gamma}$ contains a characteristic subgroup $\Gamma'$ of finite index which is a Mostow-Wang group (see Definition 9.5.1), with $^n\Gamma' = {}^n\widetilde{\Gamma}$, where $^n$ denotes the discrete nilradical. Now $\widetilde{\Gamma}$ contains a characteristic subgroup $\widehat{\Gamma}$ of finite index which is predivisible and $^n\widehat{\Gamma} = {}^n\Gamma'$. This implies that $\widehat{\Gamma}/{}^n\widehat{\Gamma}$ is free Abelian, say $\mathbb{Z}^m$.

Let $Q = \Gamma/\widehat{\Gamma}$ and $S^* = S/p(\mathcal{Z}(\Gamma))$. Note that $S^*$ is not necessarily the adjoint form of $S$. Let $K^*$ be a maximal compact subgroup of $S^*$. Note that $K^*$ is a finite quotient of $T \times K$, where $T$ is a torus generated by free Abelian factors of $p(\mathcal{Z}(\Gamma))$.

Now $\Gamma/\widetilde{\Gamma} = \Gamma/\Gamma_R \cdot \mathcal{Z}(\Gamma)$ acts on $S^*/K^*$ with compact quotient. Therefore $Q = \Gamma/\widehat{\Gamma}$ acts on $S^*/K^*$ with compact quotient via the homomorphism $\Gamma/\widehat{\Gamma} \to \Gamma/\widetilde{\Gamma}$. Let us denote ${}^n\widehat{\mathbb{T}}$ by $\Delta$. Since ${}^n\widehat{\mathbb{T}}$ is characteristic in $\widehat{\Gamma}$, it is normal in $\Gamma$. Consider the exact sequences $1 \to \Delta \to \Gamma \to \Gamma/\Delta \to 1$ and $1 \to \mathbb{Z}^m \to \Gamma/\Delta \to Q \to 1$. We get the latter exact sequence from the fact that $\widehat{\Gamma}$ is predivisible. We do the Seifert space construction with the latter exact sequence and the action of $Q$ on the space $S^*/K^*$ to obtain an action of $\Gamma/\Delta$ on $\mathbb{R}^m \times S^*/K^*$. Now we do a Seifert space construction with the first exact sequence and the action of $\Gamma/\Delta$ on $\mathbb{R}^m \times S^*/K^*$. Consequently we obtain an action of $\Gamma$ on $\Delta_{\mathbb{R}} \times \mathbb{R}^m \times S^*/K^*$. Let $\mathbb{Z}^k$ be the center of $\Gamma$. It lies in the center of $\Delta$. Since the center of $\Delta$ lies in the center of $\Delta_{\mathbb{R}}$, there is a unique subgroup $\mathbb{R}^k$ in the center of $\Delta_{\mathbb{R}}$ containing $\mathbb{Z}^k$ as a uniform lattice. The action of $\mathbb{R}^k$ on $\Delta_{\mathbb{R}} \times \mathbb{R}^m \times S^*/K^*$, by left multiplication on the first factor, commutes with the action of $\Gamma$. Therefore, it induces an action of torus $\mathbb{R}^k/\mathbb{Z}^k$ on $M = \Gamma \backslash (\Delta_{\mathbb{R}} \times \mathbb{R}^m \times S^*/K^*)$ (which has the same homotopy type as $\Gamma \backslash G/K$). Clearly, this is a smooth maximal torus action. □

REMARK 11.7.30. In [**FJ98**], Farrell and Jones, using surgery theory, show that any closed manifold $M^n$ homotopically equivalent to $\Gamma \backslash G/K$ is homeomorphic to it provided $G$ has a faithful representation into $\mathrm{GL}(m, \mathbb{R})$ for some $m$, $n \neq 3, 4$. Thus when $G$, in the theorem above, has a faithful linear representation, then $\Gamma \backslash G/K$ has a maximal torus action. Of course there are simply connected $G$ without faithful representations.

## 11.8. Toral rank of spherical space forms

11.8.1. The *toral rank* of a space $X$ is the dimension of the largest torus that acts effectively on $X$. It is also called the *toral degree of symmetry* of $X$ (by [**Hsi75**]). We have seen that the only connected Lie groups that act on admissible manifolds are tori (Theorem 3.2.2) and they must act injectively. Therefore, an upper bound on the toral rank of an admissible manifold is the rank of the center of its fundamental group. When the admissible manifold admits a maximal torus action, then this upper bound is the toral rank. In this section, we will study the toral rank of manifolds covered by the sphere, and in particular, determine the toral rank of spherical space forms.

A finite group that acts freely on $S^{n-1}$ has periodic cohomology with the minimum period dividing $n$. This can be seen, in the linear or simplicial case, by splicing copies of the chain complex for the sphere together to get a free $\mathbb{Z}G$-resolution of $\mathbb{Z}$:

$$0 \leftarrow \mathbb{Z} \leftarrow C_0(S^{n-1}) \leftarrow C_1(S^{n-1}) \leftarrow \cdots \leftarrow C_{i-1}(S^{n-1}) \leftarrow C_i(S^{n-1}) \leftarrow \cdots .$$

A finite group satisfies the *pq-condition* ($p$, $q$ are primes) if every subgroup of order $pq$ is cyclic. A finite group $F$ has periodic cohomology if and only if $F$ satisfies all the $p^2$-*conditions* [**CE56**, XII-§11]. A group $F$ that acts freely and orthogonally on a sphere $S^{2n-1}$ satisfies all $pq$-conditions ($p$ may be equal to $q$). In this case, $F \backslash S^{2n-1}$ is a spherical space form (see subsections and examples, 4.5.16–4.5.18) and $F$ has periodic cohomology a divisor of $2n$. Conversely, if $F$ is solvable and satisfies all $pq$-conditions, free linear actions do exist [**Wol77**, §6.1.11]. Though in general, all the $pq$-conditions are not sufficient for $F$ to act freely and linearly on some sphere. However, Milnor has shown that if $F$ acts freely and topologically on a sphere, it satisfies all the $2q$-*conditions* [**Mil57**].

The following, proved by surgery theory, characterizes those groups that can act freely and topologically on some sphere.

THEOREM 11.8.2 ([**MTW76**]). *A finite group $F$ can act freely on a sphere if and only if it satisfies all 2p- and $p^2$-conditions.*

For example, all the groups $SL(2,p)$, $p$ an odd prime, can act freely on a sphere. But it is only $SL(2,3)$ and $SL(2,5)$ that can act freely and orthogonally on a sphere. Note $SL(2,2) \cong \mathbb{Z}_3 \rtimes \mathbb{Z}_2$ does not satisfy the $2p$-condition but it does satisfy all the $p^2$-conditions.

In the theorem, the actions produced are not necessarily on spheres $S^{n-1}$, where $n$ is the minimal period of the cohomology of $F$. However, the discrepancy is at most a factor of 2. In the examples constructed by T. Petrie [**Pet71**], with $F = \mathbb{Z}_m \rtimes \mathbb{Z}_q$, $m,q$ odd and $\mathbb{Z}_q \to \text{Aut}(\mathbb{Z}_m)$ faithful, the minimal period is $2q$ and the free topological actions are on $S^{2q-1}$. Not all are linear; for example, $\mathbb{Z}_7 \rtimes \mathbb{Z}_3$ acts freely on the 5-sphere, does not satisfy the $pq$-condition, and does not act linearly.

The groups that act freely and orthogonally on a sphere are divided into six types I, II, ..., VI. If a group $F$ is of type I, II, III, or V and its minimal period is $2n$, then $F$ admits a free orthogonal actions on $S^{2n-1}$. But there are some in type IV and VI where the minimal sphere is $S^{4n-1}$ and the period is $2n$.

11.8.3. If a finite group $F$ acts freely on a cohomology $(n-1)$-manifold over $\mathbb{Z}$, having the $\mathbb{Z}$-cohomology of $S^{n-1}$, then the cohomology of $F$ is periodic with period dividing $n$. For an easy proof in this context, consider the Borel space $\Sigma_F = EF \times_F \Sigma$. This fibers over $F \backslash \Sigma$ with contractible fiber $EF$ and also fibers over $BG$ with fiber $\Sigma$. Now use the spectral sequence $''E$ of Chapter 10 associated with the second fibering. For facts concerning cohomology manifolds; see Section 1.8.

LEMMA 11.8.4 ([**CR69**, 4.15]). *Let $F$ act freely on an $(2n-1)$-dimensional $\mathbb{Z}$-cohomology manifold $\Sigma^{2n-1}$ having the $\mathbb{Z}$-cohomology of $S^{2n-1}$. Suppose the minimal period of the cohomology of $F$ is $2n$. If $S^1$ acts effectively on $M = F\backslash \Sigma^{2n-1}$, then $M^{S^1} = \emptyset$.*

PROOF. Suppose $x \in M^{S^1}$. Choose $\widehat{x} \in \Sigma^{2n-1}$ as a base point in $\Sigma^{2n-1}$ over $x \in M$. The $S^1$-action can be lifted to $\Sigma^{2n-1}$ and $\widehat{x}$ is fixed. Now $(\Sigma^{2n-1})^{S^1} = \Sigma'$ is a closed sphere-like cohomology manifold of odd dimension $< 2n-2$; see Subsection 1.8.1. Furthermore, if $C$ is the connected component of $M^{S^1}$ containing $x$, $\nu^{-1}(C) = \Sigma'$, where $\nu : \Sigma^{2n-1} \to M$ is the orbit mapping of the $F$-action, see Corollary 2.3.6. Since $F$ is invariant on $\Sigma'$, and dimension of $\Sigma'$ is less than $2n-2$, the period of $F$ must be less than $2n$. This is a contradiction.

Let us also observe that $M^{S^1}$ is connected (when nonempty) and therefore, it is the same as $C$. For, $M$ is rationally an odd dimensional cohomology sphere and the fixed set $M^{S^1}$ is an odd dimensional rational cohomology manifold with the rational cohomology of an odd dimensional sphere. □

PROPOSITION 11.8.5. *Let $F$ act freely and topologically on a sphere $S^{2n-1}$. Let $2d$ be the minimal period of the cohomology of $F$ and suppose there is a free topological action of $F$ on $S^{2d-1}$. Then $n = kd$ and the toral rank of $M = F\backslash S^{2n-1}$ is $\leq k$.*

PROOF. We show first that if $T^k$ acts effectively on $M$, it cannot have any fixed points. We shall use the following fact: If $K$ is normal in $G$, and $G$ acts on $X$, then $\text{Fix}(G, X) = \text{Fix}(G/K, \text{Fix}(K, X))$. Now suppose $\text{Fix}(T^k, M) \neq \emptyset$. Then $\text{Fix}(T^{k-1}, \text{Fix}(T^1, M)) \neq \emptyset$, where $T^{k-1} = T^k/T^1$.

If we lift the $T^1$-action to $S^{2kd-1}$, we get $\text{Fix}(T^1, S^{2kd-1}) = \Sigma^s$, an odd dimensional compact $\mathbb{Z}$-cohomology $s$-manifold having the integral cohomology of the $s$-sphere with $s \leq 2n-3$. Since $s$ is odd, $\Sigma^s$ is connected. As $T^1$ commutes with the covering $F$-action, the period $2d$ divides $s+1$. Thus $s = 2dr_1 - 1$, where $r_1 \leq k-1$. Thus we have $\text{Fix}(T^k, M) = \text{Fix}(T^{k-1}, F\backslash\Sigma^{2dr_1-1})$. Continuing inductively, we have $\text{Fix}(T^k, M) = \text{Fix}(T^{k-1}, F\backslash\Sigma^{2dr_1-1}) = \cdots = \text{Fix}(T^{k-s}, F\backslash\Sigma^{2dr_s-1}) \neq \emptyset$ with $r_s < r_{s-1} < \cdots < r_1 \leq k-1$. Thus $r_s \leq k-s$. To avoid a contradiction, we have $r_{k-1} = 1$. Then we have $T^1$-lifting to $\Sigma^{2d-1}$ and acting with fixed points. The fixed set is again a sphere-like cohomology manifold of dimension less than $2d-1$. But again, $F$ is invariant on this fixed set which yields a contradiction to Lemma 11.8.4. So the action of $T^k$ on $M$ is without fixed points.

Now suppose there is an effective $(T^{k+1}, M)$-action. There are a finite number of distinct isotropy groups each having rank at most $k-1$. Then there is a circle subgroup $S^1$ in $T^{k+1}$ which is not completely contained in any of the isotropy groups. Therefore, this $S^1$ acts with only finite isotropy subgroups. Now $H^*(M; \mathbb{Q}) \cong H^*(S^{2kd-1}; \mathbb{Q})$ and the cohomology of the orbit space $S^1\backslash M$ is rationally like the cohomology of $\mathbb{C}P_{kd-1}$. $T^k$ acts on $S^1\backslash M$ and $\text{Fix}(T^k, S^1\backslash M) \neq \emptyset$ because $\chi(S^1\backslash M) \neq \emptyset$. Then, each $S^1$-orbit over a fixed $w \in \text{Fix}(T^k, S^1\backslash M)$ is invariant under the $T^{k+1}$-action. Thus, there is a subgroup of rank $k$ which fixes the $S^1$-orbit. This contradicts Lemma 11.8.4. $\square$

11.8.6. If $G$ is a connected Lie group acting effectively on $X$ and $\pi_1(X)$ is finite, then the finite covering group $G'$ of $G$ corresponding to the kernel of $\text{ev}_*^x : \pi_1(G, 1) \to \pi_1(X, x)$ lifts to an effective action on the universal covering of $X$.

In Subsection 4.5.18, we saw that for a spherical space form, the group of diagonal unitary matrices $D$ in $U(n)$ with constant entries descends to an effective *unitary* action of $D/(D \cap F)$ on $M$.

PROPOSITION 11.8.7 ([**Kah70**]). *If $F$ acts freely and unitarily on $S^{2n-1}$, where $2n$ is the minimal period of the cohomology of $F$, then the descent of $D$ to $M = F\backslash S^{2n-1}$ is the only effective "unitary" circle action on $M$.*

PROOF. This is a corollary of Proposition 11.8.5, where $k = 1$. Suppose there is an effective circle action $C$ on $M$. Then a covering $C'$ of $C$ lifts to an effective action of $C'$ on $S^{2n-1}$ which commutes with $F$. Suppose, in addition, that $C'$ commutes with $D$. This will be the case if $C'$ is a unitary action on $S^{2n-1}$. If $C' \neq D$, then $C'$ and $D$ generate a 2-dimensional torus $T$ which commutes with $F$. This torus then descends to an effective torus, modulo a finite subgroup, acting on $M$. This contradicts Proposition 11.8.5. $\square$

11.8.8. Let $F$ act freely and unitarily on some sphere $S^{2n-1}$. The representation then can be conjugated, modulo an automorphism of $F$, into $U(n)$. This representation splits into $k$ irreducible complex representations $V_1 \oplus \cdots \oplus V_k$ of constant degree $d$, independent of $n$. It follows that each $V_i$ is isomorphic to $\mathbb{C}^d$, and the unit sphere in $V_i$ is homeomorphic to $S_i^{2d-1}$. Thus, $F$ acts freely on the join $S^{2d-1} \circ S^{2d-1} \circ \cdots \circ S^{2d-1}$ by direct sum representations. Therefore $n = kd$ for some $k$. Consequently, by the comment in Subsection 4.5.18 and [**Wol77**, §7.4],

$2d-1$ is the minimal dimension of a sphere for which there is a free unitary and, hence, free orthogonal action of $F$,

On each summand $\mathbb{C}_i^d$, there is the circle action $D$ which commutes with the action of $F$. We can extend this action to act trivially on the other factors. In this way, we get a $k$-torus acting on $\sum_{i=1}^{k} \mathbb{C}_i^d$ and by restriction on $S^{2n-1}$ which commutes with the action of $F$ on $S^{2n-1}$. This $T^k$-action descends to give us an effective unitary $T^k$-action on $F\backslash S^{2n-1}$. Therefore, we have

THEOREM 11.8.9 (cf. [**Kah70**]). *If $M = F\backslash S^{2n-1}$ is a spherical space form and $d$ is the (constant) minimal degree of any free irreducible unitary representation of $F$, then $M$ has an effective unitary action of $T^k$, where $kd = n$. Moreover, if the minimal period of the cohomology of $F$ is $2d$, then the toral rank of $M$ is exactly $k$.*

REMARK 11.8.10. (1) Proposition 11.8.5 has much wider applicability than just spherical space forms $M$. C.T.C. Wall and others have shown that if $M$ is a spherical space form of dimension greater than 4, then there is an infinite number of topologically distinct $M'$ $h$-cobordant to $M$ all covered by the sphere. Thus for each $M'$, the toral rank of $M'$ is less than or equal to $k$, where $M$, a spherical space form, satisfies the hypothesis of Proposition 11.8.5.

(2) In [**MTW83**], it is shown that if a finite group $F$ satisfies all $p^2$ and all $2p$-conditions, then there exists a free and smooth action on the sphere $S^{2kd-1}$ with the standard differential structure. Here $2d$ is the period of $F$ and $k$ can be taken to be 1 or 2. Even when one must choose 2, this result, except for one class of groups, is, geometrically the best possible result. Proposition 11.8.5 gives us an upper bound for the toral ranks of these manifolds. It would be interesting to have better bounds on the torus ranks of those manifolds that do not satisfy the conditions of Theorem 11.8.9.

Note that Proposition 11.8.5 can also be formulated in terms of sphere-like integral cohomology manifolds. The proof is essentially as given in Proposition 11.8.5.

(3) Note for a $(2n-1)$-dimensional lens space $F\backslash S^{2n-1}$, the period of $F$ is 2 and consequently the toral rank is exactly $n$. Using the join construction as in Subsection 11.8.8, it is easy to construct an infinite number of distinct linear $T^n$-actions on each lens space. The infinite number of distinct effective $T^2$-actions on the 3-sphere are all weakly equivalent. The $T^2$-actions on 3-dimensional lens spaces are classified in [**OR70a**]. All the other 3-dimensional space forms have period 4 and admit a unique $S^1$-action up to equivalence. This action is the descent of the unique unitary action of Proposition 11.8.7 (cf. Chapters 14 and 15).

(4) Kahn in Proposition 11.8.7 assumed that $2n$ was the minimal degree for which $F$ has a free unitary representation instead of our slightly different assumption of $2n$ being the minimal period of the cohomology of $F$. His proof of the proposition is explicit and computational and differs from what we gave. As a by-product, he showed that $D \cap \pi_1(M) = D \cap F = \text{Center}(\pi_1(M))$ and the action of $D/D \cap F$ is locally injective and effective on $M$. Consequently, we may conclude that the *linear $D/D \cap F$-action* lifts to a principal $S^1$-bundle over $\mathbb{C}P_{n-1}$ whose first Chern class or Euler class is $\pm |D \cap F|$ with $S^1 = D/D \cap F$.

CHAPTER 12

# Seifert Fiberings with Compact Connected $Q$

## 12.1. Introduction

In Chapter 10, we studied Seifert fiberings modeled on principal $T^k$-bundle over a paracompact space $W$. On $W$, there was a proper action of a discrete group $Q$, and we characterized the liftings of $Q$ to the group of weak bundle automorphisms of $P$, $\text{TOP}_{T^k}(P)$. The image of $\text{TOP}_{T^k}(P)$ in $\text{Aut}(T^k) \times \text{TOP}(W)$ contains $(\varphi \times \rho)(Q)$ if and only if $[P] \in H^2(W; \mathbb{Z}^k)^Q$. This is, of course, a necessary condition for lifting. Then $Q$ lifts to $P$, as a group of weak bundle automorphisms, if and only if $P$ is the pullback of a $T^k$-bundle over the Borel space, $W_Q = EQ \times_Q W$, where the structure group of the bundle is $T^k \rtimes \text{Aut}(T^k)$. We gave two approaches to this problem in Sections 10.2 and 10.3. It is the method of Section 10.2 adapted to $Q$, a connected compact Lie group acting on $W$, that is pursued in this chapter.

This lifting problem has had a distinguished history: [**Con68**], [**CR69**], [**Ste61**], [**Su63**], [**HY76**], [**Got77**], [**Las79**], [**LMS83**] among others. In this chapter, we present the methods and results of Hattori and Yoshida for connected Lie groups $Q$ and principal $T^k$-bundles. To accommodate the connectedness of $Q$, the cohomology of the group used in Section 10.2 needs to be replaced by the continuous cohomology of the group $Q$. The vanishing of $H^p(Q; \text{M}(W, \mathbb{R}^k))$ is crucial for discrete $Q$ and it will also be significant for compact connected $Q$.

The claim is similar to that in Section 10.2. Namely, if $P$ is a principal $T^k$-bundle over $W$ and $Q$ is a connected, compact Lie group acting on $W$ (connected, locally connected and locally compact), then the action of $Q$ lifts to a group of bundle automorphisms of $P$ if and only if $P$ is equivalent to the pullback of a principal $T^k$-bundle $P'$ over the Borel space $W_Q = EQ \times_Q W$ via $\pi \circ i : W \xrightarrow{i} EQ \times W \xrightarrow{\pi} EQ \times_Q W$. The spectral sequence, associated with the fibering of the Borel space over the classifying space $BQ$, will be used to determine when $P$ is the pullback of a principal $T^k$-bundle over $W_Q$. This will provide liftings and consequent Seifert fiberings in many interesting situations. Of course, one of the reasons for lifting $Q$-actions on $W$ to bundle automorphisms on $P$ is to consider the induced action of $T^k$ on $Q \backslash P$ and the consequent Seifert fibering $Q \backslash P \to Q \backslash W$.

## 12.2. Lifting $Q$-actions

12.2.1. Let $Q$ be a Lie group, and let $\varphi : Q \times W \to W$ be a *proper* action. Since $Q$ has a topology, we shall assume that $W$ is locally compact, Hausdorff, connected and locally connected. Then $\text{TOP}(W)$ is a topological group under the compact-open topology and $\widetilde{\varphi} : Q \to \text{TOP}(W)$ is a homomorphism of topological groups and $\widetilde{\varphi}(Q)$ is a closed (since the action is proper) subgroup of $\text{TOP}(W)$. The topology of $\widetilde{\varphi}(Q)$ is the Lie group topology of $Q/Q_0$, where $Q_0$ is the kernel of $\widetilde{\varphi}$; see Subsection 1.2.7. We shall repeatedly use the following

LEMMA 12.2.2. *Let $f : X \to Y$ and $p : Z \to Y$ be $Q$-maps between $Q$-spaces. Then the pullback $f^*(Z)$ has an induced $Q$-action and the maps in the induced commutative diagram*

$$\begin{array}{ccc} f^*(Z) & \xrightarrow{\tilde{f}} & Z \\ p^* \downarrow & & \downarrow p \\ X & \xrightarrow{f} & Y \end{array}$$

*are $Q$-maps. Moreover, if $f$ and $p$ are also $G$-maps with the action of $G$ and $Q$ commuting, then the induced maps are also $G$-maps and the induced actions commute.*

PROOF. Recall that

$$f^*(Z) = \{(x, z) \in X \times Z \mid f(x) = p(z)\}.$$

Let $(x, z) \in f^*(Z)$. Define $q \cdot (x, z) = (qx, qz)$ for all $q \in Q$ and $(x, z) \in f^*(Z)$. This gives a well-defined $Q$-action and makes the induced maps $Q$-equivariant. The rest of the lemma follows easily. □

EXERCISE 12.2.3. Let $f : P' \to P$ be a $G$-bundle equivalence between principal $G$-bundles (over a fixed base $W$) with a proper action of $Q$ on $W$. If $Q$ lifts to a group of bundle automorphisms on $P'$ (respectively, group of weak bundle automorphisms on $P'$), then $Q$ lifts similarly to $P$.

12.2.4. Let $P$ be a principal $G$-bundle over $W$ and $Q$ a Lie groups acting properly on $W$. Form the Borel space $W_Q = EQ \times_Q W$. Embed $W$ into $W_Q$ via the map $\pi \circ i : W \to W_Q$, where

$i : W \to EQ \times W$, $w \mapsto (e_0, w)$, for some fixed $e_0 \in EQ$,

$\pi : EQ \times W \to EQ \times_Q W$, the natural projection,

$\pi_2 : EQ \times W \to W$, the natural projection onto $W$.

This embeds $W$ as the fiber over the image of $e_0$ in $BQ$, the classifying space for principal $Q$-bundles, and $W \hookrightarrow W_Q \to BQ$ is a fiber bundle. Note that $\pi_2$ is $Q$-equivariant, where the $Q$-action on $EQ \times W$ is the diagonal $Q$-action given by $q \cdot (e, w) = (eq^{-1}, qw)$.

12.2.5. Suppose $Q$ lifts to a group of bundle automorphisms of $P$. Then $Q$ lifts to a group of bundle automorphisms acting freely on $\pi_2^*(P)$. Furthermore, $Q \backslash \pi_2^*(P)$ is a principal $G$-bundle over $W_Q$ and $P$ is equivalent to $i^* \circ \pi^*(Q \backslash \pi_2^*(P))$.

PROOF. Take the trivial $G$-action on $EQ \times W$ and $W$. Then $\pi_2$ is both a $G$-map and a $Q$-map with $Q$ commuting with $G$. Therefore by Lemma 12.2.2, the $Q$-action on $EQ \times W$ lifts to a group of bundle automorphisms on $\pi_2^*(P)$. Since the $Q$-action on $EQ \times W$ is free, the action of $Q$ on $\pi_2^*(P)$ is also free. Therefore we have the commuting diagram of orbit mappings

$$\begin{array}{ccc} \pi_2^*(P) & \xrightarrow{Q \backslash} & Q \backslash \pi_2^*(P) \\ G \backslash \downarrow & & \downarrow G \backslash \\ EQ \times W & \xrightarrow[Q \backslash]{\pi} & EQ \times_Q W \end{array}$$

yielding $Q\backslash\pi_2^*(P)$ a principal $G$-bundle over $W_Q$. Clearly, $\pi^*(Q\backslash\pi_2^*(P)) = \pi_2^*(P)$. Now $i^*(\pi_2^*(P)) \cong P$, hence $i^* \circ \pi^*(Q\backslash\pi_2^*(P)) \cong P$. $\square$

12.2.6. We just have shown that a *necessary* condition for $P$ to admit a lifting of a proper $Q$-action on $W$ to a group of a bundle automorphism of $P$ is that $P$ *must be the pullback of some principal $G$-bundle over $W_Q$ via the map $\pi \circ i$*. This condition is also sufficient when $Q$ is discrete and $G = T^k$ as seen in Chapter 10. Hattori and Yoshida have shown sufficiency for $G = T^k$ and $Q$ compact Lie group which generalized earlier results of Stewart and Su. However, Gottlieb gave an example, attributed to Bredon, that sufficiency fails for $G = \mathrm{Spin}(1)$. Gottlieb [**Got77**] also observed the following:

PROPOSITION 12.2.7. *If $Q$ is a Lie group and acts freely and properly on $W$, then the condition is both necessary and sufficient for lifting $Q$ to a group of bundle automorphisms on $P$.*

Consider the diagram of maps

$$\begin{array}{ccccc} W & \underset{\pi_2}{\overset{i}{\rightleftarrows}} & EQ \times W & & \\ \bar{\pi}\downarrow & & \downarrow \pi & & \\ Q\backslash W & \underset{\overline{\pi_2}}{\overset{\bar{i}}{\rightleftarrows}} & EQ \times_Q W & \xrightarrow{f} & BG. \end{array}$$

We have $\bar{\pi} \circ \pi_2 = \overline{\pi_2} \circ \pi$ since $\pi_2$ is a $Q$-map. As $Q$ acts freely, $\overline{\pi_2}$ is a fiber bundle map with fiber $EQ$. Hence $\overline{\pi_2}$ is a homotopy equivalence. Let $\bar{i}$ be a homotopy inverse to $\overline{\pi_2}$. Then we have $\pi \circ i \simeq \bar{i} \circ \bar{\pi}$ and $\overline{\pi_2} \circ \pi = \bar{\pi} \circ \pi_2$ and all the horizontal maps $i, \pi_2, \bar{i}, \overline{\pi_2}$ are homotopy equivalences. For a map $f : W_Q = EQ \times_Q W \to BG$, define $\bar{f} : Q\backslash W \to BG$ by $\bar{f} = f \circ \bar{i}$. Then we have $\bar{f} \circ \bar{\pi} \simeq f \circ \pi \circ i$. Thus if $P \in i^* \circ \pi^*(P')$, where $P'$ is a principal $G$-bundle over $W_Q$, then $P \cong \bar{\pi}^* \circ \bar{f}^*(\xi)$, with $\xi$ being the universal $G$-bundle over $BG$. Since $\bar{f} \circ \bar{\pi}$ is a $Q$-map, the pullback of the universal $G$-bundle has $G \times Q$-action lifting the $Q$-action on $W$. Since $\bar{\pi}^* \circ \bar{f}^* \xi \cong P$, we have shown what we wanted to prove. (We can define a lift of $Q$ to $P$ via $h : \bar{\pi}^* \circ \bar{f}^* \xi \to P$, the $G$-bundle equivalence by setting $q \circ h(r) = h(q(r))$. One can also use Exercise 12.2.3.)

## 12.3. Lifting $Q$-actions (for connected $Q$)

12.3.1. We will now explain the argument of Hattori and Yoshida in the case when $G = T^k$, and $Q$ and $W$ are connected and $Q$ is compact. Their argument treats $Q$ compact and not necessarily connected; see [**HY76**]. By different methods, Lashof, May and Segal [**LMS83**] treat the same $Q$ with $G$ a compact Abelian Lie group.

In the discrete case, the proof relied on the cohomology of $Q$. In the continuous case, the *continuous cohomology* of $Q$ will be used. The cohomology groups are defined analogously as in the discrete case but maps

$$Q^q = Q \times \cdots \times Q \longrightarrow A,$$

where $A$ is a topological Abelian $Q$-module, are taken continuously. Both methods, discrete and continuous, rely on the vanishing of certain cohomology groups when the coefficients are taken in the maps of $W$ into $\mathbb{R}^k$. In both cases, this is a nontrivial fact and, in the continuous case, was first proved by G.D. Mostow [**Mos61**].

Since we are only considering lifting $Q$-actions to bundle automorphisms, we can simplify our universal group $\text{TOP}_{T^k}(P)$ to

$$\text{TOP}^0_{T^k}(P) = \{f \in \text{TOP}_{T^k}(P) : f(au) = af(u)\}.$$

Then the sequence

$$0 \to \text{M}(W, T^k) \to \text{TOP}^0_{T^k}(P) \xrightarrow{j} \text{TOP}(W)$$

is exact. The image of $\rho : Q \to \text{TOP}(W)$ is contained in the image of $j : \text{TOP}^0_{T^k}(P) \to \text{TOP}(W)$ because $\rho(Q)$ is in the connected path component of the identity of $\text{TOP}(W)$; see Lemma 4.2.15. Therefore, we can form the pullback of the above sequence via the maps $j$ and $\rho$ to get the following diagram.

$$\begin{array}{ccccccccc}
1 & \longrightarrow & \text{M}(W,T^k) & \longrightarrow & \rho^*(Q) = E(P,Q) & \xrightarrow{\bar{j}} & Q & \longrightarrow & 1 \\
 & & \| & & \downarrow\epsilon & & \downarrow\rho & & \\
1 & \longrightarrow & \text{M}(W,T^k) & \longrightarrow & j^{-1}(\rho(Q)) & \xrightarrow{j} & \rho(Q) & \longrightarrow & 1 \\
 & & \| & & \downarrow\cap & & \downarrow\cap & & \\
1 & \longrightarrow & \text{M}(W,T^k) & \longrightarrow & \text{TOP}^0_{T^k}(P) & \xrightarrow{j} & \text{TOP}(W) & &
\end{array}$$

These are exact sequences of topological groups; see [**HY76**] and compare with [**Par89**]. We want to describe the extensions in terms of factor sets. In general, this is not possible, for to do so we need a *continuous* map $s : Q \to E(P,Q)$ so that $\bar{j} \circ s$ is the identity. That is, we need a cross section for the top (i.e., first) bundle. (In [**HY76**], $s$ is called a *pseudo-lifting*); see the Example 12.4.5.

LEMMA 12.3.2 ([**HY76**, Lemma 2.3]). *If $P$ is equivalent to the pullback of a principal $T^k$-bundle $S$ over $W_Q$ via $i^* \circ \pi^*(S) \cong P$, then the top sequence has a continuous section.*

PROOF. From $P \cong i^* \circ \pi^*(S)$, we have $\pi_2^*(P) \cong \pi^*(S)$ over $EQ \times W$. We may identify $P$ with $\pi^*S|_{e_0 \times W}$. There is a contraction $r_t : EQ \to EQ$ such that $r_0 = e_0$, and $r_1$ is the identity. Consequently there is a covering homotopy

$$\bar{r}_t : \pi^*S \longrightarrow \pi^*S$$

of $r_t \times 1 : EQ \times W \to EQ \times W$ so that $\bar{r}_1$ is the identity and $\bar{r}_0$ maps $\pi^*S$ onto $\pi^*S|_{e_0 \times W} = P$. The $Q$-action on $\pi^*(S)$ is given by $q \cdot (u,(e,x)) \mapsto (u,(eq^{-1},qx))$, where $u \in S$ which projects to $\langle e, x \rangle = \langle eq^{-1}, qx \rangle \in W_Q = EQ \times_Q W$.

Consider the composite map

$$Q \times P \subset Q \times \pi^*(S) \longrightarrow \pi^*(S) \xrightarrow{\bar{r}_0} P,$$

which covers the $\varphi : Q \times W \to W$ action. Thus each $q \in Q$ is an automorphism of $P$ and we get a continuous map $s : Q \to E(P,Q)$ and $\rho(Q) \to j^{-1}(\rho(Q)) \subset \text{TOP}^0_{T^k}(P)$ so that $\bar{j} \circ s = \text{id}_Q$ or $j \circ s = \text{id}_{\rho(Q)}$. $\square$

12.3.3. For an extension of topological groups

$$1 \longrightarrow A \longrightarrow E \longrightarrow Q \longrightarrow 1$$

with $A$ Abelian, we can describe the congruence classes of extensions in terms of $H^2(Q; A)$ where we use cohomology arising from continuous cochains provided that the extension when regarded as a principal $A$-bundle has a cross section; see for example, [**Hu52**, pp.11–59] and [**Mos61**]. That is, we take a continuous section $s : Q \to E$ and construct a factor set by

$$f(\alpha, \beta) = s(\beta)s(\alpha)s(\alpha\beta)^{-1}.$$

This is a 2-cocycle just as in the discrete case. In our case, the values of $f$ lie in $\mathrm{M}(W, T^k)$ with $\mathrm{M}(W, T^k)$ a left $Q$-module in the usual sense; namely,

$$\alpha \circ \lambda = \lambda \circ s(\alpha)^{-1}$$

for $\lambda \in \mathrm{M}(W, T^k)$, and $\alpha \in Q$. Our cocycle is in

$$\mathrm{M}(Q \times Q; \mathrm{M}(W, T^k)) = \mathrm{M}(Q^2; \mathrm{M}(W, T^k)) = C^2(Q; \mathrm{M}(W, T^k)).$$

Its cohomology class is denoted by $o(P)$ and it is the obstruction for splitting. The class $o(P)$ vanishes if and only if the sequence in Subsection 12.3.1 splits, that is, whenever $Q$ lifts to $P$ as a group of bundle automorphisms. What will be shown is the following.

THEOREM 12.3.4 ([**HY76**]). *If $Q$ is a compact connected Lie group, then $H^2(Q; \mathrm{M}(W, T^k)) = 0$.*

COROLLARY 12.3.5. *If $Q$ is a compact connected Lie group acting properly on $W$, then $Q$ lifts to a group of bundle automorphisms on $P$ if and only if $P$ is equivalent to the pullback $i^* \circ \pi^*(S)$ of a principal $T^k$-bundle $S$ over $W_Q$.*

Actually, [**HY76**] and [**LMS83**] prove a stronger result. The same conclusion holds for $Q$ a compact Lie group and $T^k$ replaced by a compact Abelian Lie group. They prove that the necessary condition is sufficient to show that the obstruction $o(P)$ vanishes.

We have shown in Lemma 12.3.2 that the necessary condition for lifting is sufficient for a cross section. Therefore we can construct our factor set $f$. The vanishing of $H^2(Q; \mathrm{M}(W, T^k))$ gives us a splitting and hence a lifting. Therefore, our necessary condition is also a sufficient condition when $Q$ is a connected compact Lie group and our principal bundle $P$ is a principal $T^k$-bundle. It remains to verify Theorem 12.3.4.

12.3.6. First let us recall some facts about the cohomology of topological groups with coefficients in the Abelian topological group $A$; see [**Hu52**, §4], [**Mos61**].

Let $Q$ and $A$ be topological groups with $A$ Abelian. Define the $Q$-module of continuous $p$-cochains of $Q$ into $A$ by $C^p(Q; A) = \mathrm{M}(Q^p, A)$ for $q > 0$. That is, continuous maps of $Q^p = Q \times Q \times \cdots \times Q$, the Cartesian product of $p$ copies of $Q$, into the Abelian topological group $A$. These maps forms a group where addition of maps is given by addition of functional values. We assume that $Q$ acts on $A$ as a group of transformations and so this extends to an action on $C^p(Q; A)$. We define $C^0(Q; A) = A$. For each $p \geq 0$, we define the inhomogeneous coboundary operation

$$\delta^p : C^p(Q; A) \longrightarrow C^{p+1}(Q; A),$$

which will satisfy $\delta^{p+1} \delta^p = 0$. For $p = 0$, $\delta^0 : C^0(Q; A) = A \longrightarrow C^1(Q; A) = \mathrm{M}(Q, A)$ is given by

$$(\delta^0)a(x) = xa - a$$

for $x \in Q$, $a \in A$. For $p > 0$,

$$\delta^p f(x_1, \ldots, x_{p+1}) = \; x_1 \cdot f(x_2, \ldots, x_{p+1}) + \cdots$$
$$+ \sum_{i=1}^{i=p}(-1)^i f(x_1, \ldots, x_i \cdot x_{i+1}, \ldots, x_{p+1})$$
$$+ (-1)^{p+1} f(x_1, \ldots, x_p).$$

Define $H^p(Q; A) = \ker(\delta^p)/\operatorname{Im}(\delta^{p-1})$. Then $H^p$ has the obvious functorial properties. Clearly, if $Q$ is discrete, this continuous cohomology is the same as the ordinary cohomology of the discrete group $Q$ with coefficients in the group $A$ because every map from $Q$ is continuous.

Our definition of coboundary differs slightly from that of Hattori and Yoshida because we require $Q$ to act on $\operatorname{M}(Q, A)$ as a left action instead of their right action. This is consistent with our preceding chapters.

If $0 \to A' \xrightarrow{i} A \xrightarrow{i} A'' \to 0$ is an exact sequence of Abelian topological $Q$-modules, then

$$0 \longrightarrow C^p(Q; A') \xrightarrow{i_*} C^p(Q; A) \xrightarrow{j_*} C^p(Q; A'')$$

is exact but, unfortunately, the last homomorphism is not necessarily onto. However, if $A''$ admits a continuous cross section in $A$, then $j_* : C^p(Q; A) \to C^p(Q; A'')$ is surjective as a cochain map. Consequently, under this restriction, we get a long exact sequence of cohomology. This differs from the discrete case since there a cross section always exists because all maps from a discrete $Q$ are continuous. This point is a serious one and the general lack of exactness makes calculations of continuous cohomology difficult.

If $W$ is locally compact Hausdorff space, $\operatorname{M}(W, \mathbb{R}^k)$ and $\operatorname{M}(W, T^k)$ are Abelian topological groups. We will eventually need the vanishing theorem of Mostow:

PROPOSITION 12.3.7 ([**Mos61**]). *If $Q$ is a compact Lie group and $W$ is a locally compact Hausdorff space, then $H^p(Q; \operatorname{M}(W, \mathbb{R}^k)) = 0$ for all $p \geq 1$.*

12.3.8 (Proof of Theorem 12.3.4). From the exact sequence of Abelian groups

$$0 \longrightarrow \operatorname{M}(W, \mathbb{Z}^k) \longrightarrow \operatorname{M}(W, \mathbb{R}^k) \xrightarrow{\exp} \operatorname{M}(W, T^k) \longrightarrow H^1(W; \mathbb{Z}^k) \longrightarrow 0,$$

we obtain the exact sequences

$$0 \longrightarrow \operatorname{M}(W, \mathbb{Z}^k) \longrightarrow \operatorname{M}(W, \mathbb{R}^k) \xrightarrow{\exp} \operatorname{M}_0(W, T^k) \longrightarrow 0$$

and

$$0 \longrightarrow \operatorname{M}_0(W, T^k) \longrightarrow \operatorname{M}(W, T^k) \longrightarrow H^1(W; \mathbb{Z}^k) \longrightarrow 0,$$

where $\operatorname{M}_0(W, T^k)$ is the subgroup of maps of $W$ into $T^k$ which are homotopic to a constant. Since $W$ is connected, $\operatorname{M}(W, \mathbb{Z}^k) \cong \mathbb{Z}^k$ and $\operatorname{M}(W, \mathbb{R}^k)$ is the universal covering of $\operatorname{M}_0(W, T^k)$. Also $\operatorname{M}(W, \mathbb{R}^k)$ is contractible and $\operatorname{M}_0(W, T^k)$ is a $K(\mathbb{Z}^k, 1)$. These are all $Q$-modules and so we have the resulting cochain complexes,

$$0 \longrightarrow C^*(Q; \operatorname{M}_0(W, T^k)) \longrightarrow C^*(Q; \operatorname{M}(W, T^k)) \longrightarrow C^*(Q; H^1(W; \mathbb{Z}^k)).$$

In the last term, $H^1(W; \mathbb{Z}^k)$ is discrete and therefore the last homomorphism is onto. In particular, we have

$$H^q(Q; \operatorname{M}_0(W, T^k)) \cong H^q(Q; \operatorname{M}(W, T^k)), \quad \text{for } q \geq 2$$

(we are using that $H^q(Q; \text{discrete}) = 0$ for $q \geq 1$, when $Q$ is connected) and

(12.3.1)
$$0 \longrightarrow \operatorname{M}_0(W, T^k)^Q \longrightarrow \operatorname{M}(W, T^k)^Q \longrightarrow H^1(W; \mathbb{Z}^k)$$
$$\longrightarrow H^1(Q; \operatorname{M}_0(W, T^k)) \longrightarrow H^1(Q; \operatorname{M}(W, T^k)) \longrightarrow 0$$

is exact.

It remains to show $H^2(Q; \mathrm{M}_0(W, T^k)) = 0$. The cochain complex

$$0 \longrightarrow C^*(Q; \mathbb{Z}^k) \longrightarrow C^*(Q; \mathrm{M}(W, \mathbb{R}^k)) \longrightarrow C^*(Q; \mathrm{M}_0(W, T^k))$$

is exact but the last homomorphism is not onto. Now,

$$f \in C^p(Q; \mathrm{M}_0(W, T^k)) = M(Q^p, \mathrm{M}_0(W, T^k))$$

is, by definition, the space of all continuous maps of $Q^p$ into $\mathrm{M}_0(W, T^k)$. Then $f$ is the image of an element of $C(Q^p; \mathrm{M}(W, \mathbb{R}^k))$ if and only if $f: Q^p \to \mathrm{M}_0(W, T^k)$ is trivial on $\pi_1$. ($\mathrm{M}(W, \mathbb{R}^k)$ is the contractible universal covering of $\mathrm{M}_0(W, T^k)$, and $\mathrm{M}_0(W, T^k)$ is a $K(\mathbb{Z}^k, 1)$.) Therefore, we can assign to $f$ the homomorphism $f_*: \pi_1(Q^p) \to \mathbb{Z}^k$. We need not concern ourselves with base points as all our groups are Abelian and $Q^p$ is simple. Therefore, we obtain an epimorphism

$$M(Q^p, \mathrm{M}_0(W, T^k)) \longrightarrow \mathrm{Hom}(\pi_1(Q^p), \mathbb{Z}^k) = \mathrm{Hom}^p(\pi_1(Q), \mathbb{Z}^k) \text{ (by definition)}.$$

We shall show the last map is a cochain mapping. Let

$$\partial_j: Q^{p+1} \longrightarrow Q^p, \quad j = 0, 1, \ldots, p+1$$

be defined by

$$\partial_0(u_1, \ldots, u_{p+1}) = (u_2, u_3, \ldots, u_{p+1})$$
$$\partial_i(u_1, \ldots, u_{p+1}) = (u_1, \cdots, u_i \cdot u_{i+1}, \ldots, u_{p+1}), \ 1 \leq i \leq p$$
$$\partial_{p+1}(u_1, \ldots, u_{p+1}) = (u_1, u_2, \ldots, u_p).$$

The coboundary $\delta: \mathrm{Hom}^p(\pi_1(Q), \mathbb{Z}^k) \longrightarrow \mathrm{Hom}^{p+1}(\pi_1(Q), \mathbb{Z}^k)$ is defined by $\delta = \sum_{i=1}^{p+1}(-1)^i \delta_i$, where $\delta_i$ is the transpose of the induced homomorphism $\partial_{i*}: \pi_1(Q^{p+1}) \to \pi_1(Q^p)$.

For $f \in C^p(Q; \mathrm{M}_0(W, T^k))$, with $f \mapsto f_* \in \mathrm{Hom}^p(\pi_1(Q), \mathbb{Z}^k)$, the first term of $\delta f$, $x_1 \cdot f(x_2, \ldots, x_{p+1})$, maps to $f_*(x_2, \ldots, x_{p+1})$ because the operation of $x_1$ becomes trivial on the homotopy level. Thus $(\delta f)_* = \delta f_*$ and we get a cochain mapping for $p \geq 1$. Now, define $\mathrm{Hom}^0(\pi_1(Q), \mathbb{Z}^k) = 0$. Then, the composition

$$\begin{array}{ccc}
C^0(Q; \mathrm{M}_0(W, T^k)) = \mathrm{M}_0(W, T^k) & \longrightarrow & \mathrm{Hom}^0(\pi_1(Q), \mathbb{Z}^k) = 0 \\
\delta^0 \downarrow & & \downarrow \delta \\
C^1(Q; \mathrm{M}_0(W, T^k)) & \longrightarrow & \mathrm{Hom}^1(\pi_1(Q), \mathbb{Z}^k) = \mathrm{Hom}(\pi_1(Q), \mathbb{Z}^k)
\end{array}$$

is the trivial homomorphism ensuring that $C^*(Q; \mathrm{M}_0(W, T^k)) \longrightarrow \mathrm{Hom}^*(\pi_1(Q), \mathbb{Z}^k)$ is also a cochain mapping at the 0 and 1-level. Thus we get the following exact sequence of cochain complexes

$$0 \longrightarrow C^*(Q; \mathbb{Z}^k) \longrightarrow C^*(Q; \mathrm{M}(W, \mathbb{R}^k))$$
$$\longrightarrow C^*(Q; \mathrm{M}_0(W, T^k)) \longrightarrow \mathrm{Hom}^*(\pi_1(Q), \mathbb{Z}^k) \longrightarrow 0.$$

Let $K^*$ denote the image of $C^*(Q; \mathrm{M}(W, \mathbb{R}^k)) \xrightarrow{\exp_*} C^*(Q; \mathrm{M}_0(W, T^k))$. Then we have two short exact sequences,

$$0 \longrightarrow C^*(Q; \mathbb{Z}^k) \longrightarrow C^*(Q; \mathrm{M}(W, \mathbb{R}^k)) \longrightarrow K^* \longrightarrow 0,$$
$$0 \longrightarrow K^* \longrightarrow C^*(Q; \mathrm{M}_0(W, T^k)) \longrightarrow \mathrm{Hom}^*(\pi_1(Q), \mathbb{Z}^k) \longrightarrow 0.$$

Passing the first exact sequence to cohomology, we get the long exact sequence

$$\longrightarrow H^p(Q; \mathbb{Z}^k) \longrightarrow H^p(Q; \mathrm{M}(W, \mathbb{R}^k)) \longrightarrow H^p(Q; K^*) \longrightarrow H^{p+1}(Q; \mathbb{Z}^k) \longrightarrow .$$

Now $H^p(Q; \mathbb{Z}^k) = 0$ for $p > 0$ since $Q$ is connected, so $H^p(Q; \mathrm{M}(W, \mathbb{R}^k)) \cong H^p(Q; K^*)$, for $p \geq 1$. But $H^p(Q; \mathrm{M}(W, \mathbb{R}^k)) = 0$ for all $p \geq 1$ [**Mos61**] so that $H^p(Q; K^*) = 0$ for all $p \geq 1$.

Passing the second exact sequence to cohomology, we get the long exact sequence

$$\longrightarrow H^p(Q; K^*) \longrightarrow H^p(Q; \mathrm{M}_0(W, T^k))$$
$$\longrightarrow H^p(\mathrm{Hom}^*(\pi_1(Q), \mathbb{Z}^k)) \longrightarrow H^{p+1}(Q; K^*) \longrightarrow .$$

Since $H^p(Q; K^*) = 0$ for all $p \geq 1$,

$$H^p(Q; \mathrm{M}_0(W, T^k)) \cong H^p(\mathrm{Hom}^*(\pi_1(Q), \mathbb{Z}^k))$$

for $p \geq 1$.

We now calculate $H^p(Q; \mathrm{M}_0(W, T^k))$ for $p \leq 2$. This reduces to computing $H^p(\mathrm{Hom}^*(\pi_1(Q), \mathbb{Z}^k))$ for $p = 1$ and 2. Now $\mathrm{Hom}^1(\pi_1(Q), \mathbb{Z}^k) = \mathrm{Hom}(\pi_1(Q), \mathbb{Z}^k)$. If $f$ is a 1-cochain, then

$$\delta f(\alpha, \beta) = f\partial_{0*}(\alpha, \beta) - f\partial_{1*}(\alpha, \beta) + f\partial_{2*}(\alpha, \beta)$$
$$= f(\beta) - f(\alpha + \beta) + f(\alpha)$$

since $\partial_1(\alpha, \beta) = \alpha\beta$ induces the addition in $\pi_1(Q)$. However, $f \in \mathrm{Hom}(\pi_1(Q), \mathbb{Z}^k)$, hence $\delta f(\alpha, \beta) = 0$. Since $\mathrm{Hom}^0(\pi_1(Q), \mathbb{Z}^k) = 0$,

$$H^1(\mathrm{Hom}^*(\pi_1(Q), \mathbb{Z}^k)) = \mathrm{Hom}(\pi_1(Q), \mathbb{Z}^k).$$

Similarly, for

$$f \in \mathrm{Hom}^2(\pi_1(Q), \mathbb{Z}^k) = \mathrm{Hom}(\pi_1(Q^2), \mathbb{Z}^k) = \mathrm{Hom}(\pi_1(Q), \mathbb{Z}^k) \times \mathrm{Hom}(\pi_1(Q), \mathbb{Z}^k),$$

we have

$$\delta f(\alpha, \beta, \gamma) = f(\beta, \gamma) - f(\alpha + \beta, \gamma) + f(\alpha, \beta + \gamma) - f(\alpha, \beta)$$
$$= f(\beta, \gamma) - (f(\alpha, 0) + f(\beta, \gamma)) + (f(\alpha, \beta) + f(0, \gamma)) - f(\alpha, \beta)$$
$$= f(0, \gamma) - f(\alpha, 0)$$
$$= f(-\alpha, \gamma).$$

Thus if $\delta f = 0$, then $f = 0$. Therefore,

$$H^2(\mathrm{Hom}^*(\pi_1(Q), \mathbb{Z}^k)) = 0.$$

This completes the proof of Theorem 12.3.4. □

12.3.9. To apply the theorem, we need to know when a principal $T^k$-bundle $P$ over $W$ is equivalent to a pullback of a principal $T^k$-bundle $P'$ over $W_Q$. Clearly the characteristic class $[P] \in H^2(W; \mathbb{Z}^k)$ must be an element of $H^2(W; \mathbb{Z}^k)^Q$ if there is an $E(P, Q)$ that maps into $\mathrm{TOP}^0_{T^k}(P)$ and onto $Q$. These groups are equal since the $Q$-action on the cohomology is trivial because $Q$ is connected. Therefore, $P$ is the pullback of some principal $T^k$-bundle $P'$ over $W_Q$ if and only if the image of $i^* \circ \pi^* : H^2(W_Q; \mathbb{Z}^k) \to H^2(W; \mathbb{Z}^k)$ contains $[P]$. If $[P]$ is in the image, then the number of distinct liftings of $Q$ up to strict equivalence is naturally in one-to-one correspondence with the elements of the kernel $(\pi \circ i)^*$. (Recall that given a lifting of $Q$ to $P$, then $\pi_2^*(P)$ also has a lifting and $Q\backslash\pi_2^*(P)$ is the desired $P'$. By virtue of Exercise 12.2.3, this establishes the correspondence.)

Let us now consider the spectral sequence of the fibering $W_Q \to BQ$, where $E_2^{p,q} = H^q(BQ; H^p(W; \mathbb{Z}^k))$.

PROPOSITION 12.3.10. *If $H^1(W; \mathbb{Z}^k) = 0$, we get the complete analogue of the exact sequence of Subsection 10.2.3; namely,*

$$0 \to H^2(BQ; \mathbb{Z}^k) \xrightarrow{\ell} H^2(W_Q; \mathbb{Z}^k)$$
$$\xrightarrow{e} H^2(W; \mathbb{Z}^k)^Q \longrightarrow H^3(BQ; \mathbb{Z}^k) \to H^3(W_Q; \mathbb{Z}^k).$$

Notice that $H^2(W; \mathbb{Z}^k)^Q = H^2(W; \mathbb{Z}^k)$ as $Q$ is connected.

PROOF. We consider the fibering $EQ \times_Q W = W_Q \xrightarrow{\ell} BQ$, where $W$ is the fiber and the structure group is $Q$. Since $Q$ is connected, the coefficient system in the $E_2^{p,q} = H^p(BQ; H^q(W; \mathbb{Z}^k))$ is simple (i.e., not twisted) and the sequence converges to $H^*(W_Q; \mathbb{Z}^k)$. In dimension 2, we have $H^2(W_Q; \mathbb{Z}^k) = [W_Q, B_{T^k}]$ and the edge homomorphism $H^2(W_Q; \mathbb{Z}^k) \xrightarrow{e} H^2(W; \mathbb{Z}^k)$ coincides with $i^* \circ \pi^* : H^2(W_Q; \mathbb{Z}^k) \to H^2(W; \mathbb{Z}^k)$. We obtain the terms of low degree for this spectral sequence

$$0 \to H^1(BQ; \mathbb{Z}^k) \longrightarrow H^1(W_Q; \mathbb{Z}^k)$$
$$\longrightarrow H^1(W; \mathbb{Z}^k) \longrightarrow H^2(BQ; \mathbb{Z}^k) \to H^2(W_Q; \mathbb{Z}^k).$$

If $H^1(W; \mathbb{Z}^k) = 0$, the exact sequence continues with

$$0 \to H^2(BQ; \mathbb{Z}^k) \longrightarrow H^2(W_Q; \mathbb{Z}^k)$$
$$\longrightarrow H^2(W; \mathbb{Z}^k) \longrightarrow H^3(BQ; \mathbb{Z}^k) \to H^3(W_Q; \mathbb{Z}^k).$$

Therefore the elements $[P] \in H^2(W; \mathbb{Z}^k)$, which are the characteristic classes of the bundles for which $Q$ lifts to a bundle of automorphisms of $P$, are precisely the image of $e$. For each such $P$, the group $H^2(BQ; \mathbb{Z}^k)$ classifies the distinct liftings of $Q$ to $P$ up to strict equivalence. In particular, it classifies the liftings for the product bundle. □

If $H^1(W; \mathbb{Z}^k) \ne 0$, then using the spectral sequence again, we may characterize the image of $e$ as those elements $c \in H^2(W; \mathbb{Z}^k) = E_2^{0,2}$ such that $d^2(c) = 0$ and $d^3(c) = 0$. (Here, $d^3 : E_3^{0,2}(= \ker d^2) \longrightarrow E_3^{3,0} = H^3(BQ; \mathbb{Z}^k)$ since $BQ$ is simply connected.)

Also, note that $H^2(BQ; \mathbb{Z}^k)$ is naturally isomorphic to $\mathrm{Hom}(\pi_1(Q), \mathbb{Z}^k)$. For $H^2(BQ; \mathbb{Z}^k) \cong \mathrm{Hom}(H_2(BQ), \mathbb{Z}^k)$ since $BQ$ is simply connected and $H_2(BQ) \cong \pi_2(BQ) \cong \pi_1(Q)$. We have the following corollaries.

COROLLARY 12.3.11 (cf. [**Ste61, HY76**]). *If $Q$ is a simply connected, compact Lie group acting properly on $W$, then $Q$ lifts to any principal $T^k$-bundle over $W$. Furthermore, this lifting is unique up to conjugation by elements of $\mathrm{M}(W, T^k)$.*

PROOF. As observed above, the map $H^2(W_Q; \mathbb{Z}^k) \xrightarrow{e} H^2(W; \mathbb{Z}^k)$ arises from the edge homomorphism in the spectral sequence associated with the fibering $W_Q \to BQ$. But as $\pi_1(BQ) = \pi_2(Q) = 0$ together with exact sequence

$$\pi_3(EQ) = 0 \to \pi_3(BQ) \to \pi_2(Q) = 0 \to \pi_2(EQ) = 0 \to$$

(which implies $\pi_3(BQ) = 0$), we have $BQ$ 3-connected and so $H^p(BQ, H^q(W; \mathbb{Z}^k)) = E_2^{p,q}$ is 0 for $p = 1, 2, 3$. This implies that $H^1(W; \mathbb{Z}^k) \cong H^1(W_Q; \mathbb{Z}^k)$ and $H^2(W_Q; \mathbb{Z}^k) \xrightarrow{e} H^2(W; \mathbb{Z}^k)$ is an isomorphism and so on each principal $T^k$-bundle, $Q$ lifts to a group of bundle automorphisms.

Just as in the discrete case, the liftings are in one-to-one correspondence with the group $H^1(Q, \mathrm{M}(W, T^k))$ up to conjugation by elements of $\mathrm{M}(W, T^k)$. From the proof of the theorem, we found that

$$H^1(Q; \mathrm{M}_0(W, T^k)) \cong \mathrm{Hom}(\pi_1(Q), \mathbb{Z}^k) = 0$$

and $H^1(Q; \mathrm{M}_0(W, T^k))$ mapped onto $H^1(Q; \mathrm{M}(W, T^k))$, and thus 0. $\square$

COROLLARY 12.3.12 (cf. [**Su63, HY76**]). *If $H^1(W; \mathbb{Z}^k) = 0$ and $T^n$ acts on $W$, then $T^n$ lifts to every principal $T^k$-bundle over $W$.*

PROOF. From the exact sequence arising from the spectral sequence, we have $H^3(BT^n, \mathbb{Z}^k) = 0$ and so $e$ is onto. $\square$

If $Q$ is a compact connected Lie group, then there exists a *finite central covering* $\widetilde{Q}$ of $Q$ with

$$\widetilde{Q} = T^n \times \widetilde{Q}_1 \times \cdots \times \widetilde{Q}_m$$

for some $m \geq 0$, $n \geq 0$, and where each $\widetilde{Q}_i$ is a simply connected simple Lie group. If $Q$ acts on $W$, then the homomorphism $\widetilde{Q} \to Q$ defines an action of $\widetilde{Q}$ on $W$.

COROLLARY 12.3.13. *If $Q$ acts on $W$ with $H^1(W; \mathbb{Z}^k) = 0$, then the action of $\widetilde{Q}$ lifts to every principal $T^k$-bundle over $W$.*

PROOF. $B\widetilde{Q} = BT^n \times B\widetilde{Q}_1 \times \cdots \times B\widetilde{Q}_m$. Therefore, $H^3(BQ; \mathbb{Z}^k) = 0$, and consequently, $H^2(W_{\widetilde{Q}}; \mathbb{Z}^k) \xrightarrow{j} H^2(W; \mathbb{Z}^k)$ is onto. $\square$

COROLLARY 12.3.14. *If $Q$ is connected and $H^1(W; \mathbb{Z}^k) = 0$, then the lifts of $Q$ to $P$ up to conjugation by elements of $\mathrm{M}(W, T^k)$, are in one-to-one correspondence with $\mathrm{Hom}(\pi_1(Q), \mathbb{Z}^k)$.*

PROOF. By the exact sequence (12.3.1), we have that $H^1(Q; \mathrm{M}_0(W, T^k))$ is isomorphic to $H^1(Q; \mathrm{M}(W, T^k))$ because $H^1(W; \mathbb{Z}^k) = 0$. Moreover, $H^1(Q; \mathrm{M}_0(W, T^k))$ was seen to be isomorphic to $\mathrm{Hom}(\pi_1(Q), \mathbb{Z}^k)$. $\square$

## 12.4. Examples

12.4.1. Let us work out the liftings of $S^1$ to bundle automorphisms on $S^1$-bundles over $S^2$. There is a unique, up to $S^1$-equivalence, effective $S^1$-action on $S^2$, namely, rotating about the poles $N$ and $S$. Since $S^2$ is simply connected, we have the exact sequence in Proposition 12.3.10:

$$0 \to H^2(BS^1; \mathbb{Z}) \xrightarrow{\ell} H^2(S^2{}_{S^1}; \mathbb{Z})$$
$$\xrightarrow{j} H^2(S^2; \mathbb{Z}) \longrightarrow H^3(BS^1; \mathbb{Z}) \to H^3(S^2{}_{S^1}; \mathbb{Z}).$$

But, $H^2(BS^1; \mathbb{Z}) = \mathbb{Z}$, $H^2(S^2; \mathbb{Z}) = \mathbb{Z}$, $H^3(BS^1; \mathbb{Z}) = 0$, so we get

$$0 \longrightarrow \mathbb{Z} \longrightarrow H^2(S^2{}_{S^1}; \mathbb{Z}) \xrightarrow{j} \mathbb{Z} \longrightarrow 0.$$

So, $H^2(S^2{}_{S^1}; \mathbb{Z}) = \mathbb{Z} \oplus \mathbb{Z}$.

For $S^3$, i.e., $c_1(P) = +1$, we can take the following liftings of $S^1$, a different lifting for each $n \in \mathbb{Z}$:
$$z \times (z_1, z_2) \mapsto (z^n z_1, z^{n-1} z_2).$$
These are liftings of $S^1$ because of the commutative diagram,

$$\begin{array}{ccccc}
(z_1, z_2) & \xrightarrow{S^1\backslash} & z_1/z_2 & \in S^2 = \mathbb{C}P_1 & z_1\bar{z}_1 + z_2\bar{z}_2 = 1 \\
{\cdot z}\downarrow & & {\cdot z}\downarrow & & \\
(z^n z_1, z^{n-1} z_2) & \xrightarrow{S^1\backslash} & z \cdot \left(\frac{z_1}{z_2}\right) & \text{rotation about the poles of } S^2.
\end{array}$$

For $L(m, 1)$, i.e., $c_1(P) = m$, we define $\langle z_1, z_2 \rangle \in L(m, 1)$ by taking the orbit space of the diagonal $\mathbb{Z}_m$-action on $S^3$ given by
$$e^{2\pi i \frac{k}{m}} \times (z_1, z_2) \mapsto (e^{2\pi i \frac{k}{m}} z_1, e^{2\pi i \frac{k}{m}} z_2), \quad k = 0, 1, \ldots, m-1.$$
Thus we get a different lifting of $S^1$ for each $n \in \mathbb{Z}$:
$$z \times \langle z_1, z_2 \rangle \mapsto \langle z^n z_1, z^{n-1} z_2 \rangle.$$

EXERCISE 12.4.2. If we take the ineffective $S^1$-action on $S^2$ given by $z \times \frac{z_1}{z_2} \mapsto z^p \frac{z_1}{z_2}$, then determine the liftings of $S^1$ to the Hopf fibering $S^3 \to S^2$ up to equivalence, which of these liftings of $S^1$ will be effective?

What about the product case, i.e., on $S^1 \times S^2$? We can take
$$\begin{array}{ccc}
e^{2\pi i \theta} \times (z, w) & \longrightarrow & (z, e^{2\pi i \theta} w) \\
\downarrow & & \downarrow \\
e^{2\pi i \theta} \times w & \longrightarrow & e^{2\pi i \theta} w.
\end{array}$$
Then
$$\begin{array}{ccc}
e^{2\pi i \theta} \times (z, w) & \longrightarrow & (e^{2\pi i n \theta} z, e^{2\pi i \theta} w) \\
\downarrow & & \downarrow \\
e^{2\pi i \theta} \times w & \longrightarrow & e^{2\pi i \theta} w
\end{array}$$
would give us an infinite number of spaces covering the original. These are all inequivalent liftings of $S^1$.

EXERCISE 12.4.3. Points of $\mathbb{C}P_n$ are represented by $[z_1 : z_2 : \cdots : z_{n+1}]$, the homogeneous coordinates which is the $S^1$-orbit of $(z_1, z_2, \ldots, z_{n+1}) \in S^{2n+1}$ under the Hopf map. We can take an $S^1$-action on $\mathbb{C}P_n$, say
$$z \times [z_1 : z_2 : z_3 : \cdots : z_{n+1}] \mapsto [z^m z_1 : z^{m-1} z_2 : z_3 : \cdots : z_{n+1}]$$
for definiteness. Describe the liftings of $S^1$ to $S^{2n+1}$.

EXERCISE 12.4.4. Let $Q$ be a connected Lie group acting properly on $W$, and let $P$ be a principal $G$-bundle over $W$ with $G$ connected. Let
$$H = \text{Im}(\text{ev}_*^w : \pi_1(Q, 1) \to \pi_1(W, w)),$$
and let $K$ be the kernel of $\pi_1(Q, 1) \to H$. Corresponding to $K$, there is a unique connected covering group $Q_K$ of $Q$ (whose fundamental group is $K$). Let $E$ be the image of $\text{ev}_*^u : \pi_1(G, 1) \to \pi_1(P, u)$, where $u \mapsto w$ under the bundle projection map. Then the action of $G$ lifts to $P'$, the covering space of $P$ corresponding to the subgroup $E$. This is a principal $G$-bundle over $\widetilde{W}$, the universal covering of $W$.

Show that the action of $Q$ is liftable to $P$ over $W$ if and only if the action of $Q_K$ is liftable to $P'$ over $\widetilde{W}$.

EXAMPLE 12.4.5. Consider the exact sequence of topological groups
$$1 \longrightarrow S^1 \longrightarrow U(2) \longrightarrow SO(3) \longrightarrow 1,$$
where $S^1 = \begin{bmatrix} e^{i\theta} & 0 \\ 0 & e^{i\theta} \end{bmatrix}$ is the center of $U(2)$. The question is, does there exist a global section to this principal $S^1$-bundle? Now $SO(3)$ is homeomorphic to $\mathbb{R}P^3$. We would need a global section to use factor sets to determine if there is a splitting of this group extension. Suppose there is a global section. Then $\pi_1(U(2))$ contains $\pi_1(SO(3)) = \mathbb{Z}_2$. On the other hand,
$$1 \longrightarrow SU(2) \longrightarrow U(2) \xrightarrow{\det} U(1) \longrightarrow 1$$
is also exact and splits by $z \mapsto \begin{bmatrix} z & 0 \\ 0 & 1 \end{bmatrix}$. Now $\pi_1(SU(2)) = 1$ and so $\pi_1(U(2)) = \mathbb{Z}$. Therefore, there does not exist a global section $SO(3) \to U(2)$.

EXERCISE 12.4.6. For the principal $S^1$-bundle $U(2) \to SO(3)$, let $SU(2)$ act on $W = SO(3)$ transitively. The Corollary 12.3.11 says that $Q = SU(2)$ lifts to a group of bundle automorphisms. Show directly that $(\pi \circ i)^* : H^2(W_{SO(3)}; \mathbb{Z}) \to H^2(SO(3); \mathbb{Z})$ is an isomorphism.

On the other hand, show that if $Q = SO(3)$ acting transitively on $W = SO(3)$, then this action does not lift back to a group of bundle automorphisms on $SU(2)$.

Note that we can write $U(2)$ as $U(1) \times_{\mathbb{Z}_2} SU(2)$. There are exactly two Lie group extensions of $U(1)$ by $PU(2) = SO(3)$.

EXAMPLE 12.4.7. Let $P$ be a principal $S^1$-bundle over the 2-torus so that $0 \neq [P] \in H^2(T^2; \mathbb{Z})$. Let $S^1$ be an effective $S^1$-action on the torus. This action cannot be lifted to a group of bundle automorphisms on $P$. For, a lift would generate an action of the 2-torus on the 3-dimensional nilmanifold $P$. The center of $\pi_1(P)$ is $\mathbb{Z}$ and $P$ is aspherical. This contradicts Corollary 3.1.12.

EXERCISE 12.4.8 (Generalization of Example 12.4.7). Let $P$ be a compact nilmanifold. Let $T^k$ be the maximal torus action on $P$ and $W = T^k \backslash P$. $W$ is a compact nilmanifold and if $P \neq T^k$, $\dim(W) > 0$ and center of $\pi_1(W)$ has rank greater than 0. Let $T^s$ be a torus acting effectively on $W$. Then $T^s$ cannot be lifted to a group of bundle automorphisms of $P$.

EXAMPLE 12.4.9. Let $N$ be a principal $T^k$-bundle over $M$, where $M$ is a closed aspherical manifold. Assume the rank of center $\pi_1(N) = k$. Then no nontrivial compact connected Lie group acting on $M$ can lift to bundle automorphisms on $N$.

EXAMPLE 12.4.10. Let $Q$ be a compact connected Lie group which acts smoothly, effectively preserving orientation on an oriented, connected smooth $n$-manifold $W$. For any Riemannian metric on $W$, average the metric with respect to the group $Q$ and obtain a new Riemannian metric so that $Q$ now acts as isometries on $W$. Let $P \xrightarrow{\tau} W$ be the bundle of oriented orthonormal frames of $W$. Thus, $P$ is a principal $SO(n)$-bundle over $W$. The action of $Q$ lifts to a group of automorphisms of $P$ just as in the Example 2.8.6. The lifted action is free and commutes with the $SO(n)$-action on $P$. The map $\nu : P \to Q \backslash P$ is a principal $Q$-fibering and $Q \backslash P$ inherits an $SO(n)$-action. We view the orbit mapping $\tau : Q \backslash P \xrightarrow{SO(n) \backslash} Q \backslash W$ as a Seifert fibering as in Section 4.3.

For $x \in Q\backslash P$, choose $u \in P$ and $w \in W$ such that $\nu(u) = x$, and $\tilde{\tau}(u) = w$ as in the notation of Section 4. Let us determine the fiber over $w^* = \bar{\nu}(w)$. Let $Q_w$ be the isotropy at $w$. Then $\tilde{\tau}^{-1}(w) = \mathrm{SO}(n)$, and $Q_w$ acts freely on it. Thus, the fiber over $w^*$ is isomorphic to $\mathrm{SO}(n)/Q_w$.

If $Q$ acts freely on $W$, $Q\backslash W$ is a smooth Riemannian manifold with $Q\backslash P$ isomorphic to the oriented normal frame bundle over $Q\backslash W$. If $Q$ does not act freely on $W$, it is not unreasonable to call $Q\backslash P$ and the orbit mapping $Q\backslash P \to Q\backslash W$ the oriented frame bundle over the quotient space $Q\backslash W$.

We have seen in Subsection 12.2.5 how to construct a principal $\mathrm{SO}(n)$-bundle $P'$ over $W_Q$ so that $P$ is the pullback of $P'$ via the map

$$W \xrightarrow{i} EQ \times W \xrightarrow{\pi} EQ \times_Q W = W_Q,$$

with $P'$ being $Q\backslash \pi_2^*(P)$. As mentioned in Subsection 12.3.1 and Section 10.2, this necessary condition of writing $P$ as the pullback of a principal $G$-bundle over the Borel space $W_Q$ is also sufficient to lift $Q$ to a group of bundle automorphisms of $P$ whenever $G = T^k$ and $Q$ is discrete acting properly on $W$. For what we have described above, where $n = 2$, $G = \mathrm{SO}(2) = T^1$, the principal oriented orthonormal frame bundle coincides with the unit tangent bundle over 2-manifolds. We have already extensively described these liftings in Chapters 14 and 15, and the reader may now wish to reflect back on that.

CHAPTER 13

# Deformation Spaces

### 13.1. Uniformizing groups

In a general Seifert Construction, it may happen that $\rho : Q \to \mathrm{TOP}(W)$ has an image in a subgroup that has geometric or topological significance. This means that it is likely that the associated Seifert Constructions inherit some of these properties. Let $\mathcal{U}$ be a subgroup of $\mathrm{TOP}_G(G \times W)$ containing $\ell(G)$. Put $\widehat{\mathcal{U}} = \mathcal{U} \cap (\mathrm{M}(W, G) \rtimes \mathrm{Inn}(G))$ and $\overline{\mathcal{U}} = \mathcal{U}/\widehat{\mathcal{U}}$ so that

$$\begin{array}{ccccccccc} 1 & \longrightarrow & \widehat{\mathcal{U}} & \longrightarrow & \mathcal{U} & \longrightarrow & \overline{\mathcal{U}} & \longrightarrow & 1 \\ & & \cap \downarrow & & \cap \downarrow & & \cap \downarrow & & \\ 1 & \longrightarrow & \mathrm{M}(W, G) \rtimes \mathrm{Inn}(G) & \longrightarrow & \mathrm{TOP}_G(G \times W) & \longrightarrow & \mathrm{Out}(G) \times \mathrm{TOP}(W) & \longrightarrow & 1 \end{array}$$

commutes.

If the image of $\theta : \Pi \to \mathrm{TOP}_G(G \times W)$ lies in $\mathcal{U}$, we say that the *uniformizing group* (see Definition 4.6.1) has been reduced to $\mathcal{U}$ for $\Pi$. The group $\mathcal{U}$ can then be used to study extra structures on the Seifert fiber space $\theta(\Pi)\backslash(G \times W)$.

We have already seen in Section 7.8 that if $W$ is a smooth manifold, $\Gamma$ special and $\rho : Q \to \mathrm{Diff}(W)$, then any injective Seifert Construction can be done smoothly in

$$\mathrm{Diff}_G(G \times W) = \mathcal{C}(W, G) \rtimes (\mathrm{Aut}(G) \times \mathrm{Diff}(W)),$$

where $\mathrm{Diff}_G(G \times W) = \mathrm{TOP}_G(G \times W) \cap \mathrm{Diff}(G \times W)$.

Furthermore, existence, uniqueness, and rigidity also hold because we may use a smooth partition of unity in the proof for the vanishing of the necessary cohomology groups. To obtain existence of an injective Seifert Construction in $\mathcal{U}$, one has to verify the conditions in Definition 4.6.1.

If $\mathcal{U}$ is much smaller than $\mathrm{TOP}_G(G \times W)$, uniqueness and rigidity are not likely to hold. The set of all homomorphisms of $\Pi$ into $\mathcal{U}$ which belong to the same conjugacy class in $\mathrm{TOP}_G(G \times W)$ is a deformation space of that conjugacy class. We shall illustrate these concepts by describing in some detail the deformation spaces for some of the 3-dimensional geometries. All of the 3-dimensional Seifert manifolds admit a geometric structure. For most, this means there is a reduction from $\mathrm{TOP}_G(P)$ into $\mathcal{U}$, where $\mathcal{U}$ is the isometry group of the universal covering of the manifold. We shall show that the associated deformation spaces are themselves Seifert fiber spaces over well-known deformation spaces for 2-dimensional geometries. Some material is taken from [**KLR86**].

## 13.2. $\widetilde{\mathrm{PSL}}(2,\mathbb{R})$-geometry

13.2.1. When $W$ is homeomorphic to $\mathbb{R}^2$ in a Seifert Construction, $\rho(Q)$ is a discrete subgroup of $\mathrm{TOP}(\mathbb{R}^2)$, acting properly on $\mathbb{R}^2$, then the group $\rho(Q)$ can be conjugated in $\mathrm{TOP}(\mathbb{R}^2)$ to a group which acts as isometries on $\mathbb{R}^2$ with the usual Euclidean metric (e.g., $\rho(Q)$ is crystallographic) or as isometries on $\mathbb{H}^2$ with the usual hyperbolic metric. In the former case, we say $\rho(Q)$ is isomorphic to a (Euclidean) crystallographic group, and in the latter, $\rho(Q)$ is isomorphic to a hyperbolic group. We write $\mathbb{R}^2$ with the usual Euclidean metric (respectively, hyperbolic metric) as $\mathbb{E}^2$ (respectively, $\mathbb{H}^2$). If $Q \backslash \mathbb{R}^2$ is compact, then any two embeddings of $Q$ are conjugate in $\mathrm{TOP}(\mathbb{R}^2)$.

Thus, if we reduce the uniformizing group $\mathrm{TOP}_G(G \times \mathbb{R}^2)$ to $\mathcal{U}$, where at least $\mathrm{TOP}(\mathbb{R}^2)$ is replaced by $\mathrm{Isom}(\mathbb{E}^2)$ or $\mathrm{Isom}(\mathbb{H}^2)$, we would expect to find a rich deformation theory for Seifert Constructions modeled on $G \times \mathbb{R}^2$. We will take $G = \mathbb{R}$, and $\overline{\mathcal{U}} = \mathrm{Isom}(\mathbb{E}^2)$ or $\mathrm{Isom}(\mathbb{H}^2) \subset \mathrm{TOP}(\mathbb{R}^2)$ so that the model space is $\mathbb{R} \times \mathbb{R}^2$ or $\mathbb{R} \times \mathbb{H}^2$.

For each central extension $0 \to \mathbb{Z} \to \Pi \to Q \to 1$ with $Q$ cocompact hyperbolic, the group $\Pi$ can be embedded by $\theta : \Pi \to \mathrm{TOP}_\mathbb{R}(\mathbb{R} \times \mathbb{H}^2)$ so that it is topologically and/or smoothly rigid. If $\Pi$ is torsion free, $\Pi \backslash (\mathbb{R} \times \mathbb{H}^2)$ is a classical closed Seifert 3-manifold and with a unique $S^1$-action up to equivalence. The $S^1$-orbit space or base space is a 2-dimensional orbifold homeomorphic to $Q \backslash \mathbb{H}^2$. The product $\mathbb{R} \times \mathbb{H}^2$ carries several geometries so that the Riemannian metric induced on $\mathbb{R} \times \mathbb{H}^2 \to \mathbb{H}^2$ is the hyperbolic metric. We will examine first the Riemannian metric on $\widetilde{\mathrm{PSL}}(2,\mathbb{R})$ and then later an $\mathbb{R}$-invariant Lorentz metric on $\mathbb{R} \times \mathbb{H}^2$.

13.2.2. The 3-dimensional Lie group of all $2 \times 2$ real matrices with determinant 1 is denoted by $\mathrm{SL}(2,\mathbb{R})$. The group of real Möbius transformations is $\mathrm{PSL}(2,\mathbb{R})$, and it is doubly covered by $\mathrm{SL}(2,\mathbb{R})$. The universal covering of $\mathrm{PSL}(2,\mathbb{R})$ (and hence of $\mathrm{SL}(2,\mathbb{R})$) is $\widetilde{\mathrm{PSL}}(2,\mathbb{R})$. For simplicity we use the notation $P$ for $\mathrm{PSL}(2,\mathbb{R})$ and $\widetilde{P}$ for $\widetilde{\mathrm{PSL}}(2,\mathbb{R})$. Topologically $\widetilde{P}$ is homeomorphic to $\mathbb{R} \times \mathbb{H}$:

$$P = \mathrm{PSL}(2,\mathbb{R}) \approx S^1 \times \mathbb{H},$$
$$\widetilde{P} = \widetilde{\mathrm{PSL}}(2,\mathbb{R}) \approx \mathbb{R} \times \mathbb{H}.$$

Therefore, if we use

$$\mathrm{TOP}_\mathbb{R}(\mathbb{R} \times \mathbb{H}) = \mathrm{M}(\mathbb{H}, \mathbb{R}) \rtimes (\mathrm{GL}(1,\mathbb{R}) \times \mathrm{TOP}(\mathbb{H}))$$

as our uniformizing group, we will not be able to distinguish the geometries of $\widetilde{\mathrm{PSL}}(2,\mathbb{R})$, $\mathbb{R} \times \mathbb{H}$, $\mathbb{R} \times \mathbb{R}^2$ and Nil from that of $\widetilde{P}$.

The Lie group $\mathrm{PSL}(2,\mathbb{R})$ can be viewed as the unit tangent bundle of the hyperbolic space $\mathbb{H}$, and it has a natural Riemannian metric. This metric pulls back to a Riemannian metric on $\widetilde{P}$. It turns out that this metric is right invariant. That is, all right translations by elements of the group are isometries. (Of course, one can change all the right actions to left actions.) Furthermore, the isometry group is

$$\mathrm{Isom}(\widetilde{P}) = (\mathbb{R} \times_\mathbb{Z} \widetilde{P}) \rtimes \mathbb{Z}_2,$$

where $\mathbb{R}$ is a subgroup of $\widetilde{P}$ containing the center $\mathbb{Z}$, acting as left translations. These two actions commute with each other, and $\widehat{\ell}(z) = \widehat{r}(z^{-1})$ for $z \in \mathbb{Z}$, the center of $\widetilde{P}$. The finite group $\mathbb{Z}_2$ is generated by the reflection about the $y$-axis. In fact, any orientation reversing isometry of period 2 will do. While it reverses the orientation of the base space $\mathbb{H}$, it also reverses the orientation of the fiber $\mathbb{R}$. Consequently, it preserves the orientation of $\widetilde{P}$.

We take the subgroup $\mathbb{R}$ described above as our $G$. Smoothly, $\widetilde{P} = \mathbb{R} \times \mathbb{H}$. Therefore, we have
$$G = \mathbb{R}, \quad W = \mathbb{H}.$$
Since $\mathbb{R} \times_{\mathbb{Z}} \widetilde{P}$ commutes with the left translation $G = \mathbb{R}$ and the generator of $\mathbb{Z}_2$ is an inversion of $\mathbb{R}$, $\mathrm{Isom}(\widetilde{P}) = (\mathbb{R} \times_{\mathbb{Z}} \widetilde{P}) \rtimes \mathbb{Z}_2$ lies inside $\mathrm{TOP}_{\mathbb{R}}(\mathbb{R} \times \mathbb{H})$. To make the presentation clearer, we use only the connected component of $\mathrm{Isom}(\widetilde{P})$. So, let us take
$$\mathcal{U} = \mathrm{Isom}_0(\widetilde{P}) = \mathbb{R} \times_{\mathbb{Z}} \widetilde{P}$$
so that we have the commuting diagram

$$\begin{array}{ccccccccc}
1 & \longrightarrow & \mathbb{R} & \longrightarrow & \mathcal{U} = \mathbb{R} \times_{\mathbb{Z}} \widetilde{P} & \longrightarrow & P & \longrightarrow & 1 \\
 & & \cap & & \cap & & \cap & & \\
1 & \longrightarrow & \mathrm{M}(\mathbb{H}, \mathbb{R}) & \longrightarrow & \mathrm{TOP}_{\mathbb{R}}(\mathbb{R} \times \mathbb{H}) & \longrightarrow & \mathrm{GL}(1, \mathbb{R}) \times \mathrm{TOP}(\mathbb{H}) & \longrightarrow & 1.
\end{array}$$

For this case, $Q$ is a cocompact discrete subgroup of $P$ and has a well-known presentation
$$Q = \langle \overline{x}_1, \ldots, \overline{x}_g, \overline{y}_1, \ldots, \overline{y}_g, \overline{w}_1, \ldots, \overline{w}_p \mid \overline{w}_j{}^{\alpha_j} = 1, \ \prod_{j=1}^{p} \overline{w}_j \prod_{i=1}^{g} [\overline{x}_i, \overline{y}_i] = 1 \rangle$$
for $p \geq 0$, $g \geq 0$ and all $\alpha_j \geq 2$. It is also required that the Euler characteristic of $Q$, defined as
$$\chi(Q) = (2 - 2g) - \sum_{j=1}^{p} \left(1 - \frac{1}{\alpha_j}\right),$$
satisfies $\chi(Q) < 0$. It is our intention to characterize those $[\Pi] \in H^2(Q; \mathbb{Z})$ which embed in $\mathcal{U}$ and to determine their deformation spaces.

13.2.3. For any subgroup $Q$ of $\mathrm{Isom}_0(\mathbb{H})$, one can pullback (see Subsection 5.3.1) the above extension via $Q \hookrightarrow \mathrm{Isom}_0(\mathbb{H})$ to get $\widetilde{Q}$ so that the diagram

$$\begin{array}{ccccccccc}
1 & \longrightarrow & \mathbb{R} & \longrightarrow & \widetilde{Q} & \longrightarrow & Q & \longrightarrow & 1 \\
 & & \downarrow = & & \downarrow & & \downarrow \cap & & \\
1 & \longrightarrow & \mathbb{R} & \longrightarrow & \mathrm{Isom}_0(\widetilde{P}) & \longrightarrow & \mathrm{Isom}_0(\mathbb{H}) & \longrightarrow & 1
\end{array}$$

commutes. Thus, $\mathbb{R}$ becomes a $Q$-module.

LEMMA 13.2.4. *Let $Q$ be a cocompact discrete subgroup of $\mathrm{Isom}_0(\mathbb{H})$. Then*

(1) $H^2(Q; \mathbb{R}) = \mathbb{R}$.
(2) *The class $[\widetilde{Q}] \in H^2(Q; \mathbb{R})$ is nonzero (that is, the exact sequence does not split).*

PROOF. (1) Let $1 \to \mathbb{Z} \to \widetilde{P} \to P \to 1$ be the universal covering projection, and let $1 \to \mathbb{Z} \to \widehat{Q} \to Q \to 1$ be the pullback of this exact sequence via $Q \hookrightarrow P$. Then $\mathbb{Z} \subset \widehat{Q}$ sits in $\mathbb{R} \times_{\mathbb{Z}} \widetilde{P} = \mathrm{Isom}_0(\widetilde{P})$ as the center. Denote the inclusion of $\mathbb{Z} \hookrightarrow \mathbb{R} \subset \mathrm{Isom}_0(\widetilde{P})$ by $i$. Then

$$\begin{array}{ccccccccccc}
1 & \longrightarrow & \mathbb{Z} & \longrightarrow & \widehat{Q} & \longrightarrow & Q & \longrightarrow & 1 & & [\widehat{Q}] \in H^2(Q; \mathbb{Z}) \\
 & & \downarrow i & & \downarrow & & \downarrow = & & & & \downarrow i_* \\
1 & \longrightarrow & \mathbb{R} & \longrightarrow & \widetilde{Q} & \longrightarrow & Q & \longrightarrow & 1 & & [\widetilde{Q}] \in H^2(Q; \mathbb{R})
\end{array}$$

is commutative so that $i_*[\widehat{Q}] = [\widehat{Q}]$. By Selberg's Lemma 6.1.8, $Q$ contains a torsion-free normal subgroup $Q_0$ of finite index. Thus $H^2(Q_0; \mathbb{R}) = H^2(Q_0\backslash\mathbb{H}^2; \mathbb{R}) \cong \mathbb{R}$. Then, by transfer, $H^2(Q; \mathbb{R}) = \mathbb{R}$, since $Q/Q_0$ acts on $Q_0\backslash\mathbb{H}^2$ preserving orientation. Since $H^2(Q; \mathbb{Z})$ is finitely generated, it follows that $H^2(Q; \mathbb{Z}) = \mathbb{Z} \oplus$ Torsion by the universal coefficient theorem, and the fact that $i_* : H^2(Q; \mathbb{Z}) \to H^2(Q; \mathbb{R})$ is given by $\otimes \mathbb{R}$. Thus the elements of infinite order inject and those of finite order are in the kernel.

(2) The normal torsion-free subgroup $Q_0$ of finite index in $Q$ is the fundamental group of a closed surface of genus $g > 1$. For otherwise, $\chi(Q_0) \geq 0$. Therefore, we can assume, without loss of generality, that $Q \subset P$ is torsion free. $\widehat{Q}$ is the fundamental group of the unit tangent bundle $\widehat{Q}\backslash\widetilde{P}$ of the surface $Q\backslash\mathbb{H}$ by Subsection 14.8.7. In this case, the Euler characteristic of $Q$, $2 - 2g \neq 0$, is the characteristic class of the principal $S^1$-bundle $\widehat{Q}\backslash\widetilde{P} \to Q\backslash\mathbb{H}$, and is also equal to the cohomology class $[\widehat{Q}] \in H^2(Q; \mathbb{Z}) \cong \mathbb{Z}$ of the extension $\widehat{Q}$. Therefore, $[\widehat{Q}]$ is nonzero in $H^2(Q; \mathbb{R}) = H^2(Q; \mathbb{Z}) \otimes \mathbb{R}$. We also point out, in this case, that $[\widehat{Q}] = e(\widehat{Q})$; see below for a definition of $e(\widehat{Q})$. □

13.2.5 (Euler number). Let $i : \mathbb{Z} \hookrightarrow \mathbb{R}$ be the standard inclusion. For each central extension $0 \to \mathbb{Z} \to \Pi \to Q \to 1$, there is associated a rational invariant called the *Euler number* of $\Pi$, and it is denoted by $e(\Pi)$. Recall that it can be defined in terms of a presentation of $\Pi$. Let

$$\Pi = \langle \tilde{x}_1, \ldots, \tilde{x}_g, \quad \tilde{y}_1, \ldots, \tilde{y}_g, \tilde{w}_1, \ldots, \tilde{w}_p, \tilde{z} \mid \tilde{z} \text{ central},$$
$$\tilde{w}_j^{\alpha_j} = \tilde{z}^{-\beta_j}, \prod_{j=1}^p \tilde{w}_j \prod_{i=1}^g [\tilde{x}_i, \tilde{y}_i] = \tilde{z}^b \rangle.$$

Then $e(\Pi) = -\left(b + \Sigma \frac{\beta_j}{\alpha_j}\right)$ and $|e(\Pi)|$ is an invariant of the isomorphism class of $\Pi$. In Proposition 15.2.9 and Theorem 15.2.10, it is shown that under the homomorphism

$$i_* : H^2(Q; \mathbb{Z}) \longrightarrow H^2(Q; \mathbb{R}) \cong \mathbb{R}$$

induced by $i$, we have $i_*[\Pi] = L \cdot e(\Pi)$, where $L = \text{lcm}[\alpha_1, \ldots, \alpha_p]$ and $Q$ is infinite. Thus $[\Pi]$ has infinite order in $H^2(Q; \mathbb{Z})$ if and only if $e(\Pi) \neq 0$.

We now characterize the cocompact orbifold groups modeled on $\widetilde{P}$.

THEOREM 13.2.6. *An abstract group $\Pi$ can be embedded into $\text{Isom}_0(\widetilde{P})$ as a cocompact discrete subgroup if and only if $\Pi$ is a central extension of $\mathbb{Z}$ by a discrete cocompact orientation preserving hyperbolic group $Q$ (so that $1 \to \mathbb{Z} \to \Pi \to Q \to 1$ is exact) and $[\Pi] \in H^2(Q; \mathbb{Z})$ has infinite order. Further, if this is the case, the subgroup $\mathbb{Z}$ is the center of $\Pi$, and in any discrete embedding, the image of $\mathbb{Z}$ is $\Pi \cap \mathbb{R}$, where $\mathbb{R} \subset \mathbb{R} \rtimes_{\mathbb{Z}} \widetilde{P} = \text{Isom}_0(\widetilde{P})$.*

PROOF. We now show that $\Pi \cap \mathbb{R}$ is nontrivial. Suppose $\Pi$ is a cocompact discrete subgroup of $\text{Isom}_0(\widetilde{P})$. The subgroup $\mathbb{R}$ of $\text{Isom}_0(\widetilde{P})$ is the radical (maximal connected normal solvable subgroup) of $\text{Isom}_0(\widetilde{P})$ and the quotient $\text{Isom}_0(\mathbb{H}) = \text{PSL}(2, \mathbb{R})$ has no compact factor. A theorem of Wang (see [**Rag72**, 8.27]) says that the image $Q$ of $\Pi$ in $\text{Isom}_0(\mathbb{H})$ is a lattice so that $Q$ is a discrete cocompact orientation preserving hyperbolic group. We now show that $\Pi \cap \mathbb{R}$ is nontrivial. Suppose not. Then $\Pi$ is isomorphic to $Q$ and hence, it has $\mathbb{R}$-cohomological dimension 2. However, since $\Pi$ is cocompact in $\text{Isom}_0(\widetilde{P})$, its $\mathbb{R}$-cohomological dimension is 3. This contradiction shows that $\Pi \cap \mathbb{R} = \mathbb{Z}$. Thus $\Pi$ is of the form $1 \to \mathbb{Z} \to \Pi \to Q \to 1$. Clearly, $\mathbb{Z}$ is the center of $\Pi$ since $Q$ is centerless.

We shall now sketch why $[\Pi]$ must be of infinite order in $H^2(Q;\mathbb{Z})$. By Selberg's Lemma 6.1.8, there exists a normal subgroup $Q_0$ of $Q$ which is torsion free of finite index. Let $1 \to \mathbb{Z} \to \Pi_0 \to Q_0 \to 1$ be the pullback of the above exact sequence via $Q_0 \hookrightarrow Q$. But if $[\Pi]$ has finite order, one can take $Q_0$ so that $[\Pi_0]$ has order 0 so that $\Pi_0 = \mathbb{Z} \times Q_0$. We claim that this group does not embed discretely into $\mathrm{Isom}_0(\widetilde{P})$. Choose a standard presentation for $Q_0$:

$$Q_0 = \langle a_1, b_1, \ldots, a_g, b_g \mid \prod_{i=1}^{g}[a_i, b_i] = 1\rangle.$$

Since $Q_0 \subset \mathrm{PSL}(2,\mathbb{R})$, we may think of the $a_i$, $b_i$ as elements of $\mathrm{PSL}(2,\mathbb{R})$. In $\mathrm{Isom}_0(\widetilde{P})$, these elements lift to $\{(t_{a_i}, a'_i)\}$, $\{(t_{b_i}, b'_i)\} \in \mathbb{R} \times_\mathbb{Z} \widetilde{P}$. These are unique up to the center of $\widetilde{P}$. Since $\prod_{i=1}^{g}[(t_{a_i}, a'_i), (t_{b_i}, b'_i)] = (0, t^{2g-2})$ by Theorem 14.8.8 or [**RV81**], it is nonzero. Since $\Pi_0 = \mathbb{Z} \times Q_0 \subset \mathrm{Isom}_0(\widetilde{P})$, this relation $2g-2$ would have to be 0, since the genus $g$ of $Q_0$ is greater than 1. This gives a contradiction and so $[\Pi]$ must have infinite order.

Conversely, suppose $1 \to \mathbb{Z} \to \Pi \to Q \to 1$ is exact, where $Q$ is a cocompact discrete subgroup of $\mathrm{Isom}_0(\mathbb{H}) = \mathrm{PSL}(2,\mathbb{R})$; and $[\Pi] \in H^2(Q;\mathbb{Z})$ has infinite order. Then with the natural inclusion $i : \mathbb{Z} \hookrightarrow \mathbb{R}$ and the induced homomorphism $i_* : H^2(Q;\mathbb{Z}) \to H^2(Q;\mathbb{R})$, $i_*[\Pi]$ is the pushout $[\mathbb{R}\Pi] \in H^2(Q;\mathbb{R})$ (see Subsection 5.3.4) and is nonzero. By Lemma 13.2.4, $[\widetilde{Q}] \in H^2(Q;\mathbb{R})$ is also nonzero. Therefore, there exists $\epsilon \in \mathbb{R}$ so that $(\epsilon \circ i)_*[\Pi] = \epsilon_*[\mathbb{R}\Pi] = [\widetilde{Q}]$. This implies that there exists a homomorphism of $\Pi$ into $\widetilde{Q}$ with the diagram

$$\begin{array}{ccccccccc}
1 & \to & \mathbb{Z} & \to & \Pi & \to & Q & \to & 1 \\
& & \downarrow i & & \downarrow & & \downarrow = & & \\
1 & \to & \mathbb{R} & \to & \mathbb{R}\Pi & \to & Q & \to & 1 \\
& & \downarrow \epsilon & & \downarrow & & \downarrow = & & \\
1 & \to & \mathbb{R} & \to & \widetilde{Q} & \to & Q & \to & 1
\end{array}$$

commutative and with injective vertical maps. Since $\mathbb{Z}$ and $Q$ acts properly discontinuously with compact quotient on $\mathbb{R}$ and $\mathbb{H}$, respectively, $\Pi$ is cocompact and discrete. This completes the proof. $\square$

COROLLARY 13.2.7. *Let $\rho : Q \to \mathrm{PSL}(2,\mathbb{R})$ be a discrete cocompact subgroup. For an extension $1 \to \mathbb{Z} \to \Pi \to Q \to 1$, there exists an injective homomorphism $\theta : \Pi \to \mathrm{Isom}(\widetilde{P})$ so that the diagram*

(13.2.1)
$$\begin{array}{ccccccccc}
1 & \to & \mathbb{Z} & \to & \Pi & \to & Q & \to & 1 \\
& & \downarrow \epsilon & & \downarrow \theta & & \downarrow \rho & & \\
1 & \to & \mathbb{R} & \to & \mathrm{Isom}_0(\widetilde{P}) & \to & \mathrm{PSL}(2,\mathbb{R}) & \to & 1
\end{array}$$

*commutes if and only if $[\Pi] \in H^2(Q;\mathbb{Z})$ has infinite order.*

COROLLARY 13.2.8 (Structure). *Let $M$ be a closed orbifold modeled on $(\mathrm{Isom}_0(\widetilde{P}), \widetilde{P})$-geometry. Then $M$ is an orientable closed Seifert orbifold over a hyperbolic base with $e(M) \neq 0$.*

(Realization). *Let $M$ be a compact orientable injective Seifert orbifold with an orientable hyperbolic base orbifold. Then $M$ admits an $(\mathrm{Isom}_0(\widetilde{P}), \widetilde{P})$-geometry if and only if $e(M) \neq 0$.*

EXERCISE 13.2.9. 1. Consider the diagram in Subsection 13.2.3. With fixed $\epsilon$ and $\rho$, how many $\theta$'s are there to make the diagram commutative? Such maps are classified by $H^1(Q;\mathbb{R})$; see Theorem 5.7.2. Since the action of $Q$ on $\mathbb{R}$ is trivial,

$$H^1(Q;\mathbb{R}) = \mathbb{R}^{2g},$$

where $2g$ is the first Betti number of the group $Q$. Compare this with

$$H^i(Q;\mathrm{M}(\mathbb{H},\mathbb{R})) = 0 \quad (i \geq 1),$$

so that, for any extension $1 \to \mathbb{Z} \to \Pi \to Q \to 1$, a homomorphism $\Pi \to \mathrm{TOP}_\mathbb{R}(\mathbb{R} \times \mathbb{H})$ exists and is unique, up to conjugation by elements of $\mathrm{M}(\mathbb{H},\mathbb{R})$.

2. Let $Q \subset P$ be a surface group. Suppose $\Pi \subset \mathrm{Isom}_0(\widetilde{P}) = \mathbb{R} \times_\mathbb{Z} \widetilde{P}$ is a subgroup such that $Z = \Pi \cap \mathbb{R} \cong \mathbb{Z}$, $\Pi/Z = Q$. Then $\Pi$ is *commensurable* (in the sense that both contain subgroups of finite index which are isomorphic) with $\widehat{Q}$, the fundamental group of the unit tangent bundle of the surface $Q\backslash\mathbb{H}^2$.

3. A surface group cannot embed into $\widetilde{P}$.

## 13.3. Lorentz structures and $\widetilde{\mathrm{PSL}}(2,\mathbb{R})$-geometry

The spaces admitting $\widetilde{\mathrm{PSL}}(2,\mathbb{R})$-geometry have another interesting geometric structure. Here is a more explicit description of our problem. Let us denote $\widetilde{\mathrm{PSL}}(2,\mathbb{R})$ by $P_\infty$. The space $P_\infty$ is the universal covering group of $P_1 = \mathrm{PSL}(2,\mathbb{R})$. Topologically $P_\infty$ is homeomorphic to $\mathbb{R} \times \mathbb{H}$,

$$\begin{aligned} P_1 &= \mathrm{PSL}(2,\mathbb{R}), \\ P_\infty &= \widetilde{\mathrm{PSL}}(2,\mathbb{R}). \end{aligned}$$

Consider the *indefinite* metric of signature $++--$ on $\mathbb{R}^4$. The unit sphere of this space is

$$\begin{aligned} S^{1,2} &= \{(x,y) \mid x,y \in \mathbb{R}^2, \ |x|^2 - |y|^2 = 1\} \\ &\approx O(2,2)/O(1,2). \end{aligned}$$

The linear map of $\mathbb{R}^4$, defined by the matrix

$$\begin{bmatrix} 1 & 0 & 1 & 0 \\ 0 & 1 & 0 & 1 \\ 0 & -1 & 0 & 1 \\ 1 & 0 & -1 & 0 \end{bmatrix},$$

transforms $S^{1,2}$ to

$$\begin{aligned} P_2 &= \{(x,y,z,u) \in \mathbb{R}^4 \mid xz - yu = 1\} \\ &= \mathrm{SL}_2\mathbb{R}. \end{aligned}$$

Thus $P_2$ has a complete *Lorentz metric* of signature $+,-,-$ and *constant sectional curvature* 1.

A space is called a *Lorentz orbifold* if it is the quotient of $P_\infty$ by a discrete group of Lorentz isometries acting *properly discontinuously*. One can show that such a group contains normal subgroups of finite index which act freely. A Lorentz orbifold for which the discrete group acts freely is called a *Lorentz space form*. The Lorentz structure on a space form is nonsingular and it has $P_\infty$ as its (metric) universal covering. Then $1 \to \mathbb{Z} \to P_\infty \to P_1 \to 1$ is actually a central extension with $\mathbb{Z}$ being the entire center of $P_\infty$. It turns out that the identity component of $\mathrm{Isom}(P_\infty)$ is $(P_\infty \times P_\infty)/\mathbb{Z}$ where $\mathbb{Z}$ is the diagonal central subgroup corresponding

to the center of each of the $P_\infty$-factors. The action $P_\infty \times P_\infty$, as isometries, on $P_\infty$ is given by
$$(\alpha, \beta) \cdot x = \alpha \cdot x \cdot \beta^{-1}.$$
Moreover, $\mathrm{Isom}_0(P_\infty) = P_\infty \times_\mathbb{Z} P_\infty$ has index 4 in $\mathrm{Isom}(P_\infty)$. These Lorentz space forms are analogous to the complete spherical space forms in the Riemannian case.

Let us describe some obvious ones which turn out to be *homogeneous* in the sense that $\mathrm{Isom}(M)$ acts *transitively* on $M$. Take $\Pi \subset P_\infty \times e \subset P_\infty \times_\mathbb{Z} P_\infty$ as a discrete subgroup. Then surely the centralizer of $\Pi$ in $\mathrm{Isom}(P_\infty)$, $C_{\mathrm{Isom}(P_\infty)}(\Pi)$, contains $e \times P_\infty$ in $(P_\infty \times_\mathbb{Z} P_\infty)$. In fact, it is exactly $e \times P_\infty$ (unless $\Pi \approx \mathbb{Z}$ and sits in the center of $P_\infty \times e$ in $\mathrm{Isom}(P_\infty)$). Such groups $\Pi$ are classified in [**RV81**] (see Theorem 14.8.8) and are certain Seifert manifolds over a hyperbolic base, for there is an obvious $S^1$-action on $\Pi \backslash P_\infty$ induced from $e \times P_\infty \subset \mathrm{Isom}_0(P_\infty)$. A surprising fact is that *all homogeneous Lorentz orbifolds are* actually *homogeneous Lorentz space forms and coincide with those just described above*; see [**KR85**, §10].

If $\Pi \subset \mathrm{Isom}(P_\infty)$ so that $M = \Pi \backslash P_\infty$ is compact then it is shown in [**KR85**, §7] that *M is homeomorphic to an orientable Seifert orbifold over a hyperbolic base*. However, the connections between the Seifert structure and the Lorentz structure are unclear. This is due to the fact that $\mathrm{Isom}(P_\infty)$ does not act properly on $P_\infty$. By selecting a maximal subgroup of $\mathrm{Isom}(P_\infty)$ which acts properly on $P_\infty$, these two disparate structures can be related.

The subgroup
$$J(P_\infty) = (P_\infty \times_\mathbb{Z} \times \mathbb{R}) \rtimes \mathbb{Z}_2 \subset \mathrm{Isom}(P_\infty),$$
where $\mathbb{Z}_2$ reverses the orientation of time $(= \mathbb{R})$ and space $(= \mathbb{H}^2 = \mathbb{P}_\infty/\mathbb{R})$ at the same time is the same Lie group as the $\mathrm{Isom}(P_\infty)$ in the Riemannian case. A *Lorentz orbifold* (respectively, *space form*) $M = \Pi \backslash P_\infty$, where $\Pi \subset J(P_\infty) \subset \mathrm{Isom}(P_\infty)$ is called *standard*. Therefore the standard Lorentz orbifolds and space forms coincide with the orbifolds and space forms of the $\widetilde{\mathrm{PSL}}(2,\mathbb{R})$-geometry. A Lorentz space form is homogeneous if the full group of Lorentz isometries acts transitively. It is known that homogeneous Lorentz space forms $M$ admit nonstandard complete Lorentz structures if $H^1(M; \mathbb{R}) \neq 0$; see [**Gol85**].

The *Seifert fibering*, in the theorem below, on $M$ descends from the $\mathbb{R}$-action by the second $\mathbb{R}$-factor in $J(P_\infty)$ on $P_\infty$. The following are due to Kulkarni and Raymond and are stated here, for simplicity, in the *closed cases*.

THEOREM 13.3.1 ([**KR85**, 8.5] (Structure). *A compact standard Lorentz space form (respectively, Lorentz orbifold) is an orientable Seifert manifold $M$ (respectively, Seifert orbifold) over a hyperbolic base $B$ with $e(M) \neq 0$.*

(Realization). *Let $M$ be a compact orientable Seifert manifold (respectively, orientable Seifert orbifold) over a hyperbolic base with $e(M) \neq 0$. Then, $M$ admits a structure of a standard Lorentz space form (respectively, Lorentz orbifold).*

We remark that the orbifold part breaks into two separate cases. If all fibers are $\approx S^1$, then *all* closed orientable Seifert *manifolds* with $e(M) \neq 0$ appear as the underlying spaces of Lorentz *orbifolds* (and conversely). This is similar to having the topological sphere appear as a 2-dimensional hyperbolic orbifold. If some fibers are arcs, then the Lorentz orbifolds are homeomorphic to connected sums of lens spaces (including $S^3$ and $S^2 \times S^1$). We should also mention that the statements in Theorem 13.3.1 are for the full $J(P_\infty)$ and not the connected component of the identity, and they correspond to the cocompact discrete subgroups of $\mathrm{Isom}(\widetilde{P})$ in the Riemannian case instead of $\mathrm{Isom}_0(\widetilde{P})$ as described in the preceding section. See [**KR85**, §8,9] for details and treatment of cases other than the compact ones.

## 13.4. Deformation spaces for $\widetilde{\mathrm{PSL}}(2,\mathbb{R})$-geometry

DEFINITION 13.4.1. $\mathcal{U}$ is a uniformizing group as in Section 13.1. Let $\mathcal{R}(\Pi;\mathcal{U})$ be the space of all injective homomorphisms $\theta : \Pi \to \mathcal{U}$ such that $\theta(\Pi)$ is cocompact acting properly and is discrete in $\mathcal{U}$. We topologize $\mathcal{R}(\Pi;\mathcal{U})$ as a subset of $\mathcal{U}^\Pi$. In general, if $\mathcal{U}$ is a Lie group, then $\mathcal{R}(\Pi;\mathcal{U})$ will be a real analytic space.

The space $\mathcal{R}(\Pi;\mathcal{U})$ is called the *space of discrete representations* of $\Pi$ into $\mathcal{U}$ or the *Weil space of* $(\Pi;\mathcal{U})$. When there is no confusion likely, we denote $\mathcal{R}(\Pi;\mathcal{U})$ simply by $\mathcal{R}(\Pi)$.

Recall that $\mu$ denotes conjugation. The inner-automorphisms group $\mathrm{Inn}(\mathcal{U})$ acts on $\mathcal{R}(\Pi)$ from the left by

$$\mu(u) \cdot \theta = \mu(u) \circ \theta$$

for $u \in \mathcal{U}$ and $\theta \in \mathcal{R}(\Pi)$. Denote the orbit space of this action by

$$\mathcal{T}(\Pi) = \mathrm{Inn}(\mathcal{U}) \backslash \mathcal{R}(\Pi).$$

It is called the *Teichmüller space* of $\Pi$ (or of $M$).

$\mathrm{Aut}(\Pi)$ acts on $\mathcal{R}(\Pi)$ from the right by

$$\theta \cdot f = \theta \circ f$$

for $\theta \in \mathcal{R}(\Pi)$ and $f \in \mathrm{Aut}(\Pi)$. Denote the orbit space of this action by

$$\mathcal{S}(\Pi) = \mathcal{R}(\Pi)/\mathrm{Aut}(\Pi).$$

$\mathcal{S}(\Pi)$ is the *space of discrete subgroups* of $\mathcal{U}$ each isomorphic to $\Pi$, or the *Chabauty space*.

Since the two actions of $\mathrm{Inn}(\mathcal{U})$ and $\mathrm{Aut}(\Pi)$ commute with each other, $\mathrm{Aut}(\Pi)$ acts on $\mathcal{T}(\Pi)$, and $\mathrm{Inn}(\mathcal{U})$ acts on $\mathcal{S}(\Pi)$. $\mathrm{Aut}(\Pi)$-action has an obvious kernel $\mathrm{Inn}(\Pi)$. In general situations, the kernel contains $\mathrm{Im}(\Pi)$ but it is not the entire kernel. However, for the case we are considering here (3-manifolds), it is the obvious kernel. According to MacBeath and Singermann, Wang has a counterexample in the general case.

Consequently, we get an action of $\mathrm{Out}(\Pi)$ on $\mathcal{T}(\Pi)$. We denote the orbit space by

$$\mathcal{M}(\Pi) = \mathcal{T}(\Pi)/\mathrm{Out}(\Pi).$$

It is called the *moduli space* (or *Riemann space*) of $\Pi$. It is also obtained as the orbit space

$$\mathcal{M}(\Pi) = \mathrm{Inn}(\mathcal{U}) \backslash \mathcal{S}(\Pi).$$

Summarizing in the form of a commutative diagram of orbit mappings, we have

$$\begin{array}{ccc} (\mathrm{Inn}(\mathcal{U}), \mathcal{R}(\Pi), \mathrm{Aut}(\Pi)) & \xrightarrow{\mathrm{Inn}(\mathcal{U})\backslash} & (\mathcal{T}(\Pi), \mathrm{Out}(\Pi)) \\ {\scriptstyle /\mathrm{Aut}(\Pi)}\downarrow & & \downarrow{\scriptstyle /\mathrm{Out}(\Pi)} \\ (\mathrm{Inn}(\mathcal{U}), \mathcal{S}(\Pi)) & \xrightarrow{\mathrm{Inn}(\mathcal{U})\backslash} & \mathcal{M}(\Pi). \end{array}$$

13.4.2. We shall now describe the deformation spaces for closed $M$ which have a geometric structure modeled on $(\mathcal{U}, \widetilde{P}) = (\mathrm{Isom}_0(\widetilde{P}), \widetilde{P})$ (or equivalently the *standard Lorentz structures*). We shall see that these deformation spaces all have Seifert fiberings over well studied deformation spaces of discrete cocompact orientation preserving hyperbolic groups.

Let $\Pi$ be a cocompact discrete subgroup of $\mathcal{U} = \mathrm{Isom}_0(P_\infty)$. We have the central extension
$$1 \to \mathbb{Z} \to \Pi \to Q \to 1, \qquad [\Pi] \in H^2(Q; \mathbb{Z})$$
with $Q \subset P$, and having infinite order in $H^2(Q; \mathbb{Z})$.

Since our group $\mathcal{U}$ embeds into $\mathrm{TOP}_\mathbb{R}(\mathbb{R} \times \mathbb{H})$ we have *topological rigidity* in the *strong sense* that if $\theta_1$ and $\theta_2$ are two embeddings of cocompact $\Pi$ into $\mathrm{TOP}_\mathbb{R}(\mathbb{R} \times \mathbb{H})$, then they are conjugate in $\mathrm{TOP}_\mathbb{R}(\mathbb{R} \times \mathbb{H})$. In almost all cases they will not be conjugate in $\mathcal{U}$. The elements of $\mathcal{M}(\Pi)$ then represent the different $\mathcal{U}$-structures on $M$, and we may expect large deformation spaces.

In order to understand $\mathcal{R}(\Pi)$, we need to study $\mathrm{Aut}(\Pi)$. We would like to describe $\mathrm{Aut}(\Pi)$ in terms of $\mathrm{Aut}(Q)$. Since $\mathbb{Z}$ is characteristic, any automorphism of $\Pi$ induces an automorphism of $Q$.

DEFINITION 13.4.3. Let $\mathrm{Aut}(Q(\Pi))$ be the image of $\mathrm{Aut}(\Pi) \to \mathrm{Aut}(Q)$. That is,
$$\mathrm{Aut}(Q(\Pi)) = \{\overline{\theta} \in \mathrm{Aut}(Q) : \exists \theta \in \mathrm{Aut}(\Pi) \text{ inducing } \overline{\theta}\},$$
the group of automorphisms of $Q$ which can be lifted to an automorphism of $\Pi$.

Because $\mathbb{Z} \subset \mathbb{R}$ has UAEP, one can form a pushout (see Subsection 5.3.4) to get $\mathbb{R}\Pi$ fitting into the commuting diagram,

$$\begin{array}{ccccccccc}
1 & \longrightarrow & \mathbb{Z} & \longrightarrow & \Pi & \longrightarrow & Q & \longrightarrow & 1 \\
& & \downarrow & & \downarrow & & \downarrow & & \\
1 & \longrightarrow & \mathbb{R} & \longrightarrow & \mathbb{R}\Pi & \longrightarrow & Q & \longrightarrow & 1.
\end{array}$$

LEMMA 13.4.4. *There is a commutative diagram with exact rows and injective vertical maps:*

$$\begin{array}{ccccccccc}
1 & \longrightarrow & \mathrm{Hom}(Q, \mathbb{Z}) & \longrightarrow & \mathrm{Aut}(\Pi) & \longrightarrow & \mathrm{Aut}(Q(\Pi)) & \longrightarrow & 1 \\
& & \downarrow & & \downarrow & & \downarrow & & \\
1 & \longrightarrow & \mathrm{Hom}(Q, \mathbb{R}) & \longrightarrow & \mathrm{Aut}(\mathbb{R}\Pi) & \longrightarrow & \mathrm{Aut}(Q) & \longrightarrow & 1.
\end{array}$$

PROOF. The crucial fact for the proof is $H^2(Q; \mathbb{R}) = \mathbb{R}$. Since $[\Pi] \in H^2(Q; \mathbb{Z})$ has infinite order, $[\mathbb{R}\Pi] \in H^2(Q; \mathbb{R})$ is nonzero. Since $\mathbb{R}$ is characteristic in $\mathbb{R}\Pi$, any $f \in \mathrm{Aut}(\mathbb{R}\Pi)$ induces an automorphism $\overline{f} \in \mathrm{Aut}(Q)$. Suppose $\overline{f} = \mathrm{id}$. Let $\widehat{f} : \mathbb{R} \to \mathbb{R}$ be the restriction of $f$. Then $\widehat{f}_*[\mathbb{R}\Pi] = \overline{f}^*[\mathbb{R}\Pi] = [\mathbb{R}\Pi]$ since $\overline{f} = \mathrm{id}$. Since $[\mathbb{R}\Pi]$ is nonzero, $\widehat{f} = \mathrm{id}$. Therefore, $f$ is of the form
$$f(\alpha) = \lambda(\alpha) \cdot \alpha$$
for some map $\lambda : \Pi \to \mathbb{R}$. One easily sees that $\lambda$ factors through $Q$ and it satisfies the cocycle condition
$$\lambda(\overline{\alpha}\overline{\beta}) = \lambda(\overline{\alpha}) + \overline{\alpha}\lambda(\overline{\beta})$$
for all $\overline{\alpha}, \overline{\beta} \in Q$ so that $\lambda \in Z^1(Q, \mathbb{R})$. However, since $\mathbb{R}$ is central in $\mathbb{R}\Pi$, $\mathbb{R}$ is a trivial $Q$-module so that $Z^1(Q, \mathbb{R}) = \mathrm{Hom}(Q, \mathbb{R})$. Conversely, any such

$\lambda \in \text{Hom}(Q, \mathbb{R})$ yields an automorphism $f \in \text{Aut}(\mathbb{R}\Pi)$. Moreover, $\text{Hom}(Q, \mathbb{R}) \cap \text{Aut}(\Pi) = \text{Hom}(Q, \mathbb{Z})$ (cf. Theorem 5.7.2).

Let $\bar{g} \in \text{Aut}(Q)$. We would like to find $g \in \text{Aut}(\mathbb{R}\Pi)$ which induces $\bar{g}$ on $Q$. The automorphism $\bar{g}$ induces an automorphism $\bar{g}^* : H^2(Q; \mathbb{R}) \to H^2(Q; \mathbb{R})$. Since $H^2(Q; \mathbb{R}) = \mathbb{R}$ and $[\mathbb{R}\Pi] \neq 0$, there is a real number $\epsilon$ for which $\bar{g}^*[\mathbb{R}\Pi] = \epsilon[\mathbb{R}\Pi]$. This number $\epsilon$ is nonzero and can be viewed as an automorphism of $\mathbb{R}\Pi$ such that

$$\begin{array}{ccccccccc} 1 & \longrightarrow & \mathbb{R} & \longrightarrow & \mathbb{R}\Pi & \longrightarrow & Q & \longrightarrow & 1 \\ & & \downarrow \epsilon & & \downarrow & & \downarrow \bar{g} & & \\ 1 & \longrightarrow & \mathbb{R} & \longrightarrow & \mathbb{R}\Pi & \longrightarrow & Q & \longrightarrow & 1 \end{array}$$

is commutative. Thus we have shown that $\text{Aut}(\mathbb{R}\Pi) \to \text{Aut}(Q)$ is surjective. $\square$

The deformation spaces of $\Pi$ will be studied via those of $Q$. To this end, it is necessary to define the following:

DEFINITION 13.4.5. Put $\overline{\mathcal{U}} = \mathcal{U}/\widehat{\mathcal{U}}$, where $\widehat{\mathcal{U}} = \mathcal{U} \cap (M(W, G) \rtimes \text{Inn}(G))$, as in Section 13.1. Let $\mathcal{R}(Q; \overline{\mathcal{U}})$ be the space of all injective homomorphisms $\bar{\theta}(Q)$ such that $\bar{\theta}(Q)$ is cocompact and discrete in $\overline{\mathcal{U}}$.

There is a left action of $\text{Inn}(\overline{\mathcal{U}})$ on $\mathcal{R}(Q; \overline{\mathcal{U}})$, and also a right action of $\text{Aut}(Q(\Pi))$ on $\mathcal{R}(Q; \overline{\mathcal{U}})$. We the define

$$\begin{aligned} \mathcal{T}(Q; \overline{\mathcal{U}}) &= \text{Inn}(\overline{\mathcal{U}}) \backslash \mathcal{R}(Q; \overline{\mathcal{U}}), \\ \mathcal{S}(Q(\Pi); \overline{\mathcal{U}}) &= \mathcal{R}(Q; \overline{\mathcal{U}})/\text{Aut}(Q(\Pi)), \\ \mathcal{M}(Q(\Pi); \overline{\mathcal{U}}) &= \mathcal{T}(Q; \overline{\mathcal{U}})/\text{Out}(Q(\Pi)). \end{aligned}$$

These are the *Teichmüller space*, the *restricted Chabauty space*, and the *restricted moduli space* of $Q$, respectively.

It is known that $\text{Aut}(Q(\Pi))$ is a normal subgroup of $\text{Aut}(Q)$ of finite index; see the Appendix of [**KLR86**]. This implies that $\mathcal{S}(Q(\Pi)) = \mathcal{R}(Q)/\text{Aut}(Q(\Pi))$ is a finite regular covering of $\mathcal{S}(Q)$. Also $\mathcal{M}(Q(\Pi)) = \mathcal{T}(Q)/\text{Out}(Q(\Pi))$, where $\text{Out}(Q(\Pi)) = \text{Aut}(Q(\Pi))/\text{Inn}(Q)$.

LEMMA 13.4.6. *The action of* $\text{Aut}(\Pi)$ *on* $\mathcal{R}(\Pi; \mathcal{U})$ *extends to an action of* $\text{Aut}(\mathbb{R}\Pi)$ *on* $\mathcal{R}(\Pi; \mathcal{U})$. *The subgroup* $Z^1(Q; \mathbb{R}) = H^1(Q; \mathbb{R}) = \text{Hom}(Q, \mathbb{R})$ *acts on* $\mathcal{R}(\Pi; \mathcal{U})$ *freely and properly. Moreover,* $\mathcal{R}(\Pi; \mathcal{U})/\text{Hom}(Q, \mathbb{R}) = \mathcal{R}(Q; \overline{\mathcal{U}})$.

PROOF. Clearly, an element $\theta \in \mathcal{R}(\Pi; \mathcal{U})$ determines a homomorphism $\tilde{\theta} : \mathbb{R}\Pi \to \mathcal{U}$ uniquely. The action of $\text{Aut}(\mathbb{R}\Pi)$ on $\mathcal{R}(\Pi; \mathcal{U})$ is defined as follows. For $\theta \in \mathcal{R}(\Pi; \mathcal{U})$ and $f \in \text{Aut}(\mathbb{R}\Pi)$,

$$\theta \cdot f = \tilde{\theta} \circ f|_\Pi.$$

The image of $\Pi$ under $\tilde{\theta} \circ f$ is a discrete cocompact subgroup of $\mathcal{U}$ so that $\tilde{\theta} \circ f|_\Pi \in \mathcal{R}(\Pi; \mathcal{U})$.

Suppose $\theta, \theta' \in \mathcal{R}(\Pi; \mathcal{U})$ induce the same representation $\bar{\theta} = \bar{\theta}' \in \mathcal{R}(Q; \overline{\mathcal{U}})$. Let $\tilde{\theta}, \tilde{\theta}' : \mathbb{R}\Pi \to \mathcal{U}$ be the homomorphisms induced from $\theta, \theta'$ as described above. Since $\bar{\theta} = \bar{\theta}'$, the embeddings $\tilde{\theta}$ and $\tilde{\theta}'$ are related by $\tilde{\theta}'(\alpha) = \lambda(\alpha)\tilde{\theta}(\alpha)$ for some map $\lambda : \mathbb{R}\Pi \to \mathbb{R} \subset \mathcal{U}$. Since $\tilde{\theta}$ and $\tilde{\theta}'$ must be equal on $\mathbb{R}$, the map $\lambda$ factors through $Q$, and hence

$$\lambda : Q \longrightarrow \mathbb{R}.$$

## 13.4. DEFORMATION SPACES FOR $\widetilde{PSL}(2,\mathbb{R})$-GEOMETRY

The map $\lambda$ satisfies the cocycle condition so that $\lambda \in \text{Hom}(Q,\mathbb{R})$. Conversely, let $f \in \text{Aut}(\mathbb{R}\Pi)$ inducing the identity on $Q$. Then by reversing the order of arguments given above, one sees that $\theta$ and $\theta \circ f$ represent the same element of $\mathcal{R}(Q;\overline{\mathcal{U}})$.

Clearly, unless $f$ is the identity, $\theta$ and $\theta \circ f$ will be different, which shows that the action of $\text{Hom}(Q,\mathbb{R})$ on $\mathcal{R}(\Pi;\mathcal{U})$ is free. It is also proper since the orbit space is Hausdorff; see [**KLR86**, 1.10]. $\square$

THEOREM 13.4.7 ([**KLR86**]; Theorem 2.5). *Let $\Pi$ be a compact orbifold group with $(\text{Isom}_0(\widetilde{P}), \widetilde{P})$-geometry. Let $g$ be the genus of the base orbifold. Then,*

$$\begin{aligned}
\mathcal{R}(\Pi) &= \mathcal{R}(Q) \times \mathbb{R}^{2g} \text{ trivial principal } \mathbb{R}^{2g}\text{-bundle over } \mathcal{R}(Q), \\
\mathcal{T}(\Pi) &= \mathcal{T}(Q) \times \mathbb{R}^{2g} \text{ trivial principal } \mathbb{R}^{2g}\text{-bundle over } \mathcal{T}(Q), \\
\mathcal{S}(\Pi) &= T^{2g}\text{-bundle over } \mathcal{S}(Q(\Pi)), \\
\mathcal{M}(\Pi) &= \text{Seifert fiber space over } \mathcal{M}(Q(\Pi)) \text{ with typical fiber } T^{2g}.
\end{aligned}$$

*Furthermore, $\mathcal{S}(Q(\Pi))$ is a finite sheeted covering of $\mathcal{S}(Q)$ and $\mathcal{M}(Q(\Pi))$ is a finite sheeted branched covering of $\mathcal{M}(Q)$.*

PROOF. Let $\mathbb{Z}$ be the center of $\Pi$. Then $Q = \Pi/\mathbb{Z}$ is the base orbifold group. Hence in Lemma 13.4.4, $\text{Hom}(Q,\mathbb{R}) = H^1(Q;\mathbb{R}) = \mathbb{R}^{2g}$, and

$$1 \longrightarrow \mathbb{R}^{2g} \longrightarrow \text{Aut}(\mathbb{R}\Pi) \longrightarrow \text{Aut}(Q) \longrightarrow 1$$

is exact. By Lemma 13.4.6, the group $\mathbb{R}^{2g}$ acts on $\mathcal{R}(\Pi)$ on the *right*, freely and properly so that the orbit map becomes a principal bundle

$$\mathbb{R}^{2g} \to \mathcal{R}(\Pi) \to \mathcal{R}(Q).$$

Since $\mathbb{R}^{2g}$ is contractible, its classifying space is a point and consequently $\mathcal{R}(\Pi)$ splits as $(\mathcal{R}(\Pi), \mathbb{R}^{2g}) = (\mathcal{R}(Q) \times \mathbb{R}^{2g}, \mathbb{R}^{2g})$ equivariantly, where $\mathbb{R}^{2g}$ acts on the second factor as translations.

Because the above bundle is trivial (product), and $\mathbb{R}^{2g}$ is commutative, we shall write, from now on, $\mathcal{R}(Q) \times \mathbb{R}^{2g}$ as $\mathbb{R}^{2g} \times \mathcal{R}(Q)$ to conform with our Seifert fibering construction.

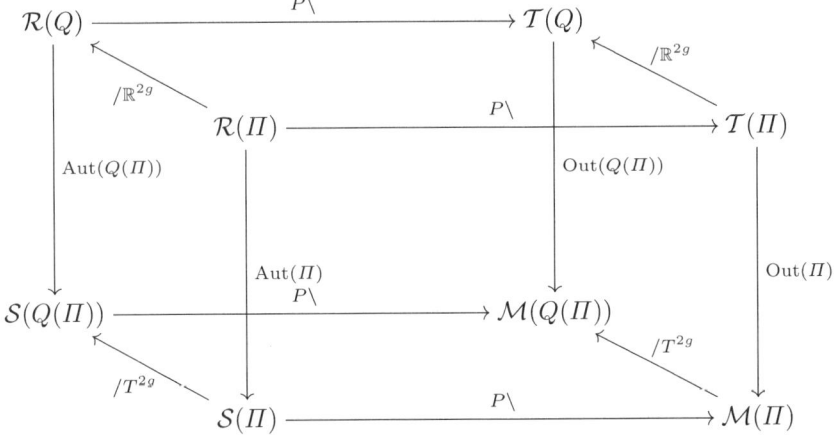

Since $\mathbb{R}$ is the center of $\text{Isom}_0(\widetilde{P})$, we have $\mu(\text{Isom}_0(\widetilde{P})) = \mu(P)$ and

$$\mathcal{T}(\Pi) = \mu(\text{Isom}_0(\widetilde{P}))\backslash\mathcal{R}(\Pi) = \mu(P)\backslash(\mathbb{R}^{2g} \times \mathcal{R}(Q)).$$

Now $\mu(P)$ acts on $\mathcal{R}(Q)$ freely and properly with quotient $\mathcal{T}(Q)$ (see [**MS75**]) which has the homotopy type of the set of two points. Therefore, $\mathcal{T}(\Pi)$ is a product

$\mathbb{R}^{2g} \times \mathcal{T}(Q)$. Moreover, it is known that $\mathcal{T}(Q)$ is diffeomorphic to two copies of $\mathbb{R}^{6g-6+2p}$, where $p$ is the number of nonfree orbit types of the $Q$-action on $\mathbb{H}$. (This is the number of distinct conjugacy classes of maximal finite subgroups of $Q$.)

For the space of subgroups
$$\mathcal{S}(\Pi) = \mathcal{R}(\Pi)/\operatorname{Aut}(\Pi) \cong (\mathbb{R}^{2g} \times \mathcal{R}(Q))/\operatorname{Aut}(\Pi),$$
note that $\operatorname{Aut}(\Pi) \cap \mathbb{R}^{2g} = \mathbb{Z}^{2g}$ and the quotient $\operatorname{Aut}(\Pi)/\mathbb{Z}^{2g}$, which we called $\operatorname{Aut}(Q(\Pi))$, is a subgroup of $\operatorname{Aut}(Q)$. By first dividing out by $\mathbb{Z}^{2g}$, we get $\mathcal{S}(\Pi) = (T^{2g} \times \mathcal{R}(Q))/\operatorname{Aut}(Q(\Pi))$. Since $\operatorname{Aut}(Q(\Pi))$ acts on $\mathcal{R}(Q)$ freely, we have a genuine fibration
$$T^{2g} \longrightarrow \mathcal{S}(\Pi) \longrightarrow \mathcal{S}(Q(\Pi)).$$
The action of $\operatorname{Aut}(\Pi)$ on $\mathbb{R}^{2g} \times \mathcal{R}(Q) = \mathcal{R}(\Pi)$ is weakly $\mathbb{R}^{2g}$-equivariant, because $\mathbb{Z}^{2g} = \operatorname{Hom}(Q;\mathbb{Z})$ is normal in $\operatorname{Aut}(\Pi)$. In other words,
$$\operatorname{Aut}(\Pi) \hookrightarrow \operatorname{TOP}_{\mathbb{R}^{2g}}(\mathbb{R}^{2g} \times \mathcal{R}(Q)) = \operatorname{M}(\mathcal{R}(Q), \mathbb{R}^{2g}) \rtimes (\operatorname{GL}(2g,\mathbb{R}) \times \operatorname{TOP}(\mathcal{R}(Q)))$$
so that
$$\begin{array}{ccccccccc}
1 & \longrightarrow & \mathbb{Z}^{2g} & \longrightarrow & \operatorname{Aut}(\Pi) & \longrightarrow & \operatorname{Aut}(Q(\Pi)) & \longrightarrow & 1 \\
& & \downarrow & & \downarrow & & \downarrow & & \\
1 & \longrightarrow & \operatorname{M}(\mathcal{R}(Q), \mathbb{R}^{2g}) & \longrightarrow & \operatorname{TOP}_{\mathbb{R}^{2g}}(\mathbb{R}^{2g} \times \mathcal{R}(Q)) & \longrightarrow & \operatorname{TOP}(\mathcal{R}(Q)) \times \operatorname{GL}(2g,\mathbb{R}) & \longrightarrow & 1
\end{array}$$
is commutative. Thus, the structure group is a subgroup of the affine group of the torus $T^{2g} \circ \operatorname{GL}(2g,\mathbb{Z})$. Let $\theta \in \mathcal{R}(\Pi)$. Then for $\alpha \in \Pi$, $\theta \cdot \mu(\alpha) = \mu(\theta(\alpha)) \circ \theta$. Therefore, on $\mathcal{T}(\Pi)$, $\operatorname{Inn}(\Pi) = \operatorname{Inn}(\mathbb{R}\Pi) = Q$ acts trivially as does $\operatorname{Inn}(Q) \cong Q$ on $\mathcal{T}(Q)$. Consequently, we have properly discontinuous actions of $\operatorname{Out}(\mathbb{R}\Pi)$ and $\operatorname{Out}(Q(\Pi))$ on $\mathcal{T}(\Pi)$ and $\mathcal{T}(Q)$.

The space of moduli $\mathcal{M}(\Pi) = \mathcal{T}(\Pi)/\operatorname{Out}(\Pi)$ requires more care. Recall that $\mathcal{R}(\Pi) = \mathbb{R}^{2g} \times \mathcal{T}(Q)$ with $\mathbb{R}^{2g}$-action by translations on the first factor. Since
$$1 \longrightarrow \mathbb{R}^{2g} \longrightarrow \operatorname{Out}(\mathbb{R}\Pi) \longrightarrow \operatorname{Out}(Q) \longrightarrow 1$$
is exact, we also have the commutative diagram,
$$\begin{array}{ccc}
(\mathcal{T}(\Pi) = \mathbb{R}^{2g} \times \mathcal{T}(Q), \operatorname{Out}(\mathbb{R}\Pi)) & \xrightarrow{/\mathbb{R}^{2g}} & (\mathcal{T}(Q), \operatorname{Out}(Q)) \\
{\scriptstyle /\operatorname{Out}(\Pi)} \downarrow & & \downarrow {\scriptstyle /\operatorname{Out}(Q(\Pi))} \\
\mathcal{M}(\Pi) = \mathcal{T}(\Pi)/\operatorname{Out}(\Pi) & \xrightarrow{q} & \mathcal{M}(Q(\Pi)).
\end{array}$$
The actions and maps arise from the embedding
$$\begin{array}{ccccccccc}
1 & \longrightarrow & \mathbb{Z}^{2g} & \longrightarrow & \operatorname{Out}(\Pi) & \longrightarrow & \operatorname{Out}(Q(\Pi)) & \longrightarrow & 1 \\
& & \downarrow & & \downarrow & & \downarrow & & \\
1 & \longrightarrow & \mathbb{R}^{2g} & \longrightarrow & \operatorname{Out}(\mathbb{R}\Pi) & \longrightarrow & \operatorname{Out}(Q) & \longrightarrow & 1
\end{array}$$
obtained from Lemma 13.4.4 by dividing out the ineffective $Q$. Note $\operatorname{Inn}(\mathbb{R}\Pi) \cap \mathbb{R}^{2g} = 1$. Now as $\operatorname{Out}(\Pi)$ normalizes $\mathbb{R}^{2g}$ and $\mathbb{Z}^{2g}$ sits in $\mathbb{R}^{2g}$ as a lattice, the mapping $q$ is a Seifert fibering with typical fiber the torus $\mathbb{R}^{2g}/\mathbb{Z}^{2g}$. In general, the fibering will not be locally trivial. In fact, if $F = (\operatorname{Out}(Q(\Pi)))_{[\overline{\theta}]}$ for some $[\overline{\theta}] \in \mathcal{T}(Q)$, then the induced extension
$$1 \longrightarrow \mathbb{Z}^{2g} \longrightarrow E \longrightarrow F \longrightarrow 1$$

acts affinely on $\mathbb{R}^{2g} \times [\bar{\theta}]$ sitting over $[\bar{\theta}]$. The orbit over $[\bar{\theta}]$ under $\text{Out}(Q(\Pi))$ determines a 2-dimensional hyperbolic orbifold up to isometry. Over this hyperbolic orbifold is the set $E\backslash\mathbb{R}^{2g} = F\backslash T^{2g}$ of metric Seifert orbifolds in $\mathcal{M}(\Pi)$ with base this hyperbolic orbifold. □

The reader should not fail to notice similarities with Theorem 14.10.3 and Subsection 14.10.4.

REMARK 13.4.8. 1. From the 3-dimensional diagram in Theorem 13.4.7, it is clear that each of the rectangular faces give rise to Seifert fiberings. In the top face, $P$ acts freely and commutes with the right $\text{Aut}(\Pi)$ and $\text{Aut}(Q(\Pi))$-actions. This makes the front and back faces into significant examples of a type of Seifert fiberings arising from what Kang and Lee [**KL97**] called $G$-mod $\Gamma$-actions. The Seifert fibering $\mathcal{S}(\Pi) \to \mathcal{M}(\Pi)$ is, in their terminology, an effective $G$-equivariant Seifert fiber space. The regular fiber is the compact homogeneous space $Q\backslash P$. For a detailed analysis of the Seifert fiberings induced by the front and back spaces; see the Appendix of [**KLR86**].

2. For another independent approach to deformation of the geometries for 3-dimensional Seifert fiberings, see [**Ohs87**].

3. Y. Kamishima pointed out, in his review [Math Review; MR0827270 (87e:57039)] of [**KLR86**], that the Teichmüller space $\mathcal{T}(\Pi, \text{Isom}(\widetilde{P}))$ may also be defined as the space of $(\text{Isom}(\widetilde{P}), \widetilde{P})$ structures on $M = \Pi\backslash\widetilde{P}$ modulo the action of the isotopy classes of fiber preserving diffeomorphisms of $M$. The group $\text{Out}(\Pi)$ is isomorphic to $\pi_0(\text{Diff}(M))$ which, in turn, is isomorphic to $\pi_0$ (isotopy classes) of Seifert automorphisms of $M$.

4. In real dimension 4 and complex dimension 2, the (complex) deformation theory for holomorphic Seifert fiberings with fiber a complex 1-torus over a Riemann surface is developed by T. Suwa [**Suw79**].

5. C.T.C. Wall [**Wal86**] has also investigated 4-dimensional geometries on closed oriented 4-manifolds. Many of these manifolds also admit compatible complex and/or Seifert structures.

6. The back face of the 3-dimensional diagram is extensively investigated in [**MS75**] and was a motivation for the rest of the diagram.

It is well known that each 2-dimensional manifold supports one of the classical 2-dimensional geometries. In [**Thu97**], Thurston developed the notion of 3-dimensional geometries and showed that there are exactly eight maximal 3-dimensional geometries. Six of the geometries manifest themselves on 3-dimensional Seifert manifolds with $S^1$ as typical fiber, the seventh one as a Seifert fibering with typical fiber a 2-torus and the last on hyperbolic manifolds. Scott's article [**Sco83a**] is also an excellent source for 3-dimensional geometries.

## 13.5. Deformation spaces for Nil-geometry

It is well known that the group of isometries of Nil is $I(\text{Nil}) = \text{Nil} \rtimes O(2)$. More precisely, the following diagram is commutative and exact:

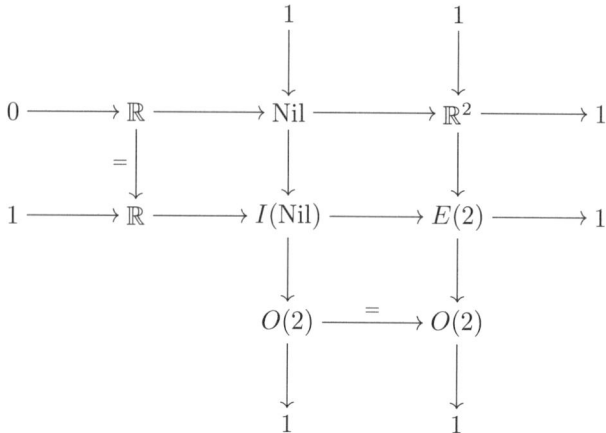

where $E(2) = \mathbb{R}^2 \rtimes O(2)$ is the group of Euclidean isometries of $\mathbb{R}^2$. The $O(2)$ factor of $I(\text{Nil})$ which reverses the orientation of $\mathbb{R}^2$ also reverses the orientation of the fiber, so that $I(\text{Nil}) = I_+(\text{Nil})$. Furthermore, all isometries preserve the fibers so that $I(\text{Nil}) = I_\mathbb{R}(\text{Nil})$.

For a subgroup $Q \subset E(2)$, we denote $Q \cap E_0(2)$ (the orientation preserving isometries) by $Q^+$. Thus, the index of $Q^+$ in $Q$ is at most 2. We have a characterization of discrete cocompact subgroups of $I(\text{Nil})$ as follows.

THEOREM 13.5.1. *An abstract group $\Pi$ can be embedded into $I(\text{Nil})$ as a compact discrete subgroup if and only if $\Pi$ is an extension of $\mathbb{Z}$ by a 2-dimensional crystallographic group $Q$ (so that $1 \to \mathbb{Z} \to \Pi \to Q \to 1$ is exact) satisfying*
  (1) $Q^+ = \ker(Q \to \text{GL}(1, \mathbb{Z}))$, *and*
  (2) $[\Pi] \in H^2(Q; \mathbb{Z})$ *has infinite order.*
*Further, if this is the case, the subgroup $\mathbb{Z}$ is characteristic in $\Pi$.*

PROOF. Suppose $\Pi \subset I(\text{Nil}) = \text{Nil} \rtimes O(2)$ is cocompact and discrete. By Auslander's theorem, Theorem 8.3.2, $\Gamma = \Pi \cap \text{Nil}$ has finite index in $\Pi$. It is also easy to see that $\Gamma$ is the maximal nilpotent normal subgroup of $\Pi$. Therefore, $\Gamma \cap \mathbb{R}$ ($\mathbb{R} = \mathcal{Z}(\text{Nil})$) is isomorphic to $\mathbb{Z}$ and $\Gamma/\mathbb{Z} = \mathbb{Z}^2$. Thus $Q = \Pi/\mathbb{Z}$ is virtually free Abelian and, in fact, is a crystallographic group. Since $1 \to \mathbb{Z} \to \Gamma \to \mathbb{Z}^2 \to 1$ does not split, $[\Gamma] \in H^2(\mathbb{Z}^2; \mathbb{Z})$ has infinite order. The sequence is induced by the inclusion $\mathbb{Z}^2 \hookrightarrow Q$, and is the pullback of the sequence $1 \to \mathbb{Z} \to \Pi \to Q$. Therefore, $[\Pi] \in H^2(Q^2; \mathbb{Z})$ has infinite order.

Conversely, if $\Pi$ satisfies the two conditions listed above, then the inverse image $\widehat{Q}$ of $Q \subset E(2)$ in $I(\text{Nil})$, $1 \to \mathbb{R} \to \widehat{Q} \to Q \to 1$ represents a nonzero element $[\widehat{Q}] \in H^2(Q; \mathbb{R}) \approx \mathbb{R}$. Since $[\Pi] \in H^2(Q; \mathbb{Z})$ has infinite order, $\mathbb{R}\Pi = i_*[\Pi]$ is nonzero, and by the same argument as in Theorem 13.2.6, $\Pi$ can be embedded in $I(\text{Nil})$ as cocompact subgroup discrete subgroup. The crystallographic group $Q$ contains a characteristic subgroup $\mathbb{Z}^2$, and the inverse image $N$ of $\mathbb{Z}^2$ in $\Pi$ is the unique maximal normal nilpotent subgroup of $\Pi$. Then $\mathbb{Z}$, which is the center of $N$, is characteristic in $\Pi$. □

COROLLARY 13.5.2 (Structure). *Let $M$ be a compact orbifold modeled on Nil-geometry. Then $M$ is a Seifert orbifold over a Euclidean base with the Euler number $e(M) \neq 0$.*

COROLLARY 13.5.3 (Realization). *Let $M$ be a compact Seifert orbifold over a Euclidean base. Then $M$ admits a geometry modeled on Nil if and only if $e(M) \neq 0$.*

Here $e(M)$ is $e(\Pi)$ where $\Pi$ is the orbifold fundamental group of the orbifold $M = \Pi\backslash\mathrm{Nil}$.

Similarly to the case of $\widetilde{\mathrm{PSL}(2,\mathbb{R})}$-geometry, $e(M) \neq 0$ is equivalent to the condition that $[\Pi] \in H^2(Q;\mathbb{Z})$ has infinite order.

COROLLARY 13.5.4. *For $\theta \in \mathcal{R}(\Pi, I(\mathrm{Nil}))$, let $\bar{\theta} : Q \to I(\mathbb{R}^2)$ be the induced map. Then $\theta \mapsto \bar{\theta}$ defines a well-defined map of $\mathcal{R}(\Pi, I(\mathrm{Nil}))$ into $\mathcal{R}(Q, I(\mathbb{R}^2))$. Furthermore, this map is injective.*

13.5.5. Let $\Pi$ be a cocompact discrete subgroup of $I(\mathrm{Nil})$. Then by Theorem 13.5.1, $\Pi$ contains a characteristic subgroup $\mathbb{Z}$ with quotient $Q$, a crystallographic group; and $1 \to \mathbb{Z} \to \Pi \to Q \to 1$ has infinite order in $H^2(Q;\mathbb{Z})$. Furthermore, $H^2(Q;\mathbb{R}) \approx \mathbb{R}$ (see the argument in Lemma 13.2.4 and use the fact that $H^2(Q;\mathbb{Z}) = \mathbb{Z} \oplus \mathrm{Torsion}$). By arguments similar to the proof of Lemma 13.4.4, one gets a commutative diagram with exact row and injective vertical maps,

$$\begin{array}{ccccccccc} 1 & \longrightarrow & Z^1(Q;\mathbb{Z}) & \longrightarrow & \mathrm{Aut}(\Pi) & \longrightarrow & \mathrm{Aut}(Q(\Pi)) & \longrightarrow & 1 \\ & & \downarrow & & \downarrow & & \downarrow & & \\ 1 & \longrightarrow & Z^1(Q;\mathbb{R}) & \longrightarrow & \mathrm{Aut}(\mathbb{R}\Pi) & \longrightarrow & \mathrm{Aut}(Q) & \longrightarrow & 1. \end{array}$$

By Corollary 13.5.4, $\mathcal{R}(\Pi) \longrightarrow \mathcal{R}(Q)$ is well defined. Moreover, it is the orbit map of the right $Z^1(Q;\mathbb{R})$-action on $\mathcal{R}(\Pi)$. Therefore, if we let $Z^1(Q;\mathbb{R}) = \mathbb{R}^k$, then $k \leq 2$ and $\mathcal{R}(\Pi) = \mathcal{R}(Q) \times \mathbb{R}^k$ with $\mathbb{R}^k$ acting freely and properly.

THEOREM 13.5.6 (Deformation spaces of $(I_0,\mathrm{Nil})$-geometry). *Let $M = \Pi\backslash\mathrm{Nil}$ be a compact Seifert orbifold with $(I_0,\mathrm{Nil})$-geometry, and let $Q = \Pi/\mathbb{Z}$ be the base orbifold group. Then*
  (1) *If $Q = \mathbb{Z}^2$ (that is, $M$ is a nilmanifold), then $\mathcal{R}(\Pi) = \mathbb{R}^2 \rtimes \mathrm{GL}(2,\mathbb{R})$, $\mathcal{T}(\Pi) = \mathrm{O}(2)\backslash\mathrm{GL}(2,\mathbb{R}) \approx \mathbb{R}^3$, $\mathcal{S}(\Pi)$ is a $T^2$-bundle over $\mathcal{S}(Q) = \mathrm{GL}(2,\mathbb{R})/\mathrm{GL}(2,\mathbb{Z})$, and $\mathcal{M}(\Pi) = \mathrm{O}(2)\backslash\mathrm{GL}(2,\mathbb{R})/\mathrm{GL}(2,\mathbb{Z})$.*
  (2) *If $Q = \mathbb{Z}^2 \rtimes \mathbb{Z}_2$, then $\mathcal{R}(\Pi) = \mathrm{GL}(2,\mathbb{R})$ and $\mathcal{T}(\Pi) = \mathrm{SO}(2)\backslash\mathrm{GL}(2,\mathbb{R}) \approx \mathbb{R}^2 \times \mathbb{R}^*$.*
  (3) *All the other cases: $\mathcal{R}(\Pi) = \mathbb{R}^2 \rtimes (\mathbb{R}^+ \times \mathrm{O}(2))$ and $\mathcal{T}(\Pi) = \mathbb{R}^*$.*

PROOF. (1) If $Q = \mathbb{Z}^2$, then $\Pi$ is a lattice of Nil, and $\mathcal{R}(\Pi) = \mathbb{R}^2 \rtimes \mathrm{GL}(2,\mathbb{R})$. Since $\mu(\mathrm{Nil} \rtimes \mathrm{O}(2)) = \mathbb{R}^2 \rtimes \mathrm{O}(2)$ and $\mathrm{Aut}(\Pi) = \mathbb{Z}^2 \rtimes \mathrm{GL}(2,\mathbb{Z})$, we have $\mathcal{T}(\Pi) = \mathrm{O}(2)\backslash\mathrm{GL}(2,\mathbb{R}) \approx \mathbb{R}^3$, $\mathcal{S}(\Pi)$ is a $T^2$-bundle over $\mathrm{GL}(2,\mathbb{R})/\mathrm{GL}(2,\mathbb{Z})$, and $\mathcal{M}(\Pi) = \mathrm{O}(2)\backslash\mathrm{GL}(2,\mathbb{R})/\mathrm{GL}(2,\mathbb{Z})$.

(2) $\mathrm{Hom}(Q,\mathbb{Z}) = 0$ so that $\mathrm{Hom}(Q,\mathbb{R}) = 0$, and hence $\mathcal{R}(\Pi) = \mathcal{R}(Q)$. Since $\mathrm{Aff}(\mathbb{R}^2) = \mathbb{R}^2 \rtimes \mathrm{GL}(2,\mathbb{R})$ acts on $\mathcal{R}(Q)$ simply transitively, $\mathcal{R}(Q) = \mathbb{R}^2 \rtimes \mathrm{GL}(2,\mathbb{R})$. Since $\mu(\mathbb{R}^2 \rtimes \mathrm{SO}(2))$ acts effectively on $\mathcal{R}(Q)$,

$$\mathcal{T}(\Pi) = \mathcal{T}(Q) = (\mathbb{R}^2 \rtimes \mathrm{SO}(2))\backslash(\mathbb{R}^2 \rtimes \mathrm{GL}(2,\mathbb{R})) = \mathrm{SO}(2)\backslash\mathrm{GL}(2,\mathbb{R}) \approx \mathbb{R}^2 \times \mathbb{R}^*.$$

(3) $\mathrm{Hom}(Q,\mathbb{Z}) = 0$ so that $\mathcal{R}(\Pi) = \mathcal{R}(Q)$. Let $\mathbb{Z}^2$ be the maximal normal Abelian subgroup and $\Phi = Q/\mathbb{Z}^2$ be the holonomy group. Fix an embedding of $\Phi$

into SO(2). Since the order of $\Phi$ is bigger than 2, an element $A \in \mathrm{GL}(2,\mathbb{R})$ satisfies $A\Phi A^{-1} \subset \mathrm{SO}(2)$ if and only if $A$ is of the form $\lambda B$, where $\lambda \in \mathbb{R}^+$ and $B \in \mathrm{O}(2)$. Thus, $\mathcal{R}(Q) = \mathbb{R}^2 \rtimes (\mathbb{R}^+ \times \mathrm{O}(2))$. Since $\mu(\mathbb{R}^2 \rtimes \mathrm{SO}(2))$ acts on $\mathcal{R}(Q)$ effectively, $\mathcal{T}(\Pi) = \mathcal{T}(Q) = (\mathbb{R}^2 \rtimes \mathrm{SO}(2))\backslash(\mathbb{R}^2 \rtimes (\mathbb{R}^+ \times \mathrm{O}(2))) \approx \mathbb{R}^+ \times \mathbb{Z}_2$. □

EXERCISE 13.5.7. Describe all the Seifert manifolds that have $(I_0, \mathrm{Nil})$-geometry in terms of Seifert invariants of Chapter 14. They are of the following form:
$M = \{o_1, g = 0, b; (\alpha_i, \beta_i)\}$ with $(\alpha_1, \alpha_2, \alpha_3) = (2,4,4), (2,3,6), (3,3,3)$, and $(\alpha_1, \alpha_2, \alpha_3, \alpha_4) = (2,2,2,2)$ all with $e(M) \neq 0$, and $(\alpha_i, \beta_i)$ relatively prime. Also, we have $\{o_1, g = 1, b \neq 0\}$, $\{n_2, g = 2, b \neq 0\}$, and $\{n_2, g = 1, b \neq -1; (2,1), (2,1)\}$.

REMARK 13.5.8. In [**KLR86**, 3.4] and [**Ohs87**], it is stated that $\mathcal{T}(\Pi) \approx \mathbb{R}^4 \times \mathbb{R}^*$ in case (1). This is a mistake.

In the cases of (2) and (3), even though $\mathcal{R}(\Pi) = \mathcal{R}(Q)$ and $\mathcal{T}(\Pi) = \mathcal{T}(Q)$, $\mathcal{S}(Q)$ and $\mathcal{M}(Q)$ are finitely covered (branched) by $\mathcal{S}(\Pi)$ and $\mathcal{M}(\Pi)$, respectively. This occurs because $\mathrm{Aut}(Q(\Pi))$ may not be equal to $\mathrm{Aut}(Q)$. However, if $\Pi$ is torsion free, i.e., $\Pi$ acts without branching and $M$ is a classical Seifert fiber space, then in case (2), and in case (3) with $\Phi = \mathbb{Z}_3$, we have $\mathrm{Aut}(Q) = \mathrm{Aut}(Q(\Pi))$. For the other holonomies, this may not hold. Even in case (2) if $\Pi$ is not torsion free, $\mathrm{Aut}(Q(\Pi))$ is a proper subgroup (of finite index) of $\mathrm{Aut}(Q)$.

CHAPTER 14

# $S^1$-actions on 3-dimensional Manifolds

## 14.1. Introduction

The reader may wish to read some of the historical remarks that we have appended to this chapter (see Section 14.14) for more of the background on the development of the classical Seifert fiberings and its interconnection with the $S^1$-actions on 3-manifolds. It was known in 1936 that $S^1$-actions on $\mathbb{R}^3$ were conjugate to linear actions [**MZ36**]. It was not until 1956, however, that R. Jacoby [**Jac56**], building on Seifert's work, showed that the fixed-point free $S^1$-actions on the 3-sphere coincided with Seifert's classification of the Seifert fiberings of the 3-sphere.

The orbit types of the $S^1$-actions are the *free orbits* (*i.e., principal orbits*), the *exceptional orbits* (*i.e., E-orbits*) where the cyclic isotropy group acts on the slice as a rotation, the *special exceptional orbits* (*i.e., SE-orbits*) where the isotropy group is $\mathbb{Z}_2$ and acts as a reflection on the slice, and the *fixed orbits* (*i.e., F-orbits*). Each connected component of an orbit type has an $S^1$-invariant neighborhood. These are attached equivariantly to the principal $S^1$-bundle of free orbits to form the 3-dimensional $S^1$-manifold. The orbit space is a 2-manifold with possible boundary consisting of the images of the components of $SE$- and $F$-orbits.

In Section 14.3, we classify when the $S^1$-action is injective, locally injective or not locally injective in terms of orbit data. In Section 14.4, we show certain manifolds with $S^1$-actions are aspherical. Suppose $M$ has an effective $S^1$-action without fixed points and $\pi_1(M)$ is infinite but not virtually $\mathbb{Z}$. Then $M$ is aspherical. They are modeled, as a Seifert fiber space, on $\mathbb{R} \times \mathbb{R}^2$ with the action of $Q$ on $\mathbb{R}^2$ topologically equivalent to a group of Euclidean or hyperbolic isometries. Furthermore, two $S^1$-manifolds of this type are shown, using the rigidity of the Seifert Construction of Chapter 7, to be weakly equivalent if and only if their fundamental groups are isomorphic. In Section 14.5, the classification is extended to $S^1$-actions with fixed points by the technique of equivariant connected sum already found in Section 2.6.

The next section, Section 14.6, makes the classification more practical by showing that the equivariant classification can be described solely in terms of numerical orbit invariants. Since there is only one $S^1$-action, up to equivalence, over each manifold by Section 14.4, these orbit invariants are also topological invariants. This procedure of utilizing Section 14.4 to obtain a topological classification in terms of numerical invariants avoids some technical 3-manifold theory or technical group theory.

Sections 14.7 through 14.11 explore interesting properties of the $(S^1, M)$ where $M$ is orientable and the action is without fixed points. Every 3-dimensional Seifert fiber space, whether it is a manifold or orbifold, lifts to a finite regular covering, $(S^1, M)$, in a fiber preserving way (possibly branched if $M$ is an orbifold), where $S^1$ acts without fixed points and $M$ is orientable. So, in a sense, this type of Seifert manifold is the most important class of Seifert fiberings. For these $S^1$-manifolds, which are generalizations of principal $S^1$-bundles over a closed orientable surface, one defines, in Section 14.7, a rational number invariant called the Euler number, $e(M)$, which is a generalization of the Euler class of a principal $S^1$-bundle. It measures the obstruction that $(S^1, M)$ has a global slice (or a multisection) for the $S^1$-action. The Euler number plays a role in Section 14.8 where it is determined which $(S^1, M)$'s are homogeneous spaces in the sense that $M$ is of the form $\Gamma \backslash G$, where $\Gamma$ is a discrete subgroup of a connected 3-dimensional Lie group $G$. In particular, every $(S^1, M)$ which is an integral homology sphere is of this type.

The locally injective $S^1$-actions on $M$ lift to a principal $S^1$-bundle $P = \tilde{M}$ over a simply connected 2-manifold $W$. The $S^1$-action commutes with the covering transformations $Q = \pi_1(P, x)/\text{ev}_*^x(\pi_1(S^1, 1))$ on $P$. The induced $Q$-action on $W$, when orientation preserving, is topologically equivalent to a holomorphic action on the Riemann sphere, the unit disk $U$ or the complex plane $\mathbb{C}$. So, in our Seifert Constructions, we may replace $W$ by $\mathbb{C}P_1$, $U$ or $\mathbb{C}$, respectively, with a holomorphic $Q$-action. The Seifert Construction, under the appropriate circumstances, can now be done holomorphically. For example, if $P$ is topologically $S^1 \times W$, replace $(S^1, S^1 \times W)$ by $(\mathbb{C}^*, \mathbb{C}^* \times W)$ with $Q$ acting holomorphically. In Subsection 14.10.1, we show how this may be done, and use this to show how neighborhoods of isolated singularities with "good" $\mathbb{C}^*$-actions are open cones over $(S^1, M)$ with $e(M) < 0$. In fact, the proper holomorphic $\mathbb{C}^*$-actions on an analytic surface over a Riemann surface are classified up to Seifert isomorphism.

In Section 14.11, some of these open cones are obtained as the complete intersection of $m - 2$ sufficiently general Brieskorn varieties in $\mathbb{C}^n$. The exponents of the polynomials determine the Seifert invariants.

The most general 3-dimensional Seifert fiberings with typical fiber $S^1$ are not necessarily the orbit space of an $S^1$-action on a 3-manifold. However, all have a 2-fold covering admitting an $S^1$-action whose involutive covering transformation is a Seifert automorphism. A precise description of these manifolds is complicated. They can be described by means of their Seifert invariants analogous to the orbit invariants used for the $S^1$-actions. This was done by Seifert in the absence of SE- and F-type fibers. In Sections 14.12 and 14.13, we recount Seifert's description and also treat the general case where SE- and F-type fibers are present. The last section is a brief historical survey of the development of the 3-dimensional Seifert fiberings with $S^1$ as typical fiber.

14.1.1 (Fibered solid torus). Let the solid torus $V = S^1 \times D^2$ be parametrized by $(z_1, \rho z_2)$ with $z_1, z_2 \in S^1 \subset \mathbb{C}$ and $0 \leq \rho \leq 1$. Define an effective $S^1$-action on $V$ by

$$z \cdot (z_1, \rho z_2) = (z^\mu z_1, z^\nu \rho z_2),$$

where $\mu$ and $\nu$ are relatively prime integers with $\mu \neq 0$. This action is free everywhere except for the core orbit $(z^\mu z_1, 0)$ where the isotropy subgroup is $\mathbb{Z}_\mu$. We call $V$ with this action a *fibered solid torus* of type $(\mu, \nu)$. The slice invariant at $(1, 0)$ on the core orbit is $(\mu, \bar{\nu})$, where $\bar{\nu} \equiv \nu \mod \mu$, and $0 < \bar{\nu} < \mu$. That is, on the slice $(1, \rho z_2)$, the action of $e^{\frac{2\pi i}{\mu}} \in S^1$ is given by

$$(1, \rho z_2) \mapsto (1, e^{\frac{2\pi i \nu}{\mu}} \rho z_2).$$

Orient $V$ by taking the product of choices orientation of $S^1$ and $D^2$, say positive for increasing angles. The orientation preserving homeomorphism $(z_1, \rho z_2) \mapsto (z_1^{-1}, \rho z_2^{-1})$ sends $(\mu, \nu)$ to $(-\mu, -\nu)$. Thus, without loss of generality, we can assume $\mu > 0$. Also, the orientation preserving homeomorphism $(z_1, \rho z_2) \mapsto (z_1, z_1^r \rho z_2)$ sends $(\mu, \nu)$ to $(\mu, r\mu + \nu)$. Thus the two actions $(S^1, V)$ and $(S^1, V')$ of type $(\mu, \nu)$ and $(\mu', \nu')$ are isomorphic via an orientation preserving homeomorphism if and only if $(\mu', \nu') = (\mu, \nu + r\mu)$ for some integer $r$. (Clearly, if $\mu \neq \mu'$ or $\nu' \not\equiv \nu$ mod $\mu$, then the slice invariants are inequivalent.) The map $(z_1, \rho, z_2) \mapsto (z_1, \rho, z_2^{-1})$ is an orientation reversing equivariant homeomorphism that send $(\mu, \nu)$ to $(\mu, -\nu)$. Therefore, $h : (S^1, V;$ with type $(\mu, \nu)) \to (S^1, V;$ with type $(\mu', \nu'))$ is an orientation preserving $S^1$-equivalence if and only if $\mu' = \mu$ and $\nu' = \nu + r\mu$ with $\mu, \mu' > 0, r \in \mathbb{Z}$. If $h$ is not required to be orientation preserving, then $\mu = \mu'$ and $\nu' = \pm\nu + r\mu$.

To visualize the orbit mapping $V \longrightarrow S^1 \backslash V$ as a Seifert fibering, it is easier to consider the action in two steps. Start with the action of $\mathbb{Z}_\mu$ on a solid torus $S^1 \times D^2$ given by

$$\lambda \cdot (z_1, \rho z_2) = (z_1 \lambda^{-1}, \lambda^\nu \rho z_2),$$

where $\lambda = e^{\frac{2\pi i}{\mu}}$. Denote the image of $(z_1, \rho z_2)$ in $S^1 \times_{\mathbb{Z}_\mu} D^2$ by $\langle z_1, \rho z_2 \rangle$. Therefore, $\langle z_1, \rho z_2 \rangle = \langle z_1 \lambda^{-s}, \lambda^{s\nu} \rho z_2 \rangle$. On the quotient space $S^1 \times_{\mathbb{Z}_\mu} D^2$, there is an $S^1$-action

$$z \cdot \langle z_1, \rho z_2 \rangle = \langle z z_1, \rho z_2 \rangle$$

induced from $z \cdot (z_1, \rho z_2) = (z z_1, \rho z_2)$ on $S^1 \times D^2$ (since this action commutes with the $\mathbb{Z}_\mu$-action).

$$\begin{array}{ccc} S^1 \times D^2 & \longrightarrow & D^2 \\ \downarrow & & \downarrow \\ S^1 \times_{\mathbb{Z}_\mu} D^2 & \longrightarrow & \mathbb{Z}_\mu \backslash D^2 \end{array} \qquad \begin{array}{ccc} (z_1, \rho z_2) & \longrightarrow & \rho z_2 \\ \downarrow & & \downarrow \\ \langle z_1, \rho z_2 \rangle & \longrightarrow & \langle \rho z_2 \rangle \end{array}$$

Here the orbit mapping

$$S^1 \times_{\mathbb{Z}_\mu} D^2 \longrightarrow S^1 \backslash (S^1 \times_{\mathbb{Z}_\mu} D^2) = \mathbb{Z}_\mu \backslash D^2$$

of this injective $S^1$-action is a Seifert fibering with typical fiber $S^1$ and one singular fiber $\mathbb{Z}_\mu \backslash S^1$. It is modeled on $(S^1, S^1 \times D^2)$ with typical fiber $S^1$ or alternatively modeled on $\mathbb{R}^1 \times D^2$ with typical fiber $\mathbb{Z} \backslash \mathbb{R}^1 = S^1$.

We claim that the $S^1$-action on a fibered solid torus $V$ of type $(\mu, \nu)$ is equivalent to the $S^1$-action $(S^1, S^1 \times_{\mathbb{Z}_\mu} D^2)$ with slice invariants $(\mu, \overline{\nu})$, where $\overline{\nu} \equiv \nu$ mod $\mu$.

$$\begin{array}{ccc} S^1 \times (S^1 \times_{\mathbb{Z}_\mu} D^2) & \longrightarrow & S^1 \times_{\mathbb{Z}_\mu} D^2 \\ 1 \times f \downarrow & & \downarrow f \\ S^1 \times (S^1 \times D^2) & \longrightarrow & S^1 \times D^2 \end{array} \qquad \begin{array}{ccc} z \langle z_1, \rho z_2 \rangle & \longrightarrow & \langle z z_1, \rho z_2 \rangle \\ 1 \times f \downarrow & & \downarrow f \\ z(z_1^\mu, z_1^\nu \rho z_2) & \longrightarrow & ((z z_1)^\mu, (z z_1)^\nu \rho z_2) \end{array}$$

For $f : \langle z_1, \rho z_2 \rangle \mapsto (z_1^\mu, z_1^\nu \rho z_2)$ is easily checked to be an $S^1$-equivalence. Thus the orbit mapping of the $S^1$-action on the fibered solid torus of type $(\mu, \nu)$ is a Seifert fibering modeled on $\mathbb{R}^1 \times D^2$ with typical fiber, $S^1 = \mathbb{Z} \backslash \mathbb{R}$, any one of the free orbits.

It is not difficult to show that any effective $S^1$-action without fixed points is equivalent to one of the $S^1$-actions on a solid fibered torus.

14.1.2 (The solid Klein bottle). On the Möbius band $M = S^1 \times_{\mathbb{Z}_2} I$,
$$(z, t) \sim (e^{\pi i} z, -t),$$
there is only one effective $S^1$-action up to equivalence. This is the action induced from the translation on the first factor of the orientable double covering $S^1 \times I$ (cf. Example 4.3.4). The orbit space is an interval with all orbits free except one end point where the isotropy is $\mathbb{Z}_2$. The solid Klein bottle is defined as $M \times I$. It has an effective $S^1$-action by acting solely on the first factor. (This is the only possible effective $S^1$-action up to equivalence.) The boundary $\partial(M \times I) = M \times \partial I \cup \partial M \times I$ is clearly a Klein bottle. (One can also envision the solid Klein bottle as an arc of Klein bottles where the last Klein bottle is collapsed to one of the circle orbits with isotropy group $\mathbb{Z}_2$.)

## 14.2. $S^1$-actions on 3-manifolds

Let $S^1$ act effectively on a connected 3-dimensional manifold $M$. We shall assume for convenience that the action is locally smooth although this is not necessary for any of the conclusions; see Remarks 14.3.6(7). Let $x \in M$, and let $S_x$ be a closed cell slice at $x$. If $x$ is a fixed point, then $S_x$ is an invariant 3-cell $D^2 \times I$, and the action is equivalent to the rotational action on $D^2$ and trivial on the second factor.

The orbit space is $(S^1 \backslash D^2) \times I$. If $x$ is not fixed, then the isotropy group is $\mathbb{Z}_\mu$, $\mu \geq 1$, and the slice is a closed 2-cell. The group $\mathbb{Z}_\mu$ acts orthogonally on the slice as either a rotation or reflection.

14.2.1 (The exceptional orbits). If $\mathbb{Z}_\mu$ acts as a rotation, then the invariant tubular neighborhood is a fibered solid torus $V$. All orbits are free except for the core orbit (assuming $\mu > 1$). We call the core orbit an *exceptional orbit* or *E-orbit* with multiplicity $\mu$. The boundary of the slice is a meridian $m$. We can choose a longitude $\ell$ which is a simple closed curve, on $\partial V$, homotopic to the E-orbit in $V$, and such that its geometric intersection with $m$ is 1. The curve gives a parametrization of the homology of $\partial V$, and if $H \sim \nu m + \mu \ell$, we have a fibered solid torus $V$ of type $(\mu, \nu)$. Here $H = S^1(1, 1)$ is a typical fiber, the $S^1$-orbit passing through the point $(1, 1)$. Of course $\ell$ can be replaced by another longitude $\ell' = \ell + sm$ and in terms of $m$ and $\ell'$, $H \sim \nu m + \mu \ell' - \mu s m = (\nu - \mu s)m + \mu \ell'$ and so $V$ becomes the equivalent type $(\mu, \nu - \mu s)$.

Note also on $V$, the curves $H$ and $m$ are already determined. We can choose a cross sectional curve $Q$ to the free action on $\partial V$. Then the geometric intersection, $Q \cdot H = 1$. The curve $m$ is unique up to isotopy, and so
$$m \sim \alpha Q + \beta H$$
for integers $\alpha$ and $\beta$. Then
$$\ell \sim \rho Q + \delta H$$
for some integers $\rho$ and $\delta$, and $\alpha \delta - \beta \rho = 1$ (since $m \cdot \ell = 1$). The last equation orients $Q$ so that $Q$ and $H$ generate the same orientation on $\partial V$ as does $m$ and $\ell$. Solving for $H$, we have
$$H \sim -\rho m + \alpha \ell \sim \nu m + \mu \ell,$$
$$Q \sim \delta m - \beta \ell.$$

Therefore, $\alpha = \mu$ and $-\rho = \nu$ and $\beta\nu \equiv 1 \bmod \alpha$. The pair $(\alpha, \beta)$ is called the *unnormalized Seifert invariant*. In the solid torus $V$, $m \sim 0$ and so in the homology of $V$, we have

$$H \sim \alpha\ell,$$
$$Q \sim -\beta\ell.$$

This says that in $V$, $H$ winds around $\ell$, $\alpha$ times and $Q$ winds around $\ell$, $-\beta$ times.

If we choose a different cross section $Q' \sim Q + tH$, then the Seifert invariant $(\alpha, \beta)$ changes to $(\alpha', \beta') = (\alpha, \beta - t\alpha)$. It is of interest to *normalize* the Seifert invariant so that $0 < \beta' < \alpha'$. This can be accomplished by choosing $t$ such that $0 < \beta - t\alpha < \alpha$. Thus we can always choose a cross section $Q$ so that the Seifert invariant $(\alpha, \beta)$ is normalized so that $0 < \beta < \alpha$. We may also choose a longitude $\ell$ so that $0 < \nu < \mu$.

The normalized Seifert invariant $(\alpha, \beta)$ is an oriented invariant. It determines the $S^1$-action up to orientation preserving equivalence. Our orientation has been determined by an orientation of the slice and typical fiber (the product of which determines the local orientation of the 3-manifold).

As a simple example, for $(\mu, \nu) = (3, 1)$, $(\alpha, \beta) = (3, 1)$, the picture looks like the following (top is identified with the bottom):

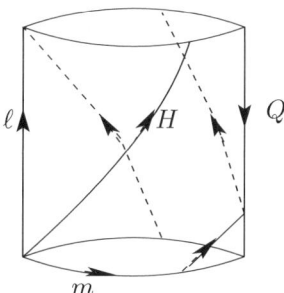

14.2.2 (The special exceptional orbits). If $\mathbb{Z}_\mu$ is $\mathbb{Z}_2$ and the action on the slice $S_x$ is a reflection, then $S^1 x$ is called a *special exceptional orbit* or an *SE-orbit*. The invariant tubular neighborhood $S^1(S_x)$ of the SE-orbit $S^1 x$ is a solid Klein bottle, $(S^1, S^1 \times_{\mathbb{Z}_2} D^2)$ with $\mathbb{Z}_2$ acting on $D^2$ by $\rho z_2 \mapsto \rho \bar{z}_2$. Each orbit passing through the axis of reflection in $S_x$ has isotropy group $\mathbb{Z}_2$ and thus is an SE-orbit. Each SE-orbit is doubly covered by nearby free orbits (those that do not intersect the axis of reflection).

PROPOSITION 14.2.3 (The orbit space). *Let $(S^1, M)$ be an effective $S^1$-action on a connected 3-manifold $M$, then the orbit space $B$ is a connected 2-manifold with possible boundary. The inverse image of a connected component of the boundary is either a connected component of fixed points or a connected component of SE-orbits. The set of E-orbits is isolated and their image in $B$ forms a discrete closed subset of $B$. All other orbits are free.*

Proof follows immediately from the proceeding.

## 14.3. $S^1$-actions on 3-manifolds as Seifert fiberings

In this section we shall assume without further notice that $S^1$ acts effectively on a connected 3-manifold without fixed points and has at most a finite number of distinct orbit types. The latter condition is immediate if $M$ is compact. Our use of the term "manifold" always assumes the boundary is empty. When a nonempty boundary is needed, we will explicitly indicate it.

LEMMA 14.3.1 ([**CR72b**, 12.9]; cf. [**Sei33**]). *Let $(S^1, M)$ be an effective action of $S^1$ without fixed points on a connected 3-manifold $M$. Then $(S^1, M)$ is injective if and only if $\pi_1(M)$ is infinite. Furthermore, if $\pi_1(M)$ is finite, then $M$ is compact and $B$ is $S^2$.*

PROOF. If $(S^1, M)$ is injective, $\pi_1(M)$ is obviously infinite. So now suppose the image of $\mathrm{ev}_*^x$ is finite. Let $K$ be the kernel of $\mathrm{ev}_*^x$, and let $'S^1$ be the covering group of $S^1$ corresponding to $K$. Then $'S^1$ has trivial evaluation homomorphism and the ineffective $('S^1, M)$-action lifts to an effective $('S^1, \widetilde{M})$-action on the universal covering $\widetilde{M}$ of $M$. Let $\{\mu_1, \ldots, \mu_s\}$ be the distinct orders of the isotropy subgroups of the action on $\widetilde{M}$ and $n$ be the least common multiple of $\{\mu_1, \ldots, \mu_s\}$. Then $'S^1/\mathbb{Z}_n$ acts freely on the simply connected $\mathbb{Z}_n \backslash \widetilde{M}$ with quotient the simply connected 2-manifold $'S^1 \backslash \widetilde{M} = W$. Since $\widetilde{M}$ is orientable, $W$ has trivial boundary (no SE-orbits) and so must be either $S^2$ or $\mathbb{R}^2$. $W$ cannot be $\mathbb{R}^2$, for then $('S^1/\mathbb{Z}_n, \mathbb{Z}_n \backslash \widetilde{M})$ would be $('S^1/\mathbb{Z}_n, 'S^1/\mathbb{Z}_n \times \mathbb{R}^2)$ and not simply connected. Therefore, $W$ must be $S^2$ if the image of $\mathrm{ev}_*^x$ is finite. Consequently, $('S^1/\mathbb{Z}_n, \mathbb{Z}_n \backslash \widetilde{M})$ is $S^3$ since it is a simply connected $S^1$-bundle over $S^2$. This means that $\widetilde{M}$ is compact, and so $M$ must be compact and $\pi_1(M)$ is finite. Thus, if $\pi_1(M)$ is infinite, $(S^1, M)$ is injective. □

COROLLARY 14.3.2. *For $(S^1, M)$ with $M$ compact and without fixed points, $M$ is aspherical if and only if $\pi_1(M)$ is infinite and not a central extension of $\mathbb{Z}$ by a finite group. For noncompact $M$, $(S^1, M)$ is always injective and $M$ is always aspherical.*

PROOF. The proof of Lemma 14.3.1 shows that $M$ must be compact and $\pi_1(M)$ finite if $(S^1, M)$ is not injective. If $(S^1, M)$ is injective, then $(S^1, M_{\mathrm{Im}(\mathrm{ev}_*^x)})$ is either $(S^1, S^1 \times S^2)$ or $(S^1, S^1 \times \mathbb{R}^2)$. So $M$ is aspherical if and only if $(S^1, M_{\mathrm{Im}(\mathrm{ev}_*^x)})$ is $(S^1, S^1 \times \mathbb{R}^2)$. □

THEOREM 14.3.3 ([**CR72b**, §12]). *Let $(S^1, M)$ be an effective action of $S^1$ without fixed points on a closed, connected 3-manifold $M$. Then $(S^1, M)$ is injective and $\tau: M \longrightarrow B = S^1 \backslash M$ is a Seifert fibering modeled on $\mathbb{R}^1 \times S^2$ or $\mathbb{R}^1 \times \mathbb{R}^2$ with typical fiber $\mathbb{Z} \backslash \mathbb{R}$ unless $B = S^2$ with fewer than four exceptional orbits.*

PROOF. Let $\{\mu_1, \ldots, \mu_s\}$ be the distinct orders of the isotropy subgroups, and let $n$ be the least common multiple of $\{\mu_1, \ldots, \mu_s\}$. Then $\mathbb{Z}_n \subset S^1$, and on $\mathbb{Z}_n \backslash M = M'$ is induced a free $\mathbb{Z}_n \backslash S^1$-action whose orbit space is again $B$. If $B$ is not $S^2$ or $\mathbb{R}P_2$, then $B$ is an aspherical 2-manifold with possible boundary. We have the following commutative diagram.

$$\begin{array}{ccccccc}
\pi_1(S^1) & \xrightarrow{\text{ev}^x_*} & \pi_1(M) & \xrightarrow{\tau_*} & \pi_1(M)/\text{Im}(\text{ev}^x_*(\pi_1(S^1))) & \longrightarrow & 1 \\
\downarrow & & \downarrow & & \downarrow & & \\
\pi_1(S^1/\mathbb{Z}_n) & \xrightarrow{\text{ev}^{x'}_*} & \pi_1(M') & \xrightarrow{\overline{\tau}_*} & \pi_1(B) & \longrightarrow & 1
\end{array}$$

The bottom sequence is part of a homotopy exact sequence of the bundle and $\tau_*$ is onto. In addition, the vertical homomorphism $\pi_1(M) \to \pi_1(M')$ is onto. If $B$ is not $\mathbb{R}P_2$ or $S^2$, then $\text{ev}_* : \pi_1(S^1/\mathbb{Z}_n) \to \pi_1(M')$ is injective, implying that $\text{ev}^x_* : \pi_1(S^1) \to \pi_1(M)$ is injective. Note also that when $\pi_1(B)$ is not finite, then $\pi_1(M)$ is infinite and not a central extension of $\mathbb{Z}$ by a finite group and hence by Corollary 14.3.2, $M$ is aspherical.

If $B$ is $\mathbb{R}P_2$, then $M'$ is a principal $S^1$-bundle over $\mathbb{R}P_2$. There are two such bundles given by the elements of $H^2(\mathbb{R}P_2; \mathbb{Z}) \cong \mathbb{Z}_2$. They are the trivial bundle $S^1 \times \mathbb{R}P_2$ and the nontrivial bundle $S^1 \times_{\mathbb{Z}_2} S^2$ which is the same as the nontrivial $S^2$-bundle over $S^1/\mathbb{Z}_2$ (note here the $S^1$-action has a global slice). In both cases, the $S^1$-action is injective. In fact, the actions are *homologically injective* (i.e., $\text{ev}^x_* : H_1(S^1, 1; \mathbb{Z}) \longrightarrow H_1(M', x; \mathbb{Z})$ is injective), a condition considerably stronger than injective. (For example, if $(S^1, M)$ is homologically injective and $H_1(M; \mathbb{Z})$ is finitely generated, then $M$ fibers equivariantly over $S^1/\mathbb{Z}_n$ for some cyclic group $\mathbb{Z}_n$. That is, $(S^1, M)$ has a global slice and $(S^1, M)$ is $S^1$-equivalent to $(S^1, S^1 \times_{\mathbb{Z}_n} N)$; see Subsections 14.7.5–14.7.10 and Sections 3.1 and 11.6 for important applications of homological injectiveness in more general contexts.) This implies, as above, that $(S^1, M)$ is homologically injective.

We also remark that *if $B$ has boundary, the action is homologically injective because $M' = S^1 \times B$*.

If $B = S^2$, we need to show that $\pi_1(M)$ is infinite whenever $M$ contains more than three exceptional orbits. In fact, in this case, *$\pi_1(M)$ will not be a central extension of $\mathbb{Z}$ by a finite group, and so $M$ will be aspherical*. Let $B = D^+ \cup D^-$ where $D^+$ and $D^-$ denote the upper and lower hemisphere. Let $b_1, \ldots, b_m$ be the images of the exceptional orbits in $B$. We can assume that $b_1$ and $b_2$ are in the interior of $D^+$ and $b_3, \ldots, b_m$ are in the interior of $D^-$. Let $M^- = \tau^{-1}(D^-)$, and $M^+ = \tau^{-1}(D^+)$. We will use the Seifert-Van Kampen theorem to show that $\pi_1(M) = \pi_1(M^+ \cup M^-) = \pi_1(M^+) *_{\pi_1(T)} \pi_1(M^-)$, where $T = \tau^{-1}(D^+ \cap D^-)$ is a trivial $S^1$-bundle over the equator $D^+ \cap D^- \subset S^2$, and thus a torus. $\pi_1(T)$ injects into both $\pi_1(M^+)$ and $\pi_1(M^-)$, and consequently, $\pi_1(M)$ is infinite and not a finite extension of $\mathbb{Z}$.

Let $x \in \tau^{-1}(b_i)$, and choose a closed cell slice $S_{x_i}$ at $x_i$. Then $D_i = \tau(S_{x_i})$ is a closed cell neighborhood of $b_i$ in $B$ and $\tau^{-1}(D_i)$ is a fibered solid torus. Let

$$B_0 = B - \bigcup_{i=1}^{m} \text{Interior}(D_i)$$

and

$$B_0^+ = D^+ \cap B_0, \qquad B_0^- = D^- \cap B_0.$$

We reconstruct $M^-$ by taking $M_0^-$ and equivariantly filling in each fibered solid torus. The action on $M_0^-$ is a product action $S^1 \times B_0^-$ and has a section which we think of as $1 \times B_0^-$. On each fibered solid torus, this section appears as a curve $Q_i$.

We orient $Q_i$ so that it is opposite to the orientation induced by the orientation of the interior of $B_0^-$. Then the meridian $m_i$ of the solid torus is attached to $\partial B_0^-$ by $m_i \sim \alpha_i Q_i + \beta_i H$. The attachment is determined up to isotopy by this homotopy relation. If we let $q_i$ denote the generator of $\pi_1(Q_i)$ and $h$ the generator of $\pi_1(H)$,

$$\pi_1(M_0^-) = \langle q_0, q_3, q_4, \ldots, q_m, h \mid q_0 q_3 q_4 \cdots q_m,\ h\ \text{central}\rangle$$

and

$$\pi_1(M_0^- \cup \tau^{-1}(D_i)) = \langle q_0, q_3, q_4, \ldots, q_m, h \mid q_0 q_3 q_4 \cdots q_m,\ q_i^{\alpha_i} h^{\beta_i},\ h\ \text{central}\rangle.$$

Continuing by attaching each fibered solid torus, we get

$$\pi_1(M^-) = \langle q_0, q_3, q_4, \ldots, q_m, h \mid q_0 q_3 q_4 \cdots q_m,\ q_i^{\alpha_i} h^{\beta_i},\ (3 \leq i \leq m),\ h\ \text{central}\rangle.$$

The quotient

$$\pi_1(M^-)/\langle h \rangle = \langle q_0, q_3, q_4, \ldots, q_m \mid q_0 q_3 q_4 \cdots q_m,\ q_i^{\alpha_i}\ (3 \leq i \leq m)\rangle$$

is the free product of $(m-2)$ finite cyclic groups.

Therefore, the group generated by $q_0$ and $h$, which is $\pi_1(\tau^{-1}(D^+ \cap D^-)) = \mathbb{Z} \times \mathbb{Z}$ is embedded in $\pi_1(M^-)$. Similarly, $q_0$ and $h$ generate $\mathbb{Z} \times \mathbb{Z}$ and is embedded in $\pi_1(M^+)$. $\pi_1(M)$ is obtained by amalgamating $\pi_1(M^+)$ and $\pi_1(M^-)$ by the embedded $\mathbb{Z} \times \mathbb{Z}$. In particular, $\pi_1(M)$ is infinite and contains $\mathbb{Z} \times \mathbb{Z}$ and is not a central extension of $\mathbb{Z}$ by a finite group. Therefore, $(S^1, M)$ is injective and $M$ is aspherical and

$$\pi_1(M) = \langle q_1, \ldots, q_m, h \mid q_1 q_2 \cdots q_m,\ q_i^{\alpha_i} h^{\beta_i},\ [q_i, h]\ (1 \leq i \leq m)\rangle$$

with $m > 3$. $\square$

PROPOSITION 14.3.4. *If $(S^1, M)$ is injective and effective with $Q = \pi_1(M, x)/\mathrm{ev}_*^x(\pi_1(S^1, 1))$ finite, $M$ compact and connected, then $(S^1, M)$ is covered by $(S^1, S^1 \times S^2)$. The orbit space $B$ is $S^2$, $\mathbb{R}P_2$ or $D^2$ with no E-orbits, $\mathbb{R}P_2$ and $D^2$ with one E-orbit, or $S^2$ with exactly two E-orbits of the same multiplicity. In fact, $(S^1, M)$ is homologically injective and $M$ is homeomorphic to $S^1 \times S^2$, $S^1 \times \mathbb{R}P_2$ or $N$, the nontrivial $S^2$-bundle over $S^1$.*

PROOF. Lift $(S^1, M)$ to the splitting action $(S^1, S^1 \times W)$. $W$ is $S^2$ since $Q$ must act effectively on $W$ and $M$ is compact. Suppose $M$ is orientable. Then $Q$ must preserve orientation on $S^2$. Let $\nu : S^1 \times S^2 \to M = Q\backslash(S^1 \times S^2)$ denote the orbit mapping, and consider the commutative diagram

$$\begin{array}{ccc} H_1(S^1; \mathbb{Q}) & \xrightarrow{i_*} & H_1(S^1 \times S^2; \mathbb{Q}) \\ \Big\| & & \Big\downarrow \nu_* \\ H_1(S^1; \mathbb{Q}) & \xrightarrow{\mathrm{ev}_*^x} & H_1(M; \mathbb{Q}). \end{array}$$

$\nu_*$ is an injection restricted to $H_1(S^1 \times S^2; \mathbb{Q})^Q$ by transfer (cf. [**Bre72**, Chapter 2, §2.4 or §7.2]). But $H_1(S^1 \times S^2; \mathbb{Q})^Q$ is all of $H_1(S^1 \times S^2; \mathbb{Q}) = \mathbb{Q}$. Therefore $\mathrm{ev}_*^x$ in $H_1$ is injective. In Chapter 11, it was shown that a general homologically injective action $(S^1, M)$ is $S^1$-equivalent to $(S^1, S^1 \times_{\mathbb{Z}_n} S)$ for some $S$ where $\mathrm{ev}_*^x(\pi_1(S^1)) \subset \pi_1(S^1 \times S) \subset \pi_1(M)$. In our spacial case, we have a tower of coverings $S^1 \times S^2 \xrightarrow{Q'\backslash} (S^1, S^1 \times S) \xrightarrow{\mathbb{Z}_n\backslash} (S^1, M)$. The group $Q$ acts freely on $S^1 \times S^2$, and there is a normal subgroup $Q'$ acting freely on $S^1 \times S^2$ commuting with the free $S^1$-action with $Q'\backslash(S^1 \times S^2) = S^1 \times S$. On $S^2$, there is induced an effective $Q'$-action with quotient $S$, an orientable 2-manifold. $S$ must be homeomorphic to $S^2$. Because the evaluation homomorphism in $\pi_1(S^1 \times S)$ projects isomorphically to the evaluation isomorphism on $\pi_1(S^1 \times S)$, the group $Q'$ must be the identity. Furthermore, $\mathbb{Z}_n$ acts effectively on both factors. For if $n \geq 1$, $\mathbb{Z}_n$ fixes two points on the sphere and so for the action to be free and diagonal on $S^1 \times S^2$, $\mathbb{Z}_n$ must be an effective rotation on $S^1$. Therefore, $S^1\backslash M$ is $S^2$ with no E-orbits ($n = 1$), or $S^2$ with two E-orbits with multiplicities $n$.

If $M$ is not orientable, let $Q'$ be the characteristic subgroup of index 2 in $Q$ that preserves the orientation of $S^2$, and let $(S^1, M')$ be the orientable double covering of $(S^1, M)$. Then $(S^1, M') = S^1 \times S^2$ or $S^1 \times_{\mathbb{Z}_n} S^2$, where $Q' = 1$ or $\mathbb{Z}_n$ and $Q$ is just an extension of 1 or $\mathbb{Z}_n$ by $\mathbb{Z}_2$. If $Q' = 1$, then $\mathbb{Z}_2$ acts on $S^2$ reversing orientation and $\mathbb{Z}_2 \backslash S^2$ is either $\mathbb{R}P_2$ or $D^2$. If $\mathbb{Z}_2$ is topologically conjugate to the antipodal map, the action on $S^1 \times S^2$ will be free since it is free on $S^2$. Therefore, the quotient is a principal $S^1$-bundle over $\mathbb{R}P_2$ and this action will be equivalent to $(S^1, S^1 \times \mathbb{R}P_2)$ or $(S^1, S^1 \times_{\mathbb{Z}_2} S^2)$, the nontrivial $S^2$-bundle over $S^1$, usually denoted by $N$. If $\mathbb{Z}_2$ is topologically conjugate to a reflection, then $\mathbb{Z}_2 \backslash S^2$ is a disk and there is a connected component of $SE$-orbits corresponding to the boundary of the disk. It is easy to see that this is also homeomorphic to $N$ and can be written as $(S^1, S^1 \times_{\mathbb{Z}_2} S^2)$. If $Q' = \mathbb{Z}_n$, then $\mathbb{Z}_2$ acts on $S^2 = \mathbb{Z}_n \backslash S^2$, up to equivalence, as either a reflection or the antipodal map. The involution on $\mathbb{Z}_n \backslash S^2$ leaves the set of two E-orbits invariant. For the reflection, they must be interchanged. Otherwise, the orbit space would have corner points along the boundary of the disk orbit space. Consequently, $(S^1, M)$ would be an orbifold and not a manifold. Therefore, $S^1 \backslash M$ is a disk with exactly one E-orbit of multiplicity $n$. It is easy to see, using Subsection 14.6.1, that $M$ is homeomorphic to $S^1 \times \mathbb{R}P_2$ if $n$ is even and $N$ if $n$ is odd. For the antipodal map, the E-orbits are again interchanged with both E-orbit images lying on a *great circle* in $\mathbb{Z}_n \backslash S^1$. The new orbit space $\mathbb{R}P_2$ with one E-orbit of multiplicity $n$. Again, using the method of Subsection 14.6.1, the manifold is homeomorphic to $N$ if $n$ is odd and $S^1 \times \mathbb{R}P_2$ if $n$ is even. All of these $(S^1, M)$'s are homologically injective because they are covered by $(S^1, S^1 \times S^2)$.

We had tacitly used the well-known fact that finite group actions on $S^2$ are conjugate to orthogonal linear actions. Another method of proof is to use Subsection 14.7.1 and Propositions 14.7.5 and 14.7.6 to deduce that $(S^1, M)$ has a global slice and is equivalent to $(S^1, S^1 \times_{\mathbb{Z}_n} S)$ when $M$ is orientable. The argument given above avoids the methods developed later in this chapter to deduce these properties. $\square$

COROLLARY 14.3.5. *If $B$ is $S^2$ and $m = 3$, then $(S^1, M)$ is aspherical if $\frac{1}{\alpha_1} + \frac{1}{\alpha_2} + \frac{1}{\alpha_3} \leq 1$, and $\pi_1(M)$ is finite if $\frac{1}{\alpha_1} + \frac{1}{\alpha_2} + \frac{1}{\alpha_3} > 1$.*

PROOF. The formula for $\pi_1(M)$ in Theorem 14.3.3 is still valid for $m \leq 3$ (the Seifert invariants used there are not normalized). Therefore, $Q = \pi_1(M)/(h)$, where $(h)$ is the cyclic central subgroup generated by $\text{Im}(\text{ev}_*^x(\pi_1(S^1)))$ is given by

$$Q = \langle q_1, q_2, q_3 \mid q_1 q_2 q_3, q_1^{\alpha_1} q_2^{\alpha_2} q_3^{\alpha_3} \rangle.$$

This quotient group is isomorphic to a subgroup of index 2 in a well-known triangle group. If $\frac{1}{\alpha_1} + \frac{1}{\alpha_2} + \frac{1}{\alpha_3} \leq 1$, $Q$ is infinite and not an extension of $\mathbb{Z}$ by a finite group. In fact, $\frac{1}{\alpha_1} + \frac{1}{\alpha_2} + \frac{1}{\alpha_3} = 1$ if and only if $(\alpha_1, \alpha_2, \alpha_3) = (2, 4, 4), (3, 3, 3), (2, 3, 6)$ and $Q$ is isomorphic to an orientation preserving Euclidean crystallographic group. If the sum is less than 1, $Q$ is isomorphic to an orientation preserving hyperbolic group. Therefore if the sum is less than or equal to 1, $(S^1, M)$ is injective and aspherical.

If $\frac{1}{\alpha_1} + \frac{1}{\alpha_2} + \frac{1}{\alpha_3} > 1$, then $(\alpha_1, \alpha_2, \alpha_3) = (2, 2, n \geq 2), (2, 3, 3), (2, 3, 4)$ or $(2, 3, 5)$ and $Q$ is finite and is isomorphic to a noncyclic subgroup of SO(3). In fact, these last groups constitute all the isomorphism classes of the noncyclic finite subgroups of SO(3). Now by Proposition 14.3.4, $(S^1, M)$ cannot be injective and so $\pi_1(M)$ must be finite. $\square$

REMARK 14.3.6. (1). All such $(S^1, M)$ in Corollary 14.3.5 where the sum is greater than 1 was shown by Seifert and Threlfall ([**ST31**] and [**ST33**]) to coincide with the 3-dimensional spherical space forms (Subsection 4.5.16) with non-Abelian fundamental group. Recall that $\pi_1(M) \subset U(2)$ and the diagonal $S^1 \subset U(2)$-action on $S^3$ descends to the $S^1$-action on $\pi_1(M) \backslash S^3$. Therefore, the action on $(S^1, M)$ is locally injective because a covering of the $S^1$-action lifts to the Hopf fibering on $S^3$.

(2). For $(S^1, M)$ with $B = S^2$ and $m \leq 2$, then we may, if there are any E-orbits, place an E-orbit over the north pole and another over the south pole in $B$. Thus over the hemispheres bounded by the equator are solid tori (singularly fibered if they contain E-orbits) and they are equivariantly sewn together along their boundaries over the equator of $B$. This amalgamation produces a lens space including $S^3$ or $S^2 \times S^1$. Whenever $S^2 \times S^1$ is produced, the $(S^1, M)$-action will be homologically injective; otherwise $\pi_1(M)$ is finite. A detailed analysis of such $(S^1, M)$ can be found in [**OR69**, §4]; see also Chapter 15. All of these lens spaces (including $S^3$ or $S^2 \times S^1$) also admit very different $S^1$-actions with fixed points and are treated in Subsection 14.5.1.

(3). It is not difficult, using [**OR69**, §4], to determine exactly when an $(S^1, M)$ fails to be locally injective. If $(S^1, M)$ is not locally injective, then $\pi_1(M)$ must be finite and not one of the spherical space forms of Corollary 14.3.5 and (1) above. Therefore the only possibilities are the lens spaces different from $S^1 \times S^2$. If $B = S^2$ with no E-orbits, the action is free and so $(S^1, M)$ is locally injective. If $B = S^2$ with one E-orbit, then $S^1$-action cannot be lifted to any covering space because $\mathrm{ev}_*^x(\pi_1(S^1, 1)) \to \pi_1(M, x)$ will be onto. Therefore this action cannot be locally injective. If $B = S^2$ with two E-orbits of multiplicities $\alpha_1$ and $\alpha_2$, then $(S^1, M)$ can be maximally lifted to a covering space whose covering transformations are isomorphic to $\mathbb{Z}_d$, where $d = \gcd(\alpha_1, \alpha_2)$. This lifted $S^1$-action is free only if $\alpha_1 = \alpha_2$; see Subsection 15.3.2.

A lifting of an effective $S^1$-action to $(S^1, M_{\mathrm{ev}_*^x(\pi_1(S^1))})$ gives rise to a prototype of a Seifert fibering. This prototype becomes a Seifert fibering modeled on a principal $S^1$-bundle if and only if the lifted action on $(S^1, M_{\mathrm{ev}_*^x(\pi_1(S^1))})$ is free, (Section 4.3). That is, a prototype is a Seifert fibering modeled on a principal $S^1$-bundle if and only if $(S^1, M)$ is locally injective. The $S^1$-action on lens spaces, cited above, are the only $S^1$-actions without fixed points on 3-manifolds which fail to be Seifert fiberings in our sense. Of course, these prototypes are not pathological and we abuse our notation by calling these orbit mappings Seifert fiberings to conform with common usage.

The $S^1$-action with fixed points are also prototypes of Seifert fiberings but of course are not modeled on principal bundles since the $S^1$-action can be lifted to the universal covering and there the action has fixed points. It is reasonable to include these actions under a notion of *generalized Seifert fiberings*, a convenient terminology that we use in Sections 14.12 and 14.13.

(4) (Exercise). Show that on any lens space, one can construct an infinite number of distinct $S^1$-actions. (Hint: Construct an infinite number of distinct $S^1$-actions on $S^3$ and let them descend to a lens space (not $S^1 \times S^2$). In fact, construct a torus action on $S^3$ that descends to a given lens space.)

(5). We have shown that when $B$ is $\mathbb{R}P^2$, the action $(S^1, M)$ is homologically injective. Over $B_0$, the action is a product action. Pass to the orientable double cover of $B_0$ and get $B'_0 = S^2 -$ (the interiors of $m$ pairs of disks). Each fibered solid torus of slice type $(\mu, \nu)$ in $(S^1, M)$ lifts to a pair of fibered solid tori of slice type $(\mu, \nu)$ and $(\mu, -\nu)$, and they are sewn onto $\tau^{-1}(B'_0)$ equivariantly to obtain $(S^1, M')$, the orientable double covering of $(S^1, M)$. Thus $S^1 \backslash M' = B' = S^2$ and contains the image exactly $2m$ exceptional orbits. Consequently, if $B = \mathbb{R}P_2$ contains the image of at least two exceptional orbits, then $(S^1, M)$ is homologically injective and $M$ is aspherical. Otherwise, $M'$ is homeomorphic to $S^1 \times S^2$.

(6). In the proof of Theorem 14.3.3, $\tau^{-1}(D^+ \cap D^-) = T$ is a *vertical* torus isomorphic to $(D^+ \cap D^-) \times H$ and $\pi_1(T)$ injects into $\pi_1(M)$. In terms of the usual terminology of 3-dimensional manifolds, $T$ is an embedded incompressible torus. When $m < 4$, $M$ has no embedded vertical tori. (There are, however, horizontal incompressible tori, embedded as global slices, if $(S^1, M)$ is homologically injective and $m = 3$.)

(7). The local smoothness assumption for analysis of $S^1$-actions on 3-manifolds is only a convenience to easily show that a slice can be chosen to be a (closed) cell and the action on slice by the isotropy group is equivalent to an orthogonal action. In general, a slice is a factor of a product neighborhood in a manifold. Thus, it is a homology manifold and because of the low dimensions, it is locally Euclidean. Furthermore, elementary facts concerning Smith theory in these low dimensions easily lead to the action of the isotropy group on the slice to be topologically equivalent to an orthogonal action. The reader can find full details in [**Ray68**].

## 14.4. The classification of $S^1$-actions on closed aspherical 3-manifolds

LEMMA 14.4.1. *Let $(S^1, M)$ be an effective $S^1$-action on a closed connected 3-manifold without fixed points. If $M$ has $t > 0$ components of SE-orbits, then $M$ is aspherical unless the orbit space is a disk and $M$ has less than two E-orbits.*

PROOF. We have seen in the proof of Theorem 14.3.3 that if $t > 0$ and $M$ has no fixed points, then $(S^1, M)$ is injective and even homologically injective. Therefore, if $Q$ is infinite, then $M$ is aspherical. If $Q$ is finite, then $B$ is a disk with less than two E-orbits. □

14.4.2. So now we know that a closed connected 3-manifold $M$ with an effective $S^1$-action without fixed points fails to be aspherical if and only if the orbit space $B$ is one of the following:

(1) $\mathbb{R}P_2$ with ($\leq 1$) E-orbits (Exercise),
(2) $S^2$ with ($\leq 2$) E-orbits,
(3) $S^2$ with three orbits such that $\frac{1}{\alpha_1} + \frac{1}{\alpha_2} + \frac{1}{\alpha_3} > 1$,
(4) $D^2$ with ($\leq 1$) E-orbit.

14.4.3. Assume now that $(S^1, M)$ is an effective action without fixed points on a connected closed aspherical 3-manifold $M$. Then $(S^1, M)$ is modeled on $(\mathbb{R}^1, \mathbb{R}^1 \times \mathbb{R}^2)$ and $(\mathbb{Z}\backslash\mathbb{R}, \pi_1(M)\backslash\mathbb{R}^1 \times \mathbb{R}^2) = (S^1, M)$, and

$$0 \longrightarrow \pi_1(S^1, 1) \xrightarrow{\mathrm{ev}^x_*} \pi_1(M) \longrightarrow Q \longrightarrow 1$$

is a central extension of $\mathbb{Z}$ by $Q$ where $Q$ acts effectively and properly on $\mathbb{R}^2 = W$ so that $S^1\backslash M = Q\backslash \mathbb{R}^2 = B$. Thus the lifting of $(S^1, M)$ to $(S^1, S^1 \times \mathbb{R}^2)$ and then that to $(\mathbb{R}^1, \mathbb{R}^1 \times \mathbb{R}^2)$ yields an example of a Seifert Construction. That is, we get an injective isomorphism $\theta : \pi_1(M) \to \mathrm{TOP}_{\mathbb{R}^1}(\mathbb{R}^1 \times \mathbb{R}^2)$ such that

$$\begin{array}{ccccccccc}
0 & \longrightarrow & \mathbb{Z} & \longrightarrow & \pi_1(M) & \longrightarrow & Q & \longrightarrow & 1 \\
& & \downarrow & & \downarrow \theta & & \downarrow \varphi \times \rho & & \\
1 & \longrightarrow & \mathrm{M}(\mathbb{R}^2, \mathbb{R}^1) & \longrightarrow & \mathrm{TOP}_{\mathbb{R}^1}(\mathbb{R}^1 \times \mathbb{R}^2) & \longrightarrow & \mathrm{Aut}(\mathbb{R}^1) \times \mathrm{TOP}(\mathbb{R}^2) & \longrightarrow & 1,
\end{array}$$

the rows are exact, the vertical homomorphisms are injective and $\varphi$ is trivial since the top row is a central extension.

We need the following important rigidity fact: Let

$$\rho_i : Q_i \longrightarrow \mathrm{TOP}(\mathbb{R}^2), \; i = 1, 2$$

be discrete groups acting properly, effectively and cocompactly on $\mathbb{R}^2$. If $\psi : Q_1 \to Q_2$ is an isomorphism, then there exists a homeomorphism $h : \mathbb{R}^2 \to \mathbb{R}^2$ so that $h \circ \rho_1 \circ h^{-1} = \rho_2 \circ \psi$; that is, the diagram

$$\begin{array}{ccc}
\mathbb{R}^2 & \xrightarrow{h} & \mathbb{R}^2 \\
\downarrow \rho_1(q) & & \downarrow \rho_2(\psi(q)) \\
\mathbb{R}^2 & \xrightarrow{h} & \mathbb{R}^2
\end{array}$$

commutes for each $q \in Q$; see [**Mac67**].

Now suppose we have $(S^1, M_1)$ and $(S^1, M_2)$, two actions satisfying the conditions above. For now, exclude those $Q$'s that have nontrivial centers. (These $Q$'s are $\mathbb{Z} \times \mathbb{Z}$, $\pi_1$(Klein bottle), $\mathbb{Z} \times (\mathbb{Z}_2 * \mathbb{Z}_2)$ and $\{v, k_1 |\; k_1 v^2 = v^2 k_1, k_1^2\}$. Here the base, $Q\backslash \mathbb{R}^2 = B$, is the torus, the Klein bottle, the annulus, and the Möbius band, respectively.) All of these $Q$'s have no E-orbits. All the remaining $Q_i$ acts properly, effectively and cocompactly on $\mathbb{R}^2$ and have trivial center. Let $\psi' : \pi_1(M_1) \longrightarrow \pi_1(M_2)$ be an isomorphism. Because the $Q_i$ are centerless, the center of $\pi_1(M_i)$ is exactly $\mathbb{Z} = \mathrm{Im}(\mathrm{ev}^x_*)$. Consequently $h$ induces an isomorphism of the exact sequences

(14.4.1)
$$\begin{array}{ccccccccc}
0 & \longrightarrow & \mathbb{Z} & \longrightarrow & \pi_1(M_1) & \longrightarrow & Q_1 & \longrightarrow & 1 \\
& & \downarrow \psi'' & & \downarrow \psi' & & \downarrow \psi & & \\
0 & \longrightarrow & \mathbb{Z} & \longrightarrow & \pi_1(M_2) & \longrightarrow & Q_2 & \longrightarrow & 1.
\end{array}$$

Since $(S^1, M_i)$ are injective and modeled on $(\mathbb{R}^1, \mathbb{R}^1 \times \mathbb{R}^2)$ and satisfy the conditions in Subsection 14.4.2, we have $\theta_i : \pi_1(M_i) \longrightarrow \mathrm{TOP}_{\mathbb{R}^1}(\mathbb{R}^1 \times \mathbb{R}^2)$. In Theorem 7.3.2 and Subsection 11.2.3, it is shown that the rigidity condition on the $Q_i$ implies that the injective homomorphisms $\theta_i$ are conjugate in $\mathrm{TOP}_{\mathbb{R}^1}(\mathbb{R}^1 \times \mathbb{R}^2)$; that is, $k\theta_1 k^{-1} = \theta_2 \circ \psi'$. This means the following.

## 14.4. THE CLASSIFICATION OF $S^1$-ACTIONS ON 3-MANIFOLDS

THEOREM 14.4.4. *Let $(S^1, M_1)$ and $(S^1, M_2)$ be effective actions on closed aspherical 3-manifolds with $M_i$ not 3-torus, or the two principal $S^1$-bundles over the Klein bottle. Then the actions $(S^1, M_1)$ and $(S^1, M_2)$ are weakly topologically equivalent if and only if $\pi_1(M_1)$ is isomorphic to $\pi_1(M_2)$.*

PROOF. Since $(S^1, M)$ is injective, there is induced a central extension $0 \to \mathbb{Z} \xrightarrow{\text{ev}^x_*} \pi_1(M, x) \to Q \to 1$. The group $Q$ acts cocompactly on $\mathbb{R}^2$ and $B = Q \backslash \mathbb{R}^2$. These groups are well known. In every case, $Q$ has trivial center except for where $Q = \mathbb{Z}^2$ and $B$ is a 2-torus; $Q$ is $\pi_1$(Klein bottle) and $B$ is the Klein bottle; $Q$ is $\mathbb{Z} \times (\mathbb{Z}_2 * \mathbb{Z}_2)$ and $B$ is an annulus; or $Q$ is $\{v, k_1, x |\ v^2 x, [v, k_1^2], [x, k_1]\}$ and $B$ is the Möbius band. In these four special cases, $B$ has no E-orbits. Thus, these $Q$'s are distinct from each other.

If $Q = \mathbb{Z} \times \mathbb{Z}$, then $(S^1, M)$ is a principal $S^1$-bundle over the 2-torus. These are the nilmanifolds discussed earlier in Example 4.5.12. The bundles are classified by their Euler classes $-b$ (i.e., their characteristic class) which is an element of $H^2(T^2, \mathbb{Z}) \cong \mathbb{Z}$. The associated $(S^1, M)$ has $H_1(M; \mathbb{Z}) \cong (\mathbb{Z} \oplus \mathbb{Z}) \oplus \mathbb{Z}_{|b|}$, where $-b$ is the Euler class of the bundle. If $b \neq 0$, the principal orbit generates the center of $\pi_1(M)$. Thus for $Q = \mathbb{Z} \times \mathbb{Z}$, $(S^1, M)$ is distinguished by $|b| \in \mathbb{Z}$, $b \neq 0$, up to weak equivalence. (The bundle over the punctured torus is a product bundle. We fill in a final fibered solid torus by attaching equivariantly $S^2 \times S^1$, where $(\partial D^2 \times 1) = m \sim Q + bH$. The curve $Q$ is matched to a cross section over the punctured torus, and $H$ is matched to a principal orbit. This corresponds to a $(1, b)$ Seifert invariant and so the extended action on the final solid torus over the puncture remains free. The integer $b$ is the negative of the obstruction to completing the cross section $Q \sim m - bH$ over the entire torus.) If $b = 0$, $M$ is the 3-torus and any circle subgroup determines a principal $S^1$-action. These actions on the 3-torus are not, in general, equivalent but of course there is an isomorphism of $\mathbb{Z}^3 = \pi_1(M_1)$ onto $\mathbb{Z}^3 = \pi_1(M_2)$ carrying $\pi_1(S^1)$ of $(S^1, M_1)$ onto $\pi_1(S^1)$ of $(S^1, M_2)$.

If $Q$ is the group $\pi_1$(Klein bottle), then there are two principal $S^1$-bundles over the Klein bottle, $S^1 \times K$ and $SK$ where $K$ represents the Klein bottle. These manifolds are not homeomorphic (check their first homology groups).

However, $S^1 \times K$ is homeomorphic to the $S^1$-manifold whose orbit space is an annulus without any E-orbits. For if $B$ is an annulus and $A$ is an arc as in the picture, then over the arc $A$ we have, in $M$, an invariant Klein bottle. This action has two components of SE-orbits and there is a global section yielding $S^1 \times K$.

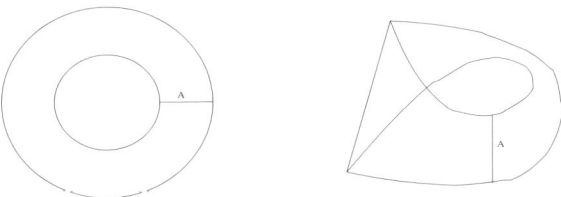

FIGURE 1. Annulus and Möbius band

When $B$ is a Möbius band and $A$ is an arc as in the above figure, we see that over each vertical line segment, such as $A$, we have invariant Klein bottle. There is a circle of these corresponding to the center circle. This bundle is twisted and homeomorphic to $SK$ which also fibers over $S^1$, with the Klein bottle as a fiber.

In all other cases except for these four exclusions, the group $Q$ has trivial center and so the fundamental group of the principal $S^1$-orbit in $(S^1, M)$ generates the image of the evaluation homomorphism, $\operatorname{ev}_*^x(\pi_1(S^1)) \subset \pi_1(M, x)$. Therefore $\psi' : \pi_1(M_1) \longrightarrow \pi_1(M_2)$ induces a diagram (14.4.1) and, because $Q$ is rigid, we have that $(S^1, M_1)$ and $(S^1, M_2)$ are weakly topologically equivalent. Note, if $\psi''$ is the *identity* map, then the actions are equivalent. □

## 14.5. The classification of $S^1$-actions with fixed points on 3-manifolds

14.5.1 (The lens spaces). Let $V_i$ ($i = 1, 2$) be two solid tori, and let $m_i$ and $\ell_i$ be meridians and longitudes on $\partial V_i$. The lens space $L(\mu, \nu)$ is formed by attaching $V_2$ to $V_1$ by an orientation reversing homeomorphism of $\partial V_2$ to $\partial V_1$, where

$$m_2 \sim \mu\ell_1 + \nu m_1.$$

It is easy to construct an infinite number of distinct $S^1$-actions without fixed points on each lens space (they all descend from an infinite number of distinct fixed point free actions on $S^3$).

Now take $V_1$ as a fibered solid torus of type $(\mu_1, \nu_1)$, and on $V_2$ take the action given by

$$z \times (z_1, \rho z_2) \mapsto (z_1, z\rho z_2).$$

This $S^1$-action is given by rotation just in the second factor of $S^1 \times D^2$. Thus the core circle is fixed and all other orbits are free. The orbit space is an annulus. On $\partial V_2$, the orbits $(z_1, z)$ are the meridians. We match $m_2$ to $H_1 \sim \mu_1\ell_1 + \nu_1 m_1$. The cross sectional curve $q_2 = (z_1, 1)$ is sent to the negative of some cross sectional curve $-q_1$ (negative in order to reverse orientation) on $\partial V_1$. We have constructed an action with exactly one E-orbit of type $(\mu_1, \nu_1)$ and a circle of fixed points on $L(\mu_1, \nu_1)$.

EXERCISE 14.5.2. Show that if a lens space $L(\mu, \nu)$ has an effective $S^1$-action with fixed point, then it has a circle of fixed points and exactly one E-orbit with isotropy group $\mathbb{Z}_\mu$. (Hint: $H_*(L(\mu, \nu); \mathbb{Q}) \cong H_*(S^3; \mathbb{Q})$ and use the Smith theorems (Section 1.8) for $S^1$-actions.)

THEOREM 14.5.3 (Classification [**Ray68**]). *Let $(S^1, M)$ be an effective $S^1$-action with fixed points on a connected closed 3-manifold $M$. Then the orbit space $B$ is a compact 2-manifold of genus $g$, with $k > 0$ boundary components $\{F_\ell : \ell = 1, \ldots, k\}$ consisting of images of the fixed set $F$, $t \geq 0$ boundary components $\{C_j : j = 1, \ldots, t\}$ consisting of the images of the components of the SE-orbits, and $\{b_1, \ldots, b_m\}$, $m \geq 0$, the images of E-orbits with respective slice invariants $(\mu_i, \nu_i)$, $i = 1, \ldots, m$. Furthermore, $M$ is homeomorphic to*

(a) $S^3 \# (S^2 \times S^1)^{2g+k-1} \# (\mathbb{R}P_2 \times S^1)^t \# L(\mu_1, \nu_1) \# \cdots \# L(\mu_m, \nu_m)$, *if $B$ is orientable and $t \geq 0$,*

(b) $N \# (S^2 \times S^1)^{g+k-2} \# L(\mu_1, \nu_1) \# \cdots \# L(\mu_m, \nu_m)$, *if $B$ is nonorientable and $t = 0$ ($N$ is the nontrivial $S^1$-bundle over $S^2$),*

(c) $(S^2 \times S^1)^{g+k-1} \# (\mathbb{R}P_2 \times S^1)^t \# L(\mu_1, \nu_1) \# \cdots \# L(\mu_m, \nu_m)$, *if $B$ is nonorientable and $t > 0$.*

The facts about the orbit space were demonstrated in Proposition 14.2.3. We need to

(1) Show that the orbit data determines the manifold $M$ and the $S^1$-action,

## 14.5. THE CLASSIFICATION OF $S^1$-ACTIONS WITH FIXED POINTS

(2) Construct, as a connected sum, a manifold with $S^1$-action to match the given data.

PROOF. (1) Let $D_i$, $i = 1, \ldots, m$ be images in $B$ of slices on E-orbits with slice invariants $(\mu_i, \nu_i)$, $0 < \nu_i < \mu_i$. Then $\tau^{-1}(D_i) = V_i$ is a fibered solid torus of type $(\mu_i, \nu_i)$. Let $N(C_j)$ be an annular collared neighborhood of $C_j$. Then $\tau^{-1}(N(C_j))$ is an invariant neighborhood of an SE-component isomorphic to $(S^1, (\text{Möbius band}) \times S^1)$ with the $S^1$-action on the first factor only, and let $N(F_\ell)$ be an annular collared neighborhood of an $F_\ell$ in $B$. Then $\tau^{-1}(N(F_\ell))$ is an $S^1$-invariant solid torus $(S^1, S^1 \times D^2)$ with the $S^1$-action solely along the second factor. Put

$$B_0 = B - \text{Interior}(\bigcup_{i=1}^{m} D_i \cup \bigcup_{j=1}^{t} N(C_j) \cup \bigcup_{\ell=1}^{k} N(F_\ell)).$$

This is a compact 2-manifold with $m+t+k > 0$ boundary components. The action is free on $M_0 = \tau^{-1}(B_0)$. On the boundaries of each fibered solid torus $\tau^{-1}(D_i)$, the fibers and the meridian curves are fixed.

The slice data $(\mu, \nu)$ depends upon how the slice has been oriented. We may suppose that the fiber $S^1$ is oriented by increasing angle. Then if $M' - (\text{SE-orbits})$ is orientable, we can orient $M'$ by a product of an orientation of the base times the orientation of the fiber. This determines an orientation of each fibered solid torus and their slice invariants. However, if $M'$ is not orientable, there is no canonical orientation for the fibered solid torus. There is an equivariant isotopy over a neighborhood of an orientation reversing curve, in the base, which reverses the local orientation on the solid fibered torus. This will change $(\mu, \nu)$ to $(\mu, -\nu)$ and the Seifert invariant $(\alpha, \beta)$ to $(\alpha, -\beta)$.

With the fixed curves $m_i$ and $H$ on $\partial V_i$, we can choose longitudes $\ell_i$ and cross sectional curves $Q_i$ so that the Seifert invariant $(\alpha_i, \beta_i)$ is normalized as $0 < \beta_i < \alpha_i$ if $M - (\text{SE-orbits})$ is oriented, and $0 < \beta_i \le \alpha_i/2$ if $M - (\text{SE-orbits})$ is not orientable. These curves $\ell_i$ and $Q_i$ are then unique up to isotopy if we normalize both $\beta_i$ and $\nu_i$. In particular, the generators of the homology on $\partial V_i$ satisfy the relations $m_i \sim \alpha_i Q_i + \beta_i H$, where $H$ denotes a free orbit. We can regard the assignment

$$s_i : \partial D_i \longrightarrow Q_i \subset \tau^{-1}(D_i)$$

as a cross section to the free action on $\partial(\tau^{-1}(D_i))$. Let $s$ be the collection of these cross sections and this gives

$$s : \bigcup_{i=1}^{m}(\partial D_i) \longrightarrow \bigcup_{i=1}^{m} \partial(\tau^{-1}(D_i))$$

a cross section to the $S^1$-action on the boundaries of the fibered solid tori. The partial section can be extended to a cross section $s_0$ to the orbit mapping over all $B_0$ since $k > 0$. Note, $\tau^{-1}(N(F_\ell))$ is a solid torus $S^1 \times D^2$ with action equivalent to $z \times (z_1, \rho z_2) \mapsto (z_1, z\rho z_2)$. The section $s_0$ intersects $\tau^{-1}(N(F_\ell))$ along a curve $P$ in $\partial(S^1 \times D^2)$ where $P \sim nS^1 + 1\partial D^2$. That is, $P$ is a $(1, n)$ torus knot which spans an annulus, in $S^1 \times D^2$, with boundary $P$ and $F_\ell$. This annulus is itself an extension of $s_0$ to $N(F_\ell) \subset B$. A similar annulus extends the cross section $s_0$ to each $N(C_j)$, $j = 1, \ldots, k$. Since a section to any of the $\tau^{-1}\partial(N(F_\ell))$ or $\tau^{-1}\partial(N(C_j))$ can be extended to a cross section of the orbit mapping of the respective $\tau^{-1}(N(F_\ell))$ or $\tau^{-1}(N(C_j))$, the section $s_0$ can be extended to a section to the orbit mapping over

$B_1 = B - \text{Interior} \bigcup_{i=1}^m D_i$. Since $m_i \sim \alpha Q_i + \beta_i H$ determines how the fibered solid tori must be sewn into $\tau^{-1}(B_1)$, the manifold and the $S^1$-action is completely determined by the given orbit data.

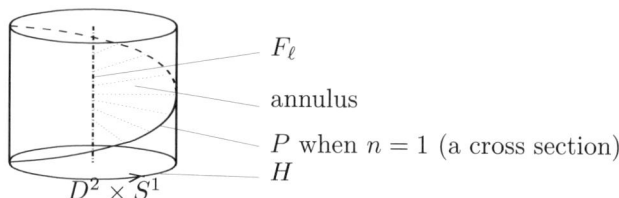

EXERCISE 14.5.4. Given two $S^1$-actions $(S^1, M)$ and $(S^1, M')$ as above with isomorphic orbit data, construct an equivariant homeomorphism between $(S^1, M)$ and $(S^1, M')$.

(2) Let $M_{\epsilon, g, k, t}$ denote an $S^1$-action on $M$ with $k$ components of F-orbits, $t$ components of SE-orbits, no E-orbits and orbit space $B = B_{\epsilon, g, k, t}$ of genus $g$ and $\epsilon = \mathfrak{o}$ or $\mathfrak{n}$, where $\epsilon = \mathfrak{o}$ represents a choice of orientation for $B$ if $B$ is orientable and $\epsilon = \mathfrak{n}$ if $B$ is not orientable. In Theorem 2.6.2, we have shown that $M_{\mathfrak{o}, g, h, 0} = S^3 \# (S^2 \times S^1)^{2g+h-1}$. $B_{\mathfrak{o}, 0, 1, 1}$ is an annulus with an $F$ and SE-boundary component.

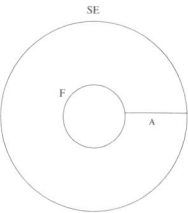

The action on $M_{\mathfrak{o}, 0, 1, 1}$ has a cross section and is $S^1$-equivalent to $\mathbb{R}P_2 \times S^1$ (note that $\tau^{-1}(\text{arc } A) = (S^1, \mathbb{R}P_2)$.) A boundary connected sum to $B_{\mathfrak{o}, g, h, 0}$ of $t$ of these annuli yields $B_{\mathfrak{o}, g, h, t}$ and corresponds to an equivariant connected sum of $t$ copies of $\mathbb{R}P_2 \times S^1$ yielding $M_{\mathfrak{o}, g, h, t}$.

$B_{\mathfrak{n}, 1, 1, 0}$ is a Möbius band and $M = N$, the nontrivial 2-sphere bundle over $S^1$. $B_{\mathfrak{n}, 2, 1, 0} = B_{\mathfrak{n}, 1, 1, 0}(\partial \#) B_{\mathfrak{n}, 1, 1, 0}$, and $M = N \# N = N \# (S^2 \times S^1)$ where $\partial \#$ means boundary connected sum. Now $M_{\epsilon, g, h, t}$ is formed by repeated equivariant connected sums with these building blocks. We finally add all the appropriate E-orbits by doing equivariant connected sums with the appropriate $L(\mu_i, \nu_i)$'s corresponding to the boundary connected sums of $B_{\epsilon, g, h, t}$ with the orbit spaces of the $(S^1, L(\mu, \nu))$'s. This completes the proof of the theorem. $\square$

## 14.6. A complete set of invariants

In a study of the actions of a group on a class of manifolds, one wishes to have both a complete equivariant classification in terms of various orbit data and also a topological identification of the manifolds on which the group acts.

The latter topological identification can take the form of classifying all the various actions on each manifold in the class. In our study of effective $S^1$-actions on closed, connected 3-manifolds, we have seen in Theorem 14.5.3 that if the action has fixed points, then the manifold is a connected sum of certain specific 3-manifolds. From the proof of Theorem 14.5.3, one easily determines all the inequivalent $S^1$-actions with fixed points on each connected sum. Thus we have both equivariant classification and topological identification when the action has fixed points.

If there are no fixed points, the classification and topological identification is more complex. So far we have seen, when $M^{S^1} = \emptyset$, that $M$ is aspherical if and only if $\pi_1(M)$ is infinite and not a central extension of $\mathbb{Z}$ by a finite group (Corollary 14.3.2). Furthermore, we can tell exactly when $M$ is aspherical by the orbit space and the Seifert invariants (Theorem 14.3.3, Lemma 14.4.1, and Subsection 14.4.2). Significantly, in Theorem 14.4.4, we showed that if $M$ is aspherical admitting an $S^1$-action, and not one of the three exceptions, then there is exactly one action on $M$ up to weak equivalence. This is a direct consequence of the resulting Seifert construction $\theta : \pi_1(M) \to \text{TOP}_\mathbb{R}(\mathbb{R}^1 \times \mathbb{R}^2)$ being unique up to conjugation in $\text{TOP}_\mathbb{R}(\mathbb{R}^1 \times \mathbb{R}^2)$. Thus actions $(S^1, M_i)$, $i = 1, 2$, with $M_i$, not one of the 3 exceptions and aspherical, are weakly equivalent if and only if $\pi_1(M_i) \cong \pi_1(M_2)$. While this is undoubtedly quite elegant, having, in addition, an explicit classification solely in terms of numerical data is helpful especially since the desired equivariant classification will not be restricted to aspherical $M$.

14.6.1 (A complete set of orbit data). Let $(S^1, M)$ be an effective action of $S^1$ on a connected closed 3-manifold. In Theorem 14.5.3, the *weighted* orbit space $B_{\epsilon,g,k,t}$ together with the normalized Seifert invariants determine an $S^1$-action up to equivalence. We also showed that if $k = 0$, but $t > 0$, the cross section $s_0 : B_0 \to M_0$ to the orbit mapping extends to $\bigcup_{j=1}^{t} N(C_j)$. Consequently, the action in this case is also completely determined by $B_{\epsilon,g,0,t}$ and the set of normalized Seifert invariants. Only the case where $k = t = 0$ remains to be analyzed. For this, we use the same notation as in Theorem 14.5.3 and adjust the argument as necessary.

As before, we have a partial cross section $s : \bigcup_{i=1}^{m} \partial D_i \to \bigcup_{i=1}^{m} \tau^{-1}(\partial D_i)$ to the orbit mapping. There is an obstruction to extending this partial cross section to $B_0$. It is an element of $H^2(B_0, \partial B_0; \mathbb{Z}) = \mathbb{Z}$, if $B$ is orientable or $\mathbb{Z}_2$ if $B$ is nonorientable; see [**Ste51**, Part III]. Remove, from $M_0$, the interior of an invariant tubular neighborhood of a typical free orbit. It is a solid torus $V$.

Let $\tau(V) = D \subset B_0$ and put $B'_0 = B_0 - \text{Interior}(D)$. The obstruction to extending the partial cross section $s$ to all of $B'_0 \subset B_0$ is an element of $H^2(B'_0, \partial B_0; \mathbb{Z}) = 0$ and so vanishes. This extension $\bar{s}$ intersects $\partial V$ in a cross sectional curve $Q$ to the free action on $\partial V = \tau^{-1}(\partial D)$. With $H$ a typical free orbit on $\partial V$, the simple closed curve $Q$ and $H$ form a homology basis for $\partial V$.

If $M$ is orientable, we choose an orientation. This orientation is the product of an orientation of the base times an orientation of the fiber. So choosing the usual orientation on $S^1$, we orient the base $B$ so that the product orientation gives the orientation of $M$. Any two extensions of the product cross section intersects $\partial V$ in homotopic curves. Thus $H$ and $Q$ are determined uniquely up to isotopy. If we parametrize the boundary of $V = D^2 \times S^1$ in terms of $Q$ and $H$, then $m$, a meridial curve on $V$, satisfies the homology $m \sim Q + bH$ or $Q \sim m - bH$. Thus the integer $-b$ is the obstruction to extending the cross section $\bar{s}$ over $D$. The integer $b$ also describes how $V$ is attached to $\tau^{-1}(B'_0) \cup \bigcup_{i=1}^{m} \tau^{-1}(D_i)$, and so, $B_{o,g,0,0}$, $b$ and the normalized Seifert invariants characterize the action on the oriented $M$ up to orientation preserving homeomorphism.

Note that it does not matter if we choose a different orientation of the fiber as long as we choose the opposite orientation for the base so that the orientation of $M$ is preserved. In fact, there is an involution on $M$ which is an orientation preserving Seifert automorphism (weak equivalence) which *flips* the fiber and base.

If $M$ is not orientable, the local orientation of a solid fibered torus of an E-orbit can be chosen so that the normalized Seifert invariant is $(\alpha, \beta)$, where $0 < \beta \leq \alpha/2$. The partial section $s$ extends to all of $B_0 - \text{Interior}(D)$. This section can be deformed to any other section keeping the section fixed (up to isotopy) on $\bigcup_{i=1}^{m} \partial D_i$ and such that the original cross sectional curve $Q$ on $\tau^{-1}(\partial D)$ is replaced by $Q'$ with $Q' \sim Q + 2nH$, where $n$ can be any integer; see [**Sei33**] for an explicit description. Thus the integer $b$ is determined uniquely up to an even integer. This corresponds to the obstruction class $b \in H^2(B_0, \bigcup_{i=1}^{m} D_i; \mathbb{Z}) = \mathbb{Z}_2$. However, if one of the $(\alpha_i, \beta_i) = (2, 1)$, another adjustment can be made, which is essentially due to $B$ being nonorientable and $-1 \equiv 1 \mod 2$. In this case, there is more than one pair $(Q, H)$ parametrizing the fibered solid torus with normalized Seifert invariant $(2, 1)$. Seifert shows then that the integer $b$ which had been reduced to 0 or 1 can now be reduced to 0. As before, the Seifert invariants $(\alpha_i, \beta_i)$ determine how the solid tori are equivariantly attached to $M_0$ to form $(S^1, M)$.

14.6.2 (Summary). Associate to each effective $S^1$-action on a connected closed 3-manifold $M$, the following symbols which encode all the data determining $(S^1, M)$:

$$(S^1, M) \longleftrightarrow \{\epsilon, g, k, t; \ b; \ (\alpha_1, \beta_1), \ldots, (\alpha_m, \beta_m)\}.$$

Here

$\epsilon = \mathfrak{o}$, an orientation of $B$ if $B$ is orientable or

$\epsilon = \mathfrak{n}$, an orientation of $B$ if $B$ is nonorientable,

$g = $ genus of $B$,

$k = $ number of connected components of fixed points,

$t = $ number of connected components of SE-orbits,

and when

$\epsilon = \mathfrak{o}$, then $0 < \beta_i < \alpha_i$ with $\gcd(\alpha_i, \beta_i) = 1$ (in addition, if $k = t = 0$, then $b \in \mathbb{Z}$; otherwise $b = 0$).

$\epsilon = \mathfrak{n}$, then $0 < \beta_i \leq \alpha_i/2$ with $\gcd(\alpha_i, \beta_i) = 1$ (in addition, if $k = t = 0$ and no $(\alpha_i, \beta_i) = (2, 1)$, then $b \in \mathbb{Z}_2$; otherwise $b = 0$).

We call the quantity $\{\epsilon, g, k, t; \ b; \ (\alpha_1, \beta_1), \ldots, (\alpha_m, \beta_m)\}$ the *Seifert presentation* of $(S^1, M)$.

We shall abuse notation and replace the symbol $\longleftrightarrow$, which associates $(S^1, M)$ with its Seifert presentation $\{\epsilon, g, k, t; \ b; (\alpha_1, \beta_1), \ldots, (\alpha_m, \beta_m)\}$, by the equality symbol. We have shown

THEOREM 14.6.3.

$$(S^1, M) = \{\epsilon, g, k, t; \ b; (\alpha_1, \beta_1), \ldots, (\alpha_m, \beta_m)\}$$
$$(S^1, M') = \{\epsilon', g', k', t'; \ b'; (\alpha_1', \beta_1'), \ldots, (\alpha_m', \beta_m')\}$$

are $S^1$-equivalent (preserving orientation if $M$ − SE-orbits and $M'$ − SE-orbits are oriented) if and only if the two sets of data are isomorphic up to permutation of the order of the normalized Seifert invariants.

REMARK 14.6.4. (1). If $k > 0$, $M$ is a specific connected sum given in Theorem 14.5.3, where $\alpha_i = \mu_i$ and $\beta_i \nu_i \equiv 1 \bmod \alpha_i$.

(2). $M$ fails to be aspherical if and only if either
 (a) $k > 0$, or
 (b) $g = t = k = 0$, $\epsilon = \mathfrak{o}$, $m \leq 2$, or
 (c) $g = t = k = 0$, $\epsilon = \mathfrak{o}$, $m = 3$ and $\frac{1}{\alpha_1} + \frac{1}{\alpha_2} + \frac{1}{\alpha_3} > 1$, or
 (d) $g = 1$, $t = k = 0$, $\epsilon = \mathfrak{n}$, $m \leq 1$,
 (e) $g = k = 0$, $t = 1$, $\epsilon = \mathfrak{o}$, $m \leq 1$.

(3). $M$ is aspherical and not one of the three exceptions ($T^3$ or aspherical $S^1$-bundle over the Klein bottle), then, by Theorem 14.4.4, the $S^1$-action is unique up to equivalence (preserving orientation on $M$ − SE-orbits if $B$ is orientable). That is, if $(S^1, M)$ and $(S^1, M')$ have distinct sets of data, $M$ and $M'$ are not homeomorphic (preserving orientation on $M$ − SE-orbits and $M'$ − SE-orbits are oriented if $B$ is orientable).

(4). The topological classification when $M$ is not aspherical can be found in [**Ray68**], [**OR68**], [**Orl72**] based upon [**Sei33**] and [**vR62**].

(5). The topological classification of *aspherical* $M$ with $t = 0$ was first done independently by [**OVZ67**] and [**Wal67c**]. Both authors worked in the larger context of classical Seifert fiberings and used methods different from those employed in Theorem 14.4.4. All three methods have proven to be important for later developments.

(6). All possible sets of data are realizable if they satisfy the conditions in Summary 14.6.2. If on the other hand, we were to allow Seifert invariants $(\alpha_i, \beta_i)$, where $\gcd(\alpha_i, \beta_i) \neq 1$, we would be describing some of the Seifert orbifolds. The latter is better handled in terms of the Seifert Construction.

14.6.5. Using the Seifert-Van Kampen Theorem and normalized Seifert invariants, we may present $\pi_1(M)$ in terms of the symbols and the conditions imposed on them in Summary 14.6.2 (compare the discussion in Theorem 14.3.3 and the notation used in Theorem 14.5.3 where $B_0 = B - \text{Interior}(\bigcup_{i=1}^m D_i \cup \bigcup_{j=1}^t N(C_j) \cup \bigcup_{\ell=1}^k N(F_\ell)))$. For $\epsilon = \mathfrak{o}$, $k = 0$,

$$\pi_1(M) = \{a_1, b_1, \ldots, a_g, b_g, y_1, \ldots, y_t, h_1, \ldots, h_t, q_1, \ldots, q_m, h \mid$$
$$h \text{ central}, [h_j, y_j], h_j^2 h^{-1}, q_i^{\alpha_i} h^{\beta_i}, \prod_{s=1}^g [a_s, b_s] \prod_{j=1}^t y_j \prod_{i=1}^m q_i h^{-b}\}.$$

Here $b = 0$ if $t > 0$, and $b \in \mathbb{Z}$ if $t = 0$.

For $\epsilon = \mathfrak{o}$, $k > 0$,

$$\pi_1(M) = \{a_1, b_1, \ldots, a_g, b_g, y_1, \ldots, y_t, x_1, \ldots, x_k, h_1, \ldots, h_t, q_1, \ldots, q_m \mid$$
$$[h_j, y_j], h_j^2, q_i^{\alpha_i}, \prod_{s=1}^g [a_s, b_s] \prod_{j=1}^t y_j \prod_{i=1}^m q_i \prod_{\ell=1}^k x_\ell\}.$$

Note this reduces to a free product because if we solve for $x_k^{-1}$ in the long relation, then $x_k$ and the long relation can be eliminated from the presentation.

For $\epsilon = \mathfrak{n}$, $k = 0$,

$$\pi_1(M) = \{v_1, \ldots, v_g, y_1, \ldots, y_t, h_1, \ldots, h_t, q_1, \ldots, q_m, h \mid$$

$$h \text{ central}, [h_j, y_j], h_j^2 h^{-1}, q_i^{\alpha_i} h^{\beta_i}, \prod_{s=1}^{g} v_s^2 \prod_{j=1}^{t} y_j \prod_{i=1}^{m} q_i h^{-b}\}.$$

Here $b = 0$ if $t > 0$, and if $t = 0$, $b = \{0 \text{ or } 1\} \in \mathbb{Z}_2$ unless some $\alpha_i = 2$ in which case $b = 0$.

For $\epsilon = \mathfrak{n}$, $k > 0$,

$$\pi_1(M) = \{v_1, \ldots, v_g, y_1, \ldots, y_t, x_1, \ldots, x_k, h_1, \ldots, h_t, q_1, \ldots, q_m \mid$$

$$[h_j, y_j], h_j^2, q_i^{\alpha_i}, \prod_{s=1}^{g} v_s^2 \prod_{j=1}^{t} y_j \prod_{i=1}^{m} q_i \prod_{\ell=1}^{k} x_\ell\}.$$

Similarly, this is a free product by eliminating the last relation, i.e., solve for $x_k$.

PROPOSITION 14.6.6 ([**Sei33**, §7]). *If* $(S^1, M) = \{\mathfrak{o}, g, 0, 0, b;\ (\alpha_1, \beta_1), \ldots, (\alpha_m, \beta_m)\}$, *then* $(S^1, -M) = \{-\mathfrak{o}, g, 0, 0;\ -b - m;\ (\alpha_1, \alpha_1 - \beta_1), \ldots, (\alpha_m, \alpha_m - \beta_m)\}$.

This follows from the fact that

$$(S^1, -M) = \{-\mathfrak{o}, g, 0, 0;\ -b;\ (\alpha_1, -\beta_1), \ldots, (\alpha_m, -\beta_m)\}$$

if we were to use unnormalized invariants. Normalizing $(\alpha_i, \beta_i)$ yields $(\alpha_i, \alpha_i - \beta_i)$, where the cross sectional curve $Q_i$ is replaced by $Q_i' = Q_i - H$. This changes the "$b$" invariant by "$-1$" for each Seifert invariant. So in the normalization, $-b$ becomes $-b - m$. For an explicit calculation, see [**Sei33**, §7] or [**JN83**, Corollary 7.3].

## 14.7. The Euler number

14.7.1. For the $S^1$-action $(S^1, M)$ with $k = t = 0$ and $\epsilon = \mathfrak{o}$, define the *Euler number* $e(M)$ ([**NR78**]) to be the rational number

$$-\left(b + \sum_{i=1}^{m} \frac{\beta_i}{\alpha_i}\right).$$

When $m = 0$, $M$ is a principal $S^1$-bundle. The sign is designed so that with our orientation convention $e(M)$ agrees with usual Euler class of a principal $S^1$-bundle. $e(M)$ is an oriented invariant and $-e(M) = e(-M)$. We are using normalized Seifert invariants in our definition but it is easily seen that one obtains the same number using unnormalized Seifert invariants; see [**LLR96**].

In fact,

$$(S^1, M) = \{\mathfrak{o}, g, 0, 0; (1, b_1), \ldots, (1, b_s); (\alpha_1, \beta_1), \ldots, (\alpha_m, \beta_m)\}$$

and

$$(S^1, M') = \{\mathfrak{o}, g', 0, 0; (1, b_1'), \ldots, (1, b_{s'}'); (\alpha_1', \beta_1'), \cdots, (\alpha_m', \beta_m')\},$$

using unnormalized invariants, are $S^1$-equivalent via an orientation preserving equivariant homeomorphism if and only if

$$g = g',\ m = m',\ \alpha_i = \alpha_i',\ \beta_i \equiv \beta_i'\ (\alpha_i)$$

(up to permutation of the unnormalized Seifert invariants), and

$$e(M) = -\left(b_1 + \cdots + b_s + \sum_{i=1}^{m} \frac{\beta_i}{\alpha_i}\right) = e(M') = -\left(b_1' + \cdots + b_{s'}' + \sum_{i=1}^{m} \frac{\beta_i'}{\alpha_i'}\right).$$

This follows from the fact cited in Proposition 14.6.6 where if $Q_i$ is replaced by $Q_i - xH$, then we must change the "$b$"-invariant to $b - x$, and $(1, b_1), (1, b_2)$ can be replaced by $(1, b_1 + b_2)$. For an explicit calculation, see [**JN83**, Corollary 7.3] or [**LLR96**].

THEOREM 14.7.2 ([**NR78**, Theorem 8.2]; cf. [**Sei33**, §9]). *Let $f : M \to M$ be an orientation reversing homotopy equivalence, where $M$ is a closed, oriented aspherical 3-manifold admitting an effective $S^1$-action. Then $M$ is of the form*

$$(S^1, M) = \{-\mathfrak{o}, g, 0, 0; m;\ (\alpha_1, \beta_1), (\alpha_1, \alpha_1 - \beta_1), \ldots, (\alpha_m, \beta_m), (\alpha_m, \alpha_m - \beta_m)\}$$

*and consequently $e(M) = 0$.*

PROOF. Since $M$ is aspherical and orientable, $k = t = 0$. Assume $M$ is not the 3-torus since it obviously satisfies the conclusion of the theorem. The homotopy equivalence $f$ induces an automorphism of $\pi_1(M)$ carrying the center $\mathbb{Z}$ onto itself and the quotient $Q = \pi_1(M)/\mathbb{Z}$ onto itself. Therefore, by rigidity, $f$ is homotopic to a Seifert automorphism $h$ which is a weak equivalence of $(S^1, M)$. This means $h(t \cdot x) = a(t) \cdot h(x)$, where $a$ is an automorphism of $S^1$, $t \in S^1$, $x \in M$. Thus $h$ is either $S^1$-equivalence $h(t \cdot x) = t \cdot h(x)$ or if $a(t) = t^{-1}$, $h(t \cdot x) = t^{-1} \cdot h(x)$. In the latter case, this is also an $S^1$-action if we define $(S^1, \diamond, M)$ by $t \diamond x = t^{-1} \cdot x$. Then, $h(t \cdot x) = t \diamond h(x)$. If $\mathfrak{o}$ denotes the orientation of $M$, change the orientation of the range space $M$ to $-\mathfrak{o}$; see Summary 14.6.2. Then the equivariant homeomorphism $h$ is orientation preserving. Consequently, the oriented normalized Seifert invariants $\{\alpha_i, \beta_i|\ i = 1, \ldots, s\}$ of $(M, \mathfrak{o})$ are sent isomorphically (up to ordering) onto some collection of Seifert invariants describing the action on the range $M$ (with respect to the orientation $-\mathfrak{o}$), $\{\alpha_i, \beta_i|\ i = 1, \ldots, s\}$. But we know that the Seifert invariants of $(M, -\mathfrak{o})$ are $\{-b-s;\ (\alpha_1, \alpha_1-\beta_1), \ldots, (\alpha_s, \alpha_s-\beta_s)\}$. Consequently, for each $(\alpha_i, \beta_i)$ in the domain $(M, \mathfrak{o})$, there must be an $(\alpha_i, \alpha_i - \beta_i)$. Furthermore, $h$ must send the "$b$" invariant $b$ of $(M, \mathfrak{o})$ to the "$b$" invariant of $(M, -\mathfrak{o})$. So, $b = -b-s$ or $s = -2b$. The number of $(\alpha, \beta)$'s where $\alpha > 2$ is even because $(\alpha, \beta) \neq (\alpha, \alpha - \beta)$. Since $s$ is even, the number of Seifert invariants of type $(2, 1)$ is also even. Consequently, as $b = -s/2$, $e(M) = 0$, and the form of $(S^1, M)$ using normalized invariants is as asserted. Note this argument works for $s = 0$ and, in this case $b$ must be 0. We remark that a stronger theorem is proved in [**NR78**]. □

EXERCISE 14.7.3. Show that
$$(S^1, M) = \{\epsilon, g, 0, t; b;\ (\alpha_1, \beta_1), \ldots, (\alpha_m, \beta_m)\}$$

lifts to $S^1$-action on the orientable double covering $(S^1, M)$ which commutes with the covering transformations where,

for $\epsilon = \mathfrak{o}$, $t > 0$, (and hence $b = 0$), we have
$$(S^1, M') = \{-\mathfrak{o}, 2g + t - 1, 0, 0; -m;$$
$$(\alpha_1, \beta_1), (\alpha_1, \alpha_1 - \beta_1), \cdots, (\alpha_m, \beta_m), (\alpha_m, \alpha_m - \beta_m)\},$$

and for $\epsilon = \mathfrak{n}$, $t > 0$, we have
$$(S^1, M') = \{-\mathfrak{o}, g + t - 1, 0, 0; -m;$$
$$(\alpha_1, \beta_1), (\alpha_1, \alpha_1 - \beta_1), \cdots, (\alpha_m, \beta_m), (\alpha_m, \alpha_m - \beta_m)\}.$$

Note the existence of these free orientation reversing involutions on $M'$ does not require aspherical assumptions.

PROPOSITION 14.7.4 ([**ST33**, §14] or [**ST80**, Appendix §14, p.409]). *Let $\mathbb{Z}_a \subset S^1$ act on*
$$(S^1, M) = \{\mathfrak{o}, g, 0, 0; b;\ (\alpha_1, \beta_1), \ldots, (\alpha_m, \beta_m)\}.$$
*Then*
$$(S^1/\mathbb{Z}_a, \mathbb{Z}_a \backslash M) = \{\mathfrak{o}, g, 0, 0; ba;\ (\alpha_1, \beta_1 a), \ldots, (\alpha_m, \beta_m a)\}.$$
*Hence if the invariants of $(S^1, M)$ are normalized, then the invariants of $\mathbb{Z}_a \backslash M$ should be interpreted as unnormalized invariants. If $\gcd(a, \alpha_i) = d \neq 1$, then the unnormalized invariant $(\alpha_i, \beta_i a)$ can be interpreted effectively as $(\frac{\alpha_i}{d}, \frac{\beta_i a}{d})$.*

PROOF. Away from each exceptional orbit, the action is free. Roughly, the idea is that on $M'_0 = B'_0 \times S^1$ ($B'_0 = B_0 - \text{Interior}(D)$) are the curves $\partial B'_0 = \{Q_0, Q_1, \ldots, Q_m\}$ which parametrize $\partial B'_0 \times S^1$ by $\{Q_0, \ldots, Q_m\} \times H$. The orbit space of the action of $\mathbb{Z}_a$ on $B_0 \times S^1$ is $B_0 \times \mathbb{Z}_a \backslash S^1$. Thus we can parametrize this orbit space by $Q_0, Q_1, \ldots, Q_m$ and $H'$ where $H' = \mathbb{Z}_0 \backslash H$. The boundary of the new solid torus for $\mathbb{Z}_a \backslash V_i$ is parametrized by $Q_i$ and $H'$. The old meridian, $m_i \sim \alpha_i Q_i + \beta_i H$ and $m_0 \sim Q_0 + bH$ are mapped to $m'_i \sim \alpha_i Q_i + \beta_i(aH')$ and $m'_0 \sim Q_0 + b(aH')$. Therefore, our Seifert invariants are changed from $\{b;\ (\alpha_1, \beta_1), \ldots, (\alpha_m, \beta_m)\}$ to $\{ba;\ (\alpha_1, \beta_1 a), \ldots, (\alpha_m, \beta_m a)\}$ on $M' = \mathbb{Z}_a \backslash M$. □

PROPOSITION 14.7.5 ([**NR78**, Theorem 7.2]). *Let $(S^1, M)$ and $(S^1, M')$ be closed orientable Seifert 3-manifolds over orientable bases $B$ and $B'$. Let $f : M \to M'$ be an orientation preserving surjective mapping such that $f(tx) = t^n f(x)$ for all $x \in M$. This induces a map $\overline{f} : B \to B'$ of degree $k$ on the base. Then*
$$e(M)n = k \cdot e(M').$$

PROOF. If $M$ and $M'$ are both principal bundles, it is easy to check that the proposition holds. For, $e(M) = \langle e(M), [B] \rangle$, where $e(M) \in H^2(B; \mathbb{Z}) = [B; K(\mathbb{Z}, 2)] = [B, \mathbb{C}P_\infty] = [B, B_{S^1}]$ is given by the classifying map of the bundle. A cohomology calculation then yields the result.

Let $a$ be a positive integer divisible by all the $\alpha_i$'s occurring in both $M$ and $M'$. Then $\mathbb{Z}/a\mathbb{Z} = \mathbb{Z}_a \subset S^1$ and $f(gx) = g^n f(x)$ for all $g \in \mathbb{Z}_a$. Therefore, we get the commutative diagram

$$\begin{array}{ccc} M & \xrightarrow{f} & M' \\ \pi \downarrow & & \downarrow \pi' \\ \mathbb{Z}_a \backslash M & \xrightarrow{f_a} & \mathbb{Z}_a \backslash M' \end{array}$$

with $f_a$ also satisfying $f_a \langle tx \rangle = t^n \langle f(x) \rangle$, where $\langle - \rangle$ denotes the image under $\pi$ and $\pi'$. Since $\mathbb{Z}_a \backslash M$ and $\mathbb{Z}_a \backslash M'$ are principal $S^1$-bundles, the proposition holds for the map $f_a$. Now by Proposition 14.7.4, the proposition holds for $\pi$ and $\pi'$. Therefore it follows for $f$. Note the proposition as stated and proved in [**NR78**] is slightly more general. □

PROPOSITION 14.7.6. *For $(S^1, M)$ as in Subsection 14.7.1, $M$ has a global slice if and only if $e(M) = 0$. Thus, the Euler number can be thought of as the obstruction for a global slice. A global slice is sometimes called a multisection.*

PROOF. Recall that $(S^1, M)$ has a global slice if and only if there is an equivariant map $(S^1, M) \xrightarrow{f} (S^1, S^1/\mathbb{Z}_n)$ for some $n \geq 1$. A global slice means we can write $(S^1, M)$ as $(S^1, S^1 \times_{\mathbb{Z}_n} S)$, where $S = f^{-1}(1)$, $1 \in S^1/\mathbb{Z}_n$, is the global slice and $\mathbb{Z}_n$ acts diagonally on $S^1 \times S$. Let $\mathbb{Z} = \text{Im}(\text{ev}_*^x) \subset \Pi' \triangleleft \pi_1(M)$. Thus, $(S^1, M)$ lifts to $(S^1, M')$, where $\pi_1(M') = \Pi'$. If $M'$ is compact, we have $e(M') = k\, e(M)$ where $k$ is the order of the branched covering $S^1 \backslash M' \to S^1 \backslash M$.

If $(S^1, M)$ has a global slice, then there must be some $k$ such that $0 = e(M') = k \cdot e(M)$, where $(S^1, M') = (S^1, S^1 \times S)$. Therefore $e(M) = 0$.

Conversely, suppose $e(M) = 0$, and let $n = \text{lcm}(\alpha_1, \alpha_2, \ldots, \alpha_m)$. Let $\mathbb{Z}_n \subset S^1$ act on $(S^1, M)$ with quotient $(S^1/\mathbb{Z}_n, M') = (S^1/\mathbb{Z}_n, \mathbb{Z}_n \backslash M)$. The quotient is a principal $S^1/\mathbb{Z}_n$-bundles over $S^1 \backslash M$, and $e(M') = e(M) \cdot n$. So $e(M') = 0$ if and only if $e(M) = 0$. If $e(M') = 0$, then $M' = S^1/\mathbb{Z}_n \times S^1 \backslash M$ and the composite $M \xrightarrow{\pi} M' \xrightarrow{\text{proj}} S^1/\mathbb{Z}_n$ is equivariant and $(\text{proj} \circ \pi)^{-1}(1)$ is a global $\mathbb{Z}_n$-slice. □

COROLLARY 14.7.7. *For $(S^1, M)$ as in Subsection 14.7.1, $e(M) = 0$ if and only if $(S^1, M)$ is homologically injective.*

PROOF. If a finite $G$ acts on a compact Hausdorff space $X$, then the orbit mapping $p : X \to G \backslash X$ induces an isomorphism of $H_n(X; \mathbb{Q})^G$ onto $H_n(G \backslash X; \mathbb{Q})$ for each $n$ (see [**Bre72**, Chapter III, 7.2]), where Čech homology is used in general. By applying this fact to the covering map $(S^1, S^1 \times S) \to (S^1, S^1 \times_{\mathbb{Z}_n} S) = (S^1, M)$, we conclude that $e(M) = 0$ implies $(S^1, M)$ is homologically injective. Conversely, if $(S^1, M)$ is homologically injective and $n = \text{lcm}(\alpha_1, \ldots, \alpha_m)$ where the $\alpha_i$ are the orders of the exceptional orbits of $(S^1, M)$, then $(S^1/\mathbb{Z}_n, \mathbb{Z}_n \backslash M) = (S^1/\mathbb{Z}_n, M')$ is a principal $S^1/\mathbb{Z}_n$ bundle over $S^1 \backslash M$. Because $H_1(S^1, \mathbb{Q}) \subset H_1(M; \mathbb{Q})^{\mathbb{Z}_n}$ is injected isomorphically into $H_1(\mathbb{Z}_n \backslash M; \mathbb{Q})$, the principal $S^1$-bundle $M'$ must be the trivial $S^1/\mathbb{Z}_n$ bundle. (One easily checks from Subsection 14.6.5 that if the Euler class of the bundle over an oriented closed surface ($e = -b$ invariant), then the image of $H_1(S^1; \mathbb{Z})$ in $H_1(\text{Bundle}; \mathbb{Z})$ is $\mathbb{Z}_{|b|}$.) Thus, $e(M') = 0$ and consequently $e(M) = 0$. □

14.7.8. Let $(S^1, M)$ be as in Subsection 14.7.1 and injective with exact

$$0 \to \pi_1(S^1, 1) \xrightarrow{\text{ev}_*^x} \pi_1(M) \longrightarrow Q \to 1,$$

the resulting central group extension. This extension can be represented by an element $a \in H^2(Q; \mathbb{Z})$, where the action of $Q$ on $\mathbb{Z}$ is trivial. With $Q$ infinite, $H^2(Q; \mathbb{Z}) = \mathbb{Z} + \text{Torsion}$. (This easily follows from $H^2(Q; \mathbb{Q}) \cong H^2(Q; \mathbb{Z}) \otimes \mathbb{Q}$ and that transfer implies $H^2(Q; \mathbb{Q}) = H^2(S^1 \backslash M; \mathbb{Q}) = \mathbb{Q}$.)

COROLLARY 14.7.9. *For an injective $(S^1, M)$ as in Subsection 14.7.1, the extension class has a finite order if and only if $e(M) = 0$.*

PROOF. Suppose $Q$ is infinite. Let $Q'$ be a finite indexed normal, torsion-free subgroup of $Q$. Thus, rationally, $H^2(Q; \mathbb{Q}) \approx H^2(Q'; \mathbb{Q})^{Q/Q'} \approx \mathbb{Q}$. Thus $(S^1, M')$, the principal bundle over $S^1 \backslash M' = Q' \backslash \mathbb{R}^2$, has $e(M') = -b = a' \in H^2(Q'; \mathbb{Z}) \approx \mathbb{Z}$. Furthermore, $a' = 0$, if and only if the bundle is trivial. This bundle is trivial if and only if $e(M) = 0$. Furthermore, $a$ is of infinite order if and only if $a'$ is not 0.

If $Q$ is finite, then $(S^1, M)$ lifts to the splitting action $(S^1, S^1 \times S^2)$ and, consequently, $e(M) = 0$. □

COROLLARY 14.7.10. *For an injective $(S^1, M)$ as in Subsection 14.7.1, a lift to a regular covering $(S^1, M')$, where $Q' = \pi_1(M', x')/\text{ev}_*^{x'}(\pi_1(S^1))$ is torsion free, is a product $(S^1, S^1 \times S)$ if and only if $e(M) = 0$.*

## 14.8. $\Gamma \backslash G$ as 3-dimensional Seifert manifolds

14.8.1. We have seen examples of Seifert fibering arising from the maps $G/K \to \Gamma \backslash G/K$, where $\Gamma$ is discrete and $K$ is a closed subgroup of a Lie group $G$. Three-dimensional examples include $\Gamma \backslash SO(3) \to \Gamma \backslash SO(3)/SO(2)$. In Example 4.5.15, the principal $SO(3) \to SO(3)/SO(2)$ fibering descends to $(\mathbb{Z}_2 \times \mathbb{Z}_2)\backslash SO(3) \to (\mathbb{Z}_2 \times \mathbb{Z}_2)\backslash SO(3)/SO(3)$, a Seifert fibering over the 2-sphere with three exceptional orbits of type $(2, 1), (2, 1), (2, 1)$. $SO(3)$ is the full group of orientation preserving isometries of $S^2$. Similarly, $PSL(2, \mathbb{R})$ and $E_0(2) = \mathbb{R}^2 \rtimes SO(2)$ are the full group of orientation preserving isometries of $\mathbb{H}^2$ (the hyperbolic plane) and of $\mathbb{R}^2$ (the Euclidean plane). These are isometries with respect to the constant curvature metrics of $S^2$, $\mathbb{H}^2$ and $\mathbb{R}^2$, respectively. In each of these cases, the maximal compact in $G$ is the circle group, $SO(2)$. Of course, $\Gamma \backslash G$ is a 3-manifold which Seifert fibers over $\Gamma \backslash G/SO(2)$. This base space is orientable and when it is closed, $M = \Gamma \backslash G$ is of the form $M = \{\mathfrak{o}, g, 0, 0; b; (\alpha_1, \beta_1), \ldots, (\alpha_m, \beta_m)\}$. In this section, we shall determine the specific Seifert invariants for $M$. This was achieved in [**RV81**], and we shall sketch the result.

14.8.2. In general, if $G$ is a connected Lie group, $K$ a compact subgroup and $\Gamma$ a discrete subgroup, we claim that the slice representation at a point $g\Gamma$ is computable from the adjoint action of $K$ on the Lie algebra $\mathfrak{g}/\mathfrak{k}$. More specifically, let $m_0 = \Gamma g_0 \in M = \Gamma \backslash G$. Consider the adjoint action of $G$ on its Lie algebra $\mathfrak{g}$. The restriction to $K$ leaves $\mathfrak{k}$ invariant. Thus, there is induced an action of $K$ on $\mathfrak{g}/\mathfrak{k}$. If $K_0$ is a subgroup of $K$, we can further restrict the adjoint action to $K_0$ on $\mathfrak{g}/\mathfrak{k}$. We can form the associated vector bundle over $K_0 \backslash K$, $(\mathfrak{g}/\mathfrak{k}) \times_{K_0} K$.

On the other hand, if $K_0$ is the isotropy subgroup of the $K$-action on $\Gamma \backslash G$ at $m_0$, a $K$-invariant tubular neighborhood is given by $(N \times_{K_0} K, K)$. The slice $N$ can be identified with the normal vectors to the orbit at $m_0$ since everything here is smooth. Moreover, the Lie algebra $\mathfrak{g}/\mathfrak{k}$ can be identified with $N$.

PROPOSITION 14.8.3 ([**RV81**, (2.1)]). *There is a $K$-equivariant diffeomorphism between $(\mathfrak{g}/\mathfrak{k}) \times_{K_0} K$ and $N \times_{K_0} K$ and the action of $K_0$ on $N \cong \mathfrak{g}/\mathfrak{k}$ is induced from the adjoint action.*

PROOF. Our task is to show that the action of $K_0$ on the linear space $N$ can be identified with the restriction to $K_0$ of the adjoint action of $K$ on $\mathfrak{g}/\mathfrak{k}$. Let $p : G \to M = \Gamma \backslash G$ be defined by $p(g) = \Gamma g_0 g$. With the $K$-action on $G$ by right multiplication, $p$ is $K$-equivariant. It is a local diffeomorphism taking $1_G \in G$ to $m_0 = \Gamma g_0$. Thus the differential of $p$ at $1_G$ identifies $\mathfrak{g} = T_{1_G}(G)$ with $T_{m_0}(M)$. Further, it identifies $\mathfrak{k}$ with the tangent space to the orbit $m_0 K$ at $m_0$; thus it identifies $\mathfrak{g}/\mathfrak{k}$ with $N$. We now find the action of $K_0$ on $N$. First, it is easily verified that $K_0 = (g_0^{-1} \Gamma g_0) \cap K$. Hence it follows that $p(k_0 g k_0^{-1}) = p(g) k_0^{-1}$ for all $k_0 \in K_0$ and all $g \in G$. Convert the right action to a left action (define $k_0 \cdot n$ to be $n \cdot k_0^{-1}$ for all $n \in N$, $k_0 \in K_0$), then the adjoint action of $K_0$ on $G$ is equivariant, via $p$, with the new left $K_0$-action on $M$. Recall that $\mathrm{Ad}(k_0) : \mathfrak{g} \to \mathfrak{g}$ is defined to be the differential at $1_G$ of the map $g \mapsto k_0 g k_0^{-1}$, $G \to G$, then the result follows. $\square$

14.8.4. Turning now to a connected 3-dimensional group $G$ with $K \cong S^1$, we have $\mathfrak{g}/\mathfrak{k} \cong \mathbb{R}^2$ and

$$S^1 \xrightarrow[\cong]{\theta} K \xrightarrow{\mathrm{Ad}} \mathrm{Aut}(\mathfrak{g}/\mathfrak{k})$$

is of the form
$$(\mathrm{Ad} \circ \theta)(z)(v) = z^r v,$$
for some nonnegative integer $r$, the cardinality of the kernel of the composition $\mathrm{Ad} \circ \theta$. Assume that $r > 0$, which will be the case for $G = \mathrm{SO}(3)$, $E_0(2)$, $\mathrm{PSL}(2, \mathbb{R})$, and any of their finite coverings. If $(S^1)_x = \mathbb{Z}_\alpha$ at $x \in \Gamma \backslash G$, the normalized Seifert invariant $(\alpha, \beta)$ is obtained from the slice invariant $(\alpha, \nu)$ where $\nu\beta \equiv 1 \bmod \alpha$ and $0 < \beta < \alpha$. Now if $\lambda = e^{\frac{2\pi i}{\alpha}}$ is the "generator" of $\mathbb{Z}_\alpha$, the action of $\lambda$ on $v \in \mathfrak{g}/\mathfrak{k} = \mathbb{R}^2 \cong \mathbb{C}$ is identified with $\lambda \times v \mapsto \lambda^{-r} v$, the minus sign coming from changing a right action to a left action. Therefore we have

PROPOSITION 14.8.5 ([**RV81**, (3.2)]). *Let $r$ be the cardinality of the kernel of the map* $\mathrm{Ad} : K \to \mathrm{Aut}(\mathfrak{g}/\mathfrak{k})$. *Let $x \in \Gamma \backslash G$ be an exceptional orbit of multiplicity $\alpha$. Its oriented normalized Seifert invariant is $(\alpha, \beta)$ where $-r\beta \equiv 1 \bmod \alpha$. In particular, $r$ and $\beta$ are relatively prime.*

COROLLARY 14.8.6. *If $G = \mathrm{SO}(3)$, $E_0(2)$ or $\mathrm{PSL}(2, \mathbb{R})$, then $r = 1$ and $(\alpha, \beta) = (\alpha, \alpha - 1)$.*

PROOF. The groups $G$ can be regarded as the unit tangent bundle to $W = G/K = \{S^2, \mathbb{R}^2, \text{ and } \mathbb{H}^2, \text{ respectively}\}$. $G$ acts transitively on $W$. If $\Gamma_w$ is the isotropy group at $w \in W$, then $\Gamma_w = K_m$, where $m \in \Gamma \backslash G$ and $w$ project to the same point in $B = \Gamma \backslash G/K$, and the actions on the slices are equivalent (see Corollary 2.7.7). The action of $\Gamma_w$ is effective and transitive on the unit tangent vectors ($\cong K$) at $w$. Therefore $r = 1$. □

We have been a little sketchy on orientation matters here but this is carefully handled in [**RV81**].

14.8.7 (The "$b$"-invariant). So far, we have not assumed $\Gamma$ is cocompact. Consequently, when $\Gamma \backslash G$ is not compact and where $G = E_0(2)$ or $\mathrm{PSL}(2, \mathbb{R})$, the Seifert manifold is completely determined by the genuses of $\Gamma \backslash G/K$ and the Seifert invariants $(\alpha_i, \alpha_i - 1)$. For example, if $\Gamma = \mathrm{PSL}(2, \mathbb{Z})$ and $G = \mathrm{PSL}(2, \mathbb{R})$, then $M = \mathrm{PSL}(2, \mathbb{Z}) \backslash \mathrm{PSL}(2, \mathbb{R}) \to \mathrm{PSL}(2, \mathbb{Z}) \backslash \mathrm{PSL}(2, \mathbb{R})/K = B$ is a Seifert fibering. Here $B$ is $S^2 -$(point) with two branch points of multiplicity 2 and 3. $M$ then fibers over the sphere minus a point with two exceptional orbits with Seifert invariants $(2, 1)$ and $(3, 2)$. This space is diffeomorphic to the complement of the $(2, 3)$ torus knot in $S^3$. We obtain the same thing, from the action $z \times (z_1, z_2) \to (z^2 z_1, z^3 z_2)$, on the 3-sphere by taking the orbit space and removing a typical fiber. Suppose $\Gamma$ is torsion free, cocompact and $G = \mathrm{PSL}(2, \mathbb{R})$ or $E_0(2)$. Then $\Gamma$ acts freely and isometrically on $W = G/\mathrm{SO}(2) = \mathbb{H}^2$ or $\mathbb{R}^2$, respectively. Then $\Gamma \backslash G$ can be identified with the unit tangent bundle of the oriented closed surface $\Gamma \backslash G/K$. The obstruction to a cross section to this unit tangent bundle is the Euler class of this bundle and is identified with the Euler characteristic $2 - 2g$ of $\Gamma \backslash G/K$. The $b$-invariant is the negative of the obstruction class and so the $b$-invariant of $\Gamma \backslash G$ is $2g - 2$. Note if $G = E_0(2)$, the group $\Gamma$ is torsion-free Euclidean crystallographic and so must be $\mathbb{Z}^2$. Then $(\mathbb{Z} \oplus \mathbb{Z}) \backslash E_0(2)$ is the unit tangent bundle of the 2-torus which is, of course, the 3-torus. Consequently its $b$-invariant is 0.

In general, $\Gamma$ is not torsion free but $\Gamma$ has a torsion-free normal subgroup $\Gamma_0$ of finite index. Then $\Gamma_0 \backslash G \to \Gamma \backslash G$ is a regular covering space and fibers over the branched covering space $\Gamma_0 \backslash G/K \to \Gamma \backslash G/K$. We calculate $e(\Gamma \backslash G)$ in terms of

$e(\Gamma_0\backslash G)$ using Proposition 14.7.5. We have

$$e(\Gamma_0\backslash G) = e(\Gamma\backslash G) \cdot \deg(\Gamma_0\backslash G/K \to \Gamma\backslash G/K)/(\text{degree on fiber})$$

which yields

$$2 - 2g_0 = -\left(b + \frac{\alpha_1 - 1}{\alpha_1} + \cdots + \frac{\alpha_m - 1}{\alpha_m}\right)\left(\frac{k}{1}\right)$$

$$= -k\left(b + m - \sum_{i=1}^{m} \frac{1}{\alpha_i}\right).$$

Now we use the Hurewitz formula for the branched covering $\Gamma_0\backslash G/K \to \Gamma\backslash G/K$ to get

$$2 - 2g_0 = k(2 - 2g) + \left(\text{correction term} = -mk + \frac{k}{\alpha_1} + \cdots + \frac{k}{\alpha_m}\right)$$

$$= k(2 - 2g) - mk + k\left(\frac{1}{\alpha_1} + \cdots + \frac{1}{\alpha_m}\right)$$

$$= -k\left((2g - 2) + m - \sum_{i=1}^{m} \frac{1}{\alpha_i}\right).$$

By comparing the two expressions for $2 - 2g_0$, we get

$$b = 2g - 2.$$

Note that SO(3) is the unit tangent bundle of $S^2$. If we let $\Gamma_0 = 1_{\text{SO}(3)} \subset \Gamma$, then we may apply the same reasoning as above to calculate the "$b$"-invariant for $\Gamma\backslash\text{SO}(3)$. Since $g = 0$, the "$b$"-invariant is $-2$ for $\Gamma\backslash\text{SO}(3)$. For the case of $E_0(2)$ also, $b = 0$ for $(\mathbb{Z} \oplus \mathbb{Z})\backslash E_0(2)$ and $b = -2$ for $\Gamma\backslash E_0(2)$ because $g < g_0 = 1$ when $m > 0$.

In summary we have

THEOREM 14.8.8 ([**RV81**]). *For $M = \Gamma\backslash G$, where $G = \text{SO}(3)$, $\text{PSL}(2, \mathbb{R})$, or $E_0(2)$, $\Gamma$ cocompact and discrete, the normalized Seifert invariants are $M = \{\mathfrak{o}, g, 0, 0; 2g - 2;\ (\alpha_1, \alpha_1 - 1), \ldots, (\alpha_m, \alpha_m - 1)\}$.*

We remark that it is reasonable to call $\Gamma\backslash G$ the unit tangent bundle of the orbifold $\Gamma\backslash G/K$.

14.8.9. Let $Q$ be a discrete group which acts properly, effectively and preserves orientation on $S^2$, $\mathbb{R}^2$ or $\mathbb{H}^2$. Then $Q$ is known to be conjugate in the group of homeomorphisms of $S^2$ or $\mathbb{R}^2$ to a subgroup of the orientation preserving isometries of the standard sphere, the Euclidean plane or the hyperbolic plane [**Mac62**]. That is, $Q$ is conjugate to a discrete subgroup $\Gamma$ of SO(3), $E_0(2)$, and $\text{PSL}(2, \mathbb{R})$, respectively. If $Q$ is cocompact and not cyclic in the spherical case, then $Q$ has a presentation of the form

$$Q = \langle a_1, \ldots, a_g, b_1, \ldots, b_g, q_1, \ldots, q_m \mid \prod_{j=1}^{m} q_j \prod_{i=1}^{g}[a_i, b_i],\ q_j^{\alpha_j}\ (j = 1, \ldots, m)\rangle.$$

The $q_j$ generate distinct conjugacy classes of maximal cyclic groups of order $\alpha_j$. They are the isotropy subgroups of points on $S^2$ or $\mathbb{R}^2$ where the action of $Q$ is not free.

The orbit space of the $Q$-action is homeomorphic, as an orbifold, to $\Gamma\backslash G/K$, and as a space, to an orientable closed surface of genus $g$. There are $m$ exceptional points with "multiplicities" $\alpha_1, \alpha_2, \cdots, \alpha_m$ corresponding to the nonfree orbits with isotropy $\mathbb{Z}_{\alpha_j}$. It is often convenient to shorten the presentation of $Q$ to

$$Q(g;\ \alpha_1, \alpha_2, \ldots, \alpha_m).$$

If $g=0$, and $m \leq 2$, the presentation of $Q$, as above, yields a finite cyclic group. This group will not correspond to an effective rotation on $S^2$ about an axis unless $\alpha_1 = \alpha_2$. (Any cyclic group $\mathbb{Z}_n \subset SO(3)$ acts freely as translations on $SO(3)$. On $S^2 = SO(3)/SO(2)$, the cyclic group acts effectively as a rotation with an antipodal pair of fixed points and with multiplicities $n$.) For cyclic $\Gamma \subset SO(3)$ or $\Gamma' \subset SU(2)$, $\Gamma\backslash SO(3)$ and $\Gamma'\backslash SU(2)$ are lens spaces. Such lens spaces have many representations as $SO(2)$-manifolds. For $\Gamma \subset SO(3)$, one can choose the right circle subgroup (they are all conjugate) and write $\Gamma\backslash SO(3)$ as $\{\mathfrak{o}, 0, 0, 0; -2; (\alpha, \alpha-1), (\alpha, \alpha-1)\}$. It is of the form $L(2\alpha, 2\alpha-1) \cong L(2\alpha, 1)$. So, *when $g=0$, we shall assume in the presentation that $m \geq 3$*. Note, if $g=0$, $m > 3$, then $M = \Gamma\backslash G$ is not spherical.

Recall that the Euler characteristic of $Q$ is defined to be

$$\chi(Q) = 2 - 2g - \sum_{i=1}^{m}\left(1 - \frac{1}{\alpha_i}\right).$$

If $\chi(Q) > 0$, $Q$ is spherical (that is, $Q$ is conjugate to a subgroup $\Gamma \subset SO(3)$). Then $g=0$, $m=3$. In fact,

$$(\alpha_1, \alpha_2, \alpha_3) = (2,2,n),\ (2,3,3),\ (2,3,4)\ \text{or}\ (2,3,5),\ \text{any}\ n \geq 2.$$

If $\chi(Q) = 0$, $Q$ is Euclidean, and $g=0$, $m=3$ or $4$; or $g=1$, $m=0$.

$$(\alpha_1, \alpha_2, \alpha_3) = (2,4,4),\ (2,3,6),\ (3,3,3)$$
$$(\alpha_1, \alpha_2, \alpha_3, \alpha_4) = (2,2,2,2)$$

and with $g=1$, there are no $\alpha$'s, and $Q \cong \mathbb{Z} \times \mathbb{Z}$.

If $\chi(Q) < 0$, $Q$ is hyperbolic. That is, all remaining possibilities in the formula for $\chi(Q)$ yield hyperbolic groups.

EXERCISE 14.8.10. If $(S^1, M)$ is an effective action of a closed orientable 3-manifold without fixed points, the presentation for $\pi_1(M)$ in Subsection 14.6.5 reduces to

$$\pi_1(M) = \langle a_1, \ldots, a_g, b_1, \ldots, b_g, q_1, \ldots, q_m,\ h \mid h\ \text{central},$$
$$\prod_{j=1}^{m} q_j \prod_{i=1}^{g} [a_i, b_i] h^{-b},\ q_j{}^{\alpha_j} h^{\beta_j}\ (j=1,\ldots,m)\rangle,$$

where $h$ is the image of the generator of $\pi_1(S^1)$ under the evaluation homomorphism $\text{ev}_*^x : \pi_1(S^1, 1) \to \pi_1(M, x)$. Then

$$\pi_1(M)/(\text{central subgroup generated by } h) = \pi_1(M)/\text{Im}(\text{ev}_*^x(\pi_1(S^1))) = Q$$

is presented by

$$Q = \langle a_1, \ldots, a_g, b_1, \ldots, b_g, q_1, \ldots, q_m,\ |\ \prod_{j=1}^{m} q_j \prod_{i=1}^{g} [a_i, b_i],\ q_j{}^{\alpha_j}\ (j=1,\ldots,m)\rangle.$$

Of course, this means that when the action is locally injective, there is an induced effective action of $Q$ on $W = \mathbb{R}^2$ or $S^2$. The action of $Q$ is topologically conjugate to a group of spherical, Euclidean, or hyperbolic isometries according to whether $\chi(Q) = 2 - 2g - m + \sum_{i=1}^{m} \frac{1}{\alpha_i}$ is greater than 0, equal to 0, or less than 0.

We have seen in Theorem 14.8.8 that $e(Q\backslash G)$ for $Q\backslash G \to Q\backslash G/K$ will be $-\left(2g - 2 + m - \sum_{i=1}^{m} \frac{1}{\alpha_i}\right) = \chi(Q)$.

Let $(S^1, M)$ be as above such that $\pi_1(M)/\langle h \rangle = Q$. Show

(1) If $\chi(Q) > 0$, then $e(M) \neq 0$. Hint: Each $\beta_i$ is relatively prime to $\alpha_i$, $i = 1, 2, 3$. Then $\frac{\beta_1}{\alpha_1} + \frac{\beta_2}{\alpha_2} + \frac{\beta_3}{\alpha_3} \equiv e(M) \bmod 1$, for $(\alpha_1, \alpha_2, \alpha_3) = (2, 2, n)$, $(2, 3, 3)$, $(2, 3, 4)$, and $(2, 3, 5)$.

(2) If $\chi(Q) = 0$ and $e(M) = 0$, then $M$ is homeomorphic to $Q\backslash E_0(2)$. If $\chi(Q) = 0$ and $e(M) \neq 0$, then $M$ is covered by a nilmanifold.

(3) If $\chi(Q) < 0$, it is possible sometimes for $e(M) = 0$. However, if there is some $j$ such that $\alpha_j$ is relatively prime to each $\alpha_i$, $i \neq j$, then $e(M) \neq 0$.

(4) If $\chi(Q) < 0$, then there is at most a finite number of $(S^1, M)$'s with $e(M) = 0$, ($Q$ fixed, of course).

14.8.11. Since we know that $b = 2g - 2$, $(\alpha, \beta) = (\alpha, \alpha - 1)$, we have completely characterized those $M$, for which $M = \Gamma\backslash G$, where $G = SO(3)$, $PSL(2, \mathbb{R})$ and $E_0(2)$.

The simply connected 3-dimensional Lie groups which admit cocompact lattices are $SU(2)$, $\widetilde{PSL(2, \mathbb{R})}$, and $\widetilde{E_0(2)} = \mathbb{R}^2 \rtimes \mathbb{R}$, the universal covering groups of $SO(3)$, $PSL(2, \mathbb{R})$, and $E_0(2)$, respectively, as well as $\mathbb{R}^3$, Nil = Heisenberg group, and a certain solvable group called Sol.

Perhaps the most interesting is $\widetilde{PSL}(2, \mathbb{R})$. The center of $\widetilde{PSL}(2, \mathbb{R})$ is $\mathbb{Z}$. Let us denote $\widetilde{PSL}(2, \mathbb{R})$ by $G_\infty$,

$$G_\infty = \widetilde{PSL}(2, \mathbb{R}),$$

and a cocompact lattice in $G_\infty$ by $\Gamma_\infty$. Then

$$\mathcal{Z}(\Gamma_\infty) = \Gamma_\infty \cap \mathcal{Z}(G_\infty) = r\mathbb{Z},$$

for some $r > 0$. If $p_m : G_\infty \to G_m = G_\infty/m\mathbb{Z}$ is the central projection, then the lattice $\Gamma_\infty$ is projected into a lattice

$$p_m(\Gamma_\infty) = \Gamma_m = \Gamma_\infty/(m\mathbb{Z} \cap \Gamma_\infty) \subset G_m.$$

Induced is the finite cyclic $(m\mathbb{Z}/\gcd(m,r)\mathbb{Z})$-covering projection $\Gamma_\infty\backslash G_\infty \to \Gamma_m\backslash G_m$. If we project $\Gamma_\infty$ to $G_r$, where $\mathcal{Z}(\Gamma_\infty) = r\mathbb{Z}$, then $\Gamma_\infty\backslash G_\infty$ is diffeomorphic to $\Gamma_r\backslash G_r$. Observe for any lattice $\Gamma \subset G_m$ that $\mathcal{Z}(\Gamma) = \mathcal{Z}(G_m) \cap \Gamma$. Let $\Gamma' = p_m^{-1}(\Gamma)$, now a lattice in $G_\infty = p_m^{-1}(G_m)$, and $\Gamma'\backslash G_\infty$ will be diffeomorphic to $\Gamma\backslash G_m$. Therefore, without any loss of generality, we can assume that

(1) $\Gamma_r$ is a lattice in $G_r$,
(2) $\Gamma_r$ is centerless.

Now project $G_r \to G_1 = PSL(2, \mathbb{R})$. Then $\Gamma_r$ is projected isomorphically onto the lattice $\Gamma_1$ in $G$ since $\mathcal{Z}(\Gamma_r) = 1$ and $\Gamma_r\backslash G_r \to \Gamma_1\backslash G_1$ is an $r$-fold cyclic covering projection. This is given by dividing out the center $\mathbb{Z}/r\mathbb{Z} \subset K_r \subset G_r$.

PROPOSITION 14.8.12. *To determine all $\Gamma_\infty\backslash G_\infty$, we need only examine lattices $\Gamma_1$ in $G_1$ and determine $\Gamma_1\backslash G_1$ as a $K_1$-space. We then obtain all possibilities by considering the finite number of cyclic $\mathbb{Z}_r$ coverings $G_r$ of $G$ for which $\Gamma_1$ lifts to $\Gamma_1 \times \mathbb{Z}_r$ and determine $\Gamma_1\backslash G$ as a $K_r$-space.*

Actually $\Gamma_\infty \backslash G_\infty \longrightarrow p_1^{-1}(p_1(\Gamma_\infty))\backslash G_\infty$ is a free cyclic covering induced by the covering $K_r \longrightarrow K$. Thus we can determine the Seifert invariants for $\Gamma_\infty \backslash G_\infty$ (and hence of $\Gamma_r \backslash G_r$) by knowing that the kernel of the map $\mathrm{Ad}: K_r \to \mathrm{Aut}(\mathfrak{g}/\mathfrak{k})$ is $\mathbb{Z}_r$. Thus, the normalized Seifert invariants for $\Gamma_\infty \backslash G_\infty$ are $(\alpha_i, \beta_i)$, where $\beta_i r \equiv -1 \bmod \alpha_i$.

The "$b$"-invariant is determined by either Proposition 14.7.5 or Subsection 14.8.7. Since $\Gamma_\infty \backslash G_\infty \longrightarrow p_1^{-1}(\Gamma_1)/G_\infty$ is a free cyclic $\mathbb{Z}_r$ covering contained in the $K_r$-action, we have

$$-(\alpha_1 \cdots \alpha_m) \cdot e(\Gamma_\infty \backslash G_\infty) = (\alpha_1 \cdots \alpha_m) \cdot \frac{1}{r} \cdot \left(2g - 2 + m - \sum_{i=1}^{m} \frac{1}{\alpha_i}\right)$$

$$= -\frac{1}{r} \cdot (\alpha_1 \cdots \alpha_m)\, e(\Gamma_1 \backslash G_1).$$

Since $r$ is prime to each $\alpha_i$, $r$ divides the integer $(\alpha_1 \cdots \alpha_m)\cdot(2g-2+m-\sum_{i=1}^m \frac{1}{\alpha_i})$. Hence, there is only a finite number of $r$'s to consider as long as $e(\Gamma_\infty \backslash G_\infty) \neq 0$. This is always the case for $G_\infty = \widetilde{\mathrm{PSL}}(2,\mathbb{R})$. (Recall that for $\Gamma \subset \mathrm{PSL}(2,\mathbb{R})$, there is a normal $\Gamma_0$, torsion free of finite index in $\Gamma$ and $e(\Gamma_0) = 2 - 2g_0 < 0$, since $\mathbb{Z} \times \mathbb{Z}$ cannot be in $\Gamma_0$.) In summary, we have

THEOREM 14.8.13 ([**RV81**]). *Let $G_\infty$ be the universal covering of $\mathrm{PSL}(2,\mathbb{R})$, and let $\Gamma_\infty$ be a discrete subgroup of $G_\infty$. Then, $\mathcal{Z}(\Gamma_\infty) = \Gamma_\infty \cap \mathcal{Z}(G_\infty) = r\mathbb{Z}$ and $M = \Gamma_\infty \backslash G_\infty$ as an $\mathrm{SO}(2)$-manifold has the form*

$$M = \{\mathfrak{o}, g, 0, 0;\; b;\; (\alpha_1, \beta_1), \ldots, (\alpha_m, \beta_m)\},$$

*where $r$ is a divisor of $(\alpha_1 \cdots \alpha_m) \cdot \left(2g-2+m-\sum_{i=1}^m \frac{1}{\alpha_i}\right)$ prime to each $\alpha_i$ so that*

$$\beta_i r \equiv -1 \bmod \alpha_i \quad \text{and} \quad rb = (2g-2) - \sum_{i=1}^{m} k_i,$$

*where $\beta_i r = \alpha_i - 1 + k_i \alpha_i$. Further, $\Gamma_\infty$ is a subgroup of index $r$ in $p^{-1}(\Gamma_1)$ with $\Gamma_\infty / \mathcal{Z}(\Gamma_\infty) \cong \Gamma_1 \subset G_1 = \mathrm{PSL}(2,\mathbb{R})$,*

$$\Gamma_1 \backslash G_1 = \{\mathfrak{o}, g, 0, 0; 2g-2;\; (\alpha_1, \alpha_1 - 1), \ldots, (\alpha_m, \alpha_m - 1)\},$$

*and $\Gamma_\infty \backslash G_\infty \longrightarrow \Gamma_1 \backslash G_1$ is an $r$-fold central covering projection. All possible solutions exist. (If we use unnormalized invariants for $M$, the conditions for $b$ is replaced by $e(M) = \frac{1}{r} e(\Gamma_1 \backslash G_1)$ with $r$ prime to each $\alpha_i$.)*

PROOF. In the previous discussion, we have shown that the condition on $r$, $b$ and the $\beta_i$ are all necessary conditions for $\Gamma_\infty \backslash G_\infty \longrightarrow \Gamma_1 \backslash G_1$ to be a free cyclic $\mathbb{Z}_r$ central covering projection. It remains to show that these conditions also guarantee the existence of $\Gamma_\infty \subset p_1^{-1}(\Gamma)$.

Let $Q(g; \alpha_1, \ldots, \alpha_m) \subset \mathrm{PSL}(2,\mathbb{R})$. Then $Q \backslash \mathrm{PSL}(2,\mathbb{R}) = M$ and

$$(S^1, M) = \{\mathfrak{o}, g, 0, 0; 2g-2;\; (\alpha_1, \alpha_1 - 1), \ldots, (\alpha_m, \alpha_m - 1)\}.$$

We may write

$$(\alpha_1 \cdots \alpha_m) \cdot \left(2g - 2 + m - \sum_{i=1}^{m} \frac{1}{\alpha_i}\right) = (\alpha_1 \cdots \alpha_m) \cdot (-e(M)).$$

Let us use unnormalized invariants and absorb the "$b$"-invariant into the $\beta$'s and so we may assume that $b = 0$, and $\beta_i' \equiv \beta_i$ mod $\alpha_i$. Then the equation may be written as

$$-(\alpha_1 \cdots \alpha_m) \cdot e(M) = \sum_{i=1}^{m} \beta_i'(\alpha_1 \cdots \widehat{\alpha}_i \cdots \alpha_m),$$

where $(\alpha_i, \beta_i')$ are the corresponding unnormalized Seifert invariants ($\widehat{\alpha}_i$ means it is omitted). As we shall see in Proposition 14.9.1, $|e(M)(\alpha_1 \cdots \alpha_m)|$ is the order of the torsion subgroup of $H_1(M; \mathbb{Z})$, but this is not significant for us now.

Let us consider the algebraic equation

$$y = \sum_{i=1}^{m} x_i \cdot (\alpha_1 \cdots \widehat{\alpha}_i \cdots \alpha_m)$$

for an integer $y$ and unknown integers $x_i$. This equation has integer solutions if and only if $y$ is a multiple of $d = \gcd(A_1, A_2, \ldots, A_m)$, where $A_i = (\alpha_1 \alpha_2 \cdots \alpha_m)/\alpha_i = (\alpha_1 \cdots \widehat{\alpha}_i \cdots \alpha_m)$. Note that $\gcd(A_1, A_2, \ldots, A_m) \cdot \text{lcm}(\alpha_1, \alpha_2, \ldots, \alpha_m) = (\alpha_1 \alpha_2 \cdots \alpha_m)$. Thus, $d = \frac{(\alpha_1 \alpha_2 \cdots \alpha_m)}{\text{lcm}(\alpha_1, \alpha_2, \ldots, \alpha_m)}$. If we divide the equation for $y$ by $A = (\alpha_1 \alpha_2 \cdots \alpha_m)$, we get

$$\frac{y}{A} = \frac{x_1}{\alpha_1} + \cdots + \frac{x_m}{\alpha_m}.$$

This equation has integer solutions if and only if $y$ is a multiple of $d$. But $\frac{d}{A} = \frac{1}{L}$ where $L = \text{lcm}(\alpha_1, \alpha_2, \ldots, \alpha_m)$, and so we have integer solutions to

$$\frac{x_1}{\alpha_1} + \cdots + \frac{x_m}{\alpha_m} = \frac{1}{L}.$$

This says that $e(M)$, for any Seifert manifold with Seifert invariants $\{(\alpha_1, \beta_1), \ldots, (\alpha_m, \beta_m)\}$ is an integral multiple of $\frac{1}{L}$.

Returning to $M = \Gamma_1 \backslash G_1$, we have

$$-e(M) \cdot (\alpha_1 \cdots \alpha_m) = \sum_{i=1}^{m} \beta_i'(\alpha_1 \cdots \widehat{\alpha}_i \cdots \alpha_m) = nd,$$

for some integer multiple of $d$ where $\beta_i' \equiv -1$ mod $\alpha_i$ and

$$-e(M) = 2g - 2 + m - \sum_{i=1}^{m} \frac{1}{\alpha_i} = \sum_{i=1}^{m} \frac{\beta_i'}{\alpha_i}.$$

Let $r$ be a divisor of $nd$ so that $r$ is prime to each $\alpha_i$. Since $d = \frac{(\alpha_1 \alpha_2 \cdots \alpha_m)}{\text{lcm}(\alpha_1, \alpha_2, \ldots, \alpha_m)}$, $d$ is either 1 or is a multiple of some $\alpha_i$. Therefore, $r$ must divide $n$. We can then find integral solutions to

$$\sum_{i=1}^{m} x_i \cdot (\alpha_1 \cdots \widehat{\alpha}_i \cdots \alpha_m) = \frac{n}{r} d,$$

say $x_i = \beta_i''$. Then

$$M'' = \{\mathfrak{o}, g, 0, 0; \ 0; \ (\alpha_1, \beta_1''), \ldots, (\alpha_m, \beta_m'')\}$$

has a $\mathbb{Z}_r$ central action as $r$ is prime to each $\alpha_i$, and

$$\mathbb{Z}_r \backslash M'' = \{\mathfrak{o}, g, 0, 0; \ 0; \ (\alpha_1, r\beta_1''), \ldots, (\alpha_m, r\beta_m'')\}.$$

It follows that $r\beta_i'' = \beta_i' \equiv -1$ mod $\alpha_i$ and $M'' = \Gamma_\infty \backslash G_\infty$, $\mathbb{Z}_r \backslash M'' = \Gamma_1 \backslash G_1$. $\square$

14.8.14 ($G_\infty \neq \widetilde{\text{PSL}(2,\mathbb{R})}$). $\widetilde{E_0(2)}$ is an infinite cyclic central covering of $E_0(2)$. Using similar notation as for $\text{PSL}(2,\mathbb{R})$, we have $\Gamma_\infty \backslash G_\infty \longrightarrow \Gamma_1 \backslash G_1$ is an $r$-fold central cyclic covering. But $e(\Gamma_1 \backslash G_1) = 0$ for each Euclidean $Q = \Gamma_1$ in Theorem 14.8.8 and Subsection 14.8.9. Therefore, $e(\Gamma_\infty \backslash G_\infty)$ must be 0. $\Gamma_\infty \backslash G_\infty$ is homeomorphic to $\Gamma_1 \backslash G_1$ even though $r \neq 1$ because the choices for the $(\alpha_i, \beta_i)$ are the same as $\Gamma_1 \backslash G_1$.

SU(2) is the 2-fold central covering of SO(3). Each $\Gamma \subset \text{SO}(3)$ lifts to $\widetilde{\Gamma} \subset \text{SU}(2)$, which is a nontrivial central extension of $\mathbb{Z}_2$ by $\Gamma$. Therefore, $\widetilde{\Gamma} \backslash \text{SU}(2) = \Gamma \backslash \text{SO}(3)$.

Nil: The Heisenberg group and its lattices are discussed in Example 4.5.12.

Sol: The solvable group Sol has no center and is diffeomorphic to $\mathbb{R}^3$. Thus $\Gamma \backslash \text{Sol}$ never has an effective $S^1$-action. However, $\Gamma \backslash \text{Sol}$ does fiber over the circle with a 2-torus as fiber and structure group infinite cyclic in $\text{GL}(2,\mathbb{Z})$. Consequently, $\Gamma \backslash \text{Sol}$ can be represented as a Seifert fibering over $S^1$ with regular fiber the torus, $(\mathbb{Z} \times \mathbb{Z}) \backslash \mathbb{R}^2$. As the group Sol is of type (R), the rigidity theorems apply to these manifolds.

## 14.9. $H_1(M; \mathbb{Z})$

PROPOSITION 14.9.1. *Let $(S^1, M)$ be an effective $S^1$-action on a closed orientable 3-manifold without fixed points. Let $T$ be the torsion subgroup of $H_1(M; \mathbb{Z})$. Then*

$$H_1(M; \mathbb{Z}) = \begin{cases} \mathbb{Z}^{2g} + T, & \text{if } e(M) \neq 0, \\ \mathbb{Z}^{2g+1} + T, & \text{if } e(M) = 0. \end{cases}$$

*Furthermore, if $e(M) \neq 0$, then $|T| = (\alpha_1 \alpha_2 \cdots \alpha_m)|e(M)|$.*

PROOF. In Subsection 14.6.5, we gave a presentation of $\pi_1(M)$. From this we see that a presentation for $H_1(M; \mathbb{Z})$ is

$$H_1(M; \mathbb{Z}) = \langle a_1, \ldots, a_g, b_1, \ldots, b_g, q_1, \ldots, q_m, h \mid q_1 + \cdots + q_m - bh = 0,$$
$$\alpha_i q_i + \beta_i h = 0 \rangle$$
$$= \mathbb{Z}^{2g} \oplus \langle q_1, \ldots, q_m, h \mid q_1 + \cdots + q_m - bh = 0, \alpha_i q_i + \beta_i h = 0 \rangle.$$

Consider the relation matrix

$$A^* = \begin{bmatrix} 1 & 1 & 1 & \cdots & 1 & -b \\ \alpha_1 & 0 & 0 & \cdots & 0 & \beta_1 \\ 0 & \alpha_2 & 0 & \cdots & 0 & \beta_2 \\ 0 & 0 & \alpha_2 & \cdots & 0 & \beta_3 \\ \cdots & . & . & \cdots & . & \cdots \\ 0 & 0 & 0 & \cdots & \alpha_m & \beta_m \end{bmatrix}$$

the cokernel of which is the second summand of $H_1(M; \mathbb{Z})$. This matrix is the same matrix as the matrix $A$ augmented by the column $(-b, \beta_1, \ldots, \beta_m)^t$ used in [**LLR96**] to calculate $H^2(Q; \mathbb{Z})$, where $Q = \pi_1(M)/(\text{central subgroup generated by}$

$h$). To calculate $\det(A^*)$, expand along the first row. One easily gets

$$\det(A^*) = (-1)^{m+1}\left(b\alpha_1\alpha_2\cdots\alpha_m + \sum_{i=1}^{m}\beta_i(\alpha_1\alpha_2\cdots\widehat{\alpha_i}\cdots\alpha_m)\right)$$

$$= (-1)^{m+1}(\alpha_1\alpha_2\cdots\alpha_m)\left(b + \sum_{i=1}^{m}\frac{\beta_i}{\alpha_i}\right)$$

$$= (-1)^m(\alpha_1\alpha_2\cdots\alpha_m)\cdot e(M).$$

If $e(M) \neq 0$, then $\operatorname{rank}(A^*) = m+1$ and $H_1(M;\mathbb{Z}) = \mathbb{Z}^{2g} \oplus T$, where $|T| = (\alpha_1\alpha_2\cdots\alpha_m)\cdot|e(M)|$. If $e(M) = 0$, then $H_1(M;\mathbb{Z}) = \mathbb{Z}^{2g+1} \oplus$ Torsion. $\square$

It is of interest to determine when the order of the torsion subgroup in Proposition 14.9.1 is trivial.

COROLLARY 14.9.2 (cf. [**Sei33**, §12]). *If $|T| = 1$ and $e(M) \neq 0$, then the $\alpha$'s are pairwise coprime. For each $M$ with $\alpha$'s pairwise coprime and fixed $g$, there exists a unique, up to orientation, $(S^1, M)$ with $|T| = 1$.*

PROOF. If $m$ has no exceptional orbits, then $m = 0$ and $H_1(M;\mathbb{Z}) = \mathbb{Z}^{2g} \oplus \mathbb{Z}_b$. Therefore, $|T| = 1$ if and only if $b = 0$ or $\pm 1$. If $b = 0$, then $(S^1, M) = (S^1, S^1 \times \Sigma_g)$ for some orientable surface $\Sigma_g$. If $b = \pm 1$, then $(S^1, M)$ is a principal $S^1$-bundle over a surface $\Sigma_g$ whose Euler class is $-b = \mp 1$.

If $m = 1$, $H_1(M;\mathbb{Z}) = \mathbb{Z}^{2g} \oplus \mathbb{Z}_{|b\alpha+\beta|}$. Now $b\alpha + \beta \neq 0$. So $|T| = 1$, if and only if $b = 0$ and $\beta = 1$ or equivalently $b = -1$ and $\beta = \alpha - 1$.

Now suppose $e(M) \neq 0$ and $m \geq 2$. If $\gcd(\alpha_i, \alpha_j) = k \neq 1$ for some $i \neq j$, then $k$ divides $\alpha_i$ and $\alpha_j$. Therefore, $k$ divides $|T|$. So, in order for $|T| = 1$, each pair $(\alpha_i, \alpha_j)$ must be coprime.

To verify the second assertion, absorb the integer $b$ into the $\beta$'s and use unnormalized Seifert invariants $\{(\alpha_i, \beta_i')\}$ with $b = 0$. Since $|T| = |\sum_{i=1}^{m}\beta_i'(\alpha_1\alpha_2\cdots\widehat{\alpha_i}\cdots\alpha_m)|$, we know by the argument used in Theorem 14.8.13, that the possible values of $|T|$ are integral multiples of

$$d = \gcd(A_1, \ldots, A_m) = (\alpha_1\cdots\alpha_m)/\operatorname{lcm}(\alpha_1, \cdots, \alpha_m).$$

Therefore, $d = \pm 1$ if and only if the $\alpha_i$ are pairwise coprime. Furthermore, there are solutions $\beta_i'$ so that $|T| = 1$. Note each $\beta_i'$ is relatively prime to $\alpha_i$ for, otherwise, $d$ cannot be $\pm 1$. Since we know $e(M) = (\pm 1/\alpha_1\cdots\alpha_m)$, the manifold $(S^1, M)$ is completely determined. $\square$

EXERCISE 14.9.3. (a) Show that there exists exactly one $(S^1, M)$, other than the 3-sphere which has finite $\pi_1(M)$ and trivial $H_1(M;\mathbb{Z})$.

(b) Find the Seifert presentations for $(S^1, M)$ where $|T| = 1$ and
  (1) $(\alpha_1, \alpha_2, \alpha_3) = (2, 3, 7)$, $g = 0$,
  (2) $(\alpha_1, \alpha_2, \alpha_3) = (2, 3, 7)$, $g = 1$,
  (3) $(\alpha_1, \alpha_2, \alpha_3) = (2, 5, 7)$, $g = 0$.

(c) Show for $(S^1, M)$, as in Subsection 14.7.1, that the order of $h$, $o(h)$ in $H_1(M;\mathbb{Z})$ is given by

$$\left|b\cdot\operatorname{lcm}(\alpha_1, \ldots, \alpha_m) + \sum_{i=1}^{m}\beta_i\cdot\operatorname{lcm}(\alpha_1, \ldots, \widehat{\alpha_i}, \ldots, \alpha_m)\right|.$$

Consequently, $o(h) = |-e(M) \cdot \mathrm{lcm}(\alpha_1, \ldots, \alpha_m)|$ and $e(M) = 0$ if and only if $o(h) = 0$ (i.e., $(S^1, M)$ is homologically injective). Furthermore, if $o(h) \neq 0$, $|T| = o(h) \cdot \gcd(A_1, \ldots, A_m)$.

COROLLARY 14.9.4. *If $(S^1, M)$ has $|T| = 1$, $\chi(Q) < 0$ and $e(M) \neq 0$, then $M = \Gamma \backslash \widetilde{\mathrm{PSL}}(2, \mathbb{R})$.*

PROOF. Let $Q \subset \mathrm{PSL}(2, \mathbb{R})$ where $Q = \pi_1(M)/$(central subgroup generated by the typical fiber $h$ in $M$). The homogeneous space $M' = Q \backslash \mathrm{PSL}(2, \mathbb{R})$ has Seifert presentation

$$M' = \{\mathfrak{o}, g, 0, 0; 2g-2; (\alpha_1, \alpha_1 - 1), \ldots, (\alpha_m, \alpha_m - 1)\}.$$

Let $r = (\alpha_1 \alpha_2 \cdots \alpha_m)|e(M')|$. It is the order of the torsion subgroup of $H_1(M', \mathbb{Z})$. The integer $r$ is $r$ times $(\alpha_1 \alpha_2 \cdots \alpha_m)|e(M)| = 1$ since $|T|$ of $H_1(M, \mathbb{Z})$ is 1. We want to show that $(S^1, M)$ satisfies the conditions of Theorem 14.8.13 so that it will be a cover of $(S^1, M')$ of the form $\Gamma \backslash \widetilde{\mathrm{PSL}}(2, \mathbb{R})$. We rewrite $(\alpha_1 \alpha_2 \cdots \alpha_m)(-e(M'))$ as $\sum_{i=1}^{m} \beta_i'(\alpha_1 \cdots \widehat{\alpha_i} \cdots \alpha_m)$, where we use unnormalized invariants with $b = 0$. Then $\beta_i' \equiv -1 \bmod \alpha_i$ for each $i$. We have then $r \equiv (-1)(\alpha_1 \cdots \widehat{\alpha_i} \cdots \alpha_m) \bmod \alpha_i$. So $r$ is prime to each $\alpha_i$. Let $u_i(\alpha_1 \cdots \widehat{\alpha_i} \cdots \alpha_m) \equiv 1 \bmod \alpha_i$. Then $u_i r \equiv -1 \bmod \alpha_i$. Put $\beta_i = u_i$. Then $\beta_i r \equiv -1 \bmod \alpha_i$ for each $i$. Now $r$ is the "$r$" of Theorem 14.8.13 and we found unnormalized $\beta_i$ so that $\beta_i r \equiv -1 \bmod \alpha_i$ for each $i$. Therefore, the $r$-fold central covering of $Q \backslash \mathrm{PSL}(2, \mathbb{R})$ is the desired $\Gamma \backslash \widetilde{\mathrm{PSL}}(2, \mathbb{R}) = (S^1, M)$. □

COROLLARY 14.9.5. *Let $(S^1, M)$ be as in Corollary 14.9.4 and, in addition, assume that $g = 0$. Then the central covering $M \longrightarrow Q \backslash \mathrm{PSL}(2, \mathbb{R})$ is the universal Abelian covering of $Q \backslash \mathrm{PSL}(2, \mathbb{R})$.*

PROOF. We have the exact sequence

$$1 \longrightarrow \pi_1(M) \longrightarrow \pi_1(Q \backslash \mathrm{PSL}(2, \mathbb{R})) \longrightarrow \mathbb{Z}_r \longrightarrow 1.$$

As the order of the torsion subgroup of $H_1(Q \backslash \mathrm{PSL}(2, \mathbb{R}))$ is $r$, $\pi_1(M)$ must be the commutator subgroup of $\pi_1(Q \backslash \mathrm{PSL}(2, \mathbb{R}))$. □

EXERCISE 14.9.6. Show that a Seifert manifold with $e(M) \neq 0$ and $\{g; \alpha_1, \alpha_2, \ldots, \alpha_m\}$ pairwise coprime is completely determined by $H_1(M; \mathbb{Z}) = \mathbb{Z}^{2g} + \mathbb{Z}/|T|$. Also, determine to what extent the analogue of Corollary 14.9.4 holds when $e(M) \neq 0$ but $\chi(Q) \geq 0$.

## 14.10. Injective holomorphic Seifert fiberings

The reader is referred to [**CR72b**] where a general and comprehensive theory of holomorphic Seifert fiberings whose universal space is a holomorphic fiber bundle over $W$ with fiber a complex torus or $\mathbb{C}^k$ is given. We shall restrict ourselves here to a special case closely related to the *classical Seifert 3-manifolds*.

14.10.1 (Seifert manifolds as links of complex analytic singularities). The 3-dimensional Seifert manifolds appear in many different guises in mathematics. We have seen some of them appearing as homogeneous spaces and of course $S^1$-actions. They also play an important role in analyzing singularities of algebraic surfaces.

Let $(S^1, M)$ be an oriented closed 3-manifold Seifert fibered over a closed orientable surface. If the action is locally injective, we lift the action to a principal $S^1$-bundle $P$ over a simply connected surface $W$. The surface $W$ is homeomorphic to $\mathbb{R}^2$ or $S^2$. The group $Q = \pi_1(M)/\text{Im}(\pi_1(S^1))$ is conjugate in TOP($W$) to a subgroup of orientation preserving isometries of the sphere, the Euclidean plane, or the hyperbolic plane. These isometry groups are closed subgroups of the respective groups of holomorphic homeomorphisms of $\mathbb{C}P_1$, $\mathbb{C}$ and $U$, the unit disk in $\mathbb{C}$. Replace the $S^1$-bundle over $W$ by a holomorphic $\mathbb{C}^*$ bundle with the same Euler class over the complex $W$. The Euler class is possibly nontrivial only for $W = \mathbb{C}P_1$. Now do the Seifert Construction holomorphically. One would expect that this is possible and is analogous to the smooth construction on a principal $\mathbb{C}^*$-bundle. Actually as we shall show, each holomorphic construction gives rise to a smooth construction on a many-to-one basis (see [**CR72b**, in particular, §5,6,7,13]).

Since $\mathbb{C}^* = S^1 \times \mathbb{R}$ smoothly, it easily follows that $M_{\mathbb{C}^*} = Q\backslash(\mathbb{C}^* \times W)$, if $W = \mathbb{C}$ or $U$, and $Q\backslash(\mathbb{R} \times P)$ if $W = S^2$, is diffeomorphic to $\mathbb{R} \times M$. The induced proper $\mathbb{C}^*$-action when restricted to the $S^1$-action is equivalent to $(S^1, \mathbb{R}^1 \times M)$ with $S^1$ acting only on the second factor. If we *fill in* each $\mathbb{C}^*$ orbit with $0 \in \mathbb{C}$, we have topologically the open mapping cylinder of the orbit mapping of the $S^1$-action on $M$. (Now all of the $\mathbb{C}^*$ orbits together with their added 0 form a holomorphic *line bundle*. A line may be singular, i.e., $\mathbb{Z}_\alpha\backslash\mathbb{C} = (\mathbb{Z}_\alpha\backslash\mathbb{C}^*)\cup 0$. Clearly, this is a generalization of a classical holomorphic line bundle which is produced whenever the $S^1$-action on $M$ is free.) This just attaches the orbit space $Q\backslash W$ to $\mathbb{R}^1 \times M$. This, too, can be done analytically yielding the Riemann surface $Q\backslash W$ as a subanalytic space. If we now collapse this added orbit space to a single point, we get topologically the open cone over $M$. Analytically, when $e(M) < 0$, the singular point becomes the link of an isolated analytic singularity with $\mathbb{C}^*$-action (cf. [**CR72b**, §13], [**NR78**, Corollary 5.3], [**Orl72**, §3], [**JN83**, Appendix]). If $e(M) > 0$, reverse the orientation. That is, if $e(M) < 0$, we can *blow down* $Q\backslash W$ to a point and get an analytic space with holomorphic $\mathbb{C}^*$-action with an isolated singularity.

14.10.2. We shall analyze this analytic formulation in terms of the Seifert Construction. We follow [**CR72b**, (especially §13)] which can be consulted for more details. In effect, we employ the techniques developed in Chapter 10 to classify these holomorphic Seifert fiberings but in a holomorphic instead of topological or smooth context. This classification will not be used elsewhere in this book. Fix $Q$ acting holomorphically on $W = U$ or $\mathbb{C}$ so that $Q\backslash W$ is compact. Any holomorphic $\mathbb{C}^*$ bundle over an open subset of $W$ is trivial; see for example [**Gun66**, p.52]. We lift the $Q$-action on $W$ to a holomorphic $Q$-action on $\mathbb{C}^* \times W$ so that it commutes with the left $\mathbb{C}^*$-action. With such a lift, the quotient space $Q\backslash\mathbb{C}^* \times W$ has an induced proper holomorphic $\mathbb{C}^*$-action.

We have the natural diffeomorphisms $Q\backslash(\mathbb{C}^* \times W) = Q\backslash((S^1 \times \mathbb{R}^1) \times W) = Q\backslash((S^1 \times W) \times \mathbb{R}^1) = M \times \mathbb{R}^1$. Therefore, the $Q$-action on $\mathbb{C}^* \times W$ is free if and only if the associated Seifert invariants for $M$, $\{(\alpha_i, \beta_i) : i = 1, \ldots, m\}$ have $\gcd(\alpha_i, \beta_i) = 1$, where $Q = Q(g; \alpha_1, \ldots, \alpha_m)$. Each lifting of the $Q$-action on $W$ to a group of holomorphic bundle automorphisms of $\mathbb{C}^* \times W$ is given, as in Subsection 10.3.1, uniquely by an element of $Z^1(Q; \mathcal{H}(W, \mathbb{C}^*))$, where $\mathcal{H}(W, \mathbb{C}^*)$ denotes the holomorphic maps of $W$ into $\mathbb{C}^*$. Cohomologous cocycles give lifted actions which differ only by conjugation by an element of $\mathcal{H}(W, \mathbb{C}^*)$. That is, the holomorphic actions are strictly equivalent. Thus these liftings are classified by the element of $H^1(Q; \mathcal{H}(W, \mathbb{C}^*))$. This group then classifies the descended holomorphic $\mathbb{C}^*$-actions on $\mathbb{Q}\backslash\mathbb{C}^* \times W$, up to strict equivalence, and simultaneously classifies the holomorphic *line bundles* over the Riemann surface $Q\backslash W$. (If the action of $Q$ is not free, then $Q\backslash \mathbb{C}^* \times W$ is a holomorphic orbifold and the classification is as in an orbifold classification, i.e., identical with the classification of the $\mathbb{C}^*$-liftings since all the $Q\backslash W$ orbifolds are "good" orbifolds.)

From the exact sequence $1 \to \mathbb{Z} \to \mathbb{C} \xrightarrow{\exp} \mathbb{C}^* \to 0$, we obtain the exact sequence

$$0 \longrightarrow \mathbb{Z} = \mathcal{H}(W, \mathbb{Z}) \longrightarrow \mathcal{H}(W, \mathbb{C}) \longrightarrow \mathcal{H}(W, \mathbb{C}^*) \longrightarrow 0$$

which gives rise to a long exact sequence of cohomology groups

$$(14.10.1) \quad \cdots \xrightarrow{\delta^{i-1}} H^i(Q; \mathbb{Z}) \longrightarrow H^i(Q; \mathcal{H}(W, \mathbb{C})) \longrightarrow H^i(Q; \mathcal{H}(W, \mathbb{C}^*)) \xrightarrow{\delta^i} \cdots,$$

and this long exact sequence is the same long exact sequence of Corollary 10.3.10 and Subsection 10.3.11 when the sheaf $\mathcal{T}^k$ is replaced by $\mathbb{C}^*$ and $W$ is $\mathbb{C}$ or $U$; cf. [**CR72b**, 6.10]. (As mentioned above, any holomorphic C*-bundle over W is holomorphically a trivial bundle. Therefore we have replaced $H^*(Q;\mathcal{T})$ by $H^*(Q;\mathcal{H}(W,\mathbb{C}^*))$ and the group $H^*(Q;\mathcal{O}(W))=H^*(Q\backslash W;\mathfrak{h}_c^0)$ by $H^*(Q;\mathcal{H}(W,\mathbb{C}))$.)

We want to calculate $H^1(Q; \mathcal{H}(W, \mathbb{C}^*))$. We point out that $\mathcal{H}(W, \mathbb{C})$ and $\mathcal{H}(W, \mathbb{C}^*)$ are nontrivial $Q$-modules. In fact, $H^i(Q; \mathcal{H}(W, \mathbb{C}))$ is isomorphic to $H^i(Q\backslash W; \mathfrak{h}_c^0)$, where $\mathfrak{h}_c^0$ is the sheaf of germs of holomorphic functions $f: \nu^{-1}(U) \to \mathbb{C}$ such that $f(qw) = f(w)$, for each $q \in Q$ and $w \in W$, $U$ open in $Q\backslash W$ and $\nu$ is the orbit mapping $W \to Q\backslash W$. The sheaf is locally free and vanishes if $i \geq 2$ and is isomorphic to $\mathbb{C}^g$, where $g$ is the genus of $Q\backslash W$ and $i = 1$.

Let us compare the smooth situation with the holomorphic one. We have the following commutative diagram of exact sequences.
(14.10.2)
$$0 \to H^1(Q;\mathbb{Z}) \to H^1(Q;\mathcal{H}(W,\mathbb{C})) \xrightarrow{e_*} H^1(Q;\mathcal{H}(W,\mathbb{C}^*)) \xrightarrow{\delta} H^2(Q;\mathbb{Z}) \to H^2(Q;\mathcal{H}(W,\mathbb{C}))$$
$$\Big\downarrow= \quad\quad \Big\downarrow \quad\quad\quad\quad \Big\downarrow \quad\quad\quad\quad \Big\downarrow=$$
$$0 \to H^1(Q;\mathbb{Z}) \to H^1(Q;\mathcal{C}(W,\mathbb{C})) \xrightarrow{e_*} H^1(Q;\mathcal{C}(W,\mathbb{C}^*)) \xrightarrow{\delta} H^2(Q;\mathbb{Z}) \to H^2(Q;\mathcal{C}(W,\mathbb{C}))$$

For the smooth case, $H^i(Q; \mathcal{C}(W, \mathbb{C})) = 0$, $i > 0$.

(For each central extension $0 \to \mathbb{Z} \to \Pi \to Q \to 0$ represented by $[\Pi] \in H^2(Q; \mathbb{Z})$, we have a smooth Seifert Construction $\theta: \Pi \to \text{Diff}_\mathbb{C}(\mathbb{C} \times W) = \text{Diff}_\mathbb{C}(\mathbb{C} \times \mathbb{R}^2)$. If we fix $i: \mathbb{Z} \to \mathbb{C}$ and $\rho: Q \to \text{Diff}(W)$, the construction is unique up to strict equivalences (Subsection 7.4.2). We have the smooth Seifert orbifold over $Q\backslash W$ with an induced injective $\mathbb{C}^*$-action and therefore an injective $S^1$-action on $\theta(\Pi)\backslash(\mathbb{C} \times W)$ ($\cong \mathbb{R}^1 \times \theta(\Pi)\backslash(\mathbb{R}^1 \times W) = \mathbb{R}^1 \times M$ since $\mathbb{C}^*$ splits smoothly as $\mathbb{R}^1 \times S^1$). The uniqueness says that for any other embedding $\theta': \Pi \to \text{Diff}_\mathbb{C}(\mathbb{C} \times W)$, keeping $i$ and $\rho$ fixed, the $\mathbb{C}^*$-action on $\theta'(\Pi)\backslash(\mathbb{C} \times W)$ is strictly smoothly equivalent to that on $\theta(\Pi)\backslash(\mathbb{C} \times W)$. This uniqueness is a consequence of $H^1(Q; \mathcal{C}(W, \mathbb{C})) = 0$.)

On the other hand, by Subsection 10.4.5, if we begin with a lift of the $Q$-action on $W$ to a group of smooth bundle automorphisms of $\mathbb{C}^* \times W$, there is a principal bundle $\mathbb{C} \times W$ covering $\mathbb{C}^* \times W$ and a unique extended lifting to a group $\Pi$ of bundle automorphisms of $\mathbb{C} \times W$. Since $\delta: H^1(Q; \mathcal{C}(W, \mathbb{C}^*)) \to H^2(Q; \mathbb{Z})$ is an isomorphism, there is a one-to-one correspondence between the smooth liftings of $Q$ and the extended smooth liftings of $Q$ to $\Pi$.

THEOREM 14.10.3 ([**CR72b**, §13]). *For each smooth action* $(\mathbb{C}^*, \Pi\backslash(\mathbb{C} \times W)) = (\mathbb{C}^*, M')$ *corresponding to the unique strict conjugacy class* $\theta(\Pi)$, *there exists a* *complex g-dimensional family of strictly holomorphically inequivalent $\mathbb{C}^*$-actions each strictly smoothly equivalent to the smooth $(\mathbb{C}^*, M')$.*

PROOF. Consider $[\Pi] \in H^2(Q;\mathbb{Z})$ in the proof of diagram (14.10.2). Since $H^2(Q; \mathcal{H}(W,\mathbb{C})) = 0$, there is $[\Pi'] \in H^1(Q; \mathcal{H}(W, \mathbb{C}^*))$ with $\delta[\Pi'] = [\Pi]$. This means that every smooth construction has a holomorphic realization. If $[\lambda] \in H^1(Q; \mathcal{H}(W, \mathbb{C}))$, then $\delta([\Pi'] + e_*[\lambda]) = \delta[\Pi']$. Using the linear structure of $H^1(Q; \mathcal{H}(W, \mathbb{C}))$, we can regard the holomorphic Seifert fibering $M_{([\Pi']+e_*[\lambda])} \to Q\backslash W$ as a deformation of the holomorphic Seifert fibering $M'_{[\Pi']} = Q\backslash(\mathbb{C}^* \times W) \to Q\backslash W$. Since $H^1(Q; \mathcal{H}(W, \mathbb{C}))$ is a complex $g$-dimensional vector space, with $g$ the genus of $Q\backslash W$, this completes the proof of the theorem. $\square$

14.10.4 (Interpreting diagram (14.10.2)). From Subsection 14.10.2, we get the exact sequence

$$0 \to H^1(Q;\mathbb{Z}) \xrightarrow{i} H^1(Q; \mathcal{H}(W,\mathbb{C})) \xrightarrow{j} H^1(Q; \mathcal{H}(W, \mathbb{C}^*)) \xrightarrow{\delta} H^2(Q;\mathbb{Z}) \to 0.$$

The elements of $H^1(Q; \mathcal{H}(W, \mathbb{C}^*))$ represent the strict equivalence class of holomorphic $\mathbb{C}^*$-actions over a fixed Riemann surface $Q\backslash W$. They are in one-to-one correspondence to the generalization of holomorphic line bundles where we have singular $\mathbb{Z}_\alpha\backslash\mathbb{C}$ fibers over the branch points of $Q\backslash W$. Formally $\mathbb{C} \times_{\mathbb{C}^*} (Q\backslash(\mathbb{C}^* \times W))$ adjoins a "0" to each $\mathbb{C}^*$-fiber and adjoins the orbit space $Q\backslash W$ as a "zero"-section to $\mathbb{C} \times_{\mathbb{C}^*} (Q\backslash(\mathbb{C}^* \times W))$. $Q\backslash(\mathbb{C}^* \times W)$, with the holomorphic $\mathbb{C}^*$-action, plays the role of the principal $\mathbb{C}^*$-bundle over $Q\backslash W$ and the "line bundle", $\mathbb{C} \times_{\mathbb{C}^*} (Q\backslash(\mathbb{C}^* \times W))$, is the associated holomorphic 1-dimensional "vector bundle" over $Q\backslash W$. Group addition in $H^1(Q; \mathcal{H}(W, \mathbb{C}^*))$ corresponds to the tensor product of "line bundles". The complex $g$-dimensional vector space $H^1(Q; \mathcal{H}(W, \mathbb{C}))$ represents the deformations of the "line" bundles keeping the complex structure on $Q\backslash W$ constant. Of course, two deformations differ by an element of $H^1(Q;\mathbb{Z})$ if they map to the same "line" bundle. Therefore, the kernel of $\delta$ is a complex $g$-torus $\mathbb{C}^g/\mathbb{Z}^{2g}$ called the *Jacobian variety* of $Q\backslash W$ and the $\delta$-image of a "line" bundle is the "Chern" (or characteristic) class of the "line" bundle. The group of holomorphic "line" bundles are then parametrized by the strict equivalence classes of holomorphic $\mathbb{C}^*$-actions over a fixed Riemann surface $Q\backslash W$. As a topological group, its connected component of the identity is the Jacobian variety. The discrete part is isomorphic to $H^2(Q;\mathbb{Z}) \cong \mathbb{Z} \oplus$ Torsion. The group $H^1(Q; \mathcal{H}(W, \mathbb{C}^*))$ may be called the *Picard variety* of $Q\backslash W$. Interpreting the exact sequence in these terms, we have [**CR72b**, §13]

$$0 \longrightarrow \text{Jacobian}(Q\backslash W) \longrightarrow \text{Picard}(Q\backslash W) \xrightarrow{\delta} H^2(Q;\mathbb{Z}) \longrightarrow 0.$$

In the discussion above, we have fixed $i : \mathbb{Z} \to \mathbb{C}$ and $\rho : Q \to \text{Hol}(W)$. If we vary these choices, we do not get anything new in the smooth case because of smooth rigidity. That is, $\theta(\Pi)$ is conjugate to $\theta'(\Pi)$ in $\text{Diff}_\mathbb{C}(\mathbb{C} \times W)$ where conjugation is taken in the whole group and not just in $\mathcal{C}(W, \mathbb{C})$ as for strict equivalence. However, in the holomorphic case, a change in $\rho : Q \to \mathcal{H}(W)$ induces a much larger deformation space than treated above.

If $Q$ is torsion free, then $H^1(Q; \mathcal{H}(W, \mathbb{C}^*))$ classifies the principal holomorphic $\mathbb{C}^*$-bundles over the surface $Q\backslash W$. In this case, the kernel of $\delta$ is the principal holomorphic $\mathbb{C}^*$-bundles which are trivial as smooth $\mathbb{C}^*$-bundles. The kernel, when interpreted as holomorphic line bundles, forms the classical Jacobian variety of line bundles over $Q\backslash W$. If $Q$ is not torsion free, those elements in the kernel smoothly correspond to the Seifert fiberings $\{\mathfrak{o}, g, 0, 0; 0; (\alpha_1, 0), \ldots, (\alpha_m, 0)\}$ of Sections 15.4 and 15.5. The topological spaces underlying these last fiberings are homeomorphic to $\mathbb{C}^* \times (Q\backslash W)$, (or equivalently $\mathbb{C} \times (Q\backslash W)$ in the "line bundle" format). Let us also note when $g = 0$, the holomorphic and smooth classification are the same. That is, $H^1(Q; \mathcal{H}(W, \mathbb{C}^*)) \to H^1(Q; \mathcal{C}(W, \mathbb{C}^*))$ is an isomorphism.

14.10.5. Recall by "filling in" a holomorphic line bundle with the Riemann surface $Q\backslash W$ obtained by adding $0 \in \mathbb{C}$ to each $\mathbb{C}^*$-fiber as described above, we may collapse the analytic subspace $Q\backslash W$ to a point. If $e(Q\backslash(\mathbb{C}^* \times W)) < 0$, this becomes a complex analytic space with an isolated singularity. The analytic justification for doing this is due to [**Gra62**].

A much more algebraic approach by Orlik and Wagreich ([**Orl72**, pp.22–65] and [**OW71**]) investigates the isolated singularities of affine 2-dimensional varieties defined by weighted homogeneous polynomials with a proper holomorphic $\mathbb{C}^*$-action. A canonical minimal resolution of such a singularity is related to an equivariant plumbing graph where the boundary of the real 4-dimensional boundary is an $(S^1, M)$ as above. Compare this also with [**NR78**, §5].

## 14.11. Brieskorn complete intersections

14.11.1. The following explicit examples are taken from [**NR78**, §2]. The reader might first review Example 1.6.9. Let $(a_1, \ldots, a_n)$ be integers, $a_i \geq 2$. An $(n-2) \times n$ matrix of complex numbers

$$C = (c_{ij})$$

is *sufficiently general* if and only if every $(n-2) \times (n-2)$ subdeterminant of $C$ is nonzero. Assume $C$ is sufficiently general. The variety

$$V_C(a_1, \ldots, a_n) = \{z \in \mathbb{C}^n \mid c_{i1} z_1^{a_1} + \cdots + c_{in} z_n^{a_n} = 0,\ i = 1, \ldots, n-2\}$$

is a complex surface which is nonsingular except perhaps at the origin, and

$$\Sigma(a_1, \ldots, a_n) = V_C(a_1, \ldots, a_n) \cap S^{2n-1}$$

is a smooth 3-manifold not depending on $C$ up to diffeomorphism.

$V_C(a_1, \ldots, a_n)$ has a $\mathbb{C}^*$-action defined by

$$z(z_1, \ldots, z_n) = (z^{q_1} z_1, \ldots, z^{q_n} z_n)$$

for $z \in \mathbb{C}^*$, where

$$q_j = \frac{\operatorname{lcm}(a_1, \ldots, a_n)}{a_j}.$$

$S^1 \subset \mathbb{C}^*$ acts without fixed points on $\Sigma(a_1, \ldots, a_n)$ and so is a Seifert manifold.

THEOREM 14.11.2 (see [**NR78**, Theorem 2.1] for a proof). $\Sigma(a_1, a_2, \ldots, a_n)$ has unnormalized Seifert invariants

$$M = \{\mathfrak{o}, g, 0, 0;\ s_1(\alpha_1, \beta_1), \ldots, s_n(\alpha_n, \beta_n)\},$$

where $s_j(\alpha_j, \beta_j)$ means $(\alpha_j, \beta_j)$ is represented $s_j$ times and

$$\alpha_j = \frac{\mathrm{lcm}(a_1, \ldots, a_n)}{\mathrm{lcm}(a_1, \ldots, \widehat{a}_j, \ldots, a_n)},$$

$$s_j = \frac{(a_1 a_2 \cdots \widehat{a}_j \cdots a_n)}{\mathrm{lcm}(a_1, \ldots, \widehat{a}_j, \ldots, a_n)},$$

$$g = \frac{1}{2}\Big(2 + (n-2)\frac{(a_1 a_2 \cdots a_n)}{\mathrm{lcm}(a_1, \ldots, a_n)} - \sum_{j=1}^{n} s_j\Big),$$

and the $\beta_i$ and $e(M)$ are determined by

$$-e(M) = \sum_{j=1}^{n} s_i \frac{\beta_j}{\alpha_j} = \frac{(a_1 a_2 \cdots a_n)}{\mathrm{lcm}(a_1, \ldots, a_n)^2}.$$

The latter equation becomes

$$\sum_{j=1}^{n} \beta_j q_j = 1$$

by dividing by the right-hand side, where $q_j = \mathrm{lcm}(a_1, \ldots, a_n)/a_j$. Clearly, $\alpha_i$ divides $q_j$ if $i \neq j$ and is prime to $q_j$ if $i = j$. Thus, the equation becomes $\beta_j q_j \equiv 1 \bmod \alpha_j$ and so determines $\beta_j \bmod \alpha_j$.

COROLLARY 14.11.3. *Let $\{a_1, \ldots, a_n\}$ be pairwise coprime. Then $\Sigma(a_1, \ldots, a_n)$ is a Seifert homology sphere and is of the form $\Gamma \backslash S^3$ or $\Gamma \backslash \widetilde{\mathrm{PSL}}(2, \mathbb{R})$. It is the universal Abelian cover of $Q \backslash \mathrm{PSL}(2, \mathbb{R})$ and centrally covers it when $Q(g = 0; a_1, \ldots, a_n)$ is infinite.*

PROOF. With $a_i, a_j$ pairwise coprime, $\alpha_j = a_j$, $s_j = 1$, $g = 0$ and $-e(M) = \frac{1}{a_1 a_2 \cdots a_n}$. Thus, by Corollary 14.9.2, $\Sigma(a_1, \ldots, a_n)$ is an integral homology sphere. By Corollary 14.9.4, $\Sigma = \Gamma \backslash \widetilde{\mathrm{PSL}}(2, \mathbb{R})$ and centrally covers $Q \backslash \mathrm{PSL}(2, \mathbb{R})$, where $Q = Q(0; a_1, \ldots, a_n)$ is hyperbolic (i.e., $Q$ is infinite). By Corollary 14.9.5, $\Sigma$ is the universal Abelian cover of $Q \backslash \mathrm{PSL}(2, \mathbb{R})$.

If $\pi_1(\Sigma(a_1, \ldots, a_n))$ is finite, then $M$ is the 3-sphere or the Poincaré homology sphere which is diffeomorphic to $\Sigma(1, 1, 1)$ or $\Sigma(2, 3, 5)$, respectively. □

COROLLARY 14.11.4. *The variety $V(a_1, a_2, a_3) = z_1^{a_1} + z_2^{a_2} + z_3^{a_3} = 0$ in $\mathbb{C}^3$, with $a_1, a_2, a_3$ pairwise coprime is the cone over a Seifert homology 3-sphere $\Sigma(a_1, a_2, a_3)$. The $\mathbb{C}^*$-action on $\mathbb{C}^3$ leaving $V$ invariant is given by $z(z_1, z_2, z_3) = (z^{a_2 a_3} z_1, z^{a_1 a_3} z_2, z^{a_1 a_2} z_3)$.*

Note that this explicitly embeds $\Sigma(a_1, a_2, a_3)$ smoothly into $S^5$.

EXERCISE 14.11.5. $\Sigma(a, a, a)$ is
(1) $S^3$ if $a = 1$ (this is the only time we allow $a_i = 1$).
(2) $\mathbb{R}P_3$ if $a = 2$.
(3) a principal $S^1$-bundle with Euler class $-a$ over a surface of genus $g = \frac{1}{2}(a-1)(a-2)$.

14.11.6. It is quite remarkable that the results for Seifert homology spheres as homogeneous spaces also apply to any $\Sigma(a_1, \ldots, a_n)$ with $g = 0$. Neumann in [**JN83**] formulates it this way:

## 14.11. BRIESKORN COMPLETE INTERSECTIONS

THEOREM 14.11.7 (Neumann). *Let $(V, p)$ be a complex analytic space with a proper $\mathbb{C}^*$-action on $V - p$ with $p$ being an isolated singularity. Suppose the Seifert manifold associated with the $\mathbb{C}^*$-action, as in Subsection 14.10.1, is*

$$(S^1, M) = \{\mathfrak{o}, 0, 0, 0; b; (\alpha_1, \beta_1), \ldots, (\alpha_n, \beta_n)\}.$$

*Let $n \geq 3$. Then the universal Abelian cover $(\widetilde{V}^{ab}, p)$ of $(V, p)$ branched at the cone point $p$ is isomorphic to $V_C((\alpha_1, \ldots, \alpha_n), 0)$ where $V_C(\alpha_1, \ldots, \alpha_n)$ is the Brieskorn complete intersection for suitable matrix $C$. In fact, one can find $C$ explicitly.*

Let us state this a little less generally and within our preceding framework.

THEOREM 14.11.8. *Let $(S^1, M) = \{\mathfrak{o}, 0, 0, 0; b; (\alpha_1, \beta_1), \ldots, (\alpha_n, \beta_n)\}$ with $e(M) \neq 0$, $n \geq 3$, $\chi(Q) < 0$. Then, $\Sigma(\alpha_1, \ldots, \alpha_n)$ is the universal Abelian cover of $M$ and $\Sigma(\alpha_1, \ldots, \alpha_n)$ is homogeneous.*

We sketch the argument. Let $M_1$, $M_2$ be two Seifert manifolds with isomorphic $Q(g = 0; \alpha_1, \ldots, \alpha_n)$. (That is, $\pi_1(M)/\text{Im}(\text{ev}^x_*(\pi_1(S^1))) = Q$.) Since $\chi(Q) < 0$, there exist embeddings $\theta_i$ with fixed $\bar{\theta}_i$:

$$\begin{array}{ccccccccc}
0 & \longrightarrow & \mathbb{Z} & \longrightarrow & \pi_1(M_i) & \longrightarrow & Q & \longrightarrow & 1 \\
& & \downarrow \theta_i|_\mathbb{Z} & & \downarrow \theta_i & & \downarrow \bar{\theta}_i & & \\
0 & \longrightarrow & \mathbb{R} & \longrightarrow & \mathbb{R} \times_\mathbb{Z} \widetilde{\text{PSL}}(2, \mathbb{R}) & \longrightarrow & \text{PSL}(2, \mathbb{R}) & \longrightarrow & 1
\end{array}$$

by Chapter 13. (The uniformizing groups for the general Seifert Construction is reduced to $\mathbb{R} \times_\mathbb{Z} \widetilde{\text{PSL}}(2, \mathbb{R})$.) Because the second line is central, it follows that the $\theta_i$ restricted to the commutator subgroup of $\pi_1(M_i)$ have the same image. If we take $M_1$ to be $\{\mathfrak{o}, g, 0, 0, 0; -2; (\alpha_1, \alpha_1 - 1), \ldots, (\alpha_n, \alpha_n - 1)\}$, i.e., $M_1 = Q \backslash \text{PSL}(2, \mathbb{R})$ with $Q = Q(g = 0; \alpha_1, \ldots, \alpha_n)$, then $\Sigma(\alpha_1, \ldots, \alpha_n)$ has the same Seifert invariants and Euler number as the universal Abelian cover of $M_1$, see [**NR78**, Example 3.3] or [**JN83**, Theorem 8.2].

$(S^1, M_2)$ with the same $Q$ but perhaps different $\beta$'s will also have $\Sigma(\alpha_1, \ldots, \alpha_n)$ as its universal Abelian cover since $\pi_1(\Sigma)$ will also be the commutator subgroup of $\pi_1(M_2)$. Since $M_1 = Q \backslash \text{PSL}(2, \mathbb{R})$, it follows that $\Sigma(\alpha_1, \ldots, \alpha_n)$ is of the form $\Gamma \backslash \widetilde{\text{PSL}}(2, \mathbb{R})$, $\Gamma = [\pi_1(M_1), \pi_1(M_1)] \subset \pi_1(M_1)$, and covers $M_i$.

EXAMPLE 14.11.9. We verify (i) and (ii).

(i) $\Sigma(2, 3, 9) = \{\mathfrak{o}, 0, 0, 0; 0; (2, 1), (2, 1), (2, 1), (3, -4)\}$ is the universal Abelian cover of

$$\begin{aligned}
Q \backslash \text{PSL}(2, \mathbb{R}) &= \{\mathfrak{o}, 0, 0, 0; -2; (2, 1), (3, 2), (9, 8)\} \\
&= \{\mathfrak{o}, 0, 0, 0; 0; (2, 1), (3, -1), (9, -1)\}.
\end{aligned}$$

The unbranched covering is induced from the branched covering of $S^1 \backslash \Sigma(2, 3, 9) \longrightarrow S^1 \backslash (Q \backslash \text{PSL}(2, \mathbb{R}))$.

The indicated Seifert invariants are normalized. We shall calculate $q_1, q_2, q_3$. Since

$$q_i = \frac{\text{lcm}(a_1, \ldots, a_n)}{a_i},$$

$$q_1 = \frac{18}{2}, \quad q_2 = 6, \quad q_3 = \frac{18}{9} = 2, \quad \alpha_2 = 1,$$

$$-e(\Sigma(2, 3, 9)) = \frac{54}{18^2} = \frac{1}{6} = 3\frac{1}{2} + \frac{1 - 4}{3}.$$

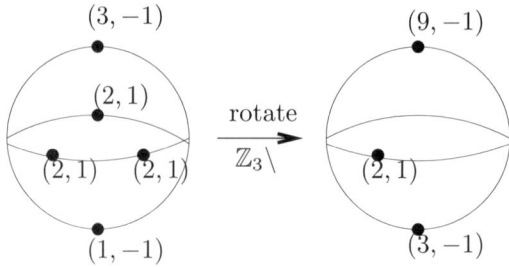

FIGURE 2. Brieskorn–1

Therefore, we have that $\Sigma(2,3,9)$ corresponds to the Seifert manifold over the first sphere in Figure 2 (Brieskorn–1). Note $\Sigma(2,3,9)$ is of the form $P_1/\Gamma$. By Proposition 14.9.1, $|H_1(Q\backslash\mathrm{PSL}(2,\mathbb{R});\mathbb{Z})| = 3$ and is isomorphic to $\mathbb{Z}_3$.

(ii) $\Sigma(2,3,9)$ is the universal Abelian cover of $-\{\mathfrak{o},0,0,0;-2;\,(2,1),(3,2),(9,5)\}$ $= \{\mathfrak{o},0,0,0;2;\,(2,-1),(3,-2),(9,-5)\} = M$ as a composite of a central $\mathbb{Z}_5$ covering followed by a $\mathbb{Z}_3$ covering as shown in Figure 3 (Brieskorn–2).

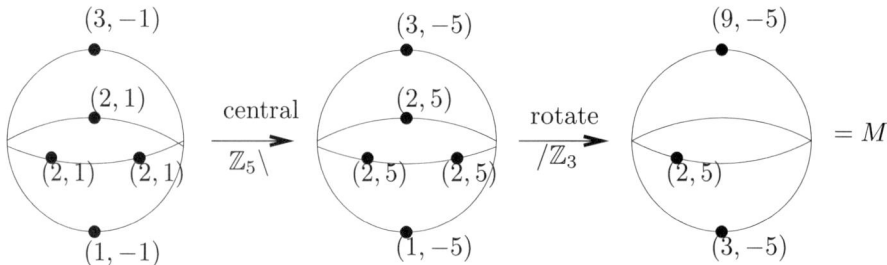

FIGURE 3. Brieskorn–2

### 14.12. Generalized Seifert 3-manifolds as local $\mathrm{SO}(2)$-actions

In this section, we classify the closed connected 3-manifolds which admit local $\mathrm{SO}(2)$-actions.

The orbits of $S^1$-actions on closed connected 3-manifolds, as seen earlier, consists of free, E, SE, and point or F orbits; see Section 14.1. These orbits are the fibers of the orbit mapping

$$\tau : (S^1, M) \longrightarrow S^1\backslash M = B.$$

Away from the E, SE, and F orbits are the free or typical orbits which form a principal $\mathrm{SO}(2)$-bundle. This bundle has a cross section provided that all orbits of $(S^1, M)$ are not free orbits. Each connected component of one of these nonfree orbits has an $S^1$-invariant neighborhood which is attached equivariantly to the principal $S^1$-bundle to form $(S^1, M)$.

A local $\mathrm{SO}(2)$-action [**OR69**] on a closed connected 3-manifold $M$ is a map onto a compact surface $B$ where the inverse image of a point in $B$ consists of a typical $S^1$-fiber, E-fiber, SE-fiber, or an F-fiber. The union of the typical fibers form an $S^1$-bundle with structure group in $\mathrm{O}(2)$. Each connected component of the nontypical fibers has an *invariant* neighborhood which is attached to the $S^1$-bundle by an O(2)-bundle isomorphism to form $M$. It can be shown that such a local SO(2)-action has at most 2- or 4-fold regular covering $(S^1, \widehat{M})$ where the fibers of $M$ lift to $S^1$-orbits of $\widehat{M}$ with $\widehat{M}$ orientable. The local SO(2)-actions are genuine Seifert fiberings if there are no point fibers. We use the terminology *generalized* Seifert fibering to indicate the possibility that the local SO(2)-action has point fibers.

The techniques used for the classification of the $S^1$-actions can be adapted to classify the local SO(2)-actions. For example, in Theorem 14.13.4 we show how the techniques of an equivariant connected sum classifies the local SO(2)-action when $F \neq \emptyset$. *So throughout Section* 14.12, *we shall assume that there are no F-fibers.*

If $F = \emptyset$ and $\pi_1(M)$ is finite, it is not hard to see that the $M$ are homeomorphic to the 3-dimensional spherical space forms of constant positive sectional curvature. All these fiberings are equivalent, as Seifert fiberings, to orbit mappings of the standard $S^1$-actions on the space forms except for the family of *prism manifolds* $(S^1, M) = \{\mathfrak{o}, 0, 0, 0; b;\ (2,1), (2,1), (\alpha_3, \beta_3)\}$, which also admit a local SO(2)-action that also fibers over $\mathbb{R}P_2$ with at most one exceptional fiber and a nonprincipal $S^1$-bundle away from the E-fiber ([**OR69**, 4(iv)], [**Orl72**, pp.112, ii and vi] or [**JN83**, p.39]).

If $F = \emptyset$ and $\pi_1(M)$ is infinite, then a local SO(2)-action is an injective Seifert fibering (Subsection 4.5.14). The idea is as follows. The base decomposition is a 2-manifold with boundary, the images of the components of SE-fibers (or F-fibers if they are present). We shall use the notation of Theorem 14.5.3.

$\tau^{-1}(B_0)$ is an O(2)-bundle, where $\tau^{-1}(B_0)$ is the closure of the complement of closed *invariant* neighborhoods of the components of the nontypical fibers for the Seifert fibering $\tau : M \to B$. If it is not a principal $S^1$-bundle, it has a double covering $\widehat{\tau^{-1}(B_0)}$ which is a principal $S^1$-bundle. If a component of the boundary $\tau^{-1}(B_0)$ is a principal $S^1$-bundle, then the $S^1$-action extends over the invariant neighborhood $V$ attached to that component. If a component of the boundary is not principal, then the invariant neighborhood $V'$ attached to it has a connected double covering $\widehat{V'}$ with an $S^1$-action covering $V'$ fiberwise. Over a boundary component of $B_0$ in $\widehat{\tau^{-1}(B_0)}$, we have either two copies of $\partial V$ or one copy of $\widehat{\partial V'}$.

Along their boundaries, we attach either two copies of $V$ equivariantly or a copy of $\widehat{V'}$ to the $S^1$-action on $\widehat{\tau^{-1}(B_0)}$. This gives us an $(S^1, \widehat{M})$ as a regular covering of $M$. The action of $\mathbb{Z}_2$ normalizes the $S^1$-action $(S^1, \widehat{M})$. Therefore, the local SO(2)-action is injective, in the sense of Subsection 4.5.14, if and only if $(S^1, \widehat{M})$ is injective. (If we allowed F-fibers in $M$, then $(S^1, \widehat{M})$ has F-fibers but of course, $(S^1, \widehat{M})$ would not be injective.)

We should also note that, in dimension 3, it suffices to only assume that the typical fibers form a locally trivial fiber bundle because TOP($S^1$) deformation retracts onto O(2). Of course we still need to attach the nontypical fibers compatibly to this bundle.

As a consequence, the space $W$ is either $\mathbb{R}^2$ or $S^2$. The universal covering of $M$ is $P = \mathbb{R}^1 \times \mathbb{R}^2$ or $\mathbb{R}^1 \times S^2$ and $B = Q \backslash W$. The manifold $M = \Pi \backslash P$ is the result of a Seifert Construction $\theta : \Pi = \pi_1(M) \to \text{TOP}_G(P)$, where

$$1 \to \mathbb{Z} \xrightarrow{i_*} \Pi \longrightarrow Q \to 1$$

is exact, $i_* : \pi_1(S^1) = \mathbb{Z} \to \pi_1(M)$ is the image of the fundamental group of a typical fiber under the inclusion $i : S^1 \to M$, and $\theta(\Pi)$ acts freely on $P$. If $W = S^2$, $Q$ is finite, while otherwise $W = \mathbb{R}^2$. The action of $Q$ on $\mathbb{R}^2$ is topologically equivalent to a hyperbolic action if $Q$ is infinite and nonsolvable and topologically equivalent to a Euclidean action if $Q$ is solvable. For the former case, the Seifert Construction $\theta : \Pi \to \text{TOP}_{\mathbb{R}^1}(\mathbb{R}^1 \times \mathbb{R}^2)$ is unique and rigid in the sense of Subsection 7.4.2 and Remark 7.4.4 and so $M$ is topologically determined by $\pi_1(M)$. We shall now verify these statements as well as sketch the orbifold case when $\theta(\Pi)$ does not act freely.

14.12.1 (The Seifert Construction; cf. Subsection 11.1.8). Let $\rho : Q \to \text{TOP}(W)$ be a homomorphism with finite kernel of a discrete group $Q$ into $\text{TOP}(W)$. Assume that the image acts properly on $W$ where $W$ is $\mathbb{R}^2$ or $S^2$ and $B = Q \backslash W$ is compact. For each extension
$$1 \to \mathbb{Z} \to \Pi \to Q \to 1,$$
there is a homomorphism $\theta : \Pi \to \text{TOP}_{\mathbb{R}^1}(\mathbb{R}^1 \times W)$ making the diagram

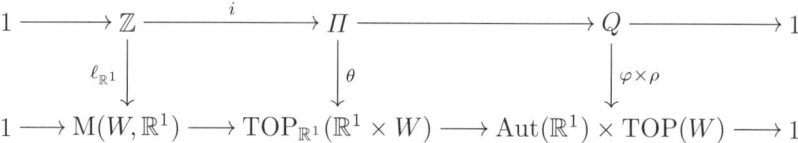

commutative and $\ell_{\mathbb{R}^1}$ injective (see Existence in Theorem 7.3.2). Without loss of generality, we can assume that $\ell_{\mathbb{R}}(\mathbb{Z}) = \Pi \cap \ell(G)$, and that $\theta$ is injective if and only if $\varphi \times \rho$ is injective (Corollary 7.7.4). $\theta$ is an *injective Seifert Construction* (see Definition 4.6.1) and it yields an effective proper action of $\Pi$ on $\mathbb{R} \times W$ which normalizes the left translational $\mathbb{R}^1$-action on $\mathbb{R}^1 \times W$. The induced mapping $\tau : M = \theta(\Pi) \backslash (\mathbb{R}^1 \times W) \to Q \backslash W$ is an injective Seifert fibering with typical fiber $S^1 = \ell(\mathbb{Z}) \backslash \mathbb{R}^1$, see Subsection 4.5.14.

If $\Pi$ does not act freely, $M$ will be an orbifold. If $W = \mathbb{R}^2$, then $M$ is aspherical if $\Pi$ acts freely. When $W = \mathbb{R}^2$, $\Pi$ acts freely if and only if $\Pi$ is torsion free; see Subsection 1.8.1 or Corollary 3.1.9. Because we assume $\varphi \times \rho$ is injective, $\rho : Q \to \text{TOP}(\mathbb{R}^2)$ has kernel $K_\rho = 1$ or $\mathbb{Z}_2 = \text{Aut}(\mathbb{Z}) \subset \text{Aut}(\mathbb{R})$. In particular, if $\Pi$ is torsion free, $K = 1$. For, if $K_\rho = \mathbb{Z}_2$, then the pullback of $1 \to \mathbb{Z} \to \Pi \to Q \to 1$ to $K_\rho$ yields the exact sequence $1 \to \mathbb{Z} \to \mathbb{Z} \rtimes \mathbb{Z}_2 \to K_\rho \to 1$, and therefore, $\mathbb{Z}_2 \subset \mathbb{Z} \rtimes \mathbb{Z}_2 \subset \Pi$.

As in Subsection 14.8.9, if $Q$ is a discrete group which acts properly and effectively with compact quotient on $\mathbb{R}^2$, then $Q$ is conjugate in $\text{TOP}(\mathbb{R}^2)$ to a group of Euclidean isometries of $\mathbb{R}^2$ or hyperbolic isometries of the hyperbolic plane. In this situation, we shall say $Q$ is *topologically Euclidean* or *topologically hyperbolic*. Topologically Euclidean $Q$ are solvable groups (extension of a solvable group by a subgroup of $\Sigma_4$, the group of permutations of (at most) four branch points, is solvable again), while topologically hyperbolic $Q$ are not solvable.

THEOREM 14.12.2. *Let $Q$ be topologically hyperbolic and $\Pi$ torsion free. Then the Seifert Construction $\theta$, as in Subsection* 14.12.1, *is unique, and rigid in the sense of Subsections* 7.4.2, *and* 7.4.3, *and Remark* 7.4.4.

PROOF. Since $\varphi \times \rho$ is injective and $\Pi$ is torsion free, $\rho$ is injective.

The base $B = Q \backslash W$ must be a compact 2-manifold with possible boundary corresponding to the images of SE-fibers, (and F-fibers in Section 14.13). As an orbifold, $B$ can have no corner reflections, for otherwise, $\Pi$ would not be acting freely.

References for the following are [**OR69**], [**Orl72**] and [**Fin76**]. $(\rho(Q), W)$ is hyperbolic if $B$ is without corner reflections and is not one of the following (the $\alpha_i$ are the orders of the branching):

(1) $\mathbb{R}P_2$ with $\leq 1$ branch point,
(2) $S^2$ with $\leq 2$ branch points,
(3) $S^2$ with 3 branch points and $1/\alpha_1 + 1/\alpha_2 + 1/\alpha_3 \geq 1$,
(4) $D^2$ with $\leq 1$ branch point,
(5) $S^2$ with 4 branch points where $\alpha_i = 2$, $i = 1, 2, 3, 4$,
(6) an annulus with no branch point,
(7) Möbius band with no branch point,
(8) Torus or Klein bottle with no branch point,
(9) $\mathbb{R}P_2$ with 2 branch points where $\alpha_1 = \alpha_2 = 2$,
(10) Disk with 2 branch points where $\alpha_1 = \alpha_2 = 2$.

(1)–(4) are analogous to Subsection 14.4.2. But here we also rule out the $Q$ that have corner reflectors and are topologically Euclidean. This requires us to add the restrictions (5)–(10). (In (3), to rule out solvable and some finite cases; (5) added to rule out solvable cases; (6) and (7) pass to orientable double cover; (8) as in (5); (9) and (10) lift to case (5).)

As $Q$ is now cocompact, acts properly and effectively on $\mathbb{R}^2$ and is isomorphic to a hyperbolic group of isometries, the action $(Q, \mathbb{R}^2)$ is topologically hyperbolic. Furthermore, $Q$ contains no normal Abelian subgroup other than the identity. Consequently, $i(\mathbb{Z}) \subset \Pi$ is the maximal normal Abelian subgroup of $\Pi$. Thus, $\mathbb{Z}$ is characteristic in $\Pi$. Therefore, under the hypothesis of the theorem, the existence, uniqueness and rigidity of Section 7.4 must hold. $\square$

COROLLARY 14.12.3 (cf. Subsection 14.4.3). *Let $M_1$ and $M_2$ be closed 3-dimensional Seifert fiberings modeled on the trivial $\mathbb{R}^1$-bundle $\mathbb{R}^1 \times \mathbb{R}^2$. If $\psi : \pi_1(M_1) \to \pi_1(M_2)$ is an isomorphism and $\pi_1(M_i)$ are torsion free and not solvable, then there is an isomorphism of Seifert fiberings $M_1 \to M_2$ induced by $k\theta_1 k^{-1} = \theta_2 \circ \psi$, where $k \in \mathrm{TOP}_{\mathbb{R}^1}(\mathbb{R}^1 \times \mathbb{R}^2)$.*

PROOF. The hypothesis implies that the vertical maps in the extensions

$$\begin{array}{ccccccccc} 1 & \longrightarrow & \mathbb{Z} & \longrightarrow & \pi_1(M_1) & \longrightarrow & Q_1 & \longrightarrow & 1 \\ & & \downarrow \psi|_{\mathbb{Z}} & & \downarrow \psi & & \downarrow \overline{\psi} & & \\ 1 & \longrightarrow & \mathbb{Z} & \longrightarrow & \pi_1(M_2) & \longrightarrow & Q_2 & \longrightarrow & 1 \end{array}$$

are isomorphisms. Moreover the $Q_i$ are topologically hyperbolic, acting via $\rho_i : Q_i \to \mathrm{TOP}(\mathbb{R}^2)$. Since $\overline{\psi}$ is an isomorphism, there is a homeomorphism $h : \mathbb{R}^2 \to \mathbb{R}^2$ so that $h \circ \rho_1 \circ h^{-1} = \rho_2 \circ \overline{\psi}$ as in Subsection 14.4.3. The corollary now follows from Theorem 7.6.10. $\square$

REMARK 14.12.4. 1. For the most general of 3-dimensional fiberings with $F = \emptyset$ and $\Pi$ acting freely, the base could also be any one of excluded ten types of 2-dimensional orbifolds in Theorem 14.12.2. Even so, all of these fiberings will be locally injective Seifert fiberings, i.e., modeled on principal bundles and obtained as a Seifert Construction $\theta : \Pi \to \mathrm{TOP}_G(P)$, except possibly $B = S^2$ with less than or equal to 2 branch points (see (2) in the proof of Theorem 14.12.2). The classification of these special situations ((1) through (10)), must be done case by case. This is accomplished in [**OR69**], [**Orl72**] and [**Fin76**].

2. The Seifert Construction does not require $\theta(\Pi)$ to act freely on $\mathbb{R}^1 \times \mathbb{R}^2$. If $\Pi$ is not torsion free, the *covering* $\mathbb{R}^1 \times \mathbb{R}^2 \to \theta(\Pi)\backslash \mathbb{R}^1 \times \mathbb{R}^2$ is branched and the quotient is a 3-dimensional orbifold.

The injective local SO(2)-actions are the result of a Seifert Construction $\theta$ modeled on $\mathbb{R}^1 \times \mathbb{R}^2$ or $\mathbb{R}^1 \times S^2$ with $\Pi$ torsion free. When $\Pi$ is not torsion free, the orbifold $\theta(\Pi)\backslash \mathbb{R}^1 \times W$, as a space, may not be a 3-manifold. For example, the orbit space of the involution on $S^1 \times S^2$ which reflects in $S^1$ and rotates in $S^2$ is not a 3-manifold.

COROLLARY 14.12.5. *If $\rho_i(Q_i)$ is topologically hyperbolic and cocompact, the existence, uniqueness and strong rigidity still holds and $M_1$ and $M_2$ (as in Corollary 14.12.3) will be homeomorphic as orbifolds. That is, if $\theta(\Pi_1)$ is isomorphic to $\theta(\Pi_2)$, then the orbifolds $M_1$ and $M_2$ are homeomorphic by means of a Seifert isomorphism.*

PROOF. Let $K_\rho = \ker(\rho)$ and $\overline{Q} = Q/K_\rho$. Then, $\rho(Q) = \overline{Q}$ is topologically hyperbolic, $\overline{Q}$ contains no nontrivial normal solvable subgroups and $Q$ contains no nontrivial normal infinite solvable subgroups. If $K_\rho = 1$, $\rho$ is injective and $\mathbb{Z}$ is characteristic in $\Pi$.

If $K_\rho = \mathbb{Z}_2$, pull back the extension $1 \to \mathbb{Z} \to \Pi \to Q \to 1$ via $K_\rho \hookrightarrow Q$ and we get

$$\begin{array}{ccccccccc} 1 & \to & \mathbb{Z} & \to & \mathbb{Z} \rtimes \mathbb{Z}_2 & \to & \mathbb{Z}_2 = K_\rho & \to & 1 \\ & & = \downarrow & & \downarrow & & \downarrow & & \\ 1 & \to & \mathbb{Z} & \to & \Gamma & \to & Q & & \\ & & & & \downarrow & & \downarrow & & \\ & & & & \overline{Q} & = & \overline{Q} = Q/K_\rho. & & \end{array}$$

Observe $\mathbb{Z} \rtimes \mathbb{Z}_2$ and $\mathbb{Z}$ are characteristic in $\Pi$ and $K_\rho$ is characteristic in $Q$. If $\psi : \Pi_1 \to \Pi_2$ is an isomorphism, then $\psi$ restricts to isomorphisms on $\mathbb{Z}$ and $\mathbb{Z} \rtimes \mathbb{Z}_2$, (if $K_\rho \neq 1$). We then obtain the same commutative diagram of extensions as in Corollary 14.12.3. We also, as in Subsection 14.4.3, get the commutative diagram

$$\begin{array}{ccc} Q_1 & \xrightarrow{\rho_1} & \mathrm{TOP}(\mathbb{R}^2) \\ \overline{\psi} \downarrow & & \downarrow \mu(h) \\ Q_2 & \xrightarrow{\rho_2} & \mathrm{TOP}(\mathbb{R}^2), \end{array}$$

where $h: \mathbb{R}^2 \to \mathbb{R}^2$ is a homeomorphism whether $K = 1$ or $\mathbb{Z}_2$. So we may apply the (strong) rigidity theorem of Section 11.2 (Chapter 7 as well), and we assert that there is induced a Seifert isomorphism $\theta_1(\Pi_1)\backslash(\mathbb{R}^1 \times \mathbb{R}^2) \to \theta_2(\Pi_2)\backslash(\mathbb{R}^1 \times \mathbb{R}^2)$. □

PROPOSITION 14.12.6. *Let $1 \to \mathbb{Z} \to \Pi \to Q \to 1$ be as in Corollary 14.12.5. If $K_\rho = \mathbb{Z}_2$, there exists a closed orientable surface $S$ and an action of $\mathbb{Z}_2 \times F$ on $S^1 \times S$ where $F$ acts as Seifert automorphisms on $S^1 \times S$, $\mathbb{Z}_2$ acts only on the first factor by reflection and $M = F\backslash(I \times S) = F\backslash((\mathbb{Z}_2\backslash S^1) \times S)$.*

*If $K_\rho = 1$, there is an orientable surface $S$ and a principal $S^1$-bundle $M''$ over $S$ and a group of Seifert automorphisms $H$ such that $M'' \to H\backslash M'' = M$.*

PROOF. Since $\Pi$ acts on the characteristic subgroup $\mathbb{Z}$ by conjugation, the subgroup $\Pi'$ of $\Pi$ which acts trivially on $\mathbb{Z}$ is characteristic and of index 1 or 2 in $\Pi$. Let $Q'$ be the image of $\Pi'$ in $Q$. Then $Q' = \ker(\varphi)$ and $Q' \cap K_\rho = 1$ since $\varphi \times \rho$ is injective. From the diagram

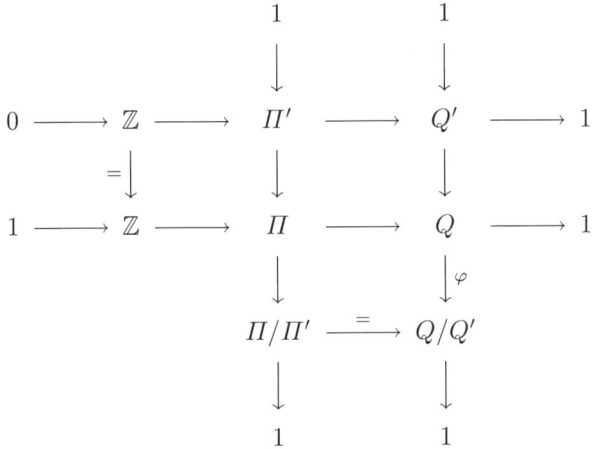

where $Q \supset K_\rho = \mathbb{Z}_2$, $\varphi$ is injective on $K_\rho$. Thus $Q = Q' \rtimes (Q/Q') = Q' \rtimes \mathbb{Z}_2$, with $\mathbb{Z}_2 \to \operatorname{Aut}(\mathbb{Z})$ nontrivial. We claim that the splitting $Q' \rtimes \mathbb{Z}_2$ is a product splitting $Q' \times \mathbb{Z}_2$. If the abstract kernel $\mathbb{Z}_2 \to \operatorname{Out}(Q')$ is nontrivial for the extension $1 \to Q' \to Q \to \mathbb{Z}_2 \to 1$, then $Q$ would be isomorphic to a subgroup of isometries of the hyperbolic plane by the Nielsen realization theorem of Zieschang [**Zie81**] or Kerckhoff [**Ker83**]. But, since we know that $Q$ is ineffective on $\mathbb{R}^2$, the abstract kernel is trivial. Furthermore, $\mathcal{Z}(Q') = 1$ which implies $H^2(\mathbb{Z}_2; \mathcal{Z}(Q')) = 0$ and therefore $Q \cong Q' \times \mathbb{Z}_2$. The action of $K_\rho = \mathbb{Z}_2$ on $Q'$ is trivial and hence, its action on $\mathbb{R}^2$ is trivial. Whether $K_\rho = 1$ or $\mathbb{Z}_2$, we always can pick $Q'' \subset Q'$ which is torsion free, of finite index and characteristic in $Q'$ as well as orientation preserving on $W = \mathbb{R}^2$. (By Selberg's Lemma 6.1.8, we get $Q''$ to be torsion free normal of finite index and orientation preserving—now take intersection with all subgroups of $Q'$ of the same index—this will be characteristic in $Q'$.) For the induced extension $1 \to \mathbb{Z} \to \Pi'' \to Q'' \to 1$, we get $M'' = \theta(\Pi'') \backslash (\mathbb{R}^1 \times \mathbb{R}^2)$. Then $M''$ is a principal $S^1$-bundle over an orientable surface $Q'' \backslash \mathbb{R}^2 = S$. Let $Q/Q''$ be $H$, then $M = H \backslash M'' = (Q/Q') \backslash (Q'/Q'') \backslash M''$.

If $K_\rho = 1$ and there is no $q \in Q$ which reverses the orientation of $S^1 \times \mathbb{R}^2$, then $Q'' \backslash (S^1 \times \mathbb{R}^2) = M''$ is a principal $S^1$-bundle over $S = Q'' \backslash W$ and $H$ acts on $M''$ as Seifert automorphisms yielding $\theta(\Pi) \backslash (\mathbb{R}^1 \times \mathbb{R}^2) = M$.

If $K_\rho = 1$ and there is $q \in Q$ reversing the orientation of $S^1 \times \mathbb{R}^2$, then the coset $qQ'' \in Q/Q'' = H$ acting on $M''$ reverses the orientation of $M''$. Consequently $e(M'') = 0$ by Theorem 14.7.2, and so $M'' = S^1 \times S$ since $Q''$ is torsion free.

If $K_\rho = \mathbb{Z}_2$, then $Q/Q'' = ('Q \times \mathbb{Z}_2)/(Q'' \times 1) = (Q'/Q'') \times \mathbb{Z}_2$. The action of $F = Q'/Q''$ and $\mathbb{Z}_2$ commute on $S^1 \times S$. The action of $\mathbb{Z}_2$ is just reflection on the first factor. Therefore if $K = \mathbb{Z}_2$, $M = (Q'/Q'') \backslash (I \times S) = F \backslash ((\mathbb{Z}_2 \backslash S^1) \times S)$. □

COROLLARY 14.12.7. *Let $1 \to \mathbb{Z} \to \Pi \to Q \to 1$ be as in Proposition 14.12.6, where $Q$ acts on $S^1 \times \mathbb{R}^2$. If some $q \in Q$ reverses the orientation of $S^1 \times \mathbb{R}^2$, then $M = \theta(\Pi) \backslash (\mathbb{R}^1 \times \mathbb{R}^2)$ is the quotient of $S^1 \times S$ for some orientable surface by a finite group $H$ of Seifert automorphisms.*

EXAMPLE 14.12.8. Let $S_g$ be the orientable double covering of the nonorientable surface $\Sigma_{g+1}$ of genus $g+1$. Form $S^1 \times_{\mathbb{Z}_2} S_g = M'$, where the diagonal $\mathbb{Z}_2$ acts as covering transformations on $S_g$ and by reflection on the $S^1$-factor. Then $M'$ is a nonprincipal $S^1$-bundle over $\Sigma_{g+1}$ with structure group $\mathbb{Z}_2$. As a manifold, $M'$ is orientable. The bundle $M' \to \Sigma_{g+1}$ has a global section by taking the image of $1 \times S_g$ in $M'$. Now reflect in each $S^1$-fiber and get an $I$-bundle $M$ over $\Sigma_{g+1}$ which is nothing but the mapping cylinder of the covering projection $S_g \to \Sigma_{g+1}$. The typical fiber for $M \to \Sigma_{g+1}$ is $S^1$ and the regular fiber is $T = \mathbb{Z}_2 \backslash S^1$. For example, if $\Sigma_1 = \mathbb{R}P_2$, then $M' = \mathbb{R}P_3 \# \mathbb{R}P_3$ and $M = \mathbb{Z}_2 \backslash M'$ is the nontrivial $I$-bundle over $\mathbb{R}P_2$. Alternatively, the space may be identified as the composition $S^1 \times S^2 \to I \times S^2 \to I \times_{\mathbb{Z}_2} S^2$. For $\Sigma_{g+1}$, the fiber space $M'$, in the terminology of the next section, is $M' = \{\mathfrak{n}_2, g+1, 0, 0; (0,0), (0,0); b = 0\}$. Note if $g = 1$, then $M'$ is homeomorphic to $\mathcal{G}_2$ but not as a Seifert isomorphism (cf. Subsection 4.1.6 and Example 4.5.10). Of course, $Q'$ is not topologically hyperbolic in this case.

REMARK 14.12.9. In Theorem 11.2.4 we obtained extensive strong rigidity, generalizations of Corollary 14.12.5 where $\mathbb{Z}$ is replaced by a special lattice and various suitable $(Q, W)$. The technique used in proving Theorem 11.2.4 is similar to that employed in proving Corollary 14.12.5.

## 14.13. A complete set of invariants for the 3-dimensional Seifert fiberings

14.13.1. Even though we have found in Theorem 14.12.2 that the local SO(2)-actions are unique and determined by $\pi_1(M)$ up to Seifert isomorphisms whenever $\pi_1(M)$ is infinite and not solvable (i.e., $Q$ is topologically hyperbolic), a characterization in terms of local data (Seifert invariants, etc.), analogous to that employed in the action case, is both practical and useful. The fibers in a local SO(2)-action are the same as in the action case. Furthermore, the action of $S^1$ on the fibers extends to an SO(2)-action on regular neighborhoods of each fiber. Therefore, there is no difference in the possible base spaces $B$. The fibers over $B_0$, *following the notation in Theorem* 14.5.3, form an $S^1$-bundle, $M_0$. But unlike the action case, this bundle may not be principal. The restriction of the bundle to a component of the boundary may be trivial ($S^1 \times S^1$) or nontrivial (a Klein bottle).

For a neighborhood of a component of SE-fibers, we see that it must be a Möbius band bundle over $S^1$. If this bundle is trivial, then the local SO(2)-action extends to a global SO(2)-action on $\tau^{-1}(N(C_j))$. This neighborhood is attached equivariantly to a boundary component of $M_0$ where the restriction to a boundary component of $M_0$ is trivial. If the Möbius band bundle is nontrivial, its boundary will be a Klein bottle and it is attached by a bundle isomorphism to a boundary component of $M_0$ which is nontrivial.

Similarly, when F-fibers are allowed, a regular neighborhood of a component of F-fibers will either be $D^2 \times S^1$ with an effective $S^1$-action of $S^1$ just along the first factor or else the nontrivial $D^2$-bundle over $S^1$, where boundary is the nontrivial $S^1$-bundle over $S^1$. Again these neighborhoods are attached by the appropriate bundle isomorphism to a boundary component of $M_0$. Neighborhoods of E-fibers are exactly the same as the SO(2) case.

$S^1$-bundles over $B_0$ with structure group O(2) are classified by the elements of $\text{Hom}(\pi_1(B_0), \mathbb{Z}_2)$. However, we want a (weaker) classification up to weak equivalence. This was accomplished by Seifert [**Sei33**, §6] when the restriction of the bundle to the boundary is trivial. He divided the bundles into six classes:

$$\mathfrak{o}_1, \mathfrak{o}_2, \mathfrak{n}_1, \mathfrak{n}_2, \mathfrak{n}_3, \mathfrak{n}_4 \text{ (or } \mathfrak{o}_0, \mathfrak{n}_0, \mathfrak{n}_n I, \mathfrak{o}_n, \mathfrak{n}_n II, \mathfrak{n}_n III)$$

with the parenthetical symbols being the respective symbols used by Seifert. The first two, $\mathfrak{o}_1, \mathfrak{o}_2$ are bundles over orientable bases, the last four are over nonorientable bases. The principal bundles are $\mathfrak{o}_1$ and $\mathfrak{n}_1$, and are trivial as long as $B_0 \neq B$. On these two types of bundles, the SO(2)-actions of $M_0$ extends to an action on $M$. The orientable $M_0$ are the types $\mathfrak{o}_1$ and $\mathfrak{n}_2$. (In type $\mathfrak{n}_2$, the $S^1$-bundle is nontrivial over each orientation reversing curve in the standard generating set.) Let $\epsilon$ denote any one of these six symbols.

We must, however, augment the meaning of $\epsilon$ if the restriction of the bundle to one of the boundary components is nontrivial. Suppose $B - B_0 = \text{Interior}((\bigcup_{i=1}^m D_i) \cup (\bigcup_{j=1}^t N(C_j)) \cup (\bigcup_{\ell=1}^k N(F_\ell)))$. For $0 \leq j \leq t' \leq t$ and $0 \leq \ell \leq k' \leq k$, let the bundle over the boundaries of $N(C_j)$ and $N(F_\ell)$ be nontrivial and trivial for $t' < j \leq t$ and $k' < \ell \leq k$. Following [**Fin76**], we denote this by the symbols $(t, t')$ and $(k, k')$, respectively.

Fintushel shows that the weak equivalence class of the bundles of type $\mathfrak{o}_1$ and $\mathfrak{o}_2$ ($B_0$ is orientable), collapses to a single class $\mathfrak{o}$ if $t' + k' > 0$ and $\mathfrak{n}_1, \mathfrak{n}_2, \mathfrak{n}_3, \mathfrak{n}_4$ collapse to a single class $\mathfrak{n}$ ($B_0$ is nonorientable) if $t' + k' > 0$. Therefore, we augment the symbols $\epsilon$ to take on the value $\mathfrak{o}$ or $\mathfrak{n}$ if $t' + k' > 0$. Furthermore, $t' + k'$ must be an even integer.

The Seifert invariants $(\alpha_i, \beta_i)$ take on exactly the same ranges as in the SO(2)-action case. The $b$-invariant is 0 if $t + k > 0$. If $t + k = 0$, then $b \in \mathbb{Z}$ when $\epsilon = \mathfrak{o}_1$ or $\mathfrak{n}_2$, and $b$ is 0 or 1 when $\epsilon = \mathfrak{o}_2, \mathfrak{n}_1, \mathfrak{n}_3$, or $\mathfrak{n}_4$ unless some $\alpha_i = 2$, in which case, $b = 0$. There are also conditions on $g$: $g \geq 1$ if $\epsilon = \mathfrak{o}_2, \mathfrak{n}_1, \mathfrak{n}_2, \mathfrak{n}$; $g \geq 2$ if $\epsilon = \mathfrak{n}_3$; and $g \geq 3$ for $\epsilon = \mathfrak{n}_4$.

THEOREM 14.13.2 (Classification up to Seifert isomorphism [**Fin76**, Theorem 2],[**OR69**, Theorem 0]). *If $M$ is a closed connected 3-manifold with local $S^1$-action, it is determined up to a Seifert isomorphism by the fiber invariants $\{\epsilon, g, k, t; (k, k'), (t, t'); b; (\alpha_1, \beta_1), \ldots, (\alpha_m, \beta_m)\}$.*

PROOF. If $M$ and $M'$ are 3-manifolds with local $S^1$-actions, then the Seifert isomorphism from $M - \text{SE}$ to $M' - \text{SE}$ is required to be orientation preserving if $M - \text{SE}$ is oriented. When $\epsilon \neq \mathfrak{o}_1$ or $\mathfrak{n}_2$, the Seifert invariants are reduced to $0 < \beta_j \leq \alpha_j/2$ since $M - \text{SE}$ will not be orientable. The bundle over $M_0$ has a cross section as long as $M_0$ is not $M$. The cross section on the boundary components of $M_0$ to which the boundaries of the neighborhoods of the SE- and F-components are attached by a bundle isomorphism extend to a cross section of the entire neighborhood of the SE or F component. The rest of the argument is very similar to the equivariant classification in Sections 14.5 and 14.6. □

14.13.3 (Topological classification). If $Q$ is topologically hyperbolic, then Theorem 14.12.2 states that the *equivariant classification*, i.e., Seifert isomorphism classification is a topological classification. If $Q$ is solvable, i.e., topologically Euclidean, or finite, then a case by case analysis can be found in [**OR69**]. Therefore, only the case where $k' + t' > 0$ remains. This is fully treated by [**Fin76**].

THEOREM 14.13.4 ([**Fin76**, Theorem 3]). *Let $M = \{\epsilon, g, k, t; (k, k'), (t, t'); 0;$ $(\alpha_1, \beta_1), \ldots, (\alpha_m, \beta_m)\}$, with $k > 0$ and $k' + t' > 0$. Then $M$ is homeomorphic to*

$$N \# (S^2 \times S^1)^{\bar{g}} \# (\mathbb{R}P_2 \times S^1)^t \# L(\alpha_1, \beta_1) \# \cdots \# L(\alpha_m, \beta_m),$$

*where $\bar{g} = 2g + k - 2$ if $\epsilon = \mathfrak{o}$, $\bar{g} = g + k - 2$ if $\epsilon = \mathfrak{n}$, and where $N$ is replaced by $S^2 \times S^1$ if $t > 0$.*

PROOF. The method of proof is analogous to Theorem 14.5.3. But first we recall from [**OR69**, Theorem 2] the result for $k' + t' = 0$ and $k > 0$. Then $M$ is homeomorphic to

$$S^3 \# (S^2 \times S^1)^{\bar{g}} \# (\mathbb{R}P_2 \times S^1)^t \# L(\alpha_1, \beta_1) \# \cdots \# L(\alpha_m, \beta_m),$$

where $\bar{g} = 2g + k - 1$ if $\epsilon = \mathfrak{o}_1$, $\bar{g} = g + k - 1$ if $\epsilon = \mathfrak{n}_2$,

$$N \# (S^2 \times S^1)^{\bar{g}} \# (\mathbb{R}P_2 \times S^1)^t \# L(\alpha_1, \beta_1) \# \cdots \# L(\alpha_m, \beta_m),$$

where $\bar{g} = 2g + k - 2$ if $\epsilon = \mathfrak{o}_2$, $\bar{g} = g + k - 2$ if $\epsilon = \mathfrak{n}_1, \mathfrak{n}_3, \mathfrak{n}_4$. Moreover, if $t > 0$, we can replace $N$ by $S^2 \times S^1$ when $\epsilon = \mathfrak{o}_2, \mathfrak{n}_1, \mathfrak{n}_3, \mathfrak{n}_4$.

The cases $\mathfrak{o}_1$ and $\mathfrak{n}_1$ were done in Theorem 14.5.3. We just need to identify the basic building blocks and apply the same method:

$$\{\mathfrak{o}_2, 1, 1, 0; (1, 0), (0, 0); 0\} = N \# N,$$

where $N$ is the nonorientable $S^2$-bundle over $S^1$. $N \# N$ can be normalized to $N \# (S^2 \times S^1)$. Likewise

$$\{\mathfrak{o}_2, g, k, 0; (k, 0), (0, 0); 0\} = N \# (S^2 \times S^1)^{2g+k-2}.$$

By sewing in components of SF-fibers where $t > 0$, $t_1 = 0$, we add copies of $(\mathbb{R}P_2 \times S^1)$.

$$\{\mathfrak{n}_2, 1, 1, 0; (1, 0), (0, 0); 0\} = S^2 \times S^1,$$

$$\{\mathfrak{n}_2, g, k, 0; (k, 0), (0, 0); 0\} = (S^2 \times S^1)^{g+k-1},$$

$$\{\mathfrak{n}_3, 2, 1, 0; (1, 0), (0, 0); 0\} = \{\mathfrak{n}_1, 1, 1, 0; (1, 0), (0, 0); 0\} \# \{\mathfrak{n}_2, 1, 1, 0; (1, 0), (0, 0); 0\}$$
$$= N \# (S^2 \times S^1),$$

$$\{\mathfrak{n}_4, 3, 1, 0; (1, 0), (0, 0); 0\} = \{\mathfrak{n}_2, 2, 1, 0; (1, 0), (0, 0); 0\} \# \{\mathfrak{n}_1, 1, 1, 0; (1, 1), (0, 0); 0\}$$
$$= N \# (S^2 \times S^1) \# (S^2 \times S^1).$$

Therefore, $M$ is a connected sum of these building blocks whenever $k' + t' = 0$ and $k > 0$.

Now all that is needed for the completion of the proof when $k' > 0$ are the additional building blocks

$$\{\mathfrak{o}, 0, 2, 0; (2, 2), (0, 0); 0\} = N \text{ and}$$
$$\{\mathfrak{o}, 0, 1, 1; (1, 1), (1, 1); 0\} = \mathbb{R}P_2 \times S^1. \qquad \square$$

LEMMA 14.13.5. *$M = \{\mathfrak{o}, 0, 0, 2; (0, 0), (2, 2); 0\} = S_* K$, one of the four possible Klein bottle bundles over $S^1$.*

PROOF. The base space $B$ is an annulus $S^1 \times I$. Over each $z \times I$ is a copy of the Klein bottle. Thus $M$ is a nontrivial Klein bottle bundle over $S^1$. An inspection of the four possible bundles shows that $M$ is $S_* K$ in the notation of [**OR69**]. (Observe that the bundle over $B_0$ is not a product.) $\qquad \square$

14.13.6 ([**Fin76**, Theorem 4]). Let $M = \{\epsilon, g, 0, t; (0,0), (t, t'); 0, (\alpha_1, \beta_1), \ldots, (\alpha_m, \beta_m)\}$ with $t' > 0$ and $M$ not as in Lemma 14.13.5, then $M$ is an injective 3-dimensional Seifert manifold with $Q$ topologically hyperbolic. $M$ is a $K(\Pi, 1)$ and the classification up to Seifert isomorphism is a topological classification.

PROOF. Granted that this manifold arises as an injective Seifert fibering with $Q$ topologically hyperbolic, the result then follows from uniqueness and rigidity as in Section 14.12. Fintushel did not use this fact and instead he observes that the orientable double cover $\widehat{M}$ is

$$\{\mathfrak{n}_2, \overline{g}, 0, 0; (0,0), (0,0); -m; (\alpha_1, \beta_1), (\alpha_1, \alpha_1 - \beta_1), \ldots, (\alpha_m, \beta_m), (\alpha_m, \alpha_m - \beta_m)\}$$

with $\overline{g} = 4g + 2t - 2$ if $\epsilon = \mathfrak{o}$ and $\overline{g} = 2g + 2t - 2$ if $\epsilon = \mathfrak{n}$.

(If we wanted to define our $M$ as a local SO(2)-action in terms of just the local fiber information and wanted to prove that it arises as an injective Seifert fibering, then the double cover $\widehat{M}$ gives us a way to do this. For $\widehat{M}$ itself has an orientable double cover $M'$ where the fibers of $\widehat{M}$ lift to an injective SO(2)-action on $M'$. Furthermore, $\pi_1(M')$ is infinite torsion free and not solvable. Therefore, $M$ itself is an injective Seifert manifold with a torsion-free nonsolvable fundamental group. From the orientable double covering, we may observe that $M$ always satisfies the conditions of Theorem 11.6.4.)

Fintushel presents the fundamental group of $M$ in terms of the fiber invariants and argues (with the help of the known classification of $\pi_1(\widehat{M})$) that all the fiber invariants determine the fundamental group of $M$ uniquely. $\square$

14.13.7 (Presentation of $\pi_1(M)$). Seifert gave presentations of $\pi_1(M)$ for $M$ when $t + k = 0$; cf. also [**OVZ67**] and [**Orl72**]. Presentations for $\pi_1(M)$ are given in [**Fin76**] for $\epsilon = \mathfrak{o}$ and $\mathfrak{n}$ when $k = 0$ and $t' > 0$.

We use the notation and the restrictions of the Subsections 14.12.1 and 14.13.1. From the cited references, we first note the presentation for $\pi_1(M)$, where $k = 0$ and $t' = 0$ (cf. Subsection 14.6.5). This corresponds to the manifolds treated in [**OR69**].

• For $\mathfrak{o}_1 : w_s = 1$, for all $s$; $\mathfrak{o}_2 : w_s = -1$, for all $s$.

$$\pi_1(M) = \{a_1, b_1, \ldots, a_g, b_g, y_1, \ldots, y_t, h_1, \ldots, h_t, q_1, \ldots, q_m, h \mid$$
$$a_s h a_s^{-1} = h^{w_s}, \; b_s h b_s^{-1} = h^{w_s}, \; q_i h q_i^{-1} = h, \; y_j h_j y_j^{-1} = h_j, \; h_j^2 = h,$$
$$q_i^{\alpha_i} = h^{-\beta_i}, \; \prod_{s=1}^{g} [a_s, b_s] \prod_{j=1}^{t} y_j \prod_{i=1}^{m} q_i = h^{-b}\}.$$

• For $\mathfrak{n}_1 : w_s = 1$, for all $s$; $\mathfrak{n}_2 : w_s = -1$, for all $s$; $\mathfrak{n}_3 : w_1 = 1, w_s = -1, s > 1$; $\mathfrak{n}_4 : w_1 = w_2 = 1, w_s = -1, s > 2$;

$$\pi_1(M) = \{v_1, \ldots, v_g, y_1, \ldots, y_t, h_1, \ldots, h_t, q_1, \ldots, q_m, h \mid$$
$$v_s h v_s^{-1} = h^{w_s}, \; q_i h q_i^{-1} = h, \; y_j h_j y_j^{-1} = h_j, \; h_j^2 = h,$$
$$q_i^{\alpha_i} = h^{-\beta_i}, \; \prod_{s=1}^{g} v_s^2 \prod_{j=1}^{t} y_j \prod_{i=1}^{m} q_i = h^{-b}\}.$$

In case $k = 0, t' \geq 1$, we have the following [**Fin76**]:

- For $\mathfrak{o}$: $\delta_j = -1$, $1 \leq j \leq t'$; $\delta_j = 1$, $t' < j \leq t$,

$$\pi_1(M) = \{a_1, b_1, \ldots, a_g, b_g, y_1, \ldots, y_t, h_1, \ldots, h_t, q_1, \ldots, q_m, h \mid$$
$$a_s h a_s^{-1} = h^{-1},\ b_s h b_s^{-1} = h^{-1},\ q_i h q_i^{-1} = h,\ y_j h_j y_j^{-1} = h_j^{\delta_j},\ h_j^2 = h,$$
$$q_i^{\alpha_i} = h^{-\beta_i},\ \prod_{s=1}^{g}[a_s, b_s] \prod_{j=1}^{t} y_j \prod_{i=1}^{m} q_i = 1\}.$$

- For $\mathfrak{n}$: $\delta_j = -1$, $1 \leq j \leq t'$; $\delta_j = 1$, $t' < j \leq t$,

$$\pi_1(M) = \{v_1, \ldots, v_g, y_1, \ldots, y_t, h_1, \ldots, h_t, q_1, \ldots, q_m, h \mid$$
$$v_s h v_s^{-1} = h^{-1},\ y_j h_j y_j^{-1} = h_j^{\delta_j},\ h_j^2 = h,$$
$$q_i h q_i^{-1} = h,\ q_i^{\alpha_i} = h^{-\beta_i},\ \prod_{s=1}^{g} v_s^2 \prod_{j=1}^{t} y_j \prod_{i=1}^{m} q_i = 1\}.$$

EXERCISE 14.13.8. If $F \neq \emptyset$, find presentations of $\pi_1(M)$ in terms of the invariants and show as in Subsection 14.6.5 that $\pi_1(M)$ is a free product of cyclic groups and $\mathbb{Z} \times \mathbb{Z}_2$'s.

14.13.9. A "normal form" for 3-dimensional Seifert orbifolds is considerably more difficult to find than the system of fiber invariants given in Theorem 14.13.2 for local SO(2)-actions. This entails a classification of all 2-dimensional orbifolds. For the good orbifolds, one has to classify the actions of all finite $Q$ on $S^2$ and all proper cocompact actions on $\mathbb{R}^2$. Presentations of spherical and Euclidean crystallographic groups were well known and these for non-Euclidean (i.e., hyperbolic) groups were given by Wilkie [**Wil66**] and Macbeath [**Mac67**]. Zieschang and Zimmerman [**ZZ82**] used these presentations to find normal forms for all the Seifert orbifolds. In contrast to the method of Corollary 14.12.5, these normal forms were used by Zieschang and Zimmerman [**ZZ82**] to show, with one class of exceptions, that two Seifert orbifolds over hyperbolic orbifolds are homeomorphic as orbifolds if and only if their orbifold fundamental groups are isomorphic. A similar analysis, but in terms of orbifold terminology and methods, was given by Bonahon and Siebenmann [**BS85**]. They too were able to show that a normal form yields an orbifold classification. These normal forms give us a "hands on" approach for working with these Seifert orbifolds much as was done with the Seifert presentations in terms of orbit invariants and the presentations using the normal forms of the local SO(2)-actions of this chapter. We also point out that the topological spaces underlying an orbifold can be unexpected such as a connected sum of lens spaces, or even fail to be locally Euclidean.

## 14.14. Historical remarks

14.14.1. In 1933, Herbert Seifert published his *tour de force* on 3-dimensional fibered spaces which have become known now as the classical Seifert 3-dimensional manifolds. This is a subclass of what are now called 3-dimensional Seifert fiberings. Seifert gave a fiberwise classification of his fiberings in terms of numerical invariants (genus of base space, type, Seifert invariants, etc.) and in some instances a topological classification such as the spherical space forms ([**ST31**] and [**ST33**]) and those fiberings which are integral homology spheres.

The general topological classification remained unknown until powerful geometric ideas introduced by students of Hirzebruch, von Randow [**vR62**] and Waldhausen ([**Wal68**], [**Wal67c**], [**Wal67a**] and [**Wal67b**]) and independently powerful group theoretic ideas of H. Zieschang as utilized in [**OVZ67**] enabled these authors to demonstrate for most (i.e., sufficiently large) classical Seifert fiberings with infinite fundamental group that the fiberwise classification agreed with the topological classification. The remaining "small" classification as well as those fiberings with nonclassical SE-type fibers were completed by Orlik-Raymond [**OR68**], [**OR69**], [**Orl72**] and Fintushel [**Fin76**].

Waldhausen's work went far beyond Seifert manifolds. He showed that the class of irreducible, sufficiently large, closed 3-manifolds were completely determined by their fundamental groups. His methods had a profound effect on the topology of 3-manifolds. Note that all aspherical Seifert 3-manifolds are irreducible (i.e., they contain no fake 3-cell) since they are universally covered by $\mathbb{R}^3$. Closed aspherical Seifert 3-manifolds are sufficiently large except for some that fiber over the 2-sphere with exactly three singular orbits with $1/\alpha_1 + 1/\alpha_2 + 1/\alpha_3 \leq 1$.

14.14.2. Seifert's work preceded the development of a global view of transformation groups, and it was not until 1956 that R. Jacoby [**Jac56**] showed that an effective $S^1$-action on a 3-sphere without fixed points coincided with the Seifert fiberings of the 3-sphere. This together with earlier work of Montgomery and Zippin [**MZ36**] on $S^1$-actions with fixed points on the 3-sphere showed that all the actions of $S^1$ on $\mathbb{R}^3$ and $S^3$ were equivalent to linear ones. The equivariant classification of $S^1$-actions on connected 3-manifolds was given in [**Ray68**]. In the presence of fixed points, the manifold is an explicit connected sum of $S^3$, lens spaces, $S^2 \times S^1$'s, $S^1 \times \mathbb{R}^2$, $S^1 \times \mathbb{R}P_2$, and $N$, the nontrivial $S^2$-bundle over $S^1$. The equivariant classification in terms of orbit invariants is the same as the Seifert fiber classification in terms of numerical Seifert invariants where relevant.

14.14.3. H. Holmann [**Hol64**] generalized the 3-dimensional Seifert fiberings to higher dimensional complex analytic singular bundles which are constructed locally in terms of complex analytic orbifolds. In [**CR72b**], sheaf-theoretic methods were used, in the spirit of Kodaira ([**Kod63**] and [**Kod64**]) and [**Con68**], for a global formulation of holomorphic Seifert fiberings where the typical fiber is a complex torus. In particular, the elliptic surfaces of Kodaira, which are closely related to 4-dimensional Seifert fiberings, have been central in the study of smooth 4-dimensional topology.

For another direction, see T. Suwa [**Suw79**] for an analysis of analytic deformation of (complex) 2-dimensional holomorphic Seifert fiber spaces.

14.14.4. In 1969, Conner and Raymond [**CR69**] introduced the technique of injective actions and the forerunner of the rigidity concept of injective fiberings. In [**CR72b**] the concept of Seifert fiberings modeled on principal $T^k$-bundles among other things was introduced enabling one to prove theorems such as Corollary 14.12.5. This has been developed much further by Lee and Raymond in a series of papers that replaces the typical fiber $T^k$ by a homogeneous space $\Gamma\backslash G$, as done in this book.

The work of von Randow [**vR62**] in describing the Seifert fiberings as the result of plumbing disk bundles over certain contractible graphs as well as the earlier work of Hirzebruch relating links of singularities in certain algebraic surfaces to Seifert manifolds led to the work of Orlik and Wagreich [**OW77**] on the resolution of good $\mathbb{C}^*$-actions and some of the work chronicled in this chapter; see also [**JN83**].

The books of Orlik [**Orl72**] and Hempel [**Hem76**] have been instrumental in presenting the Seifert manifolds as important pieces of 3-dimensional topology.

14.14.5. Important breakthroughs in 3-dimensional topology led to intense interest in the geometric aspects of Seifert manifolds.

For example, for $(S^1, M)$ with $S^1\backslash M = S^2$, arrange all the E-orbits over the equator of $S^2$. There is an orientation preserving Seifert involution on $(S^1, M)$ which reflects the fibers across the equator and reverses the orientation of each fiber. The base of the Seifert orbifold $\mathbb{Z}_2\backslash M$ is $D = \mathbb{Z}_2\backslash S^2$. The involution on $M$ fixes a link $K$ consisting of the fixed points of the involution on each of the orbits of $M$ over the equator. The link projects homeomorphically into $\mathbb{Z}_2\backslash M$ and bounds a number of bands $B$ consisting of the images in $\mathbb{Z}_2\backslash M$ of the $S^1$-orbits over the equator in $S^2$. We shall show that $\mathbb{Z}_2\backslash M$ is a 3-sphere. If we collapse each of these interval fibers of the bands in $\mathbb{Z}_2\backslash M$ to a point, we get a map $\omega : \mathbb{Z}_2\backslash M \to M'$ with $K$ doubly covering the image of the bands. The Seifert fibering $\nu : \mathbb{Z}_2\backslash M \to D$ factors through $M'$, with the fibers outside the bands, i.e., the fibers over $D - \partial D$, projecting homeomorphically into $M'$. The image $\omega(B)$ in $M'$ can be identified with $\omega(K)$ which projects homeomorphically onto $\partial D$. We have the following commutative diagram.

$$\begin{array}{ccccccc}
& & (S^1, M) & \xrightarrow{S^1\backslash} & S^1\backslash M = S^2 \\
& & \mathbb{Z}_2\backslash \downarrow & & \mathbb{Z}_2\backslash \downarrow \\
K & \xrightarrow{=} \partial B \subset B & \xrightarrow{\subset} & \mathbb{Z}_2\backslash M & \xrightarrow{\nu} & D = \mathbb{Z}_2\backslash S^2 \\
\downarrow & \downarrow & & \omega \downarrow & & \bar{\nu} \uparrow \\
\omega(K) & \xrightarrow{=} \omega(B) & \xrightarrow{\subset} & M' & \xrightarrow{=} & M'
\end{array}$$

Clearly, $\bar{\nu}^{-1} = S^1 \times (D - \partial D)$, and we can extend the obvious $S^1$-action to be fixed on $\omega(B) \approx \partial D$. Therefore, $M'$ is homeomorphic to $S^3$ and $\bar{\nu}$ is the orbit mapping of this $S^1$-action which is free on $M - \omega(K)$ and fixed on $\omega(K)$. We can shrink the bands $B$ on $\mathbb{Z}_2\backslash M$ by an isotopy $0 \le t < 1$ along each of the intervals making up $B$ while keeping the points outside of the bands fixed. At time 1, the intervals are collapsed and we see that it becomes $M'$. Therefore $\mathbb{Z}_2\backslash M$ is homeomorphic to $S^3$. (As an exercise, observe that if $e(M) \ne 0$, $\alpha_1$ is even and $\alpha_2, \ldots, \alpha_n$ are all odd, then $K$ is connected. That is, $K$ is a knot.) These links in $S^3$ are called *Montesinos links*, and they have been intensively investigated. See, for example, [**Mon73**], [**Bon79**], [**Zie84**], [**BZ87**] and [**BZ87**]. In particular, when $e(M) \ne 0$ and $\pi_1(M)$ is not solvable, then $S^3 - K$ can be shown to be a 3-dimensional hyperbolic manifold.

14.14.6. Least area technology has been very successful in addressing and solving some crucial topological questions concerning 3-dimensional Seifert fiberings. For example, Meeks and Scott [**MS86**] have shown that smooth actions on aspherical 3-dimensional Seifert manifolds are equivalent to fiber preserving actions and presumably to a group of Seifert automorphisms.

Small Seifert fiberings $(S^1, M)$ with infinite fundamental groups, i.e., Seifert fiberings over $S^2$ with exactly three orbit types do not have any embedded vertical incompressible tori. Peter Scott [**Sco83b**], again using least area technology, showed if $M'$ is a closed 3-manifold containing no fake 3-cells with $\Pi_1(M')$ isomorphic to $\Pi_1(M)$, then $M'$ is homeomorphic to M. Since Perelman's result shows that there are no fake 3-cells, the classification of closed 3-manifolds with these particular fundamental groups is complete.

Replacing a typical fiber by an exceptional or a fixed fiber is a Dehn surgery along the typical fiber. Cavicchioli and Hegenbarth [**CH96**] determine the (generalized) Seifert 3-manifold resulting from Dehn surgery, with surgery coefficient $r$, along a principal orbit of an oriented $S^1$-manifold. For example, if $r = 0$, a neighborhood of a typical fiber is replaced equivariantly by a neighborhood of a connected component of a fixed fiber. That is, $(S^1, M) = \{\mathfrak{o}, g, 0, 0; b; (\alpha_1, \beta_1), \ldots, (\alpha_m, \beta_m)\}$ is surgered to $(S^1, M') = (S^2 \times S^1)^{2g} \# L(\alpha_1, \beta_1) \# \cdots \# L(\alpha_m, \beta_m)$, as in Theorem 14.5.3. When $g = 0$, $(S^2 \times S^1)^{2g}$ is to be interpreted as $S^3$. If $r = \alpha'/\beta$, then the Seifert presentation of $(S^1, M')$ is the same as the Seifert presentation of $(S^1, M)$ except for the additional Seifert invariant $(\alpha' - \chi\beta, \beta)$, where $\chi$ is a well determined integer.

Other investigations using Dehn surgery have coupled Seifert fiberings with the latest developments of low dimensional topology; see [**Sav99**] and [**Sav02**]. The explosion in the theory of topological and smooth 4-manifolds along with developments in modern physics have strong interactions with 3-dimensional manifolds and Seifert fiberings. Many others have made dramatic contributions and advances to our knowledge of Seifert fiberings. References to some of these can be found in the bibliography.

14.14.7. Seminal independent works by Jaco-Shalen [**JS79**] and Johanson [**Joh79**] have helped to describe the structure of 3-manifolds. Essentially their work shows that an irreducible (every embedded 2-sphere bounds a ball) closed 3-manifold can be split along incompressible tori (if there are any) so that the pieces are either Seifert manifolds or "simple" pieces with no nonboundary parallel incompressible tori in them. These simple pieces if not closed, according to Thurston [**Thu97**], admit a complete hyperbolic metric. Thurston's deep and penetrating geometric and topological analysis goes far beyond this. He has conjectured that each closed 3-manifold is a connected sum of manifolds each of which can be split into a Seifert manifold or a hyperbolic manifold. The conjecture contains the Poincare conjecture. Thurston has proved many important cases of this conjecture. See [**Thu97**] and [**CHK00**].

One of the dramatic applications of these developments was the solution to the *Smith Conjecture* which stated that if a finite cyclic group acted smoothly on a closed simply connected 3-manifold with a 1-dimensional fixed point set, then the action was conjugate to a linear action on the 3-sphere; see [**MB84**].

In a different and largely analytic approach to the structure of 3-manifolds, Hamilton introduced the notion of Ricci flow. He showed in some cases that a general Riemannian metric on these primal pieces can be deformed to a locally homogeneous metric. To show that all the pieces admit a locally homogeneous metric (i.e., a geometric structure and thus a solution to Thurston's geometrization conjecture), one has to overcome formidable analytic difficulties.

G. Perelman in a recent series of papers has shown how these difficulties can be overcome to complete Hamilton's program and solve Thurston's conjectures. See J.W. Morgan [**Mor05**] for an accessible outline of these dramatic results. We have mentioned only a small part of the very large literature on 3-dimensional Seifert manifolds and 3-manifolds.

CHAPTER 15

# Classification of Seifert 3-manifolds via equivariant cohomology

## 15.1. $H^2(Q; \mathbb{Z}^k)$ and codimension-2 injective actions

In Chapter 10, we saw the Borel space $EQ \times_Q W = W_Q$ and its cohomology $H^*(W_Q; \mathbb{Z}^k) \cong H^*(Q; \mathcal{Z}^k)$ play a fundamental role in describing and classifying Seifert fiberings. This cohomology is the $Q$-equivariant cohomology of $W$ and is sometimes denoted by $H_Q^*(W; \mathbb{Z}^k)$.

In this chapter, we study some of the classical 3-dimensional Seifert manifolds by means of the technique from Sections 10.3–10.5. For necessary notations, we refer the readers to Section 10.3. This procedure is independent of Chapter 14. However, we do use results of Chapter 14 in Subsection 15.2.7, but this only serves as a convenience to avoid duplication of derivations of some properties of the classical Seifert invariants for 3-manifolds. In effect, we directly use a presentation of the group $H^2(Q; \mathcal{Z})$ to obtain numerical invariants which become the Seifert invariants over the 2-dimensional orbifold $Q \backslash W$. This sharpens and expands the tools available to study 3-manifolds.

For the classification of Seifert 3-manifolds when $W = \mathbb{R}^2$, $H^2(Q; \mathcal{Z}^1) = H^2(EQ \times_Q \mathbb{R}^2; \mathbb{Z})$ is the same as the ordinary cohomology group $H^2(Q; \mathbb{Z})$. The coefficient system is simple if $\phi : Q \to \mathrm{Aut}(\mathbb{Z})$ is trivial, otherwise it is twisted. For simplicity, we shall restrict ourselves, in this chapter, to $\phi : Q \to \mathrm{Aut}(\mathbb{Z})$ trivial and to orientation preserving homeomorphisms in the range of $\rho : Q \to \mathrm{TOP}(W)$. For $W = S^2$, $H^*(Q; \mathcal{Z}) = H^*(EQ \times_Q S^2; \mathbb{Z})$ with $\mathbb{Z}$ a simple coefficient system. All the other cases can be analyzed by a similar procedure when $\phi$ is not trivial or when $\rho$ may have orientation reversing homeomorphisms in its image.

In the last section, we return to the technique of Sections 10.1 and 10.2. We apply this to a discussion of embedding $\Pi$ in $\mathrm{TOP}_{S^1}(P_n)$, where $P_n$ is a principal $S^1$-bundle over $S^2$. Some of these results overlap the earlier sections of this chapter but are derived independently of them.

Let $Q$ act properly on $S^2$ or $\mathbb{R}^2$. Then it is known that the $Q$-action is topologically equivalent to a group of isometries on the standard sphere or a group of Euclidean isometries on the Euclidean plane, or a group of hyperbolic isometries on the hyperbolic plane. If $Q \backslash \mathbb{R}^2$ is compact and $Q$ is effective, then there is only one such action up to topological equivalence. If $Q$ is not cyclic and acts effectively on $S^2$, then the action is also unique up to equivalence on $S^2$. Presentations and properties of these groups are well known and are discussed in more detail in Chapter 14 where they are employed in a more hands-on and traditional approach to 3-dimensional Seifert fiberings. For $W = \mathbb{R}^2$, we observe the following information.

LEMMA 15.1.1 ([**CR69**, Lemma 10.1]). *Let $(Q, \mathbb{R}^2)$ be a proper and effective group of orientation preserving homeomorphisms with $Q \backslash \mathbb{R}^2$ compact. Then*
  (i) *every finite subgroup is cyclic,*
  (ii) *every nontrivial finite subgroup has a unique fixed point,*
  (iii) *every nontrivial finite subgroup lies in a maximal finite subgroup,*
  (iv) *every maximal finite subgroup is its own normalizer.*

PROOF. A finite group of orientation preserving homeomorphisms on $\mathbb{R}^2$ is topologically equivalent to a faithful representation of the group into $SO(2)$; [**Ker19**] hence (i) and (ii) follow.

Since $Q \backslash \mathbb{R}^2$ is compact, there are only a finite number of conjugacy classes of isotropy subgroups. So every finite group belongs to one of these conjugacy classes. By (ii), each of these isotropy subgroups are seen to be maximal finite groups. Let $Q_w \cap Q_{w'}$ be nontrivial. Then the intersection leaves $w$ and $w'$ fixed and by (ii), $Q_w = Q_{w'}$ so (iii) is established. For (iv), $qQ_w q^{-1} = Q_{qw}$, so if $qQ_w q^{-1} = Q_w$, then $qw = w$ and $q \in Q_w$. □

15.1.2. First we study $T^k$-bundles. In the rest of this section, we shall assume that $\varphi : Q \to \text{Aut}(\mathbb{Z}^k)$ is trivial and that $Q$ acts properly and effectively, preserving orientation of $\mathbb{R}^2$ so that $\bar{\nu} : \mathbb{R}^2 \to V = Q \backslash \mathbb{R}^2$ has compact image. Let $v_1, \ldots, v_n$ be the image of the orbits in $V$ which are not free and choose $w_\ell \in \bar{\nu}^{-1}(v_\ell)$, a representative point on each of these orbits. The group $Q$ and also its action on $\mathbb{R}^2$ are determined, up to topological conjugacy, by a set $\{g; \alpha_1, \ldots, \alpha_n\}$ where $g =$ genus of $Q \backslash \mathbb{R}^2$ and $\alpha_\ell$ is the order of $Q_{w_\ell} = \mathbb{Z}_{\alpha_\ell}$. The set $\{g; \alpha_1, \ldots, \alpha_n\}$ is called the *signature* of $Q$ and all entries are valid except for $g = 0$, $n \leq 2$ and $g = 0$, $\frac{1}{\alpha_1} + \frac{1}{\alpha_2} + \frac{1}{\alpha_3} > 1$. (For the exceptions, $Q$ is finite and acts on $S^2$.)

Instead of studying $S^1$-bundle over $\mathbb{R}^2$, we study $T^k$-bundles more generally.

PROPOSITION 15.1.3. *The lifts of $Q$ to bundle automorphisms of $T^k \times \mathbb{R}^2$ are, up to strict equivalence, in one-to-one correspondence with elements of $H^2(Q; \mathbb{Z}^k)$.*

PROOF. Since $\mathbb{R}^2$ is contractible, each $T^k$-bundle over $\mathbb{R}^2$ is trivial. Since $H^2(Q; \mathcal{Z}^k) = H^1(Q; \mathcal{T}^k)$, each 1-cocycle in $Z^1(Q; \mathcal{T}^k)$ determines a lift of $Q$ to a group of bundle automorphisms of the bundle $T^k \times \mathbb{R}^2$. A lift by a cohomologous 1-cocycle will differ from the original lift by a strict equivalence. Therefore we can say that each cohomology class represents a lift of $Q$ to a group of bundle automorphisms up to strict equivalence. Each element of $H^2(Q; \mathcal{Z}^k)$ then represents a lift of $Q$ to strict bundle automorphisms. Since $''e_1 : H^2(Q; \mathbb{Z}^k) \to H^2(Q; \mathcal{Z}^k)$ is an isomorphism (see Subsection 10.5), the distinct equivalence classes of liftings of $Q$ correspond to the elements of $H^2(Q; \mathbb{Z}^k)$ (with trivial action of $Q$ on $\mathbb{Z}^k$).

Since $Q \backslash \mathbb{R}^2$ is a closed orientable 2-manifold, $H^3(Q \backslash \mathbb{R}^2; \mathbb{Z}^k) = 0$. Therefore, by Subsection 10.5.8 and the long exact sequence of 10.5.9, we have the exact sequence

$$0 \longrightarrow H^2(Q \backslash \mathbb{R}^2; \mathbb{Z}^k) \xrightarrow{'e_1} H^2(Q; \mathbb{Z}^k) \xrightarrow{'e_2} \prod_{\ell=1}^n (\mathbb{Z}_{\alpha_\ell})^k \longrightarrow 0,$$

where the first 0 comes from $\prod_{\ell=1}^n H^1(Q_{w_\ell}; \mathbb{Z}^k) = 0 \xrightarrow{d} H^2(Q \backslash \mathbb{R}^2; \mathbb{Z}^k)$ (since $Q_{w_\ell}$ is cyclic and $Q_{w_\ell} \subset Q \xrightarrow{\varphi} \text{Aut}(\mathbb{Z}^k)$ is trivial). Here $\prod_{\ell=1}^n (\mathbb{Z}_{\alpha_\ell})^k \cong H^0(Q \backslash \mathbb{R}^2; \mathfrak{h}^2)$ is the sections of the sheaf $\mathfrak{h}^2$ over $Q \backslash \mathbb{R}^2$. The sheaf $\mathfrak{h}^2$ is 0 everywhere except for the stalk at $v_\ell$ where it is

$$H^2(Q_{w_\ell}; \mathbb{Z}^k) = H^2(\mathbb{Z}_{\alpha_\ell}; \mathbb{Z}^k) = (\mathbb{Z}_{\alpha_\ell})^k.$$

A section of this sheaf is just an element of $\prod_{\ell=1}^{n}(\mathbb{Z}_{\alpha_\ell})^k$ and each such element is a section. The map $'e_2$ assigns to $\gamma$ the class $'e_2(\gamma)$. In Subsection 10.5.5, we examined this homomorphism in terms of $'e_2 : H^1(Q;\mathcal{T}^k) \to H^0(Q\backslash W; (\mathfrak{h}^1)_{T^k})$. The cocycle $\gamma_{ji}$ is sent to a section of $\mathfrak{h}^1$ over $Q\backslash W$. Locally, this is reduced to a 1-cocycle: $Q_w \to \mathrm{M}(w, T^k) = T^k$. This describes the action of $Q_w$ on the torus $T^k$ over $w$. When $\varphi : Q_w \to \mathrm{Aut}(T^k)$ is trivial, this cocycle is just a homomorphism of $Q_w$ into $T^k$ corresponding to the action of $Q_w$ on the torus $T^k$ over $w$. That is, after dividing out by the ineffective part (the kernel of $Q_w \to T^k$), the action is just translation on $T^k$ by the subgroup, the image of $Q_w$. The canonical isomorphism $H^1(Q_w; T^k) \cong H^2(Q_w; \mathbb{Z}^k)$ establishes the connections with the integral cohomology. The exact sequence above first appears in [**CR69**, §9]. □

PROPOSITION 15.1.4. *The quotient space $Q\backslash(T^k \times \mathbb{R}^2) = M$ is an orbifold and as a space is an orientable $(k+2)$-manifold with a $T^k$-action descending from the left $T^k$-action on $T^k \times \mathbb{R}^2$. The orbit map $M \to (T^k\backslash M) = Q\backslash\mathbb{R}^2$ of the induced $T^k$-action is a Seifert fibering with typical fiber $T^k$.*

PROOF. We just need to verify that $M$, as a topological space, is an orientable $(k+2)$-closed manifold. If $w$ does not lie on a free $Q$-orbit, then choose $D$ a closed slice at $w$. $D$ can be chosen to be a closed 2-cell and the bundle over $D$ is just $T^k \times D$. The action of $Q_w \subset Q$ in the lifted $Q$-action leaves $T^k \times D$ invariant and the action is the diagonal action $T^k \times_{Q_w} D$, $Q_w$ is a cyclic group and acts effectively on the $D$ part and freely on the $T^k$ part after the ineffective part $L$ of the $Q_w$-action on $T^k$ is divided out. Then $T^k \times_{Q_w} D$ factors into $(T^k \times L\backslash D)$ and $T^k \times_{Q_w/L} (L\backslash D)$ as $L\backslash D$ is still homeomorphic to a 2-cell and $Q_w/L$ acts freely on the $T^k$-part. The quotient is an $(k+2)$-manifold. The rest of the argument is clear. □

The following is now easy.

PROPOSITION 15.1.5. *The action $Q$ on $T^k \times \mathbb{R}^2$ is free if and only if $'e_2(\gamma)|_{\bar\nu(w)}$ is an injective homomorphism for each $w \in W$.*

Each lifting of $Q$ to $T^k \times \mathbb{R}^2$ has an extended lifting to $\mathbb{R}^k \times \mathbb{R}^2$ because $\mathbb{R}^2$ is simply connected. We have the following directly from Subsection 10.4.5 and Exercise 10.4.6.

PROPOSITION 15.1.6. *If $[\gamma] \in H^1(Q; \mathcal{T}^k)$ represents a lifting of $Q$ to $T^k \times \mathbb{R}^2$, then the extended lifting of $Q$ to $\mathbb{R}^k \times \mathbb{R}^2$ is represented by the extension class $\delta[\gamma] = [\Pi] \in H^2(Q; \mathbb{Z}^k)$, where $\delta : H^1(Q; \mathcal{T}^k) \xrightarrow{\cong} H^2(Q; \mathbb{Z}^k)$ and $(T^k, M)$ is the Seifert orbifold modeled on $\mathbb{R}^k \times \mathbb{R}^2$ with typical fiber $T^k$.*

15.1.7. Note, if $[\Pi]$ represents the central extension

$$1 \to \mathbb{Z}^k \to \Pi \to Q \to 1,$$

then $\Pi$ is the orbifold fundamental group of $M = Q\backslash(T^k \times \mathbb{R}^2) = \Pi\backslash(\mathbb{R}^k \times \mathbb{R}^2)$. The fundamental group of $M$, as a space, is given by $\Pi' = \Pi/\langle F \rangle$, where $\langle F \rangle$ is the normal subgroup generated by all the finite subgroups (all cyclic) of $\Pi$ fixing some point of $\mathbb{R}^{k+2}$ [**Arm68**]. (If $A \subset \Pi$ is a finite subgroup, it is cyclic. Thus it is a direct product of prime power cyclic groups, each of which fixes some point of $\mathbb{R}^{k+2}$. Therefore, $A \subset \langle F \rangle$.)

Consequently, the $\Pi$-action on $\mathbb{R}^k \times \mathbb{R}^2$ is free (and hence the $Q$-action on $T^k \times \mathbb{R}^2$ is free) if and only if $\Pi$ is torsion free. Even if $\Pi$ is not free, there is a lift of the induced $T^k$-action on $M$ to the covering space $M_H$ of $M$, corresponding to $H = \text{Im}(\pi_1(T^k, 1) \to \pi_1(M, x))$, which commutes with the group of covering transformations $Q' = \pi_1(M)/H$. This $Q'$-action agrees with the $Q$-action on $T^k \times \mathbb{R}^2$ if and only if the original $Q$-action on $T^k \times \mathbb{R}^2$ is free. The reader may wish to recall Remark 11.2.10 and parts (5), (6) and (7) of Remark 11.6.9 for more information on these codimension-2 torus actions.

## 15.2. A presentation for $H^2(Q; \mathcal{Z})$

15.2.1. Let us, for the rest of chapter, assume that $k = 1$, $W = S^2$ or $\mathbb{R}^2$, and that $Q$ acts effectively and preserves orientation on $W$ with compact quotient $Q \backslash W$. As before, let $\varphi : Q \to \text{Aut}(S^1)$ be trivial. The group $Q$ has signature $\{g; \alpha_1, \ldots, \alpha_n\}$. We shall assume that when $g = 0$, $n \geq 3$. (Under these restrictions, $Q$ is finite and $W = S^2$ if and only if $\frac{1}{\alpha_1} + \frac{1}{\alpha_2} + \frac{1}{\alpha_3} > 1$ and $g = 0$.) We will examine $g = 0$, and $n \leq 2$ in Subsection 15.3.2. For each lifting of the $Q$-action on $W$ to a group of bundle automorphisms on a principal $S^1$-bundle $P$ over $W$, there is induced a Seifert fibering $(S^1, Q \backslash P) \xrightarrow{\tau} Q \backslash W$. If $Q$ acts freely on $P$, the quotient $Q \backslash P$ are the classical orientable Seifert manifolds over orientable bases. Because of our temporary restrictions when $g = 0$ and $n \geq 3$, the lens spaces are excluded. (We regard $S^2 \times S^1$ and $S^3$ as lens spaces.) Since we do not require $Q$ to act freely on $P$, $Q \backslash P$ may be an *orbifold* and the induced map $\tau$ has sometimes been called an *orbi-bundle* map. We shall, unless necessary, not make a distinction between free and nonfree $Q$-actions on $P$. Even if $Q$ does not act freely on $P$, we will still call $\tau$ a Seifert fibering. In fact, because of the requirements of the orientation preserving and triviality of $\varphi$, the underlying space, $|Q \backslash P|$ of $Q \backslash P$, is a classical Seifert manifold and $\tau : (S^1, |Q \backslash P|) \to Q \backslash W$ is a classical Seifert fibering. Of course, sometimes we may need to make a distinction between $Q \backslash P$ and the space $|Q \backslash P|$ when $Q$ does not act freely. In a different direction, we may also observe, by choosing $Q'$, a normal subgroup of finite index in $Q$ acting freely on $W$, that $Q \backslash P$ is the quotient of a principal $S^1$-bundle over $Q' \backslash W$ by a finite group $Q/Q'$ of Seifert automorphisms.

15.2.2. Our starting point is the exact sequence of Proposition 10.5.10 with $k = 1$ and $W = S^2$ or $\mathbb{R}^2$:

$$(15.2.1) \quad 0 \to H^2(Q \backslash W; \mathbb{Z}) \xrightarrow{'e_1} H^2(Q; \mathcal{Z}) \xrightarrow{'e_2} \mathbb{Z}_{\alpha_1} \oplus \cdots \oplus \mathbb{Z}_{\alpha_n} \to 0,$$

$$(15.2.2) \quad 0 \to H^2(Q; \mathbb{Z}) \xrightarrow{''e_1} H^2(Q; \mathcal{Z}) \xrightarrow{''e_2} H^2(W; \mathbb{Z}) \to 0.$$

The second sequence tells us that: $''e_1$ *is an isomorphism when* $W = \mathbb{R}^2$, *and* $''e_1(H^2(Q; \mathbb{Z}))$ *is the torsion of* $H^2(Q; \mathcal{Z})$ *if* $W = S^2$.

In both sequences, $H^2(Q; \mathcal{Z}) = \mathbb{Z} \oplus$ Torsion, and this torsion is a subgroup of $\prod_{i=1}^n \mathbb{Z}_{\alpha_i}$. The torsion of $H^2(Q; \mathcal{Z})$ is the same as the torsion of $H_1(Q; \mathbb{Z})$, the Abelianization of $Q$. The latter can be readily computed from a presentation of $Q$. A formula for $H^2(Q; \mathcal{Z})$ is given in [**LLR96**, section 4]:

$$H^2(Q; \mathcal{Z}) = \mathbb{Z}_{d_1} \oplus \cdots \oplus \mathbb{Z}_{d_{n-1}} \oplus \mathbb{Z}, \quad \text{when } Q \text{ is infinite,}$$

and the same calculation yields

$$H^2(Q; \mathcal{Z}) = \mathbb{Z}_{d_1} \oplus \mathbb{Z}_{d_2} \oplus \mathbb{Z}, \quad \text{when } Q \text{ is finite (hence } n \leq 3 \text{ and } W = S^2).$$

Here,

$$\gcd(1, \alpha_1) = 1, \ \gcd(\alpha_1, \alpha_2) = d_1, \ \ldots, \ \gcd(\text{lcm}(\alpha_1, \ldots, \alpha_{n-1}), \alpha_n) = d_{n-1}.$$

The order of this torsion is $\prod_{\ell=1}^{n-1} d_\ell = d$. See Exercise 15.2.11 for the claim $d = \frac{\alpha_1 \cdot \alpha_2 \cdots \alpha_n}{\text{lcm}(\alpha_1, \alpha_2, \ldots, \alpha_n)}$.

THEOREM 15.2.3 (cf. [**JN83**],[**LLR96**]). (1) *There is an epimorphism $f : \mathbb{Z}^{n+1} \to H^2(Q; \mathcal{Z})$ with kernel $K$.*
(2) *$x = (x_0, x_1, \ldots, x_n) \in K$, if and only if $x_0 = -(s_1 + \cdots + s_n)$, $x_i = s_i \alpha_i$ for fixed $(\alpha_1, \ldots, \alpha_n)$ and arbitrary $(s_1, \ldots, s_n)$, $i = 1, 2, \ldots, n$. Therefore, $f$ induces the isomorphism $f_* : \mathbb{Z}^{n+1}/K \to H^2(Q; \mathcal{Z})$.*
(3) *There is a homomorphism $e : \mathbb{Z}^{n+1} \to \mathbb{R}$, which, via $f$, factors through $H^2(Q; \mathcal{Z})$, given by*

$$e(x) = x_0 + x_1/\alpha_1 + \cdots + x_n/\alpha_n.$$

*This rational number is an integral multiple of $\frac{1}{L}$, where $L = \text{lcm}(\alpha_1, \ldots, \alpha_n)$ and all possible multiples occur.*

15.2.4. The theorem gives a way of identifying elements of $H^2(Q; \mathcal{Z})$ in terms of numerical invariants. The group $Q$ has a signature $\{g; \alpha_1, \ldots, \alpha_n\}$, where $Q$ acts on $W$ and $g$ is the genus of the orbit space $Q \backslash W$. We assume the effective action of $Q$ on $W$ is fixed. (In almost all cases, any two effective actions of $Q$ on $W$ are topologically equivalent.) For $\gamma \in H^2(Q; \mathcal{Z})$, $"e_2(\gamma)$ determines the bundle $P$ over $W$. If $W = \mathbb{R}^2$, then $Q$ is infinite and the bundle is trivial. If $Q$ is finite, then $W = S^2$ and $"e_2(\gamma) \in \mathbb{Z} \cong H^2(S^2; \mathbb{Z})$ is the Euler class of the bundle $P$ over $S^2$. Thus, $\gamma \in H^2(Q; \mathcal{Z}) \cong H^1(Q; \mathcal{T})$ determines an action of $Q$ on $P$ and $\tau : Q \backslash P \to Q \backslash W$ is a Seifert fibering. Since $Q \backslash P$ is a closed oriented Seifert manifold over an oriented base, it can also be described by another set of numerical invariants called the (classical) Seifert invariants.

We shall show in Proof 15.2.7 that for a given signature $\{g; \alpha_1, \ldots, \alpha_n\}$, the $(n+1)$-tuple $(x_0, x_1, \ldots, x_n) \in \mathbb{Z}^{n+1}$ determines the negative of the unnormalized classical Seifert invariants of Subsection 14.2.1. That is, $(x_0, x_1, \ldots, x_n) \in \mathbb{Z}^{n+1}$ determines the classical Seifert manifold of Chapter 14 whose presentation in unnormalized form is given by

$$(S^1, M) = \{g; (1, b_0); (\alpha_1, \beta_1), \ldots, (\alpha_n, \beta_n)\},$$

where $b_0 = -x_0$ and $\beta_i = -x_i$.

For example, the Seifert manifold, in normalized form, that plays the role of the unit tangent bundle of the orbifold $Q \backslash W$ of genus $g$, in Theorem 14.8.8, is

$$M = \{g; (1, 2g-2); (\alpha_1, \alpha_1 - 1), \ldots, (\alpha_n, \alpha_n - 1)\}.$$

This manifold, in terms of $(x_0, x_1, \ldots, x_n)$, is given by $(2-2g-n, 1, 1, \ldots, 1)$. For, $M$ in unnormalized form can be rewritten as

$$-\{g; (1, 2-2g); (\alpha_1, 1-\alpha_1), \ldots, (\alpha_n, 1-\alpha_n)\}$$
$$= -\{g; (1, 2-2g-n); (\alpha_1, 1), \ldots, (\alpha_n, 1)\}$$
$$= \{g; (1, -(2-2g-n)); (\alpha_1, -1), \ldots, (\alpha_n, -1)\}$$
$$= (2-2g-n, 1, 1, \ldots, 1).$$

Strictly speaking, the Seifert invariants are only defined for free actions of $Q$ on $P$. This is equivalent to requiring that the gcd of each pair $(\alpha_i, \beta_i)$ is 1 for $i = 1, 2, \ldots, n$. Nevertheless, we are interested in all $Q$-actions on $P$, free or not. The $S^1$-action on $P$, corresponding to $(x_0, x_1, \ldots, x_n)$, as we have seen, descends to an $S^1$-action on $Q\backslash P$. Consequently, $|Q\backslash P|$, the topological space underlying the orbifold $Q\backslash P$ (if the $Q$-action is not free) is still a closed orientable 3-manifold. The orbit mapping of the descended $S^1$-action on $|Q\backslash P|$ is a Seifert fibering whose *reduced* classical Seifert invariants are given by

$$\{g; -((1, x_0); (\alpha_1/\delta_1, x_1/\delta_1), \ldots, (\alpha_n/\delta_n, x_n/\delta_n))\}$$

where $\delta_i = \gcd(\alpha_i, x_i)$.

From this point of view, $\tau : |Q\backslash P| \to Q\backslash W$ can be described in terms of an element $\gamma' \in H^2(Q'; \mathcal{Z})$, where $Q'$ has signature $\{g; \alpha_1/\delta_1, \ldots, \alpha_n/\delta_n\}$, where $\alpha_i/\delta_i$ can be 1. If $Q'$ is finite, it is possible that $Q'$ is cyclic, but this will not adversely affect our arguments. In any case, $\gamma'$ and alternatively these *reduced* Seifert invariants, describe the fibering $\tau$ in terms of the *space* $|Q\backslash P|$ with the descended $S^1$-action and *not* as the *orbifold* $Q\backslash P$ resulting from the $Q$-action on $P$. For the description of the $Q$-action on $P$ and the fibering $Q\backslash P \to Q\backslash W$, $\gamma$ is appropriate.

For $Q'$ with signature $\{g; \alpha_1/\delta_1, \ldots, \alpha_n/\delta_n\}$, we note that $e(x_0, x_1/\delta_1, \ldots, x_n/\delta_n)$ has the same value as $e(x_0, x_1, \ldots, x_n)$ with signature $\{g; \alpha_1, \ldots, \alpha_n\}$. Consequently, $e(x)$ agrees with the Euler number of the Seifert fibering $\tau : Q'\backslash P' = |Q\backslash P| \to |Q\backslash W|$, with classical Seifert invariants $-\{(1, x_0); (\alpha_1/\delta_1, x_1/\delta_1), \ldots, (\alpha_n/\delta_n, x_n/\delta_n)\}$ as well as for the Seifert fibering with the *formal* unnormalized Seifert invariants $-\{(1, x_0), (\alpha_1, x_1), \ldots, (\alpha_n, x_n)\}$. (In Chapter 14, the Euler number of $\{(1, -x_0); (\alpha_1/\delta_1, -x_1/\delta_1), \ldots, (\alpha_n/\delta_n, -x_n/\delta_n)\}$ is defined to be $-(-x_0 + (-x_1/\delta_1)/(\alpha_1/\delta_1) + \cdots + (-x_n/\delta_n)/(\alpha_n/\delta_n)) = x_0 + x_1/\alpha_1 + \cdots + x_n/\alpha_n$, and so it agrees with $e(x)$.) We also point out that the genus of $Q\backslash W$ is $\frac{1}{2}$ of the rank of $H_1(Q; \mathbb{Z})$ and $H^2(Q; \mathcal{Z})$ is independent of the genus when $Q$ is infinite.

15.2.5 (Proof of Theorem 15.2.3(3)). Clearly, we have a homomorphism $e : \mathbb{Z}^{n+1} \to \mathbb{R}$ given by $e(x) = x_0 + x_1/\alpha_1 + \cdots + x_n/\alpha_n$, where $x \in \mathbb{Z}^{n+1}$. Using parts (1) and (2) of the theorem, $K$ is contained in the kernel of $e$. Thus, $e$ factors through $H^2(Q; \mathcal{Z})$, and we denote the induced homomorphism $H^2(Q; \mathbb{Z}) \to \mathbb{Q} \subset \mathbb{R}$ also by $e$. We now determine the values of this homomorphism. Let $A = \alpha_0\alpha_1 \cdots \alpha_n$ where $\alpha_0 = 1$. Let $A_i = A/\alpha_i$, $i = 0, 1, \ldots, n$, and consider the functions $\bar{e} : \mathbb{Z}^{n+1} \to \mathbb{Z}$

$$\bar{e}(x_0, x_1, \ldots, x_n) = x_0 A_0 + x_1 A_1 + \cdots + x_n A_n.$$

The image is a multiple of $d = \gcd(A_0, A_1, \ldots, A_n)$, and all possible multiple integral values of $d$ can occur. (To see this, consider the equation

(15.2.3) $$x_0 A_0 + x_1 A_1 + \cdots + x_n A_n = y.$$

It is known that this has integral solutions $x \in \mathbb{Z}^{n+1}$ if and only if $y$ is a multiple of $d$.)

Suppose
$$\bar{e}(x) = x_0 A_0 + x_1 A_1 + \cdots + x_n A_n = y = md.$$
Divide this equality by $A = A_0$ and we get
$$\bar{e}(x)/A = x_0 + x_1/\alpha_1 + \cdots + x_n/\alpha_n = md/A = m(d/A)$$
for some $m$. It is known that $L = A/d$, where $L = \mathrm{lcm}(\alpha_1, \ldots, \alpha_n)$. Therefore, $\bar{e}(x)/A = e(x) = m/L$. Note $e(f(x)) = e(x)$. Also, if $e(x) = m/L$ and $e(x') = m'/L$, then $e(x + x') = (m + m')/L$. □

15.2.6. Going back to equation (15.2.3), each solution $(x_0, x_1, \ldots, x_n)$ for a fixed $y$ can be replaced by $(x_0 - (s_1 + \cdots + s_n), x_1 + s_1\alpha_1, \ldots, x_n + s_n\alpha_n)$ with arbitrary $(s_1, \ldots, s_n)$ to get another solution. (These are not the only replacements for a solution.) We say that $x, x' \in \mathbb{Z}^{n+1}$ are strongly related if $\bar{e}(x) = \bar{e}(x')$ and $x_i = x'_i$ modulo $\alpha_i$ for each $i$. Therefore, $x$ and $x'$ are strongly related if and only if $x'_i = x_i + s_i\alpha_i$, for some integers $s_i$, $i = 1, \ldots, n$ and $x'_0 = x_0 - (s_1 + \cdots + s_n)$. Note that using (2) of Theorem 15.2.3, $x$ and $x'$ are strongly related if and only if $x - x'$ is an element of $K$. In other words, $x$ is strongly related to $x'$ if and only if $f(x) = f(x') \in H^2(Q; \mathcal{Z})$. Denote the common value of $f(x)$ by
$$f(x) = \langle x_0, x_1, \ldots, x_n \rangle.$$

In summary, an element $\gamma \in H^2(Q; \mathcal{Z})$ is a strict equivalence class (see Proposition 10.1.20) of lifts of $Q$ to bundle automorphisms of $P$, where $''e_2(\gamma)$ determines $P$. This determines the $S^1$-action on $Q\backslash P$ up to equivariant homeomorphism keeping the base fixed, i.e., up to strict equivalence (7.4.2). (To determine the $S^1$-action up to equivariant homeomorphism, i.e., $S^1$-equivalence, one must allow permutations of the set $\{\alpha_1, \ldots, \alpha_n\}$. This entails lifts of automorphisms of $Q$ and may lead to different $\gamma$.) See Proposition 15.6.2.

Furthermore, with $\{\alpha_1, \ldots, \alpha_n\}$ fixed, each of the ordered unnormalized Seifert invariants can be normalized to a unique ordered normalized Seifert invariant. The rules for this are the same as the rules of strong relation for obtaining $\langle x_0, x_1, \ldots, x_n \rangle$ from the $\{x'_0, x'_1, \ldots, x'_n\}$'s.

As we prove parts (1) and (2) of Theorem 15.2.3, we shall identify the $x_i$ with the unnormalized Seifert invariants, and thereby recapture the classification of Chapter 14. We shall also need to make sense of this when $\gcd(\alpha_i, \beta_i) \neq 1$.

15.2.7 (Proof of (1) and (2) of Theorem 15.2.3). For each $[\gamma] \in H^1(Q; \mathcal{T}) \cong H^2(Q; \mathcal{Z})$, there is a principal $S^1$-bundle $P$ over $W$, whose Euler class is $''e_2(\gamma) \in H^2(W; \mathbb{Z})$, and a lift of a $Q$-action on $W$ to a group of bundle automorphisms of $P$. This lifted action commutes with the principal $S^1$-action on $P$ and induces a Seifert fibering $(S^1, M = Q\backslash P) \xrightarrow{\tau} B = Q\backslash W$ modeled on $P$. The descended $S^1$-action is determined by a set of unnormalized Seifert invariants and uniquely by a set of normalized Seifert invariants as described in Chapter 14. All closed orientable Seifert 3-manifolds over an orientable base arise this way except when the genus $g$ of the base $B$ is 0 and $n = 1$ or $n = 2$ and $\alpha_1 \neq \alpha_2$. These bad orbifold cases are treated in Subsection 15.3.2(d).

There is an epimorphism $\bar{f} : \mathbb{Z}^{n+1} \to \mathbb{Z}_{\alpha_1} \oplus \cdots \oplus \mathbb{Z}_{\alpha_n}$ defined by $\bar{f}(x) = \bar{f}(x_0, x_1, \ldots, x_n) = (\bar{x}_1, \bar{x}_2, \ldots, \bar{x}_n)$, where $\bar{x}_i = x_i$ reduced modulo $\alpha_i$ and $\bar{x}_i \in \mathbb{Z}_{\alpha_i}$, $i = 1, 2, \ldots, n$, $x \in \mathbb{Z}^{n+1}$. Observe that if $\bar{f}(x) = 'e_2[\gamma]$ and $\bar{f}(x') = 'e_2[\gamma']$, then $\bar{f}(x + x') = 'e_2([\gamma] + [\gamma'])$. Because $\mathbb{Z}^{n+1}$ is free Abelian, we can lift $\bar{f}$ to a homomorphism $f : \mathbb{Z}^{n+1} \to H^1(Q; \mathcal{T})$ such that $'e_2 \circ f = \bar{f}$. There are many lifts but we seek one that is onto and with kernel $K$. To each 1-cocycle $\gamma \in [\gamma]$, we shall find $x = (x_0, x_1, \ldots, x_n) \in \mathbb{Z}^{n+1}$ such that $-x$ corresponds to a $(n+1)$-tuple of unnormalized Seifert invariants of Chapter 14, $\{(\alpha_0, \beta_0), (\alpha_1, \beta_1), \ldots, (\alpha_n, \beta_n)\}$, with $(\alpha_0, \beta_0) = (1, b_0)$. Moreover, it will turn out that altering the cocycle by a coboundary will change the $(n+1)$-tuples exactly as asserted in the theorem. This will yield the desired $f$.

Let $v_0$ be a point in $Q \backslash W$ which is a free $Q$-orbit. Choose $y_i \in M$, $u_i \in P$, $w_i \in W$ such that $\nu(u_i) = y_i$, $\bar{\tau}(u_i) = w_i$, and $\bar{\nu}(w_i) = v_i$. Choose a closed 2-cell $S^1$-slice $D_i$ at $y_i$. By Section 2.7, there is a closed 2-cell $S^1$-slice $\widetilde{D}_i$ at $u_i$ such that $\nu : \widetilde{D}_i \to D_i$ is a homeomorphism if $\delta_i = 1$ and $\delta_i$-branched at $u_i$ if $\delta_i > 1$. The orbit mapping $\bar{\tau}$ projects $\widetilde{D}_i$ homeomorphically onto $\bar{W}_i$, a neighborhood of $w_i$, which can be taken as a $Q_{w_i}$-slice at $w_i$. Furthermore, $\bar{W}_i / Q_{w_i} = \bar{V}_i = D_i / S^1_{y_i}$.

Let $\mathcal{V} = \{\bar{V}_0, \bar{V}_1, \ldots, \bar{V}_n, B' = B - \bigcup_{i=0}^{n} V_i\}$ be a closed covering of $B$. Then $\mathcal{W} = \{\bar{\nu}^{-1}(\bar{V}_0), \ldots, \bar{\nu}^{-1}(\bar{V}_n), \bar{\nu}^{-1}(B') = W'\}$ is a closed $Q$-invariant covering of $W$. The $S^1$-bundle over each element of $\mathcal{W}$ and over each $\partial \bar{V}_i$ and $B'$ is trivial. Now, $\{\bar{\nu}^{-1}(\bar{V}_i) \cong Q \times_{Q_{w_i}} \bar{W}_i\}$ are disjoint copies of $\bar{W}_i$ indexed by $Q/Q_{w_i}$. Thus the $S^1 \times Q$-action on $\bar{\tau}^{-1}(\bar{\nu}^{-1}(\bar{V}_i))$ is given by $(S^1 \times Q, S^1 \times (Q \times_{Q_{w_i}} \bar{W}_i))$. A typical connected component of the $S^1$-bundle over $\bar{\nu}^{-1}(\bar{V}_i)$ is the bundle over $\bar{W}_i$ which is parametrized by $(S^1, S^1 \times \widetilde{D}_i)$. The lift of $Q_{w_i} \cong \mathbb{Z}_{\alpha_i} = \langle e^{\frac{2\pi i}{\alpha_i}} = \xi_i \rangle$ acts diagonally on this solid torus and factors as

$$(S^1, S^1 \times \widetilde{D}_i) \xrightarrow{/\mathbb{Z}_{\delta_i}} (S^1, S^1 \times (\widetilde{D}_i/\mathbb{Z}_{\delta_i})) \xrightarrow{/\mathbb{Z}_{\alpha_i/\delta_i}} (S^1, S^1 \times_{\mathbb{Z}_{\alpha_i/\delta_i}} (\widetilde{D}_i/\mathbb{Z}_{\delta_i}))$$
$$= (S^1, S^1 \times_{\mathbb{Z}_{\alpha_i}} \widetilde{D}_i) = (S^1, \tau^{-1}(\bar{V}_i)).$$

More explicitly, $\xi_i \times (z_1, rz_2) \mapsto (z_1 \xi_i^{\bar{x}_i}, rz_2 \xi_i)$ for some integer $\bar{x}_i$ mod $\alpha_i$. The action on the other connected components is the same and is governed by the $Q$-equivariance of $\bar{\tau}$.

As in Subsection 10.5.5, the 1-cocycle $\gamma$ is given by $\gamma_{i,i} : Q \to M(\bar{\nu}^{-1}(\bar{V}_i), S^1)$ locally on a coordinate neighborhood of the coordinate bundle $P$ determined by $\mathcal{W}$. Since $\bar{\nu}^{-1}(\bar{V}_i) = Q \times_{Q_{w_i}} \bar{W}_i$, we have

$$H^1(Q_{w_i}; M(\bar{\nu}^{-1}(\bar{V}_i), S^1)) \cong H^1(Q; \mathrm{Hom}_{\mathbb{Z}Q_{w_i}}(\mathbb{Z}Q; M(\bar{W}_i, S^1)))$$
$$\cong H^1(Q_{w_i}; M(\bar{W}_i, S^1))$$
$$\cong H^2(Q_{w_i}; \mathbb{Z}).$$

This cocycle defines the lift of the $Q_{w_i}$-action on $\bar{W}_i$ to $S^1 \times \widetilde{D}_i$. In more detail, the edge homomorphism $'e_2 : H^1(Q; \mathcal{T}) \longrightarrow H^0(Q \backslash W; \mathfrak{h}^1_{\mathcal{T}})$ sends $\gamma_{ij}$ to a section of $\mathfrak{h}^1_{\mathcal{T}}$ over $Q \backslash W$. Locally this sends $\gamma_{i,i_{v_i}} : Q \to M(\bar{\nu}^{-1}(v_i), S^1)$ which is equivalent to the 1-cocycle $Q_{w_i} \to M(w_i, S^1)$ which generates the $\mathbb{Z}_{\alpha_i}$-action on $S^1 \times \widetilde{D}_i$ given by $\xi \times (z_1, rz_2) \mapsto (z_1, \xi^{\bar{x}_i}, rz_2 \xi)$. That is, $'e_2(\gamma_{ij}) = \bar{x}_i$. We now conclude, by Subsections 10.5.8 and 15.2.2, $'e_2(\gamma) = (\bar{x}_1, \bar{x}_2, \ldots, \bar{x}_n) \in \mathbb{Z}_{\alpha_1} \oplus \cdots \oplus \mathbb{Z}_{\alpha_n}$. Later, we will conclude that $(\alpha_i, -\bar{x}_i)$ is the unreduced normalized Seifert invariant for $Q \backslash P$, $i = 1, \ldots, n$.

The bundle $P$, with its lifted $Q$-action, is obtained by equivariantly pasting, along boundaries the trivial bundles over $\bar{\nu}^{-1}(\bar{V}_i)$ into the trivial bundle over $W'$. We get a $Q$-equivariant section $\sigma' : W' \to \bar{\tau}^{-1}(W')$ from the pullback of a section

## 15.2. A PRESENTATION FOR $H^2(Q;\mathcal{Z})$

$\sigma : B' \to \bar{\tau}^{-1}(B')$. We define a bundle coordinate chart $k_j : S^1 \times W' \to \bar{\tau}^{-1}(W')$ by $k_j(z, w) = z \cdot \sigma'(w)$, $z \in S^1$, $w \in W'$. Similarly, we have coordinate charts $k_i : S^1 \times \bar{\nu}^{-1}(\bar{V}_i) \to \bar{\tau}^{-1}\bar{\nu}^{-1}(\bar{V}_i)$. Because of $Q$-invariance, we can describe $k_i$ by restricting to a connected component. We then define $S^1 \times \widetilde{D}_i \to \bar{\tau}^{-1}(\bar{W}_i)$ by $k_i(z, u) = z \cdot k_i(1, u)$, where $z \in S^1$, $u \in \widetilde{D}_i$ and $\bar{\tau}$ restricted to $\widetilde{D}_i$ is a homeomorphism onto $\bar{W}_i$. We abbreviate $z \cdot k_i(1, u)$ as $z \cdot u$ because $\widetilde{D}_i \subset \bar{\tau}^{-1}(\bar{W}_i)$.

For $W' \cap \bar{W}_i$, we have the transition function $g_{ij} : \partial \bar{W}_i \to S^1$, given by

$$(1, w) \xrightarrow{k_j} \sigma'(w) \xrightarrow{k_i^{-1}} (\gamma_{ij}(1; w), u) = (g_{ij}(w), u),$$

where $\sigma'(w) = (1 \cdot \sigma'(w)) = (z \cdot u)$, with $z \in S^1$, $u \in \widetilde{D}_i$ and $\bar{\tau}(u) = w$; see Subsection 10.3.3. Thus $g_{ij}(w) = z$, and is a function $S^1 \to S^1$ of some homotopy class (i.e., degree) $x_i$. We can alter the section $\sigma$ near $\partial \bar{V}_i$ by a small homotopy and therefore alter $\sigma'$ near $\partial \bar{W}_i$ by an equivariant homotopy so that the map $g_{ij}$ is of the form $w \mapsto w^{x_i}$. So we may assume that our 1-cocycle $\gamma_{ij}(1, w)$ is already of this form. (The Seifert fibering is moved, keeping $\nu^{-1}(\bar{v}_0), \ldots, \nu^{-1}(\bar{v}_n)$ fixed, by an equivariant isotopy.) Consequently, we can assign the $(n+1)$-tuple $(x_0, x_1, \ldots, x_n)$ to our cocycle $\gamma_{ij}$, and this will be our $f(x)$. We now relate $f(x)$ to the Seifert invariants determined by $[\gamma]$.

We may parametrize the effective $S^1$-action on $|Q \backslash P|$ on a tubular neighborhood, $S^1 \times D^2 = S^1 \times D_i^2$, of the orbit through $y_i$, whose isotropy group is $\mathbb{Z}_{\alpha_i/\delta_i}$, by $z \times (z_1, rz_2) \mapsto (z^{\alpha_i/\delta_i} z_1, rz_2 z^{\nu_i/\delta_i})$ with $z \in S^1$, and $\gcd(\alpha_i/\delta_i, \nu_i/\delta_i) = 1$. Let $\beta'_i \nu_i/\delta_i \equiv 1 \mod \alpha_i/\delta_i = \alpha'_i$. Then, reducing $\beta'_i$ and $\nu_i/\delta_i$ modulo $\alpha_i/\delta_i$ yields the associated normalized Seifert invariant $(\alpha'_i, \beta'_i)$ as in Subsections 14.1.1 and 14.2.1.

The $S^1$-action on $|Q \backslash P|$ can be lifted to the $S^1$-action on $|Q \backslash P|_H$, the covering space associated to $H$, the image $\text{ev}_*^y(\pi_1(S^1, 1)) \subset \pi_1(|Q \backslash P|, y)$. The map $P$ to $|Q \backslash P|_H$ is a branched covering. $(S^1, S^1 \times_{\mathbb{Z}_{\alpha_i/\delta_i}} (D^2/\mathbb{Z}_{\delta_i})) = (S^1, S^1 \times_{\mathbb{Z}_{\alpha_i}} D^2)$ lifts to disjoint copies of $(S^1, S^1 \times (D^2/\mathbb{Z}_{\delta_i}))$ on $|Q \backslash P|_H$.

Let us, for clarity, temporarily drop the index $i$. Following Subsection 14.2.1, let $m$, $q$, $h$ denote meridian, cross sectional curve and typical (free) fiber on the parametrization of the boundary of the solid fibered torus in $|Q \backslash P|$ where $m \sim \alpha' q + \beta' h$. Hence $(\alpha', \beta')$ is the unnormalized Seifert invariant for this fibered torus. The inverse image in $|Q \backslash P|_H$ of this solid torus, $(S^1, S^1 \times_{\mathbb{Z}_{\alpha/\delta}} (D^2/\mathbb{Z}_\delta))$, is a disjoint collection of solid tori. On any one of these components, $m$ lifts to $\alpha' = \alpha/\delta$ copies of $m$, while $(\alpha/\delta)q$ lifts to a cross sectional curve $q'$, and $h$ lifts to $\alpha/\delta$ disjoint copies of typical fiber except for the core circle lifts to a single curve (the new core circle), where it covers the core circle $\beta/\delta$ times. If $m'$ and $h'$ denote components of the lifted $m$ and $h$, then $m' \sim q' + \beta/\delta \, h'$, where $\beta' = \beta/\delta$. On the other hand, when we lift $(S^1, S^1 \times_{\mathbb{Z}_{\alpha/\delta}} (D^2/\mathbb{Z}_\delta))$ to $P$, we get disjoint copies of $(S^1, S^1 \times D^2)$ which branch covers the $(S^1, S^1 \times (D^2/\mathbb{Z}_\delta))$ of $|Q \backslash P|_H$. The curve $\tilde{m} = 1 \times \partial D^2$ covers $m' = 1 \times \partial(D^2/\mathbb{Z}_\delta)$ $\delta$-times while $\delta q'$ lifts to $\tilde{q}$, a cross sectional curve to the trivial fibering $(S^1, S^1 \times D^2)$. The curve $h'$ lifts to $\delta$ disjoint copies of $h'$ provided that $h'$ is not the core circle. If $h'$ is the core circle, it lifts *identically*. Call a component of the lift $\tilde{h}$. Then we get $\tilde{m} \sim \tilde{q} + \beta \tilde{h} \sim \delta q' + \delta(\beta/\delta) h'$, corresponding to our action $(z_1, rz_2) \mapsto (z_1 \xi^{-\beta}, rz_2 \xi)$ on $S^1 \times D^2$. This is our explanation of the formal unnormalized Seifert invariant $(\alpha, \beta)$ when $\gcd(\alpha, \beta) = \delta$.

Let us return to our cocycle $\gamma$ and our transition functions $g_{ij}$. For each $i$, we have $g_{ij} : W' \cap \bar{W}_i \to S^1$ given by $w \mapsto w^{x_i}$, $w \in \partial \bar{W}_i \cong S^1$. The graph of

$g_{ij}$ is a cross sectional curve $c_i$ on $\partial(S^1 \times \widetilde{D}_i)$ and satisfies the homology relation $c_i \sim \widetilde{m}_i + x_i \widetilde{h}$, where $\widetilde{m}_i$ and $\widetilde{h}$ parametrize the boundary of $S^1 \times \widetilde{D}_i$. We saw, from above, that $c_i$ is the cross sectional curve $\widetilde{q}_i$ which satisfied $\widetilde{q}_i \sim \widetilde{m}_i - \beta_i \widetilde{h}$. Therefore, $x_i = -\beta_i$, where $(\alpha_i, \beta_i)$ is the formal unnormalized Seifert invariant for the Seifert fibering associated with the coordinate bundle $P$ and the cocycle $\gamma_{ij} \in [\gamma]$.

We could, at this point, complete the proof of the theorem by using, from Chapter 14, how different sets of unnormalized Seifert invariants represent the same Seifert fibering up to strict equivalence. Instead we shall reprove these facts in the present context since we have essentially used nothing from Chapter 14 other than the definitions of the Seifert invariants.

The values of the cocycle $\gamma_{ij}$ depended upon a choice of section $\sigma : B' \to \tau^{-1}(B')$. A new section $\bar{\sigma}$ results in a new lifted section $\bar{\sigma}'$ and a different cocycle $\bar{\gamma}_{ij}$. Deforming the cross section by a homotopy changes the Seifert fibering by an equivariant isotopy and so does not change the equivalence class of the new unnormalized Seifert invariants.

If we identify $\tau^{-1}(B')$ with $S^1 \times B'$, then a section $\bar{\sigma} : B' \to S^1 \times B'$ can be written as $\bar{\sigma}(v) = (\lambda(v), v)$, where $\lambda : B' \to S^1$. If $\sigma(v) = (1, v)$, then $B' \cap \bar{V}_i$ was mapped by the attaching map to $q_i$ on the boundary of $S^1 \times D_i$ and satisfied the homology relation $m_i \sim \alpha_i' q_i + \beta_i' h$ on $\partial(S^1 \times D_i)$. With $\bar{\sigma}$, $B' \cap \bar{V}_i$ is mapped to $\bar{q}_i$ and satisfies the homology relation $m_i \sim \alpha_i' \bar{q}_i + (\beta_i' - s_i \alpha_i') h$. For, $\lambda : B' \cap \bar{V}_i \to S^1$ is a map of some degree, say $s_i$, and the image of $\bar{\sigma}(B' \cap \bar{V}_i)$, when attached to $\partial(S^1 \times D_i)$, is a curve $\bar{q}_i$ such that $\bar{q}_i \sim q_i + s_i h$. Therefore, $\alpha_i' q_i \sim \alpha_i' \bar{q}_i - \alpha_i' s_i h$. Now the lift to $P$ yields a new cross sectional curve $\widetilde{\bar{q}}$ satisfying $\widetilde{m}_i \sim \widetilde{\bar{q}}_i + (\beta_i - \alpha_i s_i) \widetilde{h}$. This results in new unnormalized Seifert invariants $(\alpha_i, \beta_i - \alpha_i s_i)$ for $[\gamma]$. We examine the meaning of this for $i = 0$. Then $\alpha_0 = 1$, and we need to determine $s_0$.

Recall that we can identify $H^1(B'; \mathbb{Z})$ with the homotopy classes of maps $[B', S^1]$. From the exact sequence

$$H^1(B'; \mathbb{Z}) \xrightarrow{i^*} H^1(\partial B'; \mathbb{Z}) \xrightarrow{\delta} H^2(B', \partial B'; \mathbb{Z}) \cong \mathbb{Z},$$

we may interpret the $i$th-factor of the middle term,

$$H^1(\partial B'; \mathbb{Z}) = \prod_{i=0}^{n} H^1(B' \cap \partial \bar{V}_i; \mathbb{Z}),$$

as the homotopy class of a partial section $[B' \cap \partial \bar{V}_i] \to (B' \times S^1)$ of degree $s_i$. Then, $\delta(s_0, s_1, \ldots, s_n) = s_0 + s_1 + \cdots + s_n$ is the obstruction for these partial sections to extend to a complete section over $B'$ [**Ste51**, §3.4.2]. Since there is a section, this obstruction vanishes and so $s_0 = -(s_1 + \cdots + s_n)$ is a necessary and sufficient condition that must be satisfied with a change in section. Now $(\alpha_0, \beta_0) = (1, b_0)$ and $(\bar{\alpha}_0, \bar{\beta}_0) = (1, \bar{b}_0) = (1, b_0 - (1 \cdot s_0)) = (1, b_0 + (s_1 + \cdots + s_n))$. Since $(\alpha_i, \bar{\beta}_i) = (\alpha_i, \beta_i - \alpha_i s_i)$, the new Seifert invariants for $[\gamma]$ is

$$\{g; (1, b_0 + (s_1 + \cdots + s_n)), (\alpha_1, \beta_1 - s_1 \alpha_1), \ldots, (\alpha_n, \beta_n - s_n \alpha_n)\}.$$

In terms of the $(n+1)$-tuples, we have $f(x) = f(x') = [\gamma]$, where $x = (x_0, x_1, \ldots, x_n)$ and $x' = (x_0 - (s_1 + \cdots + s_n), x_1 + s_1 \alpha_1, \ldots, x_n + s_n \alpha_n)$ and $x_i = -\beta_i$. Clearly, $f$ is onto and the kernel is $K$. $\square$

REMARK 15.2.8. 1. By following modern conventions for orientations and obstruction theory, our cohomology class corresponds to the negative of the classical Seifert invariants.

2. $H^2(Q;\mathcal{Z})$ is an Abelian extension of $\mathbb{Z}$ by $(\mathbb{Z}_{\alpha_1} \oplus \cdots \oplus \mathbb{Z}_{\alpha_n})$ and can be represented in terms of generators and relations as follows:

$$\{X_0, X_1, \ldots, X_n | \ \alpha_i X_i = X_0, \ i = 1, \ldots, n\}$$

Let $\{u_i\}$ be the standard basis for $\mathbb{Z}^{n+1}$. That is, $u_i$ is the $(n+1)$-tuple which is 1 in the $i$th slot and 0 elsewhere, $i = 0, 1, \ldots, n$. Define $X_0 = f(u_0) = \text{Im}('e_1(1)) \in H^2(Q;\mathcal{Z})$ and $X_i = f(u_i)$. That is, $f(u_i) = \gamma \in H^2(Q;\mathcal{Z})$, where $(\alpha_j, \beta_j) = (\alpha_j, 0)$, $j = 0, 1, \ldots, \hat{i}, \ldots, n$ and $(\alpha_i, \beta_i) = (\alpha_i, 1)$. Then, $f(x) = f(x_0, x_1, \ldots, x_n) = x_0 X_0 + \cdots + x_n X_n = \langle x_0, x_1, \ldots, x_n \rangle$.

3. If $Q$ is infinite, $''e_1$ of equation (15.2.2) is an isomorphism of $H^2(Q;\mathcal{Z})$ with $H^2(Q;\mathbb{Z})$ which naturally is isomorphic to the central extensions of $\mathbb{Z}$ by $Q$. Each $\gamma \in H^2(Q;\mathcal{Z})$ determines a lift of $Q$, up to strict equivalence, to a group of bundle isomorphisms on $S^1 \times \mathbb{R}^2$. This lift has an extended lift to a group $\Pi$ of bundle isomorphisms to the principal $\mathbb{R}^1$-bundle $\mathbb{R}^1 \times \mathbb{R}^2$. This $\Pi$ is the central extension $''e_1^{-1}(\gamma)$. In [**LLR96**] and [**JN83**], Theorem 15.2.3 for $Q$ infinite is proved by showing how each central extension of $\mathbb{Z}$ by $Q$ determines a group $\Pi$ whose presentation determines the *formal* Seifert invariants of $(S^1, \Pi \backslash (\mathbb{R}^1 \times \mathbb{R}^2))$. This provides an alternative approach to Theorem 15.2.3 for $Q$ infinite.

PROPOSITION 15.2.9. *$\gamma \in H^2(Q;\mathcal{Z})$ has finite order if and only if $e(\gamma) = 0$.*

PROOF. Suppose $\gamma = \langle x_0, x_1, \ldots, x_n \rangle$ has order $a > 1$. Then $ae(\gamma) = e(a\gamma) = e(0) = 0$. Therefore $e(\gamma) = 0$. Conversely, for $\gamma = \langle x_0, x_1, \ldots, x_n \rangle$, $L\gamma = \langle Lx_0, Lx_1, \ldots, Lx_n \rangle = \langle b, 0, \ldots, 0 \rangle$ for some $b \in \mathbb{Z}$. If $e(\gamma) = 0$, then $0 = L(e(\gamma)) = e(L(\gamma)) = e\langle b, 0, \ldots, 0 \rangle$. Hence $b = 0$ and so $L\gamma = 0$. Therefore, the order of $\gamma$ divides $L$. □

THEOREM 15.2.10 (cf. [**LLR96**, Theorem 4.5]). *Tensoring $H^2(Q;\mathcal{Z})$ with $\mathbb{R}$ is a homomorphism sending $f(x)$ to $L(e(x)) \in \mathbb{R}$.*

PROOF. Let $\theta : H^2(Q;\mathcal{Z}) \to \mathbb{Z} \oplus$ Torsion be an isomorphism; see Subsection 15.2.2. For $f(x) = \gamma$, write $\theta(\gamma) = (\gamma_{_\mathbb{Z}}, \gamma_{_T})$ in terms of this isomorphism. We may assume that the positive generator of $H^2(Q\backslash W, \mathbb{Z})$ has positive image in $\gamma_{_\mathbb{Z}}$. If $\otimes 1_{_\mathbb{R}} : H^2(Q;\mathcal{Z}) \to H^2(Q;\mathcal{Z}) \otimes \mathbb{R} \approx \mathbb{R}$ is the homomorphism given by $\gamma \mapsto \gamma \otimes 1_{_\mathbb{R}}$, then $\gamma \otimes 1_{_\mathbb{R}} = \gamma_{_\mathbb{Z}} \otimes 1_{_\mathbb{R}} = \gamma_{_\mathbb{Z}}$ and $\gamma_{_T} \otimes 1_{_\mathbb{R}} = 0$. Now, only a generator $g$ of infinite order in $H^2(Q;\mathcal{Z})$ can satisfy $e(g) = \pm 1/L$. Each such $g$ is of the form $g = (g_{_\mathbb{Z}}, g_{_T}) = (\pm 1, g_{_T})$. Since $g \otimes 1_{_\mathbb{R}} = \pm 1 \in \mathbb{R}$, and $e(\gamma) = e(\gamma_{_\mathbb{Z}})$, then $\gamma \otimes 1_{_\mathbb{R}} = L(e(\gamma))$. Note, $\gamma \otimes \mathbb{R}$ does not depend upon the choice of $\theta$. □

EXERCISE 15.2.11. 1. $H^2(Q;\mathcal{Z})$ is isomorphic to $\mathbb{Z}$ if and only if $\{\alpha_1, \alpha_2, \ldots, \alpha_n\}$ are pairwise coprime.

2. The order of the torsion of $H^2(Q;\mathcal{Z})$ is $d = \frac{\alpha_1 \cdot \alpha_2 \cdots \alpha_n}{\text{lcm}(\alpha_1, \alpha_2, \ldots, \alpha_n)}$.

3. $e(\gamma) = e(\gamma')$ if and only if $\gamma - \gamma'$ is a torsion element of $H^2(Q;\mathcal{Z})$.

4. If $Q$ has signature $\{g; \alpha_1, \ldots, \alpha_n\}$ and the $\alpha_i$ are pairwise coprime, then $x \in H^2(Q;\mathcal{Z})$ is determined up to strict equivalence by $e(x)$ alone provided that $\gcd(x_i, \alpha_i) = 1$, for each $i$. (Hint: cf. with Corollary 14.9.2.) Note that $e(x)$ also characterizes $x$, and hence $|Q\backslash P|$, even when $\gcd(x_i, \alpha_i) = \delta_i > 1$.

5. In Section 15.1, $H^2(Q;\mathcal{Z})$ is given as an Abelian extension of $\mathbb{Z}$ by $\mathbb{Z}_{\alpha_1} \oplus \cdots \oplus \mathbb{Z}_{\alpha_n}$. When the set $\mathbb{Z} \times (\mathbb{Z}_{\alpha_1} \oplus \cdots \oplus \mathbb{Z}_{\alpha_n})$ is equipped with a 2-cocycle,

$$k : (\mathbb{Z}_{\alpha_1} \oplus \cdots \oplus \mathbb{Z}_{\alpha_n}) \times (\mathbb{Z}_{\alpha_1} \oplus \cdots \oplus \mathbb{Z}_{\alpha_n}) \to \mathbb{Z},$$

a multiplication can be defined on the set yielding a central group extension of $\mathbb{Z}$ by $\mathbb{Z}_{\alpha_1} \oplus \cdots \oplus \mathbb{Z}_{\alpha_n}$. (Let the integers $x_i$, $0 \leq x_i < \alpha_i$, represent a residue class of $\alpha_i \in \mathbb{Z}^+$. Similarly, for $x'_i$. Put $x = (x_1, \ldots, x_n)$, $x' = (x'_1, \ldots, x'_n)$ and define $k(x, x') = (\epsilon_1 + \epsilon_2 + \cdots + \epsilon_n)$ where $\epsilon_i = 1$ if $x_i + x'_i \geq \alpha_i$ and 0 if otherwise. Show $k$ is a suitable cocycle such that the set $\mathbb{Z} \times (\mathbb{Z}_{\alpha_1} \oplus \cdots \oplus \mathbb{Z}_{\alpha_n})$ equipped with this cocycle is congruent to $H^2(Q;\mathcal{Z})$.) Use Theorem 15.2.10 to show that $f(u_0) = 'e_1(1)$ of Remark 15.2.8(2), is $L$ times the generator of the $\mathbb{Z}$-factor in $H^2(Q;\mathcal{Z}) \cong \mathbb{Z} \oplus$ Torsion.

6. Let $x = (x_0, x_1, \ldots, x_n) \in \mathbb{Z}^{n+1}$ and define a homomorphism $\bar{f} : \mathbb{Z}^{n+1} \to \mathbb{Z}_{\alpha_1} \oplus \cdots \oplus \mathbb{Z}_{\alpha_n}$ by $\bar{f}(x) = (\bar{x}_1, \bar{x}_2, \ldots, \bar{x}_n)$ where $\bar{x}_i$ denotes the congruence class of $x_i$ mod $\alpha_i$, $i = 1, 2, \ldots, n$. Since $'e_2 : H^2(Q;\mathcal{Z}) \to \mathbb{Z}_{\alpha_1} \oplus \cdots \oplus \mathbb{Z}_{\alpha_n}$ is an epimorphism, one can lift $\bar{f}$ to a homomorphism $h : \mathbb{Z}^{n+1} \to H^2(Q;\mathcal{Z})$ so that $'e_2 \circ h = \bar{f}$. Of course, $h$ is not unique. How does one find an $h$ so that $h = f$ of Theorem 15.2.3? Of course, the key to $h$ is just to lift $x_i$ to $f(x_i)$. That is, put $x_i = \bar{x}_i + s_i \alpha_i$ and let $f(x_i) = \bar{x}_i f(u_i) + s_i f(u_0)$, where $u_i$ are as in Remark 15.2.8(2). Note from Theorem 15.2.10, $f(u_0) = (L, 0) \in \mathbb{Z} \oplus (\mathbb{Z}_{\alpha_1} + \cdots + \mathbb{Z}_{\alpha_n})$.

15.2.12. Let us fix $Q = \{g; \alpha_1, \ldots, \alpha_n\}$ and $\rho : Q \to \text{TOP}(W)$ and a principal $S^1$-bundle $P$ over $W = S^2$ or $\mathbb{R}^2$. Two lifts $\theta_1$ and $\theta_2 : Q \to \text{TOP}_{S^1}(P)$ corresponding to 1-cocycles in $Z^1(Q;\mathcal{T})$ are cohomologous if and only if the action of $\theta_1(Q)$ and $\theta_2(Q)$ are strictly equivalent. Of course this means that the $S^1$-actions on $\theta_1(Q)\backslash P$ and $\theta_2(Q)\backslash P$ are also strictly equivalent; see Proposition 10.1.20.

On the other hand, if $\gamma_1 = -\gamma_2$, then there is a diffeomorphism $M_1 = \theta_1(Q)\backslash P \xrightarrow{h} \theta_2(Q)\backslash P = M_2$ such that $h(zu) = z^{-1}h(u)$, for $u \in \theta_1(Q)\backslash P$, $z \in S^1$. Note, $h$ is not equivariant but it is weakly equivariant. For $W = \mathbb{R}^2$ and $P = S^1 \times \mathbb{R}^2$, this is most easily seen by defining $\tilde{h} : (x, w) \mapsto (x^{-1}, w)$ for $x \in S^1$, $w \in W$. The orientation reversing homeomorphism $\tilde{h}$ is $Q$-invariant, that is,

$$\theta_1(\alpha)(x, w) = (x(\lambda(\alpha w))^{-1}, \alpha w) \xrightarrow{\tilde{h}} (x^{-1}(\lambda(\alpha w)), \alpha w) = \theta_2(\alpha)(x^{-1}, w),$$

and weakly $S^1$-equivariant. For $P$ nontrivial over $S^2$, this analysis is valid for each coordinate patch. Use the cocycle conditions to see that they match together.

15.2.13. The term *strict equivalence* was introduced in Subsection 7.4.2 in the general context of an injective Seifert fibering. If $G = S^1$, $W = \mathbb{R}^2$, $\Gamma = 1$, the term there agrees with our usage in this chapter. In both cases, the $Q$-actions on $S^1 \times \mathbb{R}^2$ are strictly equivalent if the cocycles representing the $Q$-actions are cohomologous. Two induced $S^1$-actions, $(S^1, Q\backslash S^1 \times \mathbb{R}^2) \xrightarrow{S^1} Q\backslash \mathbb{R}^2$ from the same strict equivalence class can be equivariantly deformed into the other by just moving along the fibers.

If $Q$ acts freely on $S^1 \times \mathbb{R}^2$, the $S^1$-action $(S^1, M_\gamma)$ is injective. If $(S^1, M_1)$ and $(S^1, M_2)$ are $S^1$-actions on oriented closed manifolds over an orientable base of the same genus, they will be equivariantly homeomorphic by an orientation preserving homeomorphism if and only if their respective normalized Seifert invariants are the same as a set. For the strict equivalence classes of injective $S^1$-actions, the classification is similar but we will get a finer classification if we require the equivariant

## 15.2. A PRESENTATION FOR $H^2(Q;\mathcal{Z})$

homeomorphism to respect the strict equivalence classes. So the equivariant homeomorphism should send the ordered orbit with normalized Seifert invariant $(\alpha_i, \beta_i)$ of $M_1$ to the orbit with similarly ordered Seifert invariant $(\alpha_i, \beta_i)$ of $M_2$. On the other hand, an ordinary equivariant homeomorphism could send the orbit with normalized invariant $(\alpha_i, \beta_i)$ of $M_1$ to the orbit $(\alpha_j, \beta_j)$ of $M_2$ if $i \neq j$, $\alpha_i = alpha_j$, and $\beta_i = \beta_j$. What the latter requires is a homeomorphism of the base that interchanges singular orbits of the same Seifert invariants. The finer classification matches perfectly with the elements of $H^2(Q;\mathcal{Z})$.

If the $Q$-action on $S^1 \times \mathbb{R}^2$ is not free, then the classification would be as *orbibundles*. In this chapter as in Chapter 10, strict equivalence also includes locally injective actions and the concomitant $S^1$-action classifications; see Section 15.6 for further details and specific examples.

15.2.14. Since $H^2(Q;\mathcal{Z}) \cong \mathbb{Z} \oplus \text{Torsion}$, there exists a $g \in H^2(Q;\mathcal{Z})$, not necessarily unique, so that $e(g) = 1/L$. Now for an arbitrary $\gamma \in H^2(Q;\mathcal{Z})$, there is an integer $n$ so that $\gamma = ng + \sigma$, for some $\sigma \in \text{Torsion}$. Thus, by picking a convenient generator $g$ of $\mathbb{Z}$ and determining the finite number of $\sigma \in H^2(Q;\mathcal{Z})$ such that $e(\sigma) = 0$, we have a convenient way of cataloging the elements of $H^2(Q;\mathcal{Z})$ in terms of their Seifert invariants.

In the following sections, we use this procedure as a way of cataloging the elements of $H^2(Q;\mathcal{Z})$ in terms of Seifert invariants. It suffices to determine the $\gamma$ for which $\bar{e}(\gamma) = 0$ and $e(\gamma) > 0$. To obtain the results for $-\gamma$, just reverse the invariants of those found for $e(\gamma)$.

15.2.15. As before, $Q$ has signature $\{g; \alpha_1, \ldots, \alpha_n\}$. For Seifert invariants, we shall use $(\alpha, \beta)$ to mean the $(\alpha, -\beta)$ of Subsection 15.2.7 and of Chapter 14. This is convenient for, in this way

$$\gamma = \langle x_0, x_1, \ldots, x_n \rangle$$

corresponds to

$$M_\gamma = \{g; (1, x_0), (\alpha_1, x_1), \ldots, (\alpha_n, x_n)\}.$$

In fact, we combine both of these notations and write them as

$$\gamma = \langle (1, b), (\alpha_1, \beta_1), \ldots, (\alpha_n, \beta_n) \rangle,$$

where $b = x_0$, $\beta_i = x_i$, $i = 1, \ldots, n$, and with $g$ suppressed unless it is necessary to specify it. $M_\gamma$ is the same Seifert manifold whose unnormalized notation in Chapter 14 is given by $\{g; (1, -b), (\alpha_1, -\beta_1), \ldots, (\alpha_n, -\beta_n)\}$.

A lifted action of $Q$ to the principal $S^1$-bundle $P$ is free if and only if the projection $'e_2(\gamma) \in \mathbb{Z}_{\alpha_1} \oplus \cdots \oplus \mathbb{Z}_{\alpha_n}$ on each factor is a generator, i.e., if and only if $\gcd(\alpha_i, \beta_i) = 1$ for each $i$.

We denote the underlying space of $M_\gamma$ by $|M_\gamma|$. The quotient mapping $P \to Q \backslash P$ is a covering if $Q$ acts freely, and is branched if $Q$ does not act freely. In the latter case, $|M_\gamma|$, as a space, is still an oriented closed 3-manifold and the descent of the principal $S^1$-fibers still yields a Seifert fibering whose *reduced* Seifert invariants, (i.e., the Seifert invariants for $(S^1, |M_\gamma|) \to Q \backslash W$) are

$$|M_\gamma| = \{(1, b), (\alpha_1/\delta_1, \beta_1/\delta_1), \ldots, (\alpha_n/\delta_n, \beta_n/\delta_n)\}$$

with $\delta_i = \gcd(\alpha_i, \beta_i)$. Note, some of the $\alpha_i/\delta_i$ could be equal to 1. This can be combined by using $\langle (1, x_1), (1, x_2) = (1, x_1 + x_2) \rangle$. Also $S^1 \backslash |M_\gamma|$ could be a "bad" 2-dimensional orbifold.

We note that for $Q$, the *unit tangent bundle* for the orbifold $Q\backslash W$ (see Theorem 14.8.8) (i.e., $Q\backslash G$ where $G = $ SO(3), PSL(2,$\mathbb{R}$), or $E_0(2)$ and with $Q$ a lattice in $G$), is, in the notation of this section, given by $\langle g; (1,b) = (1, 2-2g-n), (\alpha_1, 1), \ldots, (\alpha_n, 1)\rangle$.

## 15.3. 3-dimensional spherical space forms

15.3.1 (The spherical space forms and orbifolds as locally injective Seifert fiber spaces). The unitary group $U(2) = S^1 \times_{\mathbb{Z}_2} SU(2)$ is the compact subgroup of complex automorphisms of $\mathbb{C}^2$. With the standard Kähler metric, it acts as isometries leaving the 3-sphere invariant. The $S^1$-factor consisting of the diagonal matrices with constant entries is the center of $U(2)$ and acts freely on $S^3$, and can be identified with the fiber of the principal $S^1$-bundle $S^1 \hookrightarrow P_1 = S^3 \to S^2$. The quotient group $S^1\backslash U(2)$ is SO(3) and its action descends to the group of orientation preserving isometries as well as the group of Kähler isometries of $S^2 = \mathbb{C}P_1$. Obviously $U(2)$ preserves the $S^1$-fibers on the 3-sphere and is an extended lifting of the isometries of $S^2$ to the principal bundle automorphisms of $P_1$.

Let $\Pi$, a finite group, act freely as isometries on the standard 3-sphere. This action, according to Subsection 4.6.3(5), is weakly equivalent to an action induced from a free unitary representation of $\Pi$ in $U(2)$ restricted to the invariant 3-sphere. So assume that $\Pi$ is a subgroup of $U(2)$ acting freely. It commutes with the $S^1$-action. Let $\Pi \cap S^1 = \mathbb{Z}_n$. Then the quotient $Q = \Pi/\mathbb{Z}_n$ is a subgroup of SO(3). The action of $\Pi$ descends to the action of $Q$ on $S^2$ and so $\Pi$ is an extended action of the group $Q$ of isometries of $S^2$ to bundle automorphisms of $P_1$. The descent of the $S^1$-action to $\Pi\backslash P_1$ exhibits $M = \Pi\backslash P_1$ as a Seifert fibering with typical $\mathbb{Z}_n\backslash S^1$-fiber over the orbifold $Q\backslash S^2$. We shall also show that each closed Seifert 3-manifold, $|M_\gamma|$, with finite fundamental group has an $S^1$-action of the type just described and is therefore a spherical space form. We assume the $Q$ acts effectively on $W = S^2$ and preserves orientation. Let $P_r$ be the principal $S^1$-bundle over $S^2$ with Euler class $r$. The bundle $P_r$ is diffeomorphic to the lens spaces $L(r, 1)$. The *positive sign* is a consequence of our choice of conventions. $\pm L(1,1)$ is the 3-sphere and $L(0, -)$ is the product bundle $S^1 \times S^2$ whose quotients are not spherical space forms. We shall use the following table.

| | Signature of $Q$ | Group | $H^2(Q;\mathbb{Z})$ | $H^3(Q;\mathbb{Z})$ |
|---|---|---|---|---|
| (1) | $(0; m, m)$ | $\mathbb{Z}_m$ cyclic | $\mathbb{Z}_m$ | 0 |
| (2) | $(0; 2, 2, 2m)$ | $D_{4m} = \mathbb{Z}_{2m} \rtimes \mathbb{Z}_2$ dihedral | $\mathbb{Z}_2 \oplus \mathbb{Z}_2$ | $\mathbb{Z}_2$ |
| (3) | $(0; 2, 2, 2m+1)$ | $D_{4m+2} = \mathbb{Z}_{2m+1} \rtimes \mathbb{Z}_2$ dihedral | $\mathbb{Z}_2$ | 0 |
| (4) | $(0; 2, 3, 3)$ | $T \cong A_4$ tetrahedral | $\mathbb{Z}_3$ | $\mathbb{Z}_2$ |
| (5) | $(0; 2, 3, 4)$ | $O \cong S_4$ octahedral | $\mathbb{Z}_2$ | $\mathbb{Z}_2$ |
| (6) | $(0; 2, 3, 5)$ | $I \cong A_5$ icosahedral | 0 | $\mathbb{Z}_2$ |

We use the following procedure to classify the spherical space forms. Write $\gamma \in H^2(Q; \mathcal{Z}) \cong \mathbb{Z} \oplus \text{Torsion}$ as $\gamma = rg_1 + g_T$, where $g_1$ is a generator of the $\mathbb{Z}$-factor and $g_T$ an element of the torsion subgroup. Then, $e(\gamma) = r/L$ and $''e_2(\gamma) = 2r$ if $H^3(Q;\mathbb{Z}) = \mathbb{Z}_2$ and $r$ if $H^3(Q;\mathbb{Z}) = 0$. Therefore, $\gamma$ represents a lift of $Q$ to $P_{''e_2(\gamma)}$. With $rg_1 + g_T$, the Seifert invariants are determined on $Q\backslash P_{''e_2(\gamma)}$. We get freeness of the lifted action of $Q$ on $P_{''e_2(\gamma)}$ if all the $\gcd(\alpha_i, \beta_i) = 1$. We then determine the extended lifting of $Q$ to $S^3 = P_1$ and get a free linear action on $S^3$.

**15.3.2** ($Q = \mathbb{Z}_m$). The action of $Q = \mathbb{Z}_m$ on $S^2$ is a rotation about a polar axis with periodicity $m$. Since $H^3(\mathbb{Z}_m; \mathbb{Z}) = 0$, $''e_2$ is onto (by the exact sequence (15.2.2)). This implies that $\mathbb{Z}_m$ on $S^2$ lifts to any one of the principal $S^1$-bundles $P_r$ over $S^2$ with $r$ denoting the Euler class of the bundle. Let $''e_2(\gamma) = r$. Since $''e_2$ splits, $Le(\gamma) = r \otimes \mathbb{R} = r$, where $L = \gcd(m,m) = m$. There are exactly $m$ distinct such $\gamma$ that lift, up to strict equivalence, to $P_r$. They are given by

(15.3.1)
$$\gamma = \langle 0, \beta, r - \beta \rangle = \langle (1,0), (m,\beta), (m, r-\beta) \rangle = \{0; (1,0), (m,\beta), (m, r-\beta)\}$$

with $0 \leq \beta < m$.

(a) For lifted $\mathbb{Z}_m$ to act freely on $P_r$, $\gcd(m, \beta) = \gcd(m, r-\beta) = 1$. If $m$ is even, $\beta$ must be odd and $r$ must be even. In particular, the lift of $\mathbb{Z}_m$ for $\gamma = \langle (1,0), (m,1), (m,1) \rangle$ acts freely on $P_2$ and $(S^1, M_\gamma) = \{0; (1,0), (m,1), (m,1)\} = (\mathrm{SO}(2), \mathbb{Z}_m \backslash P_2)$. Similarly we have $(S^1, M_\gamma) = \{o;\ (1,0), (4,9), (4,-5)\} = (S^1/\mathbb{Z}_4, \mathbb{Z}_4 \backslash P_4)$. The center $C = S^1$ of $U(2)$ descends to $C_4 = S^1/\mathbb{Z}_4$ and acts freely on $P_4$ and commutes with the $Q = \mathbb{Z}_4$-action on $P_4$. Because $C_4 \cap \mathbb{Z}_4 = 1$, the $C_4$-action descends isomorphically to an $S^1$-action on $M_\gamma$. This action yields the orbit invariants $\langle (1,0), (4,9), (4,-5) \rangle$. As a space $|M_\gamma|$ is homeomorphic to the lens space $-L(16,-3) = L(16,3)$. As an exercise, explain what happens for $\gamma = \langle (1,0), (4,9), (4,-7) \rangle$.

If $m$ is odd, the lift of $\mathbb{Z}_m$ for $\gamma = \langle (1,0), (m,2), (m,-1) \rangle$ acts freely on $P_1$, while the lift $\gamma' = \langle (1,0), (m,1), (m,1) \rangle$ acts freely on $P_2$.

(b) Observe that $(\mathrm{SO}(2), \mathrm{SO}(3)) \to S^2 = \mathrm{SO}(3)/\mathrm{SO}(2)$ is the bundle $P_2$ as well as the unit tangent bundle for $S^2$. The rotation $\mathbb{Z}_m$ of $S^2$ lifts to the unit tangent bundle by the differential and commutes with the projection of the unit tangent vectors. This generates a left-right action which we can convert to a left action by $(\lambda, z) \times x \mapsto \lambda x z^{-1}$. As verified in Theorem 14.8.8, this corresponds to $\gamma = \langle (1,0), (m,1), (m,1) \rangle$ and $\mathbb{Z}_m \backslash P_2 \to \mathbb{Z}_m \backslash \mathrm{SO}(3)/\mathrm{SO}(2)$. (Caution: Our $\beta$ here is the $-\beta$ of Chapter 14.) That is, $(S^1, M_\gamma) = (S^1, \mathbb{Z}_m \backslash P_2)$ has invariants $\{(1,0), (m,1), (m,1)\}$, where the $S^1$ is the descent of the $\mathrm{SO}(2)$ acting on $\mathrm{SO}(3) = P_2$ to $\mathbb{Z}_m \backslash P_2$. With this particular fibering, $\mathbb{Z}_m \backslash P_2$ is homeomorphic to the lens space $-L(2m, 1) = L(2m, 2m-1)$.

(c) As $H^2(\mathbb{Z}_m; \mathcal{Z}) = \mathbb{Z} \oplus \mathbb{Z}_m$, generators for $\mathbb{Z} \oplus \mathbb{Z}_m$ can be chosen to be

$$g_1 = \langle (1,0), (m,2), (m,-1) \rangle \text{ and}$$
$$g_T = \langle (1,0), (m,1), (m,-1) \rangle.$$

(To get a generator $g$, we need to find $\gamma$ such that $e(\gamma) = \frac{1}{L}$, $L = \mathrm{lcm}(m,m) = m$. So $e(g_1) = \frac{1}{L}$. The torsion elements consist of $\gamma$ such that $e(\gamma) = 0$. Now a generator for $\mathbb{Z}_m$ is $g_T$ and note that $\{ng_T : 1 \leq n \leq m\}$ spans this group.) For $\gamma$ with $''e_2(\gamma) = r$, there exists $s$ so that $rg_1 + sg_T = \gamma$, $0 \leq s < m$. For example, $\gamma = \langle (1,0), (m,7), (m,-5) \rangle = 2g_1 + 3g_T$. $M_\gamma = \mathbb{Z}_m \backslash P_2$, while $2g_1 - 3g_T = \langle (1,0), (m,1), (m,1) \rangle$. It is possible to get a Seifert fibering for $\beta g_T = \gamma = \langle (1,0), (m,\beta), (m,-\beta) \rangle$. Then $e(\gamma) = 0$ and $''e_2(\gamma) = 0$ giving a Seifert fibering modeled on $S^1 \times S^2$. Of course, this fibering exhibits $S^1 \times \mathbb{R}^2$ geometry and is not a spherical space form. In each case a free action, when $(m, \beta) = 1$, results in a Seifert fibering with $(S^1, M) = (S^1, S^1 \times_{\mathbb{Z}_m} S^2)$, homeomorphic to $S^1 \times S^2$.

(d) Consider a Seifert manifold $M$ with Seifert invariants $\{g = 0; (1,0), (\alpha_1, \beta_1), (\alpha_2, \beta_2)\}$ and with $\gcd(\alpha_i, \beta_i) = 1$. The base space of the associated Seifert fibering is a "bad" orbifold if and only if $\alpha_1 \neq \alpha_2$. That is, there is no $Q$-action on $S^2$ that realizes this orbifold. However, by taking $m = \text{lcm}(\alpha_1, \alpha_2)$, we can construct a lift of $\mathbb{Z}_m$ on $S^2$ to a bundle $P_r$ so that the underlying space of the orbifold $\mathbb{Z}_m \backslash P_r$, $|\mathbb{Z}_m \backslash P_r|$, is the original Seifert manifold $M$. To see this, let $\alpha_1 = \alpha_1' d$, $\alpha_2 = \alpha_2' d$, where $d$ is the $\gcd(\alpha_1, \alpha_2)$. Put $m = \text{lcm}(\alpha_1, \alpha_2) = \alpha_1' \alpha_2' d$. Then consider $(m, \widetilde{\beta}_1) = (\alpha_1 \alpha_2', \beta_1 \alpha_2')$, and $(m, \widetilde{\beta}_2) = (\alpha_2 \alpha_1', \beta_2 \alpha_1')$. Let $\widetilde{\gamma} = \langle (1,0), (m, \widetilde{\beta}_1), (m, \widetilde{\beta}_2) \rangle$. Note $\widetilde{\gamma}$ reduced is $\langle (1,0), (\alpha_1, \beta_1), (\alpha_2, \beta_2) \rangle$. The action of $\mathbb{Z}_m$ lifts to $P_r$, where $r = me(\widetilde{\gamma}) = \widetilde{\beta}_1 + \widetilde{\beta}_2$. Thus, $M(\widetilde{\gamma}) = \mathbb{Z}_m \backslash P_r$ and $|M(\widetilde{\gamma})| = M$. In this way, even Seifert fiberings with bad orbifold base are captured by this technique. That is, they are cyclic quotients of bundle automorphisms of principal bundles over $S^2$.

Observe that if $M$ has Seifert invariants $\{g = 0; (1,0), (\alpha, \beta)\}$, $\gcd(\alpha, \beta) = 1$, then we can choose $P_r = L(\beta, 1)$ over $S^2$ with a lift $\widetilde{\gamma}$ of the $\mathbb{Z}_\alpha$-action on $S^2$ to $P_r$. Then $\widetilde{\gamma} = \langle (1,0), (\alpha, \beta), (\alpha, 0) \rangle$ and $\mathbb{Z}_\alpha \backslash P_r = M(\widetilde{\gamma})$ with $|M(\widetilde{\gamma})| = M$.

(e) Exercise. In (d), the action of $\mathbb{Z}_m$ over $P_r$ is not free. Show

(i) $H_1(|\mathbb{Z}_m \backslash P_r|) = H_1(M) \cong \mathbb{Z}_{rd}$.
(ii) The $\text{image}(\text{ev}_*^x(H_1(S^1)) \subset H_1(M))$ is $\mathbb{Z}_r \subset \mathbb{Z}_{rd}$.
(iii) $(S^1, |M|)$ can be lifted to the unbranched $d$-fold covering space $\mathbb{Z}_{\alpha_1' \alpha_2'} \backslash P_r = M'$.
(iv) The lift of an orbit in $M$, with isotopy $\mathbb{Z}_{\alpha_i}$ is an orbit of $M'$ with isotopy $\mathbb{Z}_{\alpha_i'}$.

For (i) and (ii), see Proposition 14.9.1 and Exercise 14.9.3(c).

(f) Example. $\{(1,0), (3,2), (2,-1)\}$ represents a Seifert fibering of $S^3$ over the bad orbifold $\{g = 0; 3, 2\}$ where the typical fiber is the $(3,2)$ torus knot. For $\widetilde{\gamma}$, choose $\langle (1,0), (6,4), (6,-3) \rangle$. Then $\mathbb{Z}_6$ lifts to $P_1 = S^3$ and $\mathbb{Z}_6$ acts (not freely) commuting with the Hopf fibering. The rotation $\mathbb{Z}_6$ leaves those fibers over the poles invariant while it permutes freely the remaining fibers. $\mathbb{Z}_6 \backslash S^3$ is homeomorphic to $S^3$, and the reduced Seifert fibering has typical fiber the $(3,2)$-torus knot.

(g) Let $M$ be written using unnormalized invariants in the manner of Chapter 14. Then, $M = \{g = 0, (\alpha_1, \beta_1), (\alpha_2, \beta_2)\}$ with $(\alpha_i, \beta_i) = 1$, is a lens space, $L(p, q)$, where $p = \alpha_1 \beta_2 + \alpha_2 \beta_1 = \alpha_1 \alpha_2(-e(M))$, $q = x\alpha_2 + y\beta_2$ such that $1 = x\alpha_1 - y\beta_1$; see [**Orl72**, p.100].

Orlik writes $x\alpha_2 - y\beta_2$, but we believe the minus sign should be a plus sign. If one takes our same formula using the numerical invariants of this chapter's $(\alpha_i, \beta_i)$ instead of the equivalent $(\alpha_i, -\beta_i)$ of Chapter 14, the computation gives the correct $q$ but the negative of the correct $p$. So using the above formula for $p$ and $q$ but using this chapter's invariants gives the lens space $-L(p, q)$.

Finally, we observe that only $L(p, q)$ has an infinite number of distinct Seifert fiberings since $L(p, q)$ admits an effective linear action of the 2-dimensional torus. Each circle subgroup has a finite cover which acts linearly on $S^3$ and must commute with the free covering action of $\pi_1(L(p,q)) \cong \mathbb{Z}_p$ on $S^3$. Each such circle then gives a unique set of normalized Seifert invariants up to equivalence and orientation. Of course, $L(p, q)$ has s Seifert fibering of the form $\langle (1,0), (m, \beta), (m, \gamma - \beta) \rangle$, where $m - \gamma = p$. Note $L(1,1) = S^3$ has the form $\langle (1,0), (1, \beta), (1, 1-\beta) \rangle = \langle (1, 1) \rangle$.

EXERCISE 15.3.3. (a) Let $\gamma = \langle (1,0), (11,3), (11,4) \rangle$. Show that $Q = \mathbb{Z}_{11}$ lifts to $P_7$ such that $\mathbb{Z}_{11} \backslash P_7 = M_\gamma$ and $(\mathbb{Z}_{11}, S^2)$ has an extended lifting to $\Pi$ on $P_1$, where $\Pi = \mathbb{Z}_{77}$ and $|M_\gamma| = -L(77, -27)$.

(b) Show on $P_2$ there is an $S^1$-action with invariants $\langle(1,0),(6,-1),(4,1)\rangle$ with a double covering lifting to the 3-sphere with invariants $\langle(1,0),(3,-1),(2,1)\rangle$. In both cases, the quotient spaces are bad orbifolds.

15.3.4 ($Q = D_{4m} = \mathbb{Z}_{2m} \rtimes \mathbb{Z}_2$, dihedral group). We have

$$0 \longrightarrow \mathbb{Z}_2 \oplus \mathbb{Z}_2 \xrightarrow{''e_1} H^2(Q;\mathcal{Z}) \xrightarrow{''e_2} \mathbb{Z} \longrightarrow \mathbb{Z}_2 \longrightarrow 0$$

so $H^2(Q;\mathcal{Z}) \cong \mathbb{Z} \oplus \mathbb{Z}_2 \oplus \mathbb{Z}_2$. We choose generators

$$g_1 = \langle(1,0),(2,1),(2,-1),(2m,1)\rangle \text{ with } e(g_1) = 1/2m = 1/L,$$
$$g_{T_1} = \langle(1,0),(2,1),(2,-1),(2m,0)\rangle,$$
$$g_{T_2} = \langle(1,0),(2,-1),(2,0),(2m,m)\rangle.$$

Any $\gamma \in H^2(Q;\mathcal{Z})$ is of the form $\gamma = rg_1 + \gamma_T$ where $\gamma_T$ is a torsion element. (Taking $r \geq 0$, we get all the $\gamma$ with $e(\gamma) \geq 0$.) This $\gamma$ represents a lift of $D_{4m}$ to $P_{2r}$. If $\gamma_T \neq 0$, then $D_{4m}$ cannot act freely on $P_{2r}$ because at least one of the $\beta_i$ in $\gamma$ will be even. Consequently, $\gamma$ represents a free action of $D_{4m}$ on $P_{2r}$ if and only if $\gamma = rg_1$ and $\gcd(r,2m) = 1$.

The action of $Q$ on $P_2$ for $g_1$ is that of $D_{4m} \subset SO(3)$ acting by left translation on $SO(3) \approx P_2$ and commuting with $SO(2) \subset SO(3)$ acting as right translations. We get the following diagram for $\gamma_{g_1}$ and $\gamma_{rg_1}$.

$$\begin{array}{ccc}
(\mathbb{Z}_r \times Q, P_2 = SO(3), SO(2)) \xrightarrow{=} (\mathbb{Z}_r \times Q, P_2 = SO(3), SO(2)) \xrightarrow{/SO(2)} SO(3)/SO(2) = S^2 \\
\downarrow Q\backslash \qquad \qquad \downarrow /\mathbb{Z}_r \subset SO(2) \qquad \qquad \downarrow Q\backslash \\
(Q\backslash P_2, SO(2)) \qquad (Q, P_{2r}, SO(2)/\mathbb{Z}_r) \longrightarrow Q\backslash S^2 \\
\downarrow \mathbb{Z}_r \subset SO(2) \qquad \qquad \downarrow Q\backslash \qquad \qquad \downarrow = \\
(Q\backslash P_2/\mathbb{Z}_r, SO(2)/\mathbb{Z}_r) \xrightarrow{=} (Q\backslash P_2/\mathbb{Z}_r, SO(2)/\mathbb{Z}_r) \xrightarrow{/(SO(2)/\mathbb{Z}_r)} Q\backslash S^2
\end{array}$$

The right translations regarded as left action intersects the left $Q$-action only in the identity. Therefore, we may factor the $\mathbb{Z}_r \times Q$-action on $P_2$ either by $Q$ followed by $\mathbb{Z}_r \subset SO(2)$ or $\mathbb{Z}_r$ followed by $Q$. The $\mathbb{Z}_r$-action on $P_2$ is free and if $\gcd(r,2m) = 1$, the $\mathbb{Z}_r$-action on $Q\backslash P_{2r}$, is free. The induced $Q$-action on $P_2$, guaranteed by $''e_2(\gamma) = 2r$, is also free. Therefore, $\gamma$ gives rise to a spherical space form if and only if $\gamma = rg_1$ for some $r$ with $\gcd(r,2m) = 1$. We note

$$0 \to \mathbb{Z}_2 \to \pi_1(Q\backslash SO(3)) \to D_{4m} \to 1,$$

where $\pi_1(Q\backslash SO(3))$ is the binary dihedral group, $D^*_{2\cdot 4m}$, which is an extension of $\mathbb{Z}_2$ by $Q$. Therefore,

$$0 \to \mathbb{Z}_{2r} \to \pi_1(Q\backslash SO(3)) \to Q \to 0$$

is a central extension of $\mathbb{Z}_{2r}$ by $Q$ and is isomorphic to $\mathbb{Z}_r \times \pi_1(Q\backslash SO(3)) = \mathbb{Z}_r \times$ (binary dihedral group of order $8m$). That is, $\mathbb{Z}_r \times Q$ acts freely on $SO(3)$ and its extended lifting to $S^3$ is $\mathbb{Z}_r \times D^*_{2\cdot 4m}$. Note, if $m = 1$, we get the example of Example 4.5.15.

15.3.5 ($Q = \mathbb{Z}_{2m+1} \rtimes \mathbb{Z}_2 = D_{4m+2}$, $m \geq 1$). We have $H^2(Q;\mathbb{Z}) = \mathbb{Z}_2$ and $H^3(Q;\mathbb{Z}) = 0$, so $H^2(Q;\mathcal{Z}) = \mathbb{Z} \oplus \mathbb{Z}_2$. Therefore, there exist two distinct lifts of $Q$

to each $P_r$. They are $rg_1 + sg_T$ with $s = 0$ or $1$, where
$$g_1 = \langle (1,0), (2,0), (2,-1), (2m+1, m+1) \rangle,$$
$$g_T = \langle (1,0), (2,1), (2,-1), (2m+1, 0) \rangle.$$

Hence, $rg_1 + g_T = \langle (1,0), (2,1), (2,-(r+1)), (2m+1, r(m+1)) \rangle$. We note that the action of $Q$ on $P_r$ is free if and only if $\gamma = rg_1 + g_T$, $r \neq 0$ is even, and $\gcd(2m+1, r(m+1)) = 1$. In this case, the bundle $P_r$ supports a unique lift up to strict equivalence for which $Q$ acts freely.

Let $\gamma = rg_1 + g_T$ with $r = 2^n k$, $k$ odd and $n > 0$. We note that $\gamma$ can be rewritten as
$$\gamma = \langle (1,0), (2,1), (2,-1), (2m+1, 2^{n-1}k) \rangle, \ n > 0, \ k \text{ odd}.$$

Then $Q$ acts freely on $P_{2^n k}$ if and only if $\gcd(2m+1, k) = 1$. For, $n = 1$, $k = 1$, we get the left-right action of $(Q, \mathrm{SO}(3) = P_2, \mathrm{SO}(2))$ and $\pi_1(Q \backslash P_2) = Q^*_{2^2(2m+1)}$, the binary dihedral group. With $n = 1$, and $\gcd(2m+1, k) = 1$ $\pi_1(Q \backslash P_{2k})$ is isomorphic to $\mathbb{Z}_k \times D^*_{4(2m+1)}$.

To see this clearly for $\gamma = \langle (1,0), (2,1), (2,-1), (2m+1, 1) \rangle$, we have the actions $(Q, P_2 = \mathrm{SO}(3), \mathrm{SO}(2))$ with $Q$ a lattice acting as left translations and $\mathrm{SO}(2)$ as right translations and with $\mathrm{SO}(2) \cap Q = 1$. So by uniqueness, this must be the lift of $Q$ on $S^2$ to $P_2$ that acts freely. Now $k\gamma = \langle (1,0), (2,k), (2,-k), (2m+1, k) \rangle$. This represents the action of $\mathbb{Z}_k \subset \mathrm{SO}(2)$ on $Q \backslash P_2$ and will have Seifert invariants for $M = ((Q \backslash P_2)/\mathbb{Z}_k, \mathrm{SO}(2)/\mathbb{Z}_k = S^1)$ as $M = \{(1,0), (2,k), (2,-k), (2m+1, k)\}$. This action of $\mathbb{Z}_k$ is free if $k$ is odd and coprime to $2m+1$. If free, then $\pi_1(M)$ is a central extension of $\mathbb{Z}_k$ by $\pi_1(Q \backslash P_2) = D^*_{4(2m+1)}$. In fact, it is a product $\mathbb{Z}_k$ and $D^*_{4(2m+1)}$ since $\mathbb{Z}_k \cap Q = 1$ and hence $\mathbb{Z}_k \cap D^*_{4(2m+1)} = 1$.

On the other hand, we can factor first with $\mathbb{Z}_k \subset \mathrm{SO}(2)$; $(Q, \mathrm{SO}(3), \mathbb{Z}_k \subset \mathrm{SO}(2)) \to (Q, \mathrm{SO}(3)/\mathbb{Z}_k = P_{2k}, \mathrm{SO}(2)/\mathbb{Z}_k = S^1)$. We have a lift of $Q$ to $P_{2k}$ that acts freely if $rg_1 + g_T = \langle (1,0), (2,1), (2,-(2k+1)), (2m+1, 2k(m+1)) \rangle = \langle (1,0), (2,1), (2,-1), (2m+1, k) \rangle$. That is, on $P_{2k}$, we have $Q$ acting freely so that its Seifert invariants are the same as $M$. The fundamental group is now a central extension $0 \to \mathbb{Z}_{2k} \to \pi_1(M) \cong \mathbb{Z}_k \times D^*_{4(2m+1)} \to Q \to 1$. Note that $\mathbb{Z}_{2k}$ is the entire center of $\pi_1(M)$.

In general, for $\gamma = \langle (1,0), (2,1), (2,-1), (2m+1, 2^{n-1}k) \rangle$, $Q$ acts freely on $P_{2^n k}$, $k$ odd and $\pi_1(Q \backslash P_{2^n k})$ is a central extension of $\mathbb{Z}_{2^n k}$ by $D_{2(2m+1)}$ as in the preceding. We put $D'_{2^{n+1}(2m+1)} = \pi_1(Q \backslash P_{2^n})$, $Q$ acting freely. Then, as above, $\pi_1(Q \backslash P_{2^n k}) = \mathbb{Z}_k \times D'_{2^{n+1}(2m+1)}$. A presentation for $D'_{2^{n+1}(2m+1)}$ is
$$\{x, y \mid x^{2^{n+1}} = 1, \ y^{2m+1} = 1, \ xyx^{-1} = y^{-1}\}.$$

For example, if $4g_1 + g_T = \gamma = \langle (1,0), (2,1), (2,-1), (2m+1, 2) \rangle$, then $e(\gamma) = 2/(2m+1)$ and $Le(\gamma) = 4$. Therefore $Q$ lifts to $P_4$ and acts freely on $P_4$. However, $Q \backslash P_4$ is not $(Q \backslash P_2)/\mathbb{Z}_2 = 4g_1 = \langle (1,0), (2,0), (2,0), (2m+1, 2) \rangle$ $(\mathbb{Z}_2 \subset \mathrm{SO}(2))$, since $\mathbb{Z}_2$ does not act freely on $Q \backslash P_2$. The group $\pi_1(Q \backslash P_4)$ is a central extension of $\mathbb{Z}_4$ by $Q$ and is isomorphic to $D'_{8(2m+1)}$. Incidentally, we note that $D'_{4(2m+1)} = D^*_{4(2m+1)}$.

15.3.6 ($T \cong A_4$, the tetrahedral group). We have $H^2(T, \mathbb{Z}) = \mathbb{Z}_3$, $H^3(T; \mathbb{Z}) = \mathbb{Z}_2$. Consequently, $H^2(Q; \mathcal{Z}) = \mathbb{Z} \oplus \mathbb{Z}_3$ and $''e_2(\gamma) \in H^2(S^2; \mathbb{Z}) \cong \mathbb{Z}$ is even. There are three distinct lifts of $Q$ to each $P_{2r}$. They are $rg_1 + sg_{T_1} = rg_1 + g_{T_s}$ with $s = 0, 1$ or $2$, where

$$g_1 = \langle (1,0), (2,-1), (3,1), (3,1) \rangle = \langle (1,-1), (2,1), (3,1), (3,1) \rangle,$$
$$g_{T_1} = \langle (1,0), (2,0), (3,1), (3,-1) \rangle = \langle (1,-1), (2,0), (3,1), (3,2) \rangle,$$
$$g_{T_2} = \langle (1,0), (2,0), (3,-1), (3,1) \rangle = \langle (1,-1), (2,0), (3,2), (3,1) \rangle.$$

Consequently, $\gamma = rg_1 + g_{T_s} = \langle (1,0), (2,-r), (3, r+s), (3, r-s) \rangle$, $s = 0, 1, 2$. Note $e(\gamma)L = e(\gamma) \cdot 6 = r$. So for $\gamma$, $T = Q$ lifts to bundle isomorphisms on $P_{2r}$. This action will be free if and only if $r$ is odd and $\gcd(r+s, 3) = \gcd(r-s, 3) = 1$.

Let $r = 3^n k$, where $\gcd(k, 3) = 1$, $n \geq 0$. For $n = 0$, we have $g_r$, $g_r + g_{T_1}$ and $g_r + g_{T_2}$. $T$ acts freely on $P_{2r}$ only if $\gamma = g_r$. For $g_1$, we get the left-right action $(T, \mathrm{SO}(3) = P_2, \mathrm{SO}(2))$ and $\pi_1(T \backslash \mathrm{SO}(3)) = T^*$ the binary tetrahedral group, which is a central extension of $\mathbb{Z}_2$ by $T$. Therefore, if $\gamma = rg_1 = \langle (1,0), (2,-r), (3, r), (3, r) \rangle$, then $T \backslash P_{2r} = T \backslash \mathrm{SO}(3)/\mathbb{Z}_r$, ($\mathbb{Z}_r \subset \mathrm{SO}(2)$). This action is free if and only if $\gcd(r, 6) = 1$ and then as in (iii),

$$\pi_1(Q \backslash P_{2r}) \cong \mathbb{Z}_r \times T^*.$$

Now let $n \geq 1$. Then,

$$\gamma = rg_1 + g_{T_1} = \langle (1,0), (2, -3^n k), (3, 3^n k + 1), (3, 3^n k - 1) \rangle \text{ and}$$
$$\gamma' = rg_1 + g_{T_2} = \langle (1,0), (2, -3^n k), (3, 3^n k - 1), (3, 3^n k + 1) \rangle$$

yield strictly inequivalent actions of $T$ on $P_{2r}$. These $S^1$-actions $\gamma$ and $\gamma'$ on $Q \backslash P_{2r}$ are strictly inequivalent but are $S^1$-equivalent. For $\gamma'' = rg_1 + g_{T_0}$, the action of $T$ on $P_{2r}$ is not free.

For $k = 1$, the action of $T$ for $\gamma$ and $\gamma'$ are free and $\pi_1(|Q \backslash P_{2 \cdot 3^n}|)$ has order $(2 \cdot 3^n)(2^2 \cdot 3) = 2^3 3^{n+1} = 8 \cdot 3^{n+1}$. Put $\pi_1(|Q \backslash P_{2 \cdot 3^n}|) = T'_{8 \cdot 3^{n+1}}$. It is a central extension of $\mathbb{Z}_{2 \cdot 3^n}$ by $T$. On $(S^1, Q \backslash P_{2 \cdot 3^n})$, we may divide out by the $\mathbb{Z}_k \subset S^1$. This acts freely if $\gcd(k, 6) = 1$ and corresponds to $k\gamma'$ and $k\gamma''$. Consequently, we have $\pi_1(T \backslash P_{2 \cdot 3^n k}) \cong \pi_1((T \backslash P_{2 \cdot 3^n k})/\mathbb{Z}_k) \cong \mathbb{Z}_k \times T'_{8 \cdot 3^{n+1}}$ which is isomorphic to a central extension of $\mathbb{Z}_{2 \cdot 3^n k}$ by $T$. A presentation for $T'_{8 \cdot 3^{n+1}}$, with $n \geq 1$ is given by

$$\pi_1(T \backslash P_{2r}) = T'_{8 \cdot 3^{n+1}}$$
$$= \{x, y, z \mid x^2 = (xy)^2, zxz^{-1} = y, zyz^{-1} = xy, z^{3^{n+1}} = 1\}.$$

If $n = 0$, $T'_{8 \cdot 3} = T^*$, the binary tetrahedral group.

15.3.7 ($Q = O \cong S_4$, the octahedral group). We have $H^2(Q, \mathbb{Z}) = \mathbb{Z}_2$, $H^3(Q; \mathbb{Z}) = \mathbb{Z}_2$. Consequently, $H^2(Q; \mathcal{Z}) = \mathbb{Z} \oplus \mathbb{Z}_2$ with $''e_2(\gamma)$ even for $\gamma \in H^2(Q; \mathcal{Z})$. Since $L = 12$, we have for generators of $H^2(Q; \mathcal{Z})$

$$g_1 = \langle (1,0), (2,-1), (3,1), (4,1) \rangle,$$
$$g_T = \langle (1,0), (2,-1), (3,0), (4,2) \rangle.$$

Therefore,

$$rg_1 = \langle (1,0), (2,-r), (3,r), (4,r) \rangle,$$
$$rg_1 + g_T = \langle (1,0), (2,-(r+1)), (3,r), (4, r+2) \rangle.$$

For $\gamma = rg_1$, $O$ acts freely on $P_{2r}$ if and only if $\gcd(r, 12) = 1$. For $\gamma = rg_1 + g_T$, $O$ never acts freely on $P_{2r}$. For $g_1$, we get the left-right action of $O$, ($O$, $\mathrm{SO}(3) = P_2$, $\mathrm{SO}(2)$) on $P_2$. Consequently, for $rg_1$, the $O$-action lifts to $P_{2r}$ with $O \backslash P_{2r} =$

$(O\backslash P_2)/\mathbb{Z}_r$, $(\mathbb{Z}_r \subset \mathrm{SO}(2))$ This action of $O$ is free if and only if $\gcd(r,6) = 1$. Since $\pi_1(O\backslash \mathrm{SO}(3)) = \{x,y|\ x^4 = 1, x^2 = (xy)^3 = y^4\} = O^*$, the binary octahedral group, then $\pi_1(O\backslash P_{2r}) = O^* \times \mathbb{Z}_r$ when $\gcd(r,6) = 1$, and the action of $O$ is free on $P_{2r}$. As an example, consider $\gamma = \langle (1,0), (2,-1), (3,2), (4,1) \rangle$. Note, $e(\gamma) = 5/12$ and $\gamma = 5g_1$.

15.3.8 ($Q = I \cong A_5$, the icosahedral group). We have $H^2(Q, \mathbb{Z}) = 0$, $H^3(Q; \mathbb{Z}) = \mathbb{Z}_2$ and so $H^2(Q; \mathcal{Z}) = \mathbb{Z}$ with $''e_2(\gamma)$ even for $\gamma \in H^2(Q; \mathcal{Z})$. Since $L = 30$, we have the generator for $H^2(Q; \mathcal{Z})$,

$$g_1 = \langle (1,0), (2,-1), (3,1), (5,1) \rangle.$$

For $\gamma = rg_1 = \langle (1,0), (2,-r), (3,r), (5,r) \rangle$ $I$ acts freely on $P_{2r}$ if and only if $\gcd(r,30) = 1$. For $g_1$, the left-right action of $I$, $(I, \mathrm{SO}(3) = P_2, \mathrm{SO}(2))$ is free. Consequently, for $\gamma = rg_1$, the $I$-action lifts to bundle automorphisms on $P_{2r}$ with $I\backslash P_{2r} = (I\backslash P_2)/\mathbb{Z}_r$, $(\mathbb{Z}_r \subset \mathrm{SO}(2))$. This action of $I$ is free if and only if $\gcd(r,30) = 1$. Since $\pi_1(I\backslash \mathrm{SO}(3)) = I^* = \{x,y|\ x^4 = 1, x^2 = (xy)^3 = y^5\} \cong \{q_1, q_2, q_3, h|\ q_1 q_2 q_3 = 1, q_1^2 h = q_2^3 h = q_3^5 h = 1, h\ \text{central}\}$, then $\pi_1(I\backslash P_{2r}) = I^* \times \mathbb{Z}_r$ when $\gcd(r,30) = 1$.

15.3.9. The solution to the 3-dimensional spherical space form problem was sketched by H. Hopf [**Hop26**] and completed by H. Seifert and W. Threlfall [**ST31**]. See [**Orl72**] for a nice argument.

The space form problems go back to the nineteenth century especially in the works of Clifford and Klein. Chapter 9 discussed the flat case of 0 curvature. The solution to the spherical space form problem in higher dimensions involves, as mentioned in Chapter 4, the work of many group theorists and geometers.

The groups $G$ that act freely and unitarily on the unit sphere $S^{2n-1}$ in $\mathbb{C}^n$ are divided into six types. The first four types are solvable. Type 5 is $K \times \mathrm{SL}(2,5)$ with $K$ of type 1 with $(|K|, 30) = 1$. $G$ in type 6 is an extension of $K \times \mathrm{SL}(2,5)$ by $\mathbb{Z}_2$. Type 2 is a $\mathbb{Z}_2$ extension of type 1. Type 3 and type 4 are again extensions of type 1 groups by well-known groups. Thus, the type 1 groups play an important role. They include all the cyclic groups. In 1905, Burnside characterized these groups as follows.

THEOREM 15.3.10. [**Wol77**, (Burnside)(Zassenhaus) p.163, §5.4.1]. *Let $G$ be a finite group of order $N$ in which every Sylow subgroup is cyclic. Then $G$ is generated by two elements $A$ and $B$ such that*

$$A^m = B^n = 1,\ BAB^{-1} = A^r,\ N = mn,\ ((r-1)n, m) = 1,\ r^n \equiv 1 (\mathrm{mod}\ m).$$

*The commutator subgroup $G' = \{A\}$ and the quotient $G/G' = \{BG'\} \cong \{B\}$ are cyclic. Let $d$ be the order of $r$ in the multiplicative group of residues modulo $m$ of integers prime to $m$. Then $d$ divides $n$ and $G$ satisfies all pq-conditions if and only if $\frac{n}{d}$ is divisible by every prime divisor of $d$. Conversely, any group given by the generator and relations above has order $N$ and has every Sylow subgroup cyclic.*

In this section, we found the classification (up to topological type) of the 3-dimensional spherical space forms as Seifert fiberings using the method explored in Subsection 4.6.3(5). It is of interest to compare this procedure in dimension 3 with the classification by representations for $G$ of type 1.

For a cyclic group $\mathbb{Z}_n$, the presentation just becomes $\{B : B^n = 1\}$. Any free representation is a direct sum of irreducible free 1-dimensional unitary representations each of the form $e^{\frac{2\pi i k}{n}}$, $(k,n) = 1$. So $d = 1$, and for $U(2)$, we have representations

$$\varphi_{k,\ell} : e^{\frac{2\pi i}{n}} \mapsto \begin{bmatrix} e^{\frac{2\pi i k}{n}} & 0 \\ 0 & e^{\frac{2\pi i \ell}{n}} \end{bmatrix},$$

where $\gcd(n,k) = \gcd(n,\ell) = 1$. Choose $v$ such that $vk \equiv 1(n)$, then put $q \equiv v\ell(n)$. Then with the automorphism $\mathbb{Z}_n \to \mathbb{Z}_n$ such that $e^{\frac{2\pi i}{n}} \to e^{\frac{2\pi i v}{n}}$, the representation, modulo this automorphism, is equivalent to $\begin{bmatrix} e^{\frac{2\pi i}{n}} & 0 \\ 0 & e^{\frac{2\pi i q}{n}} \end{bmatrix}$, which is a standard representation for the lens space $L(p,q)$. The maximal torus consisting of the diagonal matrices of $U(2)$ commutes with $G = \mathbb{Z}_n$ and gives rise to an infinite number of distinct $S^1$-actions on $L(p,q)$. A finite number have fixed points while the rest give rise to Seifert fiberings with most of these over bad orbifolds.

Put $\mathbb{Z}_r = (\text{Center of } U(2)) \cap G$. Then, $G/\mathbb{Z}_r$ acts effectively on $S^2$. Consequently, one can find a locally injective Seifert fibering modeled on a principal $S^1$-bundle $P_r$ over $S^2$ with Euler class $r$ to represent $L(n,q)$. By the equality (15.3.1), $\gamma$ will need to be of the form $\langle (1,0), (\frac{n}{r}, \beta), (\frac{n}{r}, (r-\beta)) \rangle$, where $r = {}''e_2(\gamma)$ and $\beta$ can be calculated from 15.3.2(g). Note, if $r=1$, then $n$ must be odd. For example, if $L(n,q) = -L(7,5)$, then there is a locally injective Seifert fibering modeled on ${}''e_2(\gamma) = -1$ given by $\langle (1,0), (7,-2), (7,3) \rangle$.

The different representations $\varphi_{k,\ell}$ give rise to the different isometric classification of the $L(n,q)$ as $q$ varies with $(n,q) = 1$.

If $G$ is noncyclic of type 1 with $\varphi: G \to U(2)$ a free representation, then $d = 2$ and $\varphi$ is irreducible. From the presentation of $G$, we see that $n \equiv 0(4)$ and $m$ is odd. The center of $G$ is cyclic of order $\frac{n}{2}$ and is generated by $B^d$. By Subsection 11.8.10(4), $\langle B^d \rangle = G \cap \text{Center}U(2)$. Therefore, the image of $G$ under the orbit mapping $S^3 \to S^3/\text{Center}U(2) = S^2$, acts effectively as $G/\langle B^d \rangle \subset SO(3)$ on $S^2$. The image is dihedral isomorphic to $\mathbb{Z}_m \rtimes \mathbb{Z}_2$, $m$ odd, because $A$ maps isomorphically to $G/\langle B^d \rangle$ with index 2. Thus, as in Subsection 15.3.5, the action of $G$ on $S^3$ gives us a Seifert fibering modeled on the Hopf bundle. However, in this case, the action of $G$ on $S^2$ is ineffective. The effective action of $\mathbb{Z}_m \rtimes \mathbb{Z}_2$ on $S^2$ cannot be lifted to a group of free bundle automorphisms on $S^3$ since $\mathbb{Z}_m \rtimes \mathbb{Z}_2$ does not satisfy the $2p$ condition but it can be lifted to various bundles with even Euler class.

In terms of Seifert invariants, $\langle (1,b), (2,\beta_1), (2,\beta_2), (m,\beta_3) \rangle$ can be written as $\gamma = \langle (1,0), (2,1), (2,-1), (m,\beta) \rangle = (S^1, M)$. To determine $\beta$, we have from Burnside's theorem that $G/G' = G/\langle A \rangle = H_1(G; \mathbb{Z}) \cong \mathbb{Z}_n \cong \langle B \rangle$. Since $M = G \backslash S^3$, $G = \pi_1(M)$, $H_1(M; \mathbb{Z}) \cong H_1(G; \mathbb{Z}) \cong \mathbb{Z}_n$. Since $|H_1(M; \mathbb{Z})| = |(2 \cdot 2 \cdot m)e(\gamma)| = |4m\frac{\beta}{m}| = |4\beta| = n$. So $|\beta| = \frac{n}{4}$.

Put $n = 2^{a+1}k$ where $a > 0$, $k$ odd. Then $\gamma$, in the format used in Subsection 15.3.5, becomes

$$(S^1, M) = \langle (1,0), (2,1), (2,-1), (m, 2^{a-1}k) \rangle.$$

The group $Q = \mathbb{Z}_m \rtimes \mathbb{Z}_2 = D_{2m}$ acting on $S^2$ lifts to a group of bundle automorphisms on $P_{\text{even}}$. For $\gamma$ with $(k,m) = 1$, $D_{2m}$ acts freely on $\mathbb{Z}_{n/2} \backslash S^3 = P_{n/2} = P_{2^a k}$. Thus, $G = \pi_1(G \backslash S^3) = \pi_1(Q \backslash P_{n/2}) = \pi_1(M)$ and is a central extension of $\mathbb{Z}_{n/2}$ by $D_{2m}$. $G$ also splits as $\mathbb{Z}_k \times D'_{2^{a+1}m}$ where $D'_{2^{a+1}} = \pi_1(Q \backslash P_{2^a})$. The free irreducible representations of $G$ are given by

$$\varphi_{s,\ell}(A) = \begin{bmatrix} e^{\frac{2\pi i s}{m}} & 0 \\ 0 & e^{\frac{2\pi i s r}{m}} \end{bmatrix}, \quad \varphi_{s,\ell}(B) = \begin{bmatrix} 0 & 1 \\ e^{\frac{2\pi i \ell}{n/2}} & 0 \end{bmatrix},$$

where $(s, m) = 1$, $(\ell, n) = 1$, and $r$, $m$ and $n$ as in the presentations [**Wol77**, p.168].

One will notice that our association of $\pm M$ as a homeomorphism type for $\varphi_{s,\ell}(G)$ does not depend upon $s$ and $\ell$. This is because there is only one homeomorphism class up to orientation. We also ignored strict classification, by not fixing the position of the singularities of order 2, since permuting the orbits of multiplicity 2 will need an automorphism $\alpha$ of $G$ and result in $[\alpha\gamma] \neq [\gamma]$.

## 15.4. Seifert fiberings with $Q$ Euclidean crystallographic

**15.4.1.** For $Q$ orientation preserving Euclidean crystallographic, we have the following table.

|     | Signature of $Q$ | Group | $H^2(Q;\mathbb{Z}) \cong H^2(Q;\mathcal{Z})$ |
| --- | --- | --- | --- |
| (1) | $(1;)$ | $\mathbb{Z} \oplus \mathbb{Z}$ | $\mathbb{Z}$ |
| (2) | $(0; 2, 2, 2, 2)$ | $(\mathbb{Z} \oplus \mathbb{Z}) \rtimes \mathbb{Z}_2$ | $\mathbb{Z} \oplus \mathbb{Z}_2 \oplus \mathbb{Z}_2 \oplus \mathbb{Z}_2$ |
| (3) | $(0; 2, 4, 4)$ | $(\mathbb{Z} \oplus \mathbb{Z}) \rtimes \mathbb{Z}_4$ | $\mathbb{Z} \oplus \mathbb{Z}_2 \oplus \mathbb{Z}_4$ |
| (4) | $(0; 3, 3, 3)$ | $(\mathbb{Z} \oplus \mathbb{Z}) \rtimes \mathbb{Z}_3$ | $\mathbb{Z} \oplus \mathbb{Z}_3 \oplus \mathbb{Z}_3$ |
| (5) | $(0; 2, 3, 6)$ | $(\mathbb{Z} \oplus \mathbb{Z}) \rtimes \mathbb{Z}_6$ | $\mathbb{Z} \oplus \mathbb{Z}_6$ |

Since $Q$ is infinite, $W = \mathbb{R}^2$, and $P$ is the trivial bundle $S^1 \times \mathbb{R}^2$. $\mathbb{Z} \oplus \mathbb{Z}$ is a normal subgroup of $Q$ and so a lift of $Q$ to $S^1 \times \mathbb{R}^2$ restricts to a lift of $\mathbb{Z} \times \mathbb{Z}$ to $P$. Therefore cases (2) through (5) are the respective $\mathbb{Z}_2$, $\mathbb{Z}_4$, $\mathbb{Z}_3$, $\mathbb{Z}_6$ quotients of $(\mathbb{Z} \oplus \mathbb{Z}) \backslash (S^1 \times \mathbb{R}^2)$ of (1).

(1). $\gamma \in \mathbb{Z} = H^2(\mathbb{Z} \oplus \mathbb{Z}; \mathbb{Z})$. The $\gamma$ represents a lift of $\mathbb{Z} \oplus \mathbb{Z}$ acting freely on $\mathbb{R}^2$ to the product bundle $S^1 \times \mathbb{R}^2$. The quotient $(\mathbb{Z} \oplus \mathbb{Z}) \backslash S^1 \times \mathbb{R}^2$ is a principal $S^1$-bundle over $Q \backslash \mathbb{R}^2$, the 2-torus, with Euler class $r$. If $r = 0$, this is the 3-torus. Otherwise, it is a nilmanifold. See Example 4.5.12.

(2). For generators of $H^2(\mathbb{Z} \oplus \mathbb{Z} \rtimes \mathbb{Z}_2; \mathbb{Z})$, choose

$$\begin{aligned}
g_1 &= \langle (1,0), (2,1), (2,0), (2,0), (2,0) \rangle & e(g_1) &= 1/2, \\
g_{T_1} &= \langle (1,0), (2,1), (2,-1), (2,0), (2,0) \rangle & e(g_{T_1}) &= 0, \\
g_{T_2} &= \langle (1,0), (2,0), (2,0), (2,1), (2,-1) \rangle & e(g_{T_1}) &= 0, \\
g_{T_3} &= \langle (1,0), (2,1), (2,0), (2,-1), (2,0) \rangle & e(g_{T_1}) &= 0.
\end{aligned}$$

Note that $g_{T_1} + g_{T_2} = \langle (1,0), (2,1), (2,-1), (2,1), (2,-1) \rangle$. For $\gamma \in H^2(Q; \mathcal{Z})$, we have $e(\gamma) = m/2$ for some $m \in \mathbb{Z}$. Let

$$0 \to Q' \cong \mathbb{Z} \oplus \mathbb{Z} \xhookrightarrow{i} (\mathbb{Z} \oplus \mathbb{Z}) \rtimes \mathbb{Z}_2 = Q \to \mathbb{Z}_2 \to 0$$

be exact. Then $i^*(\gamma) = \gamma' \in H^2(Q'; \mathbb{Z})$ and its central extension

$$0 \to \mathbb{Z} \to N \to \mathbb{Z}^2 \to 0,$$

a torsion-free nilpotent group. Therefore, there is a possibly branched 2-fold fiber preserving covering of $M_\gamma = Q \backslash (S^1 \times \mathbb{R}^2)$ by $M_{\gamma'} = Q' \backslash (S^1 \times \mathbb{R}^2)$. Then $e(\gamma') = 2e(\gamma)$ by Section 14.7 or a simple cohomological argument. The group $\mathbb{Z}_2$ acts on $Q' \backslash (S^1 \times \mathbb{R}^2)$ as Seifert automorphisms. If $e(\gamma) = 0$, $Q \backslash (S^1 \times \mathbb{R}^2)$ is a $\mathbb{Z}_2$-quotient of the 3-torus, otherwise a $\mathbb{Z}_2$-quotient of a nilmanifold whose Euler class is $2e(\gamma)$. For example, $M_{g_1}$ is doubly branched covered by the nilmanifold whose Euler class is 1 and $M(2g_1 + g_{T_1} + g_{T_2})$ is freely doubly covered by the nilmanifold whose Euler class is 2.

If we write $\gamma = \langle (1, b), (2, \beta_1), (2, \beta_2), (2, \beta_3), (2, \beta_4) \rangle$, it is clear that all the $\beta_i$'s must be odd if $Q$ is to lift to a free action on $P$. Write $\gamma$ as $\angle (1,0), (2,1), (2,1), (2,1),$

$(2, \beta_4)\rangle$. Then $e(\gamma) = \frac{3}{2} + \frac{\beta_4}{2}$ and $Le(\gamma) = (3+\beta_4) = r$. Consequently, $r$ is even if $Q$ is to act freely. In general, $\gamma = rg_1 + g_T$ where $g_T$ is a torsion class and $Le(\gamma) = r$. It is easy to see that $Q$ acts *freely if and only if $r$ is even and $g_T = g_{T_1} + g_{T_2}$*. Consequently, for each integral value of $e(\gamma)$, there is one and only one lift of $Q$, namely that for $\gamma = rg_1 + g_{T_1} + g_{T_2}$, $r$ even, that acts freely on $S^1 \times \mathbb{R}^2$. Moreover, each of these $M_\gamma$ have distinct oriented homeomorphism type. These types are determined by $e(\gamma)$. In fact, $2^4 e(\gamma)$ is the order of the torsion of $H_1(M_\gamma; \mathbb{Z})$ provided $e(\gamma) \neq 0$. For $\gamma = g_{T_1} + g_{T_2}$, $M_\gamma$ is the flat manifold $\mathcal{G}_2 = S^1 \times_{\mathbb{Z}_2} T^2$, where $\mathbb{Z}_2$ acts by $z \mapsto -z$ on $S^1$ and $(z_1, z_2) \mapsto (\bar{z}_1, \bar{z}_2)$ on $T^2 = S^1 \times S^1$.

Now write $r = 2^s k$, $s \geq 1$, $k$ odd. Then $\gamma = rg_1 + g_{T_1} + g_{T_2} = 2^s k g_1 + (g_{T_1} + g_{T_2}) = k(2^s g_1 + (g_{T_1} + g_{T_2}))$ because $k$ is odd.

$$\gamma = 2^s k g_1 + (g_{T_1} + g_{T_2})$$
$$= \langle (1,0), (2, 2^s k + k), (2, -k), (2, k), (2, -k) \rangle$$
$$= \langle (1, 2^{s-1} k), (2, 1), (2, -1), (2, 1), (2, -1) \rangle.$$

Here we have $M_{2^s g_1 + (g_{T_1} + g_{T_2})} = \{(1, 2^{s-1}), (2,1), (2,-1), (2,1), (2,-1)\}$ and $M_\gamma = \mathbb{Z}_k \backslash M_{2^s g_1 + (g_{T_1} + g_{T_2})} = \{(1, 2^{s-1} k), (2,1), (2,-1), (2,1), (2,-1)\}$. $M_\gamma$ is doubly and freely covered by the nilmanifold whose Euler class is $2 \cdot 2^{s-1} k = 2^s k$.

Analogous to the factorizations in the spherical space forms, we have the following factorization.

$$\begin{array}{ccc}
(S^1, N_{2^s k}) & \xleftarrow{\mathbb{Z}_k \backslash} & ('S^1, N_{2^s}) \\
\downarrow /\mathbb{Z}_2 & & \downarrow /\mathbb{Z}_2 \\
(S^1, M_\gamma) & \xleftarrow{\mathbb{Z}_k \backslash} & ('S^1, M_{2^s g_1 + (g_{T_1} + g_{T_2})})
\end{array}$$

Here $N_m$ denotes the nilmanifold whose Euler class is $m$ and $'S^1$ is the group of the principal fiber of $N_{2^s}$ with $\mathbb{Z}_k \subset \, 'S^1$ and quotient $S^1 = \, 'S^1/\mathbb{Z}_k$. The action of the $\mathbb{Z}_k$ and $\mathbb{Z}_2$ are free.

(3)-(5). Analysis of the remaining three families is similar to that employed in (2). The Seifert fibering Subsection 15.4.1 are the only classical orientable Seifert fiberings over an orientable base which have an infinite solvable fundamental group (provided we exclude $S^1 \times S^2$ which arises in Subsection 15.3.1 (1) with $\gamma = \langle (1,0), (\alpha, \beta), (\alpha, -\beta) \rangle$, $\gcd(\alpha, \beta) = 1$). These manifolds with infinite solvable groups are all the 3-dimensional flat manifolds, nilmanifolds and infra-nilmanifolds.

REMARK 15.4.2. In Subsection 15.4.1, we have $Q = (\mathbb{Z} \times \mathbb{Z}) \rtimes \mathbb{Z}_m$, with $m = 1, 2, 3, 4, 6$, respectively. We may restrict any lifting of $Q$ to $S^1 \times \mathbb{R}^2$ to the normal subgroup $Q' = \mathbb{Z} \times \mathbb{Z}$. It acts freely yielding $M' = Q' \backslash (S^1 \times \mathbb{R}^2)$, a principal $S^1$-bundle over the 2-torus $T^2$. The group $\mathbb{Z}_m$ acts freely on $M'$ and is a lift of the induced $\mathbb{Z}_m$-action on $T^2$ from the $Q$-action on $\mathbb{R}^2$. So instead of analyzing the lifts of $Q$ to $S^1 \times \mathbb{R}^2$ we can just as well analyze the lifts $\mathbb{Z}_m$ to $M'$ in the spirit of Subsection 15.3.1. From the $'E$ spectral sequence of Exercise 10.5.3, we get for $m = 2$,

$$0 \to H^2(\mathbb{Z}_2 \backslash T^2; \mathbb{Z}) = \mathbb{Z} \xrightarrow{e_1} H^2(\mathbb{Z}_2; \mathcal{Z}) \xrightarrow{'e_2} \mathbb{Z}_2 \oplus \mathbb{Z}_2 \oplus \mathbb{Z}_2 \oplus \mathbb{Z}_2 \to 0,$$

and from the $''E$ spectral sequence of Theorem 10.3.12, we get for $m = 2$ the exact sequence

$$0 \to H^1(Q; \mathrm{M}(T^2, S^1)) = \mathbb{Z}_2 \oplus \mathbb{Z}_2 \oplus \mathbb{Z}_2 \xrightarrow{''e_1} H^2(\mathbb{Z}_2; \mathcal{Z}) \xrightarrow{''e_2} H^2(T^2; \mathbb{Z})^{\mathbb{Z}_2} = \mathbb{Z} \to 0.$$

In the latter computation, Lemma 10.1.7 is useful since $T^2$ is not simply connected.

Note that $H^2(\mathbb{Z}_2; \mathcal{Z}) \cong H^2((\mathbb{Z} \times \mathbb{Z}) \rtimes \mathbb{Z}_2; \mathcal{Z})$. This is not surprising since in [**CR72b**, (3.10)], it is shown that if

(1) $L \subset Q$ is normal and acts freely on $Q$ and
(2) $\varphi : L \to \mathrm{Aut}(\mathbb{Z}^k)$ is trivial,

then $H^*(Q/L; \mathcal{Z}^k) \cong H^*(Q; \mathcal{Z}^k)$.

The proof uses the identification of $H^*(Q; \mathcal{Z}^k)$ with the cohomology of the Borel space of $Q$. Therefore, we may study lifts of $\mathbb{Z}_2$ to principal $S^1$-bundles over $T^2$ instead of lifts of $(\mathbb{Z} \times \mathbb{Z}) \rtimes \mathbb{Z}_2$ to $S^1 \times \mathbb{R}$. Now our class $\gamma = 2^s k g_1 + (g_{T_1} + g_{T_2})$ represents a lift of $\mathbb{Z}_2$ to the principal $S^1$-bundle $N_{2^s k}$ over $T^2$ with Euler class $2^s k$ and $M_\gamma = \mathbb{Z}_2 \backslash P_{2^s k}$.

As an exercise, the reader may wish to work out the analysis of $H^2(Q; \mathcal{Z})$ where $Q$ satisfies (3) through (5) in Subsection 15.4.1 using the method of that section as well as the method suggested above.

## 15.5. Seifert fiberings with $Q$ hyperbolic

**15.5.1.** The signature of $Q$ is $(g; \alpha_1, \alpha_2, \ldots, \alpha_n)$ with all possibilities other than those in Subsections 15.3.1 and 15.4.1. The procedure for analysis is similar to that employed in Sections 15.3 and 15.4. In fact, $H^2(Q; \mathbb{Z})$ does not depend upon the genus of $Q \backslash W$. There are differences though. For example, $''e_2$ is always trivial as $W = \mathbb{R}^2$ and when $g = 0$, $n \leq 2$, the orbifold $Q \backslash W$ is always good. In Chapter 14, there are many applications when $Q$ is hyperbolic and acting freely. Again, we remind the reader that for the same manifold, the Seifert invariants $b$, $(\alpha_i, \beta_i)$ in Chapter 14 are opposite to the $\bar{b}$, $(\bar{\alpha}_i, \bar{\beta}_i)$ in this chapter. However, in both chapters, we use the same number for $e$.

EXERCISE 15.5.2. Let $(S^1, M_\gamma)$ and $(S^1, M_{\gamma'})$ be locally injective actions on oriented classical Seifert 3-manifolds with $\gamma = r g_1 + g_T$, $\gamma' = r' g_1' + g_T'$ where $L \cdot e(g_1) = r$ and $L' \cdot e(g_1') = r'$. Suppose $f : (S^1, M_\gamma) \to (S^1, M_{\gamma'})$ is a surjective orientation preserving map such that $f(zx) = z^n f(x)$, for all $x \in M_\gamma$ and $z \in S^1$. Let the degree of the induced map $\bar{f} : S^1 \backslash M_\gamma \to S^1 \backslash M_{\gamma'}$ be $k$. Show that $r = \frac{L}{L'} \cdot \frac{k}{n} \cdot r'$.

EXERCISE 15.5.3. For $Q' = (g = 0; 2, 3, 7)$, there exists a torsion-free normal subgroup $Q$ whose index is 168. Then $(S^1, M_\gamma) = Q \backslash (S^1 \times \mathbb{R}^2)$ is a principal $S^1$-bundle whose Euler class is $-4$ over a surface of genus 3 and $f : (S^1, M_\gamma) \to (S^1, M_{\gamma'})$ is an $S^1$-equivariant covering map, where $M_\gamma' = \{g = 0; (1, -1), (2, 1), (3, 1), (7, 1)\}$ (an integral homology 3-sphere). (Hint: Proposition 14.7.5).

## 15.6. Equivariant classification

15.6.1. Suppose we are given classical oriented Seifert orbifolds over an oriented base:

$$(S^1, M_\gamma) = \{g;\ (1, b), (\alpha_1, \beta_1), \ldots, (\alpha_n, \beta_n)\}$$
$$(S^1, M_{\gamma'}) = \{g';\ (1, b'), (\alpha'_1, \beta'_1), \ldots, (\alpha'_m, \beta'_m)\}$$

in unnormalized form with the $\gcd(\alpha, \beta)$'s all equal to 1.

PROPOSITION 15.6.2. $(S^1, M_\gamma)$ and $(S^1, M_{\gamma'})$ are $S^1$-equivariantly homeomorphic (respectively, weakly $S^1$-equivariantly homeomorphic) via an orientation preserving (respectively, reversing) homeomorphism if and only if $Q \cong Q'$, $g = g'$, $m \equiv m'$, and there exists a permutation $\sigma$ of $\{1, 2, \ldots, n\}$ such that $\alpha_j = \alpha'_{\sigma(j)}$ with $\beta_j \equiv \beta'_{\sigma(j)} \mod \alpha_j$, and $e(\gamma) = e(\gamma')$ (respectively, $\alpha_j = \alpha'_{\sigma(j)}$ with $\beta_j \equiv -\beta'_{\sigma(j)} \mod \alpha_j$, and $e(\gamma) = -e(\gamma')$).

This is Seifert's fiberwise classification. It is established in Chapter 14 and also follows from this chapter. The proposition holds in greater generality. If the $\gcd(\alpha, \beta)$'s are not 1, then $(S^1, M_\gamma)$ and $(S^1, M'_\gamma)$ will still be homeomorphic by a Seifert isomorphism. In particular, the orbifold structure is preserved.

For example, let $Q = \{g; 3, 3\}$, $g > 0$. Then $H^2(Q; \mathcal{Z}) = H^2(Q; \mathbb{Z}) \cong \mathbb{Z} \oplus \mathbb{Z}_3$. For generators, take

$$g_1 = \langle (1, -1), (3, 2), (3, 2) \rangle, \quad g_\tau = \langle (1, -1), (3, 1), (3, 2) \rangle.$$

Then $2g_\tau = \langle (1, -1), (3, 2), (3, 1) \rangle$ is strictly inequivalent to $g_\tau$, but they yield $S^1$-equivalent Seifert manifolds. The homeomorphism induces an automorphism of $H^2(Q; \mathbb{Z})$ that sends $g_\tau$ to $2g_\tau$.

If $g' = \langle (1, -1), (3, 1), (3, 1) \rangle$, then $g' = -g_1$. These strictly inequivalent manifolds are homeomorphic by a weakly $S^1$-equivariant homeomorphism which is also a Seifert isomorphism.

If we take the orbifolds $\langle (1, 1), (3, 1), (3, 0) \rangle$ and $\langle (1, 1), (3, 0), (3, 1) \rangle$, they are not strictly equivalent but are equivalent (as orbifolds) by a Seifert isomorphism.

Finally we remark that when we fixed $Q = \{g; \alpha_1, \ldots, \alpha_n\}$, we fixed the ordering of the $\{\alpha_1, \ldots, \alpha_n\}$. Of course, this choice is arbitrary. Fixing a different ordering is fixing a different reference and does not lead to anything new.

## 15.7. An Illustration

In this section, we return and explore, in a special context, some of the problems addressed in the first part Chapter 10. Sections 10.1 and 10.2 treat the embedding problem of $\Pi$ into $\mathrm{TOP}_{T^k}(P)$ from a global point of view while the approach in Sections 10.3–10.5 begins at a more local level. We have illustrated the latter approach for the concrete situation of $S^1$-actions on closed orientable 3-manifolds over an orientable base in Sections 15.2–15.6. In this section, we illustrate the global approach of Sections 10.1 and 10.2 by examining the Seifert fiberings descending from principal $S^1$-bundles over $S^2$. This analysis is independent of the first six sections of this chapter.

15.7.1. Let $P$ be a principal $S^1$-bundle over a simply connected space $W$. Then $[P] \in H^2(W; \mathbb{Z})$ determines this bundle. On the fiber $S^1$ of $P$, there is an action of

$\mathbb{Z}_n$ for every $n > 0$. We denote the quotient group $S^1/\mathbb{Z}_n$ by $S_n^1$ and the orbit space of the $\mathbb{Z}_n$-action on $P$ by $P_n = \mathbb{Z}_n\backslash P$. Then, $P_n$ is also a principal $S^1$-bundle (in fact, $S_n^1$-bundle) over $W$, whose characteristic class is given by $n[P] \in H^2(W;\mathbb{Z})$. Note that $P_1 = P$. There is a covering map of $P$ onto $P_n$.

$$\begin{array}{ccccc} S^1 & \longrightarrow & P & \longrightarrow & W \\ {\scriptstyle /\mathbb{Z}_n}\downarrow & & {\scriptstyle /\mathbb{Z}_n}\downarrow & & =\downarrow \\ S_n^1 & \longrightarrow & P_n & \longrightarrow & W \end{array}$$

Let $f \in \mathrm{TOP}_{S^1}(P)$ be a weakly $S^1$-equivariant homeomorphism. Then for every $n$, $f$ leaves $\mathbb{Z}_n \subset S^1$ invariant. Therefore $f$ induces a map $\bar{f}$ on $P_n$, giving rise to an element of $\mathrm{TOP}_{S_n^1}(P_n)$.

Conversely, let $\bar{f} \in \mathrm{TOP}_{S_n^1}(P_n)$. Since $P \to P_n$ is an $n$-fold covering, $\bar{f}$ lifts to a map $f : P \to P$, which is weakly equivariant. Because $\mathbb{Z}_n \subset S^1 \subset \mathrm{TOP}_{S^1}(P)$ induces the identity on $P_n$, we have an exact sequence

$$1 \to \mathbb{Z}_n \to \mathrm{TOP}_{S^1}(P) \to \mathrm{TOP}_{S_n^1}(P_n) \to 1.$$

which is not necessarily central. We therefore have the commutative diagram,

$$\begin{array}{ccccccc} & & 1 & & 1 & & \\ & & \downarrow & & \downarrow & & \\ & & \mathbb{Z}_n & \xrightarrow{=} & \mathbb{Z}_n & & \\ & & \downarrow & & \downarrow & & \\ 1 & \longrightarrow & \mathrm{M}(W,S^1) & \longrightarrow & \mathrm{TOP}_{S^1}(P) & \longrightarrow & \mathbb{Z}_2 \times \mathrm{TOP}(W) \\ & & \downarrow & & \downarrow & & \downarrow= \\ 1 & \longrightarrow & \mathrm{M}(W,S_n^1) & \longrightarrow & \mathrm{TOP}_{S_n^1}(P_n) & \longrightarrow & \mathbb{Z}_2 \times \mathrm{TOP}(W) \\ & & \downarrow & & \downarrow & & \\ & & 1 & & 1 & & \end{array}$$

with exact rows and columns.

Now let $\varphi \times \rho : Q \to \mathrm{id} \times \mathrm{TOP}_0(W)$, where $\mathrm{TOP}_0(W)$ is the connected component of the identity of $\mathrm{TOP}(W)$. Lemma 4.2.15 implies that $\mathrm{id} \times \mathrm{TOP}_0(W)$ is in the image of $\mathrm{TOP}_{S_n^1}(P_n)$. Let $E(P,Q)$, $E(P_n,Q)$ be the subgroups of $\mathrm{TOP}_{S^1}(P)$, $\mathrm{TOP}_{S_n^1}(P_n)$, respectively, induced from the above diagram using $\varphi \times \rho$ so that

$$\begin{array}{ccccccccc} 1 & \longrightarrow & \mathrm{M}(W,S^1) & \longrightarrow & E(P,Q) & \longrightarrow & Q & \longrightarrow & 1 \\ & & \downarrow & & \downarrow & & \downarrow= & & \\ 1 & \longrightarrow & \mathrm{M}(W,S_n^1) & \longrightarrow & E(P_n,Q) & \longrightarrow & Q & \longrightarrow & 1 \end{array}$$

are exact. Then, clearly

$$1 \to \mathbb{Z}_n \to E(P,Q) \to E(P_n,Q) \to 1$$

is exact. The covering homomorphism $p : S^1 \to S_n^1$ induces the homomorphism

$$p_* : H^2(Q, \mathrm{M}(W,S^1)) \to H^2(Q; \mathrm{M}(W,S_n^1))$$

and $p_*[E(P,Q)] = [E(P_n,Q)]$ because of the above diagram. However, $[E(P_n,Q)]$ can be regarded as $n[E(P_1,Q)]$ if we identify $S_n^1$ with $S^1$. Therefore to understand $E(P_n,Q)$ it suffices to understand $E(P_1,Q)$.

15.7.2. For a particular illustration, let us consider the principal $S^1$-bundles over $S^2$, the 2-sphere. So $G = S^1$ and $W = S^2$. Since $H^2(S^2;\mathbb{Z}) = \mathbb{Z}$, each integer $n$ corresponds to a principal $S^1$-bundle $P_n$, diffeomorphic to the lens space $L(n,1)$. For $n = \pm 1$, we have $\mp$ (the Hopf fibering); for $n = 0$ we have $S^2 \times S^1$. Every finite group $Q \subset \text{TOP}(S^2)$ which is orientation preserving is conjugate to an orientation preserving linear action of $Q$ on $S^2$, by [**Bro19**], [**Ker19**], etc. Thus, $Q$ may be topologically conjugated into $SO(3)$. So we may assume, without loss of generality, that $Q \subset SO(3)$. Therefore, we take $Q$ to be one of

$\mathbb{Z}_m$      Cyclic group,
$D_{2m} = \mathbb{Z}_m \rtimes \mathbb{Z}_2$      Dihedral group,
$T = A_4$      Tetrahedral group,
$O = S_4$      Octahedral group, and
$I = A_5$      Icosahedral group.

The universal covering of $SO(3)$ is $\text{Spin}(3) = SU(2)$ which is diffeomorphic to $S^3$. If we take the inverse image of $Q$ in $SU(2) \to SO(3)$, we get a nontrivial central extension $0 \to \mathbb{Z}_2 \to Q^* \to Q \to 1$. For $Q = D_{2m}, T, O$, or $I$, we have the *binary groups* $Q^* = (D_{2m})^*, T^*, O^*, I^*$. For $Q = \mathbb{Z}_m$, we get $Q^* = \mathbb{Z}_{2m}$, which is split if $m$ is odd and not split if $m$ is even.

Since $Q^* \subset SU(2) = S^3$, $Q^*$ acts freely on $S^3$. Thus, it is not difficult to calculate the following.

| Group | Order | $H^1(Q;S^1)$ | $H^2(Q;S^1)$ |
|---|---|---|---|
| $\mathbb{Z}_m$ | $m$ | $\mathbb{Z}_m$ | 0 |
| $D_{2(2m+1)}$ | $4m+2$ | $\mathbb{Z}_2$ | 0 |
| $D_{2(2m)}$ | $4m$ | $(\mathbb{Z}_2)^2$ | $\mathbb{Z}_2$ |
| $T$ | 12 | $\mathbb{Z}_3$ | $\mathbb{Z}_2$ |
| $O$ | 24 | $\mathbb{Z}_2$ | $\mathbb{Z}_2$ |
| $I$ | 60 | 0 | $\mathbb{Z}_2$ |

15.7.3. It is convenient to think of $S^3$ as the unit sphere in $\mathbb{C}^2$ and $U(2)$ the group of unitary transformations of $\mathbb{C}^2$ leaving $S^3$ invariant. We have $U(2) = SU(2) \times_{\mathbb{Z}_2} S^1$, where $S^1$ is the center of $U(2)$.

On $P_1 = S^3$, the group $Q^*$ acts freely and commutes with the free $S^1$-action. In $\text{TOP}_{S^1}(P_1)$, we have $Q^* \cap S^1 = \mathbb{Z}_2$. Since $S^1$ centralizes $Q^*$, the group generated by $S^1$ and $Q^*$ is isomorphic to $(S^1 \times Q^*)/\mathbb{Z}_2$, where $\mathbb{Z}_2 = \{(\tau, \tau^{-1}) | \tau \in S^1 \cap Q^*\}$. This will be denoted by $S^1 \times_{\mathbb{Z}_2} Q^*$. The action of $(S^1 \times Q^*)/\mathbb{Z}_2$ induces an action of $Q$ on $S^2 = S^1 \backslash P_1$. By dividing by $\mathbb{Z}_n$ in $S^1 \subset S^1 \times_{\mathbb{Z}_2} Q^*$, we obtain an action of $(S^1 \times_{\mathbb{Z}_2} Q^*)/\mathbb{Z}_n$ on $P_n$. This group, as we shall see, is either $S_n^1 \times Q$ or $S_n^1 \times_{\mathbb{Z}_2} Q^*$, where $S_n^1 = S^1/\mathbb{Z}_n$. The linear action of $Q$ is again induced on $S^2$.

Recall that since $Q$ is finite, from Proposition 10.1.9, $\ell_* : H^2(Q;S^1) \to H^2(Q;\text{M}(S^2,S^1))$ is an isomorphism. Therefore, there is one and only one extension class that can be embedded into the group $E(P_1,Q)$. So $S^1 \times_{\mathbb{Z}_2} Q^*$ is mapped to $E(P_1,Q)$ by this isomorphism. We may describe $E(P_1,Q)$ by the same construction that we made for $S^1 \times_{\mathbb{Z}_2} Q^*$. We have $\text{M}(S^2,S^1)$ and $Q^*$ inside $\text{TOP}_{S^1}(P)$. Since $\text{M}(S^2,S^1)$ is normalized by $Q^*$, we form the semidirect product, $\text{M}(S^2,S^1) \rtimes Q^*$.

This acts on $P_1$ with ineffective $\mathbb{Z}_2 = \{(\tau, \tau^{-1}) | \tau \in \mathrm{M}(S^2, S^1) \cap Q^*\}$. Thus $[(\mathrm{M}(S^2, S^1) \rtimes Q^*)/\mathbb{Z}_2] \in H^2(Q; \mathrm{M}(S^2, S^1))$ is the image of $[S^1 \times_{\mathbb{Z}_2} Q^*] \in H^2(Q; S^1)$. Again we denote $(\mathrm{M}(S^2, S^1) \rtimes Q^*)/\mathbb{Z}_2$ by $\mathrm{M}(S^2, S^1) \rtimes_{\mathbb{Z}_2} Q^*$. On the other hand, consider the long exact sequence of $H^*(Q; -)$ induced from the coefficient exact sequence $1 \to \mathbb{Z}_2 \to S^1 \to S^1_2 \to 1$. From the table, we have $2 \cdot H^2(Q; S^1) = 0$. Therefore, $H^2(Q; \mathbb{Z}_2) \to H^2(Q; S^1)$ is surjective. We conclude that under the homomorphisms, $[Q^*]$ corresponds as follows:

$$\begin{array}{ccccc} H^2(Q; \mathbb{Z}_2) & \to & H^2(Q; S^1) & \to & H^2(Q; \mathrm{M}(S^2, S^1)), \\ Q^* & \mapsto & S^1 \times_{\mathbb{Z}_2} Q^* & \mapsto & \mathrm{M}(S^2, S^1) \rtimes_{\mathbb{Z}_2} Q^*. \end{array}$$

An immediate consequence of the foregoing is that if $n$ is even, $S^1_n \times_{\mathbb{Z}_2} Q^*$ represents the trivial extension $S^1_n \times Q$ and $E(P_n, Q)$ then is $\mathrm{M}(S^2, S^1_n) \rtimes Q$. On the other hand, for $n$ odd, $Q \neq \mathbb{Z}_m, D_{2(2m+1)}$, we shall show that the extension $S^1_n \times_{\mathbb{Z}_2} Q^*$ does not split and so $E(P_n, Q)$ is not the trivial extension. Moreover, if $E(P, Q)$ is the nontrivial extension, then so is $E(P_n, Q)$ if $n$ is odd and the trivial extension if $n$ is even.

LEMMA 15.7.4. *For $Q = D_{2(2m)}, T, O$ and $I$, the extension $1 \to S^1 \to S^1 \times_{\mathbb{Z}_2} Q^* \to Q \to 1$ is nontrivial. For $Q = D_{2(2m+1)}$, $S^1 \times_{\mathbb{Z}_2} Q^* = S^1 \times Q$, it is the trivial extension. For $Q = \mathbb{Z}_m$, $S^1 \times \mathbb{Z}_m$ is the only central extension.*

PROOF. Let $Q = D_{2(2m)}, D_{2(2m+1)}, T, O$ or $I$. Let $E$ be any extension of $S^1$ by $Q$. Pull back the exact sequence $1 \to S^1 \to E \to Q \to 1$ via the natural homomorphism $Q^* \to Q$ to get an extension of $S^1$ by $Q^*$. Since $Q^*$ has periodic cohomology of period 4, $H_2(Q^*; \mathbb{Z}) = 0$ by [**Bro82**, p.159, Exercise 3]. By the universal coefficient theorem, $H^2(Q^*; S^1) = \mathrm{Hom}(H_2(Q^*; \mathbb{Z}), S^1) + \mathrm{Ext}(H_1(Q^*; \mathbb{Z}), S^1)$. The last term is 0 because $S^1$ is injective. Thus, $H^2(Q^*; \mathbb{Z}) = 0$. Therefore, the only extension of $S^1$ by $Q^*$ is $S^1 \times Q^*$. We get the following.

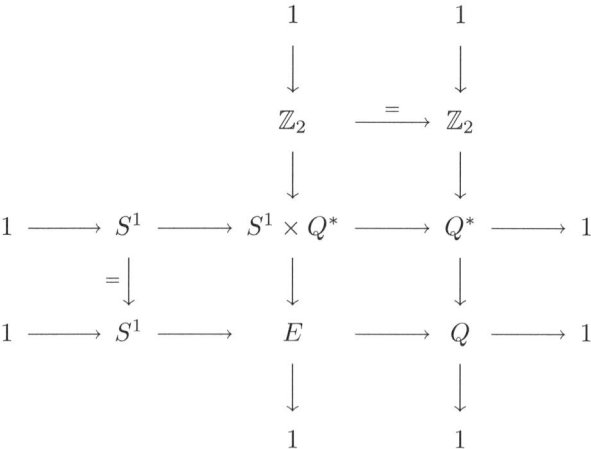

Any extension $E$ of $S^1$ by $Q$ is a quotient of $S^1 \times Q^*$ by a certain subgroup $\mathbb{Z}_2$.

It remains to identify the subgroups $\mathbb{Z}_2$ of $S^1 \times Q^*$. The inclusion $\mathbb{Z}_2 \hookrightarrow S^1 \times Q^*$ induces two homomorphisms $\mathbb{Z}_2 \to S^1$ and $\mathbb{Z}_2 \to Q^*$. There are two obvious homomorphisms of $\mathbb{Z}_2$ into $S^1$; but there is only one homomorphism $\mathbb{Z}_2 \to Q^*$ which fits into the diagram; namely, the nontrivial element of $\mathbb{Z}_2$ maps to the

central element $\tau \in \mathbb{Z}_2 \subset Q^*$. This shows that $E$ is either
$$(S^1 \times Q^*)/\langle(1,\tau)\rangle \cong S^1 \times Q$$
or
$$(S^1 \times Q^*)/\langle(e^{\pi i},\tau)\rangle = S^1 \times_{\mathbb{Z}_2} Q^*$$
In other words $\{S^1 \times Q, S^1 \times_{\mathbb{Z}_2} Q^*\}$ exhausts all the possible central extension classes of $S^1$ by $Q$. Since in the case of $Q = D_{2(2m)}, T, O$ and $I$, we know $H^2(Q; S^1) = \mathbb{Z}_2$; therefore, $S^1 \times_{\mathbb{Z}_2} Q^*$ must be the nonsplitting extension. For $Q = D_{2(2m+1)}$, we have $H^2(Q; S^1) = 0$ already so that $S^1 \times_{\mathbb{Z}_2} D^*_{2(2m+1)} = S^1 \times D_{2(2m+1)}$. For $Q = \mathbb{Z}_m$, we have $H^2(Q; S^1) = 0$, so $S^1 \times \mathbb{Z}_m$ is the only extension. □

15.7.5. Here is an explicit identification of $S^1 \times_{\mathbb{Z}_2} D^*_{2(2m+1)}$ with $S^1 \times D_{2(2m+1)}$. Recall the presentations of $D_{2(2m+1)}$ and $D^*_{2^2(2m+1)}$:

$$D^*_{2^2(2m+1)} = \langle x, y \mid x^{2m+1} = y^2, \quad yxy^{-1} = x^{-1}\rangle,$$
$$D_{2(2m+1)} = \langle \bar{x}, \bar{y} \mid \bar{x}^{2m+1} = \bar{y}^2 = 1, \quad \bar{y}\bar{x}\bar{y}^{-1} = \bar{x}^{-1}\rangle.$$

Define a map $f: D_{2(2m+1)} \to S^1 \times_{\mathbb{Z}_2} D^*_{2^2(2m+1)}$ by
$$f(\bar{x}) = (e^{\pi i}, x), \quad f(\bar{y}) = (e^{\pi i/2}, y).$$
Then $f(\bar{x}^{2m+1}) = (e^{(2m+1)\pi i}, x^{2m+1}) = (e^{\pi i}, x^{2m+1}) \in \mathbb{Z}_2$, $f(\bar{y}^2) = (e^{\pi i}, y^2) \in \mathbb{Z}_2$. So $f(\bar{x}^{2m+1}) = f(\bar{y}^2) = 1$ in $S^1 \times_{\mathbb{Z}_2} D^*_{2^2(2m+1)}$. Further, $f(\bar{y}\bar{x}\bar{y}^{-1}) = (e^{\pi i}, x^{-1}) = f(\bar{x}^{-1})$. Thus $f$ preserves the relation, and it induces an isomorphism of $S^1 \times D_{2(2m+1)}$ onto $S^1 \times_{\mathbb{Z}_2} D^*_{2^2(2m+1)}$.

PROPOSITION 15.7.6.   (1) *For $Q = D_{2(2m)}, T, O$, and $I$, $E(P_n, Q)$ is equal to $M(S^2, S^1_n) \rtimes_{\mathbb{Z}_2} Q^*$. It is a nontrivial extension of $M(S^2, S^1_n)$ by $Q$ if $n$ is odd and a trivial extension if $n$ is even.*
(2) *For $Q = D_{2(2m+1)}$ and $\mathbb{Z}_m$, $E(P_n, Q) \cong M(S^2, S^1_n) \rtimes Q$.*
(3) *For $Q = D_{2m}, T, O, I$, a central extension $E$ of $S^1$ by $Q$ can be embedded into $E(P_n, Q)$ if and only if $E$ and $E(P_n, Q)$ are both trivial or both nontrivial.*

PROOF. $H^2(Q; S^1)$ maps isomorphically onto $H^2(Q; M(S^2, S^1))$, and the class represented by $S^1 \times_{\mathbb{Z}_2} Q^*$ maps to $E(P_1, Q)$. Because $H^2(Q; M(S^2, S^1))$ is isomorphic to $\mathbb{Z}_2$ and Lemma 15.7.4, $E(P_1, Q)$ is the nontrivial extension, $S^1 \times_{\mathbb{Z}_2} Q^*$, for $D_{2m}, T, O$, and $I$. To obtain the statements about $E(P_n, Q)$ we just recall that $n[E(P_1, Q)] = [E(P_n, Q)]$ when we use the isomorphism of $S^1_n$ with $S^1$. For $Q = \mathbb{Z}_m$, $H^2(Q, S^1) \cong H^2(Q, M(S^2, S^1)) = 0$ and so the group $E(P_n, \mathbb{Z}_m)$ is the semidirect product, $M(S^2, S^1_n) \rtimes \mathbb{Z}_m$, for all $m$ and all $n$. Statement (3) is a restatement of the isomorphism $H^2(Q; S^1) \xrightarrow{\cong} H^2(Q; M(S^2, S^1))$, which is either $\mathbb{Z}_2$ or 0. □

15.7.7. Since we now know which central extension $0 \to S^1 \to E \to Q \to 1$ embeds in $\mathrm{TOP}_{S^1}(P_n)$, we can determine which group extensions $0 \to \mathbb{Z}_k \to \Pi \to Q \to 1$ embed in $E(P_n, Q)$. To embed in $E(P_n, Q)$, $\Pi$ must first embed in $S^1 \times_{\mathbb{Z}_2} Q^*$. So we need only determine the elements of $H^2(Q; \mathbb{Z}_k)$ which map to $[S^1 \times_{\mathbb{Z}_2} Q^*] \in H^2(Q; S^1)$.

THEOREM 15.7.8. *Let* $\Gamma = \mathbb{Z}_{2k+1}$ *or* $\mathbb{Z}_{2k}$; *and* $Q = \mathbb{Z}_m$, $D_{2(2m+1)}$, $D_{2(2m)}$, $T$, $O$ *or* $I$. *We can characterize the extensions* $1 \to \Gamma \to \Pi \to Q \to 1$ *which embed into* $1 \to M(S^2, S^1) \to \text{TOP}_{S^1}(S^3) \to \text{TOP}(S^2) \to 1$ *as follows:*

(1) *For* $Q = \mathbb{Z}_m$ *and* $D_{2(2m+1)}$, *every extension* $\Pi$ *embeds into* $\text{TOP}_{S^1}(S^3)$.
(2) *For* $Q = D_{2(2m)}, T, O$, *or* $I$, *there are two cases: If* $\Gamma = \mathbb{Z}_{2k+1}$, *no extension* $\Pi$ *embeds into* $\text{TOP}_{S^1}(S^3)$. *If* $\Gamma = \mathbb{Z}_{2k}$, *an extension* $\Pi$ *embeds into* $\text{TOP}_{S^1}(S^3)$ *if and only if* $[\Pi]$ *is not in the kernel of the homomorphism* $H^2(Q; \mathbb{Z}_{2k}) \to H^2(Q; S^1)$.

PROOF. As mentioned earlier, by Proposition 10.1.9, $H^2(Q; S^1) \to H^2(Q; M(S^2, S^1))$ is an isomorphism (since $Q$ is finite) and these groups are $0$ for $Q = \mathbb{Z}_{2m}$ or $D_{2(2m+1)}$; and are $\mathbb{Z}_2$ for $Q = D_{2(2m)}, T, O$, or $I$. In the later case, our target group is $M(S^2, S^1) \rtimes_{\mathbb{Z}_2} Q^* \subset \text{TOP}_{S^1}(S^3)$ which is the nontrivial extension of $M(S^2, S^1)$ by $Q$. See Tables in 15.3.1 and 15.7.2.

| Group | $H^2(Q; \mathbb{Z}_{2k+1})$ | $H^2(Q; \mathbb{Z}_{2k})$ | $H^2(Q; S^1)$ |
|---|---|---|---|
| $\mathbb{Z}_m$ | $\mathbb{Z}_{(m,2k+1)}$ | $\mathbb{Z}_{(m,2k)}$ | $0$ |
| $D_{2(2m+1)}$ | $0$ | $\mathbb{Z}_2$ | $0$ |
| $D_{2(2m)}$ | $0$ | $(\mathbb{Z}_2)^3$ | $\mathbb{Z}_2$ |
| $T$ | $\mathbb{Z}_{(3,2k+1)}$ | $\mathbb{Z}_{(3,2k)} \oplus \mathbb{Z}_2$ | $\mathbb{Z}_2$ |
| $O$ | $0$ | $(\mathbb{Z}_2)^2$ | $\mathbb{Z}_2$ |
| $I$ | $0$ | $\mathbb{Z}_2$ | $\mathbb{Z}_2$ |

In the last column under $H^2(Q; S^1)$, our target group $1 \to S^1 \to E \to Q \to 1$, which embeds into $\text{TOP}_{S^1}(S^3)$ is the zero element $E = S^1 \times Q$ when $H^2(Q; S^1) = 0$ and is the nontrivial element $S^1 \rtimes_{\mathbb{Z}_2} Q^*$ whenever $H^2(Q; S^1) = \mathbb{Z}_2$. Note that $H^2(Q; \mathbb{Z}_{2k}) \to H^2(Q; S^1)$ is surjective because it factors through $H^2(Q; \mathbb{Z}_2)$. □

REMARK 15.7.9. Recall that $U(2) = S^1 \times_{\mathbb{Z}_2} SU(2)$ is contained in $\text{TOP}_{S^1}(P_1)$. For $n \geq 0$, $U(2)/\mathbb{Z}_n \subset \text{TOP}_{S^1/\mathbb{Z}_n}(P_n)$ is a reduction of the uniformizing group. Also, $U(2) \to U(2)/\mathbb{Z}_n$ is an $n$-fold covering and $U(2)/\mathbb{Z}_n$ is isomorphic to $U(2)$ if $n$ is odd, and to $S^1/\mathbb{Z}_n \times SO(3)$ if $n$ is even, $\mathbb{Z}_n \subset S^1$.

REMARK 15.7.10. Subsections 15.7.2–15.7.9 and Section 15.3 both treat the extended liftings of $Q$ on $S^2$ to principal $S^1$-bundles over $S^2$. The results of Section 15.3 are more detailed. The local approach of Section 10.3 could also be derived from the Leray spectral sequence of a map mentioned near the end of the Subsection 10.2.3. This would make these two methods equivalent.

# Bibliography

[AD02]  A. Adem and J.F. Davis, *Topics in transformation groups*, Handbook of Topology, Elsevier Science B.V., Edited by R.J. Daverman and R.B. Sher (2002), 1–54.

[AJ76]  L. Auslander and F.E.A. Johnson, *On a conjecture of C.T.C. Wall*, J. London Math. Soc. (2) **14** (1976), 331–332.

[Are46]  Richard Arens, *Topologies for homeomorphism groups*, Amer. J. Math. **68** (1946), 593–610.

[Arm68]  M. A. Armstrong, *The fundamental group of an orbit space of a discontinuous group*, Proc. Cambridge Philos. Soc. **64** (1968), 299–301.

[Aus60]  L. Auslander, *Bieberbach's theorem on space groups and discrete uniform subgroups of Lie groups*, Ann. of Math. (2) **71** (1960), no. (3), 579–590.

[Aus61a]  ———, *Bieberbach's theorem on space groups and discrete uniform subgroups of Lie groups II*, Amer. J. Math. **83** (1961), 276–280.

[Aus61b]  ———, *Discrete uniform subgroups of solvable Lie groups*, Trans. Amer. Math. Soc. **99** (1961), 398–402.

[Aus63]  ———, *On radicals of discrete subgroups of Lie groups*, Amer. J. Math. **85** (1963), 145–150.

[Aus73]  ———, *An exposition of the structure of solvmanifolds. Part I: Algebraic theory*, Bull. Amer. Math. Soc. **79** (1973), no. 2, 227–261.

[BBN+78]  H. Brown, R. Bülow, J. Neubüser, H. Wondratscheck, and H. Zassenhaus, *Crystallographic Groups of Four-dimensional Space*, Wiley New York, 1978.

[BD02]  Yves Benoist and Karel Dekimpe, *The uniqueness of polynomial crystallographic actions*, Math. Ann. **322** (2002), no. 3, 563–571.

[BDD05]  Dietrich Burde, Karel Dekimpe, and Sandra Deschamps, *The Auslander conjecture for NIL-affine crystallographic groups*, Math. Ann. **332** (2005), no. 1, 161–176.

[Ben92]  Yves Benoist, *Une nilvariété non affine*, C. R. Acad. Sci. Paris Sér. I Math. **315** (1992), 983–986.

[Ben95]  ———, *Une nilvariété non affine.*, J. Differential Geom. **41** (1995), 21–52.

[Ber85]  J Bernstein, *On covering spaces and Lie group actions*, Contemp. Math. Amer. Math. Soc. **37** (1985), 11–13.

[BG95]  Dietrich Burde and Fritz Grunewald, *Modules for certain Lie algebras of maximal class*, J. Pure Appl. Algebra **99** (1995), 239–254.

[BH82]  W. Browder and W.C. Hsiang, *G-actions and the fundamental group*, Invent. Math. **65** (1982), 411–424.

[Blo75]  E. M. Bloomberg, *Manifolds with no periodic homeomorphism*, Trans. Amer. Math. Soc. **202** (1975), 67–78.

[Bon79]  F. Bonahon, *Involutions et fibrés de Seifert dans les variétés de dimension* 3, Université Paris-Sud XI - Orsay France, Ph.D. Dissertation, 1979.

[Bor60]  A. Borel et al., *Seminar on Transformation Groups*, Ann. of Math. Studies, vol. 46, Princeton Univ. Press, 1960.

[Bor72]  A. Borel, *Topics in the homology theory of fibre bundles*, Springer Lecture Notes in Math., vol. 36, Springer-Verlag, 1972.

[Bre72]  Glen E. Bredon, *Introduction to Compact Transformation Groups.*, Pure and Applied Mathematics, vol. 46, Academic Press, New York, 1972.

[Bre97]  ———, *Sheaf Theory, Second edition*, Graduate Texts in Mathematics, vol. 170, Springer-Verlag, New York, 1997.

[Bro19]  L. E. J. Brouwer, *Über die periodischen Transformationen der Kugel*, Math. Ann. **80** (1919), 39–41.

[Bro82]　　Kenneth S. Brown, *Cohomology of Groups*, Grad. Texts in Math., vol. 87, Springer–Verlag New York Inc., 1982.

[BS83]　　F. Bonahon and L. Siebenmann, *Seifert 3-Orbifolds and Their Role as Natural Crystalline Parts of Arbitrary Compact Irreducible 3-Orbifolds*, Universite de Paris-Sud, Orsay, 1983, (Polycopied, IHES, Orsay).

[BS85]　　———, *The classification of Seifert fibered 3-orbifolds*, Cambridge Univ. Press **95** (1985), 19–85.

[Bur96]　　Dietrich Burde, *Affine structures on nilmanifolds.*, Internat. J. Math **7** (1996), no. 5, 599–616.

[BZ87]　　Michel Boileau and Bruno Zimmermann, *Symmetries of nonelliptic Montesinos links*, Math. Ann. **277** (1987), no. 3, 563–584.

[Car72]　　J.B. Carrell, *Holomorphically injective toral actions*, Proceedings of the Second Conference on Compact Transformation Groups, Springer Lecture Notes in Math. **299** (1972), 205–236.

[CE56]　　H. Cartan and S. Eilenberg, *Homological Algebra*, Princeton University Press, 1956.

[CH96]　　Alberto Cavicchioli and Friedrich Hegenbarth, *Surgery on 3-manifolds with $S^1$-actions*, Geom. Dedicata **61** (1996), no. 3, 285–313.

[Cha65]　　Leonard S Charlap, *Compact flat Riemannian manifolds I*, Ann. of Math. (2) **31** (1965), 15–30.

[Cha86]　　Leonard S. Charlap, *Bieberbach Groups and Flat Manifolds*, Universitext, Springer–Verlag, New York Inc., 1986.

[CHK00]　　D. Cooper, C. Hodgson, and S. Kerckhoff, *Three-dimensional Orbifolds and Cone-manifold*, Memoirs, vol 5, Mathematical Society of Japan, Tokyo, 2000.

[Con68]　　Pierre E. Conner, *Lectures on the Action of a Finite Group*, Lecture Notes in Mathematics, no. 73, Springer–Verlag, New York Inc., 1968.

[CR69]　　Pierre E. Conner and Frank Raymond, *Actions of compact Lie groups on aspherical manifolds*, Topology of Manifolds, edited by J.C. Cantrell and C.H. Edwards, Markham, Chicago (1969), 227–264.

[CR71]　　———, *Injective operations of toral groups*, Topology **10** (1971), 283–296.

[CR72a]　　———, *Derived actions*, Proceedings of the Second Conference on Compact Transformation Groups, Springer Lecture Notes in Math. **299** (1972), 237–310.

[CR72b]　　———, *Holomorphic Seifert fiberings*, Proceedings of the Second Conference on Compact Transformation Groups, Springer Lecture Notes in Math. **299** (1972), 124–204.

[CR72c]　　———, *Injective operations of toral groups II*, Proceedings of the Second Conference on Compact Transformation Groups, Springer Lecture Notes in Math. **299** (1972), 109–124.

[CR72d]　　———, *Manifolds with few periodic homeomorphisms*, Proceedings of the Second Conference on Compact Transformation Groups, Springer Lecture Notes in Math. **299** (1972), 1–75.

[CR75]　　———, *Realizing finite groups of homeomorphism from homotopy classes of self-homotopy equivalences*, Manifolds—Tokyo (Proc. Internat. Conf., Tokyo, 1973) (1975), 231–238.

[CR77]　　———, *Deforming homotopy equivalences to homeomorphisms in aspherical manifolds*, Bull. Amer. Math. Soc. **83** (1977), no. (1), 36–85.

[CRW72]　　Pierre E. Conner, Frank Raymond, and Peter J. Weinberger, *Manifolds with no periodic maps*, Proceedings of the Second Conference on Compact Transformation Groups, Springer Lecture Notes in Math. **299** (1972), 81–108.

[Dav83]　　Michael Davis, *Groups generated by reflections and aspherical manifolds not covered by Euclidean space*, Ann. of Math. (2) **117** (1983), 293–324.

[DCDI03]　　Tine De Cat, Karel Dekimpe, and Paul Igodt, *Translations in simply transitive affine actions of Heisenberg type Lie groups*, Linear Algebra Appl. **359** (2003), 101–111.

[DDRM05]　　Karel Dekimpe, Bram De Rock, and Wim Malfait, *The Nielsen numbers of Anosov diffeomorphisms on flat Riemannian manifolds*, Forum Math. **17** (2005), no. 2, 325–341.

[Dek96]　　Karel Dekimpe, *Almost-Bieberbach Groups: Affine and polynomial structures*, Lect. Notes in Math., vol. 1639, Springer–Verlag, 1996.

[Dek97]　　———, *Determining the translation part of the fundamental group of an infra-solvmanifold of type (R)*, Math. Proc. Camb. Phil. Soc. **122** (1997), 515–524.

[Dek00] _____, *Polynomial structures on polycyclic groups: recent developments*, Crystallographic groups and their generalizations (Kortrijk, 1999), Contemp. Math. Amer. Math. Soc. **262** (2000), 99–120.

[Dek01] _____, *Affine and polynomial structures on virtually 2-step solvable groups*, Comm. Algebra **29** (2001), no. 11, 4965–4988.

[Dek02] _____, *Polynomial crystallographic actions on the plane*, Geom. Dedicata **93** (2002), 47–56.

[Dek03] _____, *Any virtually polycyclic group admits a NIL-affine crystallographic action*, Topology **42** (2003), no. 4, 821–832.

[DG90] G. Dula and D. Gottlieb, *Splitting off H-spaces and Conner–Raymond splitting theorem*, J. Fac. Sci. Univ. Tokyo Sect. IA Math. **37** (1990), 321–334.

[DI97] Karel Dekimpe and Paul Igodt, *Polycyclic-by-finite groups admit a bounded-degree polynomial structure*, Invent. Math. **129** (1997), no. (1), 121–140.

[DI00] _____, *Polynomial alternatives for the group of affine motions*, Math. Z. **234** (2000), 457–485.

[Die48] J. Dieudonné, *On topological groups of homeomorphisms*, Amer. J. Math. **70** (1948), 659–680.

[DIKL95] Karel Dekimpe, Paul Igodt, Suhyung Kim, and Kyung Bai Lee, *Affine structures for closed 3-dimensional manifolds with NIL–geometry*, Quart. J. Math. Oxford Ser. (2) **46** (1995), 141–167.

[DIL96] Karel Dekimpe, Paul Igodt, and Kyung Bai Lee, *Polynomial structures for nilpotent groups.*, Trans. Amer. Math. Soc. **348** (1996), 77–97.

[DIM94] Karel Dekimpe, Paul Igodt, and Wim Malfait, *There are only finitely many infra-nilmanifolds under each nilmanifold: a new proof*, Indag. Math. (N.S.) **5** (1994), no. (3), 259–266.

[DIM01] _____, *Infra-nilmanifolds and their fundamental groups*, Mathematics in the new millennium (Seoul, 2000), J. Korean Math. Soc. **38** (2001), no. 5, 883–914.

[DIP03] Karel Dekimpe, Paul Igodt, and Hannes Pouseele, *Expanding automorphisms and affine structures on nilpotent lie algebras with few generators*, Comm. Algebra **31** (2003), 5847–5874.

[Dix57] J. Dixmier, *L'application exponetielle dans les groups de Lie résolubles*, Bull. Soc. Math. France **85** (1957), 113–121.

[DK01] James F. Davis and Paul Kirk, *Lecture notes in Algebraic Topology*, American Mathematical Society, Providence, RI, 2001.

[DL03a] Karel Dekimpe and Kyung Bai Lee, *Expanding maps, Anosov diffeomorphisms and affine structures on infra-nilmanifolds*, Topology Appl. **130** (2003), no. 3, 259–269.

[DL03b] _____, *Expanding maps on infra-nilmanifolds of homogeneous type*, Trans. Amer. Math. Soc. **355** (2003), no. 3, 1067–1077.

[DLR96] Karel Dekimpe, Kyung Bai Lee, and Frank Raymond, *Bieberbach theorems for solvable Lie groups*, Transformation group theory (Taejŏn, Korea Adv. Inst. Sci. Tech., Taejŏn (1996), 28–31.

[DLR01] Karel Dekimpe, Kyung Bai Lee, and Frank Raymond, *Bieberbach theorems for solvable Lie groups*, Asian J. Math. **5** (2001), no. 3, 499–508.

[DO99] Karel Dekimpe and Veerle Ongenae, *Filiform left-symmetric algebras*, Geom. Dedicata **74** (1999), no. 2, 165–199.

[DS82] H. Donnelly and R. Schultz, *Compact group actions and maps into aspherical manifolds*, Topology **21** (1982), 443–455.

[Ebe82] P. Eberlein, *A canonical form for compact nonpositively curved manifolds whose fundamental groups have nontrivial center*, Math. Ann. **260** (1982), 23–29.

[FG83] David Fried and William M. Goldman, *Three-dimensional affine crystallographic groups*, Adv. in Math. **47** (1983), no. 1, 1–49.

[FH83] F.T. Farrell and W.C. Hsiang, *Topological characterization of flat and almost flat Riemannian manifolds $M^n (n \neq 3, 4)$*, Amer. J. Math. **105** (1983), 641–672.

[Fin76] R. Fintushel, *Local $S^1$ actions on 3-manifolds*, Pacific J. Math. **66** (1976), 111–118.

[FJ89] F.T. Farrell and L.F. Jones, *A topological analogue of Mostow's rigidity theorem*, J. Amer. Math. Soc. **2** (1989), 257–370.

[FJ93] _____, *Topological rigidity for closed non-positively curved manifolds*, Proc. Symposs. Pure Math **54, Part 3** (1993), 229–274.

[FJ98] ———, *Rigidity for aspherical manifolds with* $\pi_1 \subset GL_m(\mathbb{R})$, Asian J. Math. **2** (1998), 215–262.

[Flo55] E. Floyd, *Orbit spaces of finite transformation groups II*, Duke Math. J **22** (1955), 33–38.

[Gar66] H. Garland, *Cohomology of lattices*, Ann. of Math. (2) **84** (1966), 175–196.

[GLÖ85] D. Gottlieb, Kyung Bai Lee, and M. Özaydın, *Compact group actions and maps into $K(\pi, 1)$-spaces*, Trans. Amer. Math. Soc. **287** (1985), 419–429.

[Gol85] W. Goldman, *Nonstandard Lorentz space forms*, J. Differential Geom. **21** (1985), 301–308.

[Gor71] V. V. Gorbacevič, *Discrete subgroups of solvable Lie groups of type (E)*, Math. USSR Sbornik **14** (1971), no. N° 2, 233–251.

[Gor73] ———, *Lattices in solvable Lie groups and deformations of homogeneous spaces*, Math. USSR Sbornik **20** (1973), no. 2, 249–266.

[Got77] D. Gottlieb, *Lifting actions in fibrations*, Geometric applications of homotopy theory (Proc. Conf., Evanston, Ill.), Springer Lecture Notes in Math. **657** (1977), 217–254.

[Gra62] Hans Grauert, *Über Modifikationen und exzeptionelle analytische Mengen*, Math. Ann. **146** (1962), 331–368.

[Gro57] A. Grothendieck, *Sur quelques points d'algébre homologique*, Tohoku Math. J. (2) **9** (1957), 119–221.

[Gru70] Karl Gruenberg, *Cohomological Topics in Group Theory*, Lecture Notes in Math., vol. 143, Springer-Verlag, 1970.

[GS94] Fritz Grunewald and Dan Segal, *On affine crystallographic groups*, J. Differential Geom. **40** (1994), no. (3), 563–594.

[Gun66] R. C. Gunning, *Lectures on Riemann Surfaces*, Princeton Mathematical Notes, Princeton University Press, 1966.

[Hel62] S. Helgason, *Differential Geometry and Symmetric Spaces*, Academic Press, New York, 1962.

[Hem76] J. Hempel, *3-Manifolds*, Ann. of Math. Studies, vol. 86, Princeton Univ. Press, 1976.

[HLL05] Ku Yong Ha, Jong Bum Lee, and Kyung Bai Lee, *Maximal holonomy of infranilmanifolds with 2-dimensional quaternionic heisenberg geometry*, Trans. Amer. Math. Soc. **357** (2005), 355–383.

[Hoc65] G. Hochschild, *The Structure of Lie Groups*, Holden-Day Series in Mathematics, Holden-Day San Francisco (Calif.), 1965.

[Hol64] H. Holmann, *Seifertsche Faseraume*, Math. Ann. **157** (1964), 138–166.

[Hop26] Heinz Hopf, *Zum Clifford-Kleinschen Raumproblem*, Math. Ann. **95** (1926), 313–339.

[HR83] W. C. Hsiang and H. Rees, *Miscenko's work on Novikov's conjecture*, Contemp. Math. Amer. Math. Soc. **10** (1983), 77–98.

[Hsi75] Wu-yi Hsiang, *Cohomology Theory of Topological Transformation Groups*, Ergebnisse der Mathematik und ihrer Grenzgebiete, vol. Band 85, Springer-Verlag, New York-Heidelberg, 1975.

[Hu52] Sze-tsen Hu, *Cohomology theory in topological groups*, Michigan Math. J. **1** (1952), 11–59.

[Hu59] ———, *Homotopy Theory*, Pure and Applied Mathematics, vol. VIII, Academic Press, 1959.

[Hus94] Dale Husemoller, *Fibre Bundles* (3rd edition), Graduate Texts in Mathematics, 20, Springer-Verlag, New York, 1994.

[HY76] Akio Hattori and Tomoyoshi Yoshida, *Lifting compact group actions in fiber bundles*, Japan. J. Math. (N.S.) **2** (1976), 13–25.

[Igo84] Paul Igodt, *Generalizing a realization result of B. Zimmermann and K.B. Lee to infra-nilmanifolds*, Manuscripta Math. **47** (1984), 19–30.

[IL87] Paul Igodt and Kyung Bai Lee, *Applications of group cohomology to space constructions*, Trans. Amer. Math. Soc. **304** (1987), 69–82.

[IL90] ———, *On the uniqueness of certain affinely flat infra-nilmanifolds*, Math. Zeitschrift **204** (1990), 605–613.

[Jac56] Robb Jacoby, *One-parameter transformation groups of the three-sphere*, Proc. Amer. Math. Soc. **7** (1956), 131–142.

[Jia83] Boju Jiang, *Lectures on Nielsen Fixed Point Theory*, Contemp. Math. Amer. Math. Soc. **14** (1983).

[JN83]　M. Jenkins and W. Neuman, *Lectures on Seifert Manifolds*, Brandeis University, Waltham, Mass., 1983.

[Joh78]　F.E.A. Johnson, *On Poincaré duality groups of poly-linear type and their realisations*, Math. Z. **163** (1978), 145–148.

[Joh79]　Klaus Johannson, *Homotopy Equivalences of 3-Manifolds with Boundaries*, vol. 761, Lecture Notes in Math., Springer, Berlin-New York, 1979.

[JS79]　W. Jaco and P. Shalen, *Seifert Fibered Spaces in 3-Manifolds*, vol. 21, Mem. Amer. Math. Soc., 1979.

[Kah70]　Dost M. Kahn, *Circle and Toral Actions Covered by the Sphere*, Ph.D. disertation, The University of Michigan, 1970.

[Kam88]　Y. Kamishima, *Conformally flat 3-manifolds with compact automorphism groups*, J. London Math. Soc. (2) **38** (1988), 367–378.

[Ker19]　B. Kerékjártó, *Über die periodischen Transformationen der Kreisscheibe und der Kugelfläche*, Math. Ann. **80** (1919), 36–38.

[Ker83]　S. Kerckhoff, *The Nielsen realization problem*, Ann. of Math. (2) **117** (1983), no. 2, 235–265.

[Kim83]　Hyuk Kim, *Complete Left-Invariant Flat Affine Structures on the Simply Connected Nilpotent Lie Groups*, PhD Thesis, University of Michigan, 1983.

[Kim86]　_____, *Complete left-invariant affine structures on nilpotent Lie groups*, J. Differential Geom. **24** (1986), 373–394.

[Kim87]　_____, *Extensions of left-symmetric algebras*, Algebras, Groups and Geometries **4** (1987), 73–117.

[Kim96]　_____, *The geometry of left-symmetric algebra*, J. Korean Math. Soc. **33** (1996), 1047–1067.

[KJ03]　Hyuk Kim and Kyeonghee Jo, *Invariant measure and the Euler characteristic of projectively flat manifolds*, J. Korean Math. Soc. **40** (2003), 109–128.

[KK83]　H.T. Ku and M.C. Ku, *Group actions on aspherical $A_k(N)$-manifolds*, Trans. Amer. Math. Soc. **278** (1983), 841–859.

[KL88a]　Slawomir Kwasik and Kyung Bai Lee, *Locally linear actions on 3-manifolds*, Math. Proc. Cambridge Philos. Soc. **104** (1988), 253–260.

[KL88b]　_____, *The Nielsen numbers of homotopically periodic maps of infra-nilmanifolds*, J. London Math. Soc. (2) **38** (1988), 544–554.

[KL91]　Hyuk Kim and Hyunkoo Lee, *The Euler characteristic of a certain class of projectively flat manifolds*, Topology Appl. **40** (1991), 195–201.

[KL93]　_____, *The Euler characteristic of projectively flat manifolds with amenable fundamental groups*, Proc. Amer. Math. Soc. **118** (1993), 311–315.

[KL97]　Eunsook Kang and Kyung Bai Lee, *Fixed point theory on Seifert manifolds modeled on $S^2 \times \mathbb{R}$*, Kyungpook Math. J. **37** (1997), 123–134.

[KL98]　_____, *Injective Operations of Homogeneous Spaces*, Michigan Math. J. **45** (1998), 419–440.

[KLL05]　Seung Won Kim, Jong Bum Lee, and Kyung Bai Lee, *Averaging formula for Nielsen numbers*, Nagoya Mathematical Journal **178** (2005), 37–53.

[KLR83]　Yoshinobu Kamishima, Kyung Bai Lee, and Frank Raymond, *The Seifert construction and its applications to infra-nilmanifolds*, Quart. J. Math. Oxford Ser. (2) **34** (1983), 433–452.

[KLR86]　R. Kulkarni, Kyung Bai Lee, and Frank Raymond, *Deformation spaces for Seifert manifolds*, Geometry and Topology, University of Maryland 1983–1984, Springer Lecture Notes in Math. **1167** (1986), 180–216, Edited by J. Alexander and J. Harer.

[Kod63]　K. Kodaira, *On compact analytic surfaces III*, Ann. of Math. (2) **78** (1963), 1–40.

[Kod64]　_____, *On the structure of compact complex analytic surfaces. I*, Amer. J. Math. **86** (1964), 751–798.

[KR85]　R. Kulkarni and Frank Raymond, *3-dimensional Lorentz space-forms and Seifert fiber spaces*, J. Diff. Geometry **21** (1985), 231–268.

[KT68]　F.W. Kamber and Ph. Tondeur, *Flat manifolds with parallel torsion*, J. Differential Geom. **2** (1968), 385–389.

[Kul81]　R. Kulkarni, *Proper actions and pseudo-Reimannian space forms*, Advances in Math. **40** (1981), 10–51.

[Las79] R. Lashof, *Obstructions to equivariance*, Algebraic topology, Aarhus 1978 (Proc. Sympos., Univ. Aarhus, Aarhus, 1978), Springer Lecture Notes in Math. **763** (1979), 476–503.

[Lee82a] Kyung Bai Lee, *Geometric realization of $\pi_0\epsilon(M)$*, Proc. Amer. Math. Soc. **86** (1982), no. (2), 353–357.

[Lee82b] ———, *Geometric realization of $\pi_0\epsilon(M)$ II*, Proc. Amer. Math. Soc. **87** (1982), 175–178.

[Lee83] ———, *Aspherical manifolds with virtually 3-step nilpotent fundamental group*, Amer. J. Math. **105** (1983), 1435–1453.

[Lee88] ———, *There are only finitely many infra-nilmanifolds under each nilmanifold*, Quart. J. Math. Oxford Ser. (2) **39** (1988), 61–66.

[Lee92] ———, *Nielsen numbers of periodic maps on solvmanifolds*, Proc. Amer. Math. Soc. **116** (1992), no. 2, 575–579.

[Lee95a] ———, *Infra-solvmanifolds of type (R)*, Quart. J. Math. Oxford Ser. (2) **46** (1995), 185–195.

[Lee95b] ———, *Maps of infra-nilmanifolds*, Pacific J. Math. **168** (1995), no. 1, 157–166.

[LL02] Hyunkoo Lee and Kyung Bai Lee, *Expanding maps on homogeneous infra-nilmanifolds*, Topology Appl. **117** (2002), no. 1, 45–58.

[LL06] Jong Bum Lee and Kyung Bai Lee, *Lefschetz numbers for continuous maps, and periods for expanding maps on infra-nilmanifolds*, Journal of Geometry and Physics **56** (2006), 2011–2023.

[LL09] ———, *Averaging formula for Nielsen numbers for maps on infra-solvmanifolds of type (r)*, Nagoya Math J. **196** (2009), 117–134.

[LLR96] Jong Bum Lee, Kyung Bai Lee, and Frank Raymond, *Seifert manifolds with fiber spherical space forms*, Trans. A. M. S. **348** (1996), 3763–3798.

[LLSY07] Jong Bum Lee, Kyung Bai Lee, Joonkook Shin, and Seunghoon Yi, *Unimodular groups of type $\mathbb{R}^3 \rtimes \mathbb{R}$-geometry*, J. Korean Math. Soc. **44** (2007), 1121–1137.

[LMS83] R Lashof, P May, and G.B. Segal, *Equivariant bundles with abelian structural group*, Contemp. Math. Amer. Math. Soc. **19** (1983), 167–176.

[LP96] Kyung Bai Lee and Keun Park, *Smoothly closed geodesics in 2-step nilmanifolds*, Indiana Univ. Math. J. **45** (1996), 1–14.

[LR81] Kyung Bai Lee and Frank Raymond, *Topological, affine and isometric actions on flat Riemannian manifolds*, J. Differential Geom. **16** (1981), 255–269.

[LR82] ———, *Topological, affine and isometric actions on flat Riemannian manifolds II*, Topology and its Applications **13** (1982), 295–310.

[LR84] ———, *Geometric realization of group extensions by the Seifert construction*, Contemp. Math. Amer. Math. Soc. **33** (1984), 353–411.

[LR85a] ———, *Rigidity of almost crystallographic groups*, Contemp. Math. Amer. Math. Soc. **44** (1985), 73–78.

[LR85b] ———, *The role of Seifert fiber spaces in transformation groups*, Contemp. Math. Amer. Math. Soc. **36** (1985), 367–425.

[LR86] ———, *Some Seifert manifolds which are boundaries*, Michigan Math. J. **33** (1986), 245–251.

[LR87] ———, *Manifolds on which only tori can act*, Trans. A. M. S. **304** (1987), 487–499.

[LR89] ———, *Seifert manifolds modelled on principal bundles*, Transf. Groups (Osaka,1987); Lect. Notes in Math. **1375** (1989), 207–215.

[LR90] ———, *Examples of solvmanifolds without certain affine structure*, Illinois J. Math. **37** (1990), 69–77.

[LR91] ———, *Maximal torus actions on solvmanifolds*, Internat. J. Math. **2** (1991), 67–76.

[LR96] ———, *Seifert manifolds with $\Gamma \backslash G/K$-fiber*, Michigan Math. J **43** (1996), 437–464.

[LR02] ———, *Seifert manifolds*, Handbook of Topology, Elsevier Science B.V., Edited by R.J. Daverman and R.B. Sher (2002), 635–705.

[LS98] Kyung Bai Lee and Joonkook Shin, *Affine manifolds with dilations*, Topology and its Applications **86** (1998), no. (2), 141–150.

[LS01] Kyung Bai Lee and Andrzej Szczepański, *Maximal holonomy of almost Bieberbach groups for $Heis_5$-manifolds*, Geom. Dedicata **87** (2001), 167–180.

[LSY93] Kyung Bai Lee, Joonkook Shin, and Shoji Yokura, *Free actions of finite abelian groups on the 3-torus*, Topology and its Applications **53** (1993), no. (2), 153–175.

[LY72]   H.B. Lawson and S.T. Yau, *Compact manifolds of non-positive curvatures*, J. Diff. Geometry **7** (1972), 211–228.
[Mac62]  A. M. Macbeath, *On a theorem by J. Nielsen*, Quart. J. Math. Oxford Ser. (2) **13** (1962), 235–236.
[Mac67]  _____, *The classification of non-euclidean plane crystallographic groups*, Canad. J. Math. **19** (1967), 1192–1205.
[Mal51]  Anatolii I. Mal'cev, *On a class of homogeneous spaces*, Translations Amer. Math. Soc. **39** (1951), 1–33.
[Mal98]  Wim Malfait, *Model aspherical manifolds with no periodic maps.*, Trans. Amer. Math. Soc. **350** (1998), 4693–4708.
[MB84]   John W. Morgan and Hyman Bass, *Smith Conjecture*, Pure and Applied Mathematics, Academic Press, 1984.
[McC94]  Christopher K. McCord, *Nielsen numbers of homotopically periodic maps on infra-solvmanifolds*, Proc. Amer. Math. Soc. **120** (1994), no. 1, 311–315.
[Mil57]  J. Milnor, *Groups which act on $S^n$ without fixed points*, Amer. J. Math. **79** (1957), 623–630.
[Mil68]  _____, *Singular Points of Complex Hypersurfaces*, Ann. of Math. Studies **61** (1968), Princeton Univ. Press.
[Mil77]  John Milnor, *On fundamental groups of complete affinely flat manifolds*, Adv. Math. **25** (1977), 178–187.
[ML75]   Saunders Mac Lane, *Homology*, Die Grundlehren der Math. Wissenschaften, vol. 114, Springer–Verlag Berlin Heidelberg New York, 1975.
[Mon73]  Jose M Montesinos, *Seifert manifolds that are ramified two-sheeted cyclic coverings (Spanish)*, Bol. Soc. Mat. Mexicana (2) **18** (1973), 1–32.
[Mor05]  J. Morgan, *Recent progress on the Poincaré conjecture and the classification of 3-manifolds*, Bull. Amer. Math. Soc. (N.S.) **42** (2005), 57–78.
[Mos54]  G.D. Mostow, *Factor spaces of solvable groups*, Ann. of Math.(2) **60** (1954), 1–27.
[Mos61]  G. D. Mostow, *Cohomology of topological groups and solvmanifolds*, Ann. of Math. **73** (1961), 20–48.
[Mos73]  _____, *Strong Rigidity of Locally Symmetric Spaces*, Ann. of Math. Studies **78** (1973), Princeton Univ. Press.
[MS75]   A. M. Macbeath and D. Singleman, *Spaces of subgroups and Teichmüller spaces*, Proc. London Math. Soc. (3) **31** (1975), 211–256.
[MS86]   William H. Meeks and Peter Scott, *Finite group actions on 3-manifolds*, Invent. Math. **86** (1986), 287–346.
[MTW76] I Madsen, C.B. Thomas, and C.T.C. Wall, *The topological spherical space form problem–II*, Topology **15** (1976), 375–382.
[MTW83] _____, *The topological spherical space form problem–III. Dimensional bounds and smoothing*, Pacific J. Math. **106** (1983), 135–143.
[MY57]   D. Montgomery and Yang, *The existence of a slice*, Illinois J. Math. **65** (1957), 108–116.
[MY80]   William H. Meeks and Shing Tung Yau, *Topology of three-dimensional manifolds and the embedding problems in minimal surface theory*, Ann. of Math. (2) **112** (1980), 441–484.
[MZ36]   D. Montgomery and L. Zippin, *Periodic one-parameter groups in three-space*, Trans. Amer. Math. Soc. **40** (1936), no. 1, 24–36.
[Nie43]  J. Nielsen, *Abbildungsklassen endlicher Ordnung*, Acta Math. **75** (1943), 23–115.
[NR78]   W. Neumann and Frank Raymond, *Seifert manifolds, plumbing, μ-invariants and orientation reversing maps*, Algebraic and Geometric Topology, Springer Lecture Notes in Math. **664** (1978), 165–195.
[NS85]   A. Nicas and C Stark, *K-theory and surgery of codimension-two torus actions on aspherical manifolds*, J. London Math. Soc. (2) **31** (1985), 173–183.
[Ohs87]  K. Ohshika, *Teichmüller spaces of Seifert manifolds with infinite $\pi_1$*, Topology and Appl. **27** (1987), 75–93.
[Oli76]  R. Oliver, *A proof of the Conner conjecture*, Ann. of Math. (2) **103** (1976), 637–644.
[OR68]   Peter Orlik and Frank Raymond, *Actions of $SO(2)$ on 3-manifolds*, Proc. Conf. on Transformation Groups (New Orleans, La., 1967) Springer (1968), 297–318.

[OR69] _____, *On 3-manifolds with local SO(2) action*, Quart. J. Math. Oxford Ser. (2) (1969), 143–160.

[OR70a] _____, *Actions of the torus on 4-manifolds. I*, Trans. Amer. Math. Soc. **152** (1970), 531–559.

[OR70b] _____, *Actions of the torus on 4-manifolds. II*, Topology **13** (1970), 89–112.

[Orl72] Peter Orlik, *Seifert Manifolds*, Lect. Notes in Math., vol. 291, Springer–Verlag, 1972.

[OVZ67] P. Orlik, E. Vogt, and H. Zieschang, *Zur Topologie gefaserter dreidimensionaler Mannigfaltigkeiten*, Topology **6** (1967), 49–64.

[OW71] P. Orlik and P. Wagreich, *Isolated singularities of algebraic surfaces with $k^*$ action*, Ann. of Math. (2) **93** (1971), 205–228.

[OW77] _____, *Algebraic surfaces with $k^*$ action*, Acta. Math. **138** (1977), 43–81.

[Pal61] R. Palais, *On the existence of slices for actions of non-compact Lie groups*, Ann. of Math. (2) **73** (1961), 295–323.

[Par89] Chan-Young Park, *Homotopy Groups of Automorphisms of Some Injective Seifert Fiber Spaces*, Ph. D. Thesis, University of Michigan, 1989.

[Par91] _____, *On the weak automorphism group of a principal bundle, Product case*, Kyungpook Math. J. **31** (1991), 25–34.

[Pet71] Ted Petrie, *Free metacyclic group actions on homotopy spheres*, Ann. of Math. (2) **94** (1971), 108–124.

[Pra73] Gopal Prasad, *Strong rigidity of Q-rank 1 lattices*, Invent. Math. **21** (1973), 255–286.

[Pup95] Volker Puppe, *Simply connected 6-dimensional manifolds with little symmetry and algebras with small tangent space*, Ann. of Math. Studies **138** (1995), 283–302, Princeton Univ. Press.

[Pup07] _____, *Do manifolds have little symmetry?*, J. Fixed Point Theory Appl. (2007), 85–96.

[Rag72] M. S. Raghunathan, *Discrete Subgroups of Lie Groups*, Ergebnisse der Mathematik und ihrer Grenzgebiete, vol. 68, Springer–Verlag, 1972.

[Ray68] Frank Raymond, *Classification of actions of the circle on 3-manifolds*, Trans. Amer. Math. Soc. **131** (1968), 51–78.

[Ray79] Frank Raymond, *The Nielsen theorem for Seifert fibered spaces over locally symmetric spaces*, J. Korean Math. Soc. **16** (1979), 87–93.

[Ree83] H. Rees, *Special manifolds and Novikov's conjecture*, Topology **22** (1983), 365–378.

[RS77] Frank Raymond and Leonard L. Scott, *Failure of Nielsen's theorem in higher dimensions*, Arch. Math. **29** (1977), 643–654.

[RT76] Frank Raymond and Jeffrey L. Tollefson, *Closed 3-manifolds with no periodic maps*, Trans. Amer. Math. Soc. **221** (1976), no. 2, 403–418.

[RT82] _____, *Correction to: Closed 3-manifolds with no periodic maps*, Trans. Amer. Math. Soc. **272** (1982), no. 2, 803–807.

[RV81] Frank Raymond and A. Vasquez, *3-manifolds whose universal coverings are Lie groups*, Topology and Appl. **12** (1981), 161–179.

[RW77] Frank Raymond and D. Wigner, *Construction of aspherical manifolds*, Geometric Appl. of Homotopy Theory, Springer Lecture Notes in Math. **657** (1977), 408–422.

[Sad91a] M. Sadowski, *Equivariant splittings associated with smooth toral actions*, Algebraic Topology Poznań 1989, Springer Lecture Notes in Math. **1474** (1991), 183–192.

[Sad91b] _____, *On isomorphisms between fibrations over $T^k$ associated with a $T^k$ action*, Ann. Polon. Math. **53** (1991), 45–51.

[Sai51] M. Saito, *Sur certains groupes de Lie résoubles I, II*, Sci. Papers Coll. Gen. Ed. Univ. Tokyo **7** (1951), 157–168.

[Sav99] Nikolai Saveliev, *Lectures on the Topology of 3-Manifolds*, de Gruyter Textbook, 1999, An introduction to the Casson invariant.

[Sav02] _____, *Invariants for Homology 3-spheres*, Encyclopaedia of Mathematical Sciences, vol. 140, de Gruyter Textbook, 2002.

[Sch81a] Reinhard Schultz, *Group actions on hypertoral manifolds II*, J. Reine Angew. Math. **325** (1981), 75–86.

[Sch81b] Sol Schwartzman, *A split action associated with a compact transformation group*, Proc. Amer. Math. Soc. **83** (1981), 817–824.

[Sco83a] Peter Scott, *The geometries of 3-manifolds*, Bull. London Math. Soc. **15** (1983), 401–487.

[Sco83b] _____, *There are no fake Seifert fiber spaces with infinite $\pi_1$*, Ann. of Math. Studies **117** (1983), 35–70, Princeton Univ. Press.
[Seg83] Daniel Segal, *Polycyclic Groups*, Cambridge University Press, 1983.
[Sei33] H. Seifert, *Topologie drei-dimensionaler gefaserter Räume*, Acta. Math. **60** (1933), 147–238 [An English translation of this important paper appears in [ST80] as an Appendix].
[Sel60] A Selberg, *On discontinuous groups in higher-dimensional symmetric spaces*, 1960 Contributions to function theory (Internat. Colloq. Function Theory, Bombay), Tata Institute of Fundamental Research, Bombay (1960), 147–164.
[Spa66] E. Spanier, *Algebraic Topology*, McGraw-Hill, 1966.
[ST31] H. Seifert and W. Threlfall, *Topologische Untersuchung der Diskontinuitätsbereiche endlicher Bewegungsgruppen des dreidimensionalen sphärischen Raumes*, Math. Ann. **104** (1931), 1–70.
[ST33] _____, *Topologische Untersuchung der Diskontinuitätsbereiche endlicher Bewegungsgruppen des dreidimensionalen sphärischen Raumes (schluss)*, Math. Ann. **107** (1933), 543–586.
[ST80] _____, *A Textbook of Topology*, Academic Press, 1980, English translation.
[ST87] I.N. Stewart and D.O. Tall, *Algebraic Number Theory, Second edition*, Chapman and Hall Mathematics Series, Chapman and Hall, London, 1987.
[Ste51] N. Steenrod, *The Topology of Fibre Bundles*, Princeton University Press, 1951, Princeton Mathematical Series, vol. 14.
[Ste61] T.E. Stewart, *Lifting group actions in fiber bundles*, Ann. of Math. **18** (1961), 192–198.
[Su63] J.C. Su, *Transformation groups on cohomology projective space*, Trans. Amer. Math. Soc. **106** (1963), 305–318.
[Suw79] Tatsuo Suwa, *Deformations of holomorphic Seifert fiber spaces*, Invent. Math. **51** (1979), 77–102.
[SY79] R. Schoen and S.T. Yau, *Compact group actions and the topology of manifolds with non-positive curvature*, Topology **18** (1979), 361–380.
[Thu97] William P. Thurston, *Three-dimensional geometry and topology. Vol. 1*, Princeton Mathematical Series, 35. Princeton University Press, Princeton, NJ, 1997.
[Tol69] Jeffrey L. Tollefson, *3-manifolds fibering over $S^1$ with non-unique connected fiber*, Proc. Amer. Math. Soc. **21** (1969), 79–80.
[vR62] Rabe von Randow, *Zur Topologie von dreidimensionalen Baummannigfaltigkeiten*, Bonn. Math. Schr. No. 14, 1962.
[Wal67a] F. Waldhausen, *Eine Klasse von 3-dimensionalen Mannigfaltigkeiten. I (German)*, Invent. Math. **3** (1967), 308–333.
[Wal67b] _____, *Eine Klasse von 3-dimensionalen Mannigfaltigkeiten. II (German)*, Invent. Math. **4** (1967), 87–117.
[Wal67c] _____, *Gruppen mit Zentrum und drei-dimensionaler Mannigfaltigkeiten*, Topology **6** (1967), 505–517.
[Wal68] _____, *On irreducible 3-manifolds which are sufficiently large*, Ann. of Math. **87** (1968), 56–88.
[Wal70] C.T.C. Wall, *The topological space-form problems*, Topology of Manifolds, Markham (1970), 319–331, Proc. Inst., Univ. of Georgia, Athens, 1969.
[Wal86] _____, *Geometries and geometric structures in real dimension 4 and complex dimension 2*, Geometry and Topology, University of Maryland 1983-1984, Edited by J. Alexander and J. Harer, Springer Lecture Notes in Math. **1167** (1986), 268–292.
[Wan56] H. C. Wang, *Discrete subgroups of solvable Lie groups*, Ann. of Math. **64** (1956), 1–19.
[Wan72] _____, *Topics on totally discontinuous groups*, Pure and Appl. Math. **8** (1972), 459–487.
[Wil66] H. C. Wilkie, *On non-Euclidean crystallographic groups*, Math. Z. **91** (1966), 87–102.
[Wil79] R. L. Wilder, *Topology of Manifolds*, Colloquium Publications, vol. 32, American Mathematical Society, 1979.
[Wol77] Joseph A. Wolf, *Spaces of Constant Curvature*, Publish or Perish, Inc. Berkeley, 1977.
[Yau77] Shing Tung Yau, *Remarks on the group of isometries of a Riemannian manifold*, Topology **16** (1977), 239–247.
[Zie69] H. Zieschang, *On toric fiberings over surfaces*, Math. Notes **5** (1969), 341–345.

[Zie81]  Heiner Zieschang, *Finite Groups of Mapping Classes of Surfaces*, Lect. Notes in Math., vol. 875, Springer–Verlag, 1981.

[Zie84]  H. Zieschang, *Classification of Montesinos knots*, Topology (Leningrad, 1982), Springer Lecture Notes in Math. **1060** (1984), 378–389.

[Zim80]  Bruno Zimmermann, *Über Gruppen von Homöomorphismen Seifertscher Faserraume und flacher Mannigfaltigkeiten*, Manuscripta Math. **30** (1980), 361–373.

[Zim85]  _____, *Zür Klassifikation höherdimensionaler Seifertscher Faserräume*, Low-dimensional topology (Chelwood Gate, 1982) London Math. Soc. Lecture Note Ser., 95, Cambridge Univ. Press, Cambridge (1985), 214–255.

[ZZ79]  H. Zieschang and B. Zimmermann, *Endliche Gruppen von Abbildungsklassen gefaserter 3-Mannigfaltigkeiten*, Math. Ann. **240** (1979), 41–62.

[ZZ82]  _____, *Über Erweiterungen von $\mathbb{Z}$ und $\mathbb{Z}_2 * \mathbb{Z}_2$ durch nichteuklidische kristallographische Gruppen*, Math. Ann. **259** (1982), 29–51.

# Index

1-cocycle, 105
2-coboundary, 101
2-cocycle, 100
$2q$-conditions, 265
$A$-acyclic, 48
$C_G(K)$, 75
$E$-orbits, 299
$E(3)$, 86
$F$-orbits, 299
$G$-equivalence, 2
$G$-invariant tubular neighborhood of $Gx$, 12
$G$-isomorphism, 2
$G$-map, 1
$G$-space, 1
$G \times_{(f,\widetilde{\varphi})} Q$, 96
$H$-kernel, 11
$H$-slice, 11
$H^1(Q;C)$, 105
$H^2(Q;A)$, 100
$H^2(Q;\mathcal{Z}(G))$, 104
$K$-manifold, 56
aff$(G)$, 231
Aff$(M)$, 249
Aut$(\Pi, \Gamma)$, 221
Aut$^0(G,K)$, 160
$\mathbb{Q}$-acyclic, 48
$\mathcal{E}(M)$, 257
$\mathcal{Z}(G)$, 75
$\ell(G)$, 75
Endo$(G)$, 231
$\mathbb{L}$-group, 209
$\mathbb{Z}_p$-acyclic, 48
$\mathcal{A}$-admissible, 56
$\mathrm{M}_G(P,G)$, 74
$\mu$, 75
$\mu(a)$, 75
Fitt$(\Gamma)$, 153
Opext$(Q,G,\widetilde{\varphi})$, 97
Out$(G)$, 75
Out$(\Pi, \Gamma)$, 221
TOP$_G(P)$, 73
$p^2$-conditions, 265
$pq$-condition, 265
$r(a)$, 76

$\mathbb{L}_+$-group, 209
TOP$(X)$, 2
(S1), 122
(S2), 122
(S3), 122
(S4), 122
KAN decomposition, 117
Sol-geometry, 147
covering space
    construction, 23

abstract kernel, 95, 217
ACG, 227
action
    effective, 3
    free, 4
    ineffective, 3
    ineffective part, 3
    injective, 50
    injective torus, 66
    lifting exact sequence, 27
    locally proper, 4
    proper, 5
    simply transitive, 3
    smooth, 2
    transitive, 3
admissible group, 217
admissible space, 54, 57
affine crystallographic group, 227
affine diffeomorphism, 82
affine endomorphisms, 231
affine structure, 205
affine structure on $\Gamma$, 227
algebraic hull, 230
almost Bieberbach group, 88
almost crystallographic group, 88
almost flat, 88
aspherical, 47, 56

base space, 6
Bieberbach group, 86, 139
    almost, 88
Bieberbach Property, 115
Bieberbach Theorem

first, 139
second, 139
third, 139
Bieberbach's Theorem, 139
binary group, 379
Borel space, 10
Borel's Density Theorem, 116
Brieskorn variety, 13
bundle isomorphism, 8
bundles
    equivalent, 8

Cartan $G$-space, 5
Chabauty space, 290
classifying space, 9
closed immersion, 230
cocompact, 20
cocompact lattice, 111
cohomologically locally connected, 19
cohomologous, 101, 105
cohomology $m$-manifold, 19
commutator, 109
completely solvable, 114
congruent, 97, 101
conjugation, 75
Construction Theorem, 122
converse to general pushout, 100
covering
    order, 129
covering dimension, 129
covering projection, 23
covering space
    equivalent, 23
covering transformation, 23
crossed homomorphism, 105
crystallographic group, 85, 139
    almost, 88

decreasing central series, 109
dimension, 129
discrete nilradical, 153, 155, 208

E-orbit, 302
enantiomorphic pairs, 143
equivariant, 2
essential covering, 156
Euler class, 10
Euler number, 286, 318
evaluation map, 31
evenly covered, 23
exceptional orbit, 302
existence, 123
extended lifting, 27, 187
extension
    $G$-inner, 237
    inner, 237

factor set, 100
fiber, 6

fiber bundle, 6
fibered solid torus, 300
finitely extendable, 220
First Bieberbach Theorem, 116, 140
first cohomology, 105
Fitting subgroup, 153
fixed orbits, 299
fixed-point class, 234
    essential, 234
fixed-point set, 3
Fourth Bieberbach Theorem, 142

global $H$-slice, 11
group extension, 95

Heisenberg group, 88
Hirsh number, 229
holonomy group, 140, 156
homogeneous, 289
homogeneous space, 82
homologically injective, 236, 237, 305
homologically locally connected, 19
hyper-aspherical, 56

inflation, 225
infra-homogeneous manifold, 82
infra-homogeneous space, 82
infra-nilmanifold, 88
infra-solvmanifold, 172, 259
infra-solvmanifold of type (R), 151
injective, 66, 236
injective Seifert Construction, 340
injective Seifert fibering, 89
inverse image bundle, 8
isotropy, 3
isotropy subgroup, 3
Iwasawa decomposition, 117

Jacobian variety, 334

lattice, 110
Lefschetz number, 234
left action, 1
Levi decomposition, 110
lifting, 217
lifting exact sequence, 27
lifting of group action, 27
lifting-sequence, 27
local $T^k$-action, 202
local cohomology group, 19
local triviality condition, 6
locally injective, 43
locally smooth, 20
Lorentz metric, 288
Lorentz orbifold, 288
Lorentz space form, 288

Mal'cev completion, 259
manifold factor, 207
maximal compact subgroup, 109

maximal normal nilpotent subgroup, 259
maximal torus action, 248, 250
modeled on $G/K \times W$, 163
moduli space, 290
Montesinos links, 350
Mostow-Wang group, 173

Nielsen number, 234
Nielsen realization problem, 217
nilmanifold, 111
nilpotency, 109
nilpotent, 109
nilradical, 110, 133, 260
nonorientable, 19
normalized
  Seifert invariant, 303

orbi-bundle, 356
orbit map, 2
orbit of $G$ through $x$, 2
orbit space, 2
orbit type, 3
orientable, 19

poly-$\mathbb{L}_+$-group, 209
polynomial crystallographic group, 227
polynomial representation
  canonical type, 229
polynomial structure, 205
polynomial structure on $\Gamma$, 227
polynomially conjugated, 230
predivisible group, 173, 259
primitive, 66
principal $G$-bundle, 7
principal bundle map, 8
principal orbit, 18
principal orbits, 299
projective unitary group, 91
proper mapping, 2
Property (B), 115
prototype Seifert fibering, 69
pseudo-lifting, 272
pullback, 97
pullback bundle, 8
pushout, 98, 100
  general, 98
    converse, 100

radical, 110
regular fiber, 80
restricted Chabauty space, 292
restricted moduli space, 292
Riemann space, 290
right action, 2
rigidity, 124
rigidly related, 124, 207
rotational part, 86

SE-orbit, 299, 303

Seifert automorphism, 124
Seifert Construction, 92, 163
Seifert construction, 119
Seifert fiber space, 163
  base space, 163
  injective, 163
  smooth construction, 137
  typical fiber, 163
Seifert fibered space, 78
Seifert fibering, 79, 163
  locally injective, 89
  model space, 79
Seifert isomorphism, 124, 135
Seifert manifold, 163
Seifert presentation, 316
Selberg's lemma, 111
semiconjugate, 231
semidirect product, 97
semifree, 4
semilocally simply connected, 23
semisimple, 110
Shapiro's Lemma, 127
signature, 354
simple, 110
simply transitive, 104
singular fiber, 80
slice at $x$, 11
Smith Conjecture, 351
Smith theorem, 18
solvable, 109
solvmanifold, 172
space of discrete representations, 290
space of discrete subgroups, 290
special exceptional orbit, 299, 303
special Lie group, 122
spherical space form, 90
split Lie hull, 173
splitting theorem, 237
standard Lorentz orbifold, 289
standard Lorentz structures, 291
strictly equivalent, 123, 187, 207
strong lattice property, 144
strongly equivalent, 246
structural constants, 111
structure group, 7
sufficiently general
  matrix, 335
symmetric space, 166
  almost effective, 166
  effective, 166
  Riemannian, 166

Teichmüller space, 290, 292
topologically Euclidean, 340
topologically hyperbolic, 340
topologically realizable, 217
toral degree of symmetry, 265
toral rank, 265

total space, 6
translational part, 86
type (E), 114
type (R), 114
typical fiber, 80, 123, 163

UAEP (Unique Automorphism Extension
    Property), 98, 112
ULIEP (Unique Lattice Isomorphism
    Extension Property), 115
uniform, 85, 110, 111
uniform lattice, 111
uniformizing group, 91, 92, 119, 164, 283
unipotent, 113
unique maximal normal Abelian subgroup,
    141
uniqueness, 123
universal classifying space, 9
universal covering space, 24
universal lifting sequence, 26
unnormalized
    Seifert invariant, 303

Vietoris mapping theorem, 49
virtually $\mathcal{P}$, 1

weak $G$-equivalence, 2, 73
weak $G$-isomorphism, 2
weakly $G$-equivariant, 2
weakly $G$-equivariant bundle
    automorphism, 75
Weil space of $(\Pi; \mathcal{U})$, 290

Zariski closed, 230